MATHEMATICS FOR ELECTRICITY AND ELECTRONICS

THIRD EDITION

MATHEMATICS FOR ELECTRICITY AND ELECTRONICS

Arthur D. Kramer
NYC College of Technology

THOMSON
DELMAR LEARNING™

Australia Canada Mexico Singapore Spain United Kingdom United States

Math for Electricity and Electronics, 3rd Edition
Arthur Kramer

Vice President, Technology and Trades SBU:
Alar Elken

Editorial Director:
Sandy Clark

Senior Acquisitions Editor:
Stephen Helba

Development:
Monica Ohlinger

Marketing Director:
David Garza

Senior Channel Manager:
Dennis Williams

Marketing Coordinator:
Stacey Wiktorek

Production Director:
Mary Ellen Black

Senior Production Manager:
Larry Main

Project Editor:
Christopher Chien

Art/Design Coordinator:
Francis Hogan

Technology Project Specialist:
Linda Verde

Senior Editorial Assistant:
Dawn Daugherty

Printed in the United States of America
2 3 4 5 XX 08 07 06

For more information contact
Thomson Delmar Learning
Executive Woods
5 Maxwell Drive, PO Box 8007
Clifton Park, NY 12065-8007
Or find us on the World Wide Web at
http://www.delmarlearning.com

Library of Congress Cataloging-in-Publication Data:
ISBN: 1-4018-7096-1

NOTICE TO THE READER

Dedication

To Helen and Joseph who gave us joy and love.

Contents

Preface

Introduction

This third edition of *Mathematics for Electricity and Electronics* has been upgraded and improved from the successful second edition. The number of error boxes, practice problems, calculator solutions, examples, and exercises has been significantly increased, especially in the early chapters. The first three chapters have been expanded to enable the reader to gain a better grasp of the basic concepts. Chapter 6, Computer Number Systems, from the second edition, has been moved to chapter 25 before Boolean Algebra where it more logically belongs. A new section on hexadecimal arithmetic has been added to Computer Number Systems, and a new section on Bode plots has been added to the chapter on Applications of Logarithms.

Electronics remains one of the most important technical fields in today's world, and a special mathematics book for electricity and electronics is necessary for the following reasons:

- First and foremost, electricity and electronics use more mathematical symbols and mathematical formulas than other technical fields.
- Second, the need for intermediate and advanced mathematical ideas comes earlier in the study of electricity and electronics than in the other basic engineering technologies.
- Third, a general technical math book cannot adequately cover all of the important mathematical concepts as they apply to electricity and electronics. For some topics, such as complex numbers, the terminology and the symbolism are different from that used in general mathematics.
- Fourth, a special mathematics book can focus on the many applications to electricity and electronics, thus providing the ideal way to master the mathematical concepts.

Pedagogy

Mathematics for Electricity and Electronics is designed to serve the above needs in the best possible way. The text contains all of the mathematics required for the first and second year of electronics. Those topics necessary to prepare the student for the study of calculus are also included. The presentation and pedagogy of the text are based on the following:

- *Personal Approach.* The language is informal and geared to the student, with his or her needs in mind. Explanations directly address the reader using the second person "you" instead of the formal first person "we."
- *Clarity and Ease of Understanding.* There is a minimum of rigor with concepts presented in the most basic way. New ideas are always illustrated with concrete applications to practical problems and electrical problems. All electrical and electronic notation used is that which is generally accepted.

- *Quantitative Reasoning.* The ability to think quantitatively and predict results is an important focus of the text in addition to repetitive drill and the learning of necessary rote skills.
- *Conceptual Learning.* Learning concepts and relationships is as important as memorizing formulas and rules. The capacity to use and adapt formulas to solve problems is stressed.
- *Critical Thinking.* The ability to understand, analyze, and solve problems is emphasized throughout the text.

Organization

The chapters are arranged in a very logical and progressive order. However, care has been taken to make as many topics and chapters as independent as possible and allow for maximum flexibility in the order of presentation. Aside from the following two considerations, the chapters can be studied in almost any sound pedagogic order to fit various needs and requirements. Chapters 1 to 6 contain basic material that should be understood before studying beyond chapter 6. Chapter 16, Trigonometry of the Right Triangle, is a prerequisite for any of the subsequent chapters, 17 through 22.

Each chapter begins with a discussion of the importance of the material for electricity and electronics including a description of the chapter topics and the chapter objectives. Each topic is developed using a maximum number of examples with thorough explanation of all concepts and procedures. Each section contains many exercises which increase in difficulty with every odd exercise followed by a similar even one. Every chapter concludes with Chapter Highlights and Review Exercises. The highlights summarize the important ideas and formulas and present key examples that illustrate the most important concepts. The review exercises contain more than just additional types of problems found throughout the chapter. Where possible they are designed to combine ideas from several sections to help the student master the interrelationships. The answers to the odd exercises are given in appendix C, while the even exercises can be used for additional work or examinations.

Special Features

- Every exercise set in the non-electronic chapters includes practical problems that are designed to motivate learning by applying the concepts to such popular topics as automobiles, salary, sales price, test grades, sports, sales tax, temperature, weight, and similar topics.
- Every exercise set in the non-electronic chapters includes a special *Application to Electronics* section. Most of these problems are additional exercises using some electronics terminology but do not necessarily require an understanding of the electronics concepts. All the infomation necessary to complete the problem is given in the problem. These problems motivate learning by showing where the mathematics is used in electronics and help one to understand the mathematical ideas better when done in the context of an electronics application. Some of the applied problems reinforce concepts encountered earlier or in other classes.
- In the non-electronics chapters there are examples that "Close the Circuit" between theory and practice. They show how to apply the mathematical concepts to a problem in electricity or electronics and help one to do the *Application to Electronics* problems at the end of the section.
- Every chapter contains "Error Boxes," which point out a common misunderstanding and stress the correct approach. Each error box contains practice problems to immediately reinforce the correct procedure with the answers to the problems given on a subsequent page.

- The back-of-book CD-ROM, *Interactive Mathematics for Electricity & Electonics,* provides interactive tutorials covering the basics of electronics. Topics covered include Ohm's law, series circuits, series and parallel circuits, network theorems, magnetism, and electromagnetism.
- The book contains early chapters on arithmetic and algebra to reinforce or review basic skills and fundamental concepts. Many of the exercises in these chapters are designed to be done by hand to strengthen understanding. The calculator should only be used to check the results.
- A large number of examples explain and show calculator solutions for both direct algebraic logic (DAL) and non-DAL calculators. There are many exercises that are designed to be estimated first and then done on the calculator to foster quantitive reasoning and to develop the ability to troubleshoot for mistakes.
- There are special calculator sections at the end of the first four chapters which further explain calculator functions and can be used for self study. These sections explain the differences in calculator keystrokes and the use of an electrical engineering calculator.
- Chapter 4 is a special chapter on scientific and engineering notation, and the metric system.
- Chapter 10 contains two special sections, one on formulas and one on problem solving. These are designed to help the student develop the critical thinking skill of analyzing and solving verbal problems. The sections stress the logical steps of using a formula and setting up an equation and provide many different types of exercises to sharpen one's problem-solving skills. In addition, every set of problems throughout the book contains verbal problems to help master this ability.
- Chapters 7, 11, 14, and 15 are devoted exclusively to dc circuits, and chapters 18, 21, and 22 exclusively to ac circuits.
- There are over 1000 examples, many of which show calculator steps, and close to 500 figures to help explain and illustrate important concepts.
- There are over 5000 exercises and problems of many types and varying degrees of difficulty to develop a wide range of skills.
- Appendix A on calculator functions also includes an introduction to graphics calculators. The appendix contains references to sections in the text where each calculator function is used.
- Appendix B contains a summary of all notation, metric units, and formulas used in the text.
- Appendix C contains answers to the odd-numbered exercises.

Development

As an instructor in both mathematics and electronics for more than 40 years and an author of several mathematics and computer texts, I am very aware and sensitive to the needs of electronics students and their pitfalls in mastering mathematics. All the material in the text has been classroom tested by me and reviewed by at least twenty technical faculty from various schools. In addition, the entire book, including every example, exercise, and problem has been checked for accuracy by me and several very competent technical reviewers.

Supplements

- An updated *Instructor's Solutions Manual,* which contains solutions to all exercises and problems included in the book. (ISBN: 1-4018-7097-X)
- An **e.resource** containing an updated version of the Computerized Test Bank, PowerPoint presentations covering all 27 chapters, and a complete Image Library containing figures from the book. (ISBN: 1-4018-7098-8)

- Visit *Mathematics for Electricity & Electronics, 3e*'s **Online Companion** at www.electronictech.com for additional math problems and exercises, including Microsoft® Excel problems and examples for students.

Acknowledgments

This text reflects a considerable effort by many people at Thomson Delmar Learning and elsewhere to whom I express my sincere appreciation and gratitude. First and foremost, many thanks to my first editor at Delmar, Greg Clayton, whose hard work and vision have made this third edition possible, and to his editorial assistant Dawn Daugherty who is always there to help and guide me. Further thanks to my subsequent editors, Dave Garza and Steve Helba who guided me through this new edition, and to Michelle Ruelos Cannistraci whose gentle prodding was helpful in completing the final manuscript.

I am grateful to Monica Ohlinger and her assistant Erin Denny of Ohlinger Publishing Services for their effort and cheerful encouragement in steering the review and development of the manuscript and their "motherly" concern that I meet all the deadlines.

I am especially appreciative of the suberb work done by Sharon Green of Panache Editorial, Inc., in the composition and production of the third edition. Both Monica Ohlinger and Sharon worked on the second edition and it was my good fortune to again have them on my team.

Thanks to the following reviewers who worked hard on my behalf and supplied many helpful suggestions:

John Fitzen, Idaho State University, Pocatello, ID

Ginny Olson, Chippewa Valley Technical College, Eau Claire, WI

Satyanand Singh, NYC College of Technology, Brooklyn, NY

Dick Statham, Guilford Technical Community College, Greensboro, NC

James McDonald, Springfield Technical Community College, Springfield, MA

Stephen Vossler, Lansing Community College, Lansing, MI

Jim Scorzelli, ECPI, VA

Thanks to my colleagues at NYC College of Technology for their input, and special thanks to all my students over the years who have helped me to understand how I can best serve them in mastering the difficulties of mathematics.

Last, and above all, my deepest thanks for the patience, love, and assistance always generously given to me by my forever love, Carol.

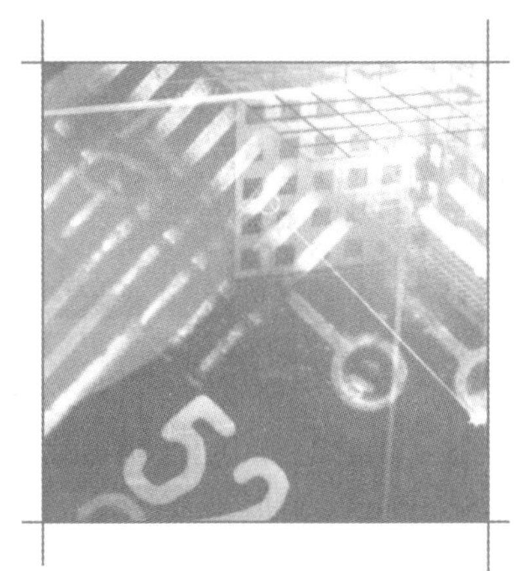

CHAPTER 1
Basic Arithmetic

CHAPTER OBJECTIVES

In this chapter, you will learn:

- The basic laws of arithmetic.
- The order of operations in arithmetic.
- The meaning of a fraction and equivalent fractions.
- How to reduce fractions to lowest terms.
- How to multiply and divide fractions.
- How to add and subtract fractions.
- How to use the calculator and its memory to perform basic arithmetic operations.

Electricity and electronics are two of the most interesting and important fields in today's technical world. They are very rewarding to study, and contain many useful concepts. However, it is necessary for you to master mathematics to gain a good understanding of these concepts. The first chapter begins by reinforcing your basic arithmetic skills and improving your understanding of fractions. It contains a review of the basic rules of arithmetic and the operations with whole numbers. Fractions occur often in electrical calculations and the meaning of a fraction is clearly explained. Understanding how to add and subtract fractions is important and is presented with thorough explanations and many clear examples to help you learn the procedures well. The last section in this chapter, and all other chapters, is called Chapter Highlights. It contains a summary of the important formulas and definitions and provides comprehensive review exercises to further reinforce your understanding of the ideas in the chapter.

• • •

1.1 Arithmetic Operations

There are four basic operations in arithmetic:

Addition, Subtraction, Multiplication, Division

Subtraction is the inverse of addition, which means, for example:

$$\text{If } 10 + 5 = 15 \text{ then } 15 - 5 = 10$$

Division is the inverse of multiplication, which means, for example:

$$\text{If } 4 \times 3 = 12 \text{ then } 12 \div 3 = 4$$

Since subtraction is the inverse of addition and division is the inverse of multiplication, the laws of arithmetic are defined for addition and multiplication. The *commutative laws* apply to switching or commuting two numbers. They state that it makes no difference which way you add or multiply two numbers a and b:

Law **Commutative Laws**

$$a + b = b + a$$
$$a \times b = b \times a \qquad (1.1)$$

For example:

$$3 + 6 = 9 \text{ and } 6 + 3 = 9 \quad \text{and} \quad 3 \times 5 = 15 \text{ and } 5 \times 3 = 15$$

Note that subtraction and division are *not* commutative. It makes a difference which number comes first when you subtract or divide:

$$7 - 3 \textit{ is not equal to } 3 - 7$$

$$10 \div 5 \textit{ is not equal to } 5 \div 10$$

The *associative laws* apply to the order in which the operations are performed on three numbers. They state that if three numbers a, b, and c are added or multiplied together, it makes no difference if you start with the first and second numbers or with the second and third numbers:

Law

Associative Laws*

$$(a + b) + c = a + (b + c)$$

$$(a \times b) \times c = a \times (b \times c) \qquad \textbf{(1.2)}$$

*Operations in parentheses are done first.

For example, suppose you want to add:

$$6 + 3 + 2$$

You can do it two ways. Using parentheses to indicate which operation is to be done first, the two ways are:

$$(6 + 3) + 2 = 9 + 2 = 11$$

$$6 + (3 + 2) = 6 + 5 = 11$$

In multiplication, suppose you want to multiply:

$$8 \times 4 \times 2$$

You can also do it two ways:

$$(8 \times 4) \times 2 = 32 \times 2 = 64$$

$$8 \times (4 \times 2) = 8 \times 8 = 64$$

Observe that subtraction and division are *not* associative. The subtraction sign and the division sign apply only to the number that immediately follows the sign as you move from left to right:

$$15 - 10 - 2 \quad \text{means} \quad (15 - 10) - 2 = 5 - 2 = 3 \quad \{\text{not } 15 - (10 - 2) = 15 - 8 = 7\}$$

$$16 \div 4 \div 2 \quad \text{means} \quad (16 \div 4) \div 2 = 4 \div 2 = 2 \quad \{\text{not } 16 \div (4 \div 2) = 16 \div 2 = 8\}$$

A third law of arithmetic that combines multiplication and addition is the *distributive law.* This law states that multiplication can be distributed over addition:

Law

Distributive Law

$$a \times (b + c) = (a \times b) + (a \times c) \qquad \textbf{(1.3)}$$

For example if you want to calculate:

$$3 \times (2 + 4)$$

You can do the operation in the parentheses first:

$$3 \times (2 + 4) = 3 \times 6 = 18$$

or you can apply the Distributive Law (1.3) and multiply each of the numbers by 3, and then add:

$$3 \times (2 + 4) = (3 \times 2) + (3 \times 4) = 6 + 12 = 18$$

The distributive law is important in algebra. The operations in arithmetic must be done in a specific order from left to right:

Procedure

Order of Operations (moving left to right)

1. Operations in parentheses.
2. Multiplication or division.
3. Addition or subtraction.

One way to help remember this order is to memorize the phrase "***P***lease ***E***xcuse ***M***y ***D***ear ***A***unt ***S***ally" (PEMDAS). The combined first letters of each word stand for: ***P***arentheses, ***E***xponents, ***M***ultiplication or ***D***ivision, ***A***ddition or ***S***ubtraction. Exponents or powers are discussed in Chapter 3.

Computers and scientific calculators are programmed to perform the arithmetic operations in the above order. It is called *algebraic logic* or the *algebraic operating system.* Study the following examples which illustrate the order of operations.

EXAMPLE 1.1

Calculate the following:

$$4 \times 19 - 36 + 6 \div 2$$

Solution Apply the order of operations moving from left to right. First perform the multiplication and division shown by each set of parentheses:

$$(4 \times 19) - 36 + (6 \div 2)$$

$$= 76 - 36 + 3$$

Then perform the subtraction and addition *moving from left to right:*

$$(76 - 36) + 3$$

$$= 40 + 3 = 43$$

You must move from left to right and subtract the 36 before adding the 3. If you add the 3 to the 36 and then subtract from 76 you will get 37, which is not the correct answer.

You can check that your calculator uses algebraic logic as follows. Enter Example 1.1 exactly as it appears above. You should get 43 when you press [=]:

$$4 \boxed{\times}\ 19 \boxed{-}\ 36 + 6 \boxed{\div}\ 2 \boxed{=} \rightarrow 43$$

EXAMPLE 1.2

Calculate by hand:

$$5 \times 17 - (12 + 4) \div 2 \times 3$$

Solution Perform the operation in parentheses first:

$$5 \times 17 - (12 + 4) \div 2 \times 3$$
$$= 5 \times 17 - (16) \div 2 \times 3$$

Do the multiplication and the division moving from left to right:

$$5 \times 17 - 16 \div 2 \times 3$$
$$= 85 - 8 \times 3$$
$$= 85 - 24 = 61$$

Observe that you divide 16 by 2 before you multiply by 3 and then do the subtraction.

You can check Example 1.2 on the calculator using parentheses keys as follows:

5 [×] 17 [−] [(] 12 [+] 4 [)] [÷] 2 [×] 3 [=] → 61

ERROR BOX

You must move from left to right when doing addition, subtraction, multiplication, or division. For example, suppose you are calculating:

$$20 - 10 \div 5 \times 3 + 15$$

You first do the division and then the multiplication:

$$20 - 2 \times 3 + 15$$
$$= 20 - 6 + 15$$

You then do the subtraction, and finally the addition:

$$= 14 + 15 = 29$$

Practice Problems: See if you can perform the following operations correctly without the calculator:

1. $30 - 12 \div 4 \times 3 + 5$ **2.** $15 \div 5 \times 6 - 4 + 3$ **3.** $8 - 2 \times 6 \div 3 + 1$

4. $7 \times 4 \div 2 - 1 + 7$ **5.** $11 - 8 \div 2 \times 2 + 5$ **6.** $20 - 10 + 15 \div 5 \times 10$

7. $8 + 16 \div 4 \times 4 - 7$ **8.** $100 \div 20 + 15 \times 10$ **9.** $5 \times 10 \div 25 + 10 - 5$

10. $500 \times 2 \div 100 \times 10$ **11.** $60 \times 4 - 80 \div 2$ **12.** $40 \div 8 \times 20 \div 25 + 6$

13. $(10 + 20) \div 3 - 5$ **14.** $10 \times (11 + 9) \div 50$ **15.** $2 \times (300 - 150) - 10$

A fraction line can be used to indicate division instead of a division sign, ÷. It is important to note that the *fraction line acts like parentheses for both the numerator and the denominator:*

$$\frac{18 + 36}{4 - 1} \textit{ is equivalent to } \frac{(18 + 36)}{(4 - 1)} \quad \textit{or} \quad (18 + 36) \div (4 - 1)$$

A forward slash, /, can also be used to indicate division as it does on a computer or some calculators:

$$(18 + 36) / (4 - 1) \textit{ is equivalent to } (18 + 36) \div (4 - 1)$$

Study the next example, which illustrates this idea.

EXAMPLE 1.3 Calculate by hand:

$$\frac{18 + 36}{4 - 1} - 5 \times 2$$

Solution You must first perform the operations in the numerator and the denominator of the fraction since the fraction line acts like parentheses:

$$\frac{18 + 36}{4 - 1} - 5 \times 2 = \frac{54}{3} - 5 \times 2$$

Then do the division and the multiplication:

$$\frac{54}{3} - 5 \times 2 = 18 - 10$$

And finally the subtraction:

$$18 - 10 = 8$$

The calculator check must place the numerator and the denominator in parenthesis:

[(] 18 [+] 36 [)] [÷] [(] 4 [−] 1 [)] [−] 5 [×] 2 [=] → 8

ERROR BOX

One of the most common errors involves division with fractions. It is important to emphasize that a fraction line, like parentheses, groups all the operations in the numerator of the fraction and all the operations in the denominator of the fraction. These operations must be done first, before any division. For example, in the fraction $\frac{6-4}{2}$, the 2 in the denominator cannot be divided into the 6 in the numerator. The subtraction must be done before the division:

$$\frac{6 - 4}{2} = \frac{2}{2} = 1$$

Practice Problems: See if you can perform the following operations correctly without the calculator:

1. $\frac{10 + 5}{5}$ **2.** $\frac{9 + 3}{3}$ **3.** $\frac{28 - 7}{7}$ **4.** $\frac{8 - 4}{4 - 2}$ **5.** $\frac{6 + 16}{6 + 5}$

6. $\frac{100 + 5}{10 - 5}$ **7.** $\frac{60}{10 + 2}$ **8.** $\frac{30}{6 - 3}$ **9.** $\frac{300 + 500}{100}$ **10.** $\frac{40 - 12}{4}$

EXAMPLE 1.4 Calculate by hand:

$$\frac{(10 - 5) \times 3}{2 + 3}$$

Solution Do the operation in parentheses first:

$$\frac{(10 - 5) \times 3}{2 + 3} = \frac{5 \times 3}{2 + 3}$$

Then calculate the numerator and denominator separately, and do the division last:

$$\frac{5 \times 3}{2 + 3} = \frac{15}{5} = 3$$

The calculator check must put the denominator in parentheses:

[(] 10 [−] 5 [)] [×] 3 [÷] [(] 2 [+] 3 [)] [=] → 3

EXAMPLE 1.5 Calculate by hand:

$$\frac{50}{15 - 5} + 5 \times (25 - 15)$$

Solution Do the operation in the denominator of the fraction and the operation in parentheses first:

$$\frac{50}{15-5} + 5 \times (25-15)$$
$$= \frac{50}{10} + 5 \times (10)$$

Then do the division and multiplication, and finally the addition:

$$\frac{50}{10} + 5 \times (10) = 5 + 50 = 55$$

The calculator check must put the denominator in parentheses:

50 [÷] [(] 15 [−] 5 [)] [×] [(] 25 [−] 15 [)] [=] → 55

EXERCISE 1.1

In exercises 1 through 40, test your understanding of arithmetic. Do each problem mentally or by hand. Use the calculator to check your results.

1. $6 + 5 - 7 + 3 + 5 + 4$
2. $8 + 2 - 3 + 9 - 1$
3. $5 \times 2 \times 3 \times 4$
4. $8 \times 5 \times 2 \times 2$
5. $12 \div 3 \times 2 \div 2$
6. $15 \times 4 - 24 \div 8$
7. $8 + 14 \div 2 \times 4$
8. $7 - 6 \div 3 + 8 \div 4$
9. $25 + 10 \times 6 \div 3 - 5$
10. $19 - 6 \times 9 \div 3 + 4$
11. $(800 + 200) \div 20$
12. $60 \div (6 + 4) \div 3$
13. $5 + (8 - 1) \times 6 \div 2$
14. $(5 - 1) \div 2 + 3 \times 4$
15. $7 \times 11 - (13 + 2) \div 5 + 3$
16. $(16 + 20) \div 18 + 12 - 5$
17. $30 + 45 \div (5 + 10) \times 5$
18. $500 - 100 \times 300 \div (50 + 50)$
19. $10 \times 20 \times 5 \div (140 - 40)$
20. $(10 + 2) \times 5 \div 6 \times 25$
21. $\frac{12 + 20}{4} - 3 \times 2$
22. $10 \times 2 + \frac{30 + 40}{7}$
23. $(5 + 35) \times \frac{13 - 4}{3}$
24. $\frac{50 - 25}{5} \times (19 - 11)$
25. $\frac{(15 - 10) \times 4}{3 + 1}$
26. $\frac{5 \times (8 + 2)}{16 - 6}$
27. $\frac{50 \times 50}{50 + 50}$
28. $\frac{10 \times 90}{10 + 90}$
29. $\frac{20 \div 4 + 5}{2 \times 15 - 20}$
30. $\frac{18 - 6 \times 2}{4 + 18 \div 9}$
31. $(27 - 9) / 6 + (12 - 5)$
32. $(20 + 7) - (13 + 3) / 2$
33. $\frac{100}{25 - 5} + 5 \times (20 - 10)$
34. $40 + (15 - 5) \times \frac{55}{30 - 25}$
35. $\frac{1000 + 500}{100} \times (350 - 250)$
36. $(1000 - 500) + \frac{2000 - 500}{150}$
37. $(400 - 200) \div (100 + 100)$
38. $(50 + 75) \div (50 - 25)$
39. $\frac{4 \times 4 + 8}{4 + 2 \times 2}$
40. $\frac{260 - 5 \times 2}{5 \times 5 + 25}$

Applied Problems

In problems 41 through 50, solve each applied problem by hand. Use the calculator to check your results.

41. One car travels 180 miles on 9 gallons of gasoline. A second car travels 210 miles on 10 gallons of gasoline. How many miles per gallon (mpg) does each car get? *Note:* mpg = (miles) / (gallons).

Answers to Error Box Problems, page 4:
1. 26 **2.** 17 **3.** 5 **4.** 20 **5.** 8 **6.** 40 **7.** 17 **8.** 155 **9.** 7 **10.** 100 **11.** 200 **12.** 10 **13.** 5 **14.** 4 **15.** 290

Answers to Error Box Problems, Page 5:
1. 3 **2.** 4 **3.** 3 **4.** 2 **5.** 2 **6.** 21 **7.** 5 **8.** 10 **9.** 8 **10.** 7

42. An electronics technician earns $640 for a 40-hour week and a math teacher earns $570 for a 30-hour week. How much does each earn per hour?

Note: Hourly earnings = (weekly income) / (hours).

43. A Mariner space probe traveling at an average rate of speed of 6000 mi/h takes 9000 hours to reach Mars. What is the total distance, in miles, traveled by the space probe?

Note: Distance = (rate of speed) × (time).

44. A bus route is 22 kilometers long in one direction. It takes the bus 50 minutes to travel the route in one direction and 70 minutes to travel the route in the other direction. What is the average rate of speed of the bus, in kilometers *per hour,* for the total trip?

Note: Average rate of speed = (*total* distance) / (*total* time).

Applications to Electronics

45. In an electric circuit, the current I in amps (A) is given by:

$$I = \frac{24}{12 + 2 + 10}$$

Calculate the value of the current.

46. In an electric circuit, the current I in amps (A) is given by:

$$I = \frac{100}{2 \times 20 + 10}$$

Calculate the value of the current.

47. In an electric circuit, the voltage V in volts (V) is given by:

$$V = 2 \times \left(\frac{40 \times 10}{40 + 10}\right)$$

Calculate the value of the voltage.

48. In an electric circuit, the voltage V in volts (V) is given by:

$$V = 3 \times \left(\frac{5 \times 20}{5 + 20}\right)$$

Calculate the value of the voltage V.

49. In Visual BASIC and other computer languages, the following keyboard symbols are used for the arithmetic operations:

Addition +
Subtraction −
Multiplication *
Division /

Apply the order of operations and calculate the following written in computer language:

$$2 + 3 * 8 / 4 - 1$$

50. As in problem 49, calculate the following expression written in computer language:

$$5 - 15 / 5 + 10 * 12$$

1.2 Multiplying and Dividing Fractions

Equivalent Fractions

Mistakes are often made when working with fractions because the concepts are not well understood. A fraction is a comparison or *ratio* between two whole numbers. For example, suppose a short distance and a long distance are compared and the ratio of the short distance to the long distance is said to be $\frac{2}{3}$. The actual distances are not necessarily 2 and 3 units. However, the ratio of the actual distances can be reduced to $\frac{2}{3}$ by dividing the numerator and denominator of the fraction by the same divisor. For example, if the short distance is 100 units and the long distance is 150 units, you can divide the numerator and denominator of the ratio by 50 reducing it to $\frac{2}{3}$:

$$\frac{100}{150} = \frac{100 / (50)}{150 / (50)} = \frac{2}{3}$$

The fraction $\frac{100}{150}$ is *equivalent* to the fraction $\frac{2}{3}$, which tells us we would need *three* of the short distances to equal *two* of the long distances. When there are common divisors, called *factors,* in the numerator and denominator of a fraction, they can be divided out and the fraction reduced to lowest terms.

Rule **Equivalent Fractions**

Two fractions are equivalent if they can each be reduced to the same fraction in lowest terms.

The following examples with fractions are designed to be done *without* the calculator to reinforce your skills in arithmetic. Use your calculator to check the results you get by hand.

EXAMPLE 1.6 Reduce to lowest terms:

$$\frac{12}{20}$$

Solution Divide the numerator and denominator by the largest common factor of 12 and 20 which is 4:

$$\frac{12}{20} = \frac{12/(4)}{20/(4)} = \frac{3}{5}$$

EXAMPLE 1.7 Simplify (reduce to lowest terms):

$$\frac{28}{42}$$

Solution Both of the numbers 28 and 42 have several factors. Before dividing, break down each number into its smallest factors:

$$\frac{28}{42} = \frac{(2)(2)(7)}{(2)(3)(7)}$$

Note that parentheses are used here instead of the multiplication sign ×. Parentheses or a dot (·) are also used for multiplication, especially in algebra, so as not to confuse the letter x with the multiplication sign ×. Now divide out the common factors 2 and 7:

$$\frac{28}{42} = \frac{\overset{1}{\cancel{(2)}}(2)\overset{1}{\cancel{(7)}}}{\underset{1}{\cancel{(2)}}(3)\underset{1}{\cancel{(7)}}} = \frac{2}{3}$$

Example 1.7 shows the smallest factors of 28 and 42. The smallest factors of a number are called prime numbers:

Definition **Prime Number**

A number greater than one that has no factors except one and the number itself.

The prime numbers are 2, 3, 5, 7, 11, 13, 17, 19, 23, 29, etc. There are simple divisibility rules to check if a number is divisible by the first three prime numbers 2, 3 and 5:

Rule **Divisibility Rules**

1. A number is divisible by two if the last digit is an even number: 0, 2, 4, 6 or 8.
2. A number is divisible by three if the sum of the digits is divisible by 3.
3. A number is divisible by five if the last digit is 0 or 5.

For example, the number 34 is divisible by two because the last digit, 4, is an even number. The number 48 is divisible by two and three because the last digit, 8, is an even number and the sum of the digits, 8 + 4 = 12 is divisible by three. The number 105 is divisible by three and five because the sum of the digits, 5 + 1 = 6, is divisible by three and the last digit is 5.

To multiply fractions, you multiply the numerators and the denominators:

Rule

Multiplying Fractions

$$\frac{A}{B} \times \frac{C}{D} = \frac{A \times C}{B \times D} \tag{1.4}$$

However, before multiplying you should divide out any common factors that are in the numerator and denominator of *either* fraction. This simplifies the multiplication. Study the following examples.

EXAMPLE 1.8

Multiply the fractions:

$$\frac{3}{10} \times \frac{4}{9}$$

Solution Factor the numbers into primes and divide out the factors 3 and 2 that both occur in a numerator and a denominator:

$$\frac{3}{10} \times \frac{4}{9} = \frac{\overset{1}{\cancel{3}}}{\underset{1}{(\cancel{2})}(5)} \times \frac{\overset{1}{(\cancel{2})}(2)}{\underset{1}{(\cancel{3})}(3)}$$

Then multiply numerators and denominators:

$$= \frac{1}{5} \times \frac{2}{3} = \frac{(1)(2)}{(5)(3)} = \frac{2}{15}$$

You can check Example 1.8 on a calculator that has a fraction key [a b/c]:

3 [a b/c] 10 [×] 4 [a b/c] 9 [=] → 2/15

or on a calculator that has a fraction conversion operation, [>Frac], such as a graphics calculator:

3 [÷] 10 [×] 4 [÷] 9 [>Frac] [=] → 2/15

EXAMPLE 1.9

Multiply the fractions:

$$3 \times \frac{3}{14} \times \frac{7}{15}$$

Solution Before multiplying, express the whole number 3 as a fraction with a denominator of 1. Then factor into primes and divide out the common factors that occur in *any* numerator and denominator:

$$\frac{3}{1} \times \frac{\overset{1}{\cancel{3}}}{(2)\underset{1}{(\cancel{7})}} \times \frac{\overset{1}{\cancel{7}}}{\underset{1}{(\cancel{3})}(5)} = \frac{3}{10}$$

ERROR BOX

A whole number is a fraction with a denominator of 1. Therefore when you multiply a whole number by a fraction, you can simply multiply the numerator by the whole number:

$$2 \times \frac{3}{7} = \frac{(2)(3)}{7} = \frac{6}{7}$$

You can also divide any common factors of the whole number with any in the denominator:

$$9 \times \frac{1}{15} = \overset{1}{\cancel{(3)}}(3) \times \frac{1}{\underset{1}{\cancel{(3)}}(5)} = \frac{(3)1}{5} = \frac{3}{5}$$

Practice Problems: See if you can correctly multiply the following:

1. $4 \times \frac{1}{6}$ **2.** $\frac{3}{20} \times 5$ **3.** $3 \times \frac{4}{5} \times \frac{1}{8}$ **4.** $6 \times \frac{5}{28} \times \frac{2}{3}$

5. $\frac{7}{50} \times 15 \times \frac{1}{6}$ **6.** $4 \times 3 \times \frac{1}{60}$ **7.** $10 \times 10 \times \frac{9}{300}$ **8.** $60 \times \frac{1}{36} \times 3$

9. $33 \times \frac{1}{2} \times \frac{6}{11}$ **10.** $400 \times \frac{3}{8} \times \frac{2}{25}$

To *divide fractions,* invert the divisor (the fraction after the division sign), and change the operation to multiplication:

Rule **Dividing Fractions**

$$\frac{A}{B} \div \frac{C}{D} = \frac{A}{B} \times \frac{D}{C} = \frac{A \times D}{B \times C} \quad (1.5)$$

When you invert a fraction you are changing it to the *reciprocal.* For example, the following numbers and their reciprocals are:

Number	Reciprocal
$\frac{3}{4}$	$\frac{4}{3}$
$\frac{1}{5}$	5
10	$\frac{1}{10}$
$\frac{11}{12}$	$\frac{12}{11}$

EXAMPLE 1.10 Divide the fractions by hand:

$$\frac{5}{12} \div \frac{15}{16}$$

Solution Invert or write the reciprocal of the fraction after the division sign and change the operation to multiplication:

$$\frac{5}{12} \div \frac{15}{16} = \frac{5}{12} \times \frac{16}{15}$$

Then divide out common factors and multiply:

$$= \frac{\overset{1}{\cancel{5}}}{\underset{1}{(\cancel{2})}\underset{1}{(\cancel{2})}(3)} \times \frac{\overset{1}{(\cancel{2})}\overset{1}{(\cancel{2})}(2)(2)}{(3)\underset{1}{(\cancel{5})}} = \frac{4}{9}$$

Study the next example, which combines multiplication and division of fractions.

EXAMPLE 1.11 Calculate by hand:

$$2 \div \frac{8}{15} \times \frac{1}{5}$$

Solution Do the operations moving from left to right. The division is done first. Change 2 to $\frac{2}{1}$, change the division to multiplication and change $\frac{8}{15}$ to its reciprocal:

$$2 \div \frac{8}{15} \times \frac{1}{5} = \frac{2}{1} \times \frac{15}{8} \times \frac{1}{5}$$

Then divide out common factors and multiply:

$$= \frac{\overset{1}{\cancel{2}}}{1} \times \frac{(3)\overset{1}{(\cancel{5})}}{\underset{1}{(\cancel{2})}(2)(2)} \times \frac{1}{\underset{1}{\cancel{5}}} = \frac{3}{(2)(2)} = \frac{3}{4}$$

You can check Example 1.11 on a calculator that has a fraction key [a b/c] :

2 [÷] [(] 8 [a b/c] 15 [)] [×] 1 [a b/c] 5 [=] → 3/4

or on a calculator that has a fraction conversion operation, [>Frac] [·]

2 [÷] [(] 8 [÷] 15 [)] [×] 1 [÷] 5 [>Frac] [=] → 3/4

EXAMPLE 1.12 Calculate by hand:

$$\frac{(5)(14)}{(7)(15)(6)}$$

Solution This example is the same as multiplying the three fractions $\frac{5}{7}$, $\frac{14}{15}$, and $\frac{1}{6}$. All the numbers in the numerator are multiplied together and all the numbers in the denominator are multiplied together. Therefore, they are all factors and you can divide common factors before multiplying:

$$\frac{(5)(14)}{(7)(15)(6)} = \frac{\overset{1}{(\cancel{5})}\overset{1}{(\cancel{2})}\overset{1}{(\cancel{7})}}{\underset{1}{(\cancel{7})}(3)\underset{1}{(\cancel{5})}\underset{1}{(\cancel{2})}(3)} = \frac{1}{9}$$

The calculator check using a fraction key and entering the numbers as three fractions multiplied together is:

5 [a b/c] 7 [×] 14 [a b/c] 15 [×] 1 [a b/c] 6 [=] → 1/9

or using the fraction conversion operation, [>Frac]:

5 [×] 14 [÷] 7 [÷] 15 [÷] 6 [=] → 1/9

EXERCISE 1.2

In exercises 1 through 12, reduce each fraction to lowest terms.

1. $\frac{6}{10}$

2. $\frac{12}{36}$

3. $\frac{28}{35}$

4. $\frac{27}{54}$

5. $\frac{39}{52}$

6. $\frac{34}{51}$

7. $\frac{50}{225}$

8. $\frac{150}{350}$

9. $\frac{1000}{5500}$

10. $\frac{3000}{6500}$

11. $\frac{27}{105}$

12. $\frac{55}{385}$

In exercises 13 through 44, calculate each problem by hand and express the answer in terms of a fraction or a whole number.

13. $\frac{5}{9} \times \frac{6}{25}$

14. $\frac{2}{21} \times \frac{7}{16}$

15. $\frac{10}{25} \times \frac{50}{75}$

16. $\frac{30}{100} \times \frac{550}{900}$

17. $\frac{8}{9} \div \frac{2}{3}$

18. $\frac{3}{11} \div \frac{1}{22}$

19. $1 \div \frac{1}{2}$

20. $2 \div \frac{2}{5}$

21. $\frac{4}{5} \div 2$

22. $\frac{5}{6} \div 10$

23. $\frac{35}{75} \div \frac{45}{60}$

24. $\frac{400}{660} \div \frac{220}{330}$

25. $\frac{3}{5} \times \frac{15}{7} \times \frac{14}{9}$

26. $\frac{3}{25} \times \frac{4}{9} \times \frac{5}{12}$

27. $4 \times \frac{3}{16} \times \frac{5}{12}$

28. $5 \times \frac{33}{50} \times \frac{25}{90}$

29. $3 \times \frac{2}{15} \div \frac{12}{25}$

30. $10 \times \frac{3}{150} \div \frac{63}{75}$

31. $\frac{4}{5} \div 6 \times \frac{5}{7}$

32. $\frac{5}{6} \div 2 \times \frac{3}{10}$

33. $\frac{(12)(15)}{(5)(3)(2)}$

34. $\frac{(8)(7)(6)}{(4)(28)}$

35. $\frac{(9)(2)(15)}{(6)(27)(5)}$

36. $\frac{(6)(4)(14)}{(7)(10)(24)}$

37. $\frac{3}{10} \div \left(12 \times \frac{1}{20}\right)$

38. $\frac{18}{25} \div \left(10 \times \frac{9}{50}\right)$

39. $\frac{8}{9} \div \frac{3}{4} \div \frac{4}{27}$

40. $\frac{1}{50} \div \frac{35}{100} \div \frac{4}{25}$

41. $5 \div \frac{25}{36} \times \frac{5}{6}$

42. $4 \div \frac{16}{21} \times \frac{2}{7}$

43. $\frac{1}{2} \div \frac{3}{4} \div \frac{2}{3}$

44. $\frac{1}{5} \div \frac{3}{10} \div \frac{1}{6}$

Applied Problems

In problems 45 through 54, solve each applied problem by hand and express the answer in terms of a fraction or a whole number.

45. A model car is to be $\frac{1}{50}$ the size of the original. If the original is 15 ft long, how long should the model be, in feet?

46. A boat model is $\frac{1}{100}$ the size of the actual boat. If the model is $\frac{3}{4}$ ft long, how long should the actual boat be, in feet?

47. Patricia decides to give $1000 of her savings to charity. She gives two-fifths to the heart fund, and divides the balance evenly between the cancer fund and an environmental fund. How much does she give to the environmental fund?

48. A $40,000 inheritance is distributed as follows: half to the spouse, three-fourths of what is left to the children, and the remainder to a private nurse. How much does the nurse inherit?

Applications to Electronics

49. In an electric circuit, the voltage V in volts (V) is given by:

$$V = \frac{3}{10} \times 90$$

Calculate the value of the voltage.

50. In an electric circuit, the voltage V in volts (V) is given by:

$$V = \frac{4}{5} \times 75$$

Calculate the value of the voltage.

51. In an electric circuit, the current I in amps (A) is given by:

$$I = \frac{3}{4} \div 12$$

Calculate the value of the current I.

52. In an electric circuit, the current I in amps (A) is given by:

$$I = \frac{3}{10} \div 15$$

Calculate the value of the current I.

53. In an electric circuit, the power P in watts (W) is given by:

$$P = \frac{3}{20} \times \frac{3}{20} \times 16$$

Calculate the value of the power P.

Answers to Error Box Problems, page 10:

1. $\frac{2}{3}$ **2.** $\frac{3}{4}$ **3.** $\frac{3}{10}$ **4.** $\frac{5}{7}$ **5.** $\frac{7}{20}$ **6.** $\frac{1}{5}$ **7.** 3 **8.** 5 **9.** 9 **10.** 12

54. In an electric circuit, the power P in watts (W) is given by:

$$P = \frac{1}{100} \times \frac{1}{100} \times 500$$

Calculate the value of the power P.

1.3 Adding Fractions

Adding fractions requires several steps. Calculators can add fractions as decimals, and some can add them as fractions. However, it is necessary for you to understand how to add fractions so that you can estimate results, recognize an incorrect answer, and be able to troubleshoot for the error. More important, in order to add fractions in algebra you need to understand how to add fractions in arithmetic.

Fractions can be added or subtracted *only if their denominators are the same.* When the denominators are the same, you add the fractions by adding the numerators over the common denominator. For example:

$$\frac{1}{8} + \frac{5}{8} = \frac{1+5}{8}$$
$$= \frac{6}{8} = \frac{3}{4}$$

Observe that you reduce the result if possible.

Rule **Adding Fractions**

$$\frac{A}{D} + \frac{C}{D} = \frac{A+C}{D} \quad (1.6)$$

To add fractions with different denominators, it is necessary to change each fraction to an equivalent fraction so that all fractions have the same common denominator. A fraction can be changed to an equivalent fraction by dividing out common factors or by the inverse process of *multiplying the numerator and denominator by the same factor.* For example, the following fractions are all equivalent:

$$\frac{30}{100} = \frac{9}{30} = \frac{6}{20} = \frac{3}{10}$$

The fraction $\frac{3}{10}$ expresses the equivalent fractions $\frac{6}{20}$, $\frac{9}{30}$, and $\frac{30}{100}$ in lowest terms. Each fraction can be obtained from $\frac{3}{10}$ by multiplying the numerator and denominator by the same factor. For example, by multiplying the numerator and denominator of $\frac{3}{10}$ by 3, the fraction is changed to $\frac{9}{30}$. Study the following examples, which show the step-by-step procedure of adding two fractions whose denominators are different.

EXAMPLE 1.13 Add the fractions:

$$\frac{2}{3} + \frac{1}{6}$$

Solution Since the denominators are not the same, you must change one or both fractions to equivalent fractions with a common denominator. You look for the Lowest Common Denominator (LCD), which is the smallest number that contains all the factors of each denominator. The denominator 6 = (2)(3) and contains the first denominator 3 as a factor. Therefore the LCD equals 6. Change the first fraction to an equivalent fraction with a denominator of 6 by multiplying the top and bottom by 2:

$$\frac{2(2)}{3(2)} + \frac{1}{6} = \frac{4}{6} + \frac{1}{6}$$

Then add the fractions over the LCD = 6:

$$= \frac{4+1}{6} = \frac{5}{6}$$

This example can be checked on a calculator with a fraction key or fraction conversion operation, [>Frac]:

2 [a b/c] 3 [+] 1 [a b/c] 6 [=] → 5/6

2 [÷] 3 [+] 1 [÷] 6 [>Frac] [=] → 5/6

EXAMPLE 1.14 Add the fractions:

$$\frac{1}{3} + \frac{2}{5}$$

Solution First find the LCD of 3 and 5. One way is to try multiples of the larger denominator until you find a multiple that the other denominator divides into. Three does not divide into (2)(5) = 10 but it does divide into (3)(5) = 15. The LCD is therefore 15. Multiply the top and bottom of the first fraction by 5 and the second fraction by 3 to change both denominators to 15:

$$\frac{1(5)}{3(5)} + \frac{2(3)}{5(3)} = \frac{5}{15} + \frac{6}{15}$$

Then add the fractions over the LCD:

$$\frac{5}{15} + \frac{6}{15} = \frac{5+6}{15} = \frac{11}{15}$$

This example can be checked on a calculator with a fraction key or fraction conversion operation:

1 [a b/c] 3 [+] 2 [a b/c] 5 [=] → 11/15

1 [÷] 3 [+] 2 [÷] 5 [>Frac] [=] → 11/15

A more direct way to find the LCD of two or more fractions is to completely factor each denominator into its prime factors, as shown in the next example.

EXAMPLE 1.15 Add the fractions:

$$\frac{1}{6} + \frac{5}{8}$$

Solution First find the LCD of 6 and 8. You can try multiples of the larger denominator until the other denominator divides into one of the multiples. Six does not divide into 8(2) = 16. Six does divide into 8(3) = 24, so 24 is the LCD. Another method, which is important to know because it is necessary in algebra, is to completely factor each denominator into its prime factors. The prime factors of each denominator are:

$$6 = (2)(3)$$
$$8 = (2)(2)(2)$$

Make up the LCD so that it contains the factors of every denominator by placing the same prime factors above each other as follows and then "bringing down" the necessary factors:

$$\begin{array}{rcccccl} 6 & = & & & (2) & (3) & \\ 8 & = & (2) & (2) & (2) & \downarrow & \\ & & \downarrow & \downarrow & \downarrow & \downarrow & \\ \hline \text{LCD} & = & (2) & (2) & (2) & (3) & = 24 \end{array}$$

If a prime factor appears more than once in a denominator, it must appear the same number of times in the LCD. Eight contains three 2's as prime factors, and therefore the LCD must contain three 2's. To determine what to multiply each fraction by to change to the LCD, see what factors are missing in the denominator to make up the LCD. Six is missing the factors (2)(2) and therefore needs to be multiplied by 4. Eight needs to be multiplied by (3):

$$\frac{1(4)}{6(4)} + \frac{5(3)}{8(3)} = \frac{4}{24} + \frac{15}{24} = \frac{19}{24}$$

You can also divide each denominator into the LCD and multiply by the result. Since 6 divides into 24 four times, the numerator and denominator of the first fraction are multiplied by 4. Since 8 divides into 24 three times, the numerator and denominator of the second fraction are multiplied by 3.

The calculator checks are:

1 [a b/c] 6 [+] 5 [a b/c] 8 [=] → 19/24

1 [÷] 6 [+] 5 [÷] 8 [>Frac] [=] → 19/24

EXAMPLE 1.16 Combine the fractions:

$$\frac{19}{20} + \frac{2}{15} - \frac{1}{12}$$

Solution Notice that the last fraction is subtracted. The procedure is the same as adding fractions, with the numerator of the last fraction being subtracted instead of added. Use the method of prime factors to find the LCD. Place the prime factors above each other and bring down the factors for the LCD:

$$\begin{array}{rcccccl} 20 & = & (2) & (2) & & (5) & \\ 15 & = & & & (3) & (5) & \\ 12 & = & (2) & (2) & (3) & \downarrow & \\ & & & \downarrow & \downarrow & \downarrow & \\ \hline \text{LCD} & = & (2) & (2) & (3) & (5) & = 60 \end{array}$$

Note that the factor of (2) must appear twice in the LCD because it appears twice in two of the denominators. The prime factors of each denominator will then be able to divide into the factors of the LCD. Now multiply the numerator and denominator of each fraction by the necessary factor to change each denominator into 60:

$$\frac{19}{20} + \frac{2}{15} - \frac{1}{12} = \frac{19(3)}{20(3)} + \frac{2(4)}{15(4)} - \frac{1(5)}{12(5)}$$

$$= \frac{57}{60} + \frac{8}{60} - \frac{5}{60}$$

Then combine the fractions:

$$= \frac{57 + 8 - 5}{60} = \frac{60}{60} = 1$$

The last numerator is subtracted because of the minus sign in front of the fraction. The answer is then reduced to lowest terms. The calculator check is the same for any calculator:

19 [÷] 20 [+] 2 [÷] 15 [−] 1 [÷] 12 [=] → 1

ERROR BOX

It is important to emphasize that when adding fractions *you can only divide out factors* in the numerator and denominator of a fraction. If any numbers are separated by a + or − sign, they are not factors. For example, in the following you cannot divide out the 5's in the next to the last step:

$$1 - \frac{3}{5} = \frac{5}{5} - \frac{3}{5} = \frac{5-3}{5} = \frac{2}{5}$$

This is because 5 is *not* a factor in the numerator even though it is a factor in the denominator. *Numbers are factors only when they are **all** separated by multiplication signs.*

Practice Problems: See if you can correctly simplify the following fractions:

1. $\frac{10-8}{8}$ **2.** $\frac{400}{400 \times 100}$ **3.** $\frac{6-4}{6+4}$ **4.** $\frac{50+20}{5+2}$

5. $\frac{6 \times 1 \times 3}{2}$ **6.** $\frac{20 - 5 \times 3}{5}$ **7.** $\frac{10 \times 5 + 2}{10 + 16}$ **8.** $\frac{4+5-3}{4 \times 5 \times 3}$

9. $\frac{3 \times 2 + 6}{20 - 5}$ **10.** $\frac{200 - 25 \times 5}{5 \times 25}$ **11.** $\frac{5 + 10 + 5}{5}$ **12.** $\frac{60 - 30 + 20}{10}$

EXAMPLE 1.17 Calculate by hand:

$$\frac{4}{5} \times \frac{1}{2} + \frac{3}{4}$$

Solution Apply the order of operations and do the multiplication before the addition:

$$\frac{4}{5} \times \frac{1}{2} + \frac{3}{4} = \frac{\cancel{(2)}^{1}(2)}{5} \times \frac{\cancel{1}}{\cancel{2}_{1}} + \frac{3}{4} = \frac{2}{5} + \frac{3}{4}$$

Find the LCD of 5 and 4:

$$\begin{array}{rcccc} 5 & = & & & (5) \\ 4 & = & (2) & (2) & \downarrow \\ & & \downarrow & \downarrow & \downarrow \\ \hline \text{LCD} & = & (2) & (2) & (5) = 20 \end{array}$$

Multiply the fractions by the necessary factor and add the numerators over the LCD:

$$\frac{2}{5} + \frac{3}{4} = \frac{2(4)}{5(4)} + \frac{3(5)}{4(5)} = \frac{8+15}{20} = \frac{23}{20}$$

The answer is a fraction that is greater than one. It is equal to:

$$\frac{23}{20} = \frac{20+3}{20} = \frac{20}{20} + \frac{3}{20} = 1 + \frac{3}{20}$$

This is called an *improper fraction.*

Definition

Improper Fraction

A fraction greater than one whose numerator is larger than its denominator.

EXAMPLE 1.18

Calculate by hand:

$$\frac{3}{25} - \frac{1}{10} \div 5$$

Solution Do the division first. Change 5 to its reciprocal and multiply:

$$\frac{3}{25} - \frac{1}{10} \div 5 = \frac{3}{25} - \left(\frac{1}{10} \times \frac{1}{5}\right) = \frac{3}{25} - \frac{1}{50}$$

Find the LCD:

$$\begin{array}{rcccc} 25 & = & & (5) & (5) \\ 50 & = & (2) & (5) & (5) \\ & & \downarrow & \downarrow & \downarrow \\ \hline \text{LCD} & = & (2) & (5) & (5) \end{array} = 50$$

Change the denominators to the LCD:

$$\frac{3}{25} - \frac{1}{50} = \frac{3(2)}{25(2)} - \frac{1}{50} = \frac{6}{50} - \frac{1}{50}$$

Subtract the fractions and reduce the answer to lowest terms:

$$= \frac{6-1}{50} = \frac{5}{50} = \frac{1}{10}$$

Two calculator checks are:

3 [a b/c] 25 [−] 1 [a b/c] 10 [÷] 5 [=] → 1/10

3 [÷] 25 [−] 1 [÷] 10 [÷] 5 [>Frac] [=] → 1/10

EXAMPLE 1.19

Calculate by hand:

$$\frac{7}{10} - \frac{2}{5} + \frac{8}{15} \times \frac{1}{2}$$

Solution Do the multiplication first:

$$\frac{7}{10} - \frac{2}{5} + \frac{8}{15} \times \frac{1}{2} = \frac{7}{10} - \frac{2}{5} + \left(\frac{\overset{4}{\cancel{(2)}}(2)(2)}{15} \times \frac{1}{\underset{1}{\cancel{2}}}\right)$$

$$= \frac{7}{10} - \frac{2}{5} + \frac{4}{15}$$

Find the LCD:

$$\begin{array}{rcccc} 10 & = & (2) & & (5) \\ 5 & = & | & & (5) \\ 15 & = & | & (3) & (5) \\ & & \downarrow & \downarrow & \downarrow \\ \hline \text{LCD} & = & (2) & (3) & (5) \end{array} = 30$$

Multiply each fraction by the necessary factor to change each denominator to 30:

$$\frac{7(3)}{10(3)} - \frac{2(6)}{5(6)} + \frac{4(2)}{15(2)} = \frac{21}{30} - \frac{12}{30} + \frac{8}{30}$$

Combine the numerators over the LCD:

$$\frac{21 - 12 + 8}{30} = \frac{17}{30}$$

The next example serves to "close the circuit" between theory and practice. It shows how to apply the preceding mathematical ideas to a problem in electronics. "Close the circuit" examples are an important theme throughout the text. They are marked with the circuit icon shown and help you to see how mathematics is used in electronics.

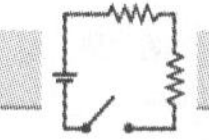

EXAMPLE 1.20

Close the Circuit

The resistance R of a circuit is given by:

$$\frac{1}{R} = \frac{1}{15} + \frac{1}{10} + \frac{1}{75}$$

The units for resistance are ohms and the symbol is a capital Greek omega: Ω. Find the value of the resistance in fraction form by adding the three fractions and inverting the result.

Solution Find the LCD by placing the prime factors of each denominator above each other:

$$\begin{array}{lcccccc}
15 & = & & (3) & (5) & & \\
10 & = & (2) & & (5) & & \\
75 & = & \downarrow & (3) & (5) & (5) & \\
 & & \downarrow & \downarrow & \downarrow & \downarrow & \\
\hline
\text{LCD} & = & (2) & (3) & (5) & (5) & = 150
\end{array}$$

The LCD must contain all the factors that are in each denominator. Multiply each fraction by the necessary factor to change each denominator to the LCD = 150:

$$\frac{1}{R} = \frac{1(10)}{15(10)} + \frac{1(15)}{10(15)} + \frac{1(2)}{75(2)} = \frac{10}{150} + \frac{15}{150} + \frac{2}{150}$$

Add the numerators over the LCD and reduce the result:

$$\frac{10 + 15 + 2}{150} = \frac{27}{150} = \frac{9}{50}$$

The result is equal to the reciprocal of R. To find the value of R invert the result:

$$R = \frac{50}{9}\ \Omega\ (\approx 5.6\ \Omega)$$

Two calculator checks, using the reciprocal key [1/x] or [x⁻¹] are:

1 [a b/c] 15 [+] 1 [a b/c] 10 [+] 1 [a b/c] 75 [=] → 9/50 [1/x] → 50/9 Ω

1 [÷] 15 [+] 1 [÷] 10 [+] 1 [÷] 75 [>Frac] [=] → 9/50 [x⁻¹] [>Frac] [=] → 50/9 Ω

Answers to Error Box Problems, page 16:

1. $\frac{1}{4}$ **2.** $\frac{1}{100}$ **3.** $\frac{1}{5}$ **4.** 10 **5.** 9 **6.** 1 **7.** 2 **8.** $\frac{1}{10}$ **9.** $\frac{4}{5}$ **10.** $\frac{3}{5}$ **11.** 4 **12.** 5

EXERCISE 1.3

In exercises 1 through 52, calculate the problems by hand, and express the answer in terms of a fraction or a whole number.

1. $\frac{6}{7}+\frac{1}{7}$
2. $\frac{3}{10}+\frac{5}{10}$
3. $\frac{3}{8}+\frac{1}{4}$
4. $\frac{1}{6}+\frac{2}{3}$
5. $\frac{2}{5}+\frac{1}{4}$
6. $\frac{2}{7}+\frac{1}{2}$
7. $\frac{5}{12}+\frac{3}{8}$
8. $\frac{1}{6}+\frac{3}{4}$
9. $\frac{4}{15}+\frac{2}{3}$
10. $\frac{1}{4}+\frac{3}{20}$
11. $\frac{1}{10}+\frac{5}{12}$
12. $\frac{4}{9}+\frac{5}{12}$
13. $\frac{21}{100}+\frac{3}{40}$
14. $\frac{9}{50}+\frac{11}{30}$
15. $\frac{9}{20}-\frac{1}{5}$
16. $\frac{3}{25}-\frac{1}{50}$
17. $\frac{5}{14}-\frac{4}{21}$
18. $\frac{7}{15}-\frac{9}{20}$
19. $\frac{3}{4}+\frac{3}{10}-\frac{1}{2}$
20. $\frac{1}{6}+\frac{11}{20}-\frac{2}{3}$
21. $2+\frac{7}{8}+\frac{2}{3}$
22. $1+\frac{2}{5}+\frac{4}{7}$
23. $\frac{5}{2}+\frac{5}{3}+\frac{5}{6}$
24. $\frac{27}{100}+\frac{12}{50}+\frac{6}{25}$
25. $\frac{5}{20}+\frac{7}{80}+\frac{3}{40}$
26. $\frac{6}{1000}+\frac{35}{100}+\frac{7}{500}$
27. $\frac{11}{40}+\frac{3}{50}+\frac{6}{20}$
28. $\frac{3}{20}+\frac{4}{30}+\frac{5}{40}$
29. $\frac{1}{5}+\frac{1}{10}+\frac{1}{20}$
30. $\frac{1}{50}+\frac{1}{100}+\frac{1}{25}$
31. $1-\frac{27}{50}+\frac{17}{100}$
32. $4-\frac{6}{15}+\frac{3}{20}$
33. $\frac{8}{15}\times\frac{1}{2}+\frac{1}{4}$
34. $\frac{5}{16}\times\frac{2}{5}+\frac{1}{6}$
35. $\frac{1}{6}+\frac{3}{8}\div\frac{3}{4}$
36. $\frac{1}{2}+\frac{2}{9}\div\frac{2}{3}$
37. $\frac{9}{20}+\frac{4}{15}\times\frac{1}{2}$
38. $\frac{7}{50}+\frac{6}{25}\times\frac{5}{12}$
39. $\frac{3}{40}-\frac{5}{16}\div 5$
40. $\frac{16}{25}-\frac{3}{10}\div 6$
41. $\left(\frac{3}{8}+\frac{1}{6}\right)\times 2$
42. $\left(\frac{1}{5}+\frac{1}{4}\right)\div 3$
43. $\frac{5}{6}+\frac{2}{5}-\frac{3}{7}\times\frac{7}{10}$
44. $\frac{3}{10}+\frac{3}{20}-\frac{3}{4}\times\frac{1}{10}$
45. $\frac{35}{100}+\frac{3}{25}-\frac{1}{2}\times\frac{6}{50}$
46. $\frac{7}{200}+\frac{3}{50}-\frac{1}{3}\times\frac{9}{40}$
47. $\frac{3}{4}-\frac{3}{5}+\frac{4}{5}\div 2$
48. $\frac{6}{10}-\frac{1}{20}+\frac{9}{15}\div 3$
49. $\left(\frac{1}{4}+\frac{1}{2}\right)\times\left(\frac{1}{3}+\frac{1}{6}\right)$
50. $\left(\frac{1}{2}+\frac{1}{3}\right)\times\left(\frac{4}{5}-\frac{1}{2}\right)$
51. $\left(\frac{2}{5}+\frac{1}{3}\right)\div\left(\frac{2}{3}+\frac{2}{9}\right)$
52. $\left(\frac{1}{5}+\frac{3}{10}\right)\div\left(\frac{3}{4}+\frac{1}{2}\right)$

Applied Problems

In problems 53 through 56, calculate each problem by hand and express the answer in terms of a fraction or a whole number.

53. A grocery clerk is given three orders for cheese: $\frac{1}{3}$ lb, $\frac{1}{2}$ lb and $\frac{1}{6}$ lb. What is the total amount of cheese, in pounds, that the clerk has to prepare?
54. A person's weight varies as follows during the course of a day: up $\frac{3}{4}$ lb, down $\frac{3}{8}$ lb, then up $\frac{1}{4}$ lb. What is the total change in the person's weight?
55. A ski area receives $\frac{3}{4}$ inch of snow on January 9, $\frac{1}{2}$ inch on January 10, and 2 inches on January 11. What is the total accumulation of snow, in inches, for the three days?
56. A bookcase is to be 5 ft 3 in high and contain five equally spaced shelves and a top (6 pieces), each $\frac{1}{2}$ in thick. The bottom shelf rests on the floor. How far apart, in feet, should each shelf be?

Applications to Electronics

In problems 57 through 62, calculate by hand the value of the resistance R in ohms (Ω) of a circuit given by the arithmetic expression. Express the answer in terms of a fraction or a whole number. For problems 57 through 60, see Example 1.20.

57. $\frac{1}{R}=\frac{1}{20}+\frac{1}{60}+\frac{1}{30}$
58. $\frac{1}{R}=\frac{1}{50}+\frac{1}{100}+\frac{1}{100}$
59. $\frac{1}{R}=\frac{1}{5}-\frac{1}{10}-\frac{1}{20}$
60. $\frac{1}{R}=\frac{1}{50}-\frac{1}{100}-\frac{1}{150}$
61. $R=5+\frac{2\times 10}{2+10}$
62. $R=10+\frac{10\times 5}{10+5}$

63. In computer languages, as in arithmetic, operations in parentheses are performed first. Calculate the following expression in computer language (see Exercise 1.1, problem 49):

1/10 * 3 / (3/2 − 1) * 20/3

64. As in problem 63, calculate the following expression in computer language:

(4/5 − 1/10) / (7/10 + 1/5) * 45/100

1.4 Hand Calculator Operations

We live in an age of rapid electronic computation. Imagine a long division problem that might take you one minute to do by hand. The hand calculator takes a few milliseconds (thousandths of a second) to do such a problem. One of the world's fastest computers can do almost half a billion (500,000,000) of these problems in one second! You need to master the calculator. It will allow you to spend less time doing calculations and more time learning important concepts. However, speed alone is of little value unless you can understand and apply concepts. For electricity and electronics, you need a scientific or engineering calculator. Scientific calculators have an exponential key [y^x], [x^y], or [∧]; trigonometric keys [sin], [cos], and [tan]; and logarithmic keys [log] and [ln]. There are two types: *Direct Algebraic Logic* (DAL) and not DAL. DAL means you enter the numbers and operations exactly as you see them. Not DAL calculators sometimes require you to enter the operation after you enter the number. The steps for both types of calculators are shown throughout the text if there is a difference. Some scientific calculators have keys for metric notation: [M] for mega, [k] for kilo, [m] for milli, and [μ] (mu) for micro. These are very useful for electronics problems. Metric units are explained in Chapter 4. Graphing calculators are helpful for graphs of voltage and current waves in ac circuits, exponential graphs, and other graphs of electronic phenomena. Appendix A contains more information on scientific and graphing calculators.

This section is not designed to replace the instruction manual for your calculator. It is designed to be used with your manual. Some of your keys and operations may be a little different from the examples shown. Study your instruction manual thoroughly to understand your particular calculator. Some calculators have an enter [ENT] key instead of an equals [=] key, such as the graphing calculators produced by Casio, Sharp, and Texas Instruments. In this text, all calculator operations are shown with an equals [=] key. The keystrokes shown will work on most scientific calculators.

One of the critical things to keep in mind is that *your calculator cannot think*! Only you possess this unique ability. You must understand the problem, interpret the information, and key the information into the calculator correctly. Furthermore, and most important, you must be able to judge if the answer makes sense. If it does not, and this can happen often, you must understand the mathematical concepts well enough to troubleshoot for the error.

Arithmetic Operations and Memory

Scientific calculators are programmed to perform the basic arithmetic operations [+], [−], [×], and [÷], according to the order of operations moving from left to right: parentheses, multiplication or division, addition or subtraction. This is called the *algebraic operating system.* See Example 1.1 to test the order of operations on your calculator. If your calculator has a [MODE] key, make sure it is set for normal calculation. The following examples show how to use the basic operations on the calculator.

EXAMPLE 1.21 Calculate:

$$7 \times (15 - 12) + 4 \div 2 \times 3$$

Solution Enter the functions and numbers as they appear in the example. The calculator will perform the operations in the correct order moving from left to right:

7 [×] [(] 15 [−] 12 [)] [+] 4 [÷] 2 [×] 3 [=] → 27

EXAMPLE 1.22 Calculate:

$$\frac{10 + 5 \times 3}{7 - 2}$$

Solution The fraction line is like parentheses, and the operations in the numerator and denominator must be done before the division. Use parentheses to separate the numerator and denominator:

[(] 10 [+] 5 [×] 3 [)] [÷] [(] 7 [−] 2 [)] [=] → 5

EXAMPLE 1.23 Calculate:

$$\frac{5 \times 21 \times 12}{15 \times 7 \times 3}$$

Solution One way to calculate this problem is to first multiply the numbers in the denominator and store the result in the memory. Then multiply the numbers in the numerator and divide by the number in the memory:

15 [×] 7 [×] 3 [=] [STO] 5 [×] 21 [×] 12 [÷] [RCL] [=] → 4

(↓ 315 under [STO]; ↑ 315 under [RCL])

Observe that you have to press the [=] key after you multiply the numbers in the denominator so the calculator will display the product 315 before you store it in the memory. The [STO] key stores the result 315. The [RCL] key recalls the result and enters it into the operations being performed. Some calculators use [M_{in}] to store a number, and [RM] or [MR] to recall a number from memory. Some calculators have more than one memory and it is necessary to key in the number of the memory after [STO] and [RCL] such as [STO] 07 and [RCL] 07.

Another way to calculate Example 1.23 is to use parentheses and group the numerator and denominator since the fraction line is like parentheses:

[(] 5 [×] 21 [×] 12 [)] [÷] [(] 15 [×] 7 [×] 3 [)] [=] → 4

There are other ways to do this example on the calculator, such as alternating between multiplication and division. See if you can do it another way.

EXAMPLE 1.24 Calculate:

$$\frac{3}{4} - \frac{1}{4} \div 2 + \frac{3}{8}$$

Solution The division sign after the fraction $\frac{1}{4}$ means the entire fraction is divided by 2. Therefore, key in parentheses around the fraction to calculate the fraction before it is divided by 2:

3 [÷] 4 [−] [(] 1 [÷] 4 [)] [÷] 2 [+] 3 [÷] 8 [=] → 1

You can also write the reciprocal of 2, which is $\frac{1}{2}$, and change to multiplication without using parentheses:

3 [÷] 4 [−] 1 [÷] 4 [×] 1 [÷] 2 [+] 3 [÷] 8 [=] → 1

Another way to do this example is by using a fraction key [$a^{b/c}$] as follows:

3 [$a^{b/c}$] 4 [−] 1 [$a^{b/c}$] 4 [÷] 2 [+] 3 [$a^{b/c}$] 8 [=] → 1

EXERCISE 1.4

In exercises 1 through 24, calculate each exercise using your calculator.

1. $(82 + 68 - 86) \div 16$
2. $39 \times 27 \div 9 + 80$
3. $48 + 64 \div 32 - 3$
4. $20 \times (1580 - 250) \div 14$
5. $\dfrac{80 - 15}{13} \times (38 - 12)$
6. $\dfrac{(14 + 10) \times 4}{3 + 5}$
7. $(1000 - 600) + \dfrac{2000 - 200}{150}$
8. $(424 + 176) \div (96 + 54)$
9. $\dfrac{176 - 34 + 66}{72 + 52 - 20}$
10. $\dfrac{142 + 58 - 20}{35 + 66 - 41}$
11. $\dfrac{252 \times (86 + 61)}{(36 - 29) \times 36}$
12. $\dfrac{(55 + 85) \times 120}{(72 - 37) \times 15}$
13. $\dfrac{155 \times 90 \times 76}{62 \times 15 \times 19}$
14. $\dfrac{41 \times 108 \times 90}{36 \times 123 \times 30}$
15. $\dfrac{500 \times 2000 \times 25}{5 \times 125 \times 1000}$
16. $\dfrac{256 \times 64 \times 50}{16 \times 8 \times 32}$
17. $\dfrac{2}{3} + \dfrac{1}{2} - \dfrac{1}{6}$
18. $\dfrac{1}{3} + \dfrac{1}{4} + \dfrac{1}{2} + \dfrac{11}{12}$
19. $\dfrac{3}{8} \times 2 - \dfrac{1}{2} + \dfrac{3}{4}$
20. $2 \times \dfrac{5}{8} \div 5 + \dfrac{3}{4}$
21. $3 + \dfrac{1}{3} \div 2 + \dfrac{5}{6}$
22. $30 \times \dfrac{3}{10} \div 3 - 1$
23. $12 \div \dfrac{3}{4} - \dfrac{2}{15} \times 30$
24. $\dfrac{5}{6} \times \dfrac{9}{10} + \dfrac{3}{8} \times \dfrac{2}{3}$

Applied Problems

25. Your calculator cannot think, but it can display words. Calculate the following correctly, and find out what helps keep the world moving by reading the display upside down:

$$\frac{300 \times 8 - 270}{55 \times 2 - 107}$$

26. Here's a good calculator trick:
    (a) Tell somebody to key in six 9s and divide by 7.
    (b) Multiply the result by any number from 1 to 6.
    (c) Then have the person tell you the first digit on the display, and *you will tell all the other digits.*

    Here's how it works: The number you get when you divide by 7 is a "cyclic" number of six digits. If you multiply by any number from 1 to 6, you get the same order of six digits starting with a different digit each time. Memorize the cyclic number, and you can do the trick.

Applications to Electronics

27. Calculate the resistance R in ohms (Ω) of a circuit given by:

$$R = \frac{(1200)(1500)}{1200 + 1500}$$

28. Calculate the resistance R in ohms (Ω) of a circuit given by:

$$R = \frac{(150)(240)}{150 + (2)(240)}$$

CHAPTER HIGHLIGHTS

1.1 ARITHMETIC OPERATIONS

Commutative Laws

$$a + b = b + a$$
$$a \times b = b \times a \qquad (1.1)$$

• • •

Associative Laws

$$(a + b) + c = a + (b + c)$$
$$(a \times b) \times c = a \times (b \times c) \qquad (1.2)$$

• • •

Distributive Law

$$a \times (b + c) = (a \times b) + (a \times c) \qquad (1.3)$$

• • •

Order of Operations (moving left to right)

1. Operations in parentheses.
2. Multiplication or division.
3. Addition or subtraction.

One way to remember this is to memorize the phrase "***P**lease **E**xcuse **M**y **D**ear **A**unt **S**ally*" (PEMDAS). The first letters of each word stand for: ***P**arentheses, **E**xponents, **M**ultiplication, **D**ivision, **A**ddition, **S**ubtraction.*

Key Example:

$$15 + 10 \times 6 \div (5 - 3) = 15 + 10 \times 6 \div (2)$$
$$= 15 + 10 \times (3) = 15 + (30) = 45$$

1.2 MULTIPLYING AND DIVIDING FRACTIONS

Equivalent Fractions

Two fractions are equivalent if they can each be reduced to the same fraction in lowest terms.

Key Example:

$$\frac{80}{100} = \frac{40}{50} = \frac{20}{25} = \frac{4}{5}$$

Prime Number

A number greater than one that has no factors except one and the number itself.

Key Example:

$$\frac{30}{42} = \frac{\overset{1}{\cancel{(2)}}\overset{1}{\cancel{(3)}}(5)}{\underset{1}{\cancel{(2)}}\underset{1}{\cancel{(3)}}(7)} = \frac{5}{7}$$

Divisibility Rules

1. A number is divisible by two if the last digit is an even number: 0, 2, 4, 6 or 8.
2. A number is divisible by three if the sum of the digits is divisible by 3.
3. A number is divisible by five if the last digit is 0 or 5.

• • •

Multiplying Fractions

$$\frac{A}{B} \times \frac{C}{D} = \frac{A \times C}{B \times D} \tag{1.4}$$

Key Example:

$$\frac{9}{20} \times \frac{5}{12} = \frac{\overset{1}{\cancel{(3)}}(3)}{(2)(2)\underset{1}{\cancel{(5)}}} \times \frac{\overset{1}{\cancel{5}}}{(2)(2)\underset{1}{\cancel{(3)}}} = \frac{3}{4} \times \frac{1}{4} = \frac{3}{16}$$

Dividing Fractions

$$\frac{A}{B} \div \frac{C}{D} = \frac{A}{B} \times \frac{D}{C} = \frac{A \times D}{B \times C} \tag{1.5}$$

Key Example:

$$\frac{3}{10} \div \frac{2}{5} = \frac{3}{10} \times \frac{5}{2} = \frac{3}{(2)\underset{1}{\cancel{(5)}}} \times \frac{\overset{1}{\cancel{5}}}{2} = \frac{3}{2} \times \frac{1}{2} = \frac{3}{4}$$

1.3 ADDING FRACTIONS

Adding Fractions

$$\frac{A}{D} + \frac{C}{D} = \frac{A + C}{D} \tag{1.6}$$

To *add fractions,* change each fraction to an equivalent fraction having the same Lowest Common Denominator (LCD) by multiplying the numerator and denominator of each fraction by the necessary factor. One way to find the LCD is to factor each denominator into the smallest factors or primes and make up the LCD so that it contains all the factors in each denominator.

Key Example: To add $\frac{1}{10} + \frac{5}{6}$, find the LCD by factoring each denominator into primes:

$$\begin{array}{rcccccl} 10 & = & (2) & & (5) & & \\ 6 & = & (2) & (3) & \downarrow & & \\ & & \downarrow & \downarrow & \downarrow & & \\ \hline \text{LCD} & = & (2) & (3) & (5) & = & 30 \end{array}$$

Add over the LCD:

$$\frac{1}{10} + \frac{5}{6} = \frac{1(3)}{10(3)} + \frac{5(5)}{6(5)} = \frac{3 + 25}{30} = \frac{28}{30} = \frac{14}{15}$$

Improper Fraction

A fraction greater than one whose numerator is larger than its denominator.

Key Example:

$$\frac{4}{5} \times \frac{1}{2} + \frac{3}{4} = \frac{2}{5} + \frac{3}{4} = \frac{2(4)}{5(4)} + \frac{3(5)}{4(5)}$$

$$= \frac{8 + 15}{20} = \frac{23}{20} \text{ or } 1 + \frac{3}{20}$$

1.4 HAND CALCULATOR OPERATIONS

Scientific calculators perform operations according to the order of arithmetic operations: parentheses, multiplication or division, addition or subtraction. There are two types: DAL (direct algebraic logic) and not DAL. The keys [Min] or [STO] store a number in the memory. The keys [MR], [RCL], or [RM] recall a number from memory.

Key Example: $\dfrac{5 \times 21 \times 12}{15 \times 7 \times 3}$

15 [×] 7 [×] 3 [=] [STO] 5 [×] 21 [×] 12 [÷] [RCL] [=] → 4

([=] [STO] ↓ 315; [RCL] ↑ 315)

Key Example: $\frac{3}{4} - \frac{1}{4} \div 2 + \frac{3}{8}$

3 [÷] 4 [−] [(] 1 [÷] 4 [)] [÷] 2 [+] 3 [÷] 8 [=] → 1

This can also be done on a calculator with a fraction key:

3 [ab/c] 4 [−] 1 [ab/c] 4 [÷] 2 [+] 3 [ab/c] 8 [=] → 1

REVIEW EXERCISES

In exercises 1 through 30, calculate each exercise by hand. Use the calculator to check your results.

1. $8 \times 3 - 4 \div 2$
2. $20 + 8 \times 12 \div 3 - 5$
3. $(15 + 25) \div 8 + 12 - 5$
4. $1000 - 200 \times 300 \div (75 + 25)$
5. $\frac{(20 + 25) \times 3}{3 + 2}$
6. $\frac{200}{25 - 5} + 5 \times (30 - 5)$
7. $\frac{5 \times 15}{5 + 15}$
8. $\frac{5 + 4 \times (9 - 7)}{8 \times 3 + 2}$
9. $\frac{4}{5} \times \frac{15}{7} \times \frac{7}{8}$
10. $11 \times \frac{3}{22} \times \frac{4}{15}$
11. $\frac{5}{12} \times \frac{6}{7} \div \frac{1}{21}$
12. $\frac{6}{15} \div 2 \times \frac{5}{9}$
13. $\frac{1}{4} \div \frac{1}{8} \div \frac{1}{2}$
14. $\frac{3}{5} \div 6 \times 100$
15. $\frac{5 \times 6 \times 8}{4 \times 24}$
16. $\frac{18}{25} \div \left(10 \times \frac{9}{100}\right)$
17. $\frac{7}{16} + \frac{1}{4}$
18. $\frac{2}{25} - \frac{3}{10}$
19. $\frac{1}{15} + \frac{1}{30} + \frac{1}{10}$
20. $3 - \frac{6}{15} + \frac{7}{30}$
21. $\frac{5}{6} + \frac{19 + 2}{2 \times 9}$
22. $\frac{100 \times 20}{1000 - 250} - \frac{2}{3}$
23. $\frac{3}{16} \times \frac{2}{3} + \frac{1}{6}$
24. $\frac{1}{8} + \frac{2}{15} \div \frac{2}{3}$
25. $\left(\frac{1}{12} + \frac{1}{3}\right) \div 3$
26. $50 \times \left(\frac{3}{5} - \frac{1}{10}\right)$
27. $\frac{33}{100} + \frac{3}{50} - \frac{1}{4} \times \frac{12}{25}$
28. $\left(\frac{16}{77} + \frac{4}{11}\right) \div 8 \times \frac{1}{2}$
29. $\left(\frac{1}{3} + \frac{1}{5}\right) \div \left(\frac{4}{5} - \frac{1}{3}\right)$
30. $\left(1 - \frac{3}{4}\right) \times \left(1 + \frac{3}{4}\right)$

In exercises 31 through 36, do each on the calculator.

31. $10 \times (1550 - 370) \div 59$
32. $\frac{56 - 22}{17} \times (103 - 74)$
33. $\frac{252 \times (86 + 61)}{(36 - 29) \times 36}$
34. $\frac{(37)(115)(60)}{(25)(69)(74)}$
35. $\frac{5}{6} + \frac{11}{12} + \frac{1}{4}$
36. $12 \times \frac{3}{4} \div (5 - 2)$

Applied Problems

In problems 37 through 56, calculate each by hand and express the answer in terms of a fraction or a whole number.

37. Joe drives 100 mi in 3 hours but takes 5 hours to make the return trip of 100 mi. What is his average rate of speed for the entire trip in mi/h? *Note:* Average rate = (total distance)/(total time)
38. An engineer earns $40,000 a year. If she works 250 days in the year, how much does she earn per day?
39. The distance shown on a map is $\frac{1}{500}$ of the actual distance. If the distance shown on the map is $\frac{1}{2}$ foot, how long is the actual distance in feet?
40. Carlos decides to live on $\frac{4}{5}$ of his weekly salary, give $\frac{1}{20}$ of it to charity and save the rest. If he earns $500 a week, how much does he save each week?
41. The list price of a computer is $2000. During a sale the list price is reduced by $\frac{1}{2}$. When the sale is over, the sales price is increased by $\frac{1}{4}$. What is the selling price after the sale is over?
42. The enclosed cut out space for an office mail box is 7 ft long by 1 ft high. It is to contain 15 equally spaced vertical partitions for 16 mail slots. If the partitions are $\frac{4}{5}$ in thick, how wide will each slot be as a fraction of a foot?

Applications to Electronics

43. Calculate the current I in amps (A) in a circuit given by:

$$I = \frac{32}{30 + 10 + 24}$$

44. Calculate the current I in amps (A) in a circuit given by:

$$I = \frac{3}{2} \div 24$$

45. Calculate the voltage V in volts (V) of a circuit given by:

$$V = 3 \times \left(\frac{10 \times 5}{10 + 5}\right)$$

46. Calculate the voltage V in volts (V) of a circuit given by:

$$V = \frac{3}{10} \times 16$$

47. Calculate the power P in watts (W) of a circuit given by:

$$P = \frac{12}{100} \times \frac{12}{100} \times 250$$

48. Calculate the power P in watts (W) of a circuit given by:

$$P = 2 \times 2 \times \left(\frac{18 \times 9}{18 + 9}\right)$$

49. The original voltage in a circuit is $V = 36$ V (volts). If the voltage decreases to $\frac{3}{4}$ of its original value, find the new value of the voltage in volts (V).

50. The original current in a circuit is $\frac{2}{3}$ A. If the current decreases to $\frac{3}{5}$ of its original value, find the new value of the current in amps (A).

In problems 51 through 54, find the resistance R *in ohms* (Ω) *of a circuit given by the formula. See Example 1.20.*

51. $\frac{1}{R} = \frac{1}{100} + \frac{1}{50} + \frac{1}{50}$

52. $\frac{1}{R} = \frac{1}{10} - \frac{1}{20} - \frac{1}{30}$

53. $R = \frac{30 \times 12}{30 - 12}$

54. $R = 25 + \frac{30 \times 20}{30 + 20}$

55. In computer languages, as in arithmetic, operations in parentheses are performed first. Calculate the following expression in computer language:

1/5 * 5 / (8/5 – 1) * 15/9

56. As in problem 55, calculate the following expression in computer language:

35/100 * (3/5 – 1/10) / (7/10 + 1/5)

CHAPTER 2

Decimals and Percentages

CHAPTER OBJECTIVES

In this chapter, you will learn:

- The meaning of the decimal number system.
- How to add, subtract, multiply, and divide decimal fractions.
- How to convert between fractions, decimal fractions, and percentages.
- How to find the percent change in a value.
- The definition of significant digits.
- The rules for rounding off numbers.
- What is meant by precision and accuracy of numbers.
- How to check calculator results and troubleshoot for errors.

Our number system is called the *decimal number system* because it is based on ten. This comes from the Latin meaning of *dec* which means ten. Decimal numbers and decimal fractions are used in almost all calculations, especially in electronics, and it is necessary for you to gain a thorough understanding of our decimal number system. This chapter explains the decimal number system and how to perform arithmetic operations on decimal fractions. Percentages represent another way to express decimal fractions and are used in certain electronic calculations such as the tolerance of a resistor. Precision and accuracy of decimal numbers are important when working with and interpreting experimental and scientific data. These concepts are explained in connection with rounding off numbers and significant digits. The last section shows you how to check results and locate errors when using the calculator.

• • •

2.1 Decimals

Our decimal number system is based on the number ten. The number 5643 written in expanded form is:

$$5643 = 5000 + 600 + 40 + 3 = 5(1000) + 6(100) + 4(10) + 3(1)$$

From *right to left* the place values of the digits are 1, 10, 100, 1000, and so forth. The place value of any digit is ten times greater than the previous digit on the right.

Decimal fractions are proper fractions with denominators of 10, 100, 1000, etc. When a decimal point is placed to the right of the unit's digit, digits to the right of the decimal point represent decimal fractions. Each digit is the numerator of a decimal fraction whose *denominators* are 10, 100, 1000, and so on, moving from *left to right.* For example, consider the number 78.19. The digits 1 and 9 to the right of the decimal point represent decimal fractions as follows:

$$78.19 = 7(10) + 8(1) + 1\left(\frac{1}{10}\right) + 9\left(\frac{1}{100}\right) = 70 + 8 + \frac{1}{10} + \frac{9}{100} = 78 + \frac{19}{100}$$

Rule

Decimal Fraction

The number of decimal digits to the right of the decimal point (decimal places) equals the number of zeros in the denominator.

Some other examples of decimal fractions are as follows:

$$0.5 = \frac{5}{10}$$

$$0.21 = \frac{21}{100}$$

$$0.076 = \frac{76}{1000}$$

Notice in the three numbers above there are no digits to the left of the decimal point, so a zero is used to emphasize the decimal point.

$$4.7 = 4 + \frac{7}{10}$$

$$18.236 = 10 + 8 + \frac{2}{10} + \frac{3}{100} + \frac{6}{1000}$$

$$= 18 + \frac{236}{1000}$$

Notice in all the numbers above that the number of zeros in the denominator of the fraction equals the number of decimal places.

Rule

Adding Decimals

Line up the decimal points and the columns. Then add or subtract in the same way as whole numbers.

EXAMPLE 2.1

Calculate the following:

$$7.74 + 5.05 - 10.4$$

Solution Line up the decimal points and the columns. Add the first two numbers, bringing down the decimal point:

$$\begin{array}{r} 7.74 \\ +5.05 \\ \hline 12.79 \end{array}$$

Place a zero at the end of 10.4 as a placeholder to aid in the subtraction of the number:

$$\begin{array}{r} 12.79 \\ -10.40 \\ \hline 2.39 \end{array}$$

Zeros can be added to the end of a decimal fraction; they do not change the value of the decimal number.

Rule

Multiplying Decimals

Multiply as whole numbers and *add the decimal places* in the numbers. The total is the number of decimal places in the answer.

EXAMPLE 2.2

Multiply:

$$0.1 \times 0.04$$

Solution Multiply as if the numbers were 1 and 4:

$$1 \times 4 = 4$$

The number 0.1 has one decimal place and 0.04 has two decimal places. Therefore, there are three decimal places in the answer:

$$0.1 \times 0.04 = 0.004$$

EXAMPLE 2.3 Multiply:

$$0.20 \times 0.31 \times 0.5$$

Solution Multiply as if the numbers were 2, 31, and 5:

$$2 \times 31 \times 5 = 62 \times 5 = 310$$

Add the decimal places to determine the number of decimal places in the answer. *The zero at the end of a decimal fraction does not count as a decimal place for multiplication.* It is used to indicate the precision and accuracy of the number. Precision and accuracy are explained later in Section 2.3. The total of the decimal places is therefore four, so there are *four* decimal places in the answer:

$$0.20 \times 0.31 \times 0.5 = 0.062 \times 0.5 = 0.0310$$

The zero at the end of the answer can be dropped and the answer written as 0.031.

EXAMPLE 2.4 Calculate:

$$0.1 \times 0.03 + 0.013$$

Solution Multiply first, and add the decimal places:

$$0.1 \times 0.03 = 0.003$$

Then do the addition, lining up the decimal points:

$$\begin{array}{r} 0.003 \\ +0.013 \\ \hline 0.016 \end{array}$$

 **ERROR BOX**

Counting the number of decimal places for multiplication can lead to a common error. A zero may or may not count as a decimal place. A zero does not count as a decimal place when:

1. It is in front of the decimal point: 0.103 counts as three decimal places for multiplication. The zero in front of the decimal point is to emphasize the decimal point.
2. It is at the end of the decimal: 0.1030 counts as three decimal places for multiplication. The zero at the end is to indicate the precision and accuracy of the number. 0.1030 is more precise and accurate than 0.103. Precision and accuracy are explained in Section 2.3.

Practice Problems: Give the number of decimal places in each number used for multiplication:

1. 100.52 **2.** 10.052 **3.** 0.0052 **4.** 0.5020
5. 0.502 **6.** 10.05020 **7.** 0.200 **8.** 100.050
9. 100.001 **10.** 0.010 **11.** 501.0010 **12.** 5010.0510

Rule **Dividing Decimals**

Move the decimal point to the right in the divisor (the number you are dividing by) and do the same in the dividend, by as many places as there are in the divisor. Then divide by the whole number, keeping the decimal point in the moved position.

EXAMPLE 2.5 Divide:

$$0.132 \div 0.12$$

The number 0.132 is the dividend. The number 0.12 is the divisor and has two decimal places. Move the decimal point in both numbers two places to the right as follows:

$$0.132 \div 0.12 = 13.2 \div 12.$$

Moving the decimal point to the right two places does not change the quotient (the division result) of the two numbers. To better understand why, consider the division as a fraction:

$$\frac{0.132}{0.12} = \frac{0.132(100)}{0.12(100)} = \frac{13.2}{12.}$$

Notice that moving the decimal point to the right in the two numbers does not change the value of the fraction. After you move the decimal point, divide the numbers, keeping the decimal point in the same position as it is in the numerator:

$$12\overline{)13.2}\quad 1.1$$

The 12 divides into the 13 once with one remaining. Then comes the decimal point and the 12 divides into 1.2, 0.1 times.

EXAMPLE 2.6 Calculate:

$$1.2 \times \frac{0.06}{1.8}$$

Solution First do the multiplication. The number 1.2 can be multiplied directly into the numerator, counting three decimal places for the result:

$$1.2 \times \frac{0.06}{1.8} = \frac{1.2 \times 0.06}{1.8} = \frac{0.072}{1.8}$$

Then move the decimal point *one place* to the right in the top and bottom and divide:

$$\frac{0.072}{1.8} = \frac{0.72}{18.} = 0.04$$

The calculator check is:

1.2 [×] 0.06 [÷] 1.8 [=] → 0.04

ERROR BOX

One often makes mistakes in the decimal point when multiplying and dividing decimals that contain zeros. For example, multiplying 0.60 × 0.6 should result in only two decimal places because none of the zeros count as decimal places:

$$0.60 \times 0.6 = 0.36$$

Also, when dividing 0.1 ÷ 0.0020, you can drop the last zero of 0.0020 as it does not affect the size of the number. You only need to move the decimal point three places:

$$\frac{0.1}{0.0020} = \frac{100}{2} = 50$$

Answers to Error Box Problems, page 29:
1. 2 **2.** 3 **3.** 4 **4.** 3 **5.** 3 **6.** 4 **7.** 1 **8.** 2 **9.** 3 **10.** 2 **11.** 3 **12.** 3

Practice Problems: See if you can do the following correctly:

1. 0.60×0.60 **2.** 1.1×0.10 **3.** 5.0×0.5 **4.** 200.0×0.20
5. $0.20 \times 0.2 \times 0.200$ **6.** $10 \div 0.1$ **7.** $1.0 \div 0.10$ **8.** $44 \div 2.20$
9. $0.25 \div 0.050$ **10.** $90 \div 0.090$

EXAMPLE 2.7 Calculate:

$$\frac{0.0025 - 0.0015}{0.5 \times (0.03 + 0.01)}$$

Solution Study the steps carefully. Do the operation in parentheses first:

$$\frac{0.0025 - 0.0015}{0.5 \times (0.03 + 0.01)} = \frac{0.0025 - 0.0015}{0.5 \times 0.04}$$

Then, since *the fraction line is like parentheses,* you must do the operations in the numerator and denominator before dividing. Subtract the numbers in the numerator, lining up the decimal points, and multiply the numbers in the denominator adding the decimal places, which total three:

$$\frac{0.0025 - 0.0015}{0.5 \times 0.04} = \frac{0.0010}{0.020}$$

Now move the decimal point two places to the right in the top and bottom and divide:

$$\frac{0.0010}{0.020} = \frac{0.10}{2.0} = 0.05$$

The zero at the end of 0.020 does not need to be considered a decimal place of the divisor.

Example 2.7 can be checked on the calculator using three sets of parentheses as follows:

[( ] 0.0025 [−] 0.0015 [ )] [÷] [( ] 0.5 [×] [( ] 0.03 [+] 0.01 [ )] [ )] [=] → 0.05

EXERCISE 2.1

In exercises 1 through 40, calculate each exercise by hand. Use the calculator to check the results.

1. $1.05 + 8.98 + 0.06$
2. $12.004 + 3.005 + 50.120$
3. $15.64 + 4.36 - 19.09$
4. $1.780 - 32.409 + 100.013$
5. $100.23 - 56.42 - 21.10$
6. $36.36 - 28.02 - 0.95$
7. 8.1×0.5
8. 60.1×0.02
9. 0.031×0.20
10. 100.1×0.10
11. $10.0 \times 0.40 \times 3.0$
12. $2.3 \times 1.5 \times 0.02$
13. $5.0 \times 0.5 \times 2.5$
14. $7.01 \times 0.020 \times 0.20$
15. $0.1 \times 0.01 \times 0.010$
16. $0.030 \times 0.400 \times 0.005$
17. $\frac{5.1}{0.17}$
18. $\frac{1.05}{0.005}$
19. $\frac{0.036}{0.30}$
20. $\frac{0.9903}{0.03}$
21. $\frac{0.70}{0.070}$
22. $\frac{0.12}{1.20}$
23. $\frac{0.10}{4.0} \times 3.2$
24. $0.40 \times \frac{2.10}{0.30}$
25. $\frac{0.72}{3.0 \times 0.3}$
26. $\frac{8.06 \times 0.04}{0.20}$
27. $(1.3 + 2.8) \times (1.6 + 1.4)$
28. $(5.5 + 3.2) \times (0.06 + 0.04)$
29. $(0.4 + 0.1) \times (0.08 - 0.04)$
30. $(0.82 + 1.68) \times (0.095 - 0.035)$
31. $\frac{1.2 \times 1.0 \times 0.03}{0.0050 + 0.0040}$
32. $\frac{1.10 \times 0.04 \times 0.11}{0.01 + 0.03}$
33. $\frac{(10.5 - 5.5) \times 0.6}{0.30}$
34. $\frac{0.02 \times (100.1 + 99.9)}{0.004}$
35. $\frac{0.02 \times (3.0 + 5.0)}{0.004}$
36. $\frac{(8.6 - 6.4) \times 0.05}{0.20}$
37. $\frac{0.5 \times 7.0 + 6.0 \times 0.70}{2.3 - 1.2}$
38. $\frac{0.09 \times 0.3 + 3.0 \times 0.006}{0.45 - 0.30}$
39. $\frac{0.80 \times 0.60 \times 0.04}{4.00 \times 3.00 \times 0.02}$
40. $\frac{0.66 \times 0.01 \times 2.0}{0.20 \times 0.50 \times 0.33}$

Applied Problems

In problems 41 through 50, solve each problem by hand. Use the calculator to check your results.

41. The three sides of a small triangle are measured to be 0.0970 mm, 0.1020 mm, and 0.0093 mm. Find the perimeter of the triangle. Perimeter = sum of the three sides.

42. The length of a small rectangle is measured to be 1.01 cm and the width to be 0.55 cm. Find the perimeter of the rectangle in centimeters. Perimeter = 2 × (length) + 2 × (width).

43. Find the area of the rectangle in problem 42 in square centimeters (cm^2). Area = (length) (width).

44. A square measures 0.02 in on a side. Find the perimeter in inches and the area of the square in square inches. Perimeter = 4 × (side) and Area = (side) × (side).

45. A STOP sign is a regular octagon (a polygon of eight equal sides). If each side measures 1.05 ft, what is the perimeter of the STOP sign?

46. Given that a baseball "diamond" is actually a square and the distance from home to first base is 27.43 meters, how many meters does a player have to run if he hits a home run?

Applications to Electronics

47. Four batteries have the following voltages: 13.01 V (volts), 12.52 V, 6.21 V, and 8.36 V. If the batteries are connected so that the total voltage equals the sum of the voltages, find the total voltage (in volts).

48. Three currents flowing in the same direction toward a junction point in a circuit are measured as follows: 0.35 A (amps), 0.09 A, and 1.06 A. If the total current entering the junction equals the sum of the currents, find the total current entering the junction in amps.

49. A rectangular computer circuit board has a length equal to 5.15 cm and a width equal to 3.35 cm. Find the perimeter of the board in cm. Perimeter = 2 × (length) + 2 × (width).

50. A square electronics component measures 1.1 mm on a side. Find the perimeter of the component in mm and the area in mm^2. Perimeter = 4 × (side) and Area = (side) × (side).

2.2 Percentages

A percentage (or percent) is a convenient way of writing a decimal fraction whose denominator is 100. The numerator is written with the percent sign, %, which represents the denominator of 100. For example:

$$20\% = \frac{20}{100} \text{ or } 0.20$$

$$100\% = \frac{100}{100} \text{ or } 1.00$$

$$130\% = \frac{130}{100} \text{ or } 1.30$$

$$1\tfrac{1}{2}\% = 1.5\% = \frac{1.5}{100}$$

$$= \frac{1.5(10)}{100(10)} = \frac{15}{1000} \text{ or } 0.015$$

Notice above that 100% = 1.00 and that a percent can be more than 100 as shown by 130%. Observe also that a percent can contain a fraction or a decimal point such as $1\frac{1}{2}\% =$ 1.5%. When 1.5% is changed to a fraction, the decimal point is eliminated by multiplying top and bottom by ten.

Rule **To Change Percent to Decimal**

Move the decimal point two places to the left.

• • •

Answers to Error Box Problems, page 30:
1. 0.36 **2.** 0.11 **3.** 2.5 **4.** 40 **5.** 0.008 **6.** 100 **7.** 10 **8.** 20 **9.** 5 **10.** 1000

Rule

To Change Decimal to Percent

Move the decimal point two places to the right.

Study the following examples, which show how to change between percents, decimals, and fractions.

EXAMPLE 2.8

Change each fraction to a percent and to a decimal:

(a) $\frac{3}{4}$ **(b)** $\frac{5}{8}$ **(c)** $\frac{7}{5}$ **(d)** $\frac{1}{200}$ **(e)** $\frac{2}{1}$

Solution

(a) To change $\frac{3}{4}$ to a percent, first change the denominator to 100 by multiplying the numerator and denominator by 25. Then move the decimal point two places to the left to change to a decimal:

$$\frac{3(25)}{4(25)} = \frac{75}{100} = 75\% = 0.75$$

(b) To change $\frac{5}{8}$ to a percent, the denominator cannot easily be changed to 100. First change the fraction to a decimal. Add a decimal point and three zeros to the numerator. Then divide by 8. This will give you three decimal places equal to the number of zeros:

$$\frac{5}{8} = \frac{5.000}{8} = 0.625$$

Now move the decimal point to the right two places to change to a percent:

$$0.625 = 62.5\%$$

(c) The fraction $\frac{7}{5}$ is greater than one and is an improper fraction. To change $\frac{7}{5}$ to a percent multiply top and bottom by 20 to change the denominator to 100. Then move the decimal point two places to the left to change to a decimal:

$$\frac{7(20)}{5(20)} = \frac{140}{100} = 140\% = 1.40$$

Since the fraction is an improper fraction greater than one, the percent is greater than 100.

(d) First change the fraction $\frac{1}{200}$ to a decimal by adding three zeros to the numerator and dividing. Then change the decimal to a percent:

$$\frac{1}{200} = \frac{1.000}{200} = 0.005 = 0.5\%$$

(e) To change $\frac{2}{1}$ to a decimal, multiply top and bottom by 100, then move the decimal point two places to the left to change to a decimal:

$$\frac{2}{1} = \frac{200}{100} = 200\% = 2.0$$

EXAMPLE 2.9

Change each decimal to a percent and to a fraction:

(a) 0.80 **(b)** 0.012 **(c)** 2.5 **(d)** 0.125

Solution

(a) To change 0.80 to a percent, move the decimal point two places to the right. Then write the fraction by putting the percent number over a denominator of 100 and reducing the fraction:

$$0.80 = 80\% = \frac{80}{100} = \frac{4(\cancel{20})}{5(\cancel{20})} = \frac{4}{5}$$

(b) To change 0.012 to a percent, move the decimal point two places to the right. Then write the fraction with a denominator of 100:

$$0.012 = 1.2\% = \frac{1.2}{100}$$

A fraction is not in the simplest form if it contains a decimal in the numerator or denominator. Eliminate the decimal in the numerator by multiplying the numerator and denominator by 10 because there is one decimal place. Then reduce the fraction:

$$\frac{1.2(10)}{100(10)} = \frac{12}{1000} = \frac{3(\cancel{4})}{250(\cancel{4})} = \frac{3}{250}$$

(c) To change 2.5 to a percent, move the decimal point two places to the right. Then write the fraction by putting the percent number over a denominator of 100 and reducing the fraction:

$$2.5 = 250\% = \frac{250}{100} = \frac{5(\cancel{50})}{2(\cancel{50})} = \frac{5}{2}$$

Note that the fraction $\frac{5}{2}$ is greater than one and is an improper fraction.

(d) To change 0.125 to a percent, move the decimal point two places to the right. Write the fraction by putting the percent number over a denominator of 100 and simplify the fraction by multiplying numerator and denominator by 10 and reducing to lowest terms:

$$0.125 = 12.5\% = \frac{12.5}{100} = \frac{12.5(10)}{100(10)} = \frac{125}{1000} = \frac{1(\cancel{125})}{8(\cancel{125})} = \frac{1}{8}$$

EXAMPLE 2.10 Change each percent to a decimal and to a fraction:

(a) 40% **(b)** 7.5% **(c)** 110% **(d)** $8\frac{1}{4}\%$

Solution

(a) To change 40% to a decimal, move the decimal point two places to the left:

$$40\% = 0.40$$

To change 40% to a fraction, put the percent number over 100 and reduce the fraction:

$$40\% = \frac{40}{100} = \frac{2(\cancel{20})}{5(\cancel{20})} = \frac{2}{5}$$

(b) To change 7.5% to a decimal, move the decimal point two places to the left:

$$7.5\% = 0.075$$

To change 7.5% to a fraction put the percent number 7.5 over 100. Then simplify the fraction by multiplying numerator and denominator by 10 and reducing to lowest terms:

$$7.5\% = \frac{7.5}{100} = \frac{7.5(10)}{100(10)} = \frac{75}{1000} = \frac{3}{40}$$

(c) To change 110% to a decimal, move the decimal point two places to the left. Change the percent to a fraction with a denominator of 100 and reduce to lowest terms:

$$110\% = 1.10 = \frac{110}{100} = \frac{11}{10}$$

Since 100% = 1.00, a percentage greater than 100 such as 110% represents a number greater than one.

(d) The number $8\frac{1}{4}$ is called a *mixed number*. It means $8 + \frac{1}{4}$. To change $8\frac{1}{4}\%$ to a decimal, first change $\frac{1}{4}$ to a decimal:

$$\frac{1}{4} = \frac{1.00}{4} = 0.25$$

Then express $8\frac{1}{4}\%$ as 8.25% and move the decimal point two places to the left:

$$8\tfrac{1}{4}\% = 8.25\% = 0.0825$$

To change 8.25% to a fraction, put the percent number 8.25 over 100. Simplify the fraction by multiplying numerator and denominator by 100 to eliminate the decimal and reducing the fraction to lowest terms:

$$8\tfrac{1}{4}\% = \frac{8.25}{100} = \frac{8.25(100)}{100(100)} = \frac{825}{10{,}000} = \frac{33(\cancel{25})}{400(\cancel{25})} = \frac{33}{400}$$

EXAMPLE 2.11 Find each of the following:

(a) 15% of 70 **(b)** 200% of 35 **(c)** $8\frac{1}{4}\%$ of 50

Solution To find a percent of a number, first change the percent to a decimal, then multiply the decimal by the number:

(a) $15\%(70) = 0.15(70) = 10.5$

(b) $200\%(35) = 2.00(35) = 70$

(c) $8\frac{1}{4}\%(50) = 8.25\%(50) = 0.0825(50) = 4.125$

ERROR BOX

Errors can occur when working with percents that contain a fraction or a decimal point. To change such a percent to a decimal, you must move the decimal point correctly. Percent means hundredths and this means you must *add two decimal places* when changing a percent to a decimal. If the percent contains a fraction, you must first change the fraction to a decimal. Study the following, and then see if you can do the practice problems correctly:

$$55.5\% = 0.555$$

$$0.55\% = 0.0055$$

$$5\tfrac{3}{4}\% = 5.75\% = 0.0575$$

$$5\tfrac{1}{2}\% = 5.5\% = 0.055$$

Practice Problems: Change each percent to a decimal:

1. 19.9% **2.** 0.99% **3.** 1.99% **4.** $1\frac{1}{2}\%$

5. $9\frac{1}{4}\%$ **6.** $\frac{3}{4}\%$ **7.** 3.5% **8.** 8.65%

9. 1.07% **10.** $5\frac{1}{5}\%$ **11.** $\frac{1}{10}\%$ **12.** $\frac{1}{100}\%$

EXAMPLE 2.12 A store gives a 15% discount on a calculator marked \$25. What is the price of the calculator?

Solution First calculate the discount. Find 15% of \$25:

$$15\%(\$25) = 0.15(\$25) = \$3.75$$

Then subtract the discount from \$25 to find the price:

$$\$25 - \$3.75 = \$21.25$$

This example can be done in one series of steps on the calculator:

$$25 \boxed{-} 0.15 \boxed{\times} 25 \boxed{=} \rightarrow 21.25$$

EXAMPLE 2.13 A store marks up an item whose wholesale price is \$50 by 200%. Find the retail price.

Solution First calculate the amount of markup. Find 200% of \$50:

$$200\%(\$50) = (2.00)(\$50) = \$100$$

Add the markup to the wholesale price to find the retail price:

$$\$50 + \$100 = \$150$$

Study the next example, which closes the circuit and shows an electrical application of decimals and percents.

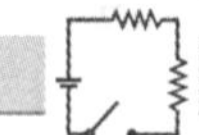

EXAMPLE 2.14
Close the Circuit

The current in a resistor is 1.5 A. The voltage across the resistor is increased by 12%, which increases the current by the same percentage. How much is the increase in the current, and what is the new current in amps?

Solution Multiply the original current by 12% to find the increase:

$$1.5(12\%) = 1.5(0.12) = 0.18$$

Then add this amount to the original current to find the new current:

$$1.5 + 0.18 = 1.68 \text{ A}$$

EXAMPLE 2.15 The list price of a computer is \$900. A discount store sells it for $\frac{1}{3}$ off list price.

(a) What is the discount price in dollars?

(b) Find the total cost, including tax of $7\frac{1}{2}\%$, in dollars.

Solution

(a) Find the discount price by subtracting $\frac{1}{3}$ of the list price from itself:

$$900 - \left(\frac{1}{3}\right)(900) = 900 - 300 = \$600$$

(b) The total cost, including a tax of $7\frac{1}{2}\%$, can be calculated by adding $7\frac{1}{2}\%$ or 7.5% of the discount price to itself:

$$600 + 600(7.5\%) = 600 + 600(0.075) = 600 + 45 = \$645$$

Answers to Error Box Problems, page 35:
1. 0.199 **2.** 0.0099 **3.** 0.0199 **4.** 0.015 **5.** 0.0925 **6.** 0.0075 **7.** 0.035 **8.** 0.0865 **9.** 0.0107 **10.** 0.052 **11.** 0.001 **12.** 0.0001

It is often necessary to calculate the percent increase or decrease in a value when you know the initial and final values:

Formula **Percent Change (increase or decrease)**

$$\text{Percent Increase} = \frac{(\text{Final Value} - \text{Initial Value}) \times 100\%}{\text{Initial Value}} \quad (2.1)$$

$$\text{Percent Decrease} = \frac{(\text{Initial Value} - \text{Final Value}) \times 100\%}{\text{Initial Value}}$$

The next example, which closes the circuit, shows an electrical application of percent change involving the resistance of a wire.

EXAMPLE 2.16

Close the Circuit

The resistance of a copper wire increases when the temperature increases. A copper wire whose resistance is 2.0 Ω (ohms) increases to 2.1 Ω when heated. What is the percent increase in the resistance?

Solution Apply formula (2.1) for percent increase:

$$\text{Percent Increase} = \frac{(2.1 - 2.0)(100\%)}{2.0} = \frac{(0.1)(100\%)}{2.0} = \frac{10\%}{2.0} = 5\%$$

This can be done on the calculator as follows, but you must add the percent sign:

[( ] 2.1 [−] 2 [ )] [×] 100 [÷] 2 [=] → 5 (5%)

EXERCISE 2.2

In exercises 1 through 44, express each number as a fraction in lowest terms, a decimal, and a percent.

1. $\frac{3}{4}$
2. $\frac{1}{2}$
3. $\frac{2}{5}$
4. $\frac{3}{10}$
5. $\frac{3}{8}$
6. $\frac{7}{8}$
7. $\frac{1}{25}$
8. $\frac{9}{100}$
9. 0.15
10. 0.25
11. 0.1
12. 0.6
13. 0.12
14. 0.16
15. 0.28
16. 0.36
17. 20%
18. 80%
19. 15%
20. 35%
21. 86%
22. 55%
23. 99%
24. 49%
25. $\frac{1}{20}$
26. $\frac{6}{50}$
27. $\frac{5}{4}$
28. $\frac{12}{5}$
29. 0.05
30. 0.08
31. 1.5
32. 1.0
33. 120%
34. 200%
35. 5.6%
36. 50.5%
37. $\frac{1}{3}$
38. $\frac{1}{16}$
39. 0.1875
40. 0.875
41. $8\frac{1}{2}\%$
42. $6\frac{1}{4}\%$
43. 0.001
44. $\frac{1}{500}$

In exercises 45 through 56, find the value of each percent.

45. 10% of 30
46. 20% of 80
47. 25% of 60
48. 75% of 40

49. 5% of 27

50. 7% of 17

51. 150% of 50

52. 125% of 10

53. $7\frac{1}{2}$% of 22

54. $3\frac{1}{4}$% of 15

55. 2.5% of 100

56. 7.25% of 50

Applied Problems

In problems 57 through 78, solve each applied problem by hand. *Use the calculator to check your results.*

57. A store gives a 12% discount on a calculator marked \$15. Find the price of the calculator.

58. A store marks up an item whose wholesale price is \$100 by 50%. Find the retail price.

59. The atmosphere at sea level is approximately 19% oxygen. How many cubic feet of oxygen are contained in 5 ft^3 of the atmosphere?

60. Approximately 0.003% of seawater is salt. How many grams (g) of salt are contained in 1,000,000 g of seawater?

61. One discount technical supply store sells a digital multimeter for one-third off its list price of \$57.00. Another discount store sells the same digital multimeter for 25% off its list price. Find the two prices for the digital multimeter in dollars and cents.

62. The list price for a textbook is \$55. A retailer pays the publisher 20% less than the list price for the text. How much does the retailer pay the publisher for the text in dollars and cents?

63. The list price of a graphing calculator is \$80.00. A discount store sells it at a 20% discount.

(a) What is the discount price in dollars and cents?

(b) What is the total cost, including a tax of $7\frac{1}{2}$%, in dollars and cents? See Example 2.15.

64. The price of a multimeter is \$30.00. A discount store sells it at $\frac{1}{3}$ off.

(a) What is the discount price in dollars and cents?

(b) What is the total cost, including tax of $8\frac{1}{4}$%, to the nearest cent? See Example 2.15.

65. Your palm pilot computer costs \$40.00 today. At a price increase of 5% a year:

(a) How much will it cost to replace it one year from now in dollars and cents?

(b) How much will it cost to replace it two years from now in dollars and cents?

66. The rent on an apartment is \$600 per month. A new lease increases the rent by 7%. What is the new rent in dollars?

67. The price of a certain model Honda is \$30,000. The hybrid version is \$33,600. What is the percent increase for the hybrid version?

68. The price of an old model laptop computer is reduced from \$1500 to \$1200. What is the percent decrease in the price?

69. Arrange the following in order of increasing size:

$$1.19,\ 110\%,\ \frac{9}{8},\ 1\frac{1}{6},\ \frac{1120}{1000}$$

70. Arrange the following in order of decreasing size:

$$0.03,\ 0.50\%,\ \frac{1}{40},\ \frac{0.5}{25}$$

Applications to Electronics

71. The tolerance band of a 150-Ω resistor is silver, which means the highest possible value could be 10% greater than its marked or nominal value. Find the highest possible value of the resistor.

72. The tolerance band of a 560-Ω resistor is gold, which means its lowest possible value could be 5% less than its marked or nominal value. Find the lowest possible value of the resistor.

73. The current in a resistor is initially 1.1 A (amps). The voltage across the resistor is increased by 10%, which increases the current by the same percentage. How much is the increase in the current, and what is the new current? See Example 2.14.

74. The current in a resistor is initially 1.1 A (amps). The voltage across the resistor is decreased by 20%, which decreases the current by the same percentage. How much is the decrease in the current, and what is the new current? See Example 2.14.

75. The current in a circuit decreases from 1.0 A to 0.6 A. What is the percent decrease in the current? See Example 2.16.

76. The voltage in a circuit increases from 80 V to 120 V. What is the percent increase in the voltage? See Example 2.16.

77. The resistance of a wire increases from 0.5 Ω (ohms) to 0.6 Ω when heated. What is the percent increase in the resistance? See Example 2.16.

78. An electronics company produces 200 microprocessors in a week. However, one week several employees are sick and the company only produces 150 microprocessors. What is the percent decrease in normal production for this week?

2.3 Precision and Accuracy

The results predicted by electrical laws are not always easy to produce experimentally. In the laboratory, measurements taken with instruments such as multimeters are only approximate and subject to certain degrees of precision and accuracy. In addition, values of electrical and electronic components tend to vary under certain conditions. The numbers obtained are usually not exact. For example, suppose you are measuring the voltage of a battery using a voltmeter and the scale reads 1.5 V. This number 1.5 is considered an *approximate* number. A more accurate voltmeter with an expanded scale might enable you to read the voltage to be 1.53 V. However, this number 1.53 is still considered to be an approximate number because a measurement can usually always be taken more accurately. Measurements are almost always approximate. One of the only times a measurement is an exact number is when it is obtained by counting. For example, if you count the number of resistances in a circuit to be 3, then this number, 3, is considered an exact number. It is therefore necessary to understand what we mean by the precision and accuracy of approximate numbers. However, before looking at precision and accuracy, you need to understand significant digits and the rules for rounding a number.

Definition

Significant Digits

1. All nonzero digits 1 to 9 are significant.
2. Zeros are significant when they are *not the first* digits in a decimal number or *not the last* digits in a whole number unless marked by a bar.

A zero or zeros at the end of a whole number are significant when they have a bar over them such as the number written $5\overline{00}0$. This number has three significant digits, 5, 0, and 0. The last zero is not significant. Study the next example which shows further how to identify significant digits.

EXAMPLE 2.17

Tell how many significant digits are in each number:

(a) 0.0045 **(b)** 1.023 **(c)** 5.80 **(d)** 0.903 **(e)** $4\bar{0}0$ **(f)** 1205

Solution

(a) 0.0045 has two significant digits, 4 and 5. The three zeros are the first digits and are used only as locators for the decimal point.

(b) 1.023 has four significant digits, 1, 0, 2, and 3. The zero is not the first digit and contributes to the accuracy of the decimal number.

(c) 5.80 has three significant digits, 5, 8, and 0. The zero is not used as a locator for the decimal point and contributes to the accuracy of the decimal number.

(d) 0.903 has three significant digits, 9, 0, and 3. The first zero is used only as a locator for the decimal point. The second zero contributes to the accuracy of the decimal number.

(e) $4\bar{0}0$ has two significant digits, 4 and 0. The bar over the first zero marks it as a significant digit. The second zero is the last digit of a whole number and is not significant.

(f) 1205 has four significant digits, 1, 2, 0, and 5. The zero is not a last digit and contributes to the accuracy of the number.

Rule

Rounding Decimals

1. If the digit to the right of the last rounded digit is *less than 5,* do not change the last rounded digit.
2. If the digit to the right of the last rounded digit is *5 or more than 5,* increase the last rounded digit by one.

For example, consider the following fractions and their decimal equivalents:

$$\tfrac{1}{9} = 0.1111\ldots \qquad \tfrac{2}{3} = 0.6666\ldots$$

These decimals go on forever. To express the fractions to two decimal places, round as follows:

$$0.1111\ldots \approx 0.11 \qquad 0.6666\ldots \approx 0.67$$

Study the following examples, which show how to round off to significant digits.

EXAMPLE 2.18 Round off each decimal to one, two, and three significant digits:

(a) 3.516 **(b)** 0.5505 **(c)** 0.07345 **(d)** 4.999 **(e)** 309.9

Solution

	one digit	*two digits*	*three digits*
(a)	$3.516 \approx 4$	$3.516 \approx 3.5$	$3.516 \approx 3.52$
(b)	$0.5505 \approx 0.6$	$0.5505 \approx 0.55$	$0.5505 \approx 0.551$
(c)	$0.07345 \approx 0.07$	$0.07345 \approx 0.073$	$0.07345 \approx 0.0735$
(d)	$4.999 \approx 5$	$4.999 \approx 5.\bar{0}$	$4.999 \approx 5.\overline{00}$
(e)	$309.3 \approx 300$	$309.3 \approx 310$	$309.3 \approx 309$

Observe in (d) that 4.999, when rounded to two, and three significant digits, has zeros that are marked as significant digits.

Rule

Rounding Whole Numbers

1. If the digit to the right of the last rounded digit is *less than 5,* change all the digits to the right of the last rounded digit to zero.
2. If the digit to the right of the last rounded digit is *5 or more than 5,* increase the last rounded digit by one and change all the digits to the right of the last rounded digit to zero.

EXAMPLE 2.19 Round off each number to one, two, and three significant digits:

(a) 3653 **(b)** 34,097 **(c)** 569 **(d)** 999 **(e)** 179,560

Solution

	one digit	*two digits*	*three digits*
(a)	$3653 \approx 4000$	$3653 \approx 3700$	$3653 \approx 3650$
(b)	$34{,}097 \approx 30{,}000$	$34{,}097 \approx 34{,}000$	$34{,}097 \approx 34{,}100$
(c)	$569 \approx 600$	$569 \approx 570$	$569 \approx 569$
(d)	$999 \approx 1000$	$999 \approx 1\bar{0}00$	$999 \approx 1\overline{00}0$
(e)	$179{,}560 \approx 200{,}000$	$179{,}560 \approx 180{,}000$	$179{,}560 \approx 18\bar{0}{,}000$

Notice in (d) and (e) that the rounded numbers have zeros that are significant.

The concepts of *accuracy* and *precision* are not the same when applied to approximate numbers as follows:

Definition

Accuracy

The accuracy of an approximate number is measured by the *number of significant digits.*

 **ERROR BOX**

It is possible to make an error when trying to determine if a zero is a significant digit. These are the three situations:

1. Zeros at the beginning of a decimal number are never significant.
2. Zeros between two nonzero digits, or at the end of a decimal number, are always significant.
3. Zeros at the end of a whole number are not significant unless indicated by a bar as shown in Example 2.17(e).

Practice Problems: Tell the number of significant digits in each number:

1. 0.0040	**2.** 0.404	**3.** 440,400	**4.** $4\bar{0},000$	**5.** 40
6. 40.0	**7.** 0.440	**8.** 4.04	**9.** 400.4	**10.** $\overline{4000}$

The number 10.23 has four significant digits. It is more accurate than 0.123, which has only three significant digits.

Definition

Precision

The precision of an approximate number is measured by the *decimal place of the last digit.*

The number 0.123 is precise to three decimal places. It is more precise than 10.23, which is precise to only two decimal places.

The concept of accuracy and significant digits is used more often than precision, but both concepts are important in experimental calculations. The results of calculations with experimental data *can not be more accurate than the data.* That is, if the data is accurate to three significant digits, the results can, at best, be accurate to three significant digits. Study Table 2-1, which compares the precision and accuracy of six numbers.

TABLE 2-1 PRECISION AND ACCURACY OF SIX NUMBERS

Number	Significant Digits	Explanation	Comment
2313	Four	All nonzero digits are significant	Precise to the nearest unit
560	Two	Last zero is not significant	Precise to the nearest ten units
$8\bar{0}00$	Two	Only one zero is marked	Least precise to the nearest hundred units
0.703	Three	The first zero is not significant	Most precise to the nearest thousandth
0.04	One	The first two zeros are not significant	Least accurate
90.2	Three	The zero is not the first digit	Precise to nearest tenth
30.450	Five	The zeros are not the first digits	Most accurate

Computers and calculators can only work with a finite number of digits and must either round off or truncate numbers. *Truncate* means to drop all the digits to the right of the last rounded digit. For example, the number 2.0175 truncated to three digits is just 2.01. Most calculators can be set to round off to a certain number of decimal places using the [FIX] key, [MODE] key, or a similar key. After pressing the [FIX] or [MODE] key, press the number key for the number of decimal places you want to round off to. See Example 2.21.

It is important to note that results calculated from data cannot be more accurate than the least accurate measurement. When doing calculations with approximate numbers, *do not round off any of the numbers.* Calculate with them as given or as measured. Then round off the final result to the desired degree of accuracy.

Study the following examples, which serve to close the circuit and show how to apply the idea of accuracy to problems in electronics.

EXAMPLE 2.20

Close the Circuit

A student measures three voltages with a voltmeter to be 2.1 V, 2.4 V, and 1.85 V. Find the total of the three voltages to *two* significant digits.

Solution Add the three voltages *as measured and then* round off the answer to two significant digits:

$$2.1 + 2.4 + 1.85 = 6.35 \text{ V} \approx 6.4 \text{ V}$$

EXAMPLE 2.21

Close the Circuit

Three students measure the current in a circuit with their ammeters to be 14 mA, 12.5 mA, and 11 mA. Find the average current to two significant digits.

Solution Find the average current by adding up the three currents *as measured* and dividing by 3:

$$\frac{14 + 12.5 + 11}{3} = 12.5$$

Then round the answer to two significant digits:

$$12.5 \approx 13 \text{ mA}$$

Example 2.21 is done on the calculator by first setting the number of decimal places using the [FIX] key, [MODE] key, or a similar key:

[FIX] 2 [(] 14 [+] 12.5 [+] 11 [)] [÷] 3 [=] → 13

EXERCISE 2.3

In exercises 1 through 20, tell how many significant digits are in each number.

1. 21.8
2. 30.2
3. 0.15
4. 0.030
5. 5.20
6. 10.01
7. 3210
8. 5000
9. 6.060
10. 0.005
11. 200.1
12. 10.4
13. 100,500
14. 0.003
15. 999.0
16. 9090
17. 0.1010
18. 0.0090
19. $71\overline{0}0$
20. $56\overline{0}0$

In exercises 21 through 40, round off each number to one, two, and three significant digits.

21. 5.152
22. 3.100
23. 318.2
24. 121.3
25. 0.4445
26. 0.4399
27. 81,872
28. 1067
29. 100.90
30. 28.999
31. 18.8625
32. 0.0899
33. 19.99
34. 39.09
35. 336
36. 454
37. 260.0
38. 678.2
39. 999
40. 8888

Answers to Error Box Problems, page 41:
1. 2 **2.** 3 **3.** 4 **4.** 2 **5.** 1 **6.** 3 **7.** 3 **8.** 3 **9.** 4 **10.** 4

Applied Problems

In problems 41 through 54, calculate each result, using the calculator if necessary. Round off the answer to three significant digits unless specified.

41. A rectangular garden is measured to be 12.5 m (meters) by 36.9 m. Find the area of the garden in square meters (m^2). Area = (length)(width).

42. The largest scientific building in the world is the Vehicle Assembly Building in Complex 39 at the John F. Kennedy Space Center, Cape Canaveral, Florida. It is 219 m (meters) long, 159 m wide and 160 m high. Assuming the building has a rectangular shape, find the volume it occupies in cubic meters (m^3). Volume of a rectangular box = (length)(width)(height).

43. The weights of four people are measured to be: 125 lb, 147 lb, 180 lb, and 155 lb. Find the average weight in pounds. See Example 2.21.

44. The fastest runner in baseball was Ernest Swanson, who took only 13.3 seconds to circle the bases, a distance of 360 ft, in 1932. What was his average rate of speed in feet per second and in miles per hour? Average rate = (distance)/(time).

Note: 1 ft/s = 0.6818 mi/h.

Applications to Electronics

45. A student measures three voltages with a voltmeter to be 6.1 V, 6.3 V, and 5.85 V. Find the total of the three voltages to two significant digits. See Example 2.20.

46. Three students measure the current in a circuit with their ammeters to be 29.5 mA, 31 mA, and 29 mA. Find the average current to two significant digits. See Example 2.21.

47. The screen of a rectangular computer monitor is measured to be 16.2 in long and 12.5 in wide. Find the area of the screen in square inches (in^2) to three significant digits. Area = (length)(width).

48. A microprocessor chip measured with a ruler is found to be 2.2 cm (centimeters) long and 1.8 cm wide. The thickness measured with a micrometer, which is a precision instrument for measuring small distances, is found to be 0.0115 cm. Find the volume of the chip, in cubic centimeters (cm^3), to two significant digits.

Note: Volume = (length)(width)(thickness).

49. Calculate the voltage V in volts (V) to two significant digits given by:

$$V = 0.14(330 + 560)$$

50. Calculate the current I in amps (A) to two significant digits given by:

$$I = \frac{13}{220} + \frac{13}{270}$$

51. The electric power consumption in a house is measured each day for four days. The measurements are: 1230 W (watts), 2004 W, 1576 W, and 1109 W. Find the average power consumption for the four days to four significant digits. See Example 2.21.

52. Sophia precisely measures the resistance of a computer circuit at five equal intervals during a one hour period of operation. Her readings are 1.28 Ω (ohms), 1.30 Ω, 1.25 Ω, 1.2 Ω, and 1.35 Ω. Find the average value of the resistance to (a) two significant digits and (b) three significant digits.

Note: The average value is only as accurate as the *least* accurate measurement. See Example 2.21.

53. The costs of three items are \$12.82, \$13.98, and \$18.76. Suppose that a computer is programmed to truncate all numbers to whole numbers before calculating.

(a) What is the total cost of the items as calculated by the computer?

(b) If the numbers are not truncated, what is the total cost rounded off to two significant digits?

54. Given the costs of four items: \$38.60, \$22.36, \$101.25, \$9.98. Do the same as in problem 53 (a) and (b) for these four costs.

2.4 Hand Calculator Operations

It is easy to make a mistake when keying in numbers or operations on the calculator. You should try to estimate or approximate answers, before calculating, so you have some idea about the size of the answer. This not only provides a check on the results but helps you to understand the concepts better. The following calculator examples show you how to estimate the answer to provide a check on calculator results.

EXAMPLE 2.22

Estimate the answer *by hand* and then calculate:

$$\frac{(64 + 57) \times 320}{840 - 50 \times 8}$$

Solution An estimate of the actual answer can readily be done without the calculator by rounding off each number to one significant digit. Use the rules for rounding whole numbers in Section 2.3. Change all the digits to zero except the first and, if the second digit is 5 or more, increase the first digit by one:

$$\frac{(64 + 57) \times 320}{840 - 50 \times 8} \approx \frac{(60 + 60) \times 300}{800 - 50 \times 8}$$

Note that 64 and 57 rounded off to one digit both become 60. The estimation, which can be done quickly by hand (or mentally), is

$$\frac{(60 + 60) \times 300}{800 - 50 \times 8} = \frac{120 \times 300}{800 - 400} = \frac{36{,}000}{400} = 90$$

This estimate gives you an idea of the size of the actual answer. The calculator solution using the original numbers and the parentheses keys is:

[(] 64 [+] 57 [)] [×] 320 [÷] [(] 840 [−] 50 [×] 8 [)] [=] → 88

The estimate of 90 is close to the calculator answer of 88, which checks that 88 is the correct answer. Suppose your calculator answer was not close to the estimate. If it was 880 or 8.80 this would indicate that there is an error somewhere and that you need to troubleshoot for the error.

EXAMPLE 2.23 Estimate the answer by hand and then calculate rounding the answer to three significant digits:

$$\frac{(2.06)(0.0340 - 0.0163)}{(3.12)(0.0560)}$$

Solution To estimate, round the numbers to one digit:

$$\frac{(2.06)(0.0340 - 0.0163)}{(3.12)(0.0560)} \approx \frac{(2)(0.03 - 0.02)}{(3)(0.06)}$$

The estimate, calculated by hand, is:

$$\frac{(2)(0.03 - 0.02)}{(3)(0.06)} = \frac{(2)(0.01)}{0.18} = \frac{0.02}{0.18} = \frac{2}{18} = \frac{1}{9} \approx 0.11$$

One calculator solution, using the [FIX] or [MODE] key to set the calculator to three decimal places is:

[FIX] 3 2.06 [×] [(] 0.0340 [−] 0.0163 [)] [÷] [(] 3.12 [×] 0.0560 [)] [=] → 0.209

The estimate of 0.11 is close to the calculator answer of 0.209 verifying the result.

Long calculations can be checked in stages by recording and estimating intermediate results. You can also check an operation by doing the inverse operation and see if you get back to the original number. In the following chapters, use your calculator to help you with calculations, but remember your calculator cannot replace mathematical understanding and reasoning. *You should always check an answer and ask yourself if it makes sense.*

EXERCISE 2.4

In exercises 1 through 18 estimate the answer first by rounding off the numbers to one digit as shown in Examples 2.22 and 2.23. Then choose what you think is the correct answer from the four choices. Finally, do the exercise on the calculator, and see if you are correct.

1. $(82 + 68 - 86) \div 32$ [2, 12, 20, 22]
2. $930 \times 81 \div 9 - 90$ [82, 820, 828, 8280]
3. $4.85 + 60.8 \div 3.20 - 2.75$ [0.211, 2.11, 21.1, 211]
4. $20 \times (1560 - 230) \div 28$ [9.5, 95, 650, 950]

5. $4 \times \frac{5}{8} \div 5 + \frac{3}{2}$ [2.5, 2.0, 1.5, 1.0]

6. $80 \times \frac{3}{10} \div 6 - 1$ [3, 5, 6, 7]

7. $\frac{72 + 52 - 20}{176 - 34 + 66}$ [0.1, 0.5, 0.9, 1. 5]

8. $\frac{152 + 23 - 30}{35 + 64 - 41}$ [1.1, 1.5, 2.0, 2.5]

9. $\frac{252 \times (86 + 61)}{(36 - 29) \times 36}$ [14.7, 47, 147, 1470]

10. $\frac{(55 + 85) \times 121}{(72 - 47) \times 14}$ [44, 44.4, 48, 48.4]

11. $\frac{159 \times 91 \times 76}{247 \times 53 \times 35}$ [0.24, 2.4, 8.2, 24]

12. $\frac{31 \times 176 \times 90}{186 \times 120 \times 110}$ [0.20, 1.2, 2.0, 12]

13. $\frac{(0.0068)(0.48 - 0.31)}{0.0017}$ [0.068, 0.68, 6.8, 68]

14. $\frac{(0.0112)(0.0211)}{(0.0400)(0.700)}$ [0.00414, 0.00844, 0.0414, 0.0844)

15. $\frac{(0.301 + 0.271)(0.504)}{(1.26)(1.60)}$ [3.66, 1.43, 0.366, 0.143]

16. $\frac{(9.09 - 7.59)(3.30)}{(0.42 + 0.18)(11.0)}$ [0.55, 0.75, 1.05, 1.55]

17. $\frac{3.03}{2.02} - \frac{27.6 - 8.40}{30.0}$ [86.0, 8.60, 0.860, 0.0860]

18. $\frac{10.5}{2.50} - \frac{8.70 + 21.5}{20}$ [1.69, 2.09, 2.69, 3.69]

In exercises 19 through 30, do the same as in exercises 1–18, except round off the calculator answers to three significant digits.

19. $3.01 + 23.2 \div 0.715$ [35.0, 35.5, 355, 305]

20. $100.1 - 7.02 + 5.132 \times 6.02$ [124, 588, 58.8, 12.4]

21. $\frac{0.1014}{(1.485)(0.0360)}$ [0.190, 1.09, 1.90, 1.9]

22. $\frac{(10.0)(5.00)(100.0)}{0.0300}$ [16,700, 170,000, 167,000, 1,670,000]

23. $\frac{(5.60)(20.6)}{(1.02)(1.03)}$ [110, 110.0, 10.0, 100]

24. $\frac{(103)(343.3)}{(99.4)(0.123)(5.73)}$ [5.05, 50.4, 50.5, 505]

25. $\frac{(6710)(0.0780)(0.890)}{51.2 + 5.15}$ [0.0827, 0.827, 8.27, 82.7]

26. $\frac{(25.1 - 12.4)(17.0)}{(232)(34.5)}$ [0.00269, 0.00270, 0.0270, 0.270]

27. $\frac{1}{820} + \frac{1}{680} + \frac{1}{470}$ [0.0480, 0.0482, 0.00482, 0.00842]

28. $\frac{1}{910} - \frac{1}{8200} - \frac{1}{13000}$ [0.0090, 0.00900, 0.00090, 0.000900]

29. $\frac{0.101 + 0.202}{52.1} - \frac{47.3}{15,000}$ [2.66, 0.266, 0.0266, 0.00266]

30. $\frac{333}{22.1 + 11.2} + \frac{5.99}{0.559}$ [2.07, 2.071, 20.7, 20.71]

Applied Problems

In problems 31 through 38, use the calculator to solve each applied problem. Round answers to three significant digits unless specified.

31. The area of the earth covered by water is 140,000,000 mi^2 (square miles) or 70.98% of the total surface. How many square miles is the entire surface of the earth?

32. The circumference of a circle is given by:

$$C = 2(\pi)(r)$$

where $\pi \approx 3.142$ (four significant digits) and r = radius. If $r = 2.37$ cm, find the circumference of the circle in cm.

33. In 2004, out of a total of 337,069 students enrolled in technology programs, 15,742 were electrical technology students, 64,440 were electronics students, 20,113 were civil technology students, and 26,395 were mechanical technology students. What percentage of the total were enrolled in:

(a) Electrical technology?
(b) Electronics?

34. The Social Security tax on self-employment income in 2003 was 15.3% of gross earnings up to $87,000. If your income exceeds $87,000, you pay only the Medicare portion of the Social Security tax, which is 2.9%, on the rest of your income. Carmen Velez earned $94,850 in her business in 2003. How much Social Security tax did she pay? Find the answer to the nearest cent.

Applications to Electronics

35. Calculate the voltage V in volts (V) of a battery given by:

$$V = 0.250\left(\frac{(220)(330)}{220 + 330}\right)$$

36. Calculate the voltage V in volts (V) of a source given by:

$$V = 0.085\left(\frac{(9100)(5600)}{9100 + 5600}\right)$$

37. Calculate the current I in amps (A) in a circuit given by:

$$I = \frac{12.6(75 + 51)}{(75)(51)}$$

38. Calculate the current I in amps (A) in a circuit given by:

$$I = \frac{220(1100 + 1600)}{(1100)(1600)}$$

CHAPTER HIGHLIGHTS

2.1 DECIMALS

A decimal number written in expanded form is:

$$753.28 = 7(100) + 5(10) + 3(1) + 2\left(\frac{1}{10}\right) + 8\left(\frac{1}{100}\right)$$

$$= 753 + \frac{28}{100}$$

Decimal Fraction

The number of decimal digits to the right of the decimal point (decimal places) equals the number of zeros in the denominator.

• • •

Adding Decimals

Line up the decimal points and the columns. Then add or subtract in the same way as whole numbers.

Key Example:

$$\begin{array}{r} 12.56 \\ -8.37 \\ \hline 4.19 \end{array}$$

Multiplying Decimals

Multiply as whole numbers and add the decimal places in the numbers. The total is the number of decimal places in the answer.

Key Example:

$$0.5 \times 0.021 \times 1.2 = 0.01260 = 0.0126$$

Dividing Decimals

Move the decimal point to the right in the divisor (the number you are dividing by) and in the dividend, by as many places as there are in the divisor. Then divide, keeping the decimal point in the moved position.

Key Example: $\frac{3.195}{0.15} = \frac{319.5}{15.} = 21.3$

2.2 PERCENTAGES

Percent means *hundredths:*

$$100\% = \frac{100}{100} = 1.00$$
$$25\% = \frac{25}{100} = 0.25$$
$$140\% = \frac{140}{100} = 1.40$$

To Change Percent to Decimal

Move the decimal point two places to the left.

• • •

To Change Decimal to Percent

Move the decimal point two places to the right.

Key Example:

$$\frac{3}{5} = \frac{3(20)}{5(20)} = \frac{60}{100} = 60\% = 0.6$$
$$0.45 = 45\% = \frac{45}{100} = \frac{9}{20}$$
$$8\tfrac{1}{4}\% = 8.25\% = 0.0825 = \frac{8.25}{100} = \frac{825}{10{,}000} = \frac{33}{400}$$

$$\text{Percent Increase} = \frac{(\text{Final Value} - \text{Initial Value}) \times 100\%}{\text{Initial Value}} \qquad \textbf{(2.1)}$$

$$\text{Percent Decrease} = \frac{(\text{Initial Value} - \text{Final Value}) \times 100\%}{\text{Initial Value}}$$

Key Example: A length decreases from 10.5 to 10.1. The percent decrease is:

$$\text{Percent decrease} = \frac{(10.5 - 10.1)(100\%)}{10.5} = \frac{(0.4)(100\%)}{10.5}$$
$$= \frac{40\%}{10.5} = 3.8\%$$

2.3 PRECISION AND ACCURACY

Significant Digits

1. All nonzero digits 1 to 9 are significant.
2. Zeros are significant when they are *not the first* digits in a decimal number or *not the last* digits in a whole number unless marked by a bar.

Key Example:

10.02 (4 significant digits)
0.120 (3 significant digits)
$5\bar{0}00$ (2 significant digits)

Rounding Decimals

1. If the digit to the right of the last rounded digit is *less than 5,* do not change the last rounded digit.
2. If the digit to the right of the last rounded digit is *5 or more than 5,* increase the last rounded digit by one.

Key Example:

$0.6505 \approx 0.7$ (1 significant digit)
$3.516 \approx 3.5$ (2 significant digits)
$8.999 \approx 9.00$ (3 significant digits)

Rounding Whole Numbers

1. If the digit to the right of the last rounded digit is *less than 5,* change all the digits to the right of the last rounded digit to zero.
2. If the digit to the right of the last rounded digit is *5 or more than 5,* increase the last rounded digit by one and change all the digits to the right of the last rounded digit to zero.

Key Example:

$29{,}582 \approx 30{,}000$ (1 significant digit)
$5674 \approx 5700$ (2 significant digits)
$10{,}949 \approx 10{,}900$ (3 significant digits)

Accuracy

The accuracy of an approximate number is measured by the *number of significant digits.*

• • •

Precision

The precision of an approximate number is measured by the *decimal place of the last digit.*

Key Example: 501 is more accurate than 500; 5.00 is more precise than 5.0.

2.4 HAND CALCULATOR OPERATIONS

Check calculator results by rounding off all numbers to one significant digit and estimating the answer by hand.

Key Example:

Estimate:

$$\frac{(3.56)(0.724-0.546)}{(0.812)(0.0390)} \approx \frac{(4)(0.7-0.5)}{(0.8)(0.04)} = \frac{(4)(0.2)}{(0.8)(0.04)}$$

$$= \frac{0.8}{0.032} = \frac{800}{32} = 25$$

Calculation:

3.56 [×] [(] 0.724 [−] 0.546 [)] [÷] [(] 0.812 [×] 0.0390 [)] = → 20.0 (3 significant digits)

REVIEW EXERCISES

In exercises 1 through 14, calculate each exercise by hand. Use the calculator to check the results.

1. $28.88 + 9.45 - 7.07$

2. $0.11 \times 0.050 \times 0.20$

3. $3.2 + 2.1 \times 0.4$

4. $8.6 - 6.8 \times 0.02$

5. $\frac{0.8804}{0.004}$

6. $\frac{9.9}{3.3 \times 0.1}$

7. $5.0 \div 0.20 \times 0.01$

8. $8.8 \times 0.5 \div 1.1$

9. $(0.10 - 0.09) \times (0.31 + 0.29)$

10. $(1.23 - 0.03) \times (0.05 + 0.07)$

11. $\frac{1.2 + 3.0 \times 0.40}{0.02 \times 6.0}$

12. $\frac{0.03 \times 0.1}{0.02 \times 6.0}$

13. $\frac{(5.2 - 3.0) \times 0.40}{0.020}$

14. $\frac{0.07 \times 0.30 \times 9.0}{2.1 \times 1.5 \times 6.0}$

In exercises 15 through 26, express each number as a fraction in lowest terms, a decimal, and a percent.

15. $\frac{4}{5}$

16. $\frac{1}{8}$

17. $\frac{3}{20}$

18. 0.3

19. 0.16

20. 2.0

21. 35%

22. 12.5%

23. $5\frac{1}{2}\%$

24. 0.009

25. $\frac{1}{200}$

26. 150%

In exercises 27 through 34, find the value of each percent.

27. 30% of 28

28. 35% of 70

29. 26% of 150

30. 6% of 15

31. 110% of 75

32. $9\frac{1}{2}\%$ of 90

33. $\frac{1}{2}\%$ of 100

34. 10.2% of 500

In exercises 35 through 38, estimate each answer by hand. Then choose the correct answer and check with the calculator.

35. $\frac{105 + 9 \times 25}{63 - 53}$ [3, 33, 60, 93]

36. $\frac{17 \times 2.0 \times 3.5}{2.5 \times 14}$ [3.4, 8.4, 34, 84]

37. $\frac{(0.08)(0.37 - 0.18)}{0.016}$ [0.19, 0.59, 0.95, 5.9]

38. $\frac{0.33}{5.5} - \frac{0.20}{28.3 + 71.7}$ [0.058, 0.098, 0.58, 0.98]

In exercises 39 through 42, estimate each answer by hand. Then choose the correct answer to three significant digits and check with the calculator.

39. $(3.70)(8.12) + 1.002$ [31, 31.0, 25, 25.0]

40. $1.15 - \frac{242}{856}$ [0.0867, 0.86, 0.867, 8.67]

41. $\frac{(39.03)(86.1)}{(10.0)(5.02)}$ [6.69, 66.9, 669, 6.70]

42. $\frac{0.707 + 0.808}{0.0102}$ [14.8, 14.9, 149, 148.5]

Applied Problems

In problems 43 through 58, round answers to two significant digits unless specified.

43. Find the area, in square inches, and the perimeter, in inches, of a right triangle whose base is 6.33 inches, height is 8.44 inches, and hypotenuse is 10.55 in. Round the answer to three significant digits. Perimeter = (sum of 3 sides). Area = $\frac{1}{2}$(base)(height).

44. The list price of a computer is $1550. One discount store gives 15% off list price and another discount store gives $\frac{1}{5}$ off list price. Find the two discount prices to the nearest cent.

45. A school is given a grant of $40,000 for its computer laboratory. After spending 60% of the grant on new hardware and software, it spends $\frac{1}{4}$ of what is left on new furniture. How much is left of the grant?

46. The earth's surface receives 317 Btu's (British thermal units) per hour per square foot of solar energy on a clear day. How many Btu's per hour are received by a solar panel whose total area is 10 ft^2 but only 75% of it is covered with solar cells? Round the answer to three significant digits.

47. A car battery is advertised for $68. A discount store sells it at a 12% discount. What is the price and the total cost, including tax of $7\frac{1}{2}$%? Give the answer to the nearest cent.

48. A store marks up an item whose wholesale price is $50 by 120%. What is the markup price?

49. The grades of four students on a test are 86, 66, 78, and 75. Find the average test score.

50. A lightning discharge from a cloud 2.5 mi high is measured at a speed of 550 mi/s (miles per second) for the downstroke and 10,000 mi/s for the powerful return stroke (ground to cloud). What is the total time, in seconds, for both strokes? Time = (distance) ÷ (rate of speed).

Applications to Electronics

51. The current in a circuit is initially 2.05 A (amps). The voltage is increased by 20%, which increases the current also by 20%.
(a) How much is the increase in the current in amps?
(b) What is the new current in amps?

52. Three students measure a resistance with their ohmmeters to be 225 Ω, 200 Ω, and 210 Ω. Find the average value of the resistance in ohms.

53. The current in a circuit decreases from 2.0 A to 1.8 A when the voltage is decreased. What is the percent decrease in the current?

54. The voltage across a resistor increases from 24 V to 32 V when the current is increased. What is the percent increase in the voltage?

55. A resistor in a circuit is 4700 Ω. The resistance of a second resistor R in the circuit is found to be 25% less than one-third of the first. What is the resistance of R?

56. The original value of the current in a circuit is 0.62 A. The resistance in the circuit changes in such a way that the current decreases to $\frac{2}{3}$ of its original value. Find the new value of the current in amps.

57. The current through a resistor is measured at five equal intervals to be 20 mA (milliamps), 21 mA, 20 mA, 21.5 mA, and 22 mA. Find the average current, in mA, through the resistor.

58. The length of a wire is shortened from 15 cm to 10 cm, causing its resistance to decrease by the same percentage.
(a) What is the percent decrease in the length and the resistance?
(b) If the resistance of the original wire was 2.5 Ω, what is the resistance in ohms of the shortened wire?

CHAPTER 3

Powers and Roots

CHAPTER OBJECTIVES

In this chapter, you will learn:

- How to raise a number to a power.
- The rules for working with powers of ten.
- How to find the square root of a number.
- How to find the cube root of a number.
- How to raise powers of ten to a positive power.
- How to find roots of powers of ten.
- How to simplify expressions with radical signs.
- How to do powers and roots on the calculator.

In scientific and technical work, numbers are often expressed in terms of powers or exponents. This is especially important in electricity and electronics where powers of ten and the metric system are used to express units and dimensions. Powers of ten simplify working with very large and very small numbers that occur in electronic measurements. The inverse process of raising to a power is finding the root of a number. Square roots and cube roots are used in electrical formulas, and roots can contain numbers times powers of ten. The calculator is designed to be used with powers of ten, and examples are shown throughout the chapter. Section 3.4 is a special section on the calculator that explains further examples with powers and roots and shows how to estimate results.

• • •

3.1 Powers

A power or exponent means repeated multiplication:

Definition

Power or Exponent *n*

$$x^n = \underbrace{(x)(x)(x)\cdots(x)}_{n \text{ times}} \qquad (x = \text{Base}) \qquad (n = \text{Positive whole number}) \tag{3.1}$$

EXAMPLE 3.1

Raise the whole numbers to the powers shown:

(a) 2^5 **(b)** 3^4 **(c)** 5^3 **(d)** 10^6 **(e)** 15^2

Solution

(a) $2^5 = (2)(2)(2)(2)(2) = 32$

(b) $3^4 = (3)(3)(3)(3) = 81$

(c) $5^3 = (5)(5)(5) = 125$

(d) $10^6 = (10)(10)(10)(10)(10)(10) = 1{,}000{,}000$

(e) $15^2 = (15)(15) = 225$

On a calculator, raising to a power is done with either the [x^y], [y^x], or [∧] key. To raise a number to the second power, just press the x squared [x^2] key. The examples above are done on the calculator as follows:

(a) 2 [x^y] 5 [=] → 32
(b) 3 [x^y] 4 [=] → 81
(c) 5 [y^x] 3 [=] → 125
(d) 10 [∧] 6 [=] → 1,000,000
(e) 15 [x^2] → 225

EXAMPLE 3.2 Raise the fractions to the powers shown and express the answers as both decimals and fractions:

(a) $\left(\frac{3}{10}\right)^3$ **(b)** $\left(\frac{2}{5}\right)^4$ **(c)** $\left(\frac{9}{40}\right)^2$

Solution To raise a fraction to a power raise the numerator and the denominator to the power:

(a) $\left(\frac{3}{10}\right)^3 = \frac{3^3}{10^3} = \frac{27}{1000} = 0.027$

(b) $\left(\frac{2}{5}\right)^4 = \frac{2^4}{5^4} = \frac{16}{625} = 0.0256$

(c) $\left(\frac{9}{40}\right)^2 = \frac{9^2}{40^2} = \frac{81}{1600} = 0.050625$

These examples can be done on the calculator using parentheses:
(a) [(] 3 [÷] 10 [)] [x^y] 3 [=] → 0.027
(b) [(] 2 [÷] 5 [)] [y^x] 4 [=] → 0.0256
(c) [(] 9 [÷] 40 [)] [x^2] [=] → 0.050625

EXAMPLE 3.3 Raise the decimals to the powers shown:

(a) $(2.1)^2$ **(b)** $(0.5)^3$ **(c)** $(0.02)^3$ **(d)** $(0.1)^4$

Solution

(a) When you raise a decimal to a power you have to carefully count the decimal places: Since there is one decimal place in 2.1, when you square the number, there will be two decimal places in the result:

$$(2.1)^2 = (2.1)(2.1) = 4.41$$

To square a number on the calculator, press the square key:

2.1 [x^2] → 4.41

(b) When you cube a number with one decimal place, there will be three decimal places in the result:

$$(0.5)^3 = (0.5)(0.5)(0.5) = 0.125$$

This is done on a calculator:

0.5 [y^x] 3 [=] → 0.125

(c) When you cube a number with two decimal places, there will be six decimal places in the result:

$$(0.02)^3 = (0.02)(0.02)(0.02) = 0.000008$$

This is done on a calculator:

0.02 [y^x] 3 [=] → 0.000008

(d) When you raise a number with one decimal place to the fourth power, there will be four decimal places in the answer:

$$(0.3)^4 = 0.0081$$

Raising to the fourth power can be done on a calculator by squaring twice:

0.3 [x^2] [x^2] → 0.0081

In the order of operations, *raising to a power, or exponent,* is done before multiplication or division:

Rule

Order of Operations

1. Operations in parentheses.
2. Exponents.
3. Multiplication or division.
4. Addition or subtraction.

To remember the order of operations: ***P***arentheses, ***E***xponents, ***M***ultiplication, ***D***ivision, ***A***ddition, ***S***ubtraction, memorize the phrase "***P***lease ***E***xcuse ***M***y ***D***ear ***A***unt ***S***ally" (PEMDAS).

ERROR BOX

Errors can occur when raising a fraction or a decimal to a power by not paying attention to the fraction line or the number of decimal places. A fraction raised to a power means the numerator *and* the denominator are both raised to the same power:

$$\left(\frac{3}{10}\right)^4 = \frac{3^4}{10^4} = \frac{81}{10{,}000} \text{ or } 0.0081$$

When raising a decimal to a power, multiply the exponent by the number of decimal places to obtain the correct number of decimal places in the answer:

$$(0.05)^3 = 0.000125 \qquad (3 \times 2 = 6 \text{ decimal places})$$

See if you can do the practice problems without the calculator. Use the calculator to check your answer.

Practice Problems: In 1 through 8, express each answer as a fraction and a decimal.

1. $\left(\frac{1}{2}\right)^4$ **2.** $\left(\frac{2}{5}\right)^3$ **3.** $\left(\frac{12}{5}\right)^2$ **4.** $\left(\frac{1}{20}\right)^3$

5. $\left(\frac{7}{10}\right)^3$ **6.** $\left(\frac{3}{25}\right)^2$ **7.** $\left(\frac{3}{2}\right)^4$ **8.** $\left(\frac{9}{50}\right)^3$

9. $(0.9)^3$ **10.** $(0.11)^2$ **11.** $(1.2)^3$ **12.** $(0.1)^4$

13. $(2.5)^2$ **14.** $(0.2)^5$ **15.** $(0.01)^3$ **16.** $(0.5)^3$

EXAMPLE 3.4 Evaluate the powers and simplify:

$$\frac{(2)^6(7)^2}{(4)^3(5)^2}$$

Solution Raise each number to the power first, then simplify:

$$\frac{(2)^6(7)^2}{(4)^3(5)^2} = \frac{(\cancel{64})(49)}{(\cancel{64})(25)} = \frac{49}{25} \text{ or } 1.96$$

One calculator solution is:

2 [yˣ] 6 [×] 7 [x²] [÷] [(] 4 [yˣ] 3 [×] 5 [x²] [)] [=] → 1.96

EXAMPLE 3.5 Perform the operations:

$$\left(\frac{3}{2}\right)^2 - (6)\left(\frac{1}{2}\right)^3$$

Solution First raise the fractions to the powers:

$$\left(\frac{3}{2}\right)^2 - (6)\left(\frac{1}{2}\right)^3 = \frac{9}{4} - (6)\left(\frac{1}{8}\right)$$

Then do the multiplication and subtract the fractions, reducing the result:

$$\frac{9}{4} - (6)\left(\frac{1}{8}\right) = \frac{9}{4} - \frac{(\cancel{2})(3)}{(\cancel{2})(2)(2)} = \frac{9}{4} - \frac{3}{4} = \frac{6}{4} = \frac{3}{2} \text{ or } 1.5$$

This can be done on a calculator:

[(] 3 [÷] 2 [)] [x²] [−] 6 [×] [(] 1 [÷] 2 [)] [yˣ] 3 [=] → 1.5

This can also be done on a calculator with a fraction key:

3 [a b/c] 2 [x²] [−] 6 [×] 1 [a b/c] 2 [yˣ] 3 [=] → 3/2

EXAMPLE 3.6 Perform the operations:

$$\frac{(0.1)^2 + (0.3)^3}{(0.3)^2 - (0.2)^2}$$

Solution Raise to the powers first:

$$\frac{(0.1)^2 + (0.3)^3}{(0.3)^2 - (0.2)^2} = \frac{0.01 + 0.027}{0.09 - 0.04}$$

Since the fraction line is like parentheses, you must do the operations in the numerator and denominator before you divide:

$$\frac{0.01 + 0.027}{0.09 - 0.04} = \frac{0.037}{0.05} = \frac{0.037(1000)}{0.05(1000)} = \frac{3.7}{50} = 0.74$$

One calculator solution is:

0.1 [x²] [+] 0.3 [yˣ] 3 [=] [÷] [(] 0.3 [x²] [−] 0.2 [x²] [)] [=] → 0.74

Answers to Error Box Problems, page 53:

1. $\frac{1}{16} = 0.0625$ **2.** $\frac{8}{125} = 0.064$ **3.** $\frac{144}{25} = 5.76$ **4.** $\frac{1}{8000} = 0.000125$ **5.** $\frac{343}{1000} = 0.343$ **6.** $\frac{9}{625} = 0.0144$ **7.** $\frac{81}{16} = 5.0625$ **8.** $\frac{729}{125{,}000} = 0.005832$ **9.** 0.729 **10.** 0.0121 **11.** 1.728 **12.** 0.0001 **13.** 6.25 **14.** 0.00032 **15.** 0.000001 **16.** 0.125

EXAMPLE 3.7 Perform the operations:

$$\left(\frac{1}{2}-\frac{1}{3}\right)^2\left(\frac{1}{2}+\frac{1}{10}\right)^2$$

Solution Do the operations in parentheses first:

$$\left(\frac{1}{2}-\frac{1}{3}\right)^2\left(\frac{1}{2}+\frac{1}{10}\right)^2=\left(\frac{3-2}{6}\right)^2\left(\frac{5+1}{10}\right)^2=\left(\frac{1}{6}\right)^2\left(\frac{6}{10}\right)^2=\left(\frac{1}{6}\right)^2\left(\frac{3}{5}\right)^2$$

Notice that the fraction $\frac{6}{10}$ is reduced to $\frac{3}{5}$. Square the fractions and multiply:

$$\left(\frac{1}{6}\right)^2\left(\frac{3}{5}\right)^2=\left(\frac{1}{36}\right)\left(\frac{9}{25}\right)=\left(\frac{1}{(2)(2)(\cancel{3})(\cancel{3})}\right)\left(\frac{(\cancel{3})(\cancel{3})}{25}\right)=\frac{1}{100}\text{ or }0.01$$

One calculator solution is:

1 [÷] 2 [−] 1 [÷] 3 [=] [x^2] [×] [(] 1 [÷] 2 [+] 1 [÷] 10 [)] [x^2] [=] → 0.01

EXERCISE 3.1

In exercises 1 through 50, evaluate each exercise by hand. Express each answer as either: a whole number, or as both a fraction and a decimal. Use the calculator to check your answers.

1. 3^4
2. 4^3
3. 2^7
4. 5^3
5. 10^5
6. 25^2
7. 11^2
8. 20^3
9. $\left(\frac{1}{2}\right)^4$
10. $\left(\frac{3}{10}\right)^3$
11. $\left(\frac{3}{5}\right)^2$
12. $\left(\frac{1}{5}\right)^3$
13. $\left(\frac{1}{10}\right)^4$
14. $\left(\frac{7}{10}\right)^2$
15. $\left(\frac{5}{4}\right)^2$
16. $\left(\frac{3}{5}\right)^4$
17. $(0.7)^2$
18. $(1.2)^2$
19. $(0.4)^3$
20. $(0.5)^3$
21. $(0.1)^4$
22. $(0.2)^4$
23. $(0.02)^3$
24. $(0.4)^3$
25. $\dfrac{(4)^4}{(8)^2(6)^2}$
26. $\dfrac{(10)^4}{(5)^2(2)^3}$
27. $\dfrac{5^2-4^2}{3^2}$
28. $\dfrac{2^3+3^3}{5^2}$
29. $\dfrac{2^4+3^2}{10^3}$
30. $\dfrac{12^2}{4^3-2^2}$
31. $\left(\frac{1}{5}\right)^2+\left(\frac{4}{5}\right)\left(\frac{1}{10}\right)^2$
32. $\left(\frac{5}{6}\right)^2-\left(\frac{2}{3}\right)^2$
33. $(6)\left(\frac{1}{4}\right)^2-(9)\left(\frac{1}{6}\right)^2$
34. $\left(\frac{3}{10}\right)^3+(4)\left(\frac{1}{10}\right)^2$
35. $\dfrac{(0.2)^3}{(0.5)^2-(0.3)^2}$
36. $\dfrac{(0.1)^3+(0.01)^2}{(0.1)^2}$
37. $\dfrac{(1.5)^2-(0.5)^2}{(0.8)^2}$
38. $\dfrac{(1.2)^2}{(0.3)^2+(0.4)^2}$
39. $(0.02)\left(\dfrac{0.4+0.2}{0.2}\right)^2$
40. $\left(\dfrac{0.9}{0.5-0.4}\right)^2(0.3)^2$
41. $\dfrac{(0.7)^2(0.3)^3}{(2.1)^2}-(0.1)^3$
42. $(0.8)^2\left(\dfrac{(0.2)^3(2.0)^2}{(0.4)^3}\right)$
43. $\dfrac{2^5}{8^2}-\left(\frac{3}{4}\right)^3$
44. $\left(\frac{1}{5}\right)^3+\dfrac{2^4}{10^3}$
45. $\dfrac{(0.3)^3-(0.1)^2}{(0.3)^2+(0.4)^2}$
46. $\dfrac{(0.4)^2-(0.2)^3}{(0.5)^2+(0.3)^2}$
47. $\left(\dfrac{0.5}{0.1}\right)^2-\left(\frac{5}{2}\right)^2$
48. $\dfrac{2^3}{0.4^2}+\left(\frac{3}{2}\right)^2$
49. $\left(\frac{1}{2}+\frac{1}{3}\right)^2\left(\frac{1}{2}-\frac{1}{10}\right)^2$
50. $\left(\frac{1}{2}-\frac{1}{6}\right)^2\left(\frac{1}{4}+\frac{1}{2}\right)^2$

Applied Problems

In problems 51 through 62, solve each problem by hand. Check with the calculator.

51. The volume of a cube $V = s^3$, where s = side. Find the volume of a small cubic container in cubic centimeters (cm^3), when $s = 0.8$ cm.

52. A baseball "diamond" is actually a square whose side s = 90 ft. Find the area A of a baseball diamond in square feet (ft^2), where $A = s^2$.

53. At an inflation rate of 10% per year, the cost C of a $20,000 car will increase to approximately $C = 20{,}000(1.10)^3$ in 3 years. How much will this approximate cost be?

54. An investment of $100 in a bank certificate of deposit (CD) whose annual percentage rate is 7% will amount to $A = 100(1.07)^5$ in five years. Find the amount A to the nearest cent.

55. The side of a square circuit board $s = 1.5$ mm. Find the area A of the board in square millimeters (mm^2), where $A = s^2$.

56. A computer cabinet has a square base whose side s = 0.6 ft. The height of the cabinet is $h = 1.1$ ft. Find the volume V of the cabinet in cubic feet (ft^3), where $V = (s^2)(h)$.

Applications to Electronics

57. Find the resistance R in ohms (Ω) of a circuit given by:

$$R = \frac{45}{1.5^2}$$

58. Find the resistance R in ohms (Ω) of a circuit given by:

$$R = \frac{50^2}{20}$$

59. The power in watts (W) dissipated by a resistor in a circuit is given by the power formula:

$$P = (I^2)(R)$$

where the current $I = 0.50$ A (amps) and the resistance $R = 68\ \Omega$ (ohms). Substitute the values and calculate P in watts.

60. In problem 59, find P in watts (W) when $I = 1.3$ A and $R = 30\ \Omega$.

61. The power in watts (W) dissipated by a resistor in a circuit is given by the power formula:

$$P = \frac{V^2}{R}$$

where the voltage $V = 1.2$ V (volts) and the resistance $R = 24\ \Omega$. Substitute the values and calculate P in watts (W).

62. In problem 61, find P in watts (W) when $V = 60$ V and $R = 200\ \Omega$.

3.2 Powers of Ten

Our number system is called the decimal system because it is based on 10 and the powers of ten. A number in the decimal system such as 2546 can be expressed in terms of powers of ten as follows:

$$\begin{aligned} 2546 &= 2(1000) + 5(100) + 4(10) + 6(1) \\ &= (2 \times 10^3) + (5 \times 10^2) + (4 \times 10^1) + (6 \times 10^0) \end{aligned}$$

where

$$\begin{aligned} 10^3 &= 1000 \\ 10^2 &= 100 \\ 10^1 &= 10 \end{aligned}$$

and the *zero power* of 10 is defined as:

Definition **Zero Power**

$$10^0 = 1 \tag{3.2}$$

A number that is a decimal fraction such as 8.723 can be expressed in terms of powers of ten as follows:

$$8.723 = 8(1) + 7(0.1) + 2(0.01) + 3(0.001)$$
$$= (8 \times 10^0) + (7 \times 10^{-1}) + (2 \times 10^{-2}) + (3 \times 10^{-3})$$

where the negative powers of ten are decimals between 0 and 1 and mean *division by positive powers* of ten:

$$10^{-1} = 0.1 = \frac{1}{10^1} = \frac{1}{10}$$
$$10^{-2} = 0.01 = \frac{1}{10^2} = \frac{1}{100}$$
$$10^{-3} = 0.001 = \frac{1}{10^3} = \frac{1}{1000}$$

Study Table 3-1, which shows more positive, zero, and negative powers of ten.

TABLE 3-1 POWERS OF TEN

Power	Fraction or Whole Number	Decimal
$10^{-9} = \frac{1}{10^9}$	$\frac{1}{10^9} = \frac{1}{1{,}000{,}000{,}000}$	0.000000001
$10^{-6} = \frac{1}{10^6}$	$\frac{1}{10^6} = \frac{1}{1{,}000{,}000}$	0.000001
$10^{-5} = \frac{1}{10^5}$	$\frac{1}{10^5} = \frac{1}{100{,}000}$	0.00001
$10^{-4} = \frac{1}{10^4}$	$\frac{1}{10^4} = \frac{1}{10{,}000}$	0.0001
$10^{-3} = \frac{1}{10^3}$	$\frac{1}{10^3} = \frac{1}{1000}$	0.001
$10^{-2} = \frac{1}{10^2}$	$\frac{1}{10^2} = \frac{1}{100}$	0.01
$10^{-1} = \frac{1}{10^1}$	$\frac{1}{10^1} = \frac{1}{10}$	0.1
10^0	1	1.0
10^1	10	10.0
10^2	100	100.0
10^3	1000	1000.0
10^4	10,000	10,000.0
10^5	100,000	100,000.0
10^6	1,000,000	1,000,000.0
10^9	1,000,000,000	1,000,000,000.0

Observe in the Decimal column in Table 3-1 that *the power of ten determines the place of the decimal point.* For every increase of one in the power of ten, the decimal point moves one place to the right. For every decrease of one in the power of ten, the decimal point moves one place to the left.

Rule **Positive Power of Ten**

The power of ten is equal to the number of zeros:

$$10^1 = 10, \quad 10^2 = 100, \quad 10^3 = 1000, \quad 10^6 = 1{,}000{,}000, \text{ etc.}$$

When the power of ten is negative, the concept of absolute value is helpful in understanding the powers of ten:

Definition

Absolute Value

The value of a number without the sign (positive value).

The absolute value of -1 is 1, the absolute value of 2 is 2, the absolute value of -3 is 3, the absolute value of 0 is 0, etc.

Definition

Negative Power of Ten

$$10^{-n} = \frac{1}{10^n} \quad (n = \text{positive number}) \tag{3.3}$$

The absolute value of the power is equal to the number of decimal places:

$$10^{-1} = 0.1 \text{ (1 dec. place)} \quad 10^{-2} = 0.01 \text{ (2 dec. places)}$$
$$10^{-6} = 0.000001 \text{ (6 dec. places), etc.}$$

EXAMPLE 3.8 Express the negative powers of ten as fractions and decimals:

(a) 10^{-1} **(b)** 10^{-3} **(c)** 10^{-6} **(d)** 10^{-8}

Solution Apply definition (3.3):

(a) $10^{-1} = \frac{1}{10^1} = \frac{1}{10} = 0.1$

(b) $10^{-3} = \frac{1}{10^3} = \frac{1}{1000} = 0.001$

(c) $10^{-6} = \frac{1}{10^6} = \frac{1}{1{,}000{,}000} = 0.000001$

(d) $10^{-8} = \frac{1}{10^8} = \frac{1}{100{,}000{,}000} = 0.00000001$

EXAMPLE 3.9 Express the decimals as powers of ten and fractions:

(a) 0.1 **(b)** 0.01 **(c)** 0.0001 **(d)** 0.0000001

Solution Apply definition (3.3):

(a) $0.1 = 10^{-1} = \frac{1}{10}$

(b) $0.01 = 10^{-2} = \frac{1}{10^2} = \frac{1}{100}$

(c) $0.0001 = 10^{-4} = \frac{1}{10^4} = \frac{1}{10{,}000}$

(d) $0.0000001 = 10^{-7} = \frac{1}{10^7} = \frac{1}{10{,}000{,}000}$

Multiplying and Dividing Powers of Ten

The three *rules for multiplying with powers of ten* are as follows:

Rule **Multiplying Positive Powers of Ten**

Add the exponents only. Do not change the base 10.

$$10^3 \times 10^2 = 10^{3+2} = 10^5$$

• • •

Rule **Multiplying Negative Powers of Ten**

Add the absolute values of the exponents and make the result negative.

$$10^{-4} \times 10^{-2} = 10^{-(4+2)} = 10^{-6}$$

• • •

Rule **Multiplying Positive and Negative Powers of Ten**

Subtract the smaller absolute value from the larger absolute value and use the sign of the larger absolute value.

$$10^6 \times 10^{-2} = 10^{+(6-2)} = 10^4 \qquad 10^{-6} \times 10^2 = 10^{-(6-2)} = 10^{-4}$$

Study the following examples, which show further how to multiply with powers of ten.

EXAMPLE 3.10 Multiply:

$$10^3 \times 10^9 \times 10^{-2}$$

Solution First apply the rule for multiplying positive powers of ten. Add 3 to 9:

$$10^3 \times 10^9 \times 10^{-2} = 10^{3+9} \times 10^{-2} = 10^{12} \times 10^{-2}$$

Then apply the rule for multiplying positive and negative powers of ten. Subtract 2 from 12 and make the result positive:

$$10^{12} \times 10^{-2} = 10^{+(12-2)} = 10^{10}$$

EXAMPLE 3.11 Multiply:

$$10^4 \times 10^{-6} \times 10^{-1}$$

Solution First apply the rule for multiplying positive and negative powers of ten. Subtract 4 from 6 and make the result negative:

$$\begin{aligned} 10^4 \times 10^{-6} \times 10^{-1} &= 10^{-(6-4)} \times 10^{-1} \\ &= 10^{-2} \times 10^{-1} \end{aligned}$$

Then apply the rule for multiplying negative powers of ten. Add 2 to 1 and make the result negative:

$$10^{-2} \times 10^{-1} = 10^{-(2+1)} = 10^{-3}$$

Rule

Dividing Powers of Ten

Change the sign of the power in the divisor and *apply the rules for multiplying* with powers of ten.

$$\frac{10^6}{10^3} = 10^6 \times 10^{-3} = 10^{+(6-3)} = 10^3$$

EXAMPLE 3.12

Divide:

$$\frac{10^5}{10^{-2}}$$

Solution Change the sign of –2 to 2 and apply the rule for multiplying positive powers of ten:

$$\frac{10^5}{10^{-2}} = 10^5 \times 10^2 = 10^{5+2} = 10^7$$

EXAMPLE 3.13

Divide:

$$\frac{10^{-3}}{10^6}$$

Solution Change the power of 6 to –6 and apply the rule for multiplying negative powers of ten:

$$\frac{10^{-3}}{10^6} = 10^{-3} \times 10^{-6} = 10^{-(3+6)} = 10^{-9}$$

EXAMPLE 3.14

Perform the operations:

$$10^{-1} \times 10^{-3} \times \frac{10^4}{10^{-6}}$$

Solution Multiply the first two powers of ten and divide the last two powers of ten:

$$10^{-1} \times 10^{-3} \times \frac{10^4}{10^{-6}} = 10^{-(1+3)} \times 10^4 \times 10^6$$

$$= 10^{-4} \times 10^{10}$$

Then multiply again:

$$10^{-4} \times 10^{10} = 10^{(10-4)} = 10^6$$

Numbers Times Powers of Ten

A very large or a very small number is often expressed in terms of a number times a power of ten as follows:

$$72{,}000{,}000 = 72 \times 1{,}000{,}000 = 72 \times 10^6$$

$$0.0045 = 4.5 \times 0.001 = 4.5 \times 10^{-3}$$

Numbers times powers of ten can be multiplied and divided as follows:

Procedure

Multiplying and Dividing Numbers Times Powers of Ten

1. Multiply or divide the numbers to obtain the number part of the answer.
2. Apply the rules for multiplying or dividing to the powers of ten.

EXAMPLE 3.15 Multiply and express the answer as a power of ten:

$$(2 \times 10^2)(6 \times 10^4)$$

Solution Multiply the numbers 2 and 6, and apply the rule for multiplying to the powers of ten:

$$(2 \times 10^2)(6 \times 10^4) = (2)(6) \times 10^{2+4} = 12 \times 10^6$$

EXAMPLE 3.16 Divide and express the answer as a power of ten:

$$\frac{8 \times 10^{-3}}{4 \times 10^{-1}}$$

Solution Divide the number 8 by 4, and apply the rule for dividing to the powers of ten:

$$\frac{8 \times 10^{-3}}{4 \times 10^{-1}} = \frac{8}{4} \times (10^{-3})(10^{1}) = 2 \times 10^{-(3-1)} = 2 \times 10^{-2}$$

EXAMPLE 3.17 Perform the operations and express the answer as a power of ten:

$$\frac{(4 \times 10^{-6})(6 \times 10^{3})}{8 \times 10^{-3}}$$

Solution First calculate the numerator. Multiply the numbers 4 and 6 and apply the rule for multiplying to the powers of ten:

$$\frac{(4 \times 10^{-6})(6 \times 10^{3})}{8 \times 10^{-3}} = \frac{(4)(6) \times 10^{-(6-3)}}{8 \times 10^{-3}} = \frac{24 \times 10^{-3}}{8 \times 10^{-3}}$$

Then divide the number 24 by 8, and apply the division rule to the powers:

$$\frac{24 \times 10^{-3}}{8 \times 10^{-3}} = \frac{24}{8} \times 10^{-3} \times 10^{3}$$

$$= 3 \times 10^{(3-3)} = 3 \times 10^{0} = 3 \times 1 = 3$$

Notice that multiplying by 10^0 is the same as multiplying by 1, and it does not change the value of the number.

This example can be done on a calculator using the [EXP] or [EE] key, which allows you to enter the power of ten after the number.

DAL: 4 [EE] [(-)] 6 [×] 6 [EE] 3 [÷] 8 [EE] [(-)] 3 [=] → 3

Not DAL: 4 [EE] 6 [+/-] [×] 6 [EE] 3 [÷] 8 [EE] 3 [+/-] [=] → 3

Note that *you do not key in the 10*, only the power.

Rule

Adding Numbers Times Powers of Ten

Numbers times powers of ten can be added or subtracted *only when the powers are the same.* Add or subtract the numbers only. *Do not change* the power of ten:

$$(5 \times 10^6) + (7 \times 10^6) = (7 + 5) \times 10^6 = 12 \times 10^6$$

If the powers are different, the powers must either be changed so they are the same, or the numbers can be changed to ordinary notation before adding or subtracting. The rule for changing the power of ten is:

Rule

Changing the Power of Ten

- To increase the power of ten, move the decimal point to the left and increase the power of ten by the number of places you moved the decimal point.

$$42 \times 10^5 = 4.2 \times 10^6 \qquad 525 \times 10^{-9} = 52.5 \times 10^{-8}$$

- To decrease the power of ten, move the decimal point to the right and decrease the power of ten by the number of places you moved the decimal point.

$$2.13 \times 10^4 = 213 \times 10^2 \qquad 7.34 \times 10^{-3} = 73.4 \times 10^{-4}$$

Observe that for the number 525×10^{-9} above, the decimal point is moved to the left one place and the power of ten is *increased* from –9 to –8. For the number 7.34×10^{-3} above, the decimal point is moved to the right one place and the power of ten is *decreased* by one from –3 to –4.

The rule for changing to ordinary notation is:

Rule

Changing from Power of Ten to Ordinary Notation

- Move the decimal point to the right (positive) or to the left (negative) the number of places indicated by the exponent.

$$0.879 \times 10^6 = 879{,}000, \quad 5.6 \times 10^3 = 5600, \quad 23.6 \times 10^{-3} = 0.0236, \quad 458 \times 10^{-6} = 0.000458$$

ERROR BOX

To avoid making an error when changing the power of ten, think of it as balancing the two factors of the number. One factor is the power of ten, and the other factor is the number that is multiplied by the power of ten. If one increases, the other must decrease by the same factor. For example, here are four ways of writing two different numbers:

$$0.08 \times 10^9 = 0.8 \times 10^8 = 8 \times 10^7 = 800 \times 10^5$$

$$300 \times 10^{-9} = 30 \times 10^{-8} = 3 \times 10^{-7} = 0.3 \times 10^{-6}$$

Notice that the number of places the decimal point is moved equals the difference in the exponents. If the decimal point is moved to the right increasing the number, the exponent decreases, and vice versa. See if you can apply these ideas and do the practice problems correctly.

Practice Problems: Write each number three ways using the given powers of ten:

1. 40×10^2 (10^6, 10^3, 10^{-1})
2. 6000×10^{-1} (10^0, 10^2, 10^3)
3. 0.002×10^6 (10^3, 10^4, 10^5)
4. 7×10^{-3} (10^{-6}, 10^{-4}, 10^{-2})
5. $50{,}000 \times 10^{-6}$ (10^{-5}, 10^{-3}, 10^0)
6. 15×10^9 (10^8, 10^{10}, 10^6)
7. 34×10^2 (10^1, 10^0, 10^{-1})
8. 9.09×10^{-1} (10^{-3}, 10^{-2}, 10^0)

Study the following examples, which show several ways that you can add numbers with different powers of ten. They also show the relationship between power of ten notation and ordinary notation.

EXAMPLE 3.18 Add the numbers and express the answer as a power of ten and in ordinary notation:

$$(9 \times 10^3) + (6 \times 10^2)$$

Solution Change 10^3 to 10^2 for the first number, so both powers are the same. Move the decimal point one place to the right, changing 9 to 90, and decrease its power of ten by 1 at the same time. This balances the entire number by multiplying part of it by ten and dividing the other part by ten, which does not change the value of the number:

$$9 \times 10^3 = 90 \times 10^2$$

You can then add the numbers as changed because the powers of ten are the same:

$$(90 \times 10^2) + (6 \times 10^2) = (90 + 6) \times 10^2 = 96 \times 10^2 = 9600$$

You can also add the numbers by changing 6 to 0.6 and 10^2 to 10^3 for the second number:

$$(9 \times 10^3) + (6 \times 10^2) = (9 \times 10^3) + (0.6 \times 10^3)$$
$$= (9 + 0.6) \times 10^3 = 9.6 \times 10^3 = 9600$$

A third way is to change to ordinary notation and add:

$$(9 \times 10^3) + (6 \times 10^2) = 9000 + 600 = 9600$$

The calculator solution is:

9 [EXP] 3 [+] 6 [EXP] 2 [=] → 9600

EXAMPLE 3.19 Subtract the numbers and express the answer as a power of ten and in ordinary notation:

$$(2 \times 10^{-2}) - (5 \times 10^{-3})$$

Solution Change to like powers of ten by changing the power of ten for the first number from −2 to −3:

$$(2 \times 10^{-2}) - (5 \times 10^{-3}) = (20 \times 10^{-3}) - (5 \times 10^{-3}) = 15 \times 10^{-3} = 0.015$$

Notice that 2×10^{-2} is changed to 20×10^{-3}. The power of 10 is *decreased by one* from −2 to −3. To balance the *decrease* in the power of ten, you move the decimal point one place to the right, changing 2 to 20, which *increases* the number by a power of ten.

You can also change the power of ten for the second number from −3 to −2:

$$(2 \times 10^{-2}) - (5 \times 10^{-3}) = (2 \times 10^{-2}) - (0.5 \times 10^{-2}) = 1.5 \times 10^{-2} = 0.015$$

Notice that 5×10^{-3} is changed to 0.5×10^{-2}. The power of ten is *increased by one* from −3 to −2. To balance the *increase* in the power of 10, you move the decimal point one place to the left, changing 5 to 0.5, which *decreases* the number by a power of ten.

You can also change to ordinary notation:

$$(2 \times 10^{-2}) - (5 \times 10^{-3}) = 0.020 - 0.005 = 0.015$$

This example is done on the calculator using the [EXP] or [EE] key for the power of ten:

DAL: 2 [EE] [(−)] 2 [−] 5 [EE] [(−)] 3 [=] → 0.015

Not DAL: 2 [EXP] 2 [+/-] [−] 5 [EXP] 3 [+/-] [=] → 0.015

Note that for a DAL calculator the negative sign is entered before the number; for a calculator that is not DAL, the negative sign is entered after the number.

Study the next example, which closes the circuit and shows an application of powers of ten to a problem involving an electric circuit and Ohm's law.

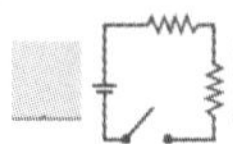

EXAMPLE 3.20

Close the Circuit

An electric circuit contains a resistance R connected to a voltage V where I is the current in the circuit.

(a) Given Ohm's law: $V = (I)(R)$ or $V = IR$. Find V in volts (V) when $I = 4.2 \times 10^{-3}$ A (amps) and $R = 1.5 \times 10^3\ \Omega$ (ohms).

(b) Given Ohm's law: $I = \frac{V}{R}$. Find I in amps when $V = 9.3$ V and $R = 1.5 \times 10^3\ \Omega$.

Solution

(a) To find V, substitute the given values for I and R in Ohm's law $V = IR$ and apply the rule for multiplying powers of ten:

$$V = IR = (4.2 \times 10^{-3})(1.5 \times 10^3) = (4.2)(1.5) \times 10^{(3-3)}$$
$$= 6.3 \times 10^0 = 6.3 \text{ V}$$

Notice in the last step that $10^0 = 1$ so it does not need to be written.

(b) To find I, substitute the given values for V and R in Ohm's law $I = \frac{V}{R}$ and apply the rule for dividing powers of ten:

$$I = \frac{V}{R} = \frac{9.3}{1.5 \times 10^3} = \frac{9.3 \times 10^0}{1.5 \times 10^3} = \frac{9.3}{1.5} \times 10^{(0-3)}$$
$$= 6.2 \times 10^{-3} = 0.0062 \text{ A}$$

EXERCISE 3.2

In exercises 1 through 12, write each number as a power of ten.

1. 1000
2. 1,000,000
3. 1
4. 1,000,000,000
5. 0.000001
6. 0.001
7. 0.0001
8. 0.1
9. $\frac{1}{100}$
10. $\frac{1}{100{,}000}$
11. $\frac{1}{10}$
12. 1

In exercises 13 through 24, write each power of ten either as a whole number or as both a decimal and a fraction.

13. 10^5
14. 10^8
15. 10^6
16. 10^0
17. 10^{-2}
18. 10^{-4}
19. 10^{-1}
20. 10^{-3}
21. $\frac{1}{10^6}$
22. $\frac{1}{10^9}$
23. $\frac{1}{10^{-1}}$
24. $\frac{1}{10^{-3}}$

In exercises 25 through 72, perform each calculation by hand and express the answer in terms of a power of ten.

25. $10^4 \times 10^2$
26. $10^3 \times 10^6$
27. $10^{-3} \times 10^{-2}$
28. $10^{-9} \times 10^{-1}$
29. $10^{-9} \times 10^3$
30. $10^5 \times 10^{-6}$
31. $10^{-3} \times 10^{12}$
32. $10^8 \times 10^{-2}$
33. $10^0 \times 10^{-9}$
34. $10^{-6} \times 10^6$
35. $\frac{10^6}{10^3}$
36. $\frac{10^5}{10^1}$
37. $\frac{10^2}{10^4}$
38. $\frac{10^7}{10^{12}}$
39. $\frac{10^{-5}}{10^{-6}}$
40. $\frac{10^{-2}}{10^{-4}}$

Answers to Error Box Problems, page 62:

1. 0.004×10^6, 4×10^3, $40{,}000 \times 10^{-1}$ **2.** 600×10^0, 6×10^2, 0.6×10^3
3. 2×10^3, 0.2×10^4, 0.02×10^5 **4.** 7000×10^{-6}, 70×10^{-4}, 0.7×10^{-2}
5. 5000×10^{-5}, 50×10^{-3}, 0.05×10^0 **6.** 150×10^8, 1.5×10^{10}, $15{,}000 \times 10^6$
7. 340×10^1, 3400×10^0, $34{,}000 \times 10^{-1}$ **8.** 909×10^{-3}, 90.9×10^{-2}, 0.909×10^0

41. $\dfrac{10^{-5}}{10^{-3}}$

42. $\dfrac{10^{-9}}{10^{-3}}$

43. $\dfrac{10^{0}}{10^{-2}}$

44. $\dfrac{10^{0}}{10^{6}}$

45. $10^3 \times 10^{-6} \times 10^2$

46. $10^{-5} \times 10^{-7} \times 10^8$

47. $10^{-3} \times 10^{10} \times 10^{-3}$

48. $10^{12} \times 10^{-9} \times 10^4$

49. $10^7 \times 10^{-7} \times 10^2$

50. $10^{-5} \times 10^7 \times 10^{-2}$

51. $10^{-1} \times 10^0 \times 10^{-2}$

52. $10^5 \times 10^{-5} \times 10^0$

53. $10^4 \times 10^{-3} \times \dfrac{10^5}{10^{-3}}$

54. $10^{12} \times 10^{-9} \times \dfrac{10^{-6}}{10^3}$

55. $\dfrac{10^8 \times 10^{-4}}{10^2 \times 10^3}$

56. $\dfrac{10^{-5} \times 10^{-4}}{10^{-9} \times 10^8}$

57. $(2 \times 10^{-4})(6 \times 10^6)$

58. $(7 \times 10^5)(7 \times 10^{-5})$

59. $(8 \times 10^5)(3 \times 10^{-4})$

60. $(9 \times 10^3)(8 \times 10^{-1})$

61. $(3 \times 10^{-6})(2 \times 10^{-6})$

62. $(5 \times 10^{-9})(9 \times 10^5)$

63. $(6 \times 10^{-6})(2 \times 10^5)$

64. $(1 \times 10^3)(4 \times 10^3)$

65. $\dfrac{15 \times 10^{-6}}{5 \times 10^{-9}}$

66. $\dfrac{20 \times 10^{-6}}{4 \times 10^{-3}}$

67. $\dfrac{10 \times 10^{4}}{4 \times 10^{-3}}$

68. $\dfrac{30 \times 10^{0}}{6 \times 10^{-7}}$

69. $\dfrac{(2 \times 10^{-3})(5 \times 10^3)}{4 \times 10^2}$

70. $\dfrac{(6 \times 10^{4})(6 \times 10^{-2})}{8 \times 10^{-3}}$

71. $\dfrac{6 \times 10^5}{(8 \times 10^2)(10^{-3})}$

72. $\dfrac{12 \times 10^{-4}}{(2 \times 10^{-3})(6 \times 10^{-6})}$

In exercises 73 through 82, calculate by hand and give the answer as a power of ten and in ordinary notation.

73. $(7 \times 10^3) + (9 \times 10^3)$

74. $(15 \times 10^{-6}) + (25 \times 10^{-6})$

75. $(18 \times 10^{-9}) - (11 \times 10^{-9})$

76. $(10 \times 10^{12}) - (5 \times 10^{12})$

77. $(5 \times 10^3) + (6 \times 10^4)$

78. $(10 \times 10^2) + (5 \times 10^3)$

79. $(2 \times 10^{-3}) + (9 \times 10^{-4})$

80. $(4 \times 10^{-1}) + (3 \times 10^{-2})$

81. $(5 \times 10^1) - (5 \times 10^0)$

82. $(10 \times 10^{-3}) - (1 \times 10^{-2})$

Applications to Electronics

In problems 83 through 90, calculate each problem by hand. Give the answer in ordinary notation.

83. Calculate the power in watts (W) of a circuit given by:

$$P = (1.2 \times 10^{-1})(4 \times 10^3)$$

84. Calculate the power in watts (W) of a circuit given by:

$$P = (0.5 \times 10^3)(7.6 \times 10^{-3})$$

85. Given Ohm's law $V = IR$, find V in volts (V) when $I = 2.5 \times 10^{-3}$ A and $R = 1.2 \times 10^3\ \Omega$.

86. Given Ohm's law $V = IR$, find V in volts (V) when $I = 3.0 \times 10^{-6}$ A and $R = 3.3 \times 10^3\ \Omega$.

87. Given Ohm's law $I = \frac{V}{R}$, find I in amps (A) when $V = 60$ V and $R = 1.2 \times 10^3\ \Omega$.

88. Given Ohm's law $I = \frac{V}{R}$, find I in amps (A) when $V = 1.1 \times 10^3$ V and $R = 10 \times 10^3\ \Omega$.

89. Given Ohm's law $R = \frac{V}{I}$, find R in ohms (Ω) when $V = 5.4$ V and $I = 2.7 \times 10^{-3}$ A.

90. Given Ohm's law $R = \frac{V}{I}$, find R in ohms when $V = 7.5$ V and $I = 25 \times 10^{-6}$ A.

3.3 Square Roots and Cube Roots

Definition

Square Root

The *square root* of a given number is a number that, when squared, exactly equals the given number. The square root or radical sign is $\sqrt{\ }$.

$$\sqrt{64} = 8 \text{ because } 8^2 = 64 \qquad \sqrt{0.09} = 0.3 \text{ because } (0.3)^2 = 0.09$$

Taking a root is the *inverse* operation of raising to a power in the same way that subtraction is the inverse of addition and division is the inverse of multiplication. Numbers whose square roots are exactly whole numbers, fractions, or finite decimals, such as 64 and 0.09 shown above, are called *perfect squares.* The perfect squares of the whole numbers 1 to 12 are:

1, 4, 9, 16, 25, 36, 49, 64, 81, 100, 121, 144

You should learn these perfect squares to help you understand the ideas better and do calculations with square roots more effectively.

EXAMPLE 3.21 Find the square roots without the calculator:

(a) $\sqrt{\frac{16}{49}}$ **(b)** $\sqrt{\frac{121}{144}}$ **(c)** $\sqrt{\frac{1}{100}}$

Solution To find the square root of a fraction, find the square root of the numerator and the denominator separately:

(a) $\sqrt{\frac{16}{49}} = \frac{\sqrt{16}}{\sqrt{49}} = \frac{4}{7}$ because $\left(\frac{4}{7}\right)^2 = \frac{4^2}{7^2} = \frac{16}{49}$

(b) $\sqrt{\frac{121}{144}} = \frac{\sqrt{121}}{\sqrt{144}} = \frac{11}{12}$ because $\left(\frac{11}{12}\right)^2 = \frac{11^2}{12^2} = \frac{121}{144}$

(c) $\sqrt{\frac{1}{100}} = \frac{\sqrt{1}}{\sqrt{100}} = \frac{1}{10}$ because $\left(\frac{1}{10}\right)^2 = \frac{1^2}{10^2} = \frac{1}{100}$

EXAMPLE 3.22 Find the square roots without the calculator:

(a) $\sqrt{1.44}$ **(b)** $\sqrt{0.0025}$ **(c)** $\sqrt{0.81}$

Solution

(a) When you square a decimal the number of decimal places doubles. Therefore, the square root of a decimal that is a perfect square will have half as many decimal places as the original number. Since 1.44 has two decimal places, its square root has one decimal place. Knowing that $12^2 = 144$, it follows that $1.2^2 = 1.44$ and therefore:

$$\sqrt{1.44} = 1.2$$

(b) The square root of a number less than one is *greater* than the number itself. This is true because, when you square a number less than one, the result is less than the original number. For example, $0.3^2 = 0.09$. Since the number in the square root, called the radicand, 0.0025, has four decimal places, its square root has two decimal places. Knowing that $5^2 = 25$, it follows that $0.05^2 = 0.0025$ and therefore:

$$\sqrt{0.0025} = 0.05$$

(c) Since the radicand 0.81 has two decimal places, the square root has one decimal place:

$$\sqrt{0.81} = 0.9$$

To find the square root on a DAL calculator, press the square root key first; otherwise, press it after the number:

DAL:	[√] 1.44 [=] → 1.2	[√] 0.0025 [=] → 0.05	[√] 0.81 [=] → 0.9
Not Dal:	1.44 [√] → 1.2	0.0025 [√] → 0.05	0.81 [√] → 0.9

Numbers that are not perfect squares, such as 2, 3, 5, 7, etc. have square roots that are infinite decimals, and it is necessary to round off the calculator result:

$$\sqrt{2} = 1.414213562\ldots \approx 1.414$$
$$\sqrt{7} = 2.645751311\ldots \approx 2.646$$

The symbol "≈" means approximately equal. That is, $\sqrt{2}$ is approximately equal to 1.414, and $\sqrt{7}$ is approximately equal to 2.646. When square roots are infinite decimals, it is sometimes easier to calculate with them in radical form instead of as decimals. The following definition of the square root for any positive number x is useful in such calculations:

Definition

Definition of Square Root ($x > 0$)

$$\sqrt{x^2} = \left(\sqrt{x}\right)^2 = x \tag{3.4}$$

The definition says that the operations of square and square root are inverse operations and "cancel each other out." When a square root and a square appear together, the radical and the square can be eliminated. For example:

$$\sqrt{9^2} = 9 \text{ and } \left(\sqrt{9}\right)^2 = 9$$

$$\sqrt{\frac{2^2}{5^2}} = \frac{2}{5} \text{ and } \left(\sqrt{\frac{2}{5}}\right)^2 = \frac{2}{5}$$

Two basic rules for roots of positive numbers x and y are:

Rules

Rules for Roots ($x, y > 0$)

$$\sqrt{xy} = \left(\sqrt{x}\right)\left(\sqrt{y}\right) \text{ or simply } \sqrt{x}\sqrt{y} \tag{3.5}$$

$$\sqrt{\frac{x}{y}} = \frac{\sqrt{x}}{\sqrt{y}} \tag{3.6}$$

Rules (3.5) and (3.6) are used to simplify radicals. You can apply them working from left to right, or from right to left. That is, products (or quotients) under a radical sign can be separated into products (or quotients) of separate radicals and vice versa.

EXAMPLE 3.23

Find the square roots without the calculator:

(a) $\sqrt{14400}$ **(b)** $\sqrt{324}$ **(c)** $\sqrt{225}$

Solution Use rule (3.5) to find the square root of a large perfect square by factoring the number into smaller perfect squares. Apply the rule from left to right, separating the radical into the product of two radicals containing perfect squares:

(a) $\sqrt{14400} = \sqrt{(144)(100)} = \sqrt{144}\,\sqrt{100} = (12)(10) = 120$

(b) $\sqrt{324} = \sqrt{(4)(81)} = \sqrt{4}\,\sqrt{81} = (2)(9) = 18$

(c) $\sqrt{225} = \sqrt{(25)(9)} = \sqrt{25}\,\sqrt{9} = (5)(3) = 15$

EXAMPLE 3.24 Simplify the products:

$$\textbf{(a)}\ \sqrt{0.2}\,\sqrt{1.8} \qquad \textbf{(b)}\ 5\sqrt{50}\,\sqrt{8}$$

Solution

(a) Apply rule (3.5) from right to left and multiply under one radical:

$$\sqrt{0.2}\,\sqrt{1.8} = \sqrt{(0.2)(1.8)} = \sqrt{0.36} = 0.6$$

(b) Multiply under one radical, and simplify:

$$5\sqrt{50}\,\sqrt{8} = 5\sqrt{(50)(8)} = 5\sqrt{400} = 5(20) = 100$$

EXAMPLE 3.25 Simplify:

$$\textbf{(a)}\ \sqrt{\frac{49}{25}} \qquad \textbf{(b)}\ \frac{\sqrt{28}}{\sqrt{7}} \qquad \textbf{(c)}\ \sqrt{\frac{0.64}{0.0256}}$$

Solution

(a) Apply rule (3.6) from left to right and separate the fraction into two radicals:

$$\sqrt{\frac{49}{25}} = \frac{\sqrt{49}}{\sqrt{25}} = \frac{7}{5} = 1.4$$

(b) Apply rule (3.6) from right to left and divide under one radical:

$$\frac{\sqrt{28}}{\sqrt{7}} = \sqrt{\frac{28}{7}} = \sqrt{4} = 2$$

(c) Apply rule (3.6) from left to right and separate the fraction into two radicals:

$$\sqrt{\frac{0.64}{0.0256}} = \frac{\sqrt{0.64}}{\sqrt{0.0256}} = \frac{0.8}{0.16} = \frac{80}{16} = 5$$

ERROR BOX

Sometimes a number may appear to be a perfect square, but it is not. The number of digits or decimal places can make the difference. Study the columns of radicals below to understand why certain numbers are or are not perfect squares. An asterisk * denotes a perfect square.

$\sqrt{400\,*} = 20$	$\sqrt{2500\,*} = 50$	$\sqrt{1440}$
$\sqrt{40}$	$\sqrt{250}$	$\sqrt{144\,*} = 12$
$\sqrt{4\,*} = 2$	$\sqrt{25\,*} = 5$	$\sqrt{14.4}$
$\sqrt{0.4}$	$\sqrt{2.5}$	$\sqrt{1.44\,*} = 1.2$
$\sqrt{0.04\,*} = 0.2$	$\sqrt{0.25\,*} = 0.5$	$\sqrt{0.144}$
$\sqrt{0.004}$	$\sqrt{0.025}$	$\sqrt{0.0144\,*} = 0.12$

Observe that for every two places you move the decimal point in a perfect square you produce another perfect square. Note also that a perfect square has an even number of decimal places (excluding significant zeros at the end).

Practice Problems: For each radical, tell whether or not it contains a perfect square. If it does, give the value of the root.

1. $\sqrt{0.16}$	**2.** $\sqrt{160}$	**3.** $\sqrt{121}$	**4.** $\sqrt{1.21}$
5. $\sqrt{0.16}$	**6.** $\sqrt{0.009}$	**7.** $\sqrt{0.9}$	**8.** $\sqrt{900}$
9. $\sqrt{4900}$	**10.** $\sqrt{4.90}$	**11.** $\sqrt{0.49}$	**12.** $\sqrt{6.25}$
13. $\sqrt{625}$	**14.** $\sqrt{1000}$	**15.** $\sqrt{10{,}000}$	

Cube Roots

A cube root is the inverse of raising to the third power. A cube root is written using the radical sign with the *index* 3 in the crook of the radical sign (for a square root the index 2 is understood):

$$\sqrt[3]{1000} = 10 \text{ because } 10^3 = 1000$$
$$\sqrt[3]{0.064} = 0.4 \text{ because } 0.4^3 = 0.064$$

The first five perfect cubes are:

$$\sqrt[3]{8} = 2, \quad \sqrt[3]{27} = 3, \quad \sqrt[3]{64} = 4, \quad \sqrt[3]{125} = 5, \quad \sqrt[3]{216} = 6$$

It is helpful to know these perfect cubes. *Rules (3.5) and (3.6) for square roots also work for all higher roots,* as shown in the next example.

EXAMPLE 3.26 Find the cube root of:

$$\sqrt[3]{\frac{8}{125}}$$

Solution Apply rule (3.6) for cube roots. Separate the fraction into two cube roots:

$$\sqrt[3]{\frac{8}{125}} = \frac{\sqrt[3]{8}}{\sqrt[3]{125}} = \frac{2}{5}$$

The definition of a cube root for a number x is similar to that for a square root:

Definition

Definition of Cube Root

$$\sqrt[3]{x^3} = \left(\sqrt[3]{x}\right)^3 = x \qquad (3.7)$$

Since raising to the third power is the inverse of taking a cube root, the two operations cancel each other. For example:

$$\sqrt[3]{8^3} = 8 \text{ and } \left(\sqrt[3]{8}\right)^3 = 8$$

The cube root of a number can be found on the calculator by using the cube root key [∛] or by applying the following rule that equates an nth root to the exponent $\frac{1}{n}$:

Formula

Root as Exponent

$$\sqrt[n]{x} = x^{1/n} \qquad \text{(Index } n = \text{Positive whole number)} \qquad (3.8)$$

For example, you can find the $\sqrt[3]{10}$ on the calculator by raising 10 to the $\frac{1}{3}$ power using the power key and the reciprocal key [1/x] or [x^{-1}]:

10 [y^x] 3 [1/x] [=] → 2.15

Similarly, you can find $\sqrt{10}$ by raising 10 to the $\frac{1}{2}$ or 0.5 power:

10 [y^x] 2 [1/x] [=] → 3.16

10 [y^x] 0.5 [=] → 3.16

See Example 3.34 for other ways to find cube roots on the calculator.

Raising to Powers and Taking Roots with Powers of Ten

The following rules apply for raising a number times a power of ten to a positive power:

Rule **Raising to Positive Powers with Powers of Ten**

Raise the number to the positive power and *multiply* the exponents to obtain the power of ten.

$$(5 \times 10^2)^3 = 5^3 \times 10^{(2)(3)} = 125 \times 10^6$$

EXAMPLE 3.27 Perform the operations:

(a) $(1.5 \times 10)^2$ **(b)** $(1.84 \times 10^3)^3$

Solution

(a) Square the 1.5 and multiply the power of ten by 2:

$$(1.5 \times 10)^2 = 1.5^2 \times 10^{1(2)} = 2.25 \times 10^2$$

(b) Raise 1.84 to the third power and multiply the power of ten by 3:

$$(1.84 \times 10^3)^3 = 1.84^3 \times 10^{3(3)} \approx 6.23 \times 10^9$$

Rule **Taking Roots with Powers of Ten**

Take the root of the number and *divide* the root index into the power of ten.

$$\sqrt{81 \times 10^6} = \sqrt{81} \times 10^{6/2} = 9 \times 10^3$$

EXAMPLE 3.28 Simplify:

(a) $\sqrt{2.25 \times 10^4}$ **(b)** $\sqrt[3]{27 \times 10^9}$

Solution

(a) Take the square root of 2.25 and divide the exponent 4 by the index 2:

$$\sqrt{2.25 \times 10^4} = \sqrt{2.25} \times 10^{4/2} = 1.5 \times 10^2$$

(b) Take the cube root of 27 and divide the exponent 9 by the index 3:

$$\sqrt[3]{27 \times 10^9} = \sqrt[3]{27} \times 10^{9/3} = 3 \times 10^3$$

EXAMPLE 3.29 Calculate and give the answer in power of ten form and in ordinary notation:

$$\frac{\sqrt{9 \times 10^6}}{(2 \times 10^3)^2}$$

Solution Apply the order of operations, and do the powers and roots first. Find the square root of the numerator and raise the number to the power in the denominator:

$$\frac{\sqrt{9 \times 10^6}}{(2 \times 10^3)^2} = \frac{\sqrt{9} \times 10^{6/2}}{2^2 \times 10^{3(2)}} = \frac{3 \times 10^3}{4 \times 10^6}$$

Answers to Error Box Problems, page 68:
1. 0.4 **2.** no **3.** 11 **4.** 1.1 **5.** no **6.** no **7.** no **8.** 30 **9.** 70 **10.** no
11. 0.7 **12.** 2.5 **13.** 25 **14.** no **15.** 100

Then do the division:

$$\frac{3 \times 10^3}{4 \times 10^6} = \frac{3}{4} \times 10^{-(6-3)} = 0.75 \times 10^{-3} = 0.00075$$

EXERCISE 3.3

In exercises 1 through 20, do each exercise by hand.

1. $\sqrt{16}$
2. $\sqrt{121}$
3. $\sqrt{\frac{25}{4}}$
4. $\sqrt{\frac{9}{100}}$
5. $\sqrt{\frac{16}{36}}$
6. $\sqrt{\frac{4}{81}}$
7. $\sqrt{0.64}$
8. $\sqrt{0.36}$
9. $\sqrt{0.01}$
10. $\sqrt{2.25}$
11. $\sqrt{40{,}000}$
12. $\sqrt{2500}$
13. $\sqrt{0.0081}$
14. $\sqrt{0.0144}$
15. $\sqrt[3]{27}$
16. $\sqrt[3]{125}$
17. $\sqrt[3]{\frac{1}{8}}$
18. $\sqrt[3]{\frac{64}{1000}}$
19. $\sqrt[3]{0.001}$
20. $\sqrt[3]{0.008}$

In exercises 21 through 38, simplify each expression by applying the rules for radicals. Do each exercise by hand and give the answers in decimal or fraction form.

21. $(\sqrt{10})^2$
22. $\sqrt{0.1^2}$
23. $\sqrt{196}$
24. $\sqrt{324}$
25. $\sqrt{784}$
26. $\sqrt{576}$
27. $\sqrt{1.96}$
28. $\sqrt{4.84}$
29. $\sqrt{4.41}$
30. $\sqrt{3.24}$
31. $\sqrt{8}\,\sqrt{18}$
32. $\sqrt{12}\,\sqrt{27}$
33. $\sqrt{0.4}\,\sqrt{0.9}$
34. $\sqrt{0.02}\,\sqrt{0.08}$
35. $\frac{\sqrt{12}}{\sqrt{3}}$
36. $\frac{\sqrt{3}}{\sqrt{75}}$
37. $\frac{\sqrt{0.09}}{\sqrt{9}}$
38. $\frac{\sqrt{0.2}}{\sqrt{0.8}}$

In exercises 39 through 66, do each exercise by hand and express the answer in terms of a power of ten.

39. $(4 \times 10^3)^3$
40. $(6 \times 10^2)^2$
41. $(2 \times 10^2)^4$
42. $(3 \times 10^4)^3$
43. $(0.8 \times 10^5)^2$
44. $(0.1 \times 10)^2$
45. $(0.1 \times 10^3)^3$
46. $(0.2 \times 10^6)^3$
47. $(1 \times 10^3)(3 \times 10^4)^2$
48. $(4 \times 10)(2 \times 10^3)^3$
49. $(1.2 \times 10^3)(5 \times 10)^3$
50. $(20 \times 10^2)(0.1 \times 10^6)^2$
51. $\sqrt{16 \times 10^6}$
52. $\sqrt{25 \times 10^4}$
53. $\sqrt{144 \times 10^2}$
54. $\sqrt{49 \times 10^{12}}$
55. $\sqrt{0.25 \times 10^4}$
56. $\sqrt{0.64 \times 10^6}$
57. $\sqrt[3]{8 \times 10^9}$
58. $\sqrt[3]{27 \times 10^3}$
59. $(\sqrt{9 \times 10^4})(\sqrt{4 \times 10^2})$
60. $(\sqrt{1 \times 10^6})(\sqrt{25 \times 10^2})$
61. $\frac{\sqrt{9 \times 10^8}}{15 \times 10^3}$
62. $\frac{\sqrt{16 \times 10^2}}{8 \times 10^2}$
63. $\frac{\sqrt{36 \times 10^2}}{(2 \times 10^3)^2}$
64. $\frac{\sqrt{64 \times 10^4}}{(5 \times 10^2)^2}$
65. $\frac{(3 \times 10^3)^2}{\sqrt{144 \times 10^2}}$
66. $\frac{(6 \times 10^2)^2}{\sqrt{81 \times 10^4}}$

Applied Problems

In problems 67 through 82, try to solve each by hand. Check with the calculator.

67. The area of a square plot of land is $A = 4900$ ft^2. Find the side of the square $s = \sqrt{A}$ in ft.
68. The area A of a square circuit board is 10,000 mm^2 (square millimeters). How long is the side $s = \sqrt{A}$ of the circuit board in mm and cm (centimeters)?

    Note: 10 mm = 1 cm.
69. Find the hypotenuse of a right triangle given by:

$$c = \sqrt{1.5^2 + 2.0^2}$$

70. The radius of a sphere is given approximately by:

$$r = \sqrt[3]{\frac{V}{4.2}}$$

where V = volume. Find r in ft, when $V = 0.0042$ ft^3.

Applications to Electronics

71. The impedance of an alternating current (ac) circuit is given by:

$$Z = \sqrt{16^2 + 12^2}$$

Impedance is measured in ohms like resistance. Find the value of Z in ohms (Ω).

72. As in problem 71, find the impedance Z in ohms of an ac circuit given by:

$$Z = \sqrt{15^2 + 20^2}$$

73. Find the voltage V in volts (V) in a circuit given by:

$$V = \sqrt{(120)(30)}$$

74. Find the current I in amps (A) in a circuit given by:

$$I = \sqrt{\frac{4}{10{,}000}}$$

75. Find the current I in amps (A) in a circuit given by:

$$I = \sqrt{\frac{10}{150 + 100}}$$

76. Find the voltage V in volts (V) in a circuit given by:

$$V = \sqrt{(16)(100 + 44)}$$

77. Find the power P in watts (W) dissipated as heat in a resistor given by:

$$P = \frac{(3.2 \times 10^3)^2}{1.6 \times 10^6}$$

78. Find the power P in watts (W) dissipated as heat in a resistor given by:

$$P = (0.12)^2(0.5 \times 10^3)$$

79. Find the current I in amps (A) in a series circuit given by:

$$I = \sqrt{\frac{24}{(5.7 \times 10^3) + (3.9 \times 10^3)}}$$

Hint: Change to ordinary notation before calculating.

80. Find the voltage drop V in volts across two resistances given by:

$$V = \sqrt{(16)(6.1 \times 10^3 + 3.9 \times 10^3)}$$

81. In Visual Basic and other computer languages, an up ↑ arrow or caret ^ denotes an exponent, and SQR(x) denotes $\sqrt{x}$. For example, $\sqrt{49} - 2^3$ is written SQR(49) – 2^3. Calculate the following expression written in computer language:

4^2/SQR(81) + 2/9

82. As in problem 81, calculate the following expression written in computer language:

SQR(32/2) * 3^2 – 11

3.4 Hand Calculator Operations

This section provides additional explanation on how to do powers and roots on the calculator, and how to estimate results and troubleshoot for errors.

Powers

Powers or exponents are done on the calculator with the power key [x^y], [y^x] or [∧]. For example, 3^5 is calculated:

3 [y^x] 5 [=] → 243

3 [∧] 5 [=] → 243

To square a number, press [x^2]:

0.7 [x^2] → 0.49

Scientific calculators are programmed to perform the order of operations correctly. Raising to a power or taking a root is done before multiplying or dividing.

EXAMPLE 3.30 Calculate:

$$\frac{(0.2)^2 + (0.5)^3}{0.5}$$

Solution One calculator solution without parentheses is:

$$0.2 \boxed{x^2} \boxed{+} 0.5 \boxed{y^x} 3 \boxed{=} \boxed{\div} 0.5 \boxed{=} \rightarrow 0.33$$

It is necessary to press the $\boxed{=}$ key after putting in the operations in the numerator. This computes the numerator before dividing. Remember that the fraction line acts like parentheses around both the numerator and the denominator.

A solution that uses parentheses corresponding to the fraction line is:

$$\boxed{(} 0.2 \boxed{x^2} \boxed{+} 0.5 \boxed{y^x} 3 \boxed{)} \boxed{\div} 0.5 \boxed{=} \rightarrow 0.33$$

EXAMPLE 3.31 Estimate and then find the solution on the calculator to three significant digits:

$$\left(\frac{1}{0.25}\right)^2 + \frac{(4.5)^3}{(2.7)^2}$$

Solution Estimate the answer by rounding off to one digit:

$$\left(\frac{1}{0.25}\right)^2 + \frac{(4.5)^3}{(2.7)^2} \approx \left(\frac{1}{0.3}\right)^2 + \frac{(5)^3}{(3)^2}$$

$$= \frac{1}{0.09} + \frac{125}{9} = \frac{100}{9} + \frac{125}{9} = \frac{225}{9} = 25$$

Note that the fraction $\frac{1}{0.09}$ is changed to $\frac{100}{9}$ before adding by multiplying the numerator and denominator by 100.

The calculator solution can be done using the reciprocal key $\boxed{1/x}$ or $\boxed{x^{-1}}$:

$$0.25 \boxed{1/x} \boxed{x^2} \boxed{+} 4.5 \boxed{y^x} 3 \boxed{\div} 2.7 \boxed{x^2} \boxed{=} \rightarrow 28.5$$

The estimate of 25 agrees closely with the calculator answer of 28.5 and verifies the calculation.

Roots

Square roots are done on the calculator by pressing the square root key $\boxed{\sqrt{\ }}$:

DAL: $\boxed{\sqrt{\ }}$ 0.36 $\boxed{=}$ $\rightarrow$0.6
Not DAL: 0.36 $\boxed{\sqrt{\ }}$ $\rightarrow$0.6

You can also use Formula (3.8) and raise to the $\frac{1}{2}$ or 0.5 power to find a square root:

$$0.36 \boxed{y^x} 0.5 \boxed{=} \rightarrow 0.6$$

EXAMPLE 3.32 Estimate and then compute with the calculator to three significant digits:

$$\sqrt{\frac{(16.5)^2(2.20)}{(6.82)(3.10)}}$$

Solution Estimate by rounding off to one digit and performing the operations under the radical:

$$\sqrt{\frac{(16.5)^2(2.20)}{(6.82)(3.10)}} \approx \sqrt{\frac{(20)^2(2)}{(7)(3)}} = \sqrt{\frac{800}{21}}$$

The estimate should be done as simply and quickly as possible *by hand* to provide a check on the calculator. To speed up the process, you should continue to round off to one digit as you estimate. Round off the denominator of 21 above to 20. Then *change the square root to the nearest perfect square* so you can do it mentally:

$$\sqrt{\frac{800}{21}} \approx \sqrt{\frac{800}{20}} = \sqrt{40} \approx \sqrt{36} = 6$$

There is little loss in accuracy by continuing to round off. The estimate will still be reliable.

The calculator solution is:

Not DAL: 16.5 [x²] [×] 2.20 [=] [÷] [(] 6.82 [×] 3.10 [)] [=] [√] → 5.32

For a DAL calculator you need extra sets of parentheses:

DAL: [√] [(] [(] 16.5 [x²] [×] 2.20 [)] [÷] [(] 6.82 [×] 3.10 [)] [)] [=] → 5.32

The estimate of 6 is close to the calculator answer 5.32 and verifies the calculation.

EXAMPLE 3.33 Estimate and then calculate to three significant digits:

$$\sqrt{0.472} + \sqrt{0.0789}$$

Solution To estimate the answer, change each decimal to the nearest perfect square. Note that the nearest perfect square to 0.472 is 0.49, and the nearest perfect square to 0.0789 is 0.09 as follows:

$$\sqrt{0.472} + \sqrt{0.0789} \approx \sqrt{0.49} + \sqrt{0.09} = 0.7 + 0.3 = 1$$

The calculator solution is:

Not DAL: 0.472 [√] [+] 0.0789 [√] [=] → 0.968

DAL: [√] [(] 0.472 [)] [+] [√] [(] 0.0789 [)] [=] → 0.968

The estimate of 1 is close to the calculator answer 0.968 and verifies the calculation.

EXAMPLE 3.34 Estimate and then calculate to three significant digits:

$$\sqrt[3]{4.86}$$

Solution You can estimate a cube root by changing it to the closest perfect cube root:

$$\sqrt[3]{4.86} \approx \sqrt[3]{8} = 2$$

Cube roots can be done in one or more ways on your calculator depending on which keys you have:

(a) Not DAL: 4.86 [$\sqrt[x]{y}$] 3 [=] → 1.69 DAL: 3 [$\sqrt[x]{y}$] 4.86 [=] → 1.69

(b) Not DAL: 4.86 [$\sqrt[3]{\ }$] → 1.69 DAL: [$\sqrt[3]{\ }$] 4.86 [=] → 1.69

(c) 4.86 [y^x] 3 [1/x] [=] → 1.69 or 4.86 [∧] 3 [x^{-1}] [=] → 1.69

(d) 4.86 [$x^{\frac{1}{y}}$] 3 [=] → 1.69

Examples (c) and (d) apply formula (3.8).

EXERCISE 3.4

In exercises 1 through 20, estimate the answer first and choose what you think is the correct answer from the four choices. Then check with the calculator, rounding off to three significant digits.

1. $(3.20)(4.26)^2$ [5.81, 58.1, 88.0, 581]
2. $(0.913)^2 (1.02)^3$ [0.185, 0.885, 8.85, 88.5]
3. $\left(\frac{0.112}{0.518}\right)^3$ [0.0101, 0.101, 0.110, 1.10]
4. $\frac{(9.87)^3}{(2.03)^4}$ [0.566, 5.66, 56.6, 566]
5. $\frac{(1.25)^4 - (0.831)^4}{3.00}$ [0.065, 0.655, 0.950, 0.955]
6. $\left(\frac{1}{8.51}\right)^2 + \left(\frac{1}{6.66}\right)^2$ [0.0364, 0.364, 3.64, 6.43]
7. $100\left(0.900 + \frac{0.180}{4.00}\right)^4$ [79.7, 7.97, 79.0, 7.90]
8. $13.5(0.281 + 0.591)^3$ [0.895, 6.63, 8.95, 66.3]
9. $\sqrt{0.0588}$ [0.242, 0.842, 2.42, 8.42]
10. $\sqrt[3]{25.6}$ [1.95. 2.94, 2.95, 6.95]
11. $\left(\sqrt{19.0}\right)\left(\sqrt{171}\right)$ [5.70, 27.0, 57.0, 570]
12. $(0.789)^2\left(\sqrt{0.105}\right)$ [0.201, 0.202, 2.02, 3.88]
13. $\sqrt{93.2} + \sqrt{29.1}$ [1.50, 15.0, 150, 650]
14. $\sqrt[3]{8.31} - \sqrt[3]{1.28}$ [0.0940, 0.940, 9.40, 94.0]
15. $\frac{\sqrt{600}}{\sqrt{5.88}}$ [1.01, 10.1, 50.1, 101]
16. $\frac{\sqrt{(0.130)(22.4)}}{\sqrt{7.92}}$ [0.606, 6.06, 9.09, 60.6]
17. $\sqrt{(47)^2 + (33)^2}$ [1.74, 5.74, 57.4, 75.4]
18. $\sqrt{(3.15)^2 - (2.26)^2}$ [0.219, 1.29, 2.19, 12.9]
19. $\frac{210}{\sqrt[3]{(3.14)^2}}$ [9.97, 99.7, 97.9, 98.0]
20. $\sqrt{\frac{(3.01)(5.05)^2}{10.2}}$ [1.47, 1.74, 2.74, 7.74]

Applied Problems

In problems 21 through 26, solve each applied problem to three significant digits.

21. The formula for the volume of a sphere is:

$$V = \frac{4}{3}\pi r^3$$

where r = radius. Find V in cm³ when r = 3.28 cm. Use π = 3.142 or the [π] key on the calculator.

22. The velocity of sound in meters per second is given by:

$$v = v_0\sqrt{1 + \frac{t}{273}}$$

where v_0 = velocity of sound at 0°C and t = temperature in degrees Celsius. Find v when v_0 = 332 m/s and t = 15°C.

Applications to Electronics

23. Find the power dissipation in watts (W) of a resistor given by:

$$P = (0.023)^2(1500)$$

24. Find the power dissipation in watts (W) of a resistor given by:

$$P = \frac{(6.6 \times 10^3)^2}{33 \times 10^3}$$

25. Find the impedance Z in ohms (Ω) of an ac circuit given by:

$$Z = \sqrt{(2200)^2 + (4300)^2}$$

26. Find the impedance Z in ohms (Ω) of an ac circuit given by:

$$Z = \sqrt{(8.2 \times 10^3)^2 + (5.1 \times 10^3)^2}$$

CHAPTER HIGHLIGHTS

3.1 POWERS

Power or Exponent *n*

$$x^n = \underbrace{(x)(x)(x)\cdots(x)}_{n \text{ times}} \quad (3.1)$$

(x = Base)
(n = Positive whole number)

Key Example:

$$0.5^3 = (0.5)(0.5)(0.5) = 0.125$$

$$\left(\frac{3}{10}\right)^4 = \frac{3^4}{10^4} = \frac{81}{10,000}$$

Order of Operations

1. Operations in parentheses.

2. Exponents.
3. Multiplication or division.
4. Addition or subtraction.

Memorize the phrase "***P***lease ***E***xcuse ***M***y ***D***ear ***A***unt ***S***ally" (PEMDAS).

Key Example:

$$\frac{(0.4)^2 + (0.3)^3}{0.5} = \frac{0.16 + 0.09}{0.5} = \frac{0.25}{0.5} = 0.5$$

3.2 POWERS OF TEN

Zero Power

$$10^0 = 1 \quad (3.2)$$

• • •

Positive Power of Ten

The power of ten is equal to the number of zeros:

$$10^1 = 10, \quad 10^2 = 100, \quad 10^3 = 1000, \quad 10^6 = 1{,}000{,}000, \text{ etc.}$$

• • •

Absolute Value

The value of a number without the sign (positive value).

• • •

Negative Power of Ten

$$10^{-n} = \frac{1}{10^n} \quad (n = \text{positive number}) \quad (3.3)$$

The absolute value of the power is equal to the number of decimal places:

$10^{-1} = 0.1$ (1 dec. place)
$10^{-2} = 0.01$ (2 dec. places)
$10^{-3} = 0.001$ (3 dec. places), etc.

See Table 3-1, page 57, which shows the powers of ten from 10^{-9} to 10^9. A number in the decimal system is expressed in powers of ten as follows:

$$5271.34 = (5 \times 10^3) + (2 \times 10^2) + (7 \times 10^1) + (1 \times 10^0) + (3 \times 10^{-1}) + (4 \times 10^{-2})$$

Multiplying Positive Powers of Ten

Add the exponents only. Do not change the base 10.

$$10^3 \times 10^2 = 10^{3+2} = 10^5$$

• • •

Multiplying Negative Powers of Ten

Add the absolute values of the exponents and make the result negative.

$$10^{-4} \times 10^{-2} = 10^{-(4+2)} = 10^{-6}$$

• • •

Multiplying Positive and Negative Powers of Ten

Subtract the smaller absolute value from the larger absolute value and use the sign of the larger absolute value.

$$10^6 \times 10^{-2} = 10^{+(6-2)} = 10^4$$
$$10^{-6} \times 10^2 = 10^{-(6-2)} = 10^{-4}$$

• • •

Dividing Powers of Ten

Change the sign of the power in the divisor and *apply the rules for multiplying with powers of ten.*

$$\frac{10^6}{10^3} = 10^6 \times 10^{-3} = 10^{+(6-3)} = 10^3$$

• • •

Multiplying and Dividing Numbers Times Powers of Ten

1. Multiply or divide the numbers to obtain the number part of the answer.
2. Apply the rules for multiplying or dividing to the powers of ten.

$$(4 \times 10^3)(8 \times 10^2) = (4)(8) \times 10^{3+2} = 32 \times 10^5$$

$$\frac{9 \times 10^{-5}}{3 \times 10^{-2}} = \frac{9}{3} \times (10^{-5})(10^2) = 3 \times 10^{-(5-2)} = 3 \times 10^{-3}$$

• • •

Adding Numbers Times Powers of Ten

Numbers times powers of ten can be added or subtracted *only when the powers are the same.* Add or subtract the numbers only. *Do not change* the power of ten:

$$(2 \times 10^3) + (4 \times 10^3) = (2 + 4) \times 10^3 = 6 \times 10^3$$

• • •

Changing the Power of Ten

- To increase the power of ten, move the decimal point to the left and increase the power of ten by the number of places you moved the decimal point.

$$42 \times 10^5 = 4.2 \times 10^6, \quad 52.5 \times 10^{-9} = 5.25 \times 10^{-8}$$

- To decrease the power of ten, move the decimal point to the right and decrease the power of ten by the number of places you moved the decimal point.

$$2.13 \times 10^4 = 213 \times 10^2, \quad 7.34 \times 10^{-3} = 73.4 \times 10^{-4}$$

• • •

Changing from Power of Ten to Ordinary Notation

Move the decimal point to the right (positive) or to the left (negative) the number of places indicated by the exponent.

To add numbers times powers of ten when the powers are different, change to ordinary notation or change to like powers:

Key Example:

$$(8 \times 10^{-4}) + (1 \times 10^{-3}) = 0.0008 + 0.001 = 0.0018$$

or

$$(8 \times 10^{-4}) + (1 \times 10^{-3}) = (0.8 \times 10^{-3}) + (1 \times 10^{-3}) = 1.8 \times 10^{-3} = 0.0018$$

3.3 SQUARE ROOTS AND CUBE ROOTS

Definition of Square Root ($x > 0$)

$$\sqrt{x^2} = \left(\sqrt{x}\right)^2 = x \qquad (3.4)$$

Key Example:

$$\sqrt{100} = \sqrt{10^2} = \left(\sqrt{10}\right)^2 = 10$$

$$\sqrt{0.04} = \sqrt{0.2^2} = \left(\sqrt{0.2}\right)^2 = 0.2$$

$$\sqrt{6.25} = \sqrt{2.5^2} = \left(\sqrt{2.5}\right)^2 = 2.5$$

Rules for Roots ($x, y > 0$)

$$\sqrt{xy} = \left(\sqrt{x}\right)\left(\sqrt{y}\right) \qquad (3.5)$$

$$\sqrt{\frac{x}{y}} = \frac{\sqrt{x}}{\sqrt{y}} \qquad (3.6)$$

Key Example:

$$\sqrt{225} = \left(\sqrt{9}\right)\left(\sqrt{25}\right) = (3)(5) = 15$$

$$\frac{\sqrt{1000}}{\sqrt{250}} = \sqrt{\frac{1000}{250}} = \sqrt{4} = 2$$

Definition of Cube Root

$$\sqrt[3]{x^3} = \left(\sqrt[3]{x}\right)^3 = x \qquad (3.7)$$

Key Example:

$$\sqrt[3]{64} = \sqrt[3]{4^3} = \left(\sqrt[3]{4}\right)^3 = 4$$

$$\sqrt[3]{0.008} = \sqrt[3]{0.2^3} = \left(\sqrt[3]{0.2}\right)^3 = 0.2$$

Root as Exponent

$$\sqrt[n]{x} = x^{1/n} \quad \text{(Index } n = \text{Positive whole number)} \qquad (3.8)$$

Key Example:

$\sqrt[3]{4.86}$: 4.86 [x^y] 3 [1/x] [=] → 1.69

Raising to Powers and Taking Roots with Powers of Ten

Raising to Positive Powers with Powers of Ten

Raise the number to the positive power and *multiply* the exponents to obtain the power of ten.

$$(5 \times 10^2)^3 = 5^3 \times 10^{(2)(3)} = 125 \times 10^6$$

• • •

Taking Roots with Powers of Ten

Take the root of the number and *divide* the root index into the power of ten.

$$\sqrt{81 \times 10^6} = \sqrt{81} \times 10^{6/2} = 9 \times 10^3$$

Key Example:

$$\frac{\sqrt{16 \times 10^4}}{(2 \times 10^2)^4} = \frac{\sqrt{16} \times 10^{4/2}}{2^4 \times 10^{(2)(4)}} = \frac{4 \times 10^2}{16 \times 10^8} = \frac{1}{4} \times 10^{-(8-2)} = 0.25 \times 10^{-6}$$

3.4 HAND CALCULATOR OPERATIONS

Powers or exponents are done on the calculator with the power key: [x^y], [y^x] or [^]. Use [x^2] to square a number and [√] to take a square root.

Key Example: The steps to calculate $\sqrt{\frac{(0.237)^2(69.4)}{(775)}}$ are:

DAL: [√] [(] [(] 0.237 [x^2] [×] 69.4 [)] [÷] 775 [)] [=] → 0.0709

Not DAL: [(] 0.237 [x^2] [×] 69.4 [)] [÷] 775 [=] [√] → 0.0709

Cube roots can be done in one or more ways on your calculator, depending on which keys you have. One of the following works on almost all calculators for $\sqrt[3]{675}$:

Key Example: 675 [y^x] 3 [1/x] [=] → 8.77
675 [^] 3 [x^{-1}] [=] → 8.77

REVIEW EXERCISES

In exercises 1 through 44, evaluate each exercise by hand and express answers as whole numbers or as both fractions and decimals.

1. 7^3

2. 11^2

3. $\left(\frac{1}{2}\right)^4$

4. $\left(\frac{2}{5}\right)^3$

5. $(0.6)^3$

6. $(0.4)^3$

7. $(1.1)^2$

8. $(0.1)^3$

9. $\frac{(3)^4}{(9)^2(10)^2}$

10. $\frac{7^2 - 3^2}{8^2}$

11. $\left(\frac{2}{5}\right)^2 + (6)\left(\frac{1}{5}\right)^2$

12. $\frac{(0.1)^3}{(1.3)^2 - (1.2)^2}$

13. $\frac{1.5^2 + 2^2}{2.5^2}$

14. $\left(\frac{1}{3} + \frac{1}{6}\right)^3$

15. $(0.3)^2 + \frac{(0.4)^2(0.1)^2}{(0.2)^3}$

16. $\left(\frac{2}{3} - \frac{1}{6}\right)^2\left(\frac{1}{3} + \frac{1}{6}\right)^2$

17. $\sqrt{49}$

18. $\sqrt{\frac{36}{25}}$

19. $\sqrt{\frac{16}{100}}$

20. $\frac{\sqrt{48}}{\sqrt{12}}$

21. $\sqrt{0.09}$

22. $\sqrt{0.0121}$

23. $\sqrt[3]{64}$

24. $\sqrt[3]{0.027}$

25. $\left(\sqrt{0.01} + \sqrt{0.81}\right)^2$

26. $\sqrt{5^2 + 12^2}$

27. $\sqrt{0.18}\,\sqrt{0.02}$

28. $\sqrt{8.1}\,\sqrt{0.1}$

29. $\left(\sqrt{13}\right)\left(\sqrt{13}\right) - \left(\sqrt{6^2}\right)$

30. $\frac{\sqrt{0.49}}{\sqrt{0.25}}$

31. 10^8

32. 10^{-5}

33. $\frac{1}{10^4}$

34. $\frac{10^0}{10^{-3}}$

35. $\frac{10^{-2}}{10^{-4}}$

36. $\frac{10^{-5}}{10^1}$

37. $10^2 \times 10^2$

38. $10^{-3} \times 10^{-1}$

39. $10^3 \times 10^{-2} \times 10^{-6}$

40. $10^{-9} \times 10^6 \times 10^2$

41. $\frac{10^2 \times 10^5}{10^{-2}}$

42. $\frac{10^{-1} \times 10^{-3}}{10^3 \times 10^{-6}}$

43. $10^3 \times 10^{-4} \times \frac{10^5}{10^{-2}}$

44. $10^9 \times 10^{-6} \times \frac{10^{-6}}{10^3}$

In exercises 45 through 66, perform each calculation by hand and express the answer in terms of a power of 10.

45. $(7 \times 10^{-4})(2 \times 10^6)$

46. $(8 \times 10^{-5})(3 \times 10^{-3})$

47. $\frac{12 \times 10^{-9}}{6 \times 10^{-6}}$

48. $\frac{(9 \times 10^4)(3 \times 10^4)}{6 \times 10^9}$

49. $\frac{2 \times 10^{-3}}{(4 \times 10^2)(5 \times 10^4)}$

50. $\frac{20 \times 10^6}{(4 \times 10^2)(2 \times 10^4)}$

51. $\frac{2(5 \times 10^6)}{10 \times 10^{-2}}$

52. $\frac{(4 \times 10^9)(5 \times 10^{-6})}{2 \times 10^3}$

53. $(8 \times 10^5) + (8 \times 10^5)$

54. $(7 \times 10^{-6}) - (6 \times 10^{-6})$

55. $(5 \times 10^2) + (6 \times 10^3)$

56. $(30 \times 10^{-7}) + (9 \times 10^{-6})$

57. $(3 \times 10^2)^4$

58. $(3 \times 10^{-6})(1.5 \times 10^6)^2$

59. $\sqrt{81 \times 10^6}$

60. $\sqrt{0.36 \times 10^2}$

61. $\left(\sqrt{1 \times 10^4}\right)\left(\sqrt{16 \times 10^2}\right)$

62. $\frac{\sqrt{9 \times 10^8}}{15 \times 10^2}$

63. $\sqrt{(7 \times 10^4)(7 \times 10^2)}$

64. $\sqrt{\frac{162 \times 10^5}{2 \times 10^1}}$

65. $\frac{\sqrt{0.81 \times 10^2}}{(3 \times 10^{-3})^2}$

66. $\frac{(2 \times 10^2)^2}{\sqrt{25 \times 10^4}}$

In exercises 67 through 74, estimate the answer by rounding off to one digit. Then choose the correct answer and check with the calculator. Round answers to three significant digits.

67. $(5.12)^3(0.0102)^2$ [0.0139, 0.0140, 1.39, 1.40]

68. $\left(\frac{1}{5.05}\right)^2 - \left(\frac{1}{10.5}\right)^2$ [0.00301, 0.0301, 0.0103, 0.301]

69. $(9.38)^2\left(\sqrt{9.91}\right)$ [1.77, 2.77, 27.7, 277]

70. $(0.101)\left(\sqrt[3]{28.3}\right)$ [0.308, 0.808, 3.08, 30.8]

71. $\sqrt{(23.1)^2 + (46.3)^2}$ [0.517, 5.17, 51.7, 71.5]

72. $\dfrac{\sqrt{66.6}}{(\sqrt{50})(\sqrt{72})}$ [0.0136, 0.136, 0.936, 1.36]

73. $\sqrt{\dfrac{(1.21)(8.13)}{(0.331)^2}}$ [0.948, 1.98, 9.48, 19.8]

74. $\sqrt{\dfrac{(1.23)^2}{(5.66)(2.51)}}$ [0.327, 0.326, 0.330, 0.3263]

Applied Problems

In problems 75 through 88, try to solve each by hand. Check with the calculator.

75. Under certain conditions, when the driver jams on the brakes of an automobile weighing 4000 lb and traveling 60 mi/h, the car will skid a distance in feet given approximately by:

$$d = \frac{4000(14)^2}{(1000)(1.4)(10)}$$

Calculate the distance d in feet.

76. The interest I earned on $1000 after two years in a savings account whose annual interest rate is 5% is given by:

$$I = 1000(0.10 + 0.05^2)$$

Find the value of I to the nearest cent.

77. The area A of a square circuit board is 2500 mm^2 (square millimeters). Find the length of the side s of the circuit board in mm and cm (centimeters), where $s = \sqrt{A}$. *Note:* 10 mm = 1 cm.

78. The volume of a cube $V = s^3$, where s = side. Find the volume of a large cubic container in m^3 (cubic meters), when $s = 0.6$ m.

Applications to Electronics

79. Find the resistance R in ohms (Ω) in a circuit given by:

$$R = \frac{32}{(0.80)^2}$$

80. Find the power in watts (W) of a circuit given by:

$$P = (1.6)^2(100)$$

81. Given Ohm's law $I = \frac{V}{R}$, find I in amps (A) when $V = 120$ V and $R = 3.0 \times 10^3$ Ω.

82. Given Ohm's law $R = \frac{V}{I}$, find R in ohms (Ω) when

$$V = 9.0 \text{ V and } I = 4.5 \times 10^{-3} \text{ A.}$$

83. Find the impedance Z in ohms (Ω) of an ac circuit given by:

$$Z = \sqrt{2.5^2 + 6.0^2}$$

84. Find the voltage in volts (V) in a circuit given by:

$$V = \sqrt{(12.1)(40)}$$

85. Find the power P in watts (W) dissipated as heat in a resistor given by:

$$P = \frac{(0.3 \times 10^3)^2}{1.2 \times 10^3}$$

86. Find the current I in amps (A) through two series resistors given by:

$$I = \sqrt{\frac{16.9}{(7.6 \times 10^4) + (2.4 \times 10^4)}}$$

Hint: Change to ordinary notation before calculating.

87. The resistance R, in ohms (Ω), of a copper wire is given by:

$$R = \frac{(5)(1.75 \times 10^{-8})}{(9 \times 10^{-6})(\pi)}$$

Calculate the value of R to two significant digits. Use $\pi = 3.14$ or the [π] key on the calculator.

88. The coupling coefficient, k, of two coils in series is given by:

$$k = \frac{45 \times 10^{-3}}{(\sqrt{45 \times 10^{-2}})(\sqrt{20 \times 10^{-2}})}$$

Calculate the value of k to two significant digits. Change to ordinary notation before calculating.

CHAPTER 4
Systems of Measurement

CHAPTER OBJECTIVES

In this chapter, you will learn:

- How to work with numbers in scientific notation.
- How to work with numbers in engineering notation.
- The basic units of the metric, or International System (SI).
- How to change between units in the metric system.
- The basic units of the U.S. customary system.
- How to convert between units in the metric system and the U.S. system.
- How to use the calculator for engineering notation and scientific notation.
- How to use the calculator to change between metric units and to convert between systems.

The metric system, also called the International System (SI), is very important in engineering and technology. Every component in an electrical or electronic system is identified by its measurement: a 1.5-volt battery, a 1.1-megohm resistor, a 60-milliamp current, a 10-microfarad capacitor, etc. This chapter will help you develop a clear understanding of the metric system and the electrical measurements that are used. It explains the basic units and shows you how to change between different metric prefixes. Two types of notation are studied, engineering notation and scientific notation. Engineering notation uses powers of ten that correspond to the metric prefixes used in electronics. Scientific notation uses powers of ten to express large and small numbers that occur in electrical applications. In addition to the metric system, the U.S. customary system is still used to a great extent in the United States. It is important to understand its relation to the metric system and to know how to convert between metric units and U.S. units. A special calculator section at the end provides additional instruction on scientific notation, engineering notation, changing metric units, and converting between metric units and U.S. units.

• • •

4.1 Scientific and Engineering Notation

Consider the following distance problem. The brightest star in the sky, Sirius, is 8.7 light-years away. This means it takes 8.7 years for light to travel from this star to the earth. If light travels at 186,000 mi/s (miles per second), how far away is Sirius? The distance, rounded to two significant digits, is calculated as follows:

$$(186{,}000 \text{ mi/sec})\ (60 \text{ sec/min})\ (60 \text{ min/hr})\ (24 \text{ hr/day})\ (365 \text{ day/yr})\ (8.7 \text{ yr})$$

$$= 51{,}031{,}500{,}000{,}000 \approx 51{,}000{,}000{,}000{,}000 \text{ miles}$$

The answer is more than 50 trillion miles. It is rounded to two significant digits because the data, 8.7 yr, is only accurate to two significant digits. Try this multiplication on the calculator. It automatically switches to *scientific notation* because the result is too large for ordinary notation:

186000 [×] 60 [×] 60 [×] 24 [×] 365 [×] 8.7 [=] → 5.10315 E13

The number on the right side of the display is the power of ten (E = exponent), and the answer to the distance problem in scientific notation to two significant digits is:

$$51{,}000{,}000{,}000{,}000 \text{ miles} = 5.1 \times 10^{13} \text{ miles}$$

Any digits after 5.1 shown on the display are not considered significant.

Definition

Scientific Notation

A number between 1 and 10 (including 1 but not 10) times a power of ten:

$$5.05 \times 10^7, \quad 6.23 \times 10^{-6}, \quad 9.99 \times 10^0, \quad 4.12 \times 10^{-2}, \quad 1.00 \times 10^{12}$$

The rule for changing from ordinary notation to scientific notation is:

Rule

Changing Ordinary Notation to Scientific Notation

Move the decimal point to the right of the first significant digit, called the *zero position.* The number of places from the zero position to the original decimal point equals the power of ten: positive if the decimal point is to the right of the zero position, negative if it is to the left.

EXAMPLE 4.1 Change to scientific notation:

(a) 532,000 **(b)** 6790 **(c)** 9.68 **(d)** 0.0702 **(e)** 0.00000410

Solution

(a) 532,000. $= 5.32 \times 10^5$

(b) 6790. $= 6.79 \times 10^3$

(c) 9.68 $= 9.68 \times 10^0$

(d) 0.0702 $= 7.02 \times 10^{-2}$

(e) 0.00000410 $= 4.10 \times 10^{-6}$

The arrows show the movement of the decimal point to the zero position. A number greater than 10 has a positive exponent such as (a) and (b) above. A number between 1 and 10 has a zero exponent, such as (c) above, and a number less than 1 has a negative exponent such as (d) and (e) above. Note that *only significant digits are shown in scientific notation.*

Engineering notation also expresses numbers with powers of ten where the powers correspond to metric units:

Definition

Engineering Notation

A number between 1 and 1000 (including 1 but not 1000) times a power of ten, which *is a multiple of three:*

$$2.22 \times 10^{-3}, \quad 65.4 \times 10^9, \quad 1.00 \times 10^{-6}, \quad 329 \times 10^0, \quad 23 \times 10^3$$

The answer to the distance problem above in engineering notation is:

$$51{,}000{,}000{,}000{,}000 \text{ miles} = 51 \times 10^{12} \text{ miles}$$

Powers of ten that are multiples of three correspond to the units used in electronics and in the metric system. For example, 10^6 corresponds to mega, 10^3 corresponds to kilo,

10^{-3} to milli, and 10^{-6} to micro. Table 4-2 in Section 4.2, page 91, shows these engineering prefixes. The rule for changing to engineering notation is:

Rule

Changing Ordinary Notation to Engineering Notation

Move the decimal point a multiple of three places: 3, 6, 9, etc., to change to a number from 1 to less than 1000. The number of places from the new position to the original decimal point equals the power of ten: positive if the original decimal point is to the right of the new position, negative if it is to the left.

EXAMPLE 4.2 Change to engineering notation:

(a) 208,000 **(b)** 92,300 **(c)** 509 **(d)** 0.0615 **(e)** 0.0000544

Solution

(a) 208,000. $= 208 \times 10^3$

(b) 92,300. $= 92.3 \times 10^3$

(c) 509. $= 509 \times 10^0$

Note in (a), (b), and (c) that a number greater than 1 has a zero or positive exponent.

(d) 0.0615 $= 61.5 \times 10^{-3}$

(e) 0.0000544 $= 54.4 \times 10^{-6}$

Note in (d) and (e) that a number less than 1 has a negative exponent.

EXAMPLE 4.3 Change each number to both scientific and engineering notation:

(a) 48,900,000 **(b)** 63,500 **(c)** 77.1 **(d)** 0.0123 **(e)** 0.00655

Solution

	Scientific Notation	*Engineering Notation*
(a) 48,900,000	4.89×10^7	48.9×10^6
(b) 63,500	6.35×10^4	63.5×10^3
(c) 77.1	7.71×10^1	77.1×10^0
(d) 0.0123	1.23×10^{-2}	12.3×10^{-3}
(e) 0.00655	6.55×10^{-3}	6.55×10^{-3}

Note that the last number, 0.00655, is the same in scientific and engineering notation.

Rule

Changing to Ordinary Notation

Move the decimal point, to the right if the exponent is positive and to the left if it is negative, by the number of places equal to the power of ten.

EXAMPLE 4.4 Change to ordinary notation:

(a) 3.40×10^3 **(b)** 663×10^{-3} **(c)** 8.23×10^{-1}

(d) 0.768×10^6 **(e)** 55.2×10^{-6}

Solution

(a) $3.40 \times 10^3 = 3400$

(b) $663 \times 10^{-3} = 0.663$

(c) $8.23 \times 10^{-1} = 0.823$

(d) $0.768 \times 10^6 = 768{,}000$

(e) $55.2 \times 10^{-6} = 0.0000552$

EXAMPLE 4.5 Given:

$$(4{,}800)(56{,}000)(0.0390)$$

(a) Express the numbers in scientific notation, then calculate and express the answer in scientific notation to three significant digits.

(b) Express the answer in ordinary and engineering notation to three significant digits.

Solution

(a) The numbers in scientific notation are:

$$(4.8 \times 10^3)(5.6 \times 10^4)(3.90 \times 10^{-2})$$

Scientific notation is useful for calculation because all the numbers are between 1 and 10. Now apply the rules for multiplying with powers of ten from Chapter 3:

$$(4.8 \times 10^3)(5.6 \times 10^4)(3.9 \times 10^{-2}) = (4.8)(5.6)(3.9) \times 10^{3+4-2} \approx 105 \times 10^5$$

Note that the answer is *not* in scientific notation because the decimal point is not to the right of the first significant digit. Move the decimal point two places to the left to the zero position and balance the number by increasing the power of ten by 2:

$$105 \times 10^5 = 1.05 \times 10^7$$

(b) Write the number in ordinary notation by applying the power of ten and moving the decimal point seven places to the right, then change to engineering notation:

$$1.05 \times 10^7 = 10{,}500{,}000 = 10.5 \times 10^6$$

ERROR BOX

To avoid making an error with the different notations you should be able to quickly recognize whether a number is in ordinary, scientific, or engineering notation. When a number is written in ordinary notation, the power of ten is actually zero: $0.345 = 0.345 \times 10^0$. When this number is written in scientific notation, the decimal point is to the right of the first digit: 3.45×10^{-1}. When this number is written in engineering notation, the power of ten is a multiple of 3 and the number is between 1 and 1000: 345×10^{-3}. It is possible for a number to be written in more than one notation, such as 5.98×10^0, which is in ordinary, scientific, *and* engineering notation.

Practice Problems: Tell which notation or notations each number is written in, or if it is not written in any notation.

1. 1339 **2.** 56.7×10^9 **3.** 2.0×10^2 **4.** 126

5. 6.91×10^{-3} **6.** 1.00×10^0 **7.** 0.555×10^5 **8.** 99.9×10^{-9}

9. 7.87×10^{-4} **10.** 10×10^{-2} **11.** 5.55 **12.** 0.872

13. 3.23×10^{-1} **14.** 999×10^9 **15.** 0.292×10^3

The mode of the calculator can be set to scientific notation or engineering notation by using the [MODE], [FSE], or a similar key. To enter any number times a power of ten on the calculator use the [EE], or [EXP] key. For example, 3.9×10^{-2} is entered as follows:

DAL: 3.9 [EE] [(-)] 2 → 3.9 E −2

Not DAL: 3.9 [EXP] 2 [+/-] → 3.9 E −2

The factor of 10 does not appear on the display. It is understood. Only the significant digits and the exponent are shown. The calculator solution for Example 4.5 above is, in scientific notation:

DAL: 4.8 [EE] 3 [×] 5.6 [EE] 4 [×] 3.9 [EXP] [(-)] 2 [=] → 1.05 E −7

Not DAL: 4.8 [EXP] 3 [×] 5.6 [EXP] 4 [×] 3.9 [EXP] 2 [+/-] [=] → 1.05 E −7

Depending on how you set your calculator, this result could be in ordinary, scientific, or engineering notation.

EXAMPLE 4.6

(a) Express the numbers in scientific notation, then calculate and express the answer in scientific notation to three significant digits.

(b) Express the answer in ordinary and engineering notation to three significant digits.

$$\frac{(328{,}000)(0.0850)}{0.00569}$$

Solution

(a) The numbers in scientific notation are:

$$\frac{(328{,}000)(0.0850)}{0.00569} = \frac{(3.28 \times 10^5)(8.50 \times 10^{-2})}{5.69 \times 10^{-3}}$$

Apply the rules for multiplying and dividing with powers of ten. Multiply the numerator first, then divide:

$$\frac{(3.28 \times 10^5)(8.50 \times 10^{-2})}{5.69 \times 10^{-3}} = \frac{(3.28)(8.50) \times 10^{5-2}}{5.69 \times 10^{-3}} = \frac{27.88 \times 10^3}{5.69 \times 10^{-3}}$$

$$= \frac{27.88}{5.69} \times 10^{3+3} = 4.8998 \times 10^6 \approx 4.90 \times 10^6$$

(b) The answer to (a) is also in engineering notation. Change to ordinary notation by applying the power of ten:

$$4.90 \times 10^6 = 4{,}900{,}000$$

The calculator solution using powers of ten and scientific notation is:

DAL: 3.28 [EE] 5 [×] 8.5 [EE] [(-)] 2 [÷] 5.69 [EE] [(-)] 3 [=] → 4.90 E6

Not DAL: 3.28 [EXP] 5 [×] 8.5 [EXP] 2 [+/-] [÷] 5.69 [EXP] 3 [+/-] [=] → 4.90 E6

EXAMPLE 4.7

Given:

$$\frac{(7{,}610{,}000)}{(0.00542)(183)}$$

(a) Write the numbers in scientific notation and estimate the answer without the calculator by rounding the numbers to one digit.

(b) Do the example on the calculator and express the answer in scientific notation to three significant digits.

Solution

(a) Scientific notation makes estimation easier because all the numbers can be rounded to a number between 1 and 10 and the calculations can be done without a calculator or even mentally:

$$\frac{7{,}610{,}000}{(0.00542)(183)} = \frac{7.61 \times 10^6}{(5.42 \times 10^{-3})(1.83 \times 10^2)} \approx \frac{8 \times 10^6}{(5 \times 10^{-3})(2 \times 10^2)}$$

Apply the rules for powers of ten to get the estimate:

$$\frac{8 \times 10^6}{(5 \times 10^{-3})(2 \times 10^2)} = \frac{8 \times 10^6}{(2)(5) \times 10^{-(3-2)}} = \frac{8 \times 10^6}{10 \times 10^{-1}} = \frac{8}{10} \times 10^{6+1}$$
$$= 0.8 \times 10^7$$

The estimate of 0.8×10^7, when changed to scientific notation, is:

$$0.8 \times 10^7 = 8 \times 10^6$$

(b) The calculator solution is:

DAL: 7.61 [EE] 6 [÷] [(] 5.42 [EE] [(-)] 3 [×] 1.83 [EE] 2 [)] [=] → 7.67 E6

Not DAL: 7.61 [EXP] 6 [÷] [(] 5.42 [EXP] 3 [+/-] [×] 1.83 [EXP] 2 [)] [=] → 7.67 E6

This answer, 7.67×10^6, agrees closely with the estimate, 8×10^6, thereby verifying the calculation.

Close the Circuit

The voltage, in volts, of a circuit is given by:

$$V = (0.000133)(470{,}000)$$

Write the numbers in engineering notation and calculate V, in volts. Express the answer in engineering notation to three significant digits.

Solution Write the numbers in engineering notation and multiply applying the rules for the powers of ten:

$$(133 \times 10^{-6})(470 \times 10^3) = (133)(470) \times 10^{-(6-3)} = 62510 \times 10^{-3}$$

Change the answer to engineering notation by moving the decimal point and changing the power of ten:

$$62510 \times 10^{-3} = 62.51 \times 10^0 \approx 62.5 \times 10^0 \text{ or } 62.5 \text{ V}$$

The calculator solution is:

DAL: 133 [EE] [(-)] 6 [×] 470 [EE] 3 [=] → 62.5 E0

Not DAL: 133 [EXP] 6 [+/-] [×] 470 [EXP] 3 [=] → 62.5 E0

The size of numbers written with powers of ten can be misleading. The increase from 10^2 to 10^3 is only 900, but the increase from 10^6 to 10^7 is 9,000,000! Figure 4-1 gives you an idea of how powers of ten relate to some of the world's important (and not so important) scientific phenomena. For example, if you could add all the words ever spoken since people first started babbling, the number would still be less than 10^{18}!

Answers to Error Box Problems, page 84:
1. Ord. **2.** Eng. **3.** Sci. **4.** Eng., Ord. **5.** Sci., Eng. **6.** All three **7.** None **8.** Eng. **9.** Sci. **10.** None **11.** All three **12.** None **13.** Sci. **14.** Eng. **15.** None

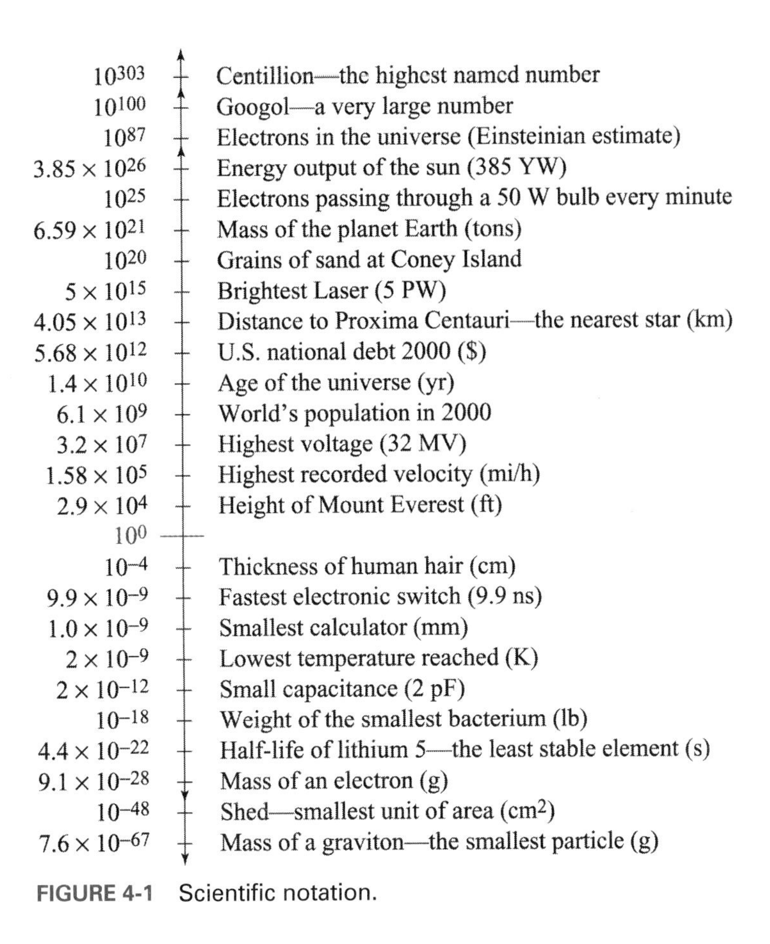

FIGURE 4-1 Scientific notation.

EXERCISE 4.1

In exercises 1 through 16, express each number in scientific notation and engineering notation.

1. 42,600,000,000
2. 11,700,000
3. 0.000930
4. 0.0000301
5. 2.35
6. 45.6
7. 0.00117
8. 0.0981
9. 334
10. 69,900
11. 0.112
12. 0.409
13. 1000
14. 10
15. 162,000
16. 7760

In exercises 17 through 50, express each number in ordinary, scientific, and engineering notation.

17. 564×10^5
18. 3.36×10^2
19. 1200×10^3
20. 14×10^6
21. 11.4×10^{-8}
22. 345×10^{-4}
23. 0.0528×10^3
24. 0.00384×10^2
25. 5690×10^{-6}
26. 1560×10^{-3}
27. 5.66×10^6
28. 9.90×10^9
29. 552×10^{-6}
30. 89.1×10^{-3}
31. 2.64×10^7
32. 4.33×10^5
33. 3.90×10^{-6}
34. 9.31×10^{-5}
35. 1.45×10^0
36. 46.7×10^0
37. 0.101×10^{-1}
38. 634×10^{-4}
39. 112×10^3
40. 87.0×10^3
41. 10.5×10^{-6}
42. 717×10^{-3}
43. 0.871×10^9
44. 4.44×10^6
45. 1230×10^{-3}
46. 1005×10^{-9}
47. 125×10^3
48. 10.1×10^{-6}
49. 5.03×10^{-1}
50. 5.50×10^{-2}

In exercises 51 through 60:
(a) Express the numbers in scientific notation, then calculate and express the answer in scientific notation to three significant digits.
(b) Express the answer in ordinary and engineering notation to three significant digits.

51. (8400)(28,000)(0.0550)
52. (0.00444)(12,800)(100,000)
53. $\dfrac{4510}{0.000334}$

54. $\dfrac{4{,}660{,}000}{0.0198}$

55. $\dfrac{8670}{(0.0139)(0.183)}$

56. $\dfrac{(0.00677)}{(0.298)(12{,}900)}$

57. $\dfrac{(0.1530)(0.00183)}{3900}$

58. $\dfrac{(1{,}560{,}000)(0.0124)}{(0.00345)}$

59. $\dfrac{(254{,}000)(2.74)}{(0.000881)(5{,}300)}$

60. $\dfrac{(0.000122)(3.34)}{(665)(0.113)}$

In exercises 61 through 80:

(a) If indicated by the instructor estimate each answer without the calculator by writing the numbers in scientific notation and rounding off to one digit. Based on the estimate choose the correct answer in engineering notation to three significant digits. See Example 4.7 for estimation procedure.

(b) Calculate the answer in engineering notation to three significant digits.

61. (0.00000101)(9,330) $[9.42 \times 10^{-6}, 9.42 \times 10^{-3}, 9.42 \times 10^{-2}, 9.42 \times 10^{0}]$

62. (4,780,000) (0.000767) $[3.66 \times 10^{2}, 3.66 \times 10^{3}, 3.67 \times 10^{3}, 36.7 \times 10^{3}]$

63. (32,200)(86,500)(2830) $[7.88 \times 10^{9}, 78.8 \times 10^{9}, 78.8 \times 10^{12}, 7.88 \times 10^{12}]$

64. (52,100)(54,500)(0.156) $[44.3 \times 10^{6}, 443 \times 10^{6}, 44.3 \times 10^{9}, 443 \times 10^{9}]$

65. $\dfrac{0.00000656}{0.000123}$ $[5.33 \times 10^{0}, 53.3 \times 10^{-6}, 533 \times 10^{-3}, 53.3 \times 10^{-3}]$

66. $\dfrac{303{,}000}{4{,}050{,}000}$ $[748 \times 10^{0}, 748 \times 10^{-3}, 7.48 \times 10^{-3}, 74.8 \times 10^{-3}]$

67. $\dfrac{(449{,}000)(0.00776)}{(0.0659)}$ $[5.29 \times 10^{4}, 5.29 \times 10^{3}, 52.9 \times 10^{3}, 52.9 \times 10^{6}]$

68. $\dfrac{3200}{(0.129)(0.00450)}$ $[5.51 \times 10^{6}, 55.1 \times 10^{3}, 55.1 \times 10^{6}, 5.51 \times 10^{3}]$

69. $\dfrac{0.877}{(45{,}100)(2670)}$ $[7.28 \times 10^{-9}, 0.728 \times 10^{-9}, 72.8 \times 10^{-9}, 728 \times 10^{-9}]$

70. $\dfrac{(0.0000210)(0.00349)}{451}$ $[163 \times 10^{-12}, 16.3 \times 10^{-9}, 1.63 \times 10^{-9}, 0.163 \times 10^{-9}]$

71. $\dfrac{(123{,}000)(0.0356)}{(1770)(0.0434)}$ $[57.0 \times 10^{0}, 570 \times 10^{-3}, 5.70 \times 10^{0}, 570 \times 10^{0}]$

72. $\dfrac{(0.00909)(0.0117)}{(0.000100)(0.514)}$ $[2.07 \times 10^{0}, 20.7 \times 10^{0}, 207 \times 10^{0}, 207 \times 10^{-3}]$

73. $\dfrac{8.89 \times 10^{12}}{9.89 \times 10^{6}}$ $[0.899 \times 10^{6}, 8.99 \times 10^{6}, 89.9 \times 10^{6}, 899 \times 10^{3}]$

74. $\dfrac{5.96 \times 10^{9}}{8.01 \times 10^{3}}$ $[74.4 \times 10^{6}, 7.44 \times 10^{6}, 74.4 \times 10^{3}, 744 \times 10^{3}]$

75. $\dfrac{3.03 \times 10^{-3}}{6.06 \times 10^{2}}$ $[5.00 \times 10^{-6}, 5.00 \times 10^{-3}, 50.0 \times 10^{-6}, 500 \times 10^{-3}]$

76. $\dfrac{92.3 \times 10^{-6}}{2.54 \times 10^{-3}}$ $[363 \times 10^{-6}, 0.363 \times 10^{-3}, 36.3 \times 10^{-3}, 3.63 \times 10^{-3}]$

77. $\dfrac{1.26 \times 10^{12}}{(35.0 \times 10^{-3})(0.900 \times 10^{6})}$ $[4.00 \times 10^{6}, 40.0 \times 10^{6}, 0.400 \times 10^{6}, 400 \times 10^{6}]$

78. $\dfrac{81.2 \times 10^{3}}{(366 \times 10^{-3})(45.5 \times 10^{3})}$ $[4.88 \times 10^{0}, 4.88 \times 10^{3}, 48.8 \times 10^{0}, 0.488 \times 10^{3}]$

79. $\dfrac{(116 \times 10^{3})(0.203 \times 10^{6})}{94.1 \times 10^{-3}}$ $[0.250 \times 10^{12}, 2.50 \times 10^{12}, 25.0 \times 10^{12}, 250 \times 10^{9}]$

80. $\dfrac{(26.5 \times 10^{-3})(69.3 \times 10^{6})}{0.0860 \times 10^{12}}$ $[21.3 \times 10^{-6}, 21.4 \times 10^{-6}, 2.13 \times 10^{-6}, 213 \times 10^{-6}]$

Applied Problems

In problems 81 through 84, solve each applied problem to three significant digits.

81. Parker Bros. Inc., manufacturer of the board game Monopoly, printed $18,500,000,000,000 of toy money in 1990 for all its games, which is more than all the real money in circulation in the world. If all this "money" were distributed equally among the world's population, estimated at 5.3 billion (5.3×10^9) in 1990, how much would each person get to the nearest cent?

82. In 1989, General Motors made one of the greatest profits ever achieved by an industrial company. Worldwide sales totaled $136,975,000,000, with its assets valued at $180,236,500,000. What percentage of the assets do the total sales represent?

83. The shortest blip of light produced at the AT&T laboratories in New Jersey lasts 8.0×10^{-15} seconds. If the speed of light is 300×10^6 meters per second, how many meters does light travel in that time? Give the answer in engineering notation. *Note:* (Rate)(Time) = (Distance)

84. The most massive living thing on earth is the giant sequoia, named the General Sherman, in Sequoia National Park, California. Its weight is estimated at 5.51×10^6 lb. The seed of such a tree weighs only 1.67×10^{-4} oz. Calculate the ratio of the weight of the mature tree to the weight of the seed. Give the answer in scientific notation. *Note:* 1 lb = 16 oz.

Applications to Electronics

In problems 85 through 92, solve each applied problem. Give the answer in engineering notation to three significant digits.

85. The world's most expensive pipeline is the Alaska pipeline, which is built to carry 2,000,000 barrels a day of crude oil. If 6.65×10^6 barrels of oil (1 million tons) can generate 4.00×10^9 kWh (kilowatt-hours) of electricity, how many kilowatt-hours could be generated in one year from the Alaska pipeline oil?

86. A CRAY scientific computer can do a simple addition in 130×10^{-12} seconds (130 picoseconds). How many additions can the CRAY computer do in 1 minute?

87. Find the voltage, in volts, of a circuit given by:

$$V = (0.00021)(870)$$

88. Find the power, in watts, of a circuit given by:

$$P = (0.550)(0.0670)$$

89. Find the current, in amps, in a circuit given by:

$$I = \frac{120 \times 10^0}{1.3 \times 10^3}$$

90. Find the current, in amps, in a circuit given by:

$$I = \frac{350 \times 10^{-3}}{3.30 \times 10^3}$$

91. The total resistance R_T of a series circuit, in ohms, is given by:

$$R_T = (3.30 \times 10^3) + (910 \times 10^0) + (1.20 \times 10^3)$$

Find R_T in ohms. *Note:* You cannot add powers of ten unless they are the same.

92. The total voltage V_T across a series circuit is given by:

$$V_T = (1.20 \times 10^0) + (430 \times 10^{-3}) + (550 \times 10^{-3})$$

Find V_T in volts. *Note:* You cannot add powers of ten unless they are the same.

4.2 Metric System (SI)

The metric system is used throughout electricity and electronics. The United States is moving "inch by 2.54 centimeters" toward using the metric system completely. In 1960 the metric system, which was first established in France in 1790, became the more complete International System, abbreviated SI, which stands for "Systeme International." Since then, almost every country has converted to the SI system except the United States. The U.S. customary system, which evolved from the old English system, is gradually being replaced by the SI system. Almost all technical and scientific work uses SI units.

The metric or SI system is based on powers of ten like our decimal system. Important SI units, their symbols, and what they measure are shown in Table 4-1. Study this table carefully. There are seven base units in the SI system. The other units are called derived units. For example, the ampere, which is named after Andre Marie Ampere, is the unit for electric current and is considered a base unit.

The other electrical units, such as coulomb, volt, ohm, and watt, are derived units. For example, the unit of charge, the coulomb, is equal to 6.25×10^{18} electrons and is the quantity of electric charge moved in one second when the current is one ampere:

$$1 \text{ coulomb} = 1 \text{ C} = (1 \text{ ampere})(1 \text{ second}) = 1 \text{ A} \cdot \text{s}$$

TABLE 4-1 SI BASE UNITS AND OTHER IMPORTANT SI UNITS

	Quantity	Unit	Symbol
Base Units	Length	meter	m
	Time	second	s
	Mass	kilogram	kg
	Electric current	ampere	A
	Thermodynamic temperature	kelvin	K
	Molecular substance	mole	mol
	Light intensity	candela	cd
	Electric charge	coulomb	C
	Electric voltage	volt	V
	Electric resistance	ohm	Ω
	Electric power	watt	W
	Electric conductance	siemens	S
	Electric capacitance	farad	F
	Electric inductance	henry	H
	Force	newton	N
	Energy	joule	J
	Frequency	hertz	Hz
	Temperature	Celsius	°C
	Pressure	kilopascal	kPa
	Angle	radian	rad
	Capacity	liter	L
	Area	square meter	m^2

The unit of energy, the joule, is used for electrical energy, mechanical energy, and heat energy. Electric power is the amount of electrical energy being used per unit of time. The unit of electric power, the watt, is equal to one joule per second. Degrees kelvin (K) is the base unit for thermodynamic or absolute temperature, but degrees Celsius is more commonly used. The conversion formula is: $°K = °C + 273.15^0$. The abbreviations for units of time are: s for seconds, min for minutes, h for hours, and d for days.

Table 4-2 shows the common metric prefixes based on powers of ten which are used to denote quantities of metric and electrical units. The powers of ten are divisible by three and correspond to engineering notation:

1 GB (gigabyte) $= 1 \times 10^9$ bytes or 1,000,000,000 bytes
1 MW (megawatt) $= 1 \times 10^6$ W or 1,000,000 W
1 kΩ (kilohm) $= 1 \times 10^3\ \Omega$ or 1000 Ω
1 mV (millivolt) $= 1 \times 10^{-3}$ V or 0.001 V
1 μA (microamp) $= 1 \times 10^{-6}$ A or 0.000001 A
1 nF (nanofarad) $= 1 \times 10^{-9}$ F or 0.000000001 F

Changes within the metric system are much easier to perform than within the U.S. system because of the powers of ten. Most changes can be done by just moving the decimal point three places to the right or left. Study the following examples which show how to change between metric units.

EXAMPLE 4.9 Change each of the following units:

(a) 1.3 kW to W **(b)** 500 mA to A **(c)** 34 V to kV
(d) 1800 kΩ to MΩ **(e)** 0.034 C to mC

TABLE 4-2 COMMON SI PREFIXES

Power of 10	Prefix	Symbol	Factor
10^{12}	tera	T	trillion (1,000,000,000,000)
10^{9}	giga	G	billion (1,000,000,000)
10^{6}	mega	M	million (1,000,000)
10^{3}	kilo	k	thousand (1000)
10^{0}	no prefix	Base Unit	one (1)
10^{-2}	centi	c	hundredth (0.01)
10^{-3}	milli	m	thousandth (0.001)
10^{-6}	micro	μ	millionth (0.000001)
10^{-9}	nano	n	billionth (0.000000001)
10^{-12}	pico	p	trillionth (0.000000000001)

Solution The procedure for changing these units is the same as that for changing powers of ten. First express the unit with the power of ten. Then change the power of ten to correspond to the new unit, and move the decimal point to balance the value.

(a) Given: $1.3\ \text{kW} = 1.3 \times 10^3\ \text{W}$. To change to the base unit W, you need to change the power of ten to 10^0. Decrease the power of ten from 3 to 0 and move the decimal point three places to the right, increasing the number by a factor of 10^3 and balancing the value:

$$1.3 \times 10^3\ \text{W} = 1300 \times 10^0\ \text{W or } 1300\ \text{W}$$

Note that changing to the 10^0 power is the same as changing to ordinary notation.

(b) Given: $500\ \text{mA} = 500 \times 10^{-3}\ \text{A}$. To change to the base unit A, increase the power of ten from −3 to 0. Move the decimal point three places to the left to decrease the number by a factor of 10^3 and balance the value:

$$500 \times 10^{-3}\ \text{A} = 0.5 \times 10^0\ \text{A or } 0.5\ \text{A}$$

(c) Given: $34\ \text{V} = 34 \times 10^0\ \text{V}$. To change to kilo, increase the power of ten from 0 to 3. Move the decimal point three places to the left to decrease the number by a factor of 10^3, balancing the value:

$$34 \times 10^0\ \text{V} = 0.034 \times 10^3\ \text{V} = 0.034\ \text{kV}$$

(d) Given: $1800\ \text{k}\Omega = 1800 \times 10^3\ \Omega$. To change to mega, increase the power of ten by 3 to 6 and move the decimal point three places to the left to balance the value.

$$1800 \times 10^3\ \Omega = 1.80 \times 10^6\ \Omega = 1.80\ \text{M}\Omega$$

(e) Given: $0.034\ \text{C} = 0.034 \times 10^0\ \text{C}$. To change to milli, decrease the power of ten from 0 to −3 and move the decimal point three places to the right to balance the value:

$$0.034 \times 10^0\ \text{C} = 34 \times 10^{-3}\ \text{C} = 34\ \text{mC}$$

It is not necessary to show every step when changing units. You should be able to do some steps mentally. Observe that in changing units you just move the decimal point three (and sometimes six) places to the left or right.

Rule

Changing Metric Units

Move the decimal point to the *left when changing to a larger unit,* and to the *right when changing to a smaller unit.* The number of places is equal to the difference in the powers of ten between the metric prefixes.

Study the following examples.

EXAMPLE 4.10 Change each of the following units:

(a) 0.085 MΩ to kΩ **(b)** 2.8 mA to μA **(c)** 100 kV to MV **(d)** 500 μC to mC

Solution

(a) Kilohms (10^3) is a smaller unit than megohms (10^6). The difference in the powers of ten is 3. Therefore, apply the rule above and move the decimal point three places to the right since you are changing to a smaller unit:

$$0.085\ \text{M}\Omega = 85\ \text{k}\Omega$$

(b) Microamps (10^{-6}) is a smaller unit than milliamps (10^{-3}). The difference in the powers of ten is 3. Move the decimal point three places to the right since you are changing to a smaller unit:

$$2.8\ \text{mA} = 2800\ \mu\text{A}$$

(c) Megavolts (10^6) is a larger unit than kilovolts (10^3). The difference in the powers of ten is 3. Move the decimal point three places to the left since you are changing to a larger unit:

$$100\ \text{kV} = 0.100\ \text{MV}$$

(d) Millicoulombs (10^{-3}) is a larger unit than microcoulombs (10^{-6}). The difference in the powers of ten is 3. Move the decimal point three places to the left since you are changing to a larger unit:

$$500\ \mu\text{C} = 0.500\ \text{mC}$$

The next example shows changes between units where the decimal point is moved six places instead of three.

EXAMPLE 4.11 Change each of the following units:

(a) 670,000 V to MV

(b) 50,000 μA to A

(c) 0.078 MΩ to Ω

(d) 0.0034 kW to mW

(e) 310 cm to m

Solution

(a) Megavolts (10^6) is a larger unit than volts (10^0). The difference in the powers of ten is 6. Therefore, move the decimal point six places to the left since you are changing to a larger unit:

$$670{,}000\ \text{V} = 0.67\ \text{MV}$$

(b) Amps (10^0) is a larger unit than microamps (10^{-6}). The difference in the powers of ten is 6. Move the decimal point six places to the left since you are changing to a larger unit:

$$50{,}000\ \mu\text{A} = 0.050\ \text{A}$$

(c) Ohms (10^0) is a smaller unit than megohms (10^6). The difference in the powers of ten is 6. Therefore, move the decimal point six places to the right:

$$0.078\ \text{M}\Omega = 78{,}000\ \Omega$$

(d) Milliwatts (10^{-3}) is a smaller unit than kilowatts (10^3). The difference in the powers of ten is 6. Therefore, move the decimal point six places to the right:

$$0.0034\ \text{kW} = 3400\ \text{mW}$$

(e) Meters (10^0) is a larger unit than centimeters (10^{-2}). The difference in the powers of ten is 2. Move the decimal point 2 places to the left since you are changing to a larger unit:

$$310\ \text{cm} = 3.10\ \text{m}$$

The next example shows some less common unit changes.

EXAMPLE 4.12 Change each of the following units:

(a) 35×10^2 Hz to kHz **(b)** 5.5×10^{-4} S to μS
(c) 76×10^5 pF to μF **(d)** 0.034 m^2 to cm^2

Solution

(a) Express 35×10^2 in ordinary notation and then move the decimal point three places to the left to change from hertz to a larger unit kilohertz:

$$35 \times 10^2\ \text{Hz} = 3500\ \text{Hz} = 3.5\ \text{kHz}$$

(b) Express 5.5×10^{-4} in ordinary notation and then move the decimal point six places to the right to change from siemens to a smaller unit microsiemens:

$$5.5 \times 10^{-4}\ \text{S} = 0.00055\ \text{S} = 550\ \mu\text{S}$$

(c) Microfarads (10^{-6}) is a larger unit than picofarads (10^{-12}) with a difference of 6 in the powers of ten. Change to ordinary notation and then move the decimal point six places to the left:

$$76 \times 10^5\ \text{pF} = 7{,}600{,}000\ \text{pF} = 7.6\ \mu\text{F}$$

(d) To change square meters to square centimeters you have to first note that 1 m = 100 cm. Since this is an equality, it can be squared on both sides and it will still be an equality:

$$(1\ \text{m})^2 = (100\ \text{cm})^2$$

which means that:

$$1 \times 10^0\ \text{m}^2 = 10{,}000\ \text{cm}^2 = 1 \times 10^4\ \text{cm}^2$$

ERROR BOX

A common error to watch out for is moving the decimal point the wrong way when changing units in the metric system. Always check that your answer makes sense. If you are changing to a *smaller* unit, then you should be *increasing* the number of units. If you are changing to a *larger* unit, then you should be *decreasing* the number of units. See if you can apply this reasoning and do the practice problems.

Practice Problems: Change each of the following units:

1. 150 mA to A **2.** 0.15 mA to μA **3.** 1.5 MΩ to kΩ

4. 1500 Ω to kΩ **5.** 0.015 V to mV **6.** 15 pF to nF

7. 110,000 V to MV **8.** 0.087 GHz to kHz **9.** 4.7 m to cm

10. 6600 μW to W **11.** 0.95 H to mH **12.** 0.5 m^2 to cm^2

Square centimeter is a smaller unit than square meter, and the difference in the powers of ten is 4. Therefore, move the decimal point four places to the right:

$$0.034 \text{ m}^2 = 340 \text{ cm}^2$$

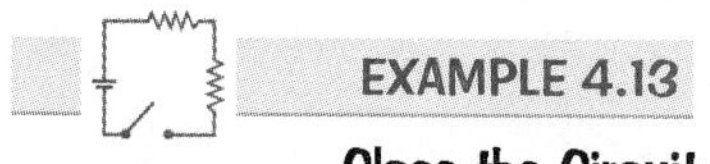

EXAMPLE 4.13

Close the Circuit

Figure 4-2 shows a basic dc circuit containing a resistor connected to a voltage source. The voltage, in volts, is given by Ohm's law $V = IR$, where I = current in amps and R = resistance in ohms. Find V when $R = 11.0$ kΩ and $I = 5.50$ mA to three significant digits.

R

I

+ V −

$V = IR$

FIGURE 4-2 Voltage in a dc circuit for Example 4.13.

Solution The units in the formula must both be base units; ohms and amps. Change milliamps to amps and kilohms to ohms before using the formula. When changing units to a base unit, you can just apply the power of ten that corresponds to the metric prefix:

$$5.50 \text{ mA} = 5.50 \times 10^{-3} \text{ A} = 0.00550 \text{ A}$$

$$11.0 \text{ k}\Omega = 11.0 \times 10^{3} \ \Omega = 11{,}000 \ \Omega$$

The voltage is then:

$$V = IR = (0.00550 \text{ A})(11{,}000 \ \Omega) = 60.5 \text{ V}$$

or using powers of ten:

$$V = IR = (5.5 \times 10^{-3} \text{ A})(11 \times 10^{3} \ \Omega) = 60.5 \times 10^{0} \text{ V} = 60.5 \text{ V}$$

The calculator solution is:

DAL: 5.5 [EE] [(−)] 3 [×] 11 [EE] 3 [=] → 60.5

Not DAL: 5.5 [EXP] 3 [+/-] [×] 11 [EXP] 3 [=] → 60.5

The solution can be done on an electrical engineering calculator that has special unit keys which take the place of entering the powers of ten. Enter the units after entering each number. Depending on your calculator, you may have to first set the mode to engineering and then press the [SHIFT] or [2nd F] key, or a special unit key, to enter the units after each number:

5.5 [SHIFT] [m] [×] 11 [SHIFT] [k] [=] → 60.5 V

EXERCISE 4.2

In exercises 1 through 50, change each of the following units. Round answers to two significant digits.

1. 2.3 kV to V
2. 0.56 kW to W
3. 31 mA to A
4. 330 Ω to kΩ
5. 1.5 kΩ to Ω
6. 5.2 W to mW
13. 7400 kV to MV

7. 1.5 MV to kV
8. 0.85 mA to μA
9. 0.0780 C to mC
10. 860 μA to mA
11. 6.8 MW to kW
12. 0.085 V to mV
21. 400 g to kg

14. 6,000 kW to MW
15. 1.30 nF to pF
16. 1.20 μS to mS
17. 3900 kHz to MHz
18. 220 μs to ms
19. 45 μH to mH
20. 4000 pF to nF
29. 3400 μA to A
30. 55,000 W to MW

22. 1500 m to km
23. 0.023 cm to mm
24. 87 mg to g
25. 0.0055 C to mC
26. 6600 mV to V
27. 40,000 V to MV
28. 120,000 Ω to MΩ
40. 55.0 mm^2 to cm^2
41. 3.5×10^2 Hz to kHz

31. 0.0075 kΩ to mΩ

32. 0.00056 A to μA

33. 0.00088 μF to pF

34. 0.00068 S to μS

35. 1000 mV to kV

36. 0.0055 C to μC

37. 4100 mm to m

38. 0.0067 km to cm

39. 0.0600 m^2 to cm^2

42. 7.6×10^4 pF to μF

43. 55×10^{-4} S to μS

44. 67×10^{-2} A to mA

45. 5.6×10^4 Ω to MΩ

46. 2100×10^2 V to MV

47. 8600 cm^2 to m^2

48. 10 cm^2 to mm^2

49. 505 nF to mF

50. 0.0013 V to μV

Applied Problems

In problems 51 through 66, round answers to two significant digits.

51. A plane climbs to an altitude of 3500 meters. How many kilometers is this altitude?

52. A human hair is 4×10^{-4} cm thick. How many millimeters is this thickness?

53. One skier beats another in a race by 0.00056 s. How many milliseconds and microseconds is this time measurement?

54. The area occupied by a snow crystal is measured to be 2×10^{-5} cm^2. How many square millimeters is this area?

Applications to Electronics

55. A large generator produces 300,000 volts. How many kilovolts and megavolts is this?

56. The frequency of a radio wave is 5.40 MHz. How many kilohertz and hertz is this frequency?

57. Using Ohm's law for voltage $V = IR$, find V in volts when $R = 5.1$ kΩ and $I = 250$ μA. See Example 4.13.

58. Using Ohm's law for voltage $V = IR$, find V in volts when $R = 2.7$ MΩ and $I = 66$ μA. See Example 4.13.

59. Using Ohm's law for current $I = \frac{V}{R}$, find I in amps *and* milliamps when $V = 240$ mV and $R = 1.5$ Ω.

60. Using Ohm's law for current $I = \frac{V}{R}$, find I in amps *and* milliamps when $V = 120$ V and $R = 1.3$ kΩ.

61. Using Ohm's law for resistance $R = \frac{V}{I}$, find R in megohms *and* kilohms when $V = 240$ V and $I = 500$ μA.

62. Using Ohm's law for resistance $R = \frac{V}{I}$, find R in ohms *and* kilohms when $V = 9.2$ V and $I = 20$ mA.

63. Using the formula for power $P = VI$, find P in watts *and* milliwatts when $V = 900$ mV and $I = 850$ mA.

64. Using the formula for power $P = VI$, find P in watts *and* kilowatts when $V = 1.2$ kV and $I = 20$ A.

65. The wave length λ (lambda) of a radio wave in meters is given by:

$$\lambda = \frac{v}{f}$$

where the velocity of the wave $v = 3.00 \times 10^8$ m/sec and f = the frequency of the wave in hertz. Find the wavelength, in meters and centimeters, of a radio wave whose frequency is 1010 kHz.

66. Using the formula for power $P = I^2R$, find P, in kilowatts and megawatts, produced by a generator when $R = 1.8$ kΩ and the current $I = 20$ A.

4.3 The U.S. Customary System

The U.S. customary system is based on numbers such as 12 and 16, which were the basis of the old English system. In part, these numbers were used because they had several divisors and were easy to work with as fractions. This is no longer important in the age of electronic computation. Some of the major differences and conversion factors between units of the U.S. customary system and the International System are shown in Table 4-3. The U.S. base units for length and mass are different from the SI units, but the other five base units are the same (see Table 4-1). In both systems, all electrical units are the same. However, the units for work energy, heat energy, and mechanical power are different in the U.S. system.

"Thinking metric," observe in Table 4-3 that:

- A meter is a little longer than a yard (3 feet).
- A kilogram is a little heavier than two pounds.
- A newton is about one-quarter of a pound force.

To convert units between systems, you can divide units like numbers and check that the answer has the correct units. Study the following examples, which show how to convert between the U.S. system and the International System.

TABLE 4-3 THE U.S. CUSTOMARY SYSTEM AND SI CONVERSIONS

Quantity	U.S. Unit	Symbol	SI Conversion Factor
Length	foot inch mile	ft in mi	1 m = 3.281 ft 1 cm = 0.3937 in 1 km = 0.6214 mi
Mass	pound	lb	1 kg = 2.205 lb
Force	pound force	lb_f	1 N = 0.2248 lb_f
Temperature	degree Fahrenheit	°F	$°C = \left(\frac{5}{9}\right)(°F - 32°)$
Velocity	feet per second	ft/s	1 m/s = 3.281 ft/s
Pressure	pounds per square foot	lb/ft^2	1 kPa = 20.89 lb/ft^2
Mechanical power	horsepower	hp	1 W = 1.341×10^{-3} hp
Work energy	foot-pound	ft · lb	1 J = 0.7376 ft · lb
Heat energy	British thermal unit	Btu	1 J = 9.485×10^{-4} Btu
Area	square foot	ft^2	1 m^2 = 10.76 ft^2 1 mm^2 = 1.550×10^{-3} in^2

EXAMPLE 4.14 Convert 34 kg to pounds. Give the answer to two significant digits.

Solution Set up the conversion factor from Table 4-3 as a unit ratio:

$$\left(\frac{2.205 \text{ lb}}{1 \text{ kg}}\right) = 1$$

Multiply by the conversion factor, which does not affect the value of a quantity since it is equal to one, and check that the units divide out:

$$34 \cancel{\text{kg}} \left(\frac{2.205 \text{ lb}}{1 \cancel{\text{kg}}}\right) = 34(2.205) \text{ lb} \approx 75 \text{ lb}$$

EXAMPLE 4.15 Convert 28 feet to meters. Give the answer to two significant digits.

Solution Obtain the conversion factor from Table 4-3 and set up the unit ratio:

$$\frac{1 \text{ m}}{3.281 \text{ ft}} = 1$$

Answers to Error Box Problems, page 93:

1. 0.15 A **2.** 150 μA **3.** 1500 kΩ **4.** 1.5 kΩ **5.** 15 mV **6.** 0.015 nF
7. 0.11 MV **8.** 87,000 kHz **9.** 470 cm **10.** 0.0066 W **11.** 950 mH **12.** 5,000 cm^2

Multiply by this ratio, which does not affect the value of a quantity since it is equal to one, and divide out units to obtain the correct units:

$$(28\ \cancel{\text{ft}})\left(\frac{1\ \text{m}}{3.281\ \cancel{\text{ft}}}\right) \approx 8.5\ \text{m}$$

EXAMPLE 4.16 Convert 25 J (joules) to foot-pounds. Give the answer to two significant digits.

Solution From Table 4-3, the unit ratio conversion factor from joules to foot-pounds is:

$$\frac{0.7376\ \text{ft}\cdot\text{lb}}{1\ \text{J}} = 1$$

Multiply by this ratio, dividing out the units, to obtain the correct units:

$$25\ \cancel{\text{J}}\left(\frac{0.7376\ \text{ft}\cdot\text{lb}}{1\ \cancel{\text{J}}}\right) \approx 18\ \text{ft}\cdot\text{lb}$$

EXAMPLE 4.17 Convert 8.60 inches to centimeters and millimeters. Give the answer to three significant digits.

Solution First change 8.60 in to cm:

$$8.60\ \cancel{\text{in}}\left(\frac{1\ \text{cm}}{0.3937\ \cancel{\text{in}}}\right) \approx 21.8\ \text{cm}$$

Move the decimal point one place to the right to change centimeters to millimeters:

$$21.8\ \text{cm} = 218\ \text{mm}$$

The calculator solution is as follows:

8.6 [÷] 0.3937 [=] → 21.8 cm [×] 10 [=] → 218 mm

EXAMPLE 4.18 Convert 37 kW to horsepower. Give the answer to two significant digits.

Solution First change to kW to watts:

$$37\ \text{kW} = 37{,}000\ \text{W}$$

Then change to horsepower. Multiply by the unit ratio conversion factor for watts-to-horsepower:

$$37{,}000\ \cancel{\text{W}}\left(\frac{1.341\times 10^{-3}\ \text{hp}}{1\ \cancel{\text{W}}}\right) \approx 50\ \text{hp}$$

EXAMPLE 4.19 Convert 66 ft/s to meters per second. Give the answer to two significant digits.

Solution Multiply by the unit ratio conversion factor from Table 4-3:

$$66\ \cancel{\text{ft/s}}\left(\frac{1.0\ \text{m/s}}{3.281\ \cancel{\text{ft/s}}}\right) = 20\ \text{m/s}$$

The one exception to the above procedure of multiplying by a unit ratio is the change from Fahrenheit to Celsius temperature. Use the formula in Table 4-3 and substitute the value given in degrees Fahrenheit, as the next example shows.

EXAMPLE 4.20 Change 50°F to Celsius to the nearest degree.

Solution Substitute 50 for °F in the formula in Table 4-3:

$$C = \left(\frac{5}{9}\right)(50 - 32) = \left(\frac{5}{9}\right)(18) = 10°C$$

EXERCISE 4.3

In exercises 1 through 40, convert each of the following units. Express answers to two significant digits.

1. 55 m to ft
2. 3.6 kg to lb
3. 120 lb to kg
4. 150 ft to m
5. 8.6 ft to m
6. 0.88 m to ft
7. 35 kg to lb
8. 185 lb to kg
9. 1000 W to hp
10. 12,000 J to Btu
11. 0.25 hp to W
12. 12 ft · lb to J
13. 1.5 hp to W
14. 0.66 Btu to J
15. 2.5 Btu to J
16. 6.4 J to ft · lb
17. 23 m^2 to ft^2
18. 29 ft^2 to m^2
19. 20 ft^2 to m^2
20. 0.78 m^2 to ft^2
21. 500 mm^2 to in^2
22. 0.92 in^2 to mm^2
23. 35 m/s to ft/s
24. 15 ft/s to m/s
25. 1.3 ft to cm
26. 5.3 cm to in
27. 5.0 km to mi
28. 6.2 mi to km
29. 5.5 in to cm
30. 54 cm to in
31. 95°F to °C
32. 32°F to °C
33. 1.5 km to ft
34. 10 in to cm
35. 120 kW to hp
36. 100 hp to kW
37. 10 MJ to Btu
38. 66 Btu to MJ
39. 0.55 lb to g
40. 5.4 hp to kW

Applied Problems

In problems 41 through 54, solve each applied problem to two significant digits unless specified.

41. The mass of a large computer system is 290 lb. How many kilograms is this mass?

42. A fast car is clocked at a speed of 55 ft/s. How many meters per second is this speed?

43. One of the thinnest calculators is manufactured by Sharp Electronics Corporation. It is 1.4 mm thick. How many inches thick is this calculator?

44. One of the tallest office buildings in the world is the Sears Tower in Chicago, which was completed in 1974. It has 110 stories and rises to 1454 ft. Its total ground area is 4,400,000 ft^2.
    (a) What is the building's height in meters to three significant digits?
    (b) What is the total ground area in square meters?

45. If one nautical mile (nmi) = 1.15 statute (land) miles, how many kilometers equal 1 nmi? *Note:* 5,280 ft = 1 mi. Round the answer to three significant digits.

46. The atmospheric pressure at sea level is 2120 lb/ft^2. How many kilopascals is this pressure?

47. The formula for work is $W = Fd$, where F = force and d = distance. When F is in newtons and d in meters, W is in joules. Find W in joules *and* foot pounds when F = 98 N and d = 55 cm.

48. Einstein's formula for atomic energy is $E = mc^2$, where m = mass in kilograms, c = speed of light in meters per second, and E = energy in joules. If $c = 3.00 \times 10^8$ m/s, calculate how many joules of atomic energy are contained in one *milligram* of mass.

Applications to Electronics

49. Wire thickness is measured in mils, where 1 mil = 1×10^{-3} inches. How many millimeters is one mil?

50. The resistance of a no. 8 copper wire whose cross section is 8.37 mm^2 is 2.06 ohms per kilometer.
    (a) What is the resistance in ohms per 1000 ft?
    (b) What is the cross section in square inches?

51. A large generator produces 10,000 kW of power. How many horsepower is this equivalent to?

52. An electrical force is equal to 19×10^{-9} N. How many pounds of force is this equal to?

53. A rectangular microprocessor has a length of 1.1 cm and a width of 0.50 cm. How many square centimeters and square inches is this area? *Note:* 1 ft^2 = 144 in^2

54. The solar energy falling on one square centimeter of the earth is 1.93 calories per minute. If 1 kW = 239 calories per second, how many kilowatts per square meter fall on the earth? Round answer to three significant digits. *Note:* 1 $= 1 \times 10^4$ cm^2.

4.4 Hand Calculator Operations

This section provides additional examples and further explanation on how to do scientific notation, engineering notation, and metric conversions on the calculator. It shows how to estimate results to help troubleshoot for mistakes.

Scientific Notation

To enter a number in scientific notation on the calculator:

1. Enter the significant digits with the decimal point to the right of the first digit.
2. Press [EE] or [EXP].
3. Enter the exponent.

For example, to enter 9.78×10^{-3}, press:

DAL: 9.78 [EE] [(-)] 3 → 9.78 E −3

Not DAL: 9.78 [EXP] 3 [+/-] → 9.78 E −3

The exponent appears to the right on the display. You do not enter the number 10; it is understood.

When a calculation in ordinary notation produces a result with too many places for the display, the calculator automatically switches to scientific notation. You can set the calculator to always display scientific notation (SCI) by pressing the [MODE], [FSE], or similar key.

Engineering Notation

You can use the same keys for engineering notation that you use for scientific notation. However, if you have an electrical engineering calculator with electrical units, you do not need to enter the exponent. When you press the units key after entering a number, the correct power of ten corresponding to the units is entered. Depending on your calculator, you may first have to set the mode to engineering (ENG) and then press the [2nd] key, [SHIFT] key, or a similar key before entering the units.

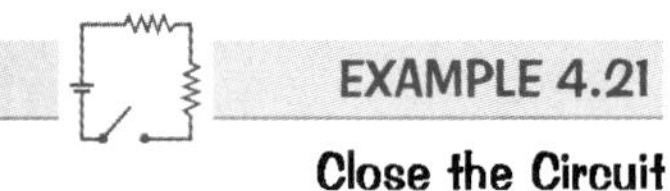

EXAMPLE 4.21

Close the Circuit

Given a resistor $R = 1.50\ \text{k}\Omega$ with a voltage drop $V = 220$ mV, find the current, in microamps, through the resistor by applying Ohm's law $I = \frac{V}{R}$. Round the answer to three significant digits.

Solution Substitute the numbers in the formula using powers of ten for the units

$$I = \frac{220 \times 10^{-3}\ \text{V}}{1.50 \times 10^{3}\ \Omega}$$

The calculator solution is:

DAL: 220 [EE] [(-)] 3 [÷] 1.5 [EE] 3 [=] → 0.000147 or 1.47 E −4

Not DAL: 220 [EXP] 3 [+/-] [÷] 1.5 [EXP] 3 [=] → 0.000147 or 1.47 E −4

The result will be in ordinary, scientific, or engineering notation, depending on how you set your calculator. To change the above answer in scientific notation to engineering notation change the power of ten from −4 to −6 and move the decimal point two places to the right to balance the number:

$$I = 1.47 \times 10^{-4} = 147 \times 10^{-6}\ \text{A} = 1.47\ \mu\text{A}$$

Using an engineering calculator, you enter each number, press [2nd], [SHIFT], or a similar key and then press the corresponding units key:

220 [2nd] [m] [÷] 1.5 [2nd] [k] [=] → 1.47 μ

The key [m] is for milli and [k] is for kilo. The calculator gives the answer with the correct units: μ = micro.

Metric Conversion

Metric conversion can be done in one series of steps on the calculator if you set up all the factors as shown in the next example.

EXAMPLE 4.22 Convert 15 inches to centimeters and millimeters. Round the answer to two significant digits.

Solution Use the conversion factor in Table 4-3, 1 cm = 0.3937 in, to change inches to centimeters. Then change centimeters to millimeters. This can be done in one series of operations as follows:

$$15 \boxed{\div} 0.3937 \rightarrow 38 \text{ cm} \boxed{\times} 10 \boxed{=} \rightarrow 380 \text{ mm}$$

Dividing by 0.3937 changes inches to centimeters, and multiplying by 10 changes centimeters to millimeters.

EXERCISE 4.4

In exercises 1 through 14, estimate the answer first and choose what you think is the correct answer (to three significant digits) from the four choices. Check the answer with the calculator.

1. $(58.6 \times 10^3)(9.19 \times 10^{-3})$ [5.39, 53.9, 539, 5390]
2. $(6.58 \times 10^6)(9.19 \times 10^{-3})$ [60.5×10^3, 6.05×10^3, 0.605×10^6, 60.5×10^6]
3. $(86.6 \times 10^3)(94.1 \times 10^3)(5.31 \times 10^3)$ [0.433×10^{12}, 4.33×10^{12}, 43.3×10^{12}, 433×10^{12}]
4. $(3.57 \times 10^6)(51.1 \times 10^{-6})(3.15 \times 10^3)$ [574×10^3, 575×10^3, 574×10^6, 575×10^6]
5. $\dfrac{55 \times 10^9}{2.2 \times 10^3}$ [2.50×10^6, 25.0×10^6, 25.0×10^9, 0.250×10^9]
6. $\dfrac{6.67 \times 10^3}{2.11 \times 10^{-6}}$ [31.6×10^6, 316×10^6, 3.16×10^9, 31.6×10^9]
7. $\dfrac{10.1 \times 10^6}{(71.1 \times 10^6)(0.813)^2}$ [0.214×10^0, 21.4×10^{-3}, 214×10^{-3}, 215×10^{-3}]
8. $\dfrac{(4.46 \times 10^6)(0.509)^3}{1.44 \times 10^3}$ [409, 409×10^{-3}, 408, 408×10^{-3}]
9. $\dfrac{(2.50 \times 10^3)^2}{(0.244)^3}$ [430×10^3, 43.0×10^6, 430×10^6, 4.30×10^9]
10. $\dfrac{(1.50 \times 10^3)^3}{(56.5)^2}$ [1.07×10^6, 1.06×10^6, 1.05×10^6, 1.06×10^3]
11. $\sqrt{4.32 \times 10^{12}}$ [2.08×10^{12}, 2.08×10^6, 20.8×10^{12}, 20.8×10^6]
12. $\sqrt{0.0893 \times 10^6}$ [2.99×10^3, 2.99×10^0, 29.9×10^0, 299×10^0]
13. $\sqrt{1.55 \times 10^{-6}}$ [1.24×10^{-6}, 1.25×10^{-3}, 1.24×10^{-3}, 1.25×10^{-3}]
14. $\sqrt{8.38 \times 10^{12}}$ [2.90×10^9, 2.90×10^6, 2.89×10^9, 2.89×10^6]

Applied Problems

In problems 15 through 22, round all answers to three significant digits.

15. The heat energy in joules radiated by a certain black body is given by:

$$E = \sigma(T^4 - T_0^4)$$

where $\sigma = 48.8 \times 10^{-9}$, $T = 291$ K (kelvin) and $T_0 = 273$ K. Calculate E in joules.

16. The formula for the radius r of a sphere is:

$$r = \sqrt[3]{\frac{3V}{4\pi}}$$

where V = volume. Find r in cm when $V = 1.13 \times 10^3$ cm^3.

Applications to Electronics

17. The current through a resistor $R = 1.20\ k\Omega$ is $I = 3.50$ mA. Using Ohm's law for voltage, $V = IR$, find the voltage drop, in volts, across the resistor.

18. The current through a resistor R is $I = 220\ \mu A$. If the voltage drop across the resistor is $V = 53.0$ mV, use Ohm's law for resistance, $R = \frac{V}{I}$, to find the value of R in ohms.

19. The current in a circuit $I = 42.5$ mA and the power $P = 1.35$ W. Using the power formula for resistance, $R = \frac{P}{I^2}$, find the resistance, in ohms, of the circuit.

20. The voltage in a circuit $V = 125$ V and the resistance $R = 1.60\ k\Omega$. Using the formula for power, $P = \frac{V^2}{R}$, find the power, in watts, in the circuit.

21. The impedance Z in ohms of an ac circuit is given by:

$$Z = \sqrt{R^2 + X^2}$$

Find Z in $k\Omega$ when $R = 14.7\ k\Omega$ and $X = 26.5\ k\Omega$.

22. In exercise 21, find Z in $k\Omega$ when $R = 140\ k\Omega$ and $X = 120\ k\Omega$.

CHAPTER HIGHLIGHTS

4.1 SCIENTIFIC AND ENGINEERING NOTATION

Scientific Notation

A number between 1 and 10 (including 1 but not 10) times a power of ten:

$$5.05 \times 10^7,\quad 6.23 \times 10^{-6},\quad 9.99 \times 10^0,\quad 4.12 \times 10^{-2},\quad 1.00 \times 10^{12}$$

• • •

Changing Ordinary Notation to Scientific Notation

Move the decimal point to the right of the first significant digit, called the *zero position.* The number of places from the zero position to the original decimal point equals the power of ten: positive if the decimal point is to the right of the zero position, negative if it is to the left.

Key Example:

$$3{,}780{,}000 = 3.78 \times 10^6 \qquad 5.03 = 5.03 \times 10^0$$

$$0.00000383 = 3.83 \times 10^{-6}$$

Engineering Notation

A number between 1 and 1000 (including 1 but not 1000) times a power of ten, which *is a multiple of three:*

$$2.22 \times 10^{-3},\quad 65.4 \times 10^9,\quad 1.00 \times 10^{-6},\quad 329 \times 10^0,\quad 23 \times 10^3$$

• • •

Changing Ordinary Notation to Engineering Notation

Move the decimal point a multiple of three places: 3, 6, 9, etc., to change to a number from 1 to less than 1000. The number of places from the new position to the original decimal point equals the power of ten: positive if the decimal point is to the right of the new position, negative if it is to the left.

Key Example:

$$3{,}560 = 3.56 \times 10^3 \qquad 87.4 = 87.4 \times 10^0$$

$$0.0000678 = 67.8 \times 10^{-6}$$

Changing to Ordinary Notation

Move the decimal point, to the right if the exponent is positive and to the left if it is negative, by the number of places equal to the power of ten.

Key Example:

$$567 \times 10^{-9} = 0.000000567 \qquad 331 \times 10^0 = 331$$

$$9.36 \times 10^3 = 9360$$

To calculate with numbers in scientific or engineering notation, apply the rules for powers of ten from Chapter 3:

Key Example:

$$\frac{7.45 \times 10^6}{(3.78 \times 10^{-3})(5.68 \times 10^3)} = \frac{7.45 \times 10^6}{(3.78)(5.68) \times 10^0} \approx 0.347 \times 10^6 = 347{,}000$$

4.2 METRIC SYSTEM (SI)

The metric system is based on ten like our decimal system. Important electrical units in the metric system are:

Voltage (V = volt)	*Current* (A = amp)	*Resistance* (Ω = ohm)
Charge (C = coulomb)	*Power* (W = watt)	*Conductance* (S = siemens)
Capacitance (F = farad)	*Inductance* (H = Henry)	*Frequency* (Hz = hertz)

Metric prefixes correspond to powers of ten in engineering notation:

M = Mega (10^6), k = kilo (10^3), m = milli (10^{-3}), μ = micro (10^{-6})

Key Example:

1 MΩ (megohm) = 1×10^6 Ω or 1,000,000 Ω.
1 kW (kilowatt) = 1×10^3 W or 1000 W
1 mA (milliamp) = 1×10^{-3} A or 0.001 A
1 μF (microfarad) = 1×10^{-6} F or 0.000001 F

Tables 4-1 and 4-2 show the SI units and metric prefixes.

Changing Metric Units

Move the decimal point to the *left when changing to a larger unit,* and to the *right when changing to a smaller unit.* The number of places is equal to the difference in the powers of ten.

Key Example:

(a) 0.21 kW = 210 W
(b) 345,000 μA = 0.345 A
(c) 0.000678 mF = 678
(d) 18 MΩ = 18,000 kΩ
(e) 4.3×10^{-5} S = 0.000043 S = 43 μS

4.3 THE U.S. CUSTOMARY SYSTEM

The electrical units in the U.S. system are the same as in the metric system. Some important differences are:

1 m= 3.281 ft
1 kg = 2.205 lb
1 W =1.341×10^{-3}
1 J = 0.7376 ft · lb

Conversions between the U.S. system and the metric system are done by multiplying by the unit ratio conversion factor given in Table 4-3. Divide units like numbers:

Key Example:

$$115 \text{ kg} = 115 \cancel{\text{kg}}\left(\frac{2.205 \text{ lb}}{1 \cancel{\text{kg}}}\right) = 254 \text{ lb}$$

$$9.82 \text{ ft} = 9.82 \cancel{\text{ft}}\left(\frac{1 \text{ m}}{3.281 \cancel{\text{ft}}}\right) = 2.99 \text{ m}$$

$$5.65 \text{ W} = 5.65 \cancel{\text{W}}\left(\frac{0.001341 \text{ hp}}{1 \cancel{\text{W}}}\right) = 0.00758 \text{ hp}$$

4.4 HAND CALCULATOR OPERATIONS

Scientific Notation and Engineering Notation

To enter 5.44×10^{-6} press:

DAL: 5.44 [EE] [(-)] 6 → 5.44 E −6
Not DAL: 5.44 [EXP] 6 [+/-] → 5.44 E −6

Key Example: Calculate $\dfrac{5.3 \times 10^{-6}}{630 \times 10^{-3}}$ as follows:

DAL: 5.3 [EE] [(-)] 6 [÷] 630 [EXP] [(-)] 3 [=] → 0.00000841 or 8.41 E −6

Not DAL: 5.3 [EXP] 6 [+/-] [÷] 630 [EXP] 3 [+/-] [=] → 0.00000841 or 8.41 E −6

Conversion Between Metric and U.S. Units

Key Example: Using the unit conversion factor 1 kg = 2.205 lb, convert 120 g to pounds on the calculator as follows:

120 [÷] 1000 [×] 2.205 [=] → 0.265 lb

REVIEW EXERCISES

In exercises 1 through 6, change each number to scientific notation and engineering notation.

1. 87,600,000
2. 0.0000560
3. 0.112
4. 1260
5. 34.7×10^4
6. 567×10^{-4}

In exercises 7 through 12, change each number to ordinary notation.

7. 33×10^7
8. 5.55×10^{-6}
9. 124×10^0
10. 971×10^3
11. 0.339×10^{-3}
12. 11.2×10^6

In exercises 13 through 20 express each number in ordinary, scientific and engineering notation.

13. 126×10^4
14. 15.2×10^{-5}
15. 0.0555×10^2
16. 48.8×10^6
17. 7.33×10^5
18. 9.48×10^{-9}
19. 3.66×10^0
20. 0.0011

In exercises 21 through 42, convert each of the following units. Express answers to three significant digits.

21. 594 to W
22. 6.50 MΩ to kΩ
23. 0.0000453 A to μA
24. 5,670,000 V to MV

25. 670 pF to
26. 0.0560 ms to μs
27. 560 mm to cm
28. 0.000593 km to cm
29. 3.20×10^4 V to kV
30. 0.000230×10^3 H to mH
31. 1040 mm^2 to cm^2
32. 45×10^3 Hz to
33. 12.0 kg to lb
34. 120 m to ft
35. 2.50 ft to m
36. 3.50 hp to kW
37. 5.30 kW to hp
38. 84.0 cm to ft
39. 6000 hp to MW
40. 30.0×10^3 J to ft · lb
41. 10 km to mi
42. 0.0343 hp to W

In exercises 43 through 48:
(a) Express the numbers in scientific notation, then calculate and express the answer in scientific notation to three significant digits.
(b) Express the answer in ordinary and engineering notation to three significant digits.

43. (0.668)(102,000)(399)
44. (1295)(32,456)(577)
45. $\frac{0.0525}{12{,}700}$
46. $\frac{7090}{(0.00246)(0.0657)}$
47. $\frac{(0.6340)(0.0721)}{9090}$
48. $\frac{(1{,}010{,}000)(55.1)}{(0.707)(2800)}$

In exercises 49 through 60:
(a) If indicated by the instructor, estimate each answer without the calculator by writing the numbers in scientific notation and rounding off to one digit. Based on the estimate choose the correct answer in engineering notation to three significant digits. See Example 4.7 for estimation procedure.
(b) Calculate the answer in engineering notation to three significant digits.

49. $(9.21 \times 10^6)(3.58 \times 10^{-3})$ [33.0×10^3, 3.30×10^3, 330×10^3, 330×10^6]
50. $(5.21 \times 10^3)(8.11 \times 10^{-6})$ [4.23×10^{-3}, 42.3×10^{-3}, 423×10^{-3}, 0.423×10^3]
51. (28,100)(32,500)(100,000) [9.13×10^{12}, 91.3×10^{12}, 913×10^{12}, 913×10^{15}]
52. $\frac{0.787}{0.000321}$ [2.45×10^4, 24.5×10^3, 2.45×10^3, 24.5×10^4]
53. $\frac{101{,}000{,}000}{505{,}000{,}000{,}000}$ [20.0×10^{-3}, 200×10^{-3}, 20.0×10^{-6}, 200×10^{-6}]
54. $\frac{3.33 \times 10^6}{7.89 \times 10^{-3}}$ [422×10^6, 42.2×10^6, 4.22×10^9, 42.2×10^9]
55. $(0.325 \times 10^6)(6.31 \times 10^{-9})$ [0.205×10^{-3}, 2.05×10^{-3}, 20.5×10^{-3}, 205×10^{-3}]
56. $\frac{(0.0000610)(0.00245)}{0.889}$ [168×10^{-9}, 16.8×10^{-9}, 1.68×10^{-9}, 0.168×10^{-9}]
57. $\frac{\sqrt{2.86 \times 10^6}}{0.532 \times 10^3}$ [0.317, 31.7, 0.318, 3.18]
58. $\left(\sqrt{9.33 \times 10^2}\right)\left(8.93 \times 10^2\right)^3$ [218×10^6, 21.8×10^6, 21.8×10^9, 2.18×10^9]
59. $\frac{(56.9 \times 10^3)(13.1 \times 10^3)}{1.49 \times 10^{-3}}$ [5.00×10^{12}, 50.0×10^{12}, 50.0×10^9, 500×10^9]
60. $\frac{12.2 \times 10^6}{(3.41 \times 10^3)(149 \times 10^{-3})}$ [25.0×10^6, 25.0×10^3, 24.0×10^6, 24.0×10^3]

Applied Problems

In problems 61 through 78, round answers to three significant digits unless specified.

61. The mean radius of the earth is 6,370,000 m. If 1 mi = 5,280 ft, what is the radius in kilometers and miles? Express the answers in ordinary, scientific, and engineering notation.

62. The mass of an electron is measured to be 9.10×10^{-28} grams. What is this weight in milligrams and pounds? Express the answers in scientific and engineering notation.

63. A "22-caliber" rifle means the bore is 0.22 in. What is the caliber of a rifle whose bore is 7.62 mm? Round the answer to two significant digits.

64. The fastest runner in baseball was Ernest Swanson, who took only 13.30 s to circle the bases, a distance of 360 ft, in 1932. What was his average speed in kilometers per hour?

Applications to Electronics

65. A power plant is rated at 150 MW. If 1 kW = 1.341 hp, how many kilowatts and horsepower is this equivalent to?

66. The world produced more than 6.82×10^{12} kWh of electricity in 1998. If 1 kWh is equivalent to 76×10^{-3} gal of oil, how many gallons of oil would it take to produce that much energy? Give the answer in scientific and engineering notation.

67. Using Ohm's law for resistance $R = \frac{V}{I}$, find R in ohms and kilohms when $V = 5.60$ V and $I = 3.80$ mA.

Note: Units must be consistent in the formula: V in volts and I in amps.

68. The total resistance R_T in ohms of a series circuit is given by:

$$R_T = (820 \times 10^3) + (1.1 \times 10^6) + (750 \times 10^3)$$

Find R_T in kilohms and megohms.

69. An ammeter measures the current I though a resistor $R = 4.70$ kΩ to be 12.0 mA. Using Ohm's law for voltage $V = IR$, find the voltage drop across the resistor.

70. In problem 69, using the power formula $P = I^2R$, find the power dissipated in the resistor in watts and milliwatts.

71. The power P dissipated by a resistor $R = 750\ \Omega$ is 90 mW. Using the power formula for current $I = \sqrt{\frac{P}{R}}$, find the current I in amps and milliamps.

72. The resistance of a parallel circuit is given by:

$$R = \frac{R_1 R_2}{R_1 + R_2}$$

If $R_1 = 820\ \Omega$ and $R_2 = 1.20 \times 10^3\ \Omega$, substitute the values and find R in ohms and kilohms.

73. The impedance Z of an ac circuit is given by:

$$Z = \sqrt{R^2 + X^2}$$

Find Z when $R = 2.2 \times 10^3\ \Omega$ and $X = 270\ \Omega$. Express the answer in ohms and kilohms.

74. In problem 73, find Z when $R = 910$ kΩ and $X = 1.50$ MΩ. Express the answer in kilohms and megohms. *Note:* Units must be consistent in the formula, both in kΩ or both in MΩ.

75. The inductance of an ac circuit is given by:

$$L = \frac{L_1 L_2}{L_1 + L_2}$$

Find L in henries when $L_1 = 50 \times 10^{-3}$ H and $L_2 = 10 \times 10^{-3}$ H.

76. Find L in problem 75 when $L_1 = 500$ μH and $L_2 = 1.50$ mH. Express the answer in microhenrys and millihenries. *Note:* Units must be consistent in the formula, both in mH or both in μH.

77. The inductance of a coil in henries is given by:

$$L = \frac{n^2 \mu A}{l}$$

Calculate L in microhenries when $n = 200$ turns, $\mu = 1.26 \times 10^{-6}$ N/A², $A = 3.00 \times 10^{-4}$ m² and $l = 0.300$ m.

78. Calculate L in microhenries in problem 77 when $n = 500$ turns, $\mu = 1.56 \times 10^{-6}$ N/A², $A = 3.50 \times 10^{-4}$ m² and $l = 0.450$ m.

CHAPTER 5

Basic Algebra

CHAPTER OBJECTIVES

In this chapter, you will learn:

- The meaning of signed numbers.
- How to add, subtract, multiply, and divide signed numbers.
- How to identify an algebraic expression.
- How to combine algebraic terms.
- The five basic rules for working with exponents.
- The meaning of negative and zero exponents.

Electricity and electronics abound with formulas and equations whose building blocks are algebraic symbols. It is essential that you learn algebra well so you can better understand and work with the ideas and concepts of electrical and electronic formulas. This chapter presents the fundamentals of algebra and serves as the foundation you need for future chapters. What makes algebra different from arithmetic is the use of positive and negative signed numbers, and literals, which are letters that are used to represent numbers. Expressions that contain a literal part and a number part are called algebraic terms. Algebraic terms can be added and multiplied, but the procedures differ from those of arithmetic. It is necessary to learn and understand these different algebraic procedures.

• • •

5.1 Signed Numbers

One of the important features of algebra that makes it different from arithmetic is the use of positive and negative signed numbers. A good way to understand signed numbers and zero is by placing them on a number line, where the numbers increase from left to right:

$$\ldots -4, -3, -2, -1, 0, +1, +2, +3, +4, \ldots$$

Therefore −3 is a larger number than −4, −2 is larger than −3, and so on. *The sign of a number indicates direction.* On the number line, negative is to the left and positive is to the right. Signed numbers are used on a thermometer where positive is up and negative is down. Up, or to the right, is usually considered positive, and down, or to the left, negative. In Chapter 3, positive and negative numbers are used for powers of ten where the number increases with the exponent:

$$10^{-3} = 0.001, \quad 10^{-1} = 0.1, \quad 10^{0} = 1, \quad 10^{3} = 1000, \text{ etc.}$$

In electricity, positive is used to indicate current flowing in one direction, and negative to indicate current flowing in the opposite direction. A current of +2 A has the same *magnitude* as a current of −2 A, but it flows in the opposite direction. For voltage, zero is used to indicate a voltage equal to ground potential. A positive voltage means a voltage above ground, and a negative voltage means a voltage below ground. A voltage of +5 V is considered to have the same magnitude as a voltage of −5 V, but −5 V is below ground potential. Two important concepts for signed numbers are:

Definition

Integer

A positive or negative whole number or zero.

• • •

Definition

Absolute Value

The value of a signed number without the sign that represents the magnitude or size of the number.

Absolute value can be thought of as equal to the positive value. It is indicated with vertical lines:

$$|-6| = 6 \quad |+9| = 9 \quad |-100| = 100 \quad |0| = 0 \quad |-3.7| = 3.7 \quad \left|+\frac{1}{2}\right| = \frac{1}{2}$$

Note that the absolute value of zero is zero, and the absolute value of any positive number N is the same as the absolute value of –N. The rules for signed numbers are similar to the rules for powers of ten given in Chapter 3:

Rule

Adding Numbers with Like Signs

Add the absolute values and apply the common sign.

EXAMPLE 5.1

Add:

(a) $(-8) + (-7)$ **(b)** $(+10) + (+21)$ **(c)** $(-6.2) + (-10.4)$

(d) $(0) + (9.6)$ **(e)** $\left(-\frac{1}{2}\right) + \left(-\frac{1}{2}\right)$

Solution Apply the rule for adding numbers with like signs:

(a) $(-8) + (-7) = -(8 + 7) = -15$

(b) $(+10) + (+21) = +(10 + 21) = +31 = 31$

(c) $(-6.2) + (-10.4) = -(6.2 + 10.4) = -16.6$

(d) $(0) + (9.6) = +(0 + 9.6) = +9.6 = 9.6$

(e) $\left(-\frac{1}{2}\right) + \left(-\frac{1}{2}\right) = -\left(\frac{1}{2} + \frac{1}{2}\right) = -1$

Notice in the answers to (b) and (d) that *when there is no sign in front, it is understood to be positive.*

Rule

Adding Numbers with Unlike Signs

Subtract the smaller absolute value from the larger absolute value and apply the sign of the larger number.

EXAMPLE 5.2

Add:

(a) $(-9) + (6)$ **(b)** $(-5) + (13)$ **(c)** $(5.6) + (-8.4)$

(d) $(1.2) + (-2.1)$ **(e)** $-\left(\frac{3}{4}\right) + \left(\frac{1}{4}\right)$

Solution Apply the rule for adding numbers with unlike signs, which is similar to subtraction in arithmetic. The parentheses above can be removed and are not necessary:

(a) $(-9) + (6) = -9 + 6 = -(9 - 6) = -3$

(b) $(-5) + (13) = -5 + 13 = +(13 - 5) = 8$

(c) $(5.6) + (-8.4) = 5.6 - 8.4 = -(8.4 - 5.6) = -2.8$

(d) $(1.2) + (-2.1) = 1.2 - 2.1 = -(2.1 - 1.2) = -0.9$

(e) $-\left(\frac{3}{4}\right) + \left(\frac{1}{4}\right) = -\frac{3}{4} + \frac{1}{4} = -\left(\frac{3}{4} - \frac{1}{4}\right) = -\frac{1}{2}$

Notice in (c) and (d) that when a plus sign is followed by a minus sign it is equivalent to just a minus sign.

EXAMPLE 5.3 Add:

$$6.6 - 2.3 - 8.8 + 1.2$$

Solution To add several signed numbers, add the numbers with like signs first, and then apply the rule for adding numbers with unlike signs:

$$\begin{aligned} 6.6 - 2.3 - 8.8 + 1.2 &= (6.6 + 1.2) - (2.3 + 8.8) \\ &= 7.8 - 11.1 = -(11.1 - 7.8) = -3.3 \end{aligned}$$

Subtraction in algebra is like addition with the sign changed:

Rule

Subtracting Signed Numbers

Change the subtraction operation to addition and change the sign of the number that follows the operation sign. Then apply the rules for adding signed numbers.

EXAMPLE 5.4 Subtract each:

(a) $10 - (-3)$ **(b)** $-8 - (2)$ **(c)** $-7.3 - (-4.6)$

(d) $22.4 - (14.2)$ **(e)** $-\frac{1}{8} - \left(\frac{1}{4}\right)$

Solution

(a) Change subtraction to addition and change -3 to 3. Then apply the rule for addition:

$$10 - (-3) = 10 + 3 = 13$$

(b) Change subtraction to addition and change 2 (which is $+2$) to -2. Apply the rule for addition:

$$-8 - (2) = -8 - 2 = -(8 + 2) = -10$$

(c) Change subtraction to addition and change -4.6 to 4.6. Apply the rule for addition:

$$-7.3 - (-4.6) = -7.3 + 4.6 = -(7.3 - 4.6) = -2.7$$

(d) Change subtraction to addition and change 14.2 to -14.2. Apply the rule for addition:

$$22.4 - (14.2) = 22.4 - 14.2 = 8.2$$

(e) Change subtraction to addition and change 1/4 to $-1/4$. Apply the rule for addition:

$$-\frac{1}{8} - \left(\frac{1}{4}\right) = -\frac{1}{8} - \frac{1}{4} = -\left(\frac{1}{8} + \frac{1}{4}\right) = -\frac{3}{8}$$

Because subtraction in algebra is just addition with the sign changed, it is helpful to think of addition and subtraction of signed numbers as a process of *algebaic combination* of signed numbers, which follows these rules:

Rules

Combining Signed Numbers

1. A plus sign followed by a plus sign equals a plus sign.
2. A plus sign followed by a minus sign equals a minus sign.
3. A minus sign followed by a plus sign equals a minus sign.
4. A minus sign followed by a minus sign equals a plus sign.

Observe in the above rules that a minus sign changes the sign that follows it, but a plus sign does not change the sign that follows it.

EXAMPLE 5.5

Perform the operations:

$$3.2 + (-4.1) - (-5.5) - (1.2)$$

Solution Apply the above rules for algebraic combination and write the example as follows:

$$3.2 + (-4.1) - (-5.5) - (1.2) = 3.2 - 4.1 + 5.5 - 1.2$$

Add the positive numbers and the negative numbers separately:

$$3.2 - 4.1 + 5.5 - 1.2 = (3.2 + 5.5) - (4.1 + 1.2) = 8.7 - 5.3$$

Add the resulting positive and negative numbers:

$$8.7 - 5.3 = 3.4$$

ERROR BOX

A common error when adding and subtracting signed numbers is to overlook a minus sign and not change the sign that follows it. Remember:

A minus followed by a minus equals a plus:

$$-(-4) = +4, \quad -(-1.5) = +1.5, \quad -(-0.63) = +0.63$$

A minus followed by a plus equals a minus:

$$-(+4) = -4, \quad -(+1.5) = -1.5, \quad -(+0.63) = -0.63$$

Practice Problems: Perform the operations:

1. $-10 - (-4) + (2)$
2. $5.6 + (-4.2) - (3.3)$
3. $50 - (+100) - (-50)$
4. $15.8 + (22.4) - (31.3)$
5. $-0.99 + (-1.01) - (-0.90)$
6. $-(-16) - (-12) + (14)$
7. $-500 -(-300) - (-100)$
8. $0.05 + (0.15) - (-0.08)$
9. $\frac{1}{5} + \frac{2}{5} - \left(\frac{3}{10}\right)$
10. $1.01 - (-1.11) + (-1.10)$

Rule

Multiplying or Dividing Two Signed Numbers

Multiply or divide the absolute values. The result is positive if the signs are alike, negative if the signs are unlike.

Multiplication of numbers in algebra is indicated by parentheses or a dot: $(-5)(3) = -5 \cdot 3$. The multiplication sign $\times$ is not used in algebra because it can be confused with the letter x.

EXAMPLE 5.6 Perform the operations:

(a) $(-5)(-3)$ **(b)** $(8)(-7)(11)$ **(c)** $(-100) \div (25)$

(d) $(-1.2)(1.4) \div (-0.8)$ **(e)** $\left(-\frac{1}{2}\right)(-4) \div (-2)$

Solution

(a) Multiply the absolute values and make the result positive:

$$(-5)(-3) = +(5)(3) = 15$$

(b) Multiply $(8)(-7)$ to get -56, and then multiply by (11) to get a negative result:

$$(8)(-7)(11) = (-56)(11) = -616$$

(c) Divide the absolute values and make the result negative:

$$(-100) \div (25) = -(100 \div 25) = -4$$

(d) Multiply the first two numbers to get -1.68, and then divide to get a positive result:

$$(-1.2)(1.4) \div (-0.8) = (-1.68) \div (-0.8) = +(1.68 \div 0.8) = 2.1$$

Exercise (d) can be done on the calculator as follows:

DAL: [(-)] 1.2 [×] 1.4 [÷] [(-)] 0.8 [=] → 2.1

Not DAL: 1.2 [+/-] [×] 1.4 [÷] 0.8 [+/-] [=] → 2.1

(e) Multiply the first two numbers and then do the division:

$$\left(-\frac{1}{2}\right)(-4) \div (-2) = +\left(\frac{1}{2}\right)(4) \div (-2) = 2 \div (-2) = -1$$

The order of operations in algebra is the same as in arithmetic:

Rule

Order of Algebraic Operations—"PEMDAS"

1. Operations in parentheses.
2. Exponents.
3. Multiplication or division.
4. Addition or subtraction.

EXAMPLE 5.7 Perform the operations:

$$-0.5(-7 + 3) + \frac{8(-3)}{6}$$

Solution Perform the operation in parentheses first:

$$-0.5(-7 + 3) + \frac{8(-3)}{6} = -0.5(-4) + \frac{8(-3)}{6}$$

Multiply and divide: $= 2 + \dfrac{-24}{6} = 2 + (-4)$

Then combine the signed numbers: $= 2 - 4 = -2$

The calculator solution is:

DAL: [(−)] 0.5 [×] [(] [(−)] 7 [+] 3 [)] [+] 8 [×] [(−)] 3 [÷] 6 [=] → −2

Not DAL: 0.5 [+/-] [×] [(] 7 [+/-] [+] 3 [)] [+] 8 [×] 3 [+/-] [÷] 6 [=] → −2

EXAMPLE 5.8 Perform the operations:

$$\frac{-1.5(0.8) - 0.2(4)}{20 + 2(-5)}$$

Solution Remember, a *fraction line is like parentheses.* Do the operations in the numerator and denominator first:

$$\frac{-1.5(0.8) - 0.2(4)}{20 + 2(-5)} = \frac{-1.2 - 0.8}{20 - 10} = \frac{-2.0}{10}$$

Then divide: $\dfrac{-2.0}{10} = -0.2$

The calculator solution is:

DAL: [(] [(−)] 1.5 [×] 0.8 [−] 0.2 [×] 4 [)] [÷] [(] 20 [+] 2 [×] [(−)] 5 [)] [=] → −0.2

Not DAL: [(] 1.5 [+/-] [×] 0.8 [−] 0.2 [×] 4 [)] [÷] [(] 20 [+] 2 [×] [5] [+/-] [)] [=] → −0.2

EXAMPLE 5.9 Perform the operations:

$$3\left(\frac{-5}{12}\right) - \left(\frac{-3}{8}\right)$$

Solution Multiply the first fraction by 3. Apply the minus sign to the second fraction, making it positive:

$$\overset{1}{\cancel{3}}\left(\frac{-5}{\underset{4}{\cancel{12}}}\right) - \left(\frac{-3}{8}\right) = \frac{-5}{4} + \frac{3}{8}$$

Note that the 3 is first divided into the 12 to simplify the multiplication. Now combine the fractions over the LCD 8:

$$\frac{-5}{4} + \frac{3}{8} = \frac{-10 + 3}{8} = \frac{-7}{8} \text{ or } -\frac{7}{8}$$

Notice the two ways the answer is written. A minus sign in a fraction can be written three ways: in the numerator, in the denominator, or in front of the fraction, and it does not change its value:

$$\frac{-7}{8} = \frac{7}{-8} = -\frac{7}{8}$$

Answers to Error Box Problems, page 108:

1. −4 **2.** −1.9 **3.** 0 **4.** 6.9 **5.** −1.10 **6.** 42 **7.** −100 **8.** 0.28 **9.** $\frac{3}{10}$ **10.** 1.02

The calculator solution is:

DAL: 3 [×] [(−)] 5 [÷] 12 [−] [(−)] 3 [÷] 8 [=] → −0.875

Not DAL: 3 [×] 5 [+/−] [÷] 12 [−] 3 [+/−] [÷] 8 [=] → −0.875

The decimal −0.875 equals the fraction $-\frac{7}{8}$.

Signs of Grouping

When it is necessary to enclose a set of parentheses in a larger grouping of operations, we use brackets [ ] to enclose parentheses and braces { } to enclose brackets.

EXAMPLE 5.10 Perform the operations:

$$2(-4) - [-3(5 - 6)]$$

Solution Work from the inside parentheses to the outside brackets. Do the operation in parentheses first:

$$2(-4) - [-3(5 - 6)] = 2(-4) - [-3(-1)]$$

Then do the operation in brackets and the remaining operations:

$$\begin{aligned} 2(-4) - [-3(-1)] &= 2(-4) - [3] \\ &= -8 - [3] = -8 - 3 = -11 \end{aligned}$$

The calculator solution using two sets of parentheses is:

DAL: 2 [×] [(−)] 4 [−] [(] [(−)] 3 [×] [(] 5 [−] 6 [)] [)] [=] → −11

Not DAL: 2 [×] 4 [+/−] [−] [(] 3 [+/−] [×] [(] 5 [−] 6 [)] [)] [=] → −11

EXAMPLE 5.11 Perform the operations:

$$15 + \{30 + [(20 - 6)(-10)] - 50\}$$

Solution Work from the inside out. Do the operation in parentheses first:

$$15 + \{30 + [(20 - 6)(-10)] - 50\} = 15 + \{30 + [14(-10)] - 50\}$$

Then do the operations in brackets and the operations in braces last:

$$\begin{aligned} &= 15 + \{30 + [-140] - 50\} = 15 + \{30 - 140 - 50\} \\ &= 15 + \{-160\} = 15 - 160 = -145 \end{aligned}$$

The calculator solution is:

DAL: 15 [+] [(] 30 [+] [(] [(] 20 [−] 6 [)] [×] [(−)] 10 [)] [−] 50 [)] [=] → −145

Not DAL: 15 [+] [(] 30 [+] [(] [(] 20 [−] 6 [)] [×] 10 [+/−] [)] [−] 50 [)] [=] → −145

EXERCISE 5.1

In exercises 1 through 58, perform the indicated operations by hand. Give answers to fraction exercises in fraction form.

1. $2 - 5$
2. $7 - 3$
3. $-2 - 8$
4. $-3 + 10$
5. $3 - (-1)$
6. $-13 - (-13)$
7. $15 - (+10)$
8. $-8 - (+5)$
9. $2.1 - (-3.2)$
10. $-0.58 - 0.12$
11. $8 - 6 - (-2) + 1$
12. $2.2 - 1.5 - (-3.1)$
13. $-1.21 + 2.21 - 1.11$
14. $5.50 + 10.05 - 5.55$
15. $-\frac{1}{3} + \frac{1}{2}$
16. $-\frac{1}{5} - \frac{3}{10}$

17. $1 - \left(-\frac{1}{4}\right) - \frac{1}{8} - \frac{1}{2}$

18. $-\frac{7}{10} + \frac{3}{5} - \left(-\frac{1}{20}\right)$

19. $(-8)(7)$

20. $(-6)(-12)$

21. $(-6)\left(-\frac{1}{2}\right)$

22. $\left(\frac{1}{6}\right)\left(-\frac{2}{7}\right)$

23. $\frac{-63}{9}$

24. $\frac{-36}{-12}$

25. $\frac{4.4}{-2.2}$

26. $\frac{-0.01}{0.1}$

27. $\left(\frac{2}{5}\right) \div (-4)$

28. $-10 \div \left(-\frac{3}{10}\right)$

29. $(-5)(-5)(-3)$

30. $(-6)(15)(2)$

31. $(-2)(0.1)(-0.3)$

32. $(-0.5)(-0.1)(-5)$

33. $-4\left(\frac{1}{2}\right)\left(-\frac{3}{4}\right)$

34. $10\left(-\frac{3}{5}\right)\left(-\frac{1}{3}\right)$

35. $\frac{6(-3)}{9}$

36. $\frac{-2(-2)}{20}$

37. $\frac{1.21}{10(-1.1)}$

38. $\frac{-6(0.5)}{-0.1}$

39. $\frac{8(0)}{-7-7}$

40. $\frac{-1-(-1)}{-1}$

41. $\frac{3-8}{-1(-2)}$

42. $\frac{-4 \cdot 3}{6 - 2 \cdot 9}$

43. $(-2)(6) - (3)(5)$

44. $(40)(2) - (-30)(-3)$

45. $0.25 - (0.5)(2) + 0.15$

46. $-2.4 + (0.6)(-2.0) - (-1.6)$

47. $\frac{5(-6)(-8)}{2(-5) + 5(3)}$

48. $\frac{(0.3)(-6)}{3.3 - 2.7}$

49. $(-3)\left(-\frac{1}{5}\right) + \frac{4(-2)}{10}$

50. $\left(\frac{1}{2}\right)\left(-\frac{1}{3}\right) - \left(\frac{2}{3}\right)\left(-\frac{1}{4}\right)$

51. $\frac{1.2 - 3.3}{6(0.7)}$

52. $\frac{10(-1.1)}{2.8 - 0.6}$

53. $5\left(\frac{-3}{-10}\right) - 4\left(\frac{7}{-12}\right)$

54. $\left(-\frac{1}{2}\right)\left(-\frac{1}{4}\right) - \left(-\frac{3}{4}\right)\left(\frac{1}{2}\right)$

55. $-6 - \frac{-16}{2} + 3\left(-\frac{1}{6}\right)$

56. $2(-0.5) + (1.2)\left(\frac{5}{12}\right)$

57. $\frac{8}{2(-0.2)} + \frac{1}{2}(-8)(0.5)$

58. $2(-0.4) + (0.6)\left(\frac{0.4}{-0.8}\right)$

Applied Problems

In problems 59 through 68, do the calculations by hand. Check with the calculator.

59. Mount Everest, the highest mountain on earth, is 8848 m or 29,028 ft above sea level. The Mariana trench, the greatest sea depth, is 11,034 m or 36,201 ft below sea level. Express these distances as signed numbers, and calculate the difference between the highest mountain and the greatest sea depth in meters and feet.

60. What is the difference between the boiling point of oxygen, −183.0°C and the melting point, −218.4°C?

61. A ship travels north (0°) for 1 hour, turns clockwise 45° and travels northeast for 1 hour, then turns counterclockwise 90° and travels northwest for 1 hour. Using positive for counterclockwise and negative for clockwise, express these angles as signed numbers and combine them to calculate the net degrees the ship has turned.

62. A football team is pushed back 8 yd on a poor play. The team then completes a pass gaining 14 yd, but incurs a penalty of 5 yd. Express these changes as signed numbers and combine them to determine the team's gain or loss.

63. What is the sign of the product of an even number of negative numbers?

64. What is the sign of the product of an odd number of negative numbers?

Applications to Electronics

65. The temperature of a light filament changes from 175°C to 17°C when the power is turned off. What signed number represents this change?

66. The voltage in a circuit with respect to ground changes from 2.5 V to −0.5 V when the ground connection is changed. What signed number represents this change?

67. In a circuit, two currents are flowing into a junction: $I_1 = 1.3$ A and $I_2 = 0.8$ A. Flowing out of the same junction are the two currents $I_3 = 1.7$ A and $I_4 = 0.4$ A. Expressing the currents flowing in as positive and the currents flowing out as negative, show that the algebraic sum of the currents at the junction is zero. This is Kirchhoff's current law.

68. Going clockwise around a closed path in a circuit the voltage drops are 10.3 V, 8.2 V, and 7.0 V. The voltage increases are 12.6 V and 12.9 V. Expressing the drops as negative and the increases as positive, show that the algebraic sum of the voltages is zero. This is Kirchhoff's voltage law.

5.2 Algebraic Terms

Algebra uses letters to represent numbers. When two letters or a number and a letter are written next to each other it is understood that they are multiplied together. Numbers, or letters that represent a fixed value, are called *constants*. Letters that assume more than one value are called *variables*. For example, in the formula for the circumference of a circle: $C = 2\pi r$, the letter π (pi) represents the infinite decimal 3.1416... and is a constant like the number 2. The number π does not vary in value. The letters C and r are variables and can assume many values. Letters and numbers make up what is called an algebraic term:

Definition

Algebraic Term

A combination of numbers and letters joined by the operations of multiplication or division.

EXAMPLE 5.12 Give six examples of algebraic terms.

Solution Each algebraic term has a *coefficient,* which is the constant in front, and a literal part, which are the letters:

Term	*Coefficient*	*Literal Part*
$5x^2y$	5	x^2y
$-abc$	-1	abc
$-3VI$	-3	VI
$\sqrt{\frac{P}{R}}$	1	$\sqrt{\frac{P}{R}}$
$-\frac{1.8}{C}$	-1.8	$\frac{1}{C}$
$4\pi r^3$	4π	r^3

If there is no number in front of a term, it is understood to be 1, as in the second and fourth terms above. In the last term, π is part of the coefficient because it is a constant value that never changes.

EXAMPLE 5.13 Given $P = 25$, $R = 9.1$ and $V = 6.3$. Find the value of each term to three significant digits:

(a) $\frac{1.5V^2}{R}$ **(b)** $0.73\sqrt{PR}$

Solution Substitute the values into the term and perform the operations on the calculator:

(a) $\frac{1.5(6.3)^2}{9.1}$ $\Rightarrow$ 1.5 [×] 6.3 [x²] [÷] 9.1 [=] $\rightarrow$ 6.54

(b) $0.73\sqrt{(25)(9.1)}$ $\Rightarrow$ *DAL:* 0.73 [×] [√] [(] 25 [×] 9.1 [)] [=] $\rightarrow$ 11.0

Not DAL: 0.73 [×] [(] 25 [×] 9.1 [=] [√] [)] [=] $\rightarrow$ 11.0

Definition

Like Terms

Terms that have exactly the same literal part.

$$4xy \text{ and } -10xy \quad -3I^2R_1 \text{ and } 5I^2R_1 \quad 5\sqrt{PR} \text{ and } -3\sqrt{PR}$$

In like terms all the letters, exponents, or roots are exactly the same. Like terms *can* be combined because they represent amounts of the same quantity.

Definition

Unlike Terms

Terms that do not have the same literal part.

$$7V \text{ and } 7V_1 \quad 4x^2 \text{ and } 5x \quad 3I^2R \text{ and } -3IR^2$$

The terms $7V$ and $7V_1$ are different because V is a *different* variable from V_1. The subscript 1 is part of the variable and cannot be separated. A variable with a subscript represents a given or specific value of a variable. V_1 could represent the voltage of battery 1 as opposed to V which could be the voltage across any component or an unknown source. Unlike terms *cannot* be added together or combined because they represent amounts of *different* quantities.

Rule

Combining Like Terms

Add or *combine the coefficients only.* Do not change the literal part, including the exponents or radicals.

EXAMPLE 5.14

Combine like terms:

(a) $-6.0y + (-2.4y)$ **(b)** $-5abc - (-10abc)$

(c) $10I^2R - (-I^2R)$ **(d)** $\dfrac{-5V}{R} + \dfrac{3V}{R}$

Solution

(a) Add the coefficients and do not change the variable:

$$-6.0y + (-2.4y) = -6.0y - 2.4y = -(6.0 + 2.4)y = -8.4y$$

(b) Subtracting a negative results in a positive term being combined with the first term:

$$-5abc - (-10abc) = -5abc + 10abc = 5abc$$

(c) Subtracting a negative results in a positive term whose coefficient is 1:

$$10I^2R - (-I^2R) = 10I^2R + I^2R = (10 + 1)\,I^2R = 11I^2R$$

(d) The coefficients are -5 and 3. They can be separated and combined as follows:

$$\frac{-5V}{R} + \frac{3V}{R} = -5\frac{V}{R} + 3\frac{V}{R} = (-5 + 3)\frac{V}{R}$$

$$= -2\frac{V}{R} \text{ or } \frac{-2V}{R}$$

EXAMPLE 5.15 Combine like terms:

$$6V_1 - 9V_2 + 7V_2 - 3V_1$$

Solution The variable V_1 is different from the variable V_2. Combine the V_1's and V_2's separately:

$$6V_1 - 9V_2 + 7V_2 - 3V_1 = (6-3)V_1 + (-9+7)V_2 = 3V_1 - 2V_2$$

The result, consisting of two terms, is called an algebraic expression.

Definition

Algebraic Expression

One or more algebraic terms connected by plus or minus signs.

EXAMPLE 5.16 Simplify the algebraic expression:

$$5R_x + R_y - (4R_y - 3R_x)$$

Solution Perform the operation in parentheses first. A minus sign in front of parentheses is like multiplying every term within the parentheses by −1. *It changes the sign of every term in the parentheses*:

$$\begin{aligned} 5R_x + R_y - (4R_y - 3R_x) &= 5R_x + R_y - (4R_y) - (-3R_x) \\ &= 5R_x + R_y - 4R_y + 3R_x \end{aligned}$$

R_x is a different variable from R_y. Combine only the like terms:

$$\begin{aligned} 5R_x + R_y - 4R_y + 3R_x &= (5+3)R_x + (1-4)R_y \\ &= 8R_x - 3R_y \end{aligned}$$

 **ERROR BOX**

When combining terms, be careful that you combine only like terms. The variables, subscripts, and the exponents must *all be the same* for like terms. If the subscripts are different, the variables are not the same. If the exponents are different, the terms are not the same. When combining like terms, work only with the coefficients; do not change the literal part or the exponents. See if you can do the practice problems correctly.

Practice Problems: Simplify each algebraic expression *if possible*:

1. $2V^2 + 2V_2$ **2.** $ab - ab^2 + ab$ **3.** $xy - yx$

4. $pqr - 6pqr + 8pqr$ **5.** $P_1 + P_2 + P_3$ **6.** $I^2R + 2IR - 2I^2R$

7. $V_0I + V - V_0I$ **8.** $x^2y - x^2y^2 + yx^2$ **9.** $\frac{W}{5} - \frac{W}{10}$

10. $0.5(X_L - X_c) + 0.5X_L - 0.5X_c$

EXAMPLE 5.17 Simplify the algebraic expression containing parentheses and brackets:

$$2V - [(2V - 2IR) - (V - IR)]$$

Solution Work from the inside out, paying careful attention to the minus signs in front of the parentheses and the brackets. First remove the two sets of parentheses, changing the signs in the second parentheses:

$$2V - [(2V - 2IR) - (V - IR)] = 2V - [2V - 2IR - V + IR]$$

A plus sign (or no sign) in front of a parenthesis does not change any signs in the parentheses. You can just remove the first parentheses. For the second parentheses, you must change every sign when you remove it because of the minus sign in front. Now combine the terms in the brackets:

$$2V - [2V - 2IR - V + IR] = 2V - [V - IR]$$

Remove the brackets changing both signs and combine like terms:

$$2V - V + IR = V + IR$$

EXAMPLE 5.18 Simplify the algebraic expression:

$$2G^2 - (G^2 + 3G - 5) - [2G - (5G + 6)]$$

Solution Remove the two sets of parentheses first, carefully applying the minus signs to every term:

$$2G^2 - (G^2 + 3G - 5) - [2G - (5G + 6)] = 2G^2 - G^2 - 3G + 5 - [2G - 5G - 6]$$

Combine like terms: $= G^2 - 3G + 5 - [-3G - 6]$

Remove the brackets and combine like terms: $= G^2 - 3G + 5 + 3G + 6 = G^2 + 11$

Notice that the terms containing G add to zero: $-3G + 3G = 0$.

EXERCISE 5.2

In exercises 1 through 10, find the value of the algebraic expression to three significant digits, given $V = 1.5$, $I = 3.8$, $R = 27$, and $P = 12$.

1. $4.4IR$
2. $0.67VI$
3. I^2R
4. $IR - 5V$
5. $\dfrac{16V^2}{R}$
6. $\dfrac{V^2}{2.3R}$
7. $7.8\sqrt{PR}$
8. $3.5\sqrt{\dfrac{P}{R}}$
9. $2.01I^2R - 18VI$
10. $8.6VI + 0.44I^2R$

In exercises 11 through 50, simplify each algebraic expression by combining like terms.

11. $9x - 11x$
12. $12f - 10f + 3f$
13. $11q + 2p - 6p$
14. $2x - 10x + 3y$
15. $5IR + 14V - 3IR + 2V$
16. $VI - 4P + 5P - 15VI$
17. $de + ed - d$
18. $xyz + yzx + zxy$
19. $2x^2 + 3x - 6x - x^2$
20. $ab^2 - a^2b + 3a^2b - ab^2$
21. $5.3V_1 - 2.8V_2 - 6.8V_1 + 0.5V_2$
22. $8.2C_1 - 4.3C_2 - 1.3C_2 - 2.4C_1$
23. $5Q_a - 2Q_b + 7Q_a$
24. $15X_L - 10X_C - 10X_L$
25. $\dfrac{1}{2}P + \dfrac{1}{5}P - \dfrac{1}{4}$
26. $\dfrac{3}{10} - \dfrac{1}{5}f + \dfrac{1}{10}f + \dfrac{1}{2}$
27. $(by - 4) - (2by - 3)$
28. $(7pr + 3p) - (-pr - 7p)$
29. $5V - (-2V + 4I) + 3I$
30. $-(C - 2F) + (2C - F)$
31. $(2x^2 + 3x - 1) - (-3x^2 + 4x - 2)$
32. $(-3z^2 - 5z - 5) - (2z^2 + 4z + 1)$
33. $3h^2 + (h^2 - 5h - 3) - (-10h - 8)$
34. $(8n^2 + 2n + 1) - (3n^2 - 3n) + 8$
35. $6rs^3 - 3rs^2 - 3rs^3$

Answers to Error Box Problems, page 115:

1. Cannot be simplified **2.** $2ab - ab^2$ **3.** 0 **4.** $3pqr$ **5.** Cannot be simplified **6.** $2IR - I^2R$ **7.** V **8.** Cannot be simplified **9.** $\dfrac{W}{10}$ **10.** $X_L - X_C$

36. $P_2^2 - 2P_1^2 + P_1^2$

37. $\frac{2R_x}{3} - \frac{R_y}{2} - \left(\frac{R_x}{3} - \frac{R_y}{2}\right)$

38. $\frac{3I_1}{4} - \left(\frac{3I_2}{5} - \frac{I_1}{8}\right) + \frac{I_2}{10}$

39. $-(3.5P_2 - 2.1P_1) + (6.8P_1 - P_2)$

40. $10.5C_1 + 3.5C_2 - (5.5C_2 - 9.5C_1)$

41. $5\pi + [\pi r - (\pi r - 1)]$

42. $x^2y + [xy^2 - (x^2y - 2)]$

43. $3V_0 - [(V_1 - V_0) - (V_0 - V_1)]$

44. $8I_T + [6 - (7I_T - 10) - 3I_T]$

45. $[n^2 - (2n + 1)] - (n - 1)$

46. $4 - [(f - g) - (f - g - 3)]$

47. $8H + \{3H - [2H + (H - 1)]\}$

48. $7 + \{5a - [2b + (3 - a)]\}$

49. $\frac{3.6v}{2} - \left(\frac{1.5r}{3} + \frac{2.4r}{4}\right) - \frac{2.5v}{5}$

50. $\left(\frac{4.2e}{3} - \frac{5.1i}{3}\right) - \left(\frac{1.2i}{4} - \frac{2.0e}{4}\right)$

Applied Problems

51. Simplify the following expression from a problem involving the volume of a solid:

$$(\pi rh + 2\pi r^2) - (4\pi r^2 - 2\pi rh)$$

52. Simplify the following expression from a problem involving the motion of a body:

$$3.26 - [\tfrac{1}{2}mv^2 - (mv^2 - 5.18)]$$

53. Two forces are acting on a body in the same direction. If one of the forces $F_1 = 3d^2 - 2d + 1$ and the other force $F_2 = 4d^2 + 3d - 5$, what is the resultant force, $F_1 + F_2$, on the body?

54. The total force produced by two forces on a body is given by $F_T = 5.5e^2 + 6.2e - 1.5$. If one force is given by $F_1 = 3.1e^2 - 7.4e - 2.9$, find the expression for the other force, $F_2 = F_T - F_1$.

Applications to Electronics

55. The voltage drop across one resistor is given by $V_1 = 1.5IR - 0.4$ and the voltage drop across a second resistor by $V_2 = 2.3IR + 0.8$. What is the total voltage drop, $V_1 + V_2$, across both resistors?

56. The power in one circuit is given by $P_2 = 8.34I^2R + 9.89$ and the power in a lower rated circuit by $P_1 = 5.76I^2R - 7.67$. What is the expression for the difference in powers, $P_2 - P_1$, between the circuits?

57. The work output of one computer is given by $\frac{1}{4}W - 7$ and the work output of a second computer by $\frac{1}{3}W - 5$. How much more work, in terms of W, is done by the computer with the greater output?

58. The total heat produced by two generators is given by $Q_T = 7Q^4 + \frac{1}{3}Q^2$. If the heat produced by one generator is given by $Q_1 = 5Q^4 + \frac{1}{6}Q^2$, what is the expression for the heat produced by the other generator $Q_2 = Q_T - Q_1$?

5.3 Rules for Exponents

As shown in Section 3.1, an exponent n defines repeated multiplication:

$$x^n = \underbrace{(x)(x)(x)\cdots(x)}_{n \text{ times}} \qquad (n = \text{Positive integer})$$

The number x is called the *base.* Observe what happens when you raise a negative base to a positive power:

$$(-3)^4 = (-3)(-3)(-3)(-3) = 81$$
$$(-3)^3 = (-3)(-3)(-3) = -27$$

When the base is negative, the result is positive for an even exponent and negative for an odd exponent. When the negative sign is *not* in the parentheses, it is not part of the base and the result is always negative:

$$-0.2^3 = -(0.2)(0.2)(0.2) = -0.008$$
$$-0.2^4 = -(0.2)(0.2)(0.2)(0.2) = -0.0016$$

When performing operations on numbers with exponents, there are five rules that simplify the calculations.

Rule **Product Rule for Exponents**

$$(x^m)(x^n) = x^{m+n} \tag{5.1}$$

Add exponents when multiplying like bases.

EXAMPLE 5.19 Multiply the following:

(a) $(x^3)(x^2)$ **(b)** $(2)^4(2)^2$ **(c)** $(-3)^2(-3)^3$ **(d)** $(5R)(R^3)$

Solution Apply the product rule (5.1) and add the exponents only; *do not change the base*:

(a) $(x^3)(x^2) = x^{3+2} = x^5$

(b) $(2)^4(2)^2 = 2^{4+2} = 2^6 = 64$

(c) $(-3)^2(-3)^3 = (-3)^{2+3} = (-3)^5 = -243$

Observe in (b) and (c) that the bases 2 and −3 are left alone and are not multiplied. Adding the exponents is the multiplication process.

(d) $(5R)(R^3) = 5R^{1+3} = 5R^4$

Observe in (d) that when you multiply $(5R)(R^3)$ you add the exponents and just carry along the coefficient of 5.

Rule **Quotient Rule for Exponents**

$$\frac{x^n}{x^m} = x^{n-m} \tag{5.2}$$

Subtract exponents when dividing like bases.

EXAMPLE 5.20 Divide the following:

(a) $\frac{P^5}{P}$ **(b)** $\frac{0.5^4}{0.5^2}$ **(c)** $\frac{(-1)^6}{(-1)^2}$ **(d)** $\frac{10^3}{10^2}$

Solution Apply the quotient rule (5.2) and subtract the exponents only; *do not change the base*:

(a) $\frac{P^5}{P} = P^{5-1} = P^4$

Observe that when there is no exponent as in (a), it is understood to be 1: $P = P^1$.

(b) $\frac{0.5^4}{0.5^2} = (0.5)^{4-2} = (0.5)^2 = 0.25$

(c) $\frac{(-1)^6}{(-1)^2} = (-1)^{6-2} = (-1)^4 = +1$

(d) $\frac{10^3}{10^2} = 10^{3-2} = 10^1 = 10$

Rule **Power Rule for Exponents**

$$(x^m)^n = x^{mn} \tag{5.3}$$

Multiply exponents when raising to a power.

EXAMPLE 5.21 Raise each to the given power:

(a) $(I^2)^2$ **(b)** $(10^2)^3$ **(c)** $4(y^3)^4$ **(d)** $[(-2)^3]^3$

Solution Apply the power rule (5.3) and multiply the exponents:

(a) $(I^2)^2 = I^{(2)(2)} = I^4$

(b) $(10^2)^3 = 10^{(2)(3)} = 10^6$

(c) $4(y^3)^4 = 4y^{(3)(4)} = 4y^{12}$

(d) $[(-2)^3]^3 = (-2)^{(3)(3)} = (-2)^9 = -512$

Observe that the exponent 3 in (c) only applies to the y, not to the coefficient of 4. If the coefficient is in the parentheses, then you apply the factor rule below.

Rule **Factor Rule for Exponents**

$$(xy)^n = x^n y^n \qquad \left(\frac{x}{y}\right)^n = \frac{x^n}{y^n} \tag{5.4}$$

Raise each factor separately to the given power.

EXAMPLE 5.22 Simplify each of the following:

(a) $(3y^2)^3$ **(b)** $\left(\frac{0.2}{R^2}\right)^2$ **(c)** $(10a^2b)^4$ **(d)** $\left(\frac{5a^2}{3b}\right)^3$

Solution Separate the product or the quotient into factors and apply the factor rule (5.4):

(a) $(3y^2)^3 = (3)^3(y^2)^3 = 27y^{(2)(3)} = 27y^6$

(b) $\left(\frac{0.2}{R^2}\right)^2 = \frac{(0.2)^2}{R^{(2)(2)}} = \frac{0.04}{R^4}$

(c) $(10a^2b)^4 = 10^4a^{(2)(4)}b^4 = 10^4a^8b^4$

(d) $\left(\frac{5a^2}{3b}\right)^3 = \frac{5^3a^{(2)(3)}}{3^3b^3} = \frac{125a^6}{27b^3}$

Rule **Radical Rule for Exponents**

$$\sqrt[n]{x^m} = x^{m/n} \tag{5.5}$$

Divide the index of the radical into the exponent of the radicand (number under the radical).

EXAMPLE 5.23 Simplify the roots:

$$\text{(a) } \sqrt{Z^4} \qquad \text{(b) } \sqrt[3]{V^6}$$

Solution Apply the radical rule (5.5) and divide the index of the radical into the exponent of the radicand. The index of a square root is understood to be 2:

$$\text{(a) } \sqrt{Z^4} = Z^{4/2} = Z^2$$

$$\text{(b) } \sqrt[3]{V^6} = V^{6/3} = V^2$$

EXAMPLE 5.24 Perform the operations and simplify:

$$\frac{10X^3 \cdot (X^2)^2}{10X^2}$$

Solution Perform the operations in the numerator by first applying the power rule (5.3) and then the product rule (5.1) for exponents. Remember, in the order of operations, *raising to a power and taking a root are done before multiplication or division:*

$$\frac{10X^3 \cdot (X^2)^2}{10X^2} = \frac{10X^3 \cdot X^4}{10X^2} = \frac{10X^7}{10X^2}$$

Observe that when you multiply $(10X^3)(X^4)$ you add the exponents and just carry along the coefficient of 10.

Now apply the quotient rule (5.2) and subtract exponents:

$$\frac{\cancel{10}X^7}{\cancel{10}X^2} = X^{7-2} = X^5$$

What happens when $n = m$ in the quotient rule? For example, suppose $m = n = 6$:

$$\frac{x^6}{x^6} = x^{6-6} = x^0$$

You obtain a zero exponent. Since the fraction $\frac{x^6}{x^6} = 1$, the zero exponent is defined as 1 for *any* number except zero:

Rule **Zero Exponent Rule**

$$x^0 = 1 \quad (x \neq 0) \tag{5.6}$$

Any number to the zero power is one.

For example:

$$100^0 = 1, \quad (-4)^0 = 1, \quad (33W)^0 = 1, \quad \left(\frac{6.5V}{R^2}\right)^0 = 1.$$

When $n < m$ (n less than m) and you apply the quotient rule, you obtain a negative exponent:

$$\frac{x^2}{x^4} = x^{2-4} = x^{-2}$$

This fraction can also be evaluated by dividing out factors:

$$\frac{x^2}{x^4} = \frac{\cancel{x} \cdot \cancel{x}}{\cancel{x} \cdot \cancel{x} \cdot x \cdot x} = \frac{1}{x^2}$$

For the quotient rule to work for negative exponents, it therefore follows that:

$$x^{-2} = \frac{1}{x^2}$$

Because of this relationship a negative exponent is defined as a reciprocal:

Rule

Negative Exponent Rule

$$x^{-n} = \left(\frac{1}{x}\right)^n = \frac{1}{x^n} \text{ and } \left(\frac{x}{y}\right)^{-n} = \left(\frac{y}{x}\right)^n \qquad (x, y \neq 0) \qquad (5.7)$$

That is, a negative power means *write the reciprocal of the base and raise it to the positive power.* A negative exponent does not mean change the sign of the base. Study the following example.

EXAMPLE 5.25

Evaluate:

(a) 4^{-3} **(b)** 5^{-1} **(c)** $\left(\frac{2}{3}\right)^{-2}$ **(d)** $(-10)^{-3}$

Solution Apply the negative exponent rule (5.7); write the reciprocal and raise to the positive power:

(a) $4^{-3} = \left(\frac{1}{4}\right)^3 = \frac{1}{4^3} = \frac{1}{64}$

(b) $5^{-1} = \frac{1}{5}$

Notice in (b) that the -1 power just means reciprocal.

(c) $\left(\frac{2}{3}\right)^{-2} = \left(\frac{3}{2}\right)^2 = \frac{3^2}{2^2} = \frac{9}{4}$

Notice in (c), when a fraction is raised to a negative exponent, you write the reciprocal of the fraction first and change the exponent to a positive number.

(d) $(-10)^{-3} = \left(\frac{1}{-10}\right)^3 = \frac{1}{(-10)^3} = \frac{1}{-1000} = -0.001$

Observe in (d) that the negative sign in front of the 10 has a very different meaning than the negative sign in the exponent.

As shown in Example 5.25(b), a special case of the negative exponent rule (5.7) is:

Rule

Reciprocal Exponent Rule

$$x^{-1} = \frac{1}{x} \text{ and } \left(\frac{x}{y}\right)^{-1} = \frac{y}{x} \qquad (x, y \neq 0) \qquad (5.8)$$

To raise a number to a negative exponent on the calculator, you use the [+/-] or [(−)] key to make the exponent negative, or you can use the reciprocal key, [1/x] or [x^{-1}], and a positive exponent. For example, four possible ways to calculate 5^{-3} are:

5 [x^{-1}] [y^x] 3 [=] → 0.008

5 [y^x] 3 [=] [x^{-1}] → 0.008

DAL: 5 [y^x] [(−)] 3 [=] → 0.008

Not DAL: 5 [y^x] 3 [+/-] [=] → 0.008

ERROR BOX

When working with negative exponents, you must apply the rules carefully to avoid an error. Remember that the negative sign in front of an exponent means reciprocal. *It does not change the sign of the base.* It means change the base to its reciprocal and change the sign of the exponent to positive:

$$\left(-\frac{1}{5}\right)^{-3} = (-5)^3 = -125 \qquad (-0.1)^{-2} = \left(\frac{1}{-0.1}\right)^2 = \frac{1}{(-0.1)^2} = \frac{1}{0.01} = 100$$

Practice Problems: Evaluate and express answers as fractions or whole numbers:

1. $(-6)^{-2}$ **2.** -6^{-2} **3.** $(0.5)^{-3}$ **4.** $(-0.5)^{-3}$

5. $\left(\frac{1}{2}\right)^{-4}$ **6.** $\left(-\frac{3}{4}\right)^{-2}$ **7.** $\frac{1}{10^{-2}}$ **8.** $(-3)^{-1}(3)^{-1}$

9. $\frac{4^{-1}}{2^{-2}}$ **10.** $(0.1)^2(-0.1)^{-2}$ **11.** $(10^{-1})^{-2}$ **12.** $\left(\frac{1}{2}\right)^{-3}\left(\frac{1}{2}\right)^2$

13. $\frac{(10)^{-3}}{(10)^{-6}}$ **14.** $(2)^2(0.2)^{-2}$

It is important to note that all the rules for exponents (5.1) to (5.5) also apply to zero and negative exponents. Also, the rules for powers of ten shown in Chapter 3, Section 3.2 are a special case of the general rules for exponents shown here. Study the following examples that apply several of the rules to zero and negative exponents.

EXAMPLE 5.26

Simplify and express the result in terms of a positive exponent:

(a) $(c^2)(c^{-3})$ **(b)** $(R_1^{-2})^{-3}$ **(c)** $\frac{v^2}{v^{-3}}$ **(d)** $\sqrt{10^{-6}}$ **(e)** $(4x^2)^{-1}$

Solution Apply the rules for exponents (5.1) to (5.5) and the rules for signed numbers:

(a) $(c^2)(c^{-3}) = c^{2+(-3)} = c^{-1} = \frac{1}{c}$

(b) $(R_1^{-2})^{-3} = R_1^{(-2)(-3)} = R_1^{6}$

(c) $\frac{v^2}{v^{-3}} = v^{2-(-3)} = v^{2+3} = v^5$

(d) $\sqrt{10^{-6}} = 10^{-6/2} = 10^{-3} = 0.001$

(e) $(4x^2)^{-1} = (4^{-1})(x^2)^{-1} = 0.25x^{-2} = \frac{1}{4x^2}$

EXAMPLE 5.27 Evaluate:

$$\left(\frac{10^{-3}}{10^{-2}}\right)^2$$

Solution Apply the rules for exponents, which are the same as the rules for powers of ten shown in Section 3.2. Simplify inside the parentheses first. Apply the quotient rule (5.2) to the fraction and subtract the exponents:

$$\left(\frac{10^{-3}}{10^{-2}}\right)^2 = (10^{-3-(-2)})^2 = (10^{-3+2})^2 = (10^{-1})^2$$

Then apply the power rule (5.3):

$$(10^{-1})^2 = 10^{(-1)(2)} = 10^{-2} = 0.01$$

EXAMPLE 5.28 Simplify:

$$(-1.1T^2)^2(T^{-1})(T^0)$$

Solution Apply the factor rule (5.4) to $(-1.1T^2)^2$and then the product rule (5.1):

$$\begin{aligned}(-1.1T^2)^2(T^{-1})(T^0) &= (-1.1)^2(T^2)^2(T^{-1})(1)\\ &= 1.21T^4(T^{-1}) = 1.21T^{4-1} = 1.21T^3\end{aligned}$$

Notice that when you apply the factor rule, you apply the exponent to the coefficient -1.1, and to T^2. Also note that T^0 becomes 1.

Multiplying and Dividing Terms

To multiply or divide algebraic terms, you work with the coefficients separately as follows:

Multiplying and Dividing Terms

Multiply or divide the coefficients separately and apply the rules for exponents to the literal parts.

EXAMPLE 5.29 Perform the operations:

(a) $(-4v^2)(3v^{-3})$ **(b)** $\dfrac{15L^3T^2}{5L^2T}$ **(c)** $\dfrac{5.6n^3}{(2n)^2(7n)}$

Solution

(a) Multiply the coefficients and apply the product rule separately to the exponents:

$$(-4v^2)(3v^{-3}) = (-4)(3)v^{2-3} = -12v^{-1} = \frac{-12}{v}$$

(b) Divide the coefficients and apply the quotient rule to the exponents of L and T:

$$\frac{15L^3T^2}{5L^2T} = \left(\frac{15}{5}\right)L^{3-2}T^{2-1} = 3LT$$

(c) Evaluate the denominator first and then apply the quotient rule:

$$\frac{5.6n^3}{(2n)^2(7n)} = \frac{5.6n^3}{4n^2(7n)} = \frac{5.6n^3}{(4)(7)n^{2+1}}$$

$$= \frac{5.6n^3}{28n^3} = \left(\frac{5.6}{28}\right)n^{3-3} = 0.20n^0 = 0.20$$

EXAMPLE 5.30 Multiply:

$$-5x(x^2 - 2x + 1)$$

Solution Each term in the parentheses must be multiplied by the factor $-5x$ on the outside, applying the product rule (5.1) each time:

$$(-5x)(x^2 - 2x + 1) = -5x(x^2) - 5x(-2x) - 5x(1) = -5x^3 + 10x^2 - 5x$$

EXAMPLE 5.31 Divide:

$$\frac{4at^3 - 6a^2t^2 - 4a^3t}{2at}$$

Solution The fraction line is like parentheses. Each term in the numerator must be divided by the denominator and the quotient rule (5.2) applied each time:

$$\frac{4at^3 - 6a^2t^2 - 4a^3t}{2at} = \frac{4at^3}{2at} - \frac{6a^2t^2}{2at} - \frac{4a^3t}{2at}$$

$$= \left(\frac{4}{2}\right)a^{1-1}t^{3-1} - \left(\frac{6}{2}\right)a^{2-1}t^{2-1} - \left(\frac{4}{2}\right)a^{3-1}t^{1-1}$$

$$= 2t^2 - 3at - 2a^2$$

Notice in the first fraction the a divides out, and in the last fraction the t divides out.

EXAMPLE 5.32 Calculate and round the answer to three significant digits in *engineering notation*:

$$\frac{\sqrt{55.6 \times 10^{-6}}}{(3.42 \times 10^3)^2}$$

Solution Powers of ten are handled like algebraic terms when performing operations. The first number is like the coefficient and the power of ten is like the literal part. Use the factor rule (5.4) and apply the exponent rules to the coefficient and the power of ten separately in the numerator and the denominator:

$$\frac{\sqrt{55.6 \times 10^{-6}}}{(3.42 \times 10^3)^2} = \frac{\left(\sqrt{55.6}\right) \times (10^{-6/2})}{(3.42)^2 \times (10^{(3)(2)})} \approx \frac{7.46 \times 10^{-3}}{11.7 \times 10^6}$$

Now divide the numbers and subtract the exponents. Express the result in engineering notation:

$$\frac{7.46 \times 10^{-3}}{11.7 \times 10^6} = \frac{7.46}{11.7} \times 10^{-3-6} \approx 0.638 \times 10^{-9} = 638 \times 10^{-12}$$

Answers to Error Box Problems, page 122:

1. 1/36 **2.** −1/36 **3.** 8 **4.** −8 **5.** 16 **6.** 16/9 **7.** 100 **8.** −1/9 **9.** 1 **10.** 1 **11.** 100 **12.** 2 **13.** 1000 **14.** 100

The calculator solution in scientific and engineering notation is:

DAL: [√] [(] 55.6 [EE] [(-)] 6 [)] [÷] 3.42 [EE] 3 [x^2] [=] → 6.38 E–10 = 638 E–12

Not DAL: 55.6 [EXP] 6 [+/-] [√] [÷] 3.42 [EXP] 3 [x^2] [=] → 6.38 E–10 = 638 E–12

Applications to DC and AC Circuits

The next three examples close the circuit and show applications to dc and ac circuits involving the rules for exponents.

EXAMPLE 5.33

Close the Circuit

Given the following values of dc circuit variables: $V = 20.0$ V, $I = 8.34$ mA, $R = 2.40$ kΩ, and $P = 167$ mW. Calculate the value of the indicated variable in each dc circuit formula to three significant digits:

(a) $P = I^2R$; P **(b)** $I = \sqrt{\frac{P}{R}}$; I **(c)** $R = \frac{V^2}{P}$; R

Solution As shown in Chapter 4, use the metric prefix and write each value in the base unit with the power of ten:

$$V = 20.0 \text{ V} = 20.0 \times 10^0 \text{ V}$$
$$I = 8.34 \text{ mA} = 8.34 \times 10^{-3} \text{ A}$$
$$R = 2.40 \text{ k}\Omega = 2.40 \times 10^3\ \Omega$$
$$P = 167 \text{ mW} = 167 \times 10^{-3} \text{ W}$$

Substitute into each formula and calculate applying the rules for exponents. The calculator solutions in engineering notation are shown after each example.

(a)
$$P = I^2R = (8.34 \times 10^{-3})^2(2.40 \times 10^3)$$
$$= (8.34^2 \times 10^{(-3)(2)})(2.40 \times 10^3)$$
$$= (69.4 \times 10^{-6})(2.40 \times 10^3)$$
$$= (69.4)(2.40) \times 10^{-6+3} \approx 167 \times 10^{-3} \text{ W} = 167 \text{ mW}$$

DAL: 8.34 [EE] [(-)] 3 [x^2] [×] 2.4 [EE] 3 [=] → 167 E–3

Not DAL: 8.34 [EXP] 3 [+/-] [x^2] [×] 2.4 [EXP] 3 [=] → 167 E–3

(b)
$$I = \sqrt{\frac{P}{R}} = \sqrt{\frac{167 \times 10^{-3}}{2.40 \times 10^3}} = \sqrt{\frac{167}{2.40} \times 10^{-3-3}} = \sqrt{69.6 \times 10^{-6}}$$
$$= \sqrt{69.6} \times 10^{-6/2} \approx 8.34 \times 10^{-3} \text{ A}$$
$$= 8.34 \text{ mA}$$

DAL: [√] [(] 167 [EE] [(-)] 3 [÷] 2.4 [EE] 3 [)] [=] → 8.34 E–3

Not DAL: 167 [EXP] 3 [+/-] [÷] 2.4 [EXP] 3 [=] [√] → 8.34 E–3

(c)
$$R = \frac{V^2}{P} = \frac{(20.0 \times 10^0)^2}{167 \times 10^{-3}} = \frac{(20.0)^2 \times 10^0}{167 \times 10^{-3}}$$
$$= \frac{400 \times 10^0}{167 \times 10^{-3}} = \frac{400}{167} \times 10^{0-(-3)} \approx 2.40 \times 10^3\ \Omega$$
$$= 2.40 \text{ k}\Omega$$

DAL: 20 [x^2] [÷] 167 [EE] [(-)] 3 [=] → 2.40 E3

Not DAL: 20 [x^2] [÷] 167 [EXP] 3 [+/-] [=] → 2.40 E3

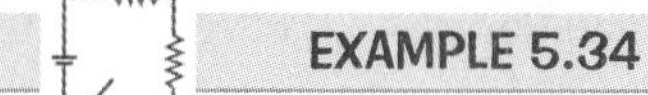

EXAMPLE 5.34

Close the Circuit

The impedance, in ohms, of an ac circuit is given by:

$$Z = \sqrt{R^2 + X^2}$$

where R is the circuit resistance and X is the circuit reactance, and both are measured in ohms. Calculate Z to three significant digits when $R = 5.1\ \text{k}\Omega$ and $X = 3.5\ \text{k}\Omega$.

Solution Express in base units with powers of ten and apply the rules for exponents:

$$Z = \sqrt{R^2 + X^2} = \sqrt{(5.1 \times 10^3)^2 + (3.5 \times 10^3)^2} = \sqrt{26.01 \times 10^6 + 12.25 \times 10^6}$$

$$= \sqrt{38.26 \times 10^6} = \sqrt{38.26} \times 10^{6/2} \approx 6.19 \times 10^3\ \Omega = 6.19\ \text{k}\Omega$$

The calculator solution is:

DAL: [√] [(] 5.1 [EE] 3 [x^2] [+] 3.5 [EE] 3 [x^2] [)] [=] → 6.19 E3

Not DAL: 5.1 [EXP] 3 [x^2] [+] 3.5 [EXP] 3 [x^2] [=] [√] → 6.19 E3

EXAMPLE 5.35

Close the Circuit

The formula for the resonant frequency of an ac circuit is given by:

$$f_r = \frac{1}{2\pi\sqrt{LC}}$$

where π is a constant ≈ 3.142, L is the circuit inductance in henrys (H), C is the circuit capacitance in farads (F), and the frequency f_r is in hertz (Hz). The resonant frequency is the frequency that produces maximum power in the circuit. Calculate f_r to three significant digits when $L = 125\ \mu\text{H}$ and $C = 245\ \text{pF}$.

Solution

Express L in henrys and C in farads using powers of ten:

$$L = 125\ \mu\text{H} = 125 \times 10^{-6}\ \text{H and } C = 245\ \text{pF} = 245 \times 10^{-12}\ \text{F}$$

Substitute in the formula, using $2\pi = 2(3.142) = 6.284$, and apply the product rule to the powers of ten under the radical:

$$f_r = \frac{1}{6.284\sqrt{LC}} = \frac{1}{6.284\sqrt{(125 \times 10^{-6})(245 \times 10^{-12})}}$$

$$= \frac{1}{6.284\sqrt{(125)(245) \times 10^{-6-12}}} = \frac{1}{6.284\sqrt{30{,}625 \times 10^{-18}}}$$

Take the square root of the first number in the radical and divide the power of ten by 2:

$$f_r = \frac{1}{6.284\left(\sqrt{30{,}625} \times 10^{-18/2}\right)} = \frac{1}{6.284(175 \times 10^{-9})}$$

$$= \frac{1}{1099.7 \times 10^{-9}} \approx 909{,}000\ \text{Hz} = 909\ \text{kHz}$$

The calculator solution in scientific and engineering notation is:

DAL: 2 [×] [π] [×] [√] [(] 125 [EE] [(-)] 6 [×] 245 [EE] [(-)] 12 [)] [=] [x^{-1}] [=] → 9.09 E5 = 909 E3

Not DAL: 125 [EXP] 6 [+/-] [×] 245 [EXP] 12 [+/-] [=] [√] [×] 2 [×] [π] [=] [1/x] [=] → 9.09 E5 = 909 E3

EXERCISE 5.3

In exercises 1 through 70, perform the operations without the calculator by applying the rules for exponents. Express variables with positive exponents.

1. $x^4 \cdot x^3$
2. $z^2 \cdot z^2$
3. $(-1)^3(-1)^2$
4. $(-2)^2(-2)^2$
5. $y^3 \cdot y^5 \cdot y$
6. $(0.1)(0.1)^2(0.1)^3$
7. $\dfrac{R^7}{R^5}$
8. $\dfrac{n^5}{n^2}$
9. $\dfrac{(-6)^3}{(-6)^2}$
10. $\dfrac{7^4}{7^2}$
11. $(v^2)^3$
12. $(P^3)^3$
13. $[(-2)^2]^3$
14. $5(C^2)^4$
15. $(3pv^2)^3$
16. $(-2c^2d)^4$
17. $(10^2 \cdot 10)^3$
18. $\dfrac{10^2}{10 \cdot 10^3}$
19. $\left(\dfrac{x}{3b^2}\right)^2$
20. $\left(\dfrac{0.1w^2}{p}\right)^3$
21. $\sqrt{I^4}$
22. $\sqrt[3]{I^6}$
23. $\dfrac{a^2(a^3)^2}{a^2}$
24. $\dfrac{(0.5^2)^3}{(0.5)^2(0.5^3)}$
25. 8^{-2}
26. 2^{-4}
27. $(7d^3)^0$
28. 100^0
29. $(-3)^{-3}$
30. 2^{-5}
31. $C^2 \cdot C^{-3} \cdot C$
32. $V_0^2 \cdot V_0^{-1} \cdot V_0^0$
33. $2^4 \cdot 2^{-3} \cdot 2^2$
34. $10^2 \cdot 10^0 \cdot 10^{-3}$
35. $\left(\dfrac{2}{9}\right)^{-1}$
36. $\left(\dfrac{3}{5}\right)^{-2}$
37. $\left(-\dfrac{1}{2}\right)^{-3}$
38. $\left(-\dfrac{3}{4}\right)^{-2}$
39. $\dfrac{9^0}{9^{-2}}$
40. $\dfrac{1}{10^{-3}}$
41. $(I^{-3})(I^5)$
42. $(p^{-1})^{-2}$
43. $(-3T)^2(T^{-2})^2$
44. $(\pi r^2)(2\pi r)^2$
45. $\dfrac{-v^3}{(-2v)^2}$
46. $\dfrac{a^2b}{(2b)^2}$
47. $\dfrac{9xy^3}{(3xy)^2}$
48. $\dfrac{(PR^3)^2}{P^2R^3}$
49. $\sqrt{10^6}$
50. $\sqrt{10^{-2}}$
51. $\sqrt{4x^4}$
52. $\sqrt{81y^6}$
53. $(0.1ab)(0.3a)(0.2b)$
54. $(-Rt)(0.2Rt^2)(0.2Rt)^2$
55. $5m(m^2 - m + 3)$
56. $-3s(8s^2 + 3s - 1)$
57. $\dfrac{1.8y^6}{0.06y^3}$
58. $\dfrac{0.28wL^2}{1.4wL}$
59. $\dfrac{6a^2b - 3ab^2}{3ab}$
60. $\dfrac{10k^3 - 5k^2 + 15k}{-5k}$
61. $\dfrac{6.6f^2 - 4.4f}{2.2f}$
62. $\dfrac{C_1^2 + C_1C_2^2}{C_1}$
63. $x^2y(x + y) - xy^2(x - y)$
64. $ab(a^2b - ab + a^2b)$
65. $I^2R(V - IR) + IR(VI + I^2R)$
66. $(-5uv)(-3uv)(uv^2 - u^2v)$
67. $(5 \times 10^{-3})^{-2}$
68. $(4 \times 10^{-6})^{-2}$
69. $\sqrt{16 \times 10^6}$
70. $\sqrt{9 \times 10^{-4}}$

In exercises 71 through 76, perform each calculation and express the answer in engineering notation to three significant digits (as numbers between 1 and 1000 times a power of ten).

71. $\left(\sqrt{5.31 \times 10^{-4}}\right)(3.15 \times 10^{-2})^3$
72. $(8.77 \times 10^{-3})^{-2}\left(\sqrt{4.40 \times 10^{-2}}\right)$
73. $\left(\sqrt{23 \times 10^{-6}}\right)\left(\sqrt{18 \times 10^{-2}}\right)$
74. $\left(\sqrt{48 \times 10^6}\right)\left(\sqrt{30 \times 10^{-4}}\right)$
75. $\dfrac{(2.23 \times 10^{-3})^2}{\sqrt{6.16 \times 10^{-6}}}$
76. $\dfrac{\sqrt[3]{8.37 \times 10^{-6}}}{(3.12 \times 10^{-3})^{-2}}$

In exercises 77 through 84, given the following values of dc circuit variables: $V = 12.0$ V, $I = 3.64$ mA, $R = 3.30$ kΩ, and $P = 43.6$ mW, calculate the value of the indicated variable in each formula to three significant digits.

77. $V = IR$; V
78. $P = VI$; P
79. $I = \dfrac{P}{V}$; I
80. $R = \dfrac{V}{I}$; R

81. $P = I^2R$; P

82. $P = \dfrac{V^2}{R}$; P

83. $I = \sqrt{\dfrac{P}{R}}$; I

84. $V = \sqrt{PR}$; V

Applications to Electronics

85. In Example 5.34, given $Z = \sqrt{R^2 + X^2}$, find Z in kilohms when $R = 8.2$ kΩ and $X = 7.3$ kΩ.

86. In Example 5.34, given $Z = \sqrt{R^2 + X^2}$, find Z in ohms when $R = 510$ Ω and $X = 680$ Ω.

87. The equivalent resistance of a parallel circuit can be expressed in terms of a negative exponent as:

$$R_{eq} = \left(\frac{1}{R_1} + \frac{1}{R_2}\right)^{-1}$$

Find R_{eq} in ohms when $R_1 = 200$ Ω and $R_2 = 300$ Ω.

88. In problem 87, find R_{eq} in ohms when $R_1 = 1.5$ kΩ and $R_2 = 1.0$ kΩ.

89. The formula for the resistance of a wire, in ohms, is:

$$R = \frac{\rho l}{A}$$

Given a copper wire where the resistivity $\rho = 17.5 \times 10^{-9}\ \Omega \cdot m$, the length $l = 5.0$ m, and the cross sectional area $A = \pi(4 \times 10^{-3})^2\ m^2$ find R in ohms and milohms to three significant digits.

90. In problem 89, find R in ohms and milohms if $A = \pi(2 \times 10^{-3})^2 m^2$ and the other values remain the same.

91. In Example 5.35, find $f_r = \dfrac{1}{2\pi\sqrt{LC}}$ when $L = 150$ μH and $C = 0.175$ nF.

92. In Example 5.35, find $f_r = \dfrac{1}{2\pi\sqrt{LC}}$ when $L = 2.50$ mH and $C = 400$ pF.

93. Multiply out the following expression in computer language and express the result in computer language:

2*X*(X^2 – 3*X + 5)

94. Do the same as problem 93 for the following expression in computer language:

(2*X^3 + 4*X^2 – 6*X)/(2*X)

CHAPTER HIGHLIGHTS

5.1 SIGNED NUMBERS

Integer

A positive or negative whole number or zero.

• • •

Absolute Value

The value of a signed number without the sign that represents the magnitude or size of the number.

Key Example:

$|-10| = 10 \quad |+25| = 25 \quad |-0.17| = 0.17 \quad |50| = 50$

Adding Numbers with Like Signs

Add the absolute values and apply the common sign.

Key Example:

(a) $(-12) + (-6) = -(12 + 6) = -18$
(b) $(+1.1) + (+5.6) = +(1.1 + 5.6) = 6.7$

Adding Numbers with Unlike Signs

Subtract the smaller absolute value from the larger absolute value and apply the sign of the larger number.

Key Example:

(a) $-16 + 8 = -(16 - 8) = -8$
(b) $-5.2 + 12.6 = +(12.6 - 5.2) = 7.4$

Subtracting Signed Numbers

Change the operation to addition and change the sign of the number that follows the operation sign. Then apply the rules for adding signed numbers.

Key Example:

(a) $5 - (-20) = 5 + 20 = 25$
(b) $5.7 - (9.6) = 5.7 - 9.6 = -3.9$

Combining Signed Numbers

1. A plus sign followed by a plus sign equals a plus sign.
2. A plus sign followed by a minus sign equals a minus sign.
3. A minus sign followed by a plus sign equals a minus sign.
4. A minus sign followed by a minus sign equals a plus sign.

Key Example:

$$1.4 + (-5.2) - (-3.5) - (6.2) = 1.4 - 5.2 + 3.5 - 6.2 = 4.9 - 11.4 = -6.5$$

Multiplying or Dividing Two Signed Numbers

Multiply or divide the absolute values. The result is positive if the signs are alike, negative if the signs are unlike.

Key Example:

(a) $(-15)(-3) = +(15)(3) = 45$

(b) $(-10) \div (0.5) = -\left(\dfrac{10}{0.5}\right) = -20$

Order of Algebraic Operations—"PEMDAS"

1. Operations in parentheses.
2. Exponents.
3. Multiplication or division.
4. Addition or subtraction.

Key Example:

$$\frac{0.1(-6) - 0.2(4)}{10 - 2(-5)} = \frac{-0.6 - 0.8}{10 + 10} = \frac{-1.4}{20} = -0.07$$

5.2 ALGEBRAIC TERMS

Algebraic Term

A combination of letters, numbers, or both, joined by the operations of multiplication or division.

Key Example:

Term	*Coefficient*	*Literal Part*
$4x^2y$	4	x^2y
$-abc$	-1	abc
$-3\sqrt{PR}$	-3	$\sqrt{PR}$

Combining Like Terms

Add or combine the coefficients only. Do not change the literal part including the exponents or radicals.

• • •

Algebraic Expression

One or more algebraic terms connected by plus or minus signs.

Key Example:

(a) $8IR - 14V + 20IR - 35V = (8 + 20)IR + (-14 - 35)V$
$= 28IR - 49V$

(b) $2x^2 - [9x^2 + (5x - 3) + 7]$
$= 2x^2 - [9x^2 + 5x - 3 + 7]$
$= 2x^2 - 9x^2 - 5x + 3 - 7 = -7x^2 - 5x - 4$

5.3 RULES FOR EXPONENTS

Product Rule for Exponents

$$(x^m)(x^n) = x^{m+n} \tag{5.1}$$

Add exponents when multiplying like bases.

Key Example:

(a) $(x^4)(x^3) = x^{4+3} = x^7$

(b) $(-2)^2(-2)^3 = (-2)^{2+3} = (-2)^5 = -32$

Quotient Rule for Exponents

$$\frac{x^n}{x^m} = x^{n-m} \tag{5.2}$$

Subtract exponents when dividing like bases.

Key Example:

(a) $\dfrac{P^3}{P} = P^{3-1} = P^2$

(b) $\dfrac{(-3)^5}{(-3)^2} = (-3)^{5-2} = (-3)^3 = -27$

Power Rule for Exponents

$$(x^m)^n = x^{mn} \tag{5.3}$$

Multiply exponents when raising to a power.

Key Example:

(a) $(V^2)^3 = V^{2 \cdot 3} = V^6$

(b) $[(-2)^3]^3 = (-2)^9 = -512$

Factor Rule for Exponents

$$(xy)^n = x^n y^n \qquad \left(\frac{x}{y}\right)^n = \frac{x^n}{y^n} \tag{5.4}$$

Raise each factor separately to the given power.

Key Example:

(a) $(5y^2)^2 = (5)^2(y^2)^2 = 25y^{(2)(2)} = 25y^4$

(b) $\left(\dfrac{2}{R}\right)^3 = \dfrac{2^3}{R^3} = \dfrac{8}{R^3}$

Radical Rule for Exponents

$$\sqrt[n]{x^m} = x^{m/n} \tag{5.5}$$

Divide the index of the radical into the exponent of the radicand (number in the radical).

Key Example:

(a) $\sqrt{r^6} = r^{6/2} = r^3$

(b) $\sqrt[3]{8V^6} = \sqrt[3]{8}V^{6/3} = 2V^2$

Zero Exponent Rule

$$x^0 = 1 \ (x \neq 0) \tag{5.6}$$

Any number to the zero power is one.

• • •

Negative Exponent Rule

$$(x,\ y \neq 0) \quad x^{-n} = \left(\frac{1}{x}\right)^n = \frac{1}{x^n}$$
$$\text{and} \left(\frac{x}{y}\right)^{-n} = \left(\frac{y}{x}\right)^n \tag{5.7}$$

Key Example:

(a) $2^{-6} = \frac{1}{2^6} = \frac{1}{64}$

(b) $(w^2)^2(w^{-3})(w^{-1}) = (w^4)(w^{-4}) = w^0 = 1$

(c) $\left(\frac{3c^2}{2d}\right)^{-2} = \left(\frac{2d}{3c^2}\right)^2 = \frac{2^2d^2}{3^2c^{2\cdot 2}} = \frac{4d^2}{9c^4}$

(d) $(-10)^{-4} = \left(\frac{1}{-10}\right)^4 = \frac{1}{(-10)^4} = \frac{1}{10{,}000} = 0.0001$

Reciprocal Exponent Rule

$$(x,\ y \neq 0) \quad x^{-1} = \frac{1}{x} \text{ and } \left(\frac{x}{y}\right)^{-1} = \frac{y}{x} \tag{5.8}$$

• • •

Multiplying and Dividing Terms

Multiply or divide the coefficients separately and apply the rules for exponents to the literal parts.

Key Example:

(a) $(-2P^3)(5P^{-4}) = (-2)(5)P^{3-4} = -10P^{-1} = \frac{-10}{P}$

(b) $\frac{5.6n^2}{(2n^2)^2} = \frac{5.6n^2}{4n^4} = \left(\frac{5.6}{4}\right)n^{2-4} = 1.4n^{-2} = \frac{1.4}{n^2}$

When multiplying or dividing an algebraic expression by an algebraic term, every term must be multiplied or divided by the algebraic term.

Key Example:

(a) $-5e^2(e^2 - e + 1) = -5e^4 + 5e^3 - 5e^2$

(b) $\frac{2.5at^3 - 1.0a^2t^2 - 1.5a^3t}{5at}$
$= \frac{2.5at^3}{5at} - \frac{1.0a^2t^2}{5at} - \frac{1.5a^3t}{5at}$
$= 0.5t^2 - 0.2at - 0.3a^2$

Powers of ten are handled like algebraic terms when working with exponents. The numbers are treated like the coefficients, and the powers of ten are treated like the literal parts.

Key Example:

$$\frac{\sqrt{9.89 \times 10^{-4}}}{(26.4 \times 10^3)^2} = \frac{(\sqrt{9.89}) \times (10^{-4/2})}{(26.4)^2 \times (10^{(3)(2)})} = \frac{3.14 \times 10^{-2}}{697 \times 10^6}$$
$$= \left(\frac{3.14}{697}\right)10^{-2-6} \approx 0.00451 \times 10^{-8}$$
$$= 45.1 \times 10^{-12}$$

REVIEW EXERCISES

In exercises 1 through 20, perform the indicated operations by hand.

1. $-9 + 11$
2. $-6.9 - (-4.2)$
3. $5.5 - 4.3 + (-9.3)$
4. $10 - (-15) + 1 - 6$
5. $\left(-\frac{3}{4}\right) + \frac{1}{8} - \frac{1}{2} + 1$
6. $\left(\frac{1}{8}\right)\left(-\frac{4}{7}\right)$
7. $\left(-\frac{1}{10}\right) \div 2$
8. $\left(\frac{1}{6}\right) \div \left(-\frac{1}{2}\right)$
9. $\frac{-49}{-7}$
10. $\frac{10.5}{-0.07}$
11. $-0.6 \cdot 10 \cdot 0.1$
12. $(-0.5)(6) + (-10)(-0.1)$
13. $(-6)\left(\frac{1}{2}\right)\left(-\frac{1}{4}\right)$
14. $\frac{15(-8)}{(-3)(-4)}$

15. $\dfrac{5.8 + 1.2}{8.8 - 9.5}$

16. $-3.6\left(\dfrac{-5.5}{-1.8}\right)$

17. $(2)\left(-\dfrac{1}{5}\right) + \dfrac{2(-2)}{10}$

18. $\left(-\dfrac{1}{3}\right)\left(-\dfrac{2}{5}\right) - \left(\dfrac{2}{3}\right)\left(\dfrac{7}{10}\right)$

19. $\left(\dfrac{1}{4}\right)(-8)(0.4) + \dfrac{2.4}{4(-0.2)}$

20. $\dfrac{10(-1.2) + (-5)(1.6)}{3.3 - 3.7}$

In exercises 21 through 62, simplify each expression.

21. $x^2 + 2x - 3x^2 + x$

22. $a^2b - ab^2 + 3a^2b + ab^2$

23. $7.3V_1 - 2.6V_2 - 6.9V_1 + 1.5V_2$

24. $2.6R_a - 3.1R_b + 8.8R_a + 1.1R_b$

25. $\dfrac{3}{5}G - \dfrac{1}{10}F + \dfrac{1}{5}F - \dfrac{1}{5}G$

26. $-\dfrac{Q}{3} + \dfrac{3}{10} + \dfrac{Q}{6} - \dfrac{7}{10}$

27. $5 - (8 + 3CV) + (2CV - 1)$

28. $(2V + IR) - (2IR - V) + 2V$

29. $X_L^2 - 2X_L^2 + 3X_C^2$

30. $9cde - 6dec - 3cde$

31. $\dfrac{R_x}{3} - \dfrac{R_y}{2} - \left(\dfrac{2R_x}{3} - \dfrac{R_y}{2}\right)$

32. $xy^2 + [xy^2 - (x^2y - 2x^2y)]$

33. $3.5I_1 - [1.6 - (0.7I_1 - 1.1) - 1.3I_1]$

34. $\left(\dfrac{5.4g}{6} - \dfrac{2.1f}{3}\right) - \left(\dfrac{2.4f}{6} - \dfrac{1.8g}{3}\right)$

35. $m^4 \cdot m^2 \cdot m^3$

36. $(kP^2)(k^2P)(kP)$

37. $(-1)^3(-1)^5(-1)$

38. $\dfrac{6^5}{6^3}$

39. $(q^3)^4$

40. $[(-4)^2]^2$

41. $(8x^3y)^2$

42. $(10^4 \cdot 10^2)^2$

43. $(-3d^2e^2)^{-1}$

44. $\left(\dfrac{x^{-2}}{y^{-1}}\right)^{-2}$

45. $\dfrac{f^4 \cdot f^2}{(f^3)^3}$

46. $\left(\dfrac{0.1p^2}{w}\right)^3$

47. $-(-303)^0$

48. $(-4)^{-3}$

49. $(3)^{-3}\left(\dfrac{2}{9}\right)^{-2}$

50. $\dfrac{5^{-2} \cdot 5^3}{5^0}$

51. $\dfrac{(6X_C^2)(-7X_C^3)}{2X_C}$

52. $\sqrt{0.36 \times 10^{-6}}$

53. $(4I^2R)(3IR)^2(5IR^2)$

54. $-6t(7t^2 + 9t - 1)$

55. $\dfrac{32x^2w}{1.6xw^2}$

56. $\dfrac{P_1^2P_2^2 + P_1P_2}{P_1P_2}$

57. $(6ab)(7b)(a^2b - ab^2)$

58. $\sqrt[3]{0.027H^{-3}}$

59. $\left(\sqrt{2.25 \times 10^6}\right)(1.4 \times 10^3)^2$

60. $(3 \times 10^2)^2(5 \times 10)^2$

61. $\dfrac{\sqrt{1.44 \times 10^{-6}}}{\sqrt{2.25 \times 10^{-2}}}$

62. $\left(\sqrt{49 \times 10^4}\right)\left(\sqrt{25 \times 10^{-4}}\right)$

In exercises 63 through 70, given the following values of dc circuit variables: $V = 6.0$ V, $I = 1.38$ mA, $R = 150\ \Omega$, and $P = 26.5$ mW. *Calculate the value of the indicated variable in each formula to three significant digits.*

63. $V = IR$; V

64. $P = VI$; P

65. $V = \dfrac{P}{I}$; V

66. $I = \dfrac{V}{R}$; I

67. $R = \dfrac{P}{I^2}$; R

68. $P = \dfrac{V^2}{R}$; P

69. $I = \sqrt{\dfrac{P}{R}}$; I

70. $V = \sqrt{PR}$; V

Applied Problems

71. What is the temperature difference between the melting point of petroleum, −70.5°C, and the boiling point, 250°C?

72. The quickest rise in temperature ever recorded was in Spearfish, South Dakota on January 22, 1943. At 7:30 AM the temperature was −4°F and at 7:32 AM it was 45°F. Express this change as a signed number.

73. The total force produced by two forces on a body is given by $F_T = 3.5e^2 + 8.2e - 2.5$. If one force is given by $F_1 = 6.1e^2 - 4.4e - 2.7$, find the expression for the other force $F_2 = F_T - F_1$.

74. The kinetic energy of one particle is given by $K_1 = \frac{1}{2}mv_1^2$. The kinetic energy of a second particle is given by $K_2 = \frac{1}{8}mv_1^2$. Simplify the expression that represents the difference in energy, $K_1 - K_2$, between the two particles.

Applications to Electronics

In applied problems 75 through 84, round all answers to three significant digits.

75. The voltage in a circuit with respect to ground changes from 2.5 V to −0.5 V when the ground connection is changed. What signed number represents this change?

76. Going clockwise around a closed path in a circuit the voltage drops are −4.3 V, −4.1 V, and −4.5 V. The voltage increases are +6.7 V and +6.2 V. Show that the algebraic sum of the voltages is zero.

77. Given the following values of dc circuit variables: V = 16.0 V, I = 48.5 mA, and R = 330 Ω, calculate the value of the indicated variable in each formula.

(a) $I = \frac{V}{R}$; I in mA

(b) $P = I^2 R$; P in mW

78. Given the following values of ac circuit variables: f = 15 kHz, L = 4.5 mH, and C = 1.5 nF, calculate the value of the indicated variable in each ac circuit formula.

(a) $X_L = 2\pi fL$; Inductive Reactance X_L in ohms.

(b) $X_C = \frac{1}{2\pi fC}$; Capacitive Reactance X_C in kilohms.

79. The total resistance of a series-parallel circuit is given by:

$$R_T = \left(\frac{1}{R_1} + \frac{1}{R_2}\right)^{-1}$$

Find R_T in ohms when R_1 = 20 Ω and R_2 = 30 Ω.

80. In problem 79, find R_T in ohms and kilohms when R_1 = 1.5 kΩ and R_2 = 750 Ω.

81. The impedance of an ac circuit is given by:

$$Z = \sqrt{R^2 + X^2}$$

Find Z in kilohms when $R = 5.6 \times 10^3$ Ω and $X = 3.2 \times 10^3$ Ω.

82. The magnitude of an electromagnetic force field is given by:

$$E = \sqrt{E_x^2 + E_y^2}$$

Find the value of E in newtons when $E_x = 5.1 \times 10^{-3}$ N and $E_y = 6.8 \times 10^{-3}$ N.

83. Calculate the current in amps required to produce a certain magnetic flux density in a circular coil given by:

$$I = \frac{(0.20)(300 \times 10^{-6})}{(4\pi) \times 10^{-6}}$$

84. Calculate the impedance of an ac circuit in kilohms given by:

$$Z = \sqrt{(1.0 \times 10^3)^2 + \left(\frac{1}{4\pi(12 \times 10^3)(20 \times 10^{-9})}\right)^2}$$

CHAPTER 6

Linear Equations

CHAPTER OBJECTIVES

In this chapter, you will learn:

- The meaning of a first-degree or linear equation.
- The step-by-step process for solving a first-degree equation.
- How to transpose terms in an equation.
- How to solve first-degree equations with fractions, decimals, and percents.

Engineering and technology use equations to express relationships and to solve all types of problems. This is especially important in electricity and electronics where equations are used in almost all applications to find unknown values of components. Solving equations is an important algebraic skill you need to master. This chapter explains the basic skills of equation solving and shows how to solve different types of first-degree equations including ones that contain fractions, decimals, and percentages. You will be able to apply many of the algebraic skills you learned in previous chapters to the solution of equations.

• • •

6.1 Solution of First-Degree Equations

A first-degree term is one that contains a variable to the first power or degree, such as:

$$2x,\ 3y,\ 5R_1,\ 2.5V,\ \tfrac{w}{2},\ 10\%I$$

A *first-degree* or *linear equation* is:

Definition

First-Degree or Linear Equation

A statement of equality that contains only first-degree terms and no higher degree terms.

Examples of linear equations in one variable are:

$$10R = 56$$

$$4.1I + 7.5I = 12.6$$

$$\frac{V_0}{20} = 5$$

$$3\%t + 16 = 7$$

$$27 = \frac{5}{9}(F - 32)$$

Notice that all the variables in the equations above have an exponent of one. Since one side of an equation is equal to the other side, you can use the following principle to simplify and solve for a variable in an equation:

Principle

Equation Principle

The following operations do not change an equality:

1. Adding or subtracting the same quantity on both sides of an equation.
2. Multiplying or dividing both sides of an equation by the same quantity. (*Division by zero is not allowed.*)

EXAMPLE 6.1 Solve and check the linear equation:

$$x - 5 = 12$$

Solution To solve the equation you need to get the unknown x by itself. You can apply the equation principle and add 5 to both sides of the equation, which leaves only x on the left side:

$$x - 5 = 12$$
$$x - 5 + 5 = 12 + 5$$
$$x = 17$$

You need to check the solution. To do the check, always substitute the solution back into the *original* equation:

$$(17) - 5 = 12 \ ?$$
$$12 = 12 \checkmark$$

EXAMPLE 6.2 Solve and check the linear equation:

$$10R = 56$$

Solution To solve the equation you need to get the unknown R by itself. Since it is multiplied by 10, you can apply the equation principle and divide both sides of the equation by the coefficient 10, which leaves you R on the left side:

Add $1.1I + 1.4I$ $\quad 10R = 56$

Divide by 10: $\quad \dfrac{\cancel{10}R}{(\cancel{10})} = \dfrac{56}{(10)}$

Simplify: $\quad R = 5.6$

You need to check the solution. To do the check, always substitute the solution back into the *original* equation:

$$10(5.6) = 56 \ ?$$
$$56 = 56 \checkmark$$

EXAMPLE 6.3 Solve and check the linear equation:

$$1.1I + 1.4I = 12.6$$

Solution Before you can isolate the unknown you need to simplify by combining the similar terms on the left side:

$$2.5I = 12.6$$

Divide both sides by 2.5: $\quad \dfrac{\cancel{2.5}I}{(\cancel{2.5})} = \dfrac{12.6}{(2.5)}$

Simplify: $\quad I = 5.04$

You need to check the solution. Substitute 5.04 for I in the original equation:

$$1.1(5.04) + 1.4(5.04) = 12.6\ ?$$
$$5.544 + 7.056 = 12.6\ ?$$
$$12.6 = 12.6\ \checkmark$$

EXAMPLE 6.4 Solve and check the equation:

$$\frac{y}{20} = 5$$

Solution Since the unknown is divided by 20, multiply both sides by 20 to solve for the unknown:

$$(\cancel{20})\frac{y}{\cancel{20}} = 5(20)$$

Simplify: $y = 100$

You need to check the solution. Substitute 100 for y in the original equation:

$$\frac{100}{20} = 5\ ?$$
$$5 = 5\ \checkmark$$

EXAMPLE 6.5 Solve and check the linear equation:

$$3t + 16 = 7$$

Solution You need to first get the term containing the unknown, $3t$, by itself on one side of the equation. Apply the equation principle and eliminate 16 by subtracting 16 from both sides of the equation:

$$3t + 16 - 16 = 7 - 16$$

Simplify both sides: $3t = -9$

Divide both sides by 3: $\frac{\cancel{3}t}{(\cancel{3})} = \frac{-9}{(3)}$

Simplify: $t = -3$

To check the solution, substitute -3 for t in the original equation:

$$3(-3) + 16 = 7\ ?$$
$$-9 + 16 = 7\ \checkmark$$

EXAMPLE 6.6 Solve and check the linear equation:

$$3H + 2(2H - 1.5) = -3$$

Solution Clear parentheses first by multiplying by 2:

$$3H + 4H - 3 = -3$$

Combine terms: $7H - 3 = -3$

Add +3 to both sides: $7H - 3 + 3 = -3 + 3$

Simplify: $7H = 0$

Divide both sides by 7: $\frac{\cancel{7}H}{(\cancel{7})} = \frac{0}{(7)}$

$$H = 0$$

The solution, H = 0, is an acceptable solution to an equation.

The check is:

$$\begin{aligned} 3(0) + 2(2(0) - 1.5) &= -3\ ? \\ 0 + 2(0 - 1.5) &= -3\ ? \\ 2(-1.5) &= -3\ ? \\ -3 &= -3\ \checkmark \end{aligned}$$

ERROR BOX

A common error when solving an equation is not to simplify each side as much as possible before doing the next step. Combine any common terms on one side and do any arithmetic operations before you add, subtract, multiply or divide both sides by some quantity.

See if you can apply this idea when doing the following problems.

Practice Problems:

1. $3a - 2a - 1 = 6$
2. $4x + 2 = 5 - (-5)$
3. $2P - 3P - 13 = 0$
4. $5c + 2.5(-8c) = 75$
5. $1.6 + 3.2(2 + R) = 9.6$
6. $7 - 11 = 2B - 4B(0.6)$
7. $6.6 - 4.4 = 2V + 8.8$
8. $3(4y - 7) + 7 = 10$
9. $5 - 7(2I + 5) = 3I - 2I$
10. $2.5 - 0.5(6Q - 4Q) = 12.5 - 2.0$

The next example shows how to handle a simple fraction in an equation. It also closes the circuit and illustrates an application of Ohm's law.

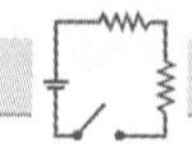

EXAMPLE 6.7

Close the Circuit

An application of Ohm's law gives the following equation for the voltage V in a circuit:

$$\frac{V}{10} - 7 = 15$$

Solve this equation for V in volts.

Solution A fraction in an equation should be eliminated as soon as possible to simplify the equation. To eliminate the fraction, multiply both sides by the lowest common denominator = 10. This means that *every term must be multiplied by* 10:

Multiply by 10: $$(\cancel{10})\frac{V}{\cancel{10}} - (10)7 = (10)15$$

Simplify: $$V - 70 = 150$$

Add 70 to both sides: $$V - 70 + 70 = 150 + 70$$

Simplify: $$V = 220$$

Check:

$$\begin{aligned} \frac{(220)}{10} - 7 &= 15\ ? \\ 22 - 7 &= 15\ ? \\ 15 &= 15\ \checkmark \end{aligned}$$

EXAMPLE 6.8

The equation to change 27°C (Celsius) to degrees Fahrenheit is:

$$27 = \frac{5}{9}(F - 32)$$

Solve this equation for F.

Solution Eliminate fractions in an equation as soon as possible to make the solution easier. Multiply both sides by the lowest common denominator = 9 to eliminate the fraction:

$$(9)27 = (\cancel{9})\frac{5}{\cancel{9}}(F - 32)$$

Note that since the fraction is multiplied by everything in the parentheses; you only multiply the factor $\frac{5}{9}$ by 9, and not the other factor $(F - 32)$:

	$243 = 5(F - 32)$
Multiply out the parentheses:	$243 = 5F - 160$
Add 160 to both sides:	$243 + 160 = 5F - 160 + 160$
Simplify:	$403 = 5F$
or	$5F = 403$

By convention the unknown term is usually written on the left. Now divide by the coefficient 5:

$$\frac{\cancel{5}F}{(\cancel{5})} = \frac{403}{(5)}$$

$$F = 80.6° \text{ F}$$

Check:

$$27 = \frac{5}{9}(80.6 - 32) \ ?$$

$$27 = \frac{5}{9}(48.6) \ ?$$

$$27 = 27 \ \checkmark$$

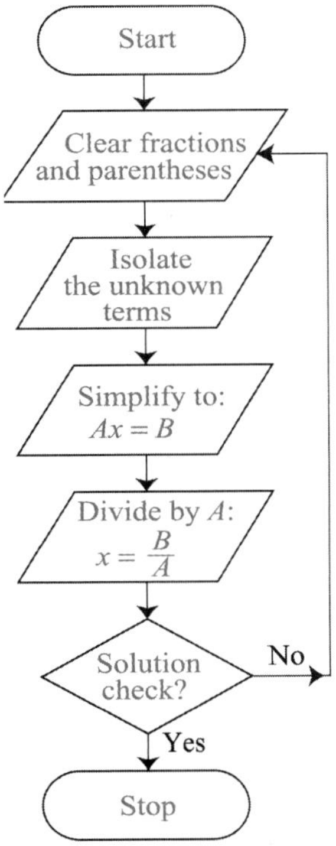

The step-by-step process or *algorithm* for solving a linear equation is as follows:

EQUATION SOLVING ALGORITHM

1. Eliminate fractions by multiplying by the LCD, and clear parentheses.
2. Combine similar terms and isolate the unknown terms on one side.
3. Simplify by putting the equation in the form $Ax = B$ where x = unknown and A and B are constants.
4. Divide both sides by the coefficient of x to obtain the solution: $x = \frac{B}{A}$
5. Check the solution in the original equation.

Start → Clear fractions and parentheses → Isolate the unknown terms → Simplify to: $Ax = B$ → Divide by A: $x = \frac{B}{A}$ → Solution check? (No → back to Clear fractions and parentheses; Yes → Stop)

FIGURE 6-1 Flowchart for solving equations.

See Figure 6-1, which illustrates the above algorithm in a flowchart.

EXERCISE 6.1

In exercises 1 through 50, solve and check each linear equation.

1. $x + 6 = 10$
2. $a - 9 = 15$
3. $R - 5 = 20$
4. $7 - y = 7$
5. $4w = 16$
6. $10p = -50$
7. $5x = -43$
8. $8x = 28$
9. $\frac{u}{4} = 13$
10. $\frac{v}{3} = 26$
11. $\frac{a}{5.5} = 10$
12. $\frac{b}{1.2} = 15$

13. $5c + 2 = 7$

14. $6d - 3 = 9$

15. $3b + 8 = 5$

16. $8 + 4y = 12$

17. $5.5 + 0.25y = 9.5$

18. $1.2z - 3.6 = 4.8$

19. $20I - 10 = 0$

20. $5 + 25V = 0$

21. $0.2X_L + 3.6 = 1.4$

22. $3.5 - 0.1X_C = 6.5$

23. $6 - \dfrac{2a}{3} = 4$

24. $8 + \dfrac{3n}{4} = 5$

25. $0.7 = 2V_0 - 2.3$

26. $1.8 + 3.7 = 3V_1 - 7.4$

27. $6 = 4x - 7x - 12$

28. $8 - 2d = 0.6(-4d)$

29. $0.28 = 0.04(w - 1.0)$

30. $3.6 - 1.2(z + 2) = 0$

31. $4 + 3(H - 0.7) - 1.9 = 0$

32. $6 - 2(G - 3.8) = 13.6$

33. $7.5I_2 + 15I_2 = 36$

34. $1.2X_C + 1.3X_C = 7.5$

35. $2x + 3(2x - 15) = -1$

36. $4y - 5(4y - 3) = 7$

37. $2a + 5 = 7(2a + 5) + 3a$

38. $2.0 - 0.5(6b - 4b) = 14.5 - 4.5$

39. $3n - \dfrac{2}{3} = 1 + \dfrac{1}{3}$

40. $\dfrac{3}{5} + 2m = \dfrac{1}{5}$

41. $\dfrac{I}{3} + \dfrac{1}{3} = 2$

42. $\dfrac{R}{2} - 2R = 6$

43. $20 = \dfrac{5}{9}(F - 32)$

44. $0 = \dfrac{5}{9}(F - 32)$

45. $42 = \dfrac{3}{10}(h + 25)$

46. $\dfrac{1}{2}(g - 10) = 9$

47. $5 - \dfrac{1}{3}(2r - 9) = 0$

48. $10 - \dfrac{2}{5}(15 - 4t) = 0$

49. $\dfrac{x}{2} + \dfrac{x}{2} = 5$

50. $\dfrac{y}{5} + \dfrac{y}{5} = 2$

Applied Problems

51. The equation to change −13°F to degrees Celsius is:

$$-13 = 1.8C + 32$$

Solve this equation for the temperature C in degrees Celsius.

52. The equation to change 22°C (Celsius) to degrees Fahrenheit is:

$$22 = \frac{5}{9}(F - 32)$$

Solve this equation for the temperature F in degrees Fahrenheit.

53. The equation to find the time it takes for a falling object to reach a velocity of 150 ft/sec is:

$$150 = 38 + 32t$$

Solve this equation for the time t in seconds.

54. The width of a rectangle is one-third its length. If the perimeter is 96 ft, the equation to find the length l is given by:

$$2l + 2\left(\frac{l}{3}\right) = 96$$

Solve this equation for the length of the rectangle in feet.

Applications to Electronics

55. An application of Ohm's law gives the following equation for the voltage V in a circuit:

$$33 = \frac{V}{20} + 28$$

Solve this equation for the voltage V in volts.

56. An application of Kirchhoff's voltage law gives the following equation for the current I in a circuit:

$$4.3I + 5.6I = 12.6$$

Solve this equation for I in amps to two significant digits.

57. An application of the power formula gives the following equation for the power P in a circuit:

$$\frac{P}{1.6} - 3.1 = 12$$

Solve this equation for P in watts to two significant digits.

58. An application of the power formula gives the following equation for the power P in a circuit:

$$50 = \frac{P}{3.4} - 12$$

Solve this equation for P in watts to two significant digits.

Answers to Error Box Problems, page 136:
1. 7 **2.** 2 **3.** −13 **4.** −5 **5.** 0.5 **6.** 10 **7.** −3.3 **8.** 2 **9.** −2 **10.** −8

6.2 Equation Solving Methods

Section 6.1 introduces the basic procedure for solving first-degree equations. This section looks at further techniques that will help you to solve equations more effectively.

Study the next two examples, which compare two ways of solving the same equation. However, the second one shows a shortcut procedure called transposing.

EXAMPLE 6.9 Solve and check the linear equation:

$$4P - 3 = 5 + 2P$$

Solution This equation has constants and the unknown on both sides of the equation. You need to put all the terms containing the unknown on one side of the equation and all the constant terms on the other side. The unknown terms are generally placed on the left side. First eliminate the 3 on the left and then the $2P$ on the right:

Add 3 to both sides: $4P - 3 + 3 = 5 + 2P + 3$

Simplify: $4P = 8 + 2P$

Subtract $2P$ from both sides: $4P - 2P = 8 + 2P - 2P$

Simplify: $2P = 8$

Divide both sides by 2: $\frac{\cancel{2}P}{(\cancel{2})} = \frac{8}{(2)}$

Simplify: $P = 4$

To check the solution, substitute $P = 4$ back into the original equation:

$$4(4) - 3 = 5 + 2(4) \;?$$

$$16 - 3 = 5 + 8 \;?$$

Check: $13 = 13$ ✓

The process of solving an equation can be simplified by observing that adding or subtracting the same quantity on both sides of an equation is equivalent to *moving that quantity to the other side of the equation and changing its sign.* For example, look at the first step in Example 6.9 above: adding 3 to both sides. This is equivalent to moving the –3 from the left side to the right side and changing the sign to +3. This process is called *transposing:*

Rule **Transposing Terms**

A term in an equation may be moved from one side to the other side by changing the sign of the term.

The next example shows how to solve Example 6.9 more efficiently by transposing terms.

EXAMPLE 6.10 Solve Example 6.9 by transposing terms:

$$4P - 3 = 5 + 2P$$

Solution Transpose (–3) to the right side:

$$4P - 3 = 5 + 2P$$

$$4P = 5 + 2P + 3$$

Simplify: $4P = 8 + 2P$

Transpose ($2P$) to the left side: $4P = 8 + 2P$

$$4P - 2P = 8$$

To simplify the process, you can transpose the –3 and the $2P$ in one step.

Simplify: $2P = 8$

Divide both sides by 2: $P = 4$

Try to use the method of transposing whenever possible to simplify solving equations.

EXAMPLE 6.11 Solve and check:

$$0.5w + 0.7 = 1.5\,(w + 1.2)$$

Solution First clear parentheses by multiplying each term in the parentheses by 1.5:

Transpose 0.7 to the right side and $1.5w$ to the left side in one step:

$$0.5w + 0.7 = 1.5w + 1.8$$

$$0.5w - 1.5w = 1.8 - 0.7$$

Combine similar terms: $-w = 1.1$

The left side is still not solved for w. You need to divide both sides by –1 to get the correct value for the unknown, which is the positive value $+w$:

$$\frac{-w}{(-1)} = \frac{1.1}{(-1)}$$

Divide right side: $w = -1.1$

Check: $0.5(-1.1) + 0.7 = 1.5(-1.1 + 1.2)$?

$$-0.55 + 0.70 = 1.5(0.1)\ ?$$

$$0.15 = 0.15 \checkmark$$

EXAMPLE 6.12 Solve and check the equation:

$$\frac{x}{3} + \frac{5x}{6} = 14$$

Solution This equation contains fractions with different denominators. You need to multiply each term by the lowest common denominator (LCD) to eliminate the fractions. Every denominator will divide into the lowest common denominator when you multiply, and every fraction will be eliminated. The LCD of 3 and 6 is 6. Multiply 6 into the numerator of each fraction and multiply the term 14 on the right side also by 6 to balance the equation:

Multiply by the LCD: $\frac{(6)x}{3} + \frac{(6)5x}{6} = (6)14$

Simplify: $\frac{\overset{2}{\cancel{(6)}}x}{\cancel{3}} + \frac{\overset{1}{\cancel{(6)}}5x}{\cancel{6}} = (6)14$

Divide by 7:

$$2x + 5x = 84$$
$$7x = 84$$
$$x = 12$$

Check:

$$\frac{(12)}{3} + \frac{5(12)}{6} = 14\ ?$$
$$4 + 10 = 14\ ✓$$

ERROR BOX

A common error when working with a fraction in an equation is not to treat the fraction line as a set of parentheses. A fraction with a minus sign in front means that *each term in the numerator is multiplied by the minus sign:*

$$-\frac{x+3}{2} = \frac{-x-3}{2}$$

Therefore, if this fraction occurs in an equation as follows:

$$5 - \frac{x+3}{2} = 2$$

You must multiply each term in the numerator by the minus sign *and* the LCD which is 2. You should first enclose the numerator in parentheses to avoid this error:

$$5 - \frac{(x+3)}{2} = 2$$

Then multiply by the LCD:
$$(2)5 - \frac{\not{2}(x+3)}{\not{2}} = (2)2$$

Simplify:
$$10 - (x+3) = 4$$

Multiply out the parentheses:
$$10 - x - 3 = 4$$

Combine terms:
$$7 - x = 4$$

Transpose 7:
$$-x = 4 - 7$$

Combine terms:
$$-x = -3$$

Divide by −1:
$$x = 3$$

Check:
$$5 - \frac{3+3}{2} = 2$$
$$5 - 3 = 2\ ✓$$

Practice Problems: Solve and check each equation:

1. $1 - \frac{x+1}{2} = 0$ **2.** $2 - \frac{y+1}{3} = 0$ **3.** $6 - \frac{R+3}{3} = 3$

4. $7 - \frac{W-2}{4} = 5$ **5.** $\frac{1}{2} - \frac{a-2}{2} = 0$ **6.** $\frac{5}{8} = -\frac{1-b}{8}$

7. $1 = -\frac{A+1}{4}$ **8.** $e - \frac{e+4}{6} = \frac{1}{6}$ **9.** $r - \frac{r-1}{5} = \frac{9}{5}$

10. $2(x-5) = -\frac{x+9}{3}$

The next two examples close the circuit and show applications of equations with fractions to voltages in a transformer and a parallel circuit.

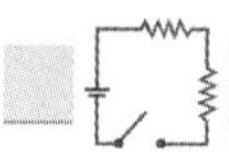

EXAMPLE 6.13

Close the Circuit

A step-down transformer reduces the voltage in a circuit based on the number of turns in the primary and secondary coils. The voltage ratio is equal to the turns ratio:

$$\frac{V_P}{V_S} = \frac{N_P}{N_S}$$

where V_P = primary voltage, V_S = secondary voltage, N_P = number of turns in the primary coil, and N_S = number of turns in the secondary coil. An equation where one ratio (fraction) equals another ratio is called a *proportion*. Find V_P to three significant figures when $V_S = 100$ V, $N_P = 1000$, and $N_S = 700$.

Solution Substitute the values in the proportion:

$$\frac{V_P}{100 \text{ V}} = \frac{1000}{700}$$

Reduce the fraction on the right:

$$\frac{V_P}{100 \text{ V}} = \frac{10}{7}$$

Multiply both sides by the LCD = 700:

$$\frac{\overset{7}{\cancel{(700)}} V_P}{\cancel{100} \text{ V}} = \frac{\overset{100}{\cancel{(700)}} 10}{\cancel{7}}$$

Simplify:

$$7V_P = 1000$$

Divide by 7:

$$V_P = \frac{1000}{7} \approx 143 \text{ V}$$

A proportion can also be solved by cross multiplication. Cross multiplication means multiplying the left numerator by the right denominator and vice versa:

Cross multiply:

$$\frac{V_P}{100} \times \frac{10}{7}$$

$$7V_P = 1000$$

Divide by 7:

$$V_P \approx 143 \text{ V}$$

Cross multiplication is equivalent to multiplying both sides of the equation by the product of the denominators. The solution check is:

$$\frac{143}{100} = \frac{1000}{700}\ ?$$

$$1.43 = 1.43 \checkmark$$

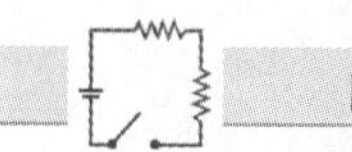

EXAMPLE 6.14

Close the Circuit

Figure 6-2 shows a circuit containing two resistors, 12 Ω and 15 Ω, in parallel. If the total current is 2.3 A, the voltage V of the battery is given by the equation:

$$\frac{V}{12} + \frac{V}{15} = 2.3$$

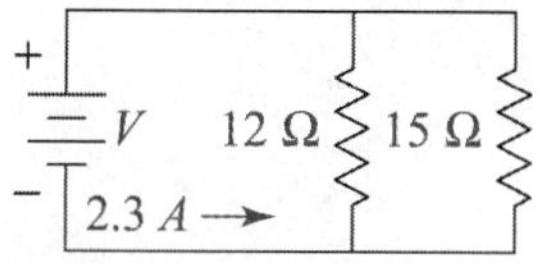

Find the V in volts to two significant digits.

FIGURE 6-2 Parallel resistors for Example 6.14.

Solution Multiply each term by LCD = 60:

$$\frac{\overset{5}{\cancel{(60)}} V}{\cancel{12}} + \frac{\overset{4}{\cancel{(60)}} V}{\cancel{15}} = (60)2.3$$

Answers to Error Box Problems, page 141:

1. −1 **2.** 5 **3.** 6 **4.** 10 **5.** 3 **6.** 6 **7.** −5 **8.** 1 **9.** 2 **10.** 3

Simplify:

$$5V + 4V = 138$$

$$9V = 138$$

Divide by 9:

$$V = \frac{138}{9} \approx 15 \text{ V}$$

Check:

$$\frac{15}{12} + \frac{15}{15} = 2.3 \ ?$$

$$1.25 + 1 = 2.25 \approx 2.3 \checkmark$$

The check result is close enough to verify the answer.

EXAMPLE 6.15 Solve and check the equation:

$$\frac{4x + 5}{6} - \frac{x + 7}{4} = 2$$

Solution This equation contains fractions that have more than one term in the numerator. See the Error Box on page 141. When you multiply each fraction by the LCD = 12, you need to *multiply every term in each numerator* since the fraction line is the same as a set of parentheses. The minus sign in front of the second fraction means that every term in the numerator must also be multiplied by the minus sign. Study the solution carefully. To avoid an error, enclose each numerator in parentheses:

Multiply by the LCD = 12:

$$\frac{\overset{2}{\cancel{(12)}}(4x + 5)}{\cancel{6}} - \frac{\overset{3}{\cancel{(12)}}(x + 7)}{\cancel{4}} = (12)2$$

$$2(4x + 5) - 3(x + 7) = (12)2$$

Multiply out parentheses:

$$8x + 10 - 3x - 21 = 24$$

Combine terms:

$$5x - 11 = 24$$

Transpose −11:

$$5x = 24 + 11 = 35$$

Divide by 5:

$$x = 7$$

Check:

$$\frac{4(7) + 5}{6} - \frac{(7) + 7}{4} = 2 \ ?$$

$$\frac{33}{6} - \frac{14}{4} = 2 \ ?$$

$$\frac{11}{2} - \frac{7}{2} = 2 \ ?$$

$$\frac{11 - 7}{2} = \frac{4}{2} = 2 \checkmark$$

EXERCISE 6.2

In exercises 1 through 42, solve and check each linear equation. Use the method of transposing to simplify the solution.

1. $4 + 5R = R - 12$

2. $6V - 11 = V - 26$

3. $7 - 3t = 13 - 2t$

4. $16 - 5w = 8 - 4w$

5. $5a + 9 = 6a + 3$

6. $3 - 7b = 3b - 2$

7. $1.5V = 4.6 - 0.5V$

8. $0.06 + 0.05I = 0.03I$

9. $2(Y - 4) = Y - 2$

10. $6 + 2P = 2(2P + 3)$

11. $5 - (f + 3) = 5f$

12. $2 - (2x - 2) = 6x$

13. $3x = 4 - 2(3 - x)$

14. $x - 3(2 + x) = 6$

15. $1.5(2.4 - t) - 7.2 = 0.3t$

16. $2.2(r - 1.0) = 6.2 + 3.4r$

17. $\frac{a}{3} - 3(a - 2) = 2$

18. $2(n - 4) = 3 - \frac{n}{2}$

19. $R - \frac{1}{2}(4R + 2) = 0$

20. $\frac{1}{3}(9K - 3) = 4K$

21. $1.5w + 0.7 = 2.5(w + 1.2)$

22. $0.4 - 0.2(t - 1.6) = 1.6t$

23. $3 - \frac{3x}{4} = 1 - x$

24. $\frac{8y}{5} + 1 = 4y - 1$

25. $E - \frac{3}{5} = \frac{4E}{5}$

26. $\frac{5f}{3} - \frac{2}{3} = 3f$

27. $\frac{V}{2} = \frac{30}{5}$

28. $\frac{140}{40} = \frac{N}{50}$

29. $\frac{V}{10} + \frac{V}{20} = 1.2$

30. $\frac{V}{6.0} = 4.5 - \frac{V}{9.0}$

31. $\frac{x}{6} + \frac{5x}{12} = 14$

32. $\frac{3x}{10} - \frac{4x}{5} = 10$

33. $\frac{a}{2} = 3 - \frac{a}{8}$

34. $\frac{3a}{4} = 15 - \frac{a}{2}$

35. $p - \frac{p-1}{2} = \frac{9}{10}$

36. $\frac{1}{4} - \frac{v-2}{6} = v$

37. $\frac{Y}{8} - \frac{Y+6}{2} = 3$

38. $\frac{Z-10}{5} + \frac{Z}{10} = 1$

39. $\frac{2x+5}{6} - \frac{x+5}{4} = 1$

40. $\frac{x+2}{3} - \frac{x+5}{6} = 4$

41. $\frac{2a+1}{5} = 2 - \frac{a+8}{10}$

42. $2 - \frac{2-3b}{2} = \frac{b-6}{4}$

Applied Problems

43. Shaquida wants to buy a calculator on sale for \$18.60. The salesman tells her that the regular price was reduced 25% for the sale. To find the regular price p, she has to solve the equation:

$$p = 18.60 + (25\%)p$$

What is the regular price p of the calculator?

44. The price of an old model computer is reduced $\frac{1}{3}$ to a sale price of \$550. To find the original price p, you need to solve the equation:

$$p - 550 = \frac{p}{3}$$

What is the original price p of the computer?

Applications to Electronics

45. In Example 6.13, given the equation for a transformer:

$$\frac{V_P}{V_S} = \frac{N_P}{N_s}$$

find V_P when $V_S = 200\ V$, $N_P = 1000$, and $N_S = 400$.

46. In Exercise 45, find N_P when $V_S = 50\ V$, $V_P = 120$ V, and $N_S = 200$.

47. In Example 6.14, if the two resistors are 10 Ω and 15 Ω and the total current is 2.5 A, the voltage V of the battery is given by the equation:

$$\frac{V}{10} + \frac{V}{15} = 2.5$$

Find V in volts to two significant digits.

48. In Example 6.14, if the two resistors are 12 Ω and 20 Ω and the total current is 1.8 A, the voltage V of the battery is given by the equation:

$$\frac{V}{12} + \frac{V}{20} = 1.8$$

Find V in volts to two significant digits.

49. An application of the parallel resistance formula to a series-parallel circuit gives the following equation for the resistance R_1:

$$120 = 2R_1 + \frac{(100)(120)}{(120)+(100)}$$

Solve the equation for R_1 in ohms to two significant digits.

50. An application of the parallel resistance formula to a series-parallel circuit gives the following equation for the resistance R_2:

$$16 = 10 + \frac{2R_2}{3}$$

Solve the equation for R_2 in ohms to two significant digits.

CHAPTER HIGHLIGHTS

6.1 SOLUTION OF FIRST-DEGREE EQUATIONS

First-Degree or Linear Equation

A statement of equality that contains only first-degree terms and no higher degree terms.

• • •

Equation Principle

The following operations do not change an equality:

1. Adding or subtracting the same quantity on both sides of an equation.
2. Multiplying or dividing both sides of an equation by the same quantity. (*Division by zero is not allowed.*)

Key Example:

	$10x - 5 - 6x = 3$
Combine terms on the left side:	$4x - 5 = 3$
Add 5 to both sides and combine terms:	$4x = 8$
Divide both sides by 4:	$x = 2$
Check:	$10(2) - 5 - 6(2) = 3?$
	$20 - 17 = 3$ ✓

Equation Solving Algorithm

1. Eliminate fractions by multiplying by the LCD, and clear parentheses.
2. Combine similar terms and isolate the unknown terms on one side.
3. Simplify by putting the equation in the form $Ax = B$ where x = unknown and A and B are constants.
4. Divide both sides by the coefficient of x to obtain the solution: $x = \frac{B}{A}$
5. Check the solution in the original equation.

Key Example:

$$\frac{2}{3}v - 4(v - 2) = 5$$

Clear parentheses: $\frac{2}{3}v - 4v + 8 = 5$

Multiply by 3: $\frac{(\cancel{3})2}{\cancel{3}}v - (3)4v + (3)8 = (3)5$

$2v - 12v + 24 = 15$

Subtract 24 from both sides and combine terms: $-10v = -9$

Divide by -10: $v = 0.9$

Check: $\frac{2}{3}(0.9) - 4(0.9 - 2) = 0.6 + 4.4 = 5$ ✓

6.2 EQUATION SOLVING METHODS

Transposing Terms

A term in an equation may be moved from one side to the other side by changing the sign of the term.

Key Example:

Transpose 8 to the right side $3z + 8 + 6z = 2 - z$

and $-z$ to the left side: $3z + 6z + z = 2 - 8$

Combine terms: $10z = -6$

Divide by 10: $z = -0.6$

Check: $3(-0.6) + 8 + 6(-0.6) = 2 - (-0.6)$

$-1.8 + 8 - 3.6 = 2 + 0.6$?

$2.6 = 2.6$ ✓

Fractions in an equation are eliminated by multiplying all terms by the lowest common denominator. Enclose each numerator in parentheses to avoid an error.

Key Example:

$$6 - \frac{x + 5}{4} = 4$$

Multiply by LCD = 4: $(4)6 - \frac{(\cancel{4})(x + 5)}{\cancel{4}} = (4)4$

$(4)6 - (x + 5) = (4)4$

Multiply out parentheses: $24 - x - 5 = 16$

Combine terms and transpose 19: $-x = 16 - 19 = -3$

Divide by -1: $x = 3$

Check: $6 - \frac{(3) + 5}{4} = 4$?

$6 - 2 = 4$ ✓

REVIEW EXERCISES

In exercises 1 through 30, solve each linear equation.

1. $2x - 2 = 10$
2. $b + 3b - 40 = 0$
3. $6q = 24 - 2q$
4. $1.2z - 3.7 = 3.5$
5. $8X_C + 4X_C = 18$
6. $\frac{d}{3} = 5.5$
7. $9 + \frac{4n}{5} = 5$
8. $\frac{5}{8} - 2w = \frac{1}{8}$
9. $1.8 + 3.7V_1 = 3V_1 - 7.3$
10. $0.22 - 0.34 = 2I_T - 0.74$
11. $4(w + 2) = 5 - w$
12. $3 - 2(x - 3) = 4x$
13. $3.5 - 2.5m = -(m - 0.50)$
14. $3.9 - 1.5(2.2 - 1.2x) = 0$
15. $3x + 8 = 4x + 2(2x + 3)$
16. $2.0 - 3.5 = 2.5 + 0.5(5a - 7a)$
17. $\frac{3p}{2} - 4(p - 2) = 3$
18. $R_x + \frac{2}{5} = \frac{4R_x}{5} + 1$
19. $10 = \frac{5}{9}(F - 32)$
20. $4 - (n - 4) = \frac{2n}{10} - 10$
21. $\frac{4}{5} - a = \frac{3a}{5}$
22. $\frac{2x}{3} - 6 = \frac{1}{3}$
23. $\frac{4w}{5} - \frac{7w}{10} = 10$
24. $\frac{V}{5} + \frac{V}{10} = 6.3$
25. $\frac{3a}{8} = 1.5 + \frac{a}{2}$
26. $\frac{P}{3.0} - 3.6 = \frac{P}{9.0}$
27. $\frac{d - 2}{2} - 2d = \frac{1}{2}$
28. $\frac{1}{5} - \frac{2g - 1}{10} = g$
29. $f - \frac{2f - 1}{3} = \frac{1}{12}$
30. $\frac{y + 7}{4} - \frac{4y + 5}{6} = 1$

Applied Problems

31. If you know that 50 mi/h = 80 km/h, you can calculate how many miles per hour 100 km/h is by solving the proportion:

$$\frac{80}{50} = \frac{100}{s}$$

Find the speed S (in mi/h) that 100 km/h is equal to.

32. Ann wants to donate 24 hours a week to three charities in the ratio of 4 to 3 to 1. To solve this problem, she needs to solve the equation:

$$4C + 3C + C = 24$$

where $4C$, $3C$, and C are the hours she works for each charity and C is the common factor. Find the value of C and how many hours she should work for each charity.

33. Bob wants to buy a television on sale for $990. The salesman tells him that the regular price was reduced 10% for the sale. Find the regular price p by solving the equation:

$$10\%p = p - 990$$

34. Jose wants to buy a computer, which sells for $1100 and is on sale for one-third off. Find the regular price P by solving the equation:

$$P - \frac{P}{3} = 1100$$

Applications to Electronics

35. An application of Ohm's law gives the following equation for the voltage V in a circuit:

$$12 - \frac{V}{10} = 9.9$$

Solve this equation for V in volts.

36. In Example 6.14, if the two resistors are 16 Ω and 20 Ω and the total current is 2.7 A, the voltage of the battery is given by the equation:

$$\frac{V}{16} + \frac{V}{20} = 2.7$$

Solve this equation for V in volts.

37. An application of Kirchhoff's voltage law gives the following equation for the resistance R in a series circuit:

$$24 = 0.80R + (0.80)(12) + 8.4$$

Solve this equation for R in ohms.

38. An application of Kirchhoff's voltage law gives the following equation for the current I in a series circuit:

$$6.2I + 8.2I - 23.6 = 0$$

Solve this equation for I in amps to two significant digits.

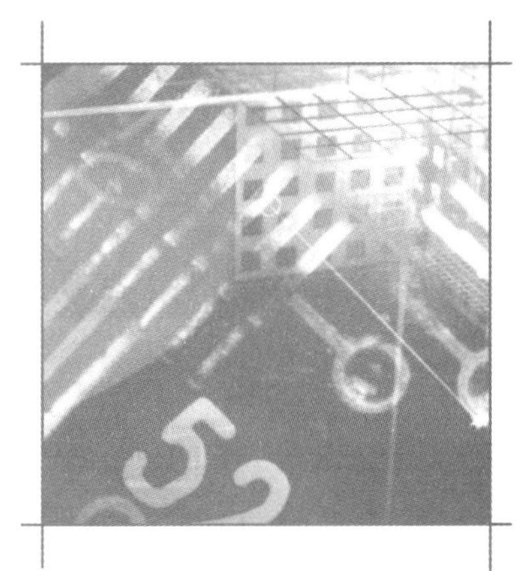

CHAPTER 7
Basics of DC Circuits

CHAPTER OBJECTIVES

In this chapter, you will learn:

- The basic concepts of dc circuits.
- Ohm's law and how to apply it.
- The meaning of direct proportion and inverse proportion.
- The concept of electric power.
- The power formula and how to derive all its forms.
- The concepts of energy and work.

A *direct current* or *dc* circuit is a circuit in which the electrons flow in one direction, from negative to positive, called electron current. Batteries are the most common source of direct current, but direct current can also be produced by a dc generator. The other type of current is alternating current. Alternating current changes direction with a certain frequency and is produced by an alternator or ac generator.

Direct current was the first type of current to be discovered. In 1828 George Simon Ohm, a German physicist, discovered one of the fundamental laws of electricity called Ohm's law. Ohm's law and another one of the basic relationships of dc circuits, the power formula, provide a sound basis for understanding all circuits. This chapter applies many of the mathematical concepts from the previous chapters to the basic theory of dc circuits.

• • •

7.1 DC Circuits and Ohm's Law

Consider the basic dc circuit in Figure 7-1 consisting of a dc voltage source V, such as a battery, connected to a resistance R, such as a light bulb or a heating element. Ohm's law explains the relationship between the current, voltage, and resistance in a dc circuit:

$I \rightarrow$, $-$, V, $+$, R

$I = \frac{V}{R}$, $R = \frac{V}{I}$, $V = IR$

FIGURE 7-1 Ohm's law and direct proportion for Examples 7.1, 7.2, and 7.3.

Formula **Ohm's Law for *I***

$$I = \frac{V}{R} \tag{7.1}$$

I = current in amps V = voltage in volts R = resistance in ohms

In Figure 7-1 the current is electron current and flows from (–) to (+). Electron flow is used throughout the text instead of conventional current, which flows from (+) to (–).

Ohm's law can be expressed using the voltage by multiplying both sides of formula (7.1) by R:

$$(R)I = \frac{V}{\cancel{R}}(\cancel{R})$$

Formula **Ohm's Law for *V***

$$V = IR \tag{7.1a}$$

Ohm's law can also be expressed using the resistance by dividing both sides of formula (7.1a) by I:

$$\frac{V}{(I)} = \frac{\cancel{I}R}{(\cancel{I})}$$

Formula **Ohm's Law for *R***

$$R = \frac{V}{I} \tag{7.1b}$$

An easy way to remember the three forms of Ohm's law is by the pie chart shown in Figure 7-2. The horizontal line across the middle is a fraction line. When you choose one of the letters, the arrangement of the other two letters represents what the chosen letter is equal to. Using the principles of equation solving, you should be able to derive any form of Ohm's law from any other form. Study the following example that illustrates how to apply Ohm's law.

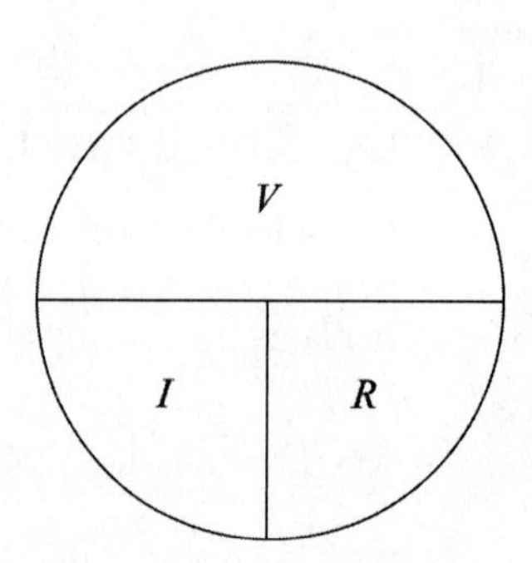

FIGURE 7-2 Pie diagram to remember the three forms of Ohm's law.

EXAMPLE 7.1 In the circuit in Figure 7-1:

(a) Find the current I in amps when $V = 24$ V and $R = 15\ \Omega$.

(b) Find the voltage V in volts and millivolts when $I = 15\ \mu$A and $R = 6.8$ kΩ.

(c) Find the resistance R in ohms and kilohms when $V = 110$ V and $I = 2.2$ mA.

Solution

(a) To find I, apply Ohm's law for I (7.1):

$$I = \frac{V}{R} = \frac{24\text{ V}}{15\ \Omega} = 1.6\text{ A}$$

(b) To find V, apply Ohm's law for V (7.1a). First change to the base units, amps and ohms, for I and R:

$$V = IR = (15\ \mu A)(6.8\ k\Omega) = (15 \times 10^{-6}\ A)(6.8 \times 10^{3}\ \Omega)$$

$$V = 102 \times 10^{-3}\ V = 102\ mV$$

The calculator solution is:

DAL: 15 [EE] [(-)] 6 [×] 6.8 [EE] 3 [=] → 0.102 V

Not DAL: 15 [EXP] 6 [+/-] [×] 6.8 [EXP] 3 [=] → 0.102 V

(c) Apply Ohm's law for R (7.1b). Change to the base units, volts and amps, for V and I:

$$R = \frac{V}{I} = \frac{110\ V}{2.2\ mA} = \frac{110\ V}{2.2 \times 10^{-3}\ A} = 50 \times 10^{3}\ \Omega = 50{,}000\ \Omega = 50\ k\Omega$$

The calculator solution is:

DAL: 110 [÷] 2.2 [EE] [(-)] 3 [=] → 50,000 Ω

Not DAL: 110 [÷] 2.2 [EXP] 3 [+/-] [=] → 50,000 Ω

ERROR BOX

A common error when applying Ohm's law is not to change all the units to the base units. If the units are not consistent, then the result will be incorrect. For example, if you are given $V = 150$ mV and $R = 10\ \Omega$, you must express V in volts before calculating the current:

$$I = \frac{0.150\ V}{10\ \Omega} = 0.015\ A = 15\ mA$$

Similarly if you are given $R = 7.5\ k\Omega$ and $I = 400\ \mu A$, you must express R in ohms and I in amps before calculating the voltage:

$$V = (7.5 \times 10^{3}\ \Omega)(400 \times 10^{-6}\ A) = 3.0\ V$$

Practice Problems: See if you can do the practice problems correctly. Apply Ohm's Law to find the indicated quantity:

1. $V = 120$ V, $R = 1.2$ kΩ; I
2. $V = 200$ mV, $R = 25$ Ω; I
3. $R = 330$ Ω, $I = 30$ mA; V
4. $R = 1.0$ kΩ, $I = 1.1$ A; V
5. $V = 12$ V, $I = 600$ μA; R
6. $V = 800$ mV, $I = 2.5$ A; R
7. $I = 35$ mA, $R = 2.0$ kΩ; V
8. $I = 700$ μA, $R = 3.0$ kΩ; V
9. $V = 2.1$ mV, $R = 300$ mΩ; I
10. $V = 750$ mV, $I = 250$ μA; R

It is important to understand the proportional changes in voltage, current, and resistance that are a consequence of Ohm's law. Study the following examples, which illustrate these proportional relationships.

EXAMPLE 7.2

For the circuit in Figure 7-1, given that the resistance is constant at $R = 100\ \Omega$, find the current I in amps and milliamps:

(a) When the voltage $V = 6.0$ V.
(b) When the voltage doubles to 12 V.
(c) Compare the current change and voltage change in (b).

Solution

(a) To find I when $V = 6.0$ V, apply Ohm's law for I (7.1):

$$I = \frac{6.0 \text{ V}}{100 \ \Omega} = 0.060 \text{ A} = 60 \text{ mA}$$

(b) When the voltage doubles to 12 V apply Ohm's law (7.1) again:

$$I = \frac{12 \text{ V}}{100 \ \Omega} = 0.120 \text{ A} = 120 \text{ mA}$$

(c) When the voltage doubles from 6.0 V to 12 V, the current also doubles from 60 mA to 120 mA. This is because Ohm's law tells us that when the resistance is constant, *the current is directly proportional to the voltage.* This means that the current and the voltage change by the same ratio or percentage. For example, if the voltage increases by 10%, then the current will increase by 10%. If the voltage decreases to half its value, then the current will decrease to half its value.

Definition

Direct Proportion

Two quantities are directly proportional if, when one changes by a certain ratio, the other changes by the same ratio.

EXAMPLE 7.3

For the circuit in Figure 7-1, the current is constant at $I = 100$ mA. Find the voltage drop V in volts across the resistance R:

(a) When $R = 100 \ \Omega$.

(b) When R increases by 10% to 110 Ω and the current remains the same.

(c) Find the percent increase in the voltage in (b) and compare with the resistance change.

Solution

(a) Apply Ohm's law for V (7.1a). Use base units for $I = 100$ mA $= 0.10$ A and $R =$ 100 Ω:

$$V = IR = (0.10 \text{ A})(100 \ \Omega) = 10 \text{ V}$$

(b) Apply Ohm's law for V (7.1a), letting $I = 0.10$ A and $R = 110 \ \Omega$:

$$V = (0.10 \text{ A})(110 \ \Omega) = 11 \text{ V}$$

(c) Applying the formula for percent increase (2.1) from Section 2.2, the percent increase in the voltage is:

$$\text{Percent Increase} = \frac{(\text{Final Value} - \text{Intial Value}) \times 100\%}{\text{Initial Value}}$$

$$= \frac{(11 - 10)(100\%)}{10} = \frac{(1)(100\%)}{10} = 10\%$$

Answers to Error Box Problems, page 149:

1. 100 mA **2.** 8.0 mA **3.** 9.9 V **4.** 1.1 kV **5.** 20 kΩ
6. 320 mΩ **7.** 70 V **8.** 2.1 V **9.** 7.0 mA **10.** 3.0 kΩ

The voltage increases in the same percentage or ratio as the resistance, 10%, when the current is constant. Therefore, the voltage is directly proportional to the resistance. Ohm's law tells us that when the current is constant, *the voltage drop across a resistance is directly proportional to the resistance.* The voltage drop across a resistance is the voltage, or potential difference, that exists between the ends of the resistance. For example, if the resistance decreases to one-half of its value, from 100 Ω to 50 Ω, and the current does not change, the voltage drop will decrease to one-half of its value from 10 V to 5 V.

Examples 7.2 and 7.3 lead to the following direct proportional relationships for Ohm's law:

Law

Ohm's Law Direct Proportional Relationships

1. The current is directly proportional to the voltage drop when the resistance is constant.
2. The voltage drop is directly proportional to the resistance when the current is constant.

The next example illustrates the inverse proportional property of Ohm's law.

EXAMPLE 7.4

Figure 7-3 shows a circuit containing a light bulb whose resistance is R_B, an ammeter to measure current, and a constant dc voltage source V_S = 24 V. Find the current I in milliamps through the light bulb:

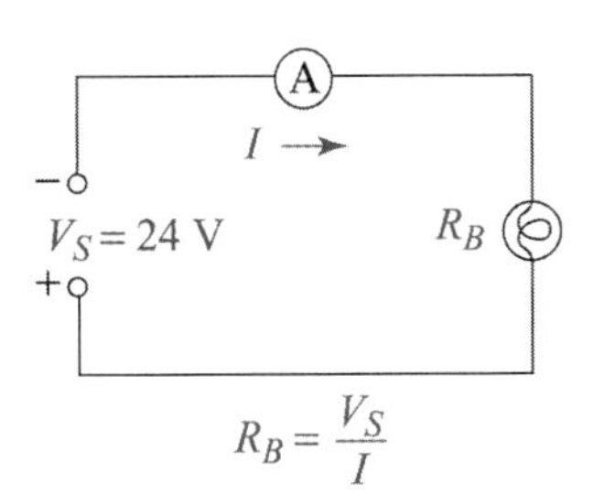

FIGURE 7-3 Ohm's law and inverse proportion for Examples 7.5 and 7.5.

(a) When R_B = 120 Ω.

(b) When R_B = 240 Ω.

(c) Compare the changes in the current and resistance in (b).

Solution

(a) Apply Ohm's law for I, (7.1), letting $V = V_S$ = 24 V and $R = R_B$ = 120 Ω:

$$I = \frac{V_S}{R_B} = \frac{24 \text{ V}}{120 \ \Omega} = 0.20 \text{ A} = 200 \text{ mA}$$

(b) When the resistance doubles to 240 Ω, the current is:

$$I = \frac{24 \text{ V}}{240 \ \Omega} = 0.10 \text{ A} = 100 \text{ mA}$$

(c) When the resistance doubles from 120 Ω to 240 Ω, the current decreases by one-half, from 200 mA to 100 mA. This is because Ohm's law tells us that when the voltage is constant, *the current is inversely proportional to the resistance.* This means that the current and resistance change in the inverse or reciprocal ratio. The resistance increases in the ratio of $\frac{2}{1}$, and the current decreases by the reciprocal ratio $\frac{1}{2}$. If the resistance were to increase three times in the ratio $\frac{3}{1}$, then the current would decrease to $\frac{1}{3}$ of its value.

Definition

Inverse Proportion

Two quantities are inversely proportional if, when one changes by a certain ratio, the other changes by the reciprocal ratio.

EXAMPLE 7.5 Suppose in Figure 7-3 that V_S is constant at 24 V and $I = 250$ mA. If I decreases by 20%, show that the current is inversely proportional to the resistance.

Solution When $V_S = 24$ V and $I = 250$ mA or 0.25 A, the resistance is:

$$R_B = \frac{24\text{ V}}{0.25\text{ A}} = 96\ \Omega$$

When I decreases by 20%, it becomes:

$$I = 250\text{ mA} - (0.20)(250\text{ mA}) = 200\text{ mA} = 0.20\text{ A}$$

The resistance is then:

$$R_B = \frac{24\text{ V}}{0.20\text{ A}} = 120\ \Omega$$

To show that the current changes are inversely proportional to the resistance, compare the ratio that the current changes to the ratio that the resistance changes. The current changes in the ratio:

$$\frac{200\text{ mA}}{250\text{ mA}} = \frac{4}{5}$$

while the resistance changes in the ratio:

$$\frac{120\ \Omega}{96\ \Omega} = \frac{5}{4}$$

The current ratio is the reciprocal of the resistance ratio. Therefore, the current is inversely proportional to the resistance.

Examples 7.4 and 7.5 lead to the following inverse proportional relationship for Ohm's law:

Rule **Ohm's Law Inverse Proportional Relationship**

3. The current is inversely proportional to the resistance when the voltage across the resistance is constant.

EXERCISE 7.1

Express answers to all exercises in engineering notation (numbers between 1 and 1000 with the appropriate units) to two significant digits.

In exercises 1 through 20, apply Ohm's law to find the indicated quantity.

1. $V = 100$ V, $R = 30\ \Omega$; I
2. $V = 9.0$ V, $R = 18\ \Omega$; I
3. $I = 2.5$ A, $R = 47\ \Omega$; V
4. $I = 50$ mA, $R = 750\ \Omega$; V
5. $V = 36$ V, $I = 4.8$ A; R
6. $V = 12$ V, $I = 100$ mA; R
7. $I = 200\ \mu$A, $R = 7.5$ kΩ; V
8. $I = 2.5$ mA, $R = 18$ kΩ; V
9. $V = 120$ V, $R = 36$ kΩ; I
10. $V = 24$ V, $R = 56$ kΩ; I
11. $I = 250$ mA, $R = 68\ \Omega$; V
12. $I = 80\ \mu$A, $R = 33$ kΩ; V
13. $V = 1.5$ kV, $I = 500$ mA; R
14. $V = 10$ kV, $I = 11$ A; R
15. $V = 3.3$ kV, $R = 1.5$ MΩ; I
16. $V = 900$ V, $R = 1.1$ MΩ; I
17. $V = 2.2$ V, $I = 90\ \mu$A; R
18. $V = 330$ mV, $I = 15\ \mu$A; R

19. V = 860 mV, I = 2.1 mA; R

20. V = 660 mV; I = 120 μA; R

21. A dc circuit contains a voltage source V_S and a constant resistor R = 56 Ω. Find the current I in amps:
(a) When V_S = 200 V.
(b) When V_S decreases to one-half of its value in (a).
(c) When V_S decreases by 10% of its value in (a).

22. A dc circuit contains a voltage source V_S and a constant resistor R = 200 Ω. Find the current I in milliamps:
(a) When V_S = 6.0 V.
(b) When V_S doubles in value from (a).
(c) When V_S increases by 50% of its value in (a).

23. In a dc circuit with a voltage source V_S and a resistor R, the current is constant at I = 200 mA. Find the voltage drop across the resistance:
(a) When R = 100 Ω.
(b) When R increases by 50% to 150 Ω.
(c) Show that the voltage drop also increases by 50% in (b). See Example 7.3.

24. In problem 23 find the voltage drop across the resistance:
(a) When R = 240 Ω.
(b) When R decreases by 50% to 120 Ω.
(c) Show that the voltage drop also decreases by 50% in (b). See Example 7.3.

25. Figure 7-4 shows a dc circuit with a constant battery voltage V_B and a resistor R. If V_B = 12 V, find the current I in milliamps:
(a) When R = 75 Ω.
(b) When R increases by one-third of its value in (a).
(c) When R increases by 100% of its value in (a).

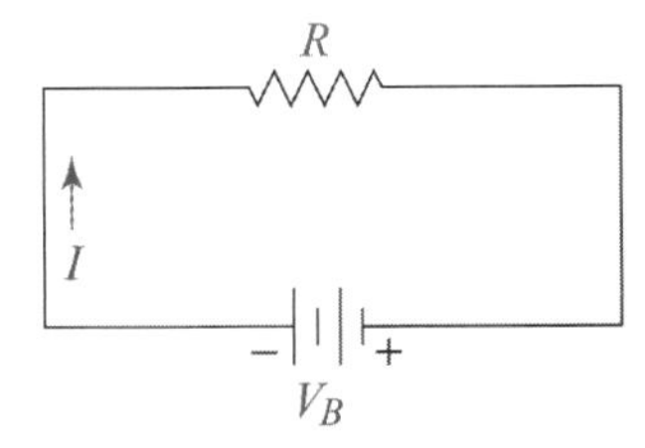

FIGURE 7-4 A dc circuit for problems 25 and 26.

26. In Figure 7-4, if V_B = 12 V, find the current I in milliamps:
(a) When R = 240 Ω.
(b) When R decreases to two-thirds of its value in (a).
(c) When R decreases by 50% of its value in (a).

27. In a dc circuit with a constant resistor, the current through the resistor I = 2.8 A. Apply the concept of direct proportion and find the current I in amps:
(a) When the voltage drop across the resistor doubles.
(b) When the voltage drop across the resistor increases by 25%.

28. In a dc circuit with a constant resistor, the current through the resistor I = 120 mA. Apply the concept of direct proportion and find the current I in milliamps:
(a) When the voltage drop across the resistor decreases to one-third of its value.
(b) When the voltage drop across the resistor decreases by 50%.

29. In a dc circuit with a constant voltage drop across a resistor R, the current through the resistor I = 50 mA. Apply the concept of inverse proportion and find the current I in milliamps:
(a) When R decreases to one-third of its value.
(b) When R decreases by 50%.

30. In a dc circuit with a constant voltage drop across a resistor R, the current through the resistor I = 3.0 A. Apply the concept of inverse proportion and find the current I in amps:
(a) When R triples in value.
(b) When R increases by 100%.

31. The resistance R of a wire is directly proportional to its length. If R = 40 mΩ for a wire 10 cm long, find the resistance R in milohms:
(a) When the length increases to 15 cm.
(b) When the length increases by 25%.

32. In problem 31, find the resistance R of the wire in milohms:
(a) When the length decreases to 6 cm.
(b) When the length decreases by 25%.

33. The resistance in a dc circuit is constant, and the current through the resistance is 300 mA. Find the percent that the voltage drop across the resistance changes:
(a) When the current through the resistance increases to 450 mA.
(b) When the current through the resistance decreases to 270 mA.

34. The resistance in a dc circuit is constant, and the voltage drop across the resistance is 3.0 V. Find the percent that the current through the resistance changes:
(a) When the voltage drop across the resistance increases to 6.0 V.
(b) When the voltage drop across the resistance decreases to 1.5 V.

35. The voltage drop across a variable resistance is constant. Find the ratio that the resistance changes:
(a) When the current through the resistance decreases to one-half of its value.
(b) When the current increases by 50%.

36. In a dc circuit, the voltage drop across a variable resistance is constant at V_R = 3 V. The current through the resistance I = 40 mA. If the current decreases by 50% when the resistance increases, show that the current is inversely proportional to the resistance. See Example 7.5.

7.2 Electric Power

Electric power is defined as the rate of using electrical energy. One watt of electric power is equal to the electrical energy produced every second by a voltage of one volt and a current of one ampere. This gives rise to the following power formula:

Formula

Power Formula

$$P = VI \tag{7.2}$$

P = power in watts V = voltage in volts I = current in amps

Formula 7.2 can be solved for I or V:

Formula

Power Formula Solved for *I*

$$I = \frac{P}{V} \tag{7.2a}$$

• • •

Formula

Power Formula Solved for *V*

$$V = \frac{P}{I} \tag{7.2b}$$

EXAMPLE 7.6

Given a dc circuit where V = source voltage, I = total current, and P = total power.

(a) Find P in watts when $I = 1.5$ A and $V = 20$ V.
(b) Find I in milliamps when $V = 100$ V and $P = 40$ W.
(c) Find V in kilovolts when $I = 20$ mA and $P = 32$ W.

Solution

(a) Apply the power formula (7.2):

$$P = VI = (20 \text{ V})(1.5 \text{ A}) = 30 \text{ W}$$

(b) Apply the power formula for I (7.2a):

$$I = \frac{P}{V} = \frac{40 \text{ W}}{100 \text{ V}} = 0.40 \text{ A} = 400 \text{ mA}$$

(c) Apply the power formula for V (7.2b). The units must be consistent in watts and amps. Use $I = 20 \text{ mA} = 0.020 \text{ A}$:

$$V = \frac{P}{I} = \frac{32 \text{ W}}{0.020 \text{ A}} = 1600 \text{ V} = 1.60 \text{ kV}$$

The power formula, like Ohm's law, contains proportional relationships as shown in the next example.

EXAMPLE 7.7

Figure 7-5 shows an electric heater connected to a 200 V power line. The heater is rated at 500 W on the low setting and 1000 W on the high setting.

(a) How much current does it use on each setting?
(b) Compare the changes in current and power.

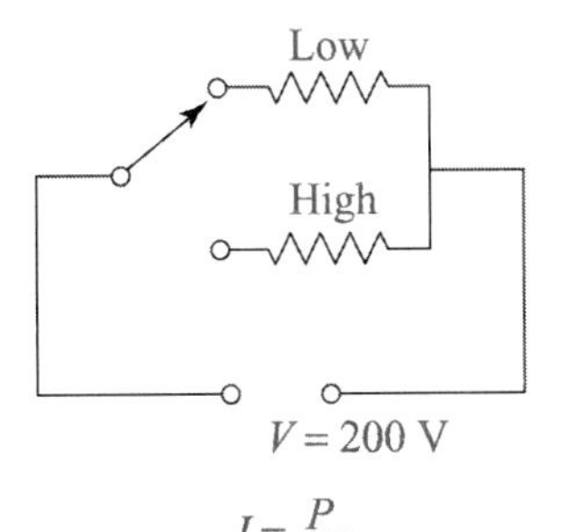

$I = \frac{P}{V}$

FIGURE 7-5 Current in an electric heater for Example 7.7.

Solution

(a) Apply the power formula for I (7.2a):

$$\text{Low: } I = \frac{P}{V} = \frac{500 \text{ W}}{200 \text{ V}} = 2.5 \text{ A}$$

$$\text{High: } I = \frac{P}{V} = \frac{1000 \text{ W}}{200 \text{ V}} = 5.0 \text{ A}$$

(b) The current doubles, from 2.5 A to 5.0 A, when the power doubles, from 500 W to 1000 W. The current and the power are directly proportional.

Example 7.7 demonstrates one of the three proportional relationships of the power law:

Rules

Power Formula Proportional Relationships

1. The power is directly proportional to the current when the voltage is constant.
2. The power is directly proportional to the voltage when the current is constant.
3. The current is inversely proportional to the voltage when the power is constant.

An example of inverse proportion in the power formula is the following: A 60 W bulb in a 120 V circuit will draw twice as much current as a 60 W bulb in a 240 V circuit. Here the voltage doubles, and the current decreases by one-half when the power is the same.

When Ohm's law is combined with the power formula, two more formulas for power arise in terms of resistance:

Formulas

Power in Terms of Resistance

$$P = I^2R \quad \textbf{(7.3a)}$$

$$P = \frac{V^2}{R} \quad \textbf{(7.3b)}$$

Formula (7.3a) is derived by using Ohm's law $V = IR$ and substituting IR for V in the power formula (7.2):

$$P = VI = (IR)I = I^2R$$

The derivation of formula (7.3b) is left as problem 45 in Exercise 7.2.

EXAMPLE 7.8

The current through a resistor $R = 560\ \Omega$ is $I = 150$ mA. Find the power dissipated in the resistor, in watts, and the voltage across the resistor, in volts, to two significant digits.

Solution Apply (7.3a) to find the power:

$$P = I^2R = (150 \times 10^{-3})^2(560) \approx 13 \text{ W}$$

Apply Ohm's law to find the voltage across the resistor:

$$V = IR = (150 \times 10^{-3})(560) = 84 \text{ V}$$

Formulas (7.3a) and (7.3b) can be solved for the other variables and expressed in two other ways:

Formulas

Resistance in Terms of Power

$$R = \frac{P}{I^2} \tag{7.4a}$$

$$R = \frac{V^2}{P} \tag{7.4b}$$

• • •

Formulas

Current and Voltage in Terms of Power

$$I = \sqrt{\frac{P}{R}} \tag{7.5a}$$

$$V = \sqrt{PR} \tag{7.5b}$$

It is not necessary to memorize all of the above formulas. If you apply some basic algebra, you can derive formulas (7.4a) and (7.5a) from formula (7.3a), and formulas (7.4b) and (7.5b) from formula (7.3b). The next example shows one of these derivations.

EXAMPLE 7.9 Derive formulas (7.4a) and (7.5a) from formula (7.3a).

Solution Divide both sides of formula (7.3a), $P = I^2R$ by I^2:

$$\frac{P}{(I^2)} = \frac{\cancel{I^2}R}{\cancel{(I^2)}}$$

You then obtain formula (7.4a):

$$R = \frac{P}{I^2}$$

Then solve this formula for I^2 to obtain:

$$I^2 = \frac{P}{R}$$

Take the positive square root of both sides of this formula. When you take a square root, you consider both the positive and negative square roots. In this case, you assume the current is positive, and it is only necessary to take the positive square root:

$$\sqrt{I^2} = \sqrt{\frac{P}{R}}$$

The square root of a squared quantity is just the quantity itself. That is, the square root cancels the square on the left side, and you have formula (7.5a):

$$I = \sqrt{\frac{P}{R}}$$

The derivation of formulas (7.4b) and (7.5b) is left as problem 46 in Exercise 7.2. The next two examples show how to apply these formulas.

EXAMPLE 7.10 Figure 7-6 shows a dc circuit containing a battery and a bulb whose resistance $R = 75\ \Omega$. If the power consumption of the bulb is 12 W:

(a) Find the current I in milliamps.

(b) Find the voltage V of the battery in volts.

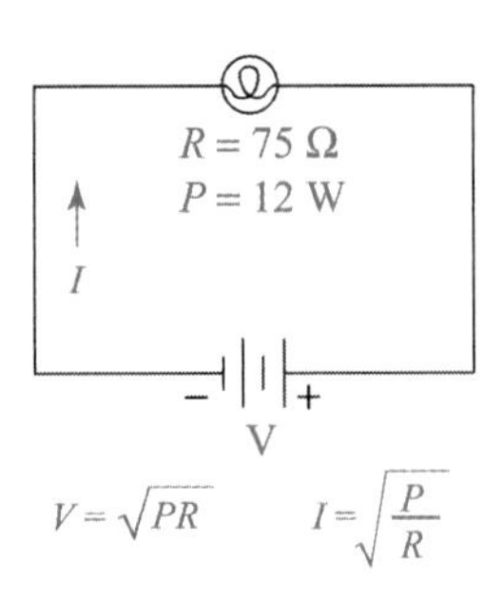

FIGURE 7-6 Power and current in a circuit for Example 7.10.

Solution

(a) Find the current in the circuit by applying formula (7.5a) with $R = 75\ \Omega$ and $P = 12$ W:

$$I = \sqrt{\frac{P}{R}} = \sqrt{\frac{12}{75}} = 0.40 \text{ A} = 400 \text{ mA}$$

(b) You can find the voltage of the battery by using formula (7.5b):

$$V = \sqrt{PR} = \sqrt{(12)(75)} = 30 \text{ V}$$

or by using Ohm's law and the current in (a):

$$V = IR = (0.40)(75) = 30 \text{ V}$$

ERROR BOX

You do not need to memorize all nine ways of writing the power formula. You can solve any problem if you know Ohm's law, the basic power formula $P = VI$, and one of the power formulas in terms of resistance, (7.3a) or (7.3b):

$$P = I^2R \quad \text{or} \quad P = \frac{V^2}{R}$$

You may have to do some algebra or an additional calculation, but there is less to remember and less chance of making an error. Here are two examples:

1. Suppose you are given $P = 30$ W and $I = 500$ mA. Using formula (7.3a) above, you can find R and V as follows. First substitute and solve for R:

$$30 = (0.5)^2(R)$$

$$R = \frac{30}{(0.5)^2} = 120\ \Omega$$

Then find V using Ohm's law:

$$V = IR = (0.5 \text{ A})(120\ \Omega) = 60 \text{ V}$$

2. Consider Example 7.10. Given $R = 75\ \Omega$, $P = 12$ W, and using formula (7.3b) above, you can find I and V as follows. First substitute and solve for V:

$$12 = \frac{V^2}{75}$$

$$V^2 = (12)(75) = 900$$

$$V = \sqrt{900} = 30 \text{ V}$$

Then find I using Ohm's law:

$$I = \frac{V}{R} = \frac{30 \text{ V}}{75\ \Omega} = 0.40 \text{ A} = 400 \text{ mA}$$

Practice Problems: Using formula (7.3a) or (7.3b) above, Ohm's law, and the power formula $P = VI$, find the indicated quantities for each to *two significant digits:*

1. $P = 48$ W, $R = 27\ \Omega$; I, V
2. $P = 200$ W, $R = 68\ \Omega$; I, V
3. $P = 5$ W, $I = 20$ mA; R, V
4. $P = 20$ W, $I = 1.5$ A; R, V
5. $P = 1.5$ kW, $V = 200$ V; R, I
6. $P = 4.5$ W, $V = 12$ V; R, I
7. $P = 250$ mW, $R = 680$ mΩ; I, V
8. $P = 3.0$ kW, $R = 3.3$ kΩ; I, V

Power and Energy

Power is defined as the amount of energy consumed per second:

Definition **Power**

$$\text{Power} = \frac{\text{Energy}}{\text{Time}}$$

From this definition, it follows that:

Formula **Energy**

$$\text{Energy} = (\text{Power})(\text{Time}) = PT$$

All mechanical, electrical, or heat energy is measured in joules (J) where:

Formula **Joule**

$$1 \text{ J} = (1 \text{ W})(1 \text{ s}) = 1 \text{ W} \cdot \text{s (watt-second)}$$

A familiar unit of electrical energy, the *kilowatt-hour* (kWh), is used to measure electrical energy consumption. One kilowatt-hour is equal to 1000 W of power consumed for 1 hour:

Formula **Kilowatt-hour**

$$1 \text{ kWh} = (1000 \text{ W})(1 \text{ h}) = 1000 \text{ W} \cdot \text{h}$$

One kilowatt-hour is also equal to an equivalent product of watts and hours that equals 1000, such as:

$$1 \text{ kWh} = (200 \text{ W})(5 \text{ h}) \quad \text{or} \quad 1 \text{ kWh} = (100 \text{ W})(10 \text{ h})$$

The relationship between joules and kilowatt-hours is:

$$1 \text{ J} = 2.778 \times 10^{-7} \text{ kWh}$$
$$1 \text{ kWh} = 3.600 \times 10^{6} \text{ J}$$

EXAMPLE 7.11 A heating coil in a dc circuit dissipates 1.5 kW of power when the current through the coil is 10 A.

(a) Find the resistance of the coil in ohms and the voltage drop in volts across the coil.

(b) Find the energy consumption of the heating coil for 8.0 h in kilowatt-hours and joules.

Answers to Error Box Problems, page 157:

1. 1.3 A, 36 V **2.** 1.7 A, 120 V **3.** 13 kΩ, 250 V **4.** 8.9 Ω, 13 V
5. 27 Ω, 7.5 A **6.** 32 Ω, 380 mA **7.** 61 mA, 410 mV **8.** 3.1 kV, 950 mA

Solution

(a) The formulas require that the units be consistent. Find the resistance by using (7.4a). Let $P = 1.5$ kW = 1500 W:

$$R = \frac{P}{I^2} = \frac{1500}{(10)^2} = 15\ \Omega$$

Now find the voltage using formula (7.5b):

$$V = \sqrt{PR} = \sqrt{(1500)(15)} = \sqrt{22{,}500} = 150\ \text{V}$$

You can also use $V = \frac{P}{I}$ or Ohm's law to find the voltage.

(b) The energy consumption for $T = 8.0$ h is:

$$\text{Energy} = PT = (1.5\ \text{kW})(8\ \text{h}) = 12\ \text{kWh}$$

The energy consumption in joules to two significant digits is:

$$12\ \cancel{\text{kWh}}\left(\frac{3.600 \times 10^6\ \text{J}}{1\ \cancel{\text{kWh}}}\right) \approx 43 \times 10^6\ \text{J} = 43\ \text{MJ}$$

EXERCISE 7.2

Express answers to all exercises in engineering notation (numbers between 1 and 1000 with the appropriate units) to two significant digits.

In exercises 1 through 30, given the values shown, use the power formulas and Ohm's law to find the indicated quantities.

1. $V = 11$ V, $I = 500$ mA; P, R
2. $V = 120$ V, $I = 1.5$ A; P, R
3. $P = 12$ W, $V = 8.0$ V; I, R
4. $P = 6.5$ W, $V = 10$ V; I, R
5. $P = 20$ W, $I = 1.5$ A; V, R
6. $P = 15$ W, $I = 200$ mA; V, R
7. $V = 12$ V, $R = 10\ \Omega$; P, I
8. $V = 20$ V, $R = 100\ \Omega$; P, I
9. $I = 200$ mA, $R = 160\ \Omega$; P, V
10. $I = 400$ mA, $R = 68\ \Omega$; P, V
11. $V = 32$ V, $P = 100$ W; R, I
12. $V = 9.0$ V, $P = 10$ W; R, I
13. $I = 800$ mA, $P = 14$ W; R, V
14. $I = 1.6$ A, $P = 1.0$ kW; R, V
15. $P = 1.2$ kW, $R = 300\ \Omega$; I, V
16. $P = 6.0$ W, $R = 20\ \Omega$; I, V
17. $P = 25$ W, $R = 56\ \Omega$; V, I
18. $P = 10$ W, $R = 33\ \Omega$; V, I
19. $V = 14$ V, $I = 120\ \mu$A; P, R
20. $V = 1.5$ V, $P = 110$ mW; R, I
21. $I = 45$ mA, $R = 1.5$ kΩ; P, V
22. $V = 720$ mV, $R = 27\ \Omega$; P, I
23. $P = 360$ W, $I = 1.5$ A; R, V
24. $P = 15$ mW, $I = 25$ mA; R, V
25. $P = 55$ mW, $R = 150\ \Omega$; I, V
26. $P = 250$ mW, $R = 3.3$ kΩ; V, I
27. $V = 5.5$ kV, $R = 9.1$ kΩ; P, I
28. $V = 240$ V, $R = 15\ \Omega$; P, I
29. $I = 2.6$ A, $R = 1.8\ \Omega$; P, V
30. $I = 750$ mA, $R = 220\ \Omega$; P, V
31. An electric stove connected to a 240-V power line is rated at 400 W on the lowest setting and 1200 W on the highest setting.
    (a) How much current does the stove use on each setting?
    (b) Find the resistance of the stove for each setting.
32. A radar unit on a boat draws 4.2 A when operating and 1.2 A when standing by. If the voltage for the unit is 12 V:
    (a) What is the power consumption when operating and when standing by?
    (b) Find the resistance of the unit for each setting.
33. An automobile lightbulb uses 440 mA in a dc circuit. How much current will the bulb use:
    (a) If the voltage is constant and the wattage of the bulb is doubled?
    (b) If the voltage is constant and the wattage of the bulb decreases by 25%?
    (c) If the voltage doubles and the wattage of the bulb is the same?

34. A 60-W bulb is replaced by a smaller wattage bulb in a constant voltage dc circuit. What should be the wattage of the smaller bulb:

(a) To decrease the current by one-third?
(b) To decrease the current by 50%?

35. The voltage in a high-power circuit is measured at 3.5 kV and the current at 4.3 A.

(a) What is the power in the circuit in kilowatts?
(b) What is the resistance of the circuit?

36. A video circuit draws 100 mA and dissipates 20 W of power.

(a) What is the voltage across the circuit?
(b) What is the resistance of the circuit?

37. In an amplifier circuit, a 7.5-Ω resistor draws 150 mA of current.

(a) How much power does the resistor dissipate?
(b) What is the voltage drop across the resistor?

38. The voltage drop over a high-voltage power line is 50 V, and the power dissipated is 4.4 kW.

(a) Find the total resistance of the power line.
(b) Find the current in the line.

39. Figure 7-7 shows a circuit containing two resistors connected in parallel where *the voltage drop across each resistor equals the voltage (V) across the circuit.* If $R_1 = 27\ \Omega$, $R_2 = 36\ \Omega$, and $V = 9.2$ V, find the power dissipated by each resistor and the total power dissipated.

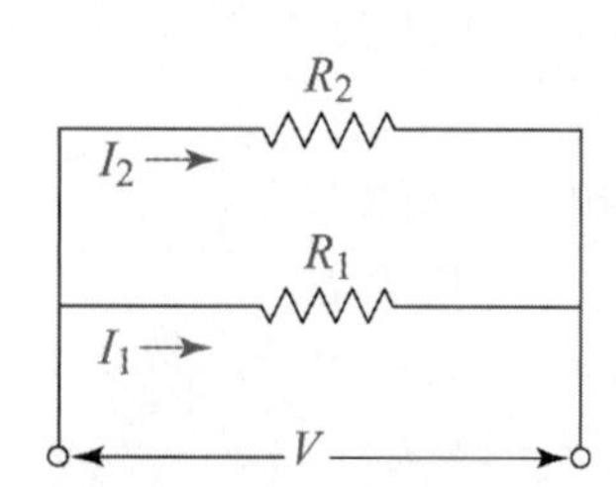

FIGURE 7-7 Power dissipation in a parallel circuit for problems 39 and 40.

40. In the parallel circuit in Figure 7-7, the currents through the resistors are $I_1 = 85$ mA and $I_2 = 58$ mA. If $R_1 = 150\ \Omega$ and $R_2 = 220\ \Omega$, find the power dissipated by each resistor and the total power dissipated.

41. A circuit whose total resistance is 130 Ω uses 50 W of power.

(a) Find the voltage across the circuit.
(b) Find the energy consumption in kilowatt-hours and joules for 24 h.

42. A 430-Ω resistor dissipates 1.5 kW of power.

(a) Find the current through the resistor.
(b) Find the energy consumption in kilowatt-hours and joules for 12 h.

43. Figure 7-8 shows a series circuit containing two resistors R_1 and R_2. The current in the circuit $I = 330$ mA. If $R_1 = 3.3$ kΩ and $R_2 = 1.5$ kΩ, find the power dissipated in each resistor and the total power dissipated.

Note: The current through each resistor equals the current in the circuit.

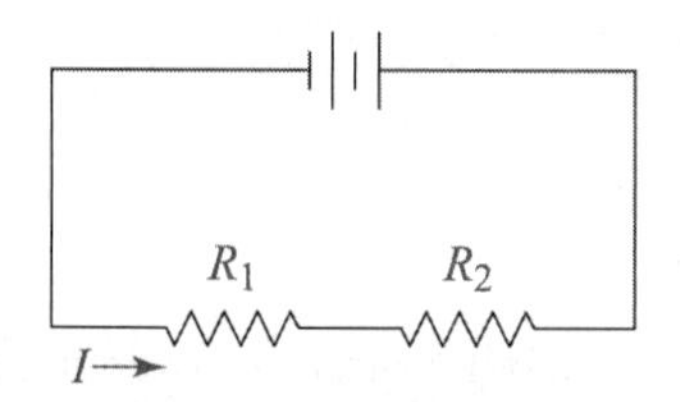

FIGURE 7-8 Power dissipation in a series circuit for problems 43 and 44.

44. In Figure 7-8, find the power dissipated in each resistor and the total power dissipated when $I = 900\ \mu$A, $R_1 = 22$ kΩ, and $R_2 = 11$ kΩ.

45. Using Ohm's law and the power formula, derive formula (7.3b):

$$P = \frac{V^2}{R}$$

46. Using formula (7.3b) in problem 45, derive formulas (7.4b) and (7.5b):

$$R = \frac{V^2}{P}$$

$$V = \sqrt{PR}$$

CHAPTER HIGHLIGHTS

7.1 OHM'S LAW

A direct current circuit is a circuit where the electrons flow in one direction, from negative to positive (electron current). Ohm's law can be expressed three ways:

Ohm's Law for *I*

$$I = \frac{V}{R} \quad (7.1)$$

• • •

Ohm's Law for *V*

$$V = IR \quad (7.1a)$$

• • •

Ohm's Law for *R*

$$R = \frac{V}{I} \quad (7.1b)$$

I = current in amps V = voltage in volts
R = resistance in ohms

An easy way to remember the three forms of Ohm's law is by using the pie chart shown in Figure 7-2.

Key Example: When $V = 240$ V and $R = 1.1$ kΩ:

$$I = \frac{V}{R} = \frac{240 \text{ V}}{1100 \ \Omega} \approx 220 \text{ mA}$$

Direct Proportion

Two quantities are directly proportional if, when one changes by a certain ratio, the other changes by the same ratio.

• • •

Ohm's Law Direct Proportional Relationships

1. The current is directly proportional to the voltage drop when the resistance is constant.
2. The voltage drop is directly proportional to the resistance when the current is constant.

Key Example:

(a) The voltage across a constant resistor is 3.0 V, and the current through the resistor is 15 mA. If the voltage triples to 9.0 V, the current will triple to 45 mA.

(b) The voltage across a variable resistor is 3.0 V when the resistor is 220 Ω. If the resistance decreases by 50% to 110 Ω and the current does not change, then the voltage across the resistor will decrease by 50% to 1.5 V.

Inverse Proportion

Two quantities are inversely proportional if, when one changes by a certain ratio, the other changes by the reciprocal ratio.

• • •

Ohm's Law Inverse Proportional Relationship

3. The current is inversely proportional to the resistance when the voltage across the resistance is constant.

Key Example:

(a) If the resistance doubles and the voltage is constant, the current decreases by half and vice versa.

(b) The voltage drop across a resistance of 100 Ω is constant. If the resistance increases by 100% to 200 Ω, then the resistance increases in the ratio: $\frac{200}{100} = \frac{2}{1}$. The current will then decrease by the reciprocal ratio $= \frac{1}{2}$.

7.2 ELECTRIC POWER

Power Formula

$$P = VI \tag{7.2}$$

P = power in watts V = voltage in volts
I = current in amps

• • •

Power Formula Solved for *I*

$$I = \frac{P}{V} \tag{7.2a}$$

• • •

Power Formula Solved for *V*

$$V = \frac{P}{I} \tag{7.2b}$$

Key Example: When $V = 10$ kV and $I = 850$ mA, the power consumption is:

$$P = (10{,}000 \text{ V})(0.850 \text{ A}) = 8500 \text{ W} = 8.5 \text{ kW}$$

Power Formula Proportional Relationships

1. The power is directly proportional to the current when the voltage is constant.
2. The power is directly proportional to the voltage when the current is constant.
3. The current is inversely proportional to the voltage when the power is constant.

Key Example:

(a) If the voltage across a resistance is constant and the current doubles, then the power dissipated will double.

(b) If the current through a resistance is constant and the voltage decreases by 50%, then the power dissipated will decrease by 50%.

(c) If the power dissipated in a resistance is constant and the voltage doubles, then the current will decrease by one-half.

Power in Terms of Resistance

$$P = I^2 R \tag{7.3a}$$

$$P = \frac{V^2}{R} \tag{7.3b}$$

• • •

Resistance in Terms of Power

$$R = \frac{P}{I^2} \tag{7.4a}$$

$$R = \frac{V^2}{P} \quad (7.4b)$$

• • •

Current and Voltage in Terms of Power

$$I = \sqrt{\frac{P}{R}} \quad (7.5a)$$

$$V = \sqrt{PR} \quad (7.5b)$$

The units for all formulas must be consistent: watts for P, volts for V, amps for I, and ohms for R.

Key Example:

(a) Given $R = 1.2\ \text{k}\Omega$ and $V = 70$ V. The power is:

$$P = \frac{(70\ \text{V})^2}{1200\ \Omega} \approx 4.1\ \text{W}$$

(b) Given $P = 60$ W and $R = 150\ \Omega$. The current is:

$$I = \sqrt{\frac{60\ \text{W}}{150\ \Omega}} \approx 0.63\ \text{A} = 630\ \text{mA}$$

It is not necessary to memorize all of the formulas (7.1) to (7.5). You can derive any formula from the two formulas, $P = VI$ and $V = IR$, and solve any problem using fewer formulas. See the Error Box in Section 7.2 which illustrates these ideas.

Power, Energy, and Work

Energy

$$\text{Energy} = (\text{Power})(\text{Time}) = PT$$

• • •

Joule

$$1\ \text{J} = (1\ \text{W})(1\ \text{s}) = 1\ \text{W} \cdot \text{s (watt-second)}$$

• • •

Kilowatt-hour

$$1\ \text{kWh} = 1000\ \text{W} \cdot \text{h} = 3.6 \times 10^6\ \text{J}$$

Key Example: The voltage across a dc circuit is 18 V, and the current through the circuit is 6 A. The energy consumption for 24 h is:

$$\text{Energy} = (\text{Power})(\text{Time}) = (VI)T = (18\ \text{V})(6\ \text{A})(24\ \text{h}) \approx 2.59\ \text{kWh} = (2.59)(3.6 \times 10^6) \approx 9.3\ \text{MJ}$$

REVIEW EXERCISES

Express answers to all exercises in engineering notation (numbers between 1 and 1000 with the appropriate units) to two significant digits.

In exercises 1 through 16, apply Ohm's law and the power formulas to find the indicated quantities.

1. $V = 36$ V, $R = 120\ \Omega$; I, P
2. $V = 1.4$ kV, $R = 510\ \Omega$; I, P
3. $I = 100$ mA, $V = 9.0$ V; R, P
4. $I = 2.6$ A, $V = 110$ V; R, P
5. $I = 2.2$ A, $R = 82\ \Omega$; V, P
6. $I = 34$ mA, $R = 1.3\ \text{k}\Omega$; V, P
7. $V = 600$ mV, $I = 7.5$ mA; R, P
8. $V = 2.2$ kV, $R = 1.2\ \text{k}\Omega$: I, P
9. $P = 100$ W; $V = 80$ V; I, R
10. $P = 4.5$ W; $V = 5.5$ V; I, R
11. $P = 15$ mW; $I = 450\ \mu$A; V, R
12. $P = 7.0$ kW, $I = 15$ A; V, R
13. $P = 6.6$ kW, $R = 750\ \Omega$; V, I
14. $P = 12$ W, $R = 7.5\ \text{k}\Omega$; V, I
15. $P = 900\ \mu$W, $I = 2.5$ mA; V, R
16. $P = 370$ mW, $I = 50$ mA; V, R

Applications to Electronics

17. A dc circuit contains a voltage source V_S and a constant resistor $R = 150\ \Omega$; find the current I:
    (a) When the voltage $V_S = 12$ V.
    (b) When V_S decreases by half.
    (c) What is the decrease in I when V_S decreases by half? Why?
18. A dc circuit contains a voltage source V_S and a constant resistor $R = 36\ \Omega$; find the current I:
    (a) When the voltage $V_S = 18$ V.
    (b) When V_S increases by 50%.
    (c) What is the percent increase in I when V_S increases by 50%? Why?
19. Given the dc circuit in Figure 7-9 with a constant voltage $V_S = 30$ V, find the current I:
    (a) When $R = 200\ \Omega$.
    (b) When R decreases by 50%.
    (c) By what percentage does I increase when R decreases by 50%? Why?

20. Given the dc circuit in Figure 7-9 with a constant voltage $V_S = 12$ V, find the current I:
(a) When $R = 20\ \Omega$.
(b) When R doubles.
(c) By what ratio does I change when R doubles? Why?

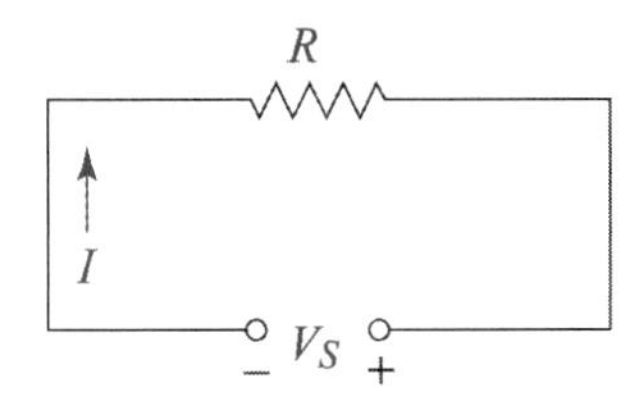

FIGURE 7-9 A dc circuit for problems 19 through 22.

21. Given the dc circuit in Figure 7-9 with $I = 1.2$ A, find V_S:
(a) When $R = 7.5\ \Omega$.
(b) When R decreases by half and the current does not change.

22. Given the circuit in Figure 7-9 with $I = 200$ mA, find I:
(a) When the voltage increases by 50% and the resistance remains constant.
(b) When the resistance doubles and the voltage remains constant.

23. The *change* in the resistance R of a tungsten bulb filament is directly proportional to the *change* in the temperature. If $R = 10\ \Omega$ at 20°C and 12 Ω at 30°C, what will be the resistance at 60°C?

24. The capacitance C of a parallel plate capacitor is *inversely proportional* to the distance d between the plates. If $C = 10\ \mu\text{F}$ when $d = 1.0$ mm, what will be the capacitance when $d = 0.50$ mm?

25. A 1.2-kΩ resistor dissipates 50 W of power.
(a) Find the current through the resistor.
(b) Find the energy consumption in kilowatt-hours and joules for 72 h.

26. A high-voltage circuit draws 560 μA and uses 100 mW of power.
(a) What is the voltage across the circuit and the resistance of the circuit?
(b) If the current doubles and the voltage remains constant, how much power does the circuit use?

27. The voltage drop over a long-distance, high-voltage power line is 40 V and the power dissipated is 2.4 kW.
(a) Find the total resistance of the power line.
(b) Find the current in the line.

28. A series circuit contains two resistors, R_1 and R_2. The current through each resistor $I = 110$ mA. If $R_1 = 4.7\ \text{k}\Omega$ and $R_2 = 1.3\ \text{k}\Omega$:
(a) Find the power dissipated in each resistor.
(b) Find the total power P_T dissipated through both resistors.

29. An electric heater is rated at 500 W on the low setting and 1500 W on the high setting. If the voltage across the heater is constant, what is the ratio of the high current to the low current?

30. A 100-W bulb is replaced by a larger wattage bulb in a constant voltage dc circuit. What should be the wattage of the larger bulb:
(a) To increase the current by 100%?
(b) To increase the current one and one-half times?

CHAPTER 8

Multiplying and Factoring Polynomials

CHAPTER OBJECTIVES

In this chapter, you will learn:

- The meaning of a monomial, binomial, trinomial, and polynomial.
- How to multiply two binomials.
- How to multiply a trinomial by a binomial.
- How to separate a monomial factor from a polynomial.
- How to factor a polynomial into two binomials.
- The meaning of a second-degree or quadratic equation.
- How to solve a second-degree equation by factoring and taking roots.
- How to solve a second-degree equation by the quadratic formula.

An algebraic expression containing two or more terms is called a polynomial. Polynomials are a basic part of equations and formulas in electricity and electronics. This chapter builds upon the ideas of basic algebra and shows you how to multiply and factor polynomials and to identify special cases. Factoring is an important technique that is used to simplify algebraic expressions, solve equations, and manipulate formulas. It is one of the methods used to solve some second-degree equations, which have important applications in ac circuits. The quadratic formula, which can be used to solve any second-degree equation, is derived using factoring methods and is also studied in this chapter.

• • •

8.1 Multiplying Polynomials

A *monomial* is an expression containing one algebraic term, and a *binomial* is an expression containing two algebraic terms. Some examples of binomials are:

$$3x + 1 \qquad V_1 - V_2 \qquad 9a^2 - \frac{b^2}{4} \qquad 3.2P + 6.4Q$$

Binomials that have like terms are called similar binomials. Two examples of similar binomials are:

$$(3x - 1) \text{ and } (x + 2) \qquad (5IR + 4V) \text{ and } (IR - 3V)$$

EXAMPLE 8.1

Multiply the similar binomials:

$$(x + 2)(3x - 1)$$

Solution When the distributive law (1.3) in chapter 1 is applied twice to this product, it shows that *each term in the first binomial has to be multiplied by each term* in the second binomial. This means there are four products. You do the products using the "FOIL" method, which is an acronym for the order in which the products are done: **F**irst, **O**utside, **I**nside, and **L**ast.

First **O**utside **I**nside **L**ast

$$(x + 2)(3x - 1) = (x)(3x) + (x)(-1) + (2)(3x) + (2)(-1)$$

$$= 3x^2 - x + 6x - 2 = 3x^2 + 5x - 2$$

(Diagram labels: O, F, I, L)

Notice that the outside product, $-x$, and the inside product, $6x$, combine to give the middle term, $5x$. This happens when the binomials are similar and the result is a polynomial containing three terms called a *trinomial.*

EXAMPLE 8.2 Multiply the similar binomials:

$$(4n - 3)(3n + 2)$$

Solution Multiply by the FOIL method and combine the outer and inner products to produce the middle term of the trinomial:

$$(4n - 3)(3n + 2) = \overset{F}{12n^2} + \overset{O}{8n} - \overset{I}{9n} - \overset{L}{6} = 12n^2 - n - 6$$

EXAMPLE 8.3 Multiply the similar binomials:

$$(2V - 3R)(3V - 2R)$$

Solution

$$(2V - 3R)(3V - 2R) = \overset{F}{6V^2} - \overset{O}{4VR} - \overset{I}{9VR} + \overset{L}{6R^2} = 6V^2 - 13VR + 6R^2$$

EXAMPLE 8.4 Multiply the similar binomials:

$$(X + Y)(X - Y)$$

Solution Notice that these binomials are almost the same but differ only by the middle sign. When you multiply this special combination, the outside and inside products add to zero and there is no middle term:

$$(X + Y)(X - Y) = X^2 - \cancel{XY} + \cancel{XY} - Y^2 = X^2 - Y^2$$

This result is called the *difference of two squares* and is an important product used to simplify more complex expressions.

Formula **Difference of Two Squares**

$$(X + Y)(X - Y) = X^2 - Y^2 \quad \textbf{(8.1)}$$

EXAMPLE 8.5 Multiply the similar binomials:

$$(4P + 5)(4P - 5)$$

Solution Since the binomials differ only by the middle sign, you can apply formula (8.1). Square the first and last terms and place a minus sign between the terms:

$$(4P + 5)(4P - 5) = (4P)^2 - (5)^2 = 16P^2 - 25$$

EXAMPLE 8.6 Multiply the identical binomials:

$$(A + B)(A + B)$$

Solution The outside and the inside products are the same, resulting in a *middle term, which is twice the product of the two terms* in the binomial:

$$(A + B)(A + B) = A^2 + AB + AB + B^2 = A^2 + 2AB + B^2$$

Since these binomials are the same, this product is called a binomial square.

Formula **Binomial Square**

$$(A + B)(A + B) = (A + B)^2 = A^2 + 2AB + B^2 \tag{8.2}$$

EXAMPLE 8.7 Square the binomial:

$$(2R - T)^2$$

Solution Apply formula (8.2). Square the first and last terms and multiply the product of the terms by 2 to obtain the middle term:

$$(2R - T)^2 = (2R)^2 + 2(2R)(-T) + (-T)^2 = 4R^2 - 4RT + T^2$$

The binomial square should not be confused with the difference of two squares. The binomial square is a trinomial, whereas the difference of two squares is a binomial.

EXAMPLE 8.8 Multiply the binomial by the trinomial:

$$(Z - 2) \text{ and } (Z^2 + 3Z + 1)$$

Solution To multiply a binomial by a trinomial, you can use a method similar to long multiplication in arithmetic. Set up the binomial below the trinomial and multiply each term in the binomial by each term in the trinomial. There are six products. Place them in two rows *lining up the similar terms:*

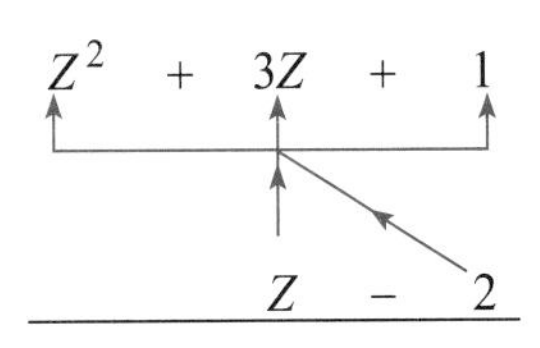

Multiply −2 by $(Z^2 + 3Z + 1)$:	$- 2Z^2 - 6Z - 2$
Multiply Z by $(Z^2 + 3Z + 1)$:	$Z^3 + 3Z^2 + Z$
Add similar terms:	$Z^3 + Z^2 - 5Z - 2$

EXERCISE 8.1

In exercises 1 through 30, multiply out each of the binomial products. Try to do some of the simpler ones mentally.

1. $(x + 2)(x + 4)$
2. $(a - 3)(a - 4)$
3. $(C - 1)(C + 4)$
4. $(R + 1)(R - 5)$
5. $(I + 10)(I - 10)$
6. $(m - 5)(m + 5)$
7. $(2V - 5)(V + 2)$
8. $(3P + 1)(P - 3)$
9. $(3x + 4)(x + 1)$
10. $(y - 6)(2y - 4)$
11. $(t - 4)^2$
12. $(G - 3)^2$

13. $(2P+4)(2P-5)$
14. $(5f-g)(2f-g)$
15. $(V_1+3)(V_1-3)$
16. $(2R_2+1)(2R_2-1)$
17. $(2X-R)(3X-2R)$
18. $(4a+7b)(a-5b)$
19. $(3x+3)^2$
20. $(2k+2)^2$
21. $(5L+2C)(5L-2C)$
22. $(d+2e)(d-2e)$
23. $(4y+3)(4y-3)$
24. $(5s+6t)(5s-6t)$
25. $3(a-b)(a+b)$
26. $5(I+R)(I-R)$
27. $(2.0p+1.5q)(0.4p+0.2q)$
28. $(1.2x+0.5y)(1.2x+3.0y)$
29. $2(C+L)^2$
30. $3(X-2)^2$

In exercises 31 through 36, multiply the binomial by the trinomial. See Example 8.8.

31. $(E+1)(E^2-2E+1)$
32. $(c-d)(c^2+2cd+d^2)$
33. $(L-2)(L^2-L+3)$
34. $(Q+1)(2Q^2+Q-4)$
35. $(T-1)(T^2+T+1)$
36. $(x+y)(x^2-xy+y^2)$

Applied Problems

37. The length of a rectangular circuit component is given by $(n+3)$ and the width by $(n-2)$. Write the expression for the area of the component and multiply out the expression. Area = (length)(width).

38. The width of a rectangular computer cabinet is given by $(2w-3)$ and the length by $(2w+3)$. Write the expression for the area of the cabinet and multiply out the expression. Area = (length)(width).

39. The length of the side of a square window is given by $(l-5)$. Write the expression for the area of the window and multiply out the expression. Area = $(\text{side})^2$.

40. The radius of a circle is given by $r = a + 1$. Write the expression for the area of the circle and multiply out the expression. Area of a circle = πr^2.

Applications to Electronics

41. The current in a circuit is given by $I = t + 4$ and the resistance by $R = 2t + 2$, where t = time. Using Ohm's law, $V = IR$, write the expression for the voltage in the circuit and multiply out the expression.

42. The current in a circuit is given by $I = 2t - 3$ and the voltage by $V = 3t + 1$, where t = time. Using the power formula, $P = VI$, write the expression for the power in the circuit and multiply out the expression.

43. In the relationship between the voltage and the temperature T of a transmission line, the algebraic expression that arises is:

$$(1+0.03T)(T-24)$$

Multiply out this expression.

44. The current in a circuit is given by $I = t + 4$ and the resistance $R = 10\ \Omega$. Using the power law, $P = I^2R$, write the expression for the power in the circuit and multiply out the expression.

8.2 Monomial Factors

To factor means to separate a number or algebraic expression into its divisors. Factoring is the reverse of multiplication and is a very important technique used in algebra to simplify expressions. Some numbers and their smallest, or prime, factors are:

$$8 = 2\cdot 2\cdot 2 = 2^3$$
$$28 = 2\cdot 2\cdot 7 = 2^2\cdot 7$$
$$36 = 2\cdot 2\cdot 3\cdot 3 = 2^2\cdot 3^2$$
$$210 = 2\cdot 3\cdot 5\cdot 7$$

EXAMPLE 8.9 Show the prime factors of the four terms:

$$6xy,\ -15pqr,\ 21a^2b^2,\ 60I_1^2R_1$$

Solution Note carefully how many prime factors each term has:

Monomial	Prime Factors	Number of Prime Factors
$6xy$	$2\cdot 3\cdot x\cdot y$	4
$-15pqr$	$-3\cdot 5\cdot p\cdot q\cdot r$	5
$21a^2b^2$	$3\cdot 7\cdot a\cdot a\cdot b\cdot b$	6
$60I_1^2R_1$	$2\cdot 2\cdot 3\cdot 5\cdot I_1\cdot I_1\cdot R_1$	7

The following example shows how to factor two or more terms by separating a common factor.

EXAMPLE 8.10 Separate the common factors in the terms of the binomial:

$$30x - 42$$

Solution The prime factors of each term are $30x = 2 \cdot 3 \cdot 5 \cdot x$ and $42 = 2 \cdot 3 \cdot 7$. Both terms have the common prime factors of $2 \cdot 3 = 6$ which is the *greatest common factor*. Separate the greatest common factor and obtain the other factor by dividing each term by 6:

$$30x - 42 = (6)\left(\frac{30x}{(6)} - \frac{42}{(6)}\right) = 6(5x - 7)$$

To check the factoring, multiply out the parentheses and see if you get back the original polynomial.

EXAMPLE 8.11 Separate the greatest common factor:

$$x^3 + 3x^2$$

Solution The prime factors of each term are $x^3 = x \cdot x \cdot x$ and $3x^2 = 3 \cdot x \cdot x$. The greatest common factor is $x \cdot x = x^2$, which is the highest power of x that divides into each term. Separate x^2 and divide each term by x^2 to obtain the other factor:

$$x^3 + 3x^2 = (x^2)\left(\frac{x^3}{(x^2)} + \frac{3x^2}{(x^2)}\right) = x^2(x + 3)$$

EXAMPLE 8.12 Separate the greatest common factor:

$$20P^3 - 15P^2 + 5P$$

Solution The prime factors of each term are:

$$\begin{aligned} 20P^3 &= 2 \cdot 2 \cdot \ \ \cdot 5 \cdot P \cdot P \cdot P \\ 15P^2 &= \qquad 3 \cdot 5 \cdot P \cdot P \\ 5P &= \qquad\quad 5 \cdot P \end{aligned}$$

The greatest common factor is $5P$. Divide $5P$ into each term to obtain the terms in the other factor:

$$20P^3 - 15P^2 + 5P = (5P)\left(\frac{20P^3}{(5P)} - \frac{15P^2}{(5P)} + \frac{5P}{(5P)}\right) = 5P(4P^2 - 3P + 1)$$

Notice that, when the entire term is the common factor, you must place a 1 in the parentheses so that when you multiply back you will get the original expression.

EXAMPLE 8.13 Separate the greatest common factor:

$$0.25IR_1 - 0.50IR_2$$

Solution The prime factors of each term are:

$$\begin{aligned} 0.25IR_1 &= 0.5 \cdot 0.5 \cdot I \cdot R_1 \\ 0.50IR_2 &= 0.5 \qquad \cdot I \cdot R_2 \end{aligned}$$

Note that R_1 is a different factor from R_2. The subscript is part of the variable and cannot be separated from it. The greatest common factor is $0.5I$, which divides into both terms as follows:

$$(0.5I)\left(\frac{0.25IR_1}{(0.5I)} - \frac{0.50IR_2}{(0.5I)}\right) = 0.5I(0.5R_1 - R_2)$$

EXAMPLE 8.14 Separate the greatest common factor:

$$24a^3b - 30a^2b^2 + 42ab^3$$

Solution The prime factors are:

$$\begin{aligned} 24a^3b &= 2 \cdot 2 \cdot 2 \cdot 3 \cdot a \cdot a \cdot a \cdot b \\ 30a^2b^2 &= 2 \cdot 3 \cdot 5 \cdot a \cdot a \cdot b \cdot b \\ 42ab^3 &= 2 \cdot 3 \cdot 7 \cdot a \cdot b \cdot b \cdot b \end{aligned}$$

The greatest common factor is $2 \cdot 3 \cdot a \cdot b = 6ab$. Divide each term by $6ab$ to obtain the other factor:

$$\begin{aligned} 24a^3b - 30a^2b^2 + 42ab^3 &= (6ab)\left(\frac{24a^3b}{(6ab)} - \frac{-30a^2b^2}{(6ab)} + \frac{42ab^3}{(6ab)}\right) \\ &= 6ab(4a^2 - 5ab + 7b^2) \end{aligned}$$

You should always check by multiplying back and see if you get the original expression.

EXERCISE 8.2

In exercises 1 through 12, show all the prime factors of each monomial.

1. 36
2. 150
3. $28xy^2$
4. $30n^2p$
5. $-68a^2b^2$
6. $-100x^2y$
7. $-39R_1R_2$
8. $57V^2I^2$
9. $50I_1^2R_1$
10. $12L_2C_2^2$
11. $\frac{6}{x}$
12. $\frac{8}{y}$

In exercises 13 through 40, factor each expression by separating the greatest common factor.

13. $5R + 15$
14. $12V - 3$
15. $3x - ax$
16. $10y + by$
17. $a^2 - 2a$
18. $2x^2 + x$
19. $3p^3 + p^2$
20. $f^3 - 10f^2$
21. $0.5VI_1 - 0.25VI_2$
22. $1.2I_1R + 1.6I_2R$
23. $0.2rs^2 + 0.4rt^2$
24. $1.5ab - 2.0ax$
25. $6.6XY^2 + 3.3XY$
26. $1.5wL - 2.5wL^2$
27. $4I^2R_1 + 2I^2R_2$
28. $pq^2 - p^2q$
29. $\frac{12a}{b} + \frac{16}{b}$
30. $\frac{x}{25} - \frac{x^2}{25}$
31. $\frac{4P_1}{I} + \frac{8P_2}{I}$
32. $\frac{10V_1}{R} - \frac{5V_2}{R}$
33. $10s^2 + 20s - 15$
34. $12m^2 - 18m + 12$
35. $24v^3 - 16v^2 + 12v$
36. $20h^4 + 50h^3 - 10h^2$
37. $24a^3b - 30a^2b^2 + 42ab^3$
38. $36cd^3 + 18c^2d^2 + 27c^3d$
39. $x^2yz + xy^2z + xyz^2$
40. $R^2S^2T - R^2ST^2 - RS^2T^2$

Applied Problems

41. The volume of a box is given by:

$$4w^3 + 28w^2 + 40w$$

Factor this expression.

42. The volume of a segment of a sphere is given by:

$$\pi rh^2 - \pi h^3$$

Factor this expression.

Applications to Electronics

43. The electric power in one circuit is given by $56I_1^2R$ and in a second circuit by $42I_2^2R$. Factor the expression that represents the total power consumed in both circuits.

44. The electric power dissipated in one resistor is given by $\frac{50V^2}{R_1}$ and in a second resistor by $\frac{25V^2}{R_2}$. Factor the expression that represents the total power dissipated in both resistors.

45. In a circuit containing a resistor R in series with a capacitor C, the current at a certain time after the switch is closed is given by:

$$I = \frac{Ve^{-1}}{R} - \frac{V_0e^{-1}}{R}$$

Factor this expression by separating the greatest common factor.

46. In a circuit containing a resistor R in series with an inductor L, the current at a certain time after the switch is closed is given by:

$$I = \frac{V}{R} - \frac{Ve^{-1}}{R}$$

Factor this expression by separating the greatest common factor.

8.3 Binomial Factors

When you multiply two similar binomials you get a trinomial, except for the special case of the difference of two squares (Example 8.4), which gives you a binomial. To factor the difference of two squares or a trinomial, you need to do the reverse of the FOIL method. This is done by trying different pairs of binomials, which contain possible factors, until the correct combination is obtained.

Difference of Two Squares

EXAMPLE 8.15 Factor the binomials:

(a) $4a^2 - 49$ **(b)** $25P_1^2 - 36P_2^2$

Solution Check that both terms are perfect squares, and there is a minus sign in between. Then apply formula (8.1) from right to left and factor into the sum and difference of the square roots:

(a) $4a^2 - 49 = (2a^2) - (7)^2 = (2a + 7)(2a - 7)$

(b) $25P_1^2 - 36P_2^2 = (5P_1)^2 - (6P_2)^2 = (5P_1 + 6P_2)(5P_1 - 6P_2)$

EXAMPLE 8.16 Factor completely into primes:

$$2\pi f^2 - 2\pi g^2$$

Solution First, separate the greatest common factor, 2π, to simplify the expression. You will then see that the other factor is the difference of two squares, and you can factor again:

$$2\pi f^2 - 2\pi g^2 = 2\pi(f^2 - g^2)$$

$$2\pi(f^2 - g^2) = 2\pi(f + g)(f - g)$$

The answer contains four prime factors: 2, π, $(f + g)$, and $(f - g)$.

Trinomials with Leading Coefficient = 1

To factor a trinomial into two binomials, you apply the FOIL method backward. Study Table 8-1, which shows the four basic sign combinations for the binomial factors of a trinomial with leading coefficient = 1.

Observe in Table 8-1 that the *sum* of the numbers in the factors is the coefficient of the middle term and the *product* of the numbers is the last term. When the last term is positive, the numbers have like signs equal to the sign of the middle term. When the last term is negative, the numbers have unlike signs and the sign of the middle term equals the sign of the number whose absolute value is greater.

TABLE 8-1 BINOMIAL FACTORS, LEADING COEFFICIENT = 1

	Sum Product	
Positive Last Term	$x^2 - 5x \oplus 6 = (x-2)(x-3)$ $x^2 + 5x \oplus 6 = (x+2)(x+3)$	**Like Signs**
Negative Last Term	$x^2 - 5x \ominus 6 = (x-6)(x+1)$ $x^2 + 5x \ominus 6 = (x+6)(x-1)$	**Unlike Signs**

EXAMPLE 8.17 Factor completely into primes:

$$a^2 + 9a + 18$$

Solution Consider the signs first. They are all positive, which means that the numbers in the factors are both positive and the binomials have the form:

$$(a + \quad)(a + \quad)$$

The product is 18 and the sum is 9. The possible factors of 18 are (1)(18), (2)(9), and (3)(6). The last combination has a sum of 9. Therefore, the factors are:

$$(a + 3)(a + 6)$$

EXAMPLE 8.18 Factor completely into primes:

$$L^2 + 4L - 12$$

Solution The last term is negative, which means the numbers have unlike signs and the binomial factors have the form:

$$(L + \quad)(L - \quad)$$

The product is −12, and the sum is + 4. The possible factors of 12 are (1)(12), (2)(6), and (3)(4). Try different pairs of factors with opposite signs until you find a sum equal to +4. The factors (+6) and (−2) have a sum of +4, and therefore the factors are:

$$L^2 + 4L - 12 = (L + 6)(L - 2)$$

EXAMPLE 8.19 Factor completely into primes:

$$f^2 - 12f - 45$$

Solution The last term is negative, which means the numbers have unlike signs and the binomial factors have the form:

$$(f + \quad)(f - \quad)$$

The possible factors of 45 are (1)(45), (3)(15), and (5)(9). The middle term is negative. Try different pairs of factors with opposite signs until you find a sum equal to −12. The factors (+3) and (−15) have a sum of −12 and therefore the factors are:

$$f^2 - 12f - 45 = (f + 3)(f - 15)$$

ERROR BOX

When factoring expressions, it is necessary to multiply back to check that the factors are correct. Some expressions cannot be factored and are prime. For example, the difference of two squares, $A^2 - B^2$, *can* be factored. The sum of two squares, $A^2 + B^2$, *cannot* be factored. You may think that the factors of $A^2 + B^2$ are $(A + B)(A + B)$, but they are not. If you multiply out this product, you will see that it has a middle term, $2AB$:

$$(A + B)(A + B) = A^2 + AB + AB + B^2 = A^2 + 2AB + B^2$$

See if you can factor the following problems correctly and tell which ones are prime.

Practice Problems: Factor each expression, or tell if it is prime.

1. $V^2 + R^2$ **2.** $V^2 - R^2$ **3.** $V^2 + 2VR + R^2$
4. $V^2 - 2R^2$ **5.** $V^2 - 4R^2$ **6.** $V^2 + 4R^2$
7. $V^2 - 2VR + R^2$ **8.** $4V^2 - R^2$ **9.** $4V^2 + 4VR + R^2$
10. $4V^2 - 4R^2$

Trinomials with Leading Coefficient ≠ 1

EXAMPLE 8.20 Factor completely into primes:

$$2y^2 - 13y + 15$$

Solution You need to consider all three terms. The last term is positive, and the middle term is negative. The binomial factors therefore have the form:

$$(\ -\ )(\ -\ )$$

The factors of the first term $2y^2$ are: $(y)(2y)$. Possible factors of the last term 15 are: (1)(15) and (3)(5). Try combinations of the possible factors. There are four combinations. Multiply out each combination, using the FOIL method, until you get the correct factors:

$$(y - 1)(2y - 15) = 2y^2 - 15y - 2y + 15 = 6y^2 - 17y + 15$$
$$(y - 15)(2y - 1) = 2y^2 - y - 30y + 15 = 2y^2 - 31y + 15$$
$$(y - 3)(2y - 5) = 2y^2 - 5y - 6y + 15 = 2y^2 - 11y + 15$$
$$(y - 5)(2y - 3) = 2y^2 - 3y - 10y + 15 = 2y^2 - 13y + 15 \quad \text{Correct}$$

Try multiplying and adding the outer and inner products mentally, and you will be able to find the correct factors more quickly.

EXAMPLE 8.21 Factor completely into primes:

$$4V^2 - 4V + 1$$

Solution The first and last term are perfect squares, which means the trinomial could be a perfect square like formula (8.2) with equal factors. Try squaring the binomial whose terms are the square roots of the first and last terms of the trinomial, and you will obtain the factors:

$$(2V - 1)^2 = (2V - 1)(2V - 1) = 4V^2 - 4V + 1$$

EXAMPLE 8.22 Factor completely into primes:

$$6RT^2 - 9RT - 15R$$

Solution Always look for a greatest common factor first. In this trinomial, the greatest common factor is $3R$, which can be separated first:

$$3R(2T^2 - 3T - 5)$$

Then factor the trinomial into two binomials. The last sign is negative, so the binomials have unlike signs. The only possible factors of $2T^2$ are $(2T)(T)$, and the only possible factors of 5 are (1)(5).

There are four combinations of binomials with unlike signs:

$(T+1)(2T-5) = 2T^2 - 5T + 2T - 5 = 2T^2 - 3T - 5$ Correct
$(T-1)(2T+5) = 2T^2 + 5T - 2T - 5 = 2T^2 + 3T - 5$
$(T+5)(2T-1) = 2T^2 - T + 10T - 5 = 2T^2 + 9T - 5$
$(T-5)(2T+1) = 2T^2 + T - 10T - 5 = 2T^2 - 9T - 5$

The first combination is correct and the complete answer has four prime factors:

$$6RT^2 - 9RT - 15R = 3R(T+1)(2T-5)$$

EXERCISE 8.3

In exercises 1 through 42, factor each expression completely into primes or tell if it is prime. Separate any common factors first.

1. $a^2 - 4b^2$
2. $16x^2 - 25y^2$
3. $X^2 - 0.16$
4. $0.04 - f^2$
5. $0.25X^2 - Y^2$
6. $1 - 0.36A^2$
7. $Z^2 + 6Z + 5$
8. $d^2 + 3d + 2$
9. $v^2 - 4v + 3$
10. $E^2 - 8E + 12$
11. $C_1^2 + 6C_1 + 9$
12. $V_T^2 + 16V_T + 64$
13. $p^2 + q^2$
14. $c^2 + 4d^2$
15. $A^2 - 8A + 16$
16. $4B^2 - 4B + 1$
17. $n^2 - 6n - 7$
18. $p^2 - 4p - 12$
19. $50m^2 - 200$
20. $3h^2 - 27$
21. $G^2 + 4G + 1$
22. $2f^2 + 7f + 3$
23. $4e^2 - 5e + 1$
24. $3Q^2 - 10Q + 3$
25. $2x^2 - 5x - 3$
26. $2y^2 - 13y - 7$
27. $5x^2 + xy - 6y^2$
28. $8a^2 - 10ab - 3b^2$
29. $M^2 - 2M - 1$
30. $d^2 + d + 1$
31. $8R^2 + 28R - 16$
32. $6t^2 + 4t - 10$
33. $2x^2 - 8y^2$
34. $5R^2 - 80$
35. $10PV^2 + 15PV - 25P$
36. $2uv^2 + 12uv + 10u$
37. $S^2 - 0.6S + 0.09$
38. $V_0^2 + V_0 + 0.25$
39. $L^4 + 2L^2 + 1$
40. $I^4 - 4I^2 + 4$
41. $\frac{T^2}{2} - \frac{25}{2}$
42. $\frac{2m^2}{n^2} - 8$

Applied Problems

43. The height of a rocket is given by $h = 4t^2 + 28t + 48$, where t = time. Factor this expression into primes.

44. The distance s that a body travels is given by $s = 2t^3 - 3t^2 - 2t$, where t = time. Factor this expression into primes.

Applications to Electronics

45. The current in a circuit is given by: $I = 0.25t^2 - 1.0t + 1.0$, where t = time. Factor this expression into primes.

46. The voltage in a circuit is given by: $V = 4t^2 - 4$, where t = time. Factor this expression into primes.

47. In a circuit, the power dissipated by one resistor is: $P_1 = I_1^2R$. The power dissipated by another resistor is: $P_2 = I_2^2R$. Factor the expression $P_1 - P_2$ that represents the difference in power dissipated by the two resistors.

48. In a circuit, the power dissipated by one resistor is: $P_1 = \frac{V_1^2}{R}$. The power dissipated by another resistor is: $P_2 = \frac{V_2^2}{R}$. Factor the expression $P_1 - P_2$ that represents the difference in power dissipated by the two resistors.

Answers to Error Box Problems, page 173:
1. prime **2.** $(V+R)(V-R)$ **3.** $(V+R)^2$ **4.** prime **5.** $(V-2R)(V+2R)$ **6.** prime **7.** $(V-R)^2$ **8.** prime **9.** $(2V+R)^2$ **10.** $4(V-R)(V+R)$

49. A variable with a subscript such as V_1 is written in computer language as V1. Change the following computer language expression to algebraic notation and factor the expression:

X^2*Y – X0^2*Y

50. As in problem 49, change the computer language expression to algebraic notation and factor the expression:

K^2*T^2 – 2*K*T*T1 + T1^2

8.4 Second-Degree Equations

A *second-degree* or *quadratic* equation is an equation that contains a second-degree term and no higher-degree term. The general form for any second-degree equation in x is:

Equation

General Second-Degree (Quadratic) Equation

$$ax^2 + bx + c = 0 \ (a, b, c \text{ constants}) \qquad (8.3)$$

The terms in the general second-degree equation (8.3) must be written in order: second-degree term ax^2, first-degree term bx, and the constant term c. Examples of some second-degree equations and their constants a, b, and c are:

$$x^2 + 2x - 8 = 0 \quad a = 1, b = 2, c = -8$$
$$2R^2 - 3R + 1 = 0 \quad a = 2, b = -3, c = 1$$
$$3V_0^2 - 5V_0 - 4 = 0 \quad a = 3, b = -5, c = -4$$
$$I^2 - 25 = 0 \quad a = 1, b = 0, c = -25$$
$$p^2 + 2p = 0 \quad a = 1, b = 2, c = 0$$

Notice in the last two equations that when the first-degree term is missing, $b = 0$, and when the constant term is missing, $c = 0$. There are several methods for solving second-degree equations, depending on the values of a, b, and c. Three methods are shown here: taking square roots, factoring, and the quadratic formula.

Solution by Taking Square Roots

EXAMPLE 8.23 Solve the quadratic equation to three significant digits:

$$2x^2 - 24 = 0$$

Solution When the first-degree term is missing ($b = 0$), you can solve the equation by isolating x^2 and taking the square root of both sides, which eliminates the square:

Transpose 24: $2x^2 = 24$

Divide by 2: $x^2 = 12$

Take the square root of both sides: $\sqrt{x^2} = \pm\sqrt{12}$

$$x \approx \pm 3.46$$

Since every positive number x has two square roots, $+\sqrt{x}$ and $-\sqrt{x}$, both square roots need to be taken and are written together as $\pm\sqrt{x}$. Note that *every second-degree equation has two solutions.*

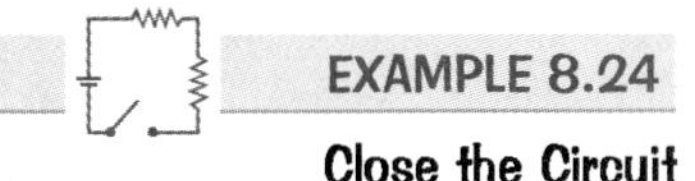

EXAMPLE 8.24

Close the Circuit

In an ac circuit containing a resistance R and a reactance X, the relationship between the resistance, reactance, and the impedance Z is given by the pythagorean relation:

$$Z^2 = R^2 + X^2$$

If $Z = 6.0$ kΩ and $X = 5.0$ kΩ, find R in ohms and kilohms to two significant digits.

Solution First substitute the given values in ohms in the formula:

$$(6.0 \times 10^3)^2 = R^2 + (5.0 \times 10^3)^2$$

Square the values: $36 \times 10^6 = R^2 + 25 \times 10^6$

Solve for R^2: $R^2 = 36 \times 10^6 - 25 \times 10^6$

$$R^2 = 11 \times 10^6$$

Since R is positive, the negative root does not apply. Take only the positive square root of both sides of the equation:

$$R = \sqrt{11 \times 10^6} \approx 3.3 \times 10^3\ \Omega = 3300\ \Omega = 3.3\ \text{k}\Omega$$

Solution by Factoring

EXAMPLE 8.25 Solve the quadratic equation:

$$w^2 + 2w = 48$$

Solution To solve by factoring, the equation must be in the form of the general second-degree equation (9.3) with all the terms except zero on the left side:

Transpose 48: $w^2 + 2w - 48 = 0$

Factor the left side: $(w + 8)(w - 6) = 0$

Each factor can now be set equal to zero because the product of two numbers can only equal zero when one or both of the numbers equals zero.

Set each factor to zero: $w + 8 = 0$ and $w - 6 = 0$

Solve each first degree equation: $w = -8$ $w = 6$

Note that both answers satisfy the original equation $w^2 + 2w = 48$:

$$w = -8 \rightarrow (-8)^2 + 2(-8) = 64 - 16 = 48$$
$$w = 6 \rightarrow (6)^2 + 2(6) = 36 + 12 = 48$$

EXAMPLE 8.26 Solve the quadratic equation:

$$2I^2 - 50 = 0$$

Solution When each term in an equation has a common factor, you can simplify the equation by dividing each term by the common factor:

Divide each term by 2: $\frac{2I^2}{2} - \frac{50}{2} = \frac{0}{2}$

$$I^2 - 25 = 0$$

Factor the left side: $(I + 5)(I - 5) = 0$

Set each factor equal to zero: $I + 5 = 0$ and $I - 5 = 0$

Solve each first-degree equation: $I = -5$ $I = 5$

This equation can also be solved by isolating I^2 and taking square roots.

EXAMPLE 8.27 Solve the quadratic equation:

$$2x^2 + 5x - 3 = 4x + 12$$

Solution First put the equation in the general form (8.3):

Transpose $4x$ and 12: $2x^2 + 5x - 3 - 4x - 12 = 0$

Combine terms: $2x^2 + x - 15 = 0$

Factor the left side: $(2x - 5)(x + 3) = 0$

Set each factor equal to zero: $2x - 5 = 0$ and $x + 3 = 0$

Solve the first-degree equations: $x = \frac{5}{2}$ or 2.5 $\quad x = -3$

EXAMPLE 8.28 Solve the quadratic equation:

$$4V^2 = 25V$$

Solution Transpose $25V$ to put it in general form (8.3):

$$4V^2 - 25V = 0$$

When the constant term $c = 0$, the equation can be solved by separating the unknown as a common factor.

Separate the common factor V: $V(4V - 25) = 0$

Set each factor equal to zero: $V = 0$ and $4V - 25 = 0$

Solve each equation: $V = 0 \quad V = \frac{25}{4} = 6.25$

Notice that one of the solutions is zero, which is an acceptable answer.

Solution by the Quadratic Formula

The quadratic formula can be used to solve any quadratic equation and is necessary when a quadratic equation cannot be solved by factoring or by taking square roots. Given the general quadratic equation (8.3):

$$ax^2 + bx + c = 0$$

The two solutions for x are given by the quadratic formula:

Formula

Quadratic Formula

$$x = \frac{-b \pm \sqrt{b^2 - 4ac}}{2a} \tag{8.4}$$

The useful and important quadratic formula (8.4) comes from the general quadratic equation by changing it to a perfect square (completing the square) and solving for x by taking square roots.

EXAMPLE 8.29 Solve the quadratic equation by the quadratic formula:

$$5x^2 + 8x = 6x + 3$$

Solution To use the quadratic formula, you must first put the equation in the general form (8.3) by moving all the nonzero terms to one side in order:

Transpose $6x$ and 3: $$5x^2 + 8x - 6x - 3 = 0$$

Simplify: $$5x^2 + 2x - 3 = 0$$

Write down the quadratic formula and the values of a, b, and c:

$$x = \frac{-b \pm \sqrt{b^2 - 4ac}}{2a}$$

$$a = 5,\ b = 2,\ c = -3$$

Substitute the values into the quadratic formula, and simplify the radical:

$$x = \frac{-(2) \pm \sqrt{(2)^2 - 4(5)(-3)}}{2(5)}$$

$$x = \frac{-2 \pm \sqrt{4 - (-60)}}{10} = \frac{-2 \pm \sqrt{64}}{10} = \frac{-2 \pm 8}{10}$$

Note that under the radical you must multiply the minus sign in front of $-4(5)(-3)$ before combining this product with $(2)^2$. Separate the solution into two answers, one using the plus sign and one using the minus sign, and calculate the two solutions:

Simplify: $$x = \frac{-2 + 8}{10} \quad \text{and} \quad x = \frac{-2 - 8}{10}$$

$$x = \frac{6}{10} = \frac{3}{5} = 0.6 \quad \text{and} \quad x = \frac{-10}{10} = -1$$

Note that this example can also be solved by factoring.

EXAMPLE 8.30 Solve the quadratic equation:

$$2P^2 - 3P = 1.$$

Express the roots in radical form and as decimals to three significant digits.

Solution Transpose 1:

$$2P^2 - 3P - 1 = 0$$

Observe that the equation cannot be factored. Write down the quadratic formula with P in place of x and write down the coefficients:

$$P = \frac{-b \pm \sqrt{b^2 - 4ac}}{2a}$$

$$a = 2,\ b = -3,\ c = -1$$

Substitute into the formula and simplify:

$$P = \frac{-(-3) \pm \sqrt{(-3)^2 - 4(2)(-1)}}{2(2)}$$

$$P = \frac{+3 \pm \sqrt{9 - (-8)}}{4} = \frac{+3 \pm \sqrt{17}}{4}$$

The two roots in radical form are:

$$P = \frac{+3 + \sqrt{17}}{4} \quad \text{and} \quad P = \frac{+3 - \sqrt{17}}{4}$$

Use the calculator to find the roots in decimal form to three significant digits. You must calculate the numerator first before you divide:

DAL: 3 [+] [√] 17 [=] [÷] 4 [=] → 1.78; 3 [−] [√] 17 [÷] 4 [=] → −0.281
Not DAL: 3 [+] 17 [√] [=] [÷] 4 [=] → 1.78; 3 [−] 17 [√] [=] [÷] 4 [=] → −0.281

ERROR BOX

A common error to watch out for when using the quadratic formula is simplifying the radical incorrectly. You must first multiply out $4ac$ and b^2 separately, and then combine these terms. You cannot divide the denominator into the radical, and you cannot divide the denominator into just the first term in the numerator. The fraction line is like a set of parentheses, and all the terms in the numerator must be divided by the denominator or none at all. For example, for the quadratic equation $x^2 - 4x + 2 = 0$, the quadratic formula yields:

$$x = \frac{+4 \pm \sqrt{8}}{2}$$

You cannot divide the 2 into the radicand 8. The radical represents the infinite decimal $\sqrt{8} = 2.8284\ldots$, which is an irrational number. You cannot divide the 2 into the 4 either, because 4 is not a factor of the numerator.

See if you can correctly find the irrational roots of the following quadratic equations.

Practice Problems: Find the roots in radical form and as decimals to three significant digits:

1. $V^2 - 2V = 1$ **2.** $R^2 + 1 = 4R$ **3.** $2I = 1 - 2I^2$ **4.** $3x^2 = 1 - 3x$
5. $y^2 + 4y = 2$

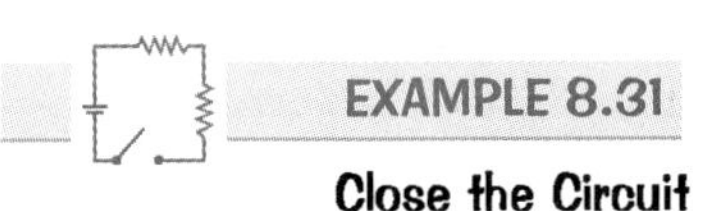

EXAMPLE 8.31

Close the Circuit

Figure 8-1 shows a load R_L connected in series to a resistance R_1. The voltage of the source $V = 80$ V and $R_1 = 10\ \Omega$. If the power used by the load $P_L = 100$ W, the two possible values of the current I in the circuit are given by the quadratic equation:

$$80I = 10I^2 + 100$$

$R_1 = 10\ \Omega$, I, 80 V, Load R_L, $P_L = 100$ W

$80I = 10I^2 + 100$

FIGURE 8-1 Series circuit with a load R_L for Example 8.31.

This equation says that the power supplied by the source, $80I$, equals the power dissipated in the resistor, $10I^2$, plus the power used by the load, 100 W. Find I in amps to two significant digits.

Solution To solve this quadratic equation, transpose $80I$ and put it in the general form with the terms in order:

$$10I^2 - 80I + 100 = 0$$

Divide each term by 10 to simplify: $I^2 - 8I + 10 = 0$

Apply the quadratic formula where I takes the place of x:

$$I = \frac{-(-8) \pm \sqrt{(-8)^2 - 4(1)(10)}}{2(1)} = \frac{8 \pm \sqrt{24}}{2}$$

There are two possible solutions for I:

$$I = \frac{8 + \sqrt{24}}{2} \approx 6.4 \text{ A} \quad \text{and} \quad I = \frac{8 - \sqrt{24}}{2} \approx 1.6 \text{ A}$$

Each solution for I corresponds to a different value for the resistance of the load R_L. Depending on the whether the load resistance needs to be high or low, only one answer may apply in a particular circuit.

EXERCISE 8.4

In exercises 1 through 36, round all decimal answers to three significant digits.

In exercises 1 through 20, solve each quadratic equation by taking square roots or factoring.

1. $x^2 - 4 = 0$
2. $2y^2 = 32$
3. $49e^2 = 9$
4. $25 - 4f^2 = 0$
5. $q^2 - 3q - 10 = 0$
6. $t^2 + 6t + 8 = 0$
7. $3x^2 = 2x + 1$
8. $2z^2 + 3 = 5z$
9. $4y = 4y^2 + 1$
10. $d^2 + 6d = -9$
11. $3G_1^2 = 8G_1 - 4$
12. $5V_T^2 + 2V_T = 7$
13. $56P^2 = 8P$
14. $10C^2 + 18C = 0$
15. $1.28 - 0.50w^2 = 0$
16. $1.21 = 2.25I^2$
17. $53Z^2 - 29 = 0$
18. $32 - 44K^2 = 0$
19. $10a^2 + 5a = 5$
20. $6a^2 - 20a + 16 = 0$

In exercises 21 through 36, solve each quadratic equation by the quadratic formula.

21. $3y^2 - y = 10$
22. $2w^2 + 3 = 7w$
23. $x^2 + 5x - 1 = 0$
24. $x^2 + 2x - 1 = 0$
25. $y^2 - y = 1$
26. $w^2 - 3 = w$
27. $V^2 - 1 - 2V = 0$
28. $p^2 - 1 + 3p = 0$
29. $2m^2 - m - 2 = 0$
30. $2R^2 - 1 = 2R$
31. $2e^2 - 6 = e^2 + 2e$
32. $4f^2 + 4f = 2 - 2f$
33. $2S^2 + 0.5 = 3S$
34. $V_T^2 - 3V_T = 2.5$
35. $6X_L = 3X_L^2 + 2$
36. $Z_T^2 - 3 = 9Z_T$

Applied Problems

In problems 37 through 46, solve each applied problem by setting up and solving a quadratic equation. Round decimal answers to two significant digits.

37. The hypotenuse of a right triangle is 8.0 ft long. If the long side is twice the length of the short side, find the two sides of the right triangle. Set up and solve a quadratic equation using the Pythagorean formula: (short side)^2 + (long side)^2 = (hypotenuse)^2.
38. The long side of a right triangle is 5.0 m long. If the hypotenuse is three times the length of the short side, find the short side and the hypotenuse of the right triangle. Set up and solve a quadratic equation using the Pythagorean formula: (short side)^2 + (long side)^2 = (hypotenuse)^2.
39. The area of a rectangular parking lot is 96 ft^2. If the length is 10 more than the width, find the length and width of the lot to the nearest foot. Set up and solve a quadratic equation using the formula: Area = (length)(width).
40. The area of a large rectangular video screen is 0.56 m^2. If the length is one meter more than the width, find the length and width of the monitor in meters and centimeters. Set up and solve a quadratic equation using the formula: Area = (length)(width).

Applications to Electronics

41. In an ac circuit containing a resistance R and a reactance X, the relationship between the resistance, reactance, and the impedance Z is given by the Pythagorean relation:

$$Z^2 = R^2 + X^2$$

If Z = 4.0 kΩ and X = 2.0 kΩ, find R in ohms and kilohms. See Example 8.24.

42. In problem 41, if Z = 12 kΩ and R = 10 kΩ, find X in ohms and kilohms.
43. The power output P of a generator is given by:

$$P = VI - RI^2$$

where V = generator voltage, R = internal resistance, and I = current. Find the two possible values of I in amps

---

Answers to Error Box Problems, page 179:

1. $\frac{2 \pm \sqrt{8}}{2} = 2.41, -0.414$ 2. $\frac{4 \pm \sqrt{12}}{2} = 3.73, 0.268$ 3. $\frac{-2 \pm \sqrt{12}}{4} = 0.366, -1.37$
4. $\frac{-3 \pm \sqrt{21}}{6} = 0.264, -1.26$ 5. $\frac{-4 \pm \sqrt{24}}{2} = 0.449, -4.45$

when $P = 40$ W, $V = 13$ V, and $R = 1.0\ \Omega$. Substitute the values and solve the quadratic equation.

44. In problem 43, find the two possible values of I in amps when $P = 100$ W, $V = 200$ V, and $R = 50\ \Omega$.

45. In Example 8.31, if $V = 100$ V, $R_1 = 100\ \Omega$, and $P_L = 10$ W, the current I is given by:

$$100I = 10 + 100I^2$$

Find the two possible values of I in amps.

46. In Example 8.31, if $V = 50$ V, $R_1 = 30\ \Omega$, and $P_L = 15$ W, the current I is given by:

$$50I = 15 + 30I^2$$

Find the two possible values of I in amps.

CHAPTER HIGHLIGHTS

8.1 MULTIPLYING POLYNOMIALS

Key Example:

First Outside Inside Last

$$(x + 2)(3x - 1) = (x)(3x) + (x)(-1) + (2)(3x) + (2)(-1)$$
$$= 3x^2 - x + 6x - 2 = 3x^2 + 5x - 2$$

Difference of Two Squares

$$(X + Y)(X - Y) = X^2 - Y^2 \quad (8.1)$$

Key Example:

$$(3p + 4q)(3p - 4q) = (3p)^2 - (4q)^2 = 9p^2 - 16q^2$$

Binomial Square

$$(A + B)(A + B) = (A + B)^2 = A^2 + 2AB + B^2 \quad (8.2)$$

Key Example:

$$(2b - 5y)^2 = (2b)^2 + 2(2b)(-5y) + (-5y)^2 = 4b^2 - 20by + 25y^2$$

To multiply a binomial by a trinomial, use a method similar to long multiplication in arithmetic.

Key Example:

$$\begin{array}{rrrrr} & 3x^2 & + 2x & - 1 & \\ & & x & - 2 & \\ \hline \text{Multiply } -2 \text{ by } (3x^2 + 2x - 1): & & -6x^2 & -4x & +2 \\ \text{Multiply } x \text{ by } (3x^2 + 2x - 1): & 3x^3 & +2x^2 & -x & \\ \hline \text{Add similar terms:} & 3x^3 & -4x^2 & -5x & +2 \end{array}$$

8.2 MONOMIAL FACTORS

Key Example:

Prime Factors:

$$28 = 2 \cdot 2 \cdot 7 = 2^2 \cdot 7$$
$$21a^2b^2 = 3 \cdot 7 \cdot a \cdot a \cdot b \cdot b$$
$$60I_1I_2 = 2 \cdot 2 \cdot 3 \cdot 5 \cdot I_1 \cdot I_2$$

When factoring, separate the greatest common factor first and divide each term by the greatest common factor to obtain the terms in the other factor.

Key Example:

$$10p^3 - 25p^2 + 5p = 5p(2p^2 - 5p + 1)$$

8.3 BINOMIAL FACTORS

Difference of Two Squares

Key Example:

$$4x^2 - 9y^2 = (2x + 3y)(2x - 3y)$$
$$3x^2 - 12 = 3(x^2 - 4) = 3(x + 2)(x - 2)$$

TABLE 8-1 BINOMIAL FACTORS, LEADING COEFFICIENT = 1

	Sum Product	
Positive Last Term	$x^2 - 5x \oplus 6 = (x - 2)(x - 3)$ $x^2 + 5x \oplus 6 = (x + 2)(x + 3)$	**Like Signs**
Negative Last Term	$x^2 - 5x \ominus 6 = (x - 6)(x + 1)$ $x^2 + 5x \ominus 6 = (x + 6)(x - 1)$	**Unlike Signs**

Trinomials with Leading Coefficient ≠ 1

Key Example:

Binomial Square:

$$4z^2 - 20z + 25 = (2z - 5)(2z - 5) = (2z - 5)^2$$

Key Example: To factor $2R^2 - 9R - 5$, try combinations of factors of the first term, $2R^2$ and the last term, -5, until you obtain the correct middle term:

$$(R + 1)(2R - 5) = 2R^2 - 3R - 5$$
$$(R - 1)(2R + 5) = 2R^2 + 3R - 5$$
$$(R + 5)(2R - 1) = 2R^2 + 9R - 5$$
$$(R - 5)(2R + 1) = 2R^2 - 9R - 5 \quad \text{Correct}$$

8.4 SECOND-DEGREE EQUATIONS

General Second-Degree (Quadratic) Equation

$$ax^2 + bx + c = 0 \ (a, b, c \text{ constants}) \tag{8.3}$$

Solution by Taking Square Roots

Key Example:

$$2x^2 + 15 = 25$$

Solve for x^2: $2x^2 = 10$

$$x^2 = 5$$

Take square roots: $\sqrt{x^2} = x = \pm\sqrt{5} \approx \pm 2.24$

Solution by Factoring

Key Example:

$$2x^2 + 6x + 1 = 5x + 7$$

Put in the general form (8.3): $2x^2 + x - 6 = 0$

Factor: $(2x - 3)(x + 2) = 0$

Set each factor to zero: $2x - 3 = 0 \quad x + 2 = 0$

Solve linear equations: $x = 1.5 \quad x = -2$

Solution by the Quadratic Formula

Quadratic Formula

$$x = \frac{-b \pm \sqrt{b^2 - 4ac}}{2a} \tag{8.4}$$

Key Example:

$$2x^2 - 5 = 4x$$

Put in the general form (8.3): $2x^2 - 4x - 5 = 0$

Write down a, b, c: $a = 2, b = -4, c = -5$

Substitute in the formula:

$$x = \frac{-(-4) \pm \sqrt{(-4)^2 - 4(2)(-5)}}{2(2)}$$

Simplify and calculate the root:

$$x = \frac{4 \pm \sqrt{16 - (-40)}}{4} = \frac{4 \pm \sqrt{56}}{4} \approx \frac{4 \pm 7.48}{4}$$

Compute two values:

$$x = \frac{4 + 7.48}{4} = 2.87 \quad \text{and} \quad x = \frac{4 - 7.48}{4} = -0.87$$

REVIEW EXERCISES

In exercises 1 through 10, multiply out each of the products.

1. $(W - 4)(W + 2)$
2. $(3E_1 + 10)(2E_1 - 1)$
3. $(0.2k - 0.1)(0.1k + 0.2)$
4. $(7f + 10)^2$
5. $(2m - 1)(2m + 1)$
6. $(5N - 4)(5N + 4)$
7. $3(X + Y)(X - Y)$
8. $5(a - b)^2$
9. $(I + 2)(I^2 - 4I + 4)$
10. $(R + 1)^2(R - 1)$

In exercises 11 through 38, factor each expression into primes or tell if it is prime.

11. $2x^2 - 4x$
12. $r^3 + 3r^2$
13. $12A^2C + 12B^2C$
14. $120I^2R + 160I^2R_0$
15. $21z^3 + 15z^2 - 27z$
16. $15v^4 + 45v^3 - 30v^2$
17. $\frac{3V_1}{R} + \frac{9V_2}{R}$
18. $\frac{5a}{x} + \frac{10b}{x}$
19. $r^2st + rs^2t + rst^2$
20. $at^3 - a^2t^2 + a^3t$
21. $x^2 + y^2$
22. $2Q^2 - 9$
23. $16x^2 - 25y^2$
24. $f^2 - 0.01$
25. $100 - \frac{A^2}{4}$
26. $I^2 - 6I + 5$
27. $P^2 + 4P - 21$
28. $16C^2 - 16C + 4$
29. $X^2 + 0.3X + 0.02$
30. $10n^2 - 40$
31. $2q^2 - 5q + 3$
32. $3R^2 + 4R - 15$

33. $6u^2 - 13uv + 6v^2$

34. $12t^2 + 30t + 12$

35. $25X^2 + 15X - 10$

36. $16a^2 - 20ab - 6b^2$

37. $\frac{s^2}{10} - \frac{9}{10}$

38. $\frac{I^2}{R} - \frac{9}{R}$

In exercises 39 through 58, solve each quadratic equation by taking square roots, factoring, or the quadratic formula. Round decimal answers to three significant digits.

39. $x^2 = 16$

40. $4 - 25y^2 = 0$

41. $m^2 - 4m - 5 = 0$

42. $3t^2 + t = 2$

43. $7P_T = 8P_T^2$

44. $2R_1^2 - 50R_1 = 0$

45. $0.10w^2 - 0.441 = 0$

46. $10c - 3c^2 - 8 = 0$

47. $Z(Z + 5) = 50$

48. $16f^2 + 9 = 24f$

49. $2k^2 + 8 = (k + 2)^2$

50. $(d + 1)^2 + 2d^2 = 2$

51. $x^2 - 4x + 1 = 0$

52. $w^2 + 2w - 1 = 0$

53. $3V_1 - 1 = V_1^2$

54. $3e - 4e^2 = e^2 - 1$

55. $2a^2 - \frac{1}{2} = a$

56. $v^2 - 0.3v = 0.5$

57. $2s^2 + 2s + 1 = 10$

58. $5C + 2 = 3C^2 - 4$

Applied Problems

In applied problems 59 through 74, round decimal answers to two significant digits.

59. The width of a rectangular cabinet is given by $(3l - 2)$ and the length by $(2l + 3)$. Write the expression for the area of the cabinet and multiply out the expression.

60. The area of the sector of an electromagnet formed by two circles of radii r_1 and r_2 whose angle is θ is given by:

$$\frac{\theta}{2}(r_1 + r_2)(r_1 - r_2)$$

Multiply out this expression.

61. The height of a rocket is given by $h = 6t^2 + 42t + 72$, where t = time. Factor this expression into primes.

62. The kinetic energy of one particle is given by $K_1 = \frac{1}{2}mv_1^2$. The kinetic energy of a second particle is given by $K_2 = \frac{1}{2}mv_2^2$. Factor the expression $K_2 - K_1$, that represents the difference in energy between the two particles.

63. The area of a rectangular video monitor is 600 cm^2. If the length is 5 cm more than the width, what are the dimensions of the monitor? Set up and solve a quadratic equation.

64. The hypotenuse of a right triangle is 4 cm long. If the short side = s and the long side = $1.5s$, find the two sides of the right triangle. Set up and solve a quadratic equation using the Pythagorean formula: (short side)^2 + (long side)^2 = (hypotenuse)^2.

Applications to Electronics

65. The current in a circuit is given by $I = 2t - 3$ and the resistance by $R = t + 5$, where t = time. Using Ohm's law, $V = IR$, write the expression for the voltage in the circuit in terms of t and multiply out the expression.

66. The current in a circuit is given by $I = t - 5$ and the resistance by $R = 3.0\ \Omega$. Using the power law, $P = I^2R$, write the expression for the power in the circuit in terms of t and multiply out the expression.

67. The voltage across one resistor is: $V_1 = 75I_1R$. The voltage across a second resistor is: $V_2 = 125I_2R$. Factor the expression, $V_1 + V_2$, that represents the total voltage across both resistors.

68. In a circuit, the power dissipated by one resistor is: $P_1 = \frac{V_1^2}{2}$. The power dissipated by a second resistor is: $P_2 = \frac{V_2^2}{8}$. Factor the expression, $P_2 - P_1$, that represents the difference in the power dissipated by both resistors.

69. The power output P of a generator is given by

$$P = VI - RI^2$$

where V = generator voltage, R = internal resistance, and I = current. Find the current I in amps when $P = 25$ W, $V = 10$ V, and $R = 1.0\ \Omega$.

70. In problem 69, find the two possible values of the current in amps when $P = 120$ W, $V = 200$ V, and $R = 25\ \Omega$.

71. In an ac circuit containing a resistor and a capacitor, the relationship between the resistance R, the capacitive reactance X_C, and the total impedance Z_T is given by the Pythagorean relation:

$$Z_T^2 = R^2 + X_C^2$$

If $Z_T = 3.3$ kΩ and $X_C = 1.2$ kΩ, find R in kilohms.

72. The total current in an ac circuit is related to the current through the resistance I_R and the current through the inductance I_L by the Pythagorean relation:

$$I_T^2 = I_R^2 + I_L^2$$

If $I_T = 4.5$ A and $I_R = 1.5\ I_L$, find I_R and I_L in amps.

73. In Example 8.31, if $V = 12$ V, $R_1 = 2.0\ \Omega$, and $P_L = 16$ W, the current I is given by:

$$12I = 16 + 2.0I^2$$

Find the two possible values of the current in amps.

74. In problem 73, the current that produces the maximum power in a load is given by:

$$12I = 18 + 2.0I^2$$

Find the value of the current, in amps, that produces this maximum power.

CHAPTER 9
Algebraic Fractions

CHAPTER OBJECTIVES

In this chapter, you will learn:

- How to reduce algebraic fractions to lowest terms.
- How to multiply and divide algebraic fractions.
- How to find the lowest common denominator.
- The meaning of equivalent algebraic fractions.
- How to add algebraic fractions.
- How to simplify and solve equations with fractions.

When working with algebraic fractions, factors play an important role. The better you can factor expressions, the easier it will be to perform operations on algebraic fractions. Algebraic fractions contain variables and therefore cannot be evaluated or changed to decimals like arithmetic fractions. You must understand the meaning of a fraction and all the rules that apply. Many formulas in electronics involve algebraic fractions, such as Ohm's law and the formulas for power and parallel resistance, which are the first formulas you learn in dc circuits. The factoring techniques learned in Chapter 8 will help you perform operations with algebraic fractions, and help you work with them in equations and formulas.

• • •

9.1 Reducing Fractions

To reduce an algebraic fraction, look for a common factor in the numerator and denominator that can be divided out, as the following example shows.

EXAMPLE 9.1

Reduce the fraction:

$$\frac{12x}{28x^2}$$

Solution Factor the numerator and denominator. This fraction has common factors of 2, 2, and x. Divide out the factors and reduce the fraction:

$$\frac{12x}{28x^2} = \frac{(\cancel{2})(\cancel{2})(\cancel{x})(3)}{(\cancel{2})(\cancel{2})(\cancel{x})(7x)} = \frac{3}{7x}$$

The result is an equivalent fraction in lowest terms.

EXAMPLE 9.2

Reduce the fraction:

$$\frac{12y^2 + 5y}{5y}$$

Solution The numerator of this fraction has two terms. When the numerator or denominator has more than one term, proceed carefully and factor first before you divide. You cannot divide the $5y$ in the denominator into the $5y$ in the numerator. The $5y$ in the numerator is a term and not a factor. *You can only divide factors, not terms.* First separate the common factor of y in the numerator and then divide out factors:

$$\frac{12y^2 + 5y}{5y} = \frac{(y)(12y + 5)}{5y} = \frac{12y + 5}{5}$$

The y is the only common factor that can be divided out to reduce the fraction. If you do not separate the factor first, you may make the common error of dividing one term in a fraction without dividing the other. For example, if you divide only the $5y$ in the top with the $5y$ in the bottom, without factoring, this will produce the wrong answer $12y^2 + 1$. Remember the fraction line is like parentheses, and every term in the numerator is divided by the denominator. See the Error Box in this section to reinforce this idea.

ERROR BOX

A very common error to watch out for is dividing out terms that are not factors. This is such a common error that it is necessary to emphasize it again. Suppose you had to simplify the following fraction:

$$\frac{3 + 5}{3 - 1}$$

You would never think to incorrectly divide out the 3's because you can simply calculate the result:

$$\frac{3 + 5}{3 - 1} = \frac{8}{2} = 4$$

But, if you had to simplify the following fraction in algebra:

$$\frac{a + b}{a - c}$$

you might be tempted to divide out the a's because there is nothing that can be done to simplify this fraction. However, when there is more than one term in the numerator or denominator of a fraction, do three things before you divide out any expression:

1. Separate all common factors from every term in the numerator and the denominator.
2. Factor more, if possible, into binomials so you have all prime factors.
3. Ask yourself: Is the expression a factor of *every* term in the numerator *and* the denominator?

Practice Problems: Apply these ideas and see if you can reduce the following fractions correctly or tell if they are not reducible.

1. $\frac{I^2 - 1}{I + 1}$
2. $\frac{I^2 + 1}{I + 1}$
3. $\frac{I^2 - 2I + 1}{I - 1}$
4. $\frac{I^2R + V}{IR + V}$
5. $\frac{I^2R + IV}{IR + V}$
6. $\frac{I^2 + 1}{I - 1}$
7. $\frac{I^2 - 1}{I - 1}$
8. $\frac{I^2 - 2I + 1}{I + 1}$
9. $\frac{R^2 - V^2}{IR - IV}$
10. $\frac{I^2R + IV}{R + IV}$

EXAMPLE 9.3 Reduce the fraction:

$$\frac{12R^2 + 4R}{2R^2 - 2R}$$

Solution Before you divide, you must factor the numerator and the denominator completely. Separate the common factors $4R$ in the numerator and $2R$ in the denominator. Then divide out common factors:

$$\frac{12R^2 + 4R}{2R^2 - 2R} = \frac{\overset{2}{\cancel{4R}}(3R+1)}{\cancel{2R}(R-1)} = \frac{2(3R+1)}{R-1}$$

Look again at this example. You might have been tempted at first to divide $2R^2$ into $12R^2$ and $-2R$ into $4R$. The result would then be $6 - 2 = 4$, which is very different from the correct answer.

Because division is the inverse of multiplication, you can only divide quantities that are multiplied together. If there is a plus or minus sign between expressions, they are not factors and you cannot divide them.

EXAMPLE 9.4 Reduce the fraction:

$$\frac{x^2 - 4t^2}{x - 2t}$$

Solution Again, you cannot divide individual terms. First factor the numerator, which is the difference of two squares. The denominator cannot be factored. Then divide out the common factor, which is the binomial $(x - 2t)$.

$$\frac{x^2 - 4t^2}{x - 2t} = \frac{(x+2t)\cancel{(x-2t)}}{\cancel{x-2t}} = x + 2t$$

EXAMPLE 9.5 Reduce the fraction:

$$\frac{a^2 + 2ab + b^2}{a^2 - ab - 2b^2}$$

Solution First factor both the numerator and denominator completely into binomials and then divide out the common binomial factor $(a + b)$:

$$\frac{a^2 + 2ab + b^2}{a^2 - ab - 2b^2} = \frac{\cancel{(a+b)}(a+b)}{\cancel{(a+b)}(a-2b)} = \frac{a+b}{a-2b}$$

EXAMPLE 9.6 Reduce the fraction:

$$\frac{P^2 - V^2}{P^2 + V^2}$$

Solution The numerator can be factored because it is the *difference* of two squares, but the denominator can *not* be factored because it is the *sum* of two squares:

$$\frac{(P+V)(P-V)}{P^2 + V^2}$$

Resist any temptation to divide out the P's or the V's. The denominator cannot be factored, and this fraction cannot be reduced. It is in lowest terms as given.

EXERCISE 9.1

In exercises 1 through 30, reduce each fraction to lowest terms if possible.

1. $\frac{30x}{12x^2}$
2. $\frac{8y^2}{20y}$
3. $\frac{210a^3}{270a}$
4. $\frac{66b^2}{165b^3}$
5. $\frac{ax + bx}{ax}$
6. $\frac{RT - R}{RT}$
7. $\frac{KL}{KM - KN}$
8. $\frac{PQ}{P^2 + P}$
9. $\frac{ab + ad}{ac}$
10. $\frac{xy + y^2}{xy}$
11. $\frac{mn + n}{m + 1}$
12. $\frac{a - ax}{1 - x}$
13. $\frac{2R_1^2 - 2R_2^2}{10R_1 + 10R_2}$
14. $\frac{3x + 3y}{6x^2 - 6y^2}$

15. $\dfrac{5.1v + 5.1w}{1.7v + 1.7w}$

16. $\dfrac{2.5m^2 + 5m}{1.5m^2 + 3m}$

17. $\dfrac{x^2 + y^2}{x + y}$

18. $\dfrac{a^2 - 2b^2}{a + 2b}$

19. $\dfrac{4V^2 - 6V - 4}{4V - 8}$

20. $\dfrac{3P + 6}{3P^2 + 3P - 6}$

21. $\dfrac{t^2 + 6t + 9}{3t^2 + 8t - 3}$

22. $\dfrac{d^2 - 4d + 4}{3d^2 - 5d - 2}$

23. $\dfrac{6I_1^{\ 2} + 6I_2^{\ 2}}{5I_1^{\ 2} - 5I_2^{\ 2}}$

24. $\dfrac{m^2 + 8m + 16}{m^2 + 4m + 4}$

25. $\dfrac{4a^2 + 4ax + x^2}{4a^2 - x^2}$

26. $\dfrac{2b^2 - 3by - 2y^2}{2b^2 - by - y^2}$

27. $\dfrac{7X_L^{\ 2} + 13X_L - 2}{9X_L^{\ 2} + 14X_L - 8}$

28. $\dfrac{6P^2 - 17P - 3}{11P^2 - 29P - 12}$

29. $\dfrac{24f^2 - 54}{6f^2 - 9f - 27}$

30. $\dfrac{16e^2 - 16}{2e^2 - 2e - 4}$

Applied Problems

31. The input force of a machine is given by $1.5F_x - 1.5F_y$ and the output force by $3F_x^{\ 2} - 3F_y^{\ 2}$. The mechanical advantage is defined as:

$$\text{M.A.} = \frac{\text{Output force}}{\text{Input force}}$$

Write and simplify the algebraic fraction for the mechanical advantage.

32. The height of a video monitor is given by $6l + 2w$ and the volume is given by: $24l^3 - 16l^2w - 8lw^2$. The area of the screen is given by:

$$\text{Screen Area} = \frac{\text{Volume}}{\text{Height}}$$

Write and simplify the algebraic fraction for the area of the screen.

Applications to Electronics

33. The voltage in a circuit $V = 8t^2 - 2$, and the current $I = 4t - 2$. Using Ohm's law for resistance:

$$R = \frac{V}{I}$$

Substitute the expressions for V and I and simplify the formula for R in terms of t.

34. A computer processes information based on the formula:

$$t = \frac{I^2 - 4}{2n + 6}$$

If $n = I - 5$, substitute the expression for n and simplify the formula for t in terms of I.

9.2 Multiplying and Dividing Fractions

The formulas for multiplying and dividing algebraic fractions are the same as formulas (1.4) and (1.5) in Chapter 1 for arithmetic fractions:

Formulas

Multiplication

$$\frac{A}{B} \cdot \frac{C}{D} = \frac{AC}{BD}$$

Multiply across the top and bottom.

• • •

Division

$$\frac{A}{B} \div \frac{C}{D} = \frac{A}{B} \cdot \frac{D}{C} = \frac{AD}{BC}$$

Multiply by the reciprocal.

Answers to Error Box Problems, page 186:

1. $I - 1$ **2.** Not reducible **3.** $I - 1$ **4.** Not reducible **5.** I **6.** Not reducible **7.** $I + 1$ **8.** Not reducible **9.** $\dfrac{R + V}{I}$ **10.** Not reducible

EXAMPLE 9.7

Multiply the fractions:

$$\frac{12y^3}{7x^2} \cdot \frac{21x}{24y^2}$$

Solution As in arithmetic, you first divide out common factors in any numerator and denominator before multiplying. Factor completely into primes and divide as follows:

$$\frac{12y^3}{7x^2} \cdot \frac{21x}{24y^2} = \frac{(\cancel{2})(\cancel{2})(\cancel{3})(\cancel{y})(\cancel{y})(y)}{(\cancel{7})(\cancel{x})(x)} \cdot \frac{(3)(\cancel{7})(\cancel{x})}{(\cancel{2})(\cancel{2})(2)(\cancel{3})(\cancel{y})(\cancel{y})} = \frac{3y}{2x}$$

You can also factor mentally and divide without separating the factors.

EXAMPLE 9.8

Multiply the fractions:

$$\frac{p+1}{2t} \cdot \frac{6t}{5p^2+5p}$$

Solution Before multiplying, factor completely into primes and divide out common factors:

$$\frac{p+1}{2t} \cdot \frac{6t}{5p^2+5p} = \frac{\cancel{p+1}}{\cancel{2t}} \cdot \frac{3(\cancel{2t})}{5p(\cancel{p+1})} = \frac{3}{5p}$$

EXAMPLE 9.9

Multiply the fractions:

$$\frac{I_1^2 - I_2^2}{4I_1} \cdot \frac{8I_1^2}{2I_1 + 2I_2}$$

Solution Factor each numerator and denominator completely into primes and divide out common factors:

$$\frac{I_1^2 - I_2^2}{4I_1} \cdot \frac{8I_1^2}{2I_1 + 2I_2} = \frac{(\cancel{I_1 + I_2})(I_1 - I_2)}{(\cancel{2})(\cancel{2})\cancel{I_1}} \cdot \frac{(\cancel{2})(\cancel{2})(\cancel{2})(\cancel{I_1})(I_1)}{\cancel{2}(\cancel{I_1 + I_2})} = I_1(I_1 - I_2)$$

All the factors in both denominators divide out, leaving a 1 in the denominator, which is understood.

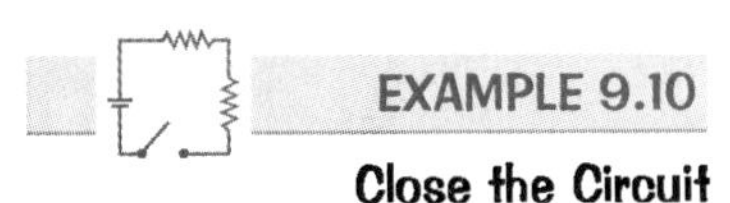

EXAMPLE 9.10

Close the Circuit

Figure 9-1 shows a circuit containing two different resistors R_1 and R_2 in parallel connected to a voltage source V. The formula for the total or equivalent resistance is given by:

$$R_T = \frac{R_1 R_2}{R_1 + R_2}$$

If $R_2 = 2R_1$, express R_T in terms of R_1.

Solution First substitute $2R_1$ for R_2 in the expression for R_T. Then simplify and reduce the fraction:

$$R_T = \frac{R_1 R_2}{R_1 + R_2} = \frac{R_1(2R_1)}{R_1 + (2R_1)}$$

$$= \frac{2R_1^{\cancel{2}}}{3\cancel{R_1}} = \frac{2R_1}{3}$$

R_1 R_2 I_T $-\ V\ +$

$$R_T = \frac{R_1 R_2}{R_1 + R_2}$$

FIGURE 9-1 Two parallel resistors for Example 9.10.

EXAMPLE 9.11 Divide the fractions:

$$\frac{mv^2 + 4mv + 4m}{3v} \div \frac{2v^2 + v - 6}{6v - 9}$$

Solution Write the reciprocal of the second fraction and factor completely into primes:

$$\frac{mv^2 + 4mv + 4m}{3v} \div \frac{2v^2 + v - 6}{6v - 9} = \frac{m(v^2 + 4v + 4)}{3v} \cdot \frac{6v - 9}{2v^2 + v - 6}$$
$$= \frac{m(v + 2)^2}{3v} \cdot \frac{3(2v - 3)}{(2v - 3)(v + 2)}$$

Notice there is a common factor of m in the first numerator that is separated first. Now divide out common factors and multiply:

$$\frac{m(v + 2)^{\cancel{2}}}{\cancel{3}v} \cdot \frac{\cancel{3}\cancel{(2v - 3)}}{\cancel{(2v - 3)}\cancel{(v + 2)}} = \frac{m(v + 2)}{v}$$

EXAMPLE 9.12 Multiply:

$$\frac{a + x}{3a^2 + 5ax + 2x^2} \cdot (3a + 2x)$$

Solution The binomial $(3a + 2x)$ has a denominator of 1 that is understood. Put $(3a + 2x)$ over 1, factor, and divide out common factors:

$$\frac{a + x}{3a^2 + 5ax + 2x^2} \cdot \frac{3a + 2x}{1} = \frac{\cancel{a + x}}{\cancel{(3a + 2x)}\cancel{(a + x)}} \cdot \frac{\cancel{3a + 2x}}{1} = \frac{1}{1} = 1$$

Note that, when all the factors divide out, the answer is 1, not 0. The factors do not "cancel" but divide into themselves once.

EXERCISE 9.2

In exercises 1 through 30, perform the indicated operations. Divide out common factors first.

1. $\frac{9x}{2} \cdot \frac{4}{3x^2}$
2. $\frac{4}{10y} \cdot \frac{5y^2}{8}$
3. $\frac{5b^2y}{4by^3} \cdot \frac{8y}{10b}$
4. $\frac{6r}{18t^3} \cdot \frac{9t}{r^2}$
5. $\frac{0.9V}{6I} \div \frac{0.3V}{I}$
6. $\frac{5.6}{0.8Z} \div \frac{14}{Z^2}$
7. $\frac{6b^2}{3a^2 + 9a} \cdot \frac{5a + 15}{2b}$
8. $\frac{r - t}{5t^3} \cdot \frac{rt^2}{r^2 - rt}$
9. $\frac{8g^2 - 8h^2}{16h} \cdot \frac{5g}{7g + 7h}$
10. $\frac{21u - 7v}{9u^2 - v^2} \cdot \frac{3u + v}{uv}$
11. $\frac{s^2 - 2st + t^2}{s} \div (s - t)$
12. $\frac{w + 10}{w - 10} \div \frac{1}{w^2 - 100}$
13. $\frac{8m - 4}{4m^2 - 4m + 1} \cdot (2m - 1)$
14. $\frac{3q^2 + 2q - 1}{q + 1} \cdot \frac{1}{3q - 1}$
15. $\frac{x^2 - y^2}{4x - 4y} \cdot \frac{2x^2 + 2y^2}{x + y}$
16. $\frac{2x^2 - 2a^2}{x^2 + ax} \cdot \frac{ax}{2x - 2a}$
17. $\frac{I_1 + I_2}{15I_2} \div \frac{I_1^2 - I_2^2}{3I_2^2}$

18. $\dfrac{V_2^2}{V_2^2 - V_1^2} \div \dfrac{V_1 V_2}{V_2 - V_1}$

19. $\dfrac{a^2 + 5a + 4}{2a + 2} \cdot \dfrac{4a + 4}{3a + 12}$

20. $\dfrac{6c^2 - 13c + 6}{4c + 1} \cdot \dfrac{8c + 2}{6c - 4}$

21. $\dfrac{5D + 10}{D^2 - 4} \div \dfrac{1}{2D^2 - 5D + 2}$

22. $\dfrac{4m^2 - 12mn + 9n^2}{2n} \div (2m - 3n)$

23. $\dfrac{R_1 - R_2}{R_1} \cdot \dfrac{R_2}{R_1^2 - R_1 R_2}$

24. $\dfrac{V_0^2 - V_0}{V_0^2} \cdot \dfrac{V_0 + 1}{V_0 - 1}$

25. $\dfrac{xy^2 + 4xy + 4x}{6y - 9} \cdot \dfrac{4y - 6}{y + 2}$

26. $\dfrac{2a + 2}{a^2 b - 2ab - 3b} \cdot \dfrac{5ab - 15b}{2b}$

27. $\dfrac{(9 \times 10^6)K}{K - 1} \cdot \dfrac{2K - 2}{(18 \times 10^3)K^2}$

28. $\dfrac{(5 \times 10^{-3})N^2}{2N + 1} \div \dfrac{(15 \times 10^{-6})N}{4N + 2}$

29. $\dfrac{3.5x}{2.5y} \cdot \dfrac{0.5x^2 y}{2.1xy^2} \div \dfrac{0.1x^2}{0.3y^2}$

30. $\dfrac{0.6c}{0.3d} \cdot \dfrac{1.8c}{1.2d} \div \dfrac{1.2c^2 d}{0.4cd^2}$

Applied Problems

31. Simplify the following algebraic expression, which occurs when calculating the center of gravity of a body:

$$\left(\frac{4\pi r^3}{3}\right)\left(\frac{2}{\pi r^2}\right) \div 2\pi$$

32. The dimensions of a rectangular computer cabinet are given by the expressions:

$$\text{height} = h, \quad \text{width} = \frac{h}{h + 2}, \quad \text{length} = \frac{3h + 6}{h}$$

Write and simplify the expression for the volume of the cabinet using the formula: Volume = (length)(width)(height).

Applications to Electronics

33. Given Ohm's law $I = \frac{V}{R}$ and the power formula $P = VI$, show that $P = \frac{V^2}{R}$ by substituting the Ohm's law expression for I in the power formula.

34. Given Ohm's law $R = \frac{V}{I}$ and the power formula $V = \frac{P}{I}$, show that $R = \frac{P}{I^2}$ by substituting the power formula expression for V in Ohm's law.

35. In Example 9.10 given:

$$R_T = \frac{R_1 R_2}{R_1 + R_2}$$

Express R_T in terms of R_1 if $R_2 = 3R_1$.

36. In problem 35, express R_T in terms of R_2 if $R_1 = 2R_2$.

9.3 Addition of Fractions

Addition of algebraic fractions requires a good grasp of factoring and a clear understanding of the process of adding fractions in arithmetic as explained in Chapter 1, Section 1.3. The example below reviews some of the ideas from Chapter 1 using the fundamental rule of fractions:

Rule

Rule of Fractions

Multiplying or dividing the numerator and the denominator by the same quantity (other than zero) does not change the value of a fraction.

EXAMPLE 9.13 Add the fractions:

$$\frac{5x}{24} + \frac{7x}{30}$$

Solution You can only add fractions if the denominators are the same. To change the fractions to a common denominator, look for the lowest common denominator (LCD) that both denominators divide into evenly. This can be done by trying multiples of the larger denominator: 30, 60, 90, and so on, until you find a multiple that both denominators divide into. The other method shown in Chapter 1, which is necessary for algebraic fractions, is to factor each denominator into its smallest factors or primes and list them as follows to obtain the LCD:

$$\begin{array}{lcccccccccccccccc}
24 & = & 2 & \cdot & 2 & \cdot & 2 & \cdot & 3 & & & = & 2^3 & \cdot & 3 & & \\
30 & = & \downarrow & & \downarrow & & 2 & \cdot & 3 & \cdot & 5 & = & 2 & \cdot & 3 & \cdot & 5 \\
 & & & & & & \downarrow & & \downarrow & & \downarrow & & & & & & \\
\hline
\text{LCD} & = & 2 & \cdot & 2 & \cdot & 2 & \cdot & 3 & \cdot & 5 & = & 2^3 & \cdot & 3 & \cdot & 5 = 120
\end{array}$$

Note: The LCD contains the highest power of each prime factor. Both the denominators divide evenly into the LCD because it contains all the prime factors of each denominator:

$$\frac{120}{24} = \frac{\cancel{2} \cdot \cancel{2} \cdot \cancel{2} \cdot \cancel{3} \cdot 5}{\cancel{2} \cdot \cancel{2} \cdot \cancel{2} \cdot \cancel{3}} = 5$$

$$\frac{120}{30} = \frac{2 \cdot 2 \cdot \cancel{2} \cdot \cancel{3} \cdot \cancel{5}}{\cancel{2} \cdot \cancel{3} \cdot \cancel{5}} = 4$$

Once you find the LCD, apply the fundamental rule of fractions: Multiply the numerator and denominator of each fraction by the factor obtained when its denominator is divided into the LCD. The top and bottom of the first fraction is multiplied by (5), and the top and bottom of the second fraction is multiplied by (4). This changes the denominator of each fraction to the LCD so you can add the fractions:

$$\frac{(5)5x}{(5)24} + \frac{(4)7x}{(4)30} = \frac{25x}{120} + \frac{28x}{120}$$

$$= \frac{25x + 28x}{120} = \frac{53x}{120}$$

EXAMPLE 9.14 Add the fractions:

$$\frac{5}{12x^2} + \frac{7}{18x}$$

Solution The prime factors of each denominator are:

$$\begin{array}{lcccccccccccccccc}
12x^2 & = & 2 & \cdot & 2 & \cdot & 3 & \cdot & & & x & \cdot & x & = & 2^2 & \cdot & 3 \cdot x^2 \\
18x & = & & & 2 & \cdot & 3 & \cdot & 3 & \cdot & x & & & = & 2 & \cdot & 3^2 \cdot x \\
 & & & & & & & & & & & & & & \downarrow & & \downarrow \quad \downarrow \\
\hline
 & & & & & & & & & & & \text{LCD} & & = & 2^2 & \cdot & 3^2 \cdot x^2 = 36x^2
\end{array}$$

The LCD contains the highest power of each prime factor. For each fraction, divide its denominator into the LCD and multiply the numerator and denominator by the result, which is simply the factors missing from each denominator that the LCD contains. The first fraction is missing (3), and the second fraction is missing $(2)(x) = 2x$:

$$\frac{(3)5}{(3)12x^2} + \frac{(2x)7}{(2x)18x} = \frac{15}{36x^2} + \frac{14x}{36x^2} = \frac{15 + 14x}{36x^2}$$

It is more convenient to *add and multiply in one step* over the LCD:

$$\frac{5}{12x^2} + \frac{7}{18x} = \frac{(3)5 + (2x)7}{36x^2} = \frac{15 + 14x}{36x^2}$$

EXAMPLE 9.15 Add the fractions:

$$\frac{5}{G} + \frac{G+1}{G-1}$$

Solution The denominator (G – 1) represents only *one* prime factor. The LCD therefore contains the two prime factors: (G) and (G – 1). Multiply the numerator and denominator of the first fraction by (G – 1) and the second fraction by (G). The multiplication and addition are shown in one step over the LCD:

$$\frac{5}{G} + \frac{G+1}{G-1} = \frac{5(G-1)+(G)(G+1)}{\underbrace{(G)(G-1)}_{\text{LCD}}}$$

To avoid an error when there is more than one term in the numerator, enclose all the terms in parentheses, as shown for (G + 1) and (G – 1) above. Now multiply the terms in the numerator and simplify:

$$\frac{5(G-1)+(G)(G+1)}{(G)(G-1)} = \frac{5G - 5 + G^2 + G}{(G)(G-1)} = \frac{G^2 + 6G - 5}{(G)(G-1)}$$

ERROR BOX

Errors in adding algebraic fractions are usually the result of not clearly understanding the meaning of a fraction or the concept of a factor. The fraction line is the same as two sets of parentheses, one enclosing all the terms in the numerator and one enclosing all the terms in the denominator. Factors are separated by multiplication or division. Plus and minus signs separate terms, not factors. Consider the following addition of two fractions:

$$\frac{2}{a-b} - \frac{1}{2a}$$

The denominator of the first fraction contains two terms, a and $-b$, which make one factor $(a - b)$. The denominator of the second fraction contains one term, $2a$, which has two factors, 2 and a. The LCD is then $(2a)(a - b)$. To add the fractions you multiply the first fraction by $(2a)$ and the second fraction by $(a - b)$. Since the fraction line is like parentheses, the minus sign applies to all terms in the numerator of the second fraction. Keep all the factors and the numerators in parentheses to avoids errors:

$$\frac{(2a)(2) - (a-b)(1)}{(2a)(a-b)}$$

Now, when you multiply out, the minus sign in front of the second fraction is correctly applied to all terms in the numerator:

$$\frac{4a - (a-b)}{(2a)(a-b)} = \frac{4a - a + b}{(2a)(a-b)}$$

Leave the denominator in factored form since sometimes you can divide out a common factor after you simplify the numerator:

$$\frac{4a - a + b}{(2a)(a-b)} = \frac{3a + b}{(2a)(a-b)}$$

Practice Problems: See if you can correctly add the fractions:

1. $\frac{b}{a} - \frac{1}{ab}$
2. $\frac{2}{ab} - \frac{a}{b}$
3. $\frac{1}{a-b} - \frac{1}{a}$
4. $\frac{2}{a+b} - \frac{1}{b}$
5. $\frac{3}{a-b} - \frac{1}{a}$
6. $\frac{a+b}{2a} - \frac{b}{a}$
7. $\frac{a-b}{ab} - \frac{1}{b}$
8. $\frac{a-b}{b} - \frac{b}{a+b}$
9. $\frac{a}{a-b} - \frac{a+b}{a}$
10. $\frac{1}{a+b} - \frac{1}{a-b}$

EXAMPLE 9.16 Add the fractions:

$$\frac{2}{T+1} - \frac{T-3}{T^2-1}$$

Factor the denominators into prime factors to find the LCD:

$$\begin{aligned} T+1 &= (T+1) \\ T^2-1 &= (T+1)\ (T-1) \\ &\quad\ \ \downarrow \qquad\ \ \downarrow \\ \hline \text{LCD} &= (T+1)\ (T-1) \end{aligned}$$

Multiply the first fraction by (T – 1). The second fraction does not need to be changed. Combine and multiply in one step:

$$\frac{2}{T+1} - \frac{T-3}{(T+1)(T-1)} = \frac{2(T-1)-(T-3)}{\underbrace{(T+1)(T-1)}_{\text{LCD}}}$$

Observe that the numerator of the second fraction is enclosed in parentheses because every term in the numerator is affected by the minus sign. Simplify the top of the combined fraction and reduce the result:

$$\frac{2T-2-T+3}{(T+1)(T-1)} = \frac{\cancel{T+1}}{(\cancel{T+1})(T-1)} = \frac{1}{T-1}$$

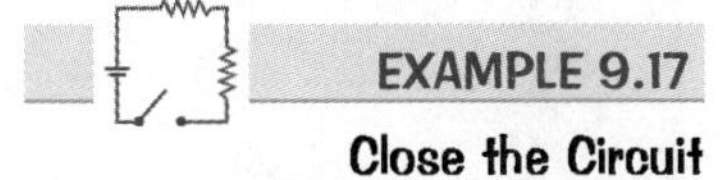

EXAMPLE 9.17

Close the Circuit

Figure 9-2 shows a series-parallel circuit containing a resistor R_1 in parallel with two series resistors R_2 and R_3. The voltage across $R_1 = 6$ V, which equals the voltage across R_2 and R_3. The total current I_T equals the sum of the currents in each branch:

$$I_T = \frac{6}{R_1} + \frac{6}{R_2 + R_3}$$

If $R_2 = R_1$, simplify the expression for I_T by adding the fractions.

Solution Let $R_2 = R_1$ in the expression for I_T:

$$I_T = \frac{6}{R_1} + \frac{6}{R_1 + R_3}$$

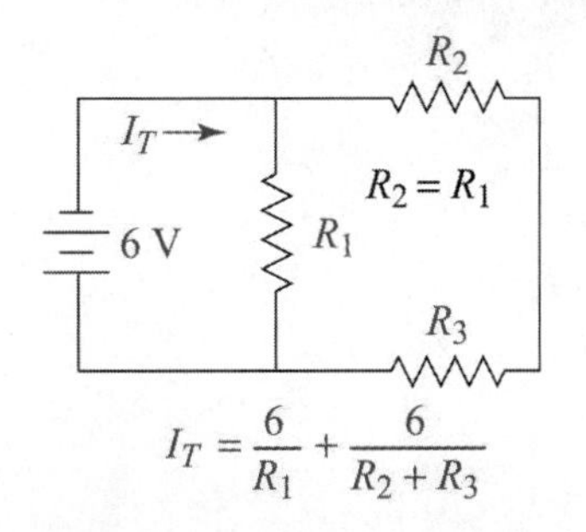

$$I_T = \frac{6}{R_1} + \frac{6}{R_2 + R_3}$$

FIGURE 9-2 Series-parallel circuit for Example 9.17.

The LCD = $R_1(R_1 + R_3)$. Multiply the first fraction by $(R_1 + R_3)$ and the second fraction by R_1 and simplify:

$$I_T = \frac{6(R_1 + R_3) + 6(R_1)}{(R_1)(R_1 + R_3)} = \frac{6R_1 + 6R_3 + 6R_1}{(R_1)(R_1 + R_3)} = \frac{12R_1 + 6R_3}{(R_1)(R_1 + R_3)}$$

EXERCISE 9.3

In exercises 1 through 50, add the fractions and simplify.

1. $\frac{7x}{10} + \frac{x}{10}$
2. $\frac{7y}{15} - \frac{4y}{15}$
3. $\frac{5}{9a} - \frac{2}{9a}$
4. $\frac{4}{3t} + \frac{8}{3t}$
5. $\frac{R}{12} + \frac{3}{20}$
6. $\frac{2}{45} + \frac{3V}{18}$
7. $\frac{7}{8} - \frac{3I}{10}$
8. $\frac{2K}{9} + \frac{4}{15}$
9. $\frac{1}{21z} + \frac{2}{9z}$
10. $\frac{14}{15m} - \frac{6}{25m}$
11. $\frac{7p}{18} + \frac{13p}{30}$
12. $\frac{11q}{21} - \frac{7q}{15}$
13. $\frac{t}{s} + \frac{3t}{2s}$
14. $\frac{3a}{I_1} + \frac{2a}{5I_1}$
15. $\frac{6}{rw^2} - \frac{8}{rw}$
16. $\frac{5}{x^2} + \frac{6}{kx}$
17. $\frac{4V_1}{V_2} + \frac{3V_2}{V_1}$
18. $\frac{6R_1}{R} - \frac{2R}{R_1}$
19. $\frac{X}{6Y^2} - \frac{3X}{12Y}$
20. $\frac{5A}{16K} + \frac{3A}{8K^2}$
21. $\frac{9}{10cd^2} + \frac{5}{6cd}$
22. $\frac{4}{5m^2k} + \frac{7}{mk^2}$
23. $\frac{5}{2 \times 10^3} - \frac{75}{4 \times 10^4}$
24. $\frac{1}{3 \times 10^5} + \frac{5}{9 \times 10^6}$
25. $\frac{1}{a} - \frac{2}{a + x}$
26. $\frac{2}{y} + \frac{1}{b - y}$
27. $\frac{3}{x + 1} + \frac{2}{x}$
28. $\frac{4}{y - 1} - \frac{3}{y}$
29. $\frac{2}{h} - \frac{h - 1}{h + 2}$
30. $\frac{f + 3}{f - 3} + \frac{3}{f}$
31. $1 + \frac{3c}{d} - \frac{5c^2}{2d^2}$
32. $\frac{3w}{x} + \frac{x}{w} - 2$
33. $\frac{3}{10y^2} - \frac{1}{2y^2} + \frac{x}{2y}$
34. $\frac{3a}{b} + \frac{4}{3b^2} - \frac{1}{2b^2}$
35. $\frac{2.3}{I} + \frac{5.8}{I + VI}$
36. $\frac{0.31}{P - nP} - \frac{0.46}{P}$
37. $\frac{3V}{2V + 2} + \frac{1}{2}$
38. $\frac{2n}{n - 2} - \frac{1}{3}$
39. $\frac{1}{e + 1} - \frac{1}{e - 1}$
40. $\frac{1}{R + 1} + \frac{5}{R - 1}$
41. $\frac{p + 1}{2} - \frac{2}{p - 1}$
42. $\frac{1}{Z + 2} - \frac{Z - 2}{2}$
43. $\frac{3}{t^2 - 1} - \frac{1}{t + 1}$
44. $\frac{1}{v - 2} - \frac{4}{v^2 - 4}$
45. $\frac{4}{3} - \frac{D + 1}{D} + \frac{D + 3}{3D}$
46. $\frac{X - 1}{X} + \frac{X + 2}{2X} - \frac{1}{2}$
47. $\frac{3r - 5}{r^2 - 1} - \frac{4}{r + 1}$
48. $\frac{2n^2}{k^2 - n^2} - \frac{k}{k + n}$
49. $\frac{3}{I_1 + I_2} - \frac{1}{I_1} + \frac{7}{I_2}$
50. $\frac{5}{V_1} - \frac{1}{V_2 - V_1} + \frac{8}{V_2}$

Applied Problems

51. Given a right triangle where $a^2 + b^2 = c^2$, show that the following expression equals $\frac{c^2}{b^2}$ by adding the terms:

$$\frac{a^2}{b^2} + 1$$

52. The force in a spring mechanism is given by:

$$F = \frac{3}{k} + \frac{2k - 3}{k + 2}$$

Simplify this expression by adding the fractions.

Applications to Electronics

53. Figure 9-3 on page 196 shows a circuit containing two resistors R_1 and R_2 in parallel. The equivalent resistance R_{eq} of this circuit is given by:

$$\frac{1}{R_{eq}} = \frac{1}{R_1} + \frac{1}{R_2}$$

Find the expression for R_{eq} by adding the two fractions and then taking the reciprocal of the sum.

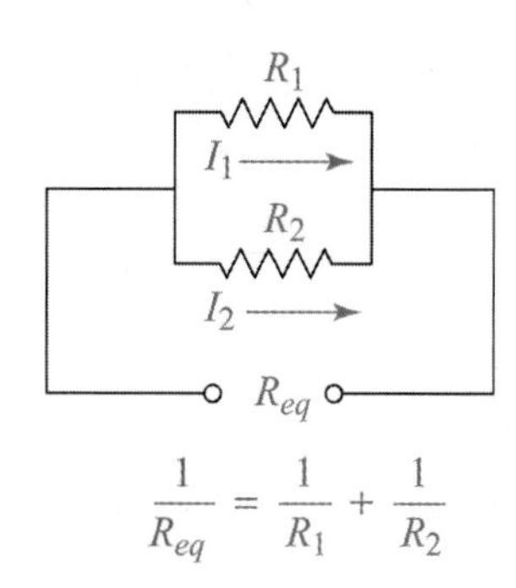

$$\frac{1}{R_{eq}} = \frac{1}{R_1} + \frac{1}{R_2}$$

FIGURE 9-3 Parallel resistors for problems 53 and 54.

54. In the circuit in Figure 9-3, if I_T = total current, then the current I_1 through R_1 and the current I_2 through R_2 are given by:

$$I_1 = \frac{I_T R_2}{R_1 + R_2} \qquad I_2 = \frac{I_T R_1}{R_1 + R_2}$$

By adding these fractions show that $I_T = I_1 + I_2$.

55. In Example 9.17 given the expression for I_T:

$$I_T = \frac{6}{R_1} + \frac{6}{R_1 + R_3}$$

Simplify the expression when $R_3 = R_2$.

56. In problem 55, simplify the expression for I_T, when $R_3 = R_1$.

57. In a circuit containing three resistors R_1, R_2 and R_3 in parallel, the equivalent resistance R_{eq} of this circuit is given by:

$$\frac{1}{R_{eq}} = \frac{1}{R_1} + \frac{1}{R_2} + \frac{1}{R_3}$$

Find the expression for R_{eq} by adding the three fractions and then taking the reciprocal of the sum.

58. A series-parallel circuit contains two series resistors R_1 and R_2 in parallel with a third resistor R_3. If $R_3 = R_1$, and $R_2 = 100\ \Omega$, the total current I_T is given by:

$$\frac{V}{R_1 + 100} + \frac{V}{R_1}$$

Simplify this expression for I_T by adding the fractions.

9.4 Equations with Fractions

Chapter 6 shows the solution of equations with fractions whose denominators are integers. This section extends these ideas to equations with fractions whose denominators contain algebraic terms. The algorithm given in Section 6.1 applies:

EQUATION SOLVING ALGORITHM

1. Eliminate fractions by multiplying by the LCD and clear parentheses.
2. Combine similar terms and isolate the unknown terms on one side.
3. Simplify by putting the equation in the form $Ax = B$, where x = unknown and A and B are constants.
4. Divide both sides by the coefficient of x to obtain the solution: $x = \frac{B}{A}$
5. Check the solution in the original equation.

EXAMPLE 9.18 Solve and check:

$$\frac{3}{2V} + \frac{2}{4V} = 1$$

Answers to Error Box Problems, page 193:

1. $\frac{b^2 - 1}{ab}$ **2.** $\frac{2 - a^2}{ab}$ **3.** $\frac{b}{a(a-b)}$ **4.** $\frac{b-a}{b(a+b)}$ **5.** $\frac{2a+b}{a(a-b)}$ **6.** $\frac{a-b}{2a}$

7. $-\frac{1}{a}$ **8.** $\frac{a^2 - 2b^2}{b(a+b)}$ **9.** $\frac{b^2}{a(a-b)}$ **10.** $\frac{-2b}{(a+b)(a-b)}$

Solution *It is not necessary to add the fractions.* Simplify the equation by eliminating the fractions first, as the algorithm states. Multiply every term in the equation by the LCD $= 4V$. The LCD can be written as a factor in the numerator of each fraction and the denominators divided into the LCD:

$$\frac{\overset{2}{\cancel{(4V)}}3}{\cancel{2V}} + \frac{\overset{1}{\cancel{(4V)}}2}{\cancel{4V}} = (4V)1$$

Observe that when each fraction is multiplied by the LCD, every denominator divides out and all fractions are eliminated. Do not confuse this procedure with addition of fractions. When you add fractions, you multiply the numerator *and* the denominator of the fraction. When you multiply an equation with fractions, you multiply *only the numerator* of each fraction. Now simplify the numerators and solve the equation without fractions:

$$6 + 2 = 4V$$
$$4V = 8$$
$$V = 2$$

The solution check is:

$$\frac{3}{2(2)} + \frac{2}{4(2)} = 1\,?$$
$$\frac{3}{4} + \frac{2}{8} = 1\,?$$
$$\frac{3}{4} + \frac{1}{4} = 1 \checkmark$$

EXAMPLE 9.19

Close the Circuit

The equivalent resistance R_{eq} of two parallel resistors is given by:

$$\frac{1}{R_{eq}} = \frac{1}{R_1} + \frac{1}{R_2}$$

If $R_1 = 1.5\ k\Omega$ and $R_2 = 2.0\ k\Omega$, find the value of R_{eq} in ohms and kilohms to two significant digits. Check the solution.

Solution You can substitute the values of R_1 and R_2 in ohms or kilohms. Using kilohms, the equation is:

$$\frac{1}{R_{eq}} = \frac{1}{1.5} + \frac{1}{2.0}$$

Eliminate the fractions by multiplying each fraction by the LCD = $(6)(R_{eq})$ and solve for R_{eq}:

$$\frac{(6)\cancel{(R_{eq})}}{\cancel{R_{eq}}} = \frac{\overset{4}{\cancel{(6)}}(R_{eq})}{\cancel{1.5}} + \frac{\overset{3}{\cancel{(6)}}(R_{eq})}{\cancel{2.0}}$$

Simplify: $6 = 4R_{eq} + 3R_{eq}$

Put unknown terms on the left side: $7R_{eq} = 6$

The solution rounded to two significant digits is:

$$R_{eq} = \frac{6}{7} \approx 0.86\ k\Omega = 860\ \Omega$$

To check the solution, use the calculator and press the reciprocal key three times:

$$\frac{1}{1.5} + \frac{1}{2.0} = \frac{1}{0.86}\,?$$

1.5 [x^{-1}] [+] 2.0 [x^{-1}] [=] [x^{-1}] → $0.857 \approx 0.86$ ✓

EXAMPLE 9.20 Solve and check:

$$\frac{2}{w-2} - \frac{9}{w} = \frac{1}{w}$$

Solution The first fraction contains a binomial denominator $(w - 2)$, which is one factor. The LCD = $(w)(w - 2)$. Multiply each term by the LCD to eliminate the denominators:

$$(w)\cancel{(w-2)}\frac{2}{\cancel{w-2}} - \cancel{(w)}(w-2)\frac{9}{\cancel{w}} = \cancel{(w)}(w-2)\frac{1}{\cancel{w}}$$

Notice that the minus sign in front of the second fraction means that −1 must be multiplied by each term as follows:

$$2w - (w - 2)(9) = w - 2$$

Clear parentheses: $2w - 9w + 18 = w - 2$

Combine terms: $-7w + 18 = w - 2$

Transpose w and 18: $-7w - w = -2 - 18$

Combine terms: $-8w = -20$

Divide by −8: $w = \frac{-20}{-8} = \frac{5}{2} = 2.5$

Check: $\frac{2}{2.5-2} - \frac{9}{2.5} = \frac{1}{2.5}$?

$4 - 3.6 = 0.4$?

$0.4 = 0.4$ ✓

EXERCISE 9.4

Round all answers to two significant digits.

In exercises 1 through 44, solve each equation with fractions.

1. $\frac{3a}{4} + \frac{a}{6} = 11$
2. $h - \frac{h}{2} = \frac{3}{4}$
3. $\frac{4V}{5} + \frac{V}{10} = V - 2$
4. $\frac{I}{15} - \frac{2I}{5} = I - 2$
5. $\frac{2x}{3} - (x - 1) = \frac{x}{6}$
6. $\frac{r}{5} = \frac{7r}{10} - (r - 2)$
7. $\frac{2}{3x} + \frac{1}{3x} = 1$
8. $\frac{5}{2y} - \frac{1}{2y} = 1$
9. $\frac{1}{10} - \frac{1}{6a} = \frac{1}{3a}$
10. $\frac{2}{b} + \frac{1}{2b} = \frac{5}{6}$
11. $\frac{3}{4P} + \frac{1}{2P} = \frac{5}{2}$
12. $\frac{5}{2R} + \frac{1}{3} = \frac{3}{R}$
13. $\frac{120}{500} = \frac{24}{N_p}$
14. $\frac{300}{10} = \frac{100}{I_p}$
15. $\frac{10}{V_S} = \frac{200}{50}$
16. $\frac{250}{N_S} = \frac{20}{4}$
17. $\frac{2}{3f} - \frac{1}{3} = \frac{3}{5f}$
18. $\frac{1}{3} + \frac{3}{4h} = \frac{5}{6h}$
19. $\frac{V_1 + 1}{2} = \frac{3V_1}{8}$
20. $\frac{2I_a + 1}{3} = \frac{3I_a + 2}{9}$
21. $\frac{1}{10} + \frac{1}{15} = \frac{1}{R}$
22. $\frac{1}{200} + \frac{1}{300} = \frac{1}{R}$
23. $\frac{1}{12} + \frac{1}{R} = \frac{1}{6}$
24. $\frac{1}{R} + \frac{1}{16} = \frac{1}{12}$
25. $\frac{3}{5} - \frac{4 - 15y}{10} = 2y$
26. $\frac{2x}{3} - \frac{3x - 1}{8} = 1$
27. $\frac{1}{2x} = \frac{3}{4x} - 1$
28. $\frac{9}{10y} = \frac{2}{5y} + 2$
29. $\frac{3}{2d} - \frac{9}{d^2} = \frac{6}{5d}$
30. $\frac{5}{4c} - \frac{7}{10c} = \frac{11}{2c^2}$
31. $\frac{3.2V_0}{15} + \frac{8.8V_0}{45} = \frac{9.2}{5.0}$
32. $\frac{0.15P_1}{10} - \frac{0.16}{6.0} = \frac{0.25P_1}{30}$
33. $\frac{1}{10} + \frac{1}{12} + \frac{1}{15} = \frac{1}{R_T}$
34. $\frac{1}{2} + \frac{1}{10} + \frac{1}{20} = \frac{1}{R_T}$
35. $\frac{1}{R_1} + \frac{1}{10} + \frac{1}{12} = \frac{1}{2}$
36. $\frac{1}{R_1} + \frac{1}{300} + \frac{1}{200} = \frac{1}{100}$
37. $\frac{4}{X} = \frac{5}{X - 1}$
38. $\frac{10}{Y + 2} = \frac{4}{Y}$
39. $\frac{5}{G + 3} - \frac{1}{G} = \frac{3}{2G}$
40. $\frac{5}{Z} + \frac{3}{Z - 4} = \frac{20}{Z}$
41. $\frac{2w - 1}{w + 3} + \frac{1}{w} = 2$
42. $\frac{3f}{f - 2} - 3 = \frac{5}{f}$

43. $\frac{6.6}{V-1} = \frac{8.8}{V+1}$

44. $\frac{0.03}{2R-1} = \frac{0.05}{3R+1}$

Applied Problems

45. If the interest earned on \$500 for 2 years is \$25, then the interest on \$800 for 5 years is given by the following equation:

$$\frac{(500)(2)}{25} = \frac{(800)(5)}{I}$$

Solve this equation for the interest I.

46. For a hydraulic lift, the ratio of the applied force F_1 to the resulting force F_2 is equal to the ratio of the squares of the diameters of the pistons:

$$\frac{F_2}{F_1} = \frac{D_2^2}{D_1^2}$$

What applied force F_1 is necessary to lift a 2400-kg automobile if $D_1 = 10$ cm and $D_2 = 80$ cm?

Applications to Electronics

47. For two resistors in parallel, the following proportion results from Ohm's law:

$$\frac{I_1}{R_2} = \frac{I_2}{R_1}$$

Find R_2 in ohms when $R_1 = 120\ \Omega$, $I_1 = 4.8$ A, and $I_2 = 2.2$ A.

48. When resistors R_1, R_2, and R_3 form the arms of a Wheatstone bridge (a device for accurately measuring the value of an unknown resistance R_x), the following proportion results when the bridge is balanced:

$$\frac{R_1}{R_x} = \frac{R_2}{R_3}$$

Find R_x in ohms when $R_1 = 110\ \Omega$, $R_2 = 33\ \Omega$, and $R_3 = 270\ \Omega$.

49. In Example 9.19 given the relationship:

$$\frac{1}{R_{eq}} = \frac{1}{R_1} + \frac{1}{R_2}$$

find R_{eq} in ohms and kilohms when $R_1 = 1.5\ \text{k}\Omega$ and $R_2 = 7.5\ \text{k}\Omega$

50. In problem 49, Find R_1 in ohms when $R_2 = 30\ \Omega$ and $R_{eq} = 12\ \Omega$

51. The change in the resistance of a conductor is proportional to the temperature change expressed by the formula:

$$\frac{R_2 - R_1}{R_4 - R_3} = \frac{T_2 - T_1}{T_4 - T_3}$$

If the resistance of an aluminum wire increases from $R_1 = 20\ \text{m}\Omega$ to $R_2 = 40\ \text{m}\Omega$ when the temperature increases from $T_1 = 0°\text{C}$ to $T_2 = 80°\text{C}$, how much will the resistance increase in milohms (mΩ) when the temperature increases from $T_3 = 0°\text{C}$ to $T_4 = 100°\text{C}$? Substitute the values in the formula and solve for the resistance increase $R_4 - R_3$ in milohms.

52. In a transformer, the voltage ratio is inversely proportional to the current ratio:

$$\frac{V_P}{V_S} = \frac{I_S}{I_P}$$

Find I_P in milliamps when $V_S = 12$ V, $V_P = 120$ V, and $I_S = 600$ mA.

CHAPTER HIGHLIGHTS

9.1 REDUCING FRACTIONS

Factor the numerator and denominator completely into primes and divide common factors.

Key Example:

$$\frac{9x+9}{3x^2-3} = \frac{\overset{3}{\cancel{9}}\cancel{(x+1)}}{\cancel{3}\cancel{(x+1)}(x-1)} = \frac{3}{(x-1)}$$

Study the Error Box in Section 9.1 to reinforce your understanding of factors and fractions.

9.2 MULTIPLYING AND DIVIDING FRACTIONS

Multiplication

Multiply across the top and bottom.

$$\frac{A}{B} \cdot \frac{C}{D} = \frac{AC}{BD}$$

• • •

Division

Multiply by the reciprocal.

$$\frac{A}{B} \div \frac{C}{D} = \frac{A}{B} \cdot \frac{D}{C} = \frac{AD}{BC}$$

Before multiplying or dividing, factor completely all numerators and denominators and divide out common factors.

Key Example:

$$\frac{2p-4}{2t}\cdot\frac{6t}{3p^2-6p}=\frac{2(\cancel{p-2})}{\cancel{2t}}\cdot\frac{\cancel{6t}}{\cancel{3}p(\cancel{p-2})}=\frac{2}{p}$$

9.3 ADDITION OF FRACTIONS

Rule of Fractions

Multiplying or dividing the numerator and the denominator by the same quantity (other than zero) does not change the value of a fraction.

Key Example:

$$\frac{7P}{30R}-\frac{P-1}{12R^2}$$

Factor each denominator into primes and take the *highest power* of each prime factor to make up the LCD:

$$\begin{array}{lcccccccl} 30R & = & 2 & \cdot & 3 & \cdot & 5 & \cdot & R \\ 12R^2 & = & 2^2 & \cdot & 3 & & \cdot & & R^2 \\ & = & \downarrow & & \downarrow & & \downarrow & & \\ \hline \text{LCD} & = & 2^2 & \cdot & 3 & \cdot & 5 & \cdot & R^2 = 60R^2 \end{array}$$

Apply the rule of fractions. Change all denominators to the LCD and add the fractions:

$$\frac{7P}{30R}-\frac{P-1}{12R^2}=\frac{(2R)7P}{(2R)30R}-\frac{(5)(P-1)}{(5)12R^2}$$

$$=\frac{(2R)7P-(5)(P-1)}{60R^2}=\frac{14PR-5P+5}{60R^2}$$

9.4 EQUATIONS WITH FRACTIONS

EQUATION SOLVING ALGORITHM

1. Eliminate fractions by multiplying by the LCD and clear parentheses.
2. Combine similar terms and isolate the unknown terms on one side and all the constant terms on the other side.
3. Simplify by putting the equation in the form $Ax = B$, where x = unknown and A and B are constants.
4. Divide both sides by the coefficient of x to obtain the solution: $x = \frac{B}{A}$
5. Check the solution in the original equation.

Key Example:

$$\frac{3}{2V}+\frac{2}{4V}=1$$

Multiply by the LCD = $4V$: $\frac{\overset{2}{\cancel{(4V)}}3}{\cancel{2V}}+\frac{\overset{1}{\cancel{(4V)}}2}{\cancel{4V}}=(4V)1$

Simplify: $6+2=4V$

Divide by 4: $4V=8$

$V=2$

The solution check is: $\frac{3}{2(2)}+\frac{2}{4(2)}=1\,?$

$$\frac{3}{4}+\frac{2}{8}=\frac{3}{4}+\frac{1}{4}=1\checkmark$$

REVIEW EXERCISES

In exercises 1 through 8, reduce each fraction to lowest terms if possible.

1. $\frac{75x^3}{15x}$
2. $\frac{8W^2-6W}{2W}$
3. $\frac{15ab}{3a^2b+12ab^2}$
4. $\frac{25R_1^2-R_2^2}{25R_1+5R_2}$
5. $\frac{P^2-1}{2P-1}$
6. $\frac{x^2+y^2}{x+y}$
7. $\frac{4f^2-4f+1}{4f^2-1}$
8. $\frac{6y^2+7y-3}{6y^2+11y+3}$

In exercises 9 through 34, perform the indicated operations.

9. $\frac{12y}{25x}\cdot\frac{20x^2}{27y}$
10. $\frac{c^2d}{cd-d^2}\cdot\frac{c-d}{3cd}$
11. $\frac{6aV^2}{35}\div\frac{4a^2V}{21a}$
12. $\frac{5g-25h}{g^2-25h^2}\div\frac{15h}{g+5h}$
13. $\frac{K}{L+M}\cdot\frac{KL+KM}{K^2}$
14. $\frac{0.25c^2-d^2}{0.3d}\cdot\frac{1.5c}{0.5c+d}$
15. $\frac{12x-3}{16x^2-1}\cdot(4x+1)$
16. $\frac{n^2-d^2}{n^2-2nd+d^2}\div(n+d)$
17. $\frac{3R_1+R_2}{2R_1+R_2}\cdot\frac{R_1+2R_2}{3R_1+R_2}$
18. $\frac{P^2-V^2}{PV-V^2}\cdot\frac{24P}{4P+4V}$
19. $\frac{2}{3}\cdot\frac{3x-3}{a+1}\cdot\frac{5a+5}{2x-2}$

20 $\frac{2y^2+y}{2y^2+9y-5}\cdot\frac{y^2+3y-10}{2y+1}$

21. $\frac{9}{5x} + \frac{12}{5x}$

22. $\frac{D}{12} - \frac{2D}{20}$

23. $\frac{5}{6r} + \frac{7}{10r}$

24. $\frac{7t}{V_1} + \frac{11t}{2V_2}$

25. $\frac{15}{14a^2b} - \frac{3}{7ab^2}$

26. $\frac{0.5}{3 \times 10^3} + \frac{1}{6 \times 10^4}$

27. $\frac{4}{x-1} + \frac{1}{2x}$

28. $\frac{2k}{3m} - \frac{3m}{5k} - 1$

29. $\frac{5}{V-5} + \frac{10}{v+5}$

30. $\frac{b}{b-y} - \frac{b+y}{b}$

31. $\frac{0.44}{t} + \frac{0.22}{t-rt}$

32. $\frac{1}{a^2 - x^2} - \frac{2}{a+x}$

33. $\frac{m-1}{2m} - \frac{m+1}{m} - \frac{3}{5}$

34. $\frac{5}{R_1 - R_2} - \frac{3}{R_2} + \frac{1}{R_1}$

In exercises 35 through 46, solve each equation with fractions. Round answers to two significant digits.

35. $\frac{x}{2} + \frac{3x}{4} = 10$

36. $\frac{9}{10X} - \frac{1}{2X} = 8$

37. $\frac{400}{100} = \frac{600}{V_s}$

38. $\frac{5}{R_x} = \frac{12}{56}$

39. $\frac{5}{3} - \frac{p+7}{21} = \frac{p}{7}$

40. $\frac{I_1 - 2}{4} = \frac{2I_1 + 3}{6}$

41. $\frac{1}{15} + \frac{1}{12} = \frac{1}{R}$

42. $\frac{1}{20} - \frac{1}{R_x} = \frac{1}{30}$

43. $\frac{3}{d^2} + \frac{2}{d} - \frac{1}{2d} = 0$

44. $\frac{1}{R} + \frac{1}{24} + \frac{1}{36} = \frac{1}{12}$

45. $\frac{5}{P+1} = \frac{2}{P} + \frac{2}{P+1}$

46. $\frac{3}{2V_1} - \frac{2}{V_1+1} = \frac{1}{6V_1}$

Applied Problems

47. The area of a rectangular microprocessor circuit is given by:

$$h^2 - 3h$$

and the width by:

$$\frac{2h-6}{9}$$

Find the expression for the length of the circuit in terms of h. Length = (area) ÷ (width).

48. The average rate for a round trip whose one way distance is D is given by:

$$R_{avg} = 2D \div \left(\frac{D}{r_1} + \frac{D}{r_2}\right)$$

where r_1 = average rate going and r_2 = average rate returning.

(a) Simplify the formula for R_{avg} by adding the fractions and then dividing.

(b) If you average $r_1 = 20$ mi/h going one way on a trip and $r_2 = 30$ mi/h returning, what is your average rate for the entire trip? Substitute in the formula from (a).

Note: The answer is less than 25 mi/h.

Applications to Electronics

49. In a circuit, the resistance as a function of time t is given by:

$$R = 0.5t^2 - 0.5t$$

and the current by:

$$I = \frac{2t}{t-1}$$

Using Ohm's law $V = IR$, find the expression for the voltage as a function of t.

50. The power in a circuit as a function of time t is given by:

$$P = \frac{4t^2}{(t-1)^2}$$

and the current by:

$$I = \frac{t^2}{t^2 - 1}$$

Using the power law $V = \frac{P}{I}$, find the expression for the voltage as a function of t.

51. The total resistance in a series-parallel circuit is given by:

$$R_T = 2R_2 + \frac{R_2 R_3}{R_2 + R_3}$$

Simplify this expression for R_T by combining all the terms into one fraction.

52. A series-parallel circuit contains two series resistors R_1 and R_2 in parallel with the resistor R_3. When $R_2 = 75\ \Omega$ and $R_3 = 15\ \Omega$, the total current is given by:

$$I_T = \frac{V}{R_1 + 75} + \frac{V}{15}$$

Simplify this expression for I_T by combining all the terms into one fraction.

53. The total capacitance C_T of two capacitors in series is given by:

$$\frac{1}{C_T} = \frac{1}{C_1} + \frac{1}{C_2}$$

Find C_1 in microfarads when $C_T = 20\ \mu F$ and $C_2 = 30\ \mu F$.

54. In problem 53, find C_2 in picofarads when $C_T = 100$ pF and $C_1 = 300$ pF.

55. Assuming no mutual inductance, the total inductance of three inductors in parallel is given by:

$$\frac{1}{L_T} = \frac{1}{L_1} + \frac{1}{L_2} + \frac{1}{L_3}$$

Find L_T in millihenries when $L_1 = 5.0$ mH, $L_2 = 5.0$ mH, and $L_3 = 10$ mH.

56. In problem 55, find L_3 in henries when $L_T = 500$ mH, $L_1 = 1.0$ H, and $L_2 = 2.0$ H. *Note:* The units must be consistent in the equation.

57. The total resistance of a series-parallel circuit containing a resistor R_1 in series with two parallel resistors R_2 and R_3 is given by:

$$R_T = R_1 + \frac{R_2 R_3}{R_2 + R_3}$$

Find R_2 in ohms when $R_T = 150\ \Omega$, $R_1 = 90\ \Omega$, and $R_3 = 150\ \Omega$.

58. In problem 57, if $R_T = 18\ \Omega$, $R_1 = 12\ \Omega$, and $R_3 = 2R_2$, find R_2 and R_3 in ohms.

CHAPTER 10

Formulas and Problem Solving

CHAPTER OBJECTIVES

In this chapter, you will learn:

- The meaning of a literal equation.
- How to solve formulas and literal equations for one of the variables.
- How to analyze and formulate an equation for a verbal problem.
- How to find and check the solution to a verbal problem.

Electricity and electronics depend on formulas and equations to express relationships and concepts, and to solve all types of problems. It is not an easy task to translate verbal problems into mathematics in order to find their solution. However, people must communicate ideas in English, Spanish, Chinese, or some other verbal language. Therefore, if you are to apply the skills you are learning, you must be able to translate verbal problems into mathematical formulas and be able to work with formulas. You need to develop the ability to apply and manipulate formulas and to solve equations with formulas. The first part of this chapter shows you how to simplify and solve formulas for different variables using the methods of equation solving. The second part emphasizes the very important and critical skill of problem solving. It shows you a step-by-step procedure designed to help you analyze a problem and set up an equation to find the solution. Many types of problems are discussed, and many different types of exercises are given to help you master the techniques.

• • •

10.1 Formulas and Literal Equations

A formula that contains an equal sign and two or more variables is also called a *literal equation.* By applying the *Equation Principle* given in Chapter 6 to a formula or a literal equation, you can solve it for different variables or put it in a more useful form:

Principle

Equation Principle

The following operations do not change an equality:

1. Adding or subtracting the same quantity on both sides of an equation.
2. Multiplying or dividing both sides of an equation by the same quantity. (*Division by zero is not allowed.*)

For example, by applying the equation principle, Ohm's law can be changed to any of its various forms so you don't have to memorize each form. This can provide an easier or more direct way to solve a problem. Study the following examples, which show how to use these concepts in electricity and electronics.

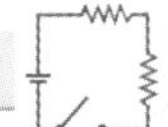

EXAMPLE 10.1

Close the Circuit

Given the power formula $P = VI$ and Ohm's law solved for I, $I = \frac{V}{R}$, derive the following formula:

$$P = \frac{V^2}{R}$$

Solution You need to combine the two formulas and eliminate I. You can apply the following substitution principle:

Principle

Substitution Principle

In an equation, any quantity may be substituted for an equal quantity.

Start with Ohm's law:

$$I = \frac{V}{R}$$

Substitute this expression for I in the formula $P = VI$ as follows:

$$P = VI = V\left(\frac{V}{R}\right)$$

Multiply out and you obtain the formula for P:

$$P = V\left(\frac{V}{R}\right) = \frac{V^2}{R}$$

EXAMPLE 10.2

Close the Circuit

Given the formula in Example 10.1:

$$P = \frac{V^2}{R}$$

Solve this formula for V.

Solution Multiply both sides by R and isolate V^2:

$$(R)P = (\cancel{R})\frac{V^2}{\cancel{R}}$$

You now have: $RP = V^2$ or $V^2 = PR$

To solve for V, take the positive square root of both sides of the formula:

$$\sqrt{V^2} = \sqrt{PR}$$
$$V = \sqrt{PR}$$

The following examples introduce some different formulas and show how to solve them for one of the variables by using different techniques. The meaning of the formula and what it represents is explained in each example. Study these examples carefully to learn the different techniques.

EXAMPLE 10.3

The formula for the velocity v of a body with constant acceleration a is:

$$v = v_0 + at$$

where v_0 = initial velocity and t = time. Solve this formula for a.

Solution First isolate the term at:

Transpose v_0: $v - v_0 = at$

Divide both sides by t: $\frac{v - v_0}{(t)} = \frac{a\cancel{t}}{(\cancel{t})}$

The formula for a is then: $a = \frac{v - v_0}{t}$

EXAMPLE 10.4 The formula for the surface area A of a cylinder is:

$$A = 2\pi rh + 2\pi r^2$$

where π is a constant ≈ 3.14, r = radius, and h = height. Solve this formula for h.

Solution

Transpose πr^2: $$A - 2\pi r^2 = 2\pi rh$$

Divide by $2\pi r$: $$A - \frac{A - 2\pi r^2}{(2\pi r)} = \frac{2\pi rh}{(2\pi r)}$$

The formula for h is then: $$h = \frac{A - 2\pi r^2}{2\pi r}$$

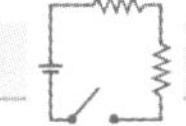

Close the Circuit

The formula for the total voltage V_T of a series circuit with two resistors R_1 and R_2 is:

$$V_T = IR_1 + IR_2$$

where I = current. Solve this formula for I.

Solution Since I appears in two terms, you must first separate I as a common factor:

$$V_T = I(R_1 + R_2)$$

Then divide both sides of the equation by the other factor in parentheses, $(R_1 + R_2)$, which is the coefficient of I:

$$\frac{V_T}{(R_1 + R_2)} = \frac{I(R_1 + R_2)}{(R_1 + R_2)}$$

Writing I on the left, the formula for I is then:

$$I = \frac{V_T}{(R_1 + R_2)}$$

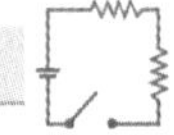

Close the Circuit

The formula for the equivalent resistance R_{eq} of two resistors R_1 and R_2 in parallel is:

$$R_{eq} = \frac{R_1 R_2}{R_1 + R_2}$$

Solve this formula for R_1.

Solution You need to isolate the terms containing R_1. First clear the fraction by multiplying both sides by the LCD, which is the binomial $(R_1 + R_2)$:

$$(R_1 + R_2)R_{eq} = (R_1 + R_2)\frac{R_1 R_2}{R_1 + R_2}$$

Multiply out the parentheses on the left:

$$R_1 R_{eq} + R_2 R_{eq} = R_1 R_2$$

Now transpose the term containing R_1 on the left side to the right side:

$$R_2 R_{eq} = R_1 R_2 - R_1 R_{eq}$$

Factor out R_1 on the right side:

$$R_2 R_{eq} = R_1(R_2 - R_{eq})$$

Divide by the coefficient of R_1, $(R_2 - R_{eq})$:

$$\frac{R_2 R_{eq}}{(R_2 - R_{eq})} = \frac{R_1(\cancel{R_2 - R_{eq}})}{(\cancel{R_2 - R_{eq}})}$$

Writing R_1 on the left, the formula for R_1 is then:

$$R_1 = \frac{R_2 R_{eq}}{R_2 - R_{eq}}$$

The next example also closes the circuit and shows the usefulness of combining formulas to produce a new electrical formula, which can be used to solve a problem in a more direct way.

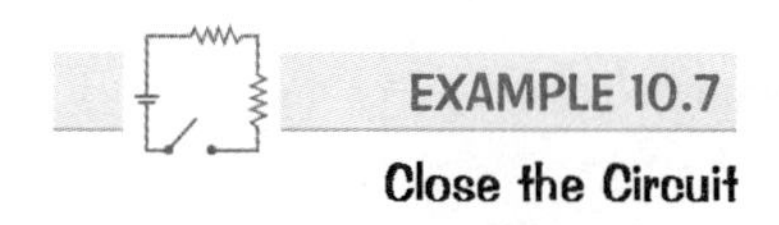

EXAMPLE 10.7

Close the Circuit

Consider the circuit in Figure 10-1, which contains a variable resistor, or rheostat, in series with a light bulb and an ammeter to measure the current. As the rheostat is varied, the total resistance in the circuit will vary. This will change the current in the circuit, causing the light to vary in brightness and the power consumption to change. Suppose the voltage is constant at 120 V and the rheostat is varied so the power changes from P_1 to P_2. When the power = P_1, the ammeter measures the current I_1 = 1.5 A. When the power = P_2, the ammeter measures the current I_2 = 2.4 A.

(a) Using the power law $P = VI$, find the change in the power of the circuit. This is represented algebraically as $\Delta P = P_2 - P_1$, where Δ = Greek letter delta for "difference."

(b) Find a formula for ΔP in terms of the change in the current: $\Delta I = I_2 - I_1$.

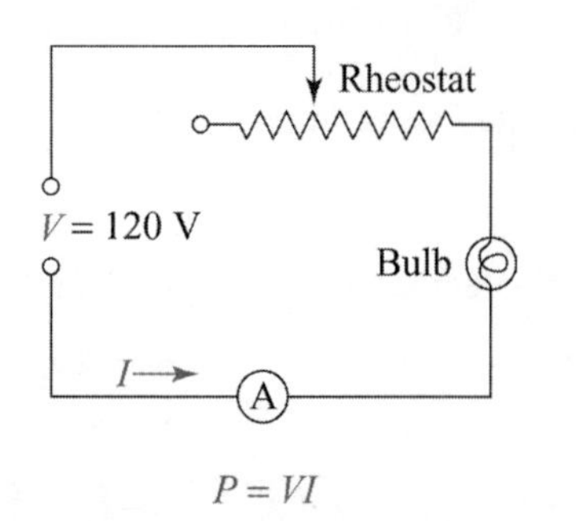

FIGURE 10-1 Rheostat circuit for Example 10.7.

Solution

(a) To find the change in the power, apply the power law and find P_1 and P_2:

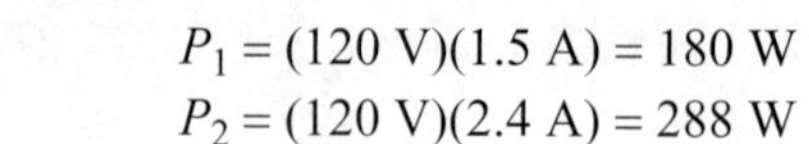

$$P_1 = (120 \text{ V})(1.5 \text{ A}) = 180 \text{ W}$$
$$P_2 = (120 \text{ V})(2.4 \text{ A}) = 288 \text{ W}$$

The change in the power is then:

$$\Delta P = P_2 - P_1 = 288 \text{ W} - 180 \text{ W} = 108 \text{ W}$$

(b) To find a formula for ΔP in terms of ΔI, start with the formulas for P_1 and P_2:

$$P_1 = VI_1$$
$$P_2 = VI_2$$

Taking the difference of these formulas, the change in the power in terms of V and I is then:

$$\Delta P = P_2 - P_1 = VI_2 - VI_1$$

Factoring out V, the formula can be written in terms of ΔI as follows:

$$\Delta P = V(I_2 - I_1) = V(\Delta I)$$

You now have a useful formula for the change in power in terms of the change in current. This formula can be used to directly obtain the answer to part (a).
Substituting V = 120 V, I_1 = 1.5 A and I_2 = 2.4 A:

$$\Delta P = V(\Delta I) = (120 \text{ V})(2.4 \text{ A} - 1.5 \text{ A}) = (120 \text{ V})(0.9 \text{ A}) = 108 \text{ W}$$

Deriving the formula for ΔP may at first appear to be more involved than the solution in part (a), however, if the change in power needs to be calculated several times for various current changes, then the derived formula provides a direct way to get the results. For example, suppose the current decreases 0.5A, that is:

$$\Delta I = -0.5 \text{ A}$$

The decrease in power is then:

$$\Delta P = V(\Delta I) = (120 \text{ V})(-0.5 \text{ A}) = -60 \text{ W}$$

This formula shows the relationship between the change in power and the change in current, which can be more useful than the actual values.

EXERCISE 10.1

In exercises 1 through 34, solve each formula for the indicated letter.

1. $P = VI; I$ — (P = power in a circuit)

2. $V = \frac{P}{I}; P$ — (V = voltage in a circuit)

3. $C = 2\pi r; r$ — (C = circumference of a circle)

4. $A = lw; w$ — (A = area of a rectangle)

5. $I^2 = \frac{P}{R}; P$ — (I = current in a circuit)

6. $R = \frac{V^2}{P}; P$ — (R = resistance in a circuit)

7. $a = \frac{F}{m}; m$ — (a = acceleration of a body)

8. $v = v_0 + at; t$ — (v = velocity of a body)

9. $X_C = \frac{1}{2\pi fC}; f$ — (X_C = capacitive reactance)

10. $C = \frac{eA}{d}; e$ — (C = capacitance of two parallel plates)

11. $R = \frac{V^2}{P}; V$ — (R = resistance in a circuit)

12. $P = I^2R; I$ — (P = power in a circuit)

13. $A = \pi r^2; r$ — (A = area of a circle)

14. $A = \frac{s^2}{2}; s$ — (A = area of an isosceles right triangle)

15. $V = I(R + r_g); r_g$ — (V = generator voltage)

16. $P = I(V_1 + V_2); V_1$ — (P = power in a series circuit)

17. $P = 2(l + w); w$ — (P = perimeter of a rectangle)

18. $A = 2w(2l + w); l$ — (A = surface area of a rectangular solid)

19. $V = V_0 + IR$; I (V = voltage in a circuit)

20. $V = B - \dfrac{V_0}{A}$; V_0 (V = transistor voltage)

21. $P = VI_1 + VI_2$; V (P = power in a parallel circuit)

22. $IR_1 + IR_2 = I_0R_0$; I (I = current in a parallel circuit)

23. $A = l_1w + l_2w$; w (A = area of rectangular border)

24. $A = \pi r_1^2 - \pi r_2^2$; π (A = area of a ring)

25. $V_T = IR_1 + IR_2 + IR_3$; I (V_T = total voltage of series circuit)

26. $P_T = VI_1 + VI_2 + VI_3$; V (P_T = total power in a parallel circuit)

27. $I_C = \dfrac{\beta I_E}{\beta + 1}$; β (I_C = collector current in transistor)

28. $I_1 = \dfrac{I_T G_1}{G_1 + G_2}$; G_1 (I_1 = current in parallel circuit)

29. $R_{eq} = \dfrac{R_1 R_2}{R_1 + R_2}$; R_2 (R_{eq} = equivalent resistance of R_1 and R_2 in parallel)

30. $L_T = \dfrac{L_1 L_2}{L_1 + L_2}$; L_1 (L_T = total inductance of L_1 and L_2 in parallel)

31. $\dfrac{1}{L_T} = \dfrac{1}{L_1} + \dfrac{1}{L_2}$; L_2 (L_T = total inductance of L_1 and L_2 in parallel)

32. $\dfrac{1}{R_{eq}} = \dfrac{1}{R_1} + \dfrac{1}{R_2}$; R_1 (R_{eq} = equivalent resistance of R_1 and R_2 in parallel)

33. $Z = \dfrac{1}{\frac{1}{\omega L} - \omega C}$; L (Z = impedance of parallel LC circuit)

34. $R_{AB} = \dfrac{R_A R_B}{R_A + R_B + R_C}$; R_A (Star to delta resistance conversion)

Applied Problems

In problems 35 through 56, do each problem by solving a literal equation. Round answers to two significant digits.

35. The boiling point of water in degrees Fahrenheit at a height h feet above sea level is given approximately by the formula:

$$T = 212 - \frac{h}{550}$$

(a) Solve this formula for h.
(b) What is the increase in height if T changes from 211°F to 209°F?

36. The velocity of a falling body is given by:

$$v = v_0 + gt$$

where v_0 = initial velocity, t = time and g = gravitational acceleration.

(a) Solve this formula for t.
(b) If v_0 = 7.5 m/s, and g = 9.8 m/s^2, how long will it take for this velocity to triple?

37. Given the temperature conversion formula:

$$F = \frac{9}{5}C + 32$$

(a) Solve this formula for C.
(b) What is the increase in degrees Celsius if the temperature doubles from 45°F to 90°F?

38. In problem 37:
(a) If F changes from F_1 to F_2 when C changes from C_1 to C_2, find the formula for the change in degrees Fahrenheit, $\Delta F = F_2 - F_1$, in terms of $\Delta C = C_2 - C_1$. (See Example 10.7.)
(b) Using the formula found in (a), if the temperature falls 50° on the Celsius scale, how many degrees does it fall on the Fahrenheit scale?

Applications to Electronics

39. Using the power formula $P = VI$ and Ohm's law $V = IR$, derive the formula:

$$P = I^2R$$

40. Using the voltage formula $V = \frac{P}{I}$ and Ohm's law $R = \frac{V}{I}$, derive the formula:

$$R = \frac{P}{I^2}$$

41. Using the power formula $P = VI$ and Ohm's law $I = \frac{V}{R}$, derive the formula:

$$R = \frac{V^2}{P}$$

42. Using the power formula $P = I^2R$ and Ohm's law $I = \frac{V}{R}$, derive the formula:

$$P = \frac{V^2}{R}$$

43. The battery voltage in the series circuit in Figure 10-2 is given by $V = V_0 + IR_1$, where V_0 is the voltage drop across R_0.
(a) Solve this formula for R_1.
(b) Find R_1 when $V = 12$ V, $V_0 = 4.8$ V, and $I = 1.2$ A.

44. In problem 43:
(a) Solve the voltage formula $V = V_0 + IR_1$ for I.
(b) Using the formula found in (a), if $V = 7.5$ V, $V_0 = 4.5$ V, and $R_1 = 150$ Ω, find I in amps and milliamps.

45. The equivalent resistance R_{eq} of the parallel circuit in Figure 10-3 is given by:

$$R_{eq} = \frac{R_1R_x}{R_1 + R_x}$$

Solve this formula for the unknown resistance R_x.

46. The equivalent resistance R_{eq} of the parallel circuit in Figure 10-3 is related to R_1 and R_x by the formula:

$$\frac{1}{R_{eq}} = \frac{1}{R_1} + \frac{1}{R_x}$$

Solve this formula for the unknown resistance R_x. *Hint:* First multiply each term by the LCD.

47. The resistance of a material is a linear function of its temperature t in degrees Celsius expressed by:

$$R_t = R_0 + R_0\alpha t$$

where R_0 = resistance at 0°C and α (alpha) = temperature coefficient of resistance.
(a) Solve this formula for α.
(b) Find α when $R_t = 25$ Ω, $R_0 = 20$ Ω, and $t = 80$°C.

48. In problem 47:
(a) Solve the resistance formula for R_0.
(b) Find R_0 if $R_t = 340$ Ω when $t = 20$°C and $\alpha = 0.0018$ per degree Celsius.

49. The resistance of a wire is directly proportional to its length l and inversely proportional to its cross sectional area A expressed by the formula:

$$R = \frac{\rho l}{A}$$

The constant ρ (rho) is called the resistivity and has the units Ω-m (ohm-meter).
(a) Solve this formula for ρ.
(b) Find ρ when $R = 3.5$ mΩ, $l = 10$ m, and $A = 60 \times 10^{-6}$ m².

Note: R must be in ohms.

50. Figure 10-4 shows a resistor R_1 in parallel with two series resistors R_2 and R_3. The total resistance R_T of the series-parallel circuit can be expressed by the formula:

$$\frac{1}{R_T} = \frac{1}{R_1} + \frac{1}{R_2 + R_3}$$

(a) If $R_2 = R_3$, solve this formula for R_T in terms of R_1 and R_2.
(b) Find R_T in kilohms when $R_1 = 20$ kΩ and $R_2 = 10$ kΩ.

51. In a circuit containing a resistance, an inductance, and a capacitance in series, resonance occurs when the inductive

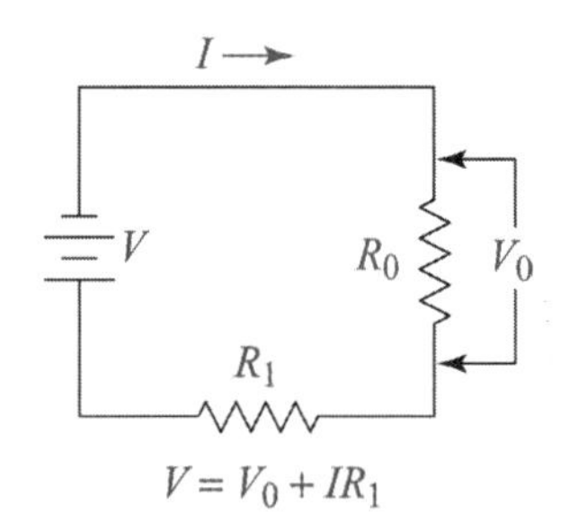

FIGURE 10-2 Series circuit for problems 43 and 44.

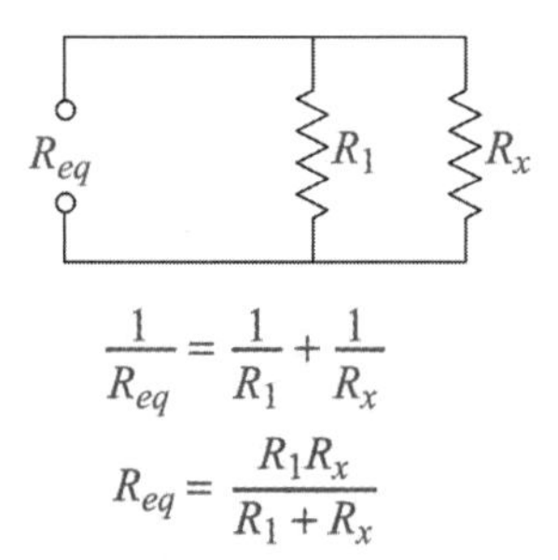

FIGURE 10-3 Parallel circuit for problems 45 and 46.

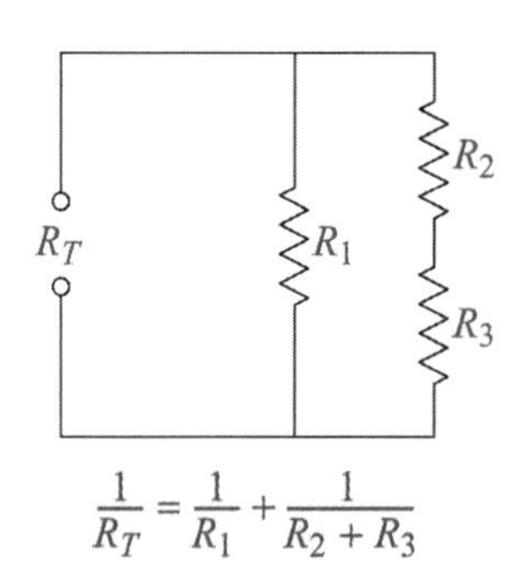

FIGURE 10-4 Series-parallel circuit for problem 50.

reactance equals the capacitive reactance given by the relationship:

$$2\pi f L = \frac{1}{2\pi f C}$$

Solve this formula for the frequency f.

52. The total power dissipation of two resistors R_1 and R_2 in series is given by:

$$P_T = I^2R_1 + I^2R_2$$

(a) Solve this formula for I.

(b) If $P_T = 100$ W, $R_1 = 2.0$ kΩ, and $R_2 = 1.6$ kΩ, calculate the current.

53. Figure 10-5 contains a rheostat in series with a light bulb whose resistance $R = 80\ \Omega$.

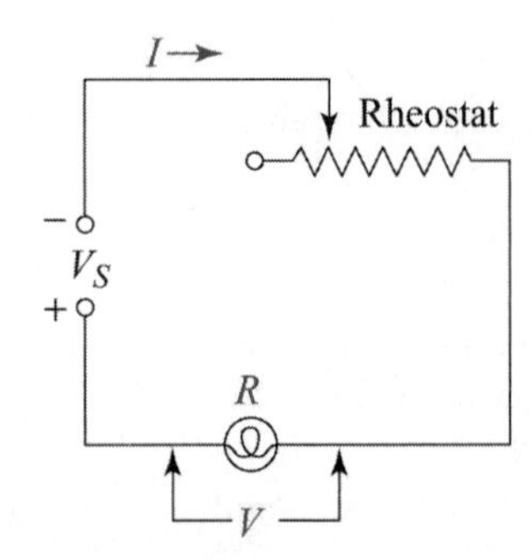

FIGURE 10-5 Rheostat with bulb for problems 53 through 56.

(a) Using Ohm's law $V = IR$, find the change in the voltage drop across the bulb when the current changes from $I_1 = 1.3$ A to $I_2 = 1.1$ A. Assume the resistance of the bulb is constant.

(b) Find the formula for the change in the voltage drop across the light bulb, $\Delta V = V_2 - V_1$, in terms of the current change $\Delta I = I_2 - I_1$.

(c) Check your answer to (a) using the formula for ΔV. See Example 10.7.

(d) Using the formula for ΔV, find ΔV in volts when $\Delta I = 400$ mA.

54. In Figure 10-5, given V_S = source voltage:

(a) Using the power formula $P = VI$, find the formula for the power change, $\Delta P = P_2 - P_1$, in the circuit in terms of the current change $\Delta I = I_2 - I_1$.

(b) Find ΔP when the current changes from $I_1 = 1.7$ A to $I_2 = 1.5$ A, and $V_S = 110$ V. See Example 10.7.

55. In problem 53:

(a) Find the formula for ΔV if the current doubles from I to $2I$.

(b) Find the formula for ΔV if the current decreases by 10%.

56. In problem 54:

(a) Find the formula for ΔP if the current increases by 50% from I to $1.5I$.

(b) Find the formula for ΔP if the current decreases by one-half its value.

10.2 Problem Solving

There are many types of verbal and applied problems. Some involve substitution into given formulas, some require knowledge of certain relationships, and some require that you formulate an equation from the given information. These problem-solving techniques require a certain skill and the best way to develop this skill is by doing many problems. You need to spend the time and apply yourself, but you should not be discouraged. Keep at it, and you will develop the skill to solve verbal and applied problems. This will not only help you to think more clearly, but you will find working with mathematics more enjoyable. Study the first example carefully. It illustrates the basic procedures of problem solving and is a useful practical problem.

EXAMPLE 10.8

Suppose the grade on your first test in this course is 76, and the grade on your second test is 78. What grade do you need on the third test if you want to raise your average for the three tests to 80?

Solution Observe the order of steps used in solving the problem. First, determine what you are trying to find and pick a representative letter to represent the unknown. For example,

$$\text{Let } g = \text{Grade needed on the third test}$$

Then ask yourself, "What should the equation say?" Here is the "verbal equation":

$$\frac{(\text{First grade}) + (\text{Second grade}) + (\text{Third grade})}{3} = \text{Average of three tests}$$

The average is the sum of the three tests divided by 3. This is what you should be thinking about in setting up the problem. You now need an algebraic expression for the average of the three tests in terms of g:

$$\text{Average} = \frac{76 + 78 + g}{3}$$

The algebraic equation follows from the verbal equation:

$$\frac{76 + 78 + g}{3} = 80$$

Multiply both sides by 3 to clear the fraction and solve for g:

$$\begin{aligned} (\cancel{3})\frac{76 + 78 + g}{\cancel{3}} &= (3)80 \\ 76 + 78 + g &= 240 \\ g &= 240 - 76 - 78 \\ g &= 86 \end{aligned}$$

Check the answer in the original equation:

$$\frac{76 + 78 + 86}{3} = \frac{240}{3} = 80 \checkmark$$

You should also *check that the answer makes logical sense.* This last step is essential. Very often you can quickly discover an incorrect answer to a verbal problem by seeing that it does not fit the facts. For example, if you obtained an answer of 70, you would know right away that something is wrong. If you want to raise your test average to 80, you need a grade higher than 80 to balance your lower grades.

Example 10.8 illustrates the following step-by-step method, or algorithm, for problem solving:

PROBLEM SOLVING ALGORITHM

1. Study the problem several times and determine what is given and what you are trying to find.
2. Choose a meaningful letter for the unknown and write down clearly what it represents. For example, t for time, p for price, d for distance, V for voltage, P for power.
3. Determine what the equation should say in words, that is, the verbal equation.
4. Write down the expressions for any other quantities, in terms of the unknown, that are needed for the equation.
5. Formulate and solve the algebraic equation.
6. Check that the answer makes logical sense and that it satisfies the equation.

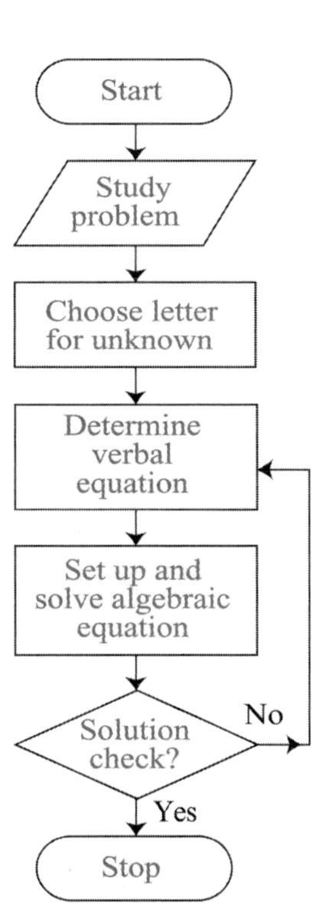

FIGURE 10-6 Flowchart for problem solving.

See Figure 10-6, which illustrates this algorithm as a flowchart.

EXAMPLE 10.9

Last week your friend bought you an electrical engineering graphics calculator, which cost $84.50. The cost included 8% sales tax. Today you find out that the calculator can be bought for the same price on the internet with no sales tax and only a 3% shipping charge.

(a) Find the price of the calculator before tax.

(b) Find the total cost over the internet.

(c) How much could your friend have saved by buying the calculator online?

Solution

(a) Choose a meaningful letter for the unknown:

Let p = Price of the calculator before tax.

Determine the verbal equation:

Price + Tax = Total cost

Set up the expression for the tax. The tax is 8% times the price. Therefore:

$$\text{Tax} = 8\%p = 0.08p$$

The algebraic equation follows from the verbal equation:

$$p + 0.08p = 84.50$$

Solve the equation. Change p to $1.00p$ to clarify the arithmetic:

$$\begin{aligned} 1.00p + 0.08p &= 84.50 \\ 1.08p &= 84.50 \\ p &= \frac{84.50}{1.08} \approx \$78.24 \text{ (nearest cent)} \end{aligned}$$

Check that the answer makes sense. Since it is several dollars less than the total cost, it does seem approximately correct. Check the answer in the equation:

$$78.24 + 0.08(78.24) = 78.24 + 6.26 = 84.50 \checkmark$$

(b) The cost of the calculator plus 3% shipping is:

$$78.24 + 0.03(78.24) = \$80.59$$

(c) Subtract this cost from the original cost to find the savings:

$$\text{Savings} = \$84.50 - \$80.59 = \$3.91$$

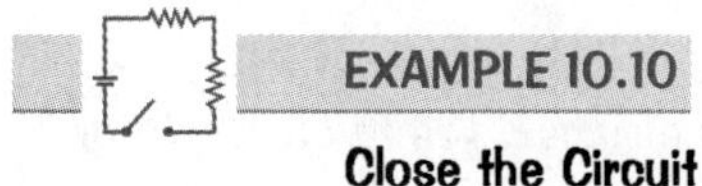

EXAMPLE 10.10

Close the Circuit

A copper wire one meter long is to be cut in such a way that the electrical resistance of one piece is two-thirds the electrical resistance of the other piece. Given that electrical resistance is directly proportional to length, how long should each piece be in *centimeters*?

Solution Since electrical resistance is directly proportional to length, the length of the short piece should be two-thirds the length of the long piece. The length of the short piece is given *in terms* of the long piece; therefore, choose the unknown to be the long piece:

Let l = length of the long piece

Think of the verbal equation:

Long piece + short piece = 1 m = 100 cm

Note that the length is changed to centimeters, which is what the problem asks for. Set up the expression for the short piece:

$$\text{Length of the short piece} = \frac{2}{3}l$$

Set up the algebraic equation from the verbal equation:

$$l + \frac{2}{3}l = 100$$

Solve the equation, clearing the fraction first:

$$(3)l + (\cancel{3})\frac{2}{\cancel{3}}l = (3)100$$

$$3l + 2l = 300$$

$$5l = 300$$

$$l = 60 \text{ cm}$$

The length of the short piece is then:

$$\frac{2}{3}l = \frac{2}{3}(60) = 40 \text{ cm}$$

The answers make sense and they check in the equation:

$$60 + \frac{2}{3}(60) = 60 + 40 = 100 \checkmark$$

As mentioned previously, skill in problem solving comes through constant practice. As in any sport, it is not difficult to understand the rules, but only through constant practice can you really play well!

EXAMPLE 10.11

Close the Circuit

Figure 10-7 shows three resistors connected in series and the voltage drops across each. The total voltage $V_T = 70$ V. The voltage drop V_2 across resistor R_2 is 50% of V_1, and the voltage drop V_3 across R_3 is 25% of V_1. In a series circuit, the total voltage equals the sum of the voltage drops. Applying this concept, what is the value of each voltage drop in volts?

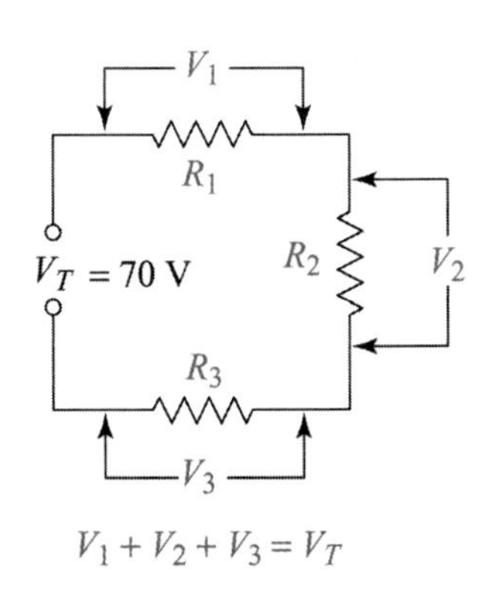

FIGURE 10-7 Resistors in series for Example 10.11.

Solution Since the voltage drops V_2 and V_3 are given in terms of V_1, start with:

Let V_1 = voltage drop across R_1

Apply the concept that the total voltage equals the sum of the voltage drops to formulate the verbal equation:

Total voltage = Voltage across R_1 + Voltage across R_2 + Voltage across R_3

Set up the expressions for the other quantities in the equation:

$$V_2 = 50\%V_1 = 0.50V_1$$

$$V_3 = 25\%V_1 = 0.25V_1$$

Set up the algebraic equation from the verbal equation:

$$V_T = V_1 + V_2 + V_3$$

Substitute $0.50V_1$ for V_2, $0.25V_1$ for V_3, and 70 for V_T in the algebraic equation:

$$V_1 + 0.50V_1 + 0.25V_1 = 70$$

Solve the equation:

$$1.75V_1 = 70$$

$$V_1 = \frac{70}{1.75} = 40\text{V}$$

Then:

$$V_2 = 0.50V_1 = 20\text{V}$$

$$V_3 = 0.25V_1 = 10\text{V}$$

Check:

$$40 \text{ V} + 20 \text{ V} + 10 \text{ V} = 70 \text{ V} \checkmark$$

ERROR BOX

A common error to watch out for is incorrectly changing a verbal phrase into an algebraic expression by applying the words directly. For example, the phrase "The voltage of the first battery is two volts more than the voltage of the second battery," may lead to the equation $V_1 + 2 = V_2$. This is not correct. The first battery has the *higher* voltage and you need to add the 2 to the smaller quantity, V_2, so the equality is correct:

$$V_1 = V_2 + 2$$

Another example is the phrase "One current I_1 is 25% less than I_2," which may lead to the equation $I_1 = I_2 - 25\%$. This is not correct. Twenty-five percent less than I_2 means: *Multiply I_2 by 25% and subtract it from I_2:*

$$I_1 = I_2 - 25\%(I_2) \quad or \quad I_1 = I_2 - 0.25I_2$$

See if you can apply these ideas and do each of the practice problems correctly.

Practice Problems: For each phrase, write an algebraic equation:

1. The first resistance R_1 is 50 less than the second resistance R_2.
2. The second voltage drop V_2 is 10 V more than the first voltage drop V_1.
3. The current in the third resistor I_3 is 50% more than the current in the second resistor I_2.
4. The power in the first circuit P_1 is 10% less than the power in the second circuit P_2.
5. The discount computer cost C_d is reduced by one-third of the retail cost C_r.
6. The car speed r_c exceeds the train speed r_t by 20 mi/h.
7. The power input P_{in} of the amplifier is 15% less than the power output P_{out}.
8. The first capacitance C_1 is 10% less than twice the second capacitance C_2.
9. If the time t_1 is 2 s more than t_2, and 3 s less than t_3, how is t_2 related to t_3?
10. If resistor R_2 is twice R_1, and R_3 is one-third of R_2, how is R_3 related to R_1?

EXAMPLE 10.12

Sharon paddles her canoe upstream in 3 hours but only takes 2 hours to return downstream the same distance. If her speed upstream is 4 mi/h, what is her speed downstream?

Solution This is a "motion" problem. The relationship between the distance traveled D, the average rate of speed R, and the time T, is given by the distance formula:

$$D = RT$$

Since you are asked to find the speed downstream:

Let r = speed of the canoe downstream.

What can be used to formulate the equation? Since the distance upstream is equal to the distance downstream, the verbal equation is:

Distance upstream = Distance downstream

Applying the distance formula $D = RT$, the distances are:

Distance upstream = (4)(3)

Distance downstream = (r)(2)

The algebraic equation and solution are then:

$$(4)(3) = (r)(2)$$
$$2r = 12$$
$$r = 6 \text{ mi/h}$$

Check: $(4)(3) = (6)(2)$ ✓

The connection between sports and mathematics mentioned before Example 10.11 can be carried further. One cannot develop much skill playing a sport by watching a lot of games. Likewise, studying a lot of text examples will not develop much skill in problem solving. You must participate and *solve as many problems yourself as possible.* Mathematics is not a spectator sport!

EXERCISE 10.2

In problems 1 through 34:

(a) Set up and solve an algebraic equation to find the answer to the problem.

(b) Check that the answer makes logical sense and that it satisfies the equation.

1. Don and Jean together earn \$96,000 a year. If Jean earns \$11,000 more than Don, how much does each earn?

2. The cost of renting a computer and a laser printer is \$600 per month. If the cost for the computer is three times that for the printer, what is the monthly cost for each?

3. Your boss is raising your salary to \$58,500 and indicates this is a 10% raise. Based on that figure, what is your present salary to the nearest dollar?

4. Stacey paid \$32.39 for a calculator, which includes 8% sales tax. What is the price of the calculator, before tax, to the nearest cent?

5. An electronics store gives one-third off list prices. If Steven paid \$28 for a voltmeter before tax, what is the list price?

6. An insurance company gives Betty \$4300 to replace her stolen car and informs her that the "book" value is only 40% of the original value. What is the original value, according to the insurance company?

7. To be eligible for a government job, you must average at least 80% on three tests. If your first two scores are 73% and 78%, what do you need on the third test to be eligible for a government job?

8. The average of your class tests is 77. The instructor indicates that your final average equals two-thirds of your class average plus one-third of your final exam grade. What do you need on the final exam to raise your average to 80?

9. A laptop computer sells for \$3000, which includes the price of a software package. This price is a 15% increase over the price of the computer without the software package.

(a) Find the cost of the computer without the software package to the nearest cent.

(b) If next week there is a sale and there will only be a 5% price increase for the computer with the software package, how much will you save if you buy the computer next week? (See Example 10.9.)

10. Mary Ellen paid \$864 for a stereo, which included 8.5% sales tax.

(a) Find the price of the stereo, before tax, to the nearest cent.

(b) How much could she have saved to the nearest cent if she bought the stereo online with no sales tax, but with a shipping cost equal to 5% of the price? (See Example 10.9.)

11. A new high-speed laser printer operates 3.5 times faster than an older model. Together, they print 2475 lines per minute. How many lines per minute does the new model print?

12. In a phone order business, one computer system completes an order in two-thirds the time of another system.

If the total time for both systems to each complete an order is exactly 4 minutes, how long does it take each system to complete its order?

13. Enrique sails his boat upsteam in 4 hours but only takes 2 hours to sail downstream the same distance. If the average speed upstream is 4 mi/h, how fast is the speed downstream? (See Example 10.12.)

14. Fernando paddles his canoe upriver in 2 hours 30 minutes and downriver the same distance in 1 hour 30 minutes. If his average speed downriver is 5 mi/h, what is his average speed upstream? (See Example 10.12.)

15. A train leaves Los Angeles, averaging 30 mi/h. A car leaves 1 hour later, averaging 50 mi/h. How many hours will the car take to catch up to the train? *Hint:* Distances are equal.

16. Gordon leaves his house, walking at a rate of 3 mi/h. Two hours later his grandson Jordan leaves the house, biking toward him at a rate of 12 mi/h. How many minutes will his grandson take to catch up to him? *Hint:* Distances are equal.

17. Howard leaves Los Angeles driving to Las Vegas, averging 50 mi/h. At the same time Karolyn leaves Las Vegas driving to Los Angeles, averaging 60 mi/h. If the distance from Los Angeles to Las Vegas is 275 mi, after how many hours will they meet?

18. A plane takes 3.0 h to fly from New York to Chicago against the wind, and 2.0 h to fly the same distance from Chicago to New York with the wind. If the air speed and the wind speed are the same for both trips, and the air speed is 500 mi/h, what is the wind speed?

Applications to Electronics

19. The total inductance L_T of two inductances in series equals the sum of the inductances $L_1 + L_2$. If $L_T = 11$ mH and L_2 is 10% greater than L_1, find each inductance in millihenries.

20. In problem 19, find L_1 and L_2 in microhenries if $L_T = 360$ μH and L_1 is 20% less than L_2.

21. Figure 10-8 shows a circuit containing three parallel resistors R_1, R_2, and R_3. The total current I_T in the circuit equals the sum of the currents through the resistors: $I_T = I_1 + I_2 + I_3$. The current in the second resistor is 1.5 times the current in the first resistor. The current in the third resistor is 3 times the current in the first resistor. If the total current is 275 mA, how much current flows through each resistor in milliamps?

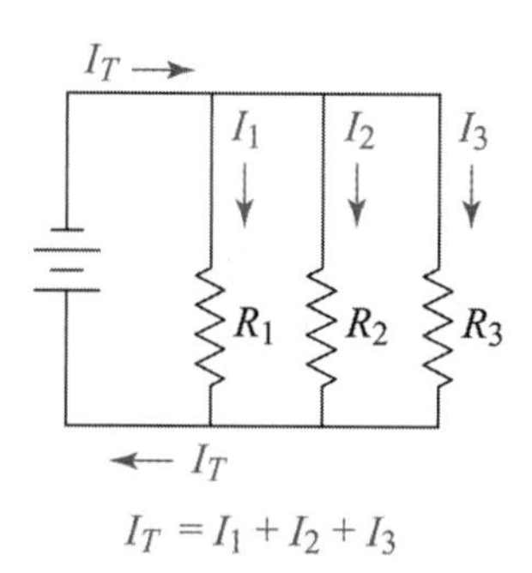

FIGURE 10-8 Three parallel resistors for problems 21 and 22.

22. In Figure 10-8, given that $I_1 = 36$ mA, I_2 equals one-third of I_T and I_3 equals one-half of I_2, find I_T, I_2, and I_3 in milliamps.

23. Figure 10-9 shows two batteries of the same type connected in series aiding so that the total voltage across both batteries equals the sum of the voltages across each battery: $V_T = V_1 + V_2$. However, it is found that the second battery does not contain as much charge as the first, and V_2 is 5% less than V_1. If $V_T = 11.7$ V, find V_1 and V_2.

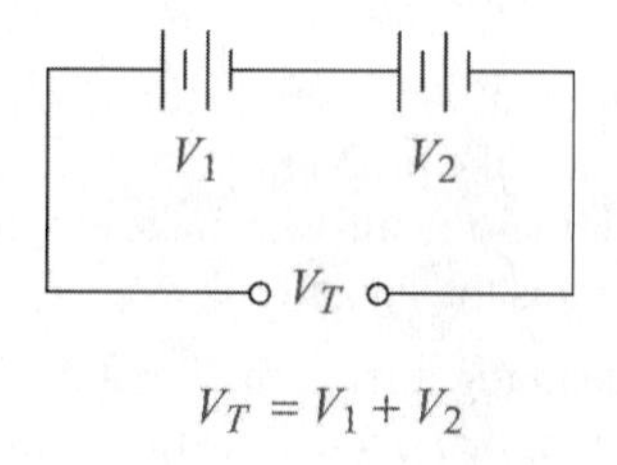

FIGURE 10-9 Series-aiding voltages for problem 23.

24. The total output of three "identical" electrical generators in a power plant is 20,000 MW. An electrical technician finds that two of the generators have the same power output but that the third generator puts out 20% more than one of the other two. What is the power output of each generator?

25. An aluminum wire 3.5 m long is to be cut so that the resistance of one piece is 75% of the resistance of the other piece. How many centimeters should each piece be? (See Example 10.10.)

Answers to Error Box Problems (may vary), page 214:
1. $R_1 + 50 = R_2$ **2.** $V_2 = V_1 + 10$ **3.** $I_3 = I_2 + 0.5I_2$ **4.** $P_1 = P_2 - 0.1P_2$
5. $C_d = C_r - (1/3)C_r$ **6.** $r_c = r_t + 20$ **7.** $P_{in} = P_{out} - 0.15(P_{out})$
8. $C_1 = 2C_2 - 0.10(2C_2)$ **9.** $t_2 + 2 = t_3 - 3$ **10.** $R_3 = (2/3)R_1$

26. A copper wire 7 feet long is to be cut into three pieces. The second piece is to have twice the resistance of the first piece and the third piece is to have twice the resistance of the second piece. How long should each piece be in inches? (See Example 10.10.)

27. In Example 10.11, find the voltage drop across each resistor in volts if V_2 is twice V_1 and V_3 is one-third of V_1.

28. In Example 10.11, find the voltage drop across each resistor in volts if V_1 is one-half of V_2 and V_3 is 50% greater than V_2.

29. Two currents, I_1 and I_2, flow into a branch point in a circuit while two other currents, I_3 and I_4, flow out of the same branch point. Kirchoff's current law says that the sum of the currents entering the branch point equals the sum of the currents leaving the branch point. If $I_1 = 4I_3$, $I_2 = 2I_3$, and $I_4 = 5$ A, find I_1, I_2, and I_3.

30. In problem 29, find I_1, I_2, and I_4 if $I_1 = 2I_2$, $I_4 = I_2 + 2$, and $I_3 = 5$ A.

31. Figure 10-10 shows a variable resistor in series with a bulb. The voltage drop across the bulb is 1.5 times the voltage drop across the resistor. If the applied voltage $V = 12$ V, what are the voltage drops across the bulb and the resistor? Apply Kirchhoff's voltage law, which says that applied voltage equals the sum of the voltage drops.

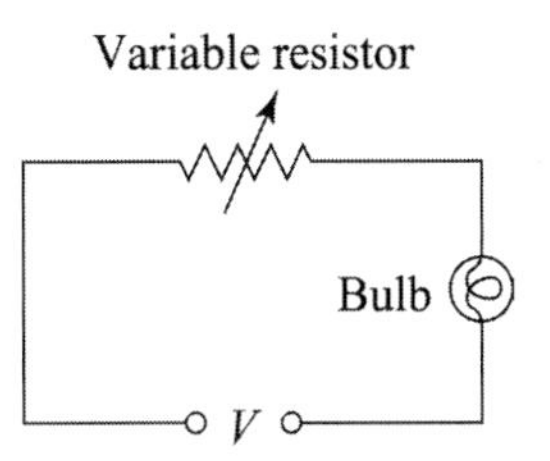

FIGURE 10-10 Variable resistor with bulb for problems 31 and 32.

32. In the circuit in Figure 10-10, the power dissipated in the variable resistor is 40% less than the power dissipated in the bulb. The total power dissipated is 9.6 W. If the total power equals the sum of the two dissipated powers, how much power is dissipated in the bulb?

33. On a local area network containing one server, 45 users sign on at the same time with the same group ID, causing an error. As you troubleshoot for the error, you find there is a server overload equal to 125% of capacity. Correct the error by finding the maximum number of users that can sign on at any one time with the same group ID.

34. In problem 33, if a second server is added to the network with 70% capacity of the first, what will be the maximum number of users that can sign on at any one time with the same group ID?

CHAPTER HIGHLIGHTS

10.1 FORMULAS AND LITERAL EQUATIONS

Equation Principle

The following operations do not change an equality:

1. Adding or subtracting the same quantity on both sides of an equation.
2. Multiplying or dividing both sides of an equation by the same quantity. (*Division by zero is not allowed.*)

Key Example: To solve $P = I^2R$ for I, solve for I^2 and take square roots:

Divide by R: $$\frac{P}{(R)} = \frac{I^2\cancel{R}}{(\cancel{R})}$$

$$\frac{P}{R} = I^2$$

Take the square root of both sides: $$I = \sqrt{\frac{P}{R}}$$

Substitution Principle

In an equation, any quantity may be substituted for an equal quantity.

Key Example: To derive the formula $R = \frac{V^2}{P}$, using the power formula $P = VI$ and Ohm's law $I = \frac{V}{R}$, substitute $\frac{V}{R}$ for I:

$$P = VI = V\left(\frac{V}{R}\right)$$

$$P = \frac{V^2}{R}$$

Multiply by R: $$(R)P = \frac{V^2}{\cancel{R}}(\cancel{R})$$

Divide by P: $$\frac{R\cancel{P}}{(\cancel{P})} = \frac{V^2}{(P)}$$

$$R = \frac{V^2}{P}$$

To solve for a variable appearing in more than one term, separate the variable on one side as a common factor before dividing by the other factor to isolate it.

Key Example:

Solve for V: $P = VI_1 + VI_2$

Factor out V: $P = V(I_1 + I_2)$

Divide by $(I_1 + I_2)$: $\frac{P}{(I_1 + I_2)} = \frac{V\cancel{(I_1 + I_2)}}{\cancel{I_1 + I_2}}$

$$V = \frac{P}{I_1 + I_2}$$

Study Examples 10.6 and 10.7 to review other techniques used in solving literal equations.

10.2 PROBLEM SOLVING

PROBLEM SOLVING ALGORITHM

1. Study the problem several times and determine what is given and what you are trying to find.
2. Choose a meaningful letter for the unknown and write down clearly what it represents. For example, t for time, p for price, d for distance, V for voltage, P for power.
3. Determine what the equation should say in words, that is, the verbal equation.
4. Write down the expressions for any other quantities, in terms of the unknown, that are needed for the equation.
5. Formulate and solve the algebraic equation.
6. Check that the answer makes logical sense and that it satisfies the equation.

See Figure 10-6, which illustrates this algorithm as a flowchart.

Key Example: A multimeter costs \$85.50, including $8\frac{1}{4}\%$ tax. Find the price before tax.

Let p = price before tax

Verbal equation: Price + Tax = Cost

Algebraic equation: $p + 0.0825p = \$85.50$

$$1.0825p = 85.50$$

$$p = \frac{85.50}{1.0825} = \$78.98$$

Key Example: The total resistance of a series circuit containing three resistors is 143 Ω. R_1 is 50 Ω less than R_2, and R_3 is 25 Ω more than R_2. If the total resistance equals the sum of the three resistances, find the three resistances in ohms. Since R_1 and R_3 are defined in terms of R_2, let R_2 = unknown.

Then: $R_1 = R_2 - 50$

$R_3 = R_2 + 50$

Algebraic equation: $R_1 + R_2 + R_3 = 143$

Substitute values: $(R_2 - 50) + R_2 + (R_2 + 25) = 143$

Solve for R_2: $3R_2 - 25 = 143$

$$R_2 = 56\ \Omega$$

$$R_1 = 56 - 50 = 6\ \Omega$$

$$R_3 = 56 + 25 = 81\ \Omega$$

Study Examples 10.10 and 10.12 for other examples of problems and solutions.

REVIEW EXERCISES

In exercises 1 through 10, solve each formula for the indicated letter.

1. $E = \frac{kQ}{r^2}$; Q (E = electric intensity)
2. $R_S = \frac{I_M R_M}{I_S}$; I_S (R_S = shunt resistance)
3. $F = \frac{Q_1 Q_2}{r^2}$; r (F = magnetic force)
4. $P = V_1 I + V_2 I$; I (P = power in a series circuit)
5. $d = v_0 t + \frac{1}{2}at^2$; a (d = distance of a falling object)

6. $d = d_0 + vt;\ v$ (d = distance traveled at constant velocity)

7. $I = \dfrac{V}{R_1 + R_2};\ R_1$ (I = total current in a series circuit)

8. $C_T = \dfrac{C_1 C_2}{C_1 + C_2};\ C_1$ (C_T = total capacitance of C_1 and C_2 in series)

9. $\dfrac{1}{L_T} = \dfrac{1}{L_1} + \dfrac{1}{L_2};\ L_1$ (L_T = total inductance of L_1 and L_2 in parallel)

10. $V_{TH} = \left[\dfrac{R_L}{R_S + R_L}\right] V_S;\ R_L$ (V_{TH} = Thevenin voltage)

11. The volume of a cylinder is given by

$$V = \pi r^2 h$$

where r is the radius of the circular base and h is the height.

(a) Solve this formula for r.

(b) If the volume increases four times from 2 in^3 to 8 in^3 and h stays the same at 4 in, by what ratio will r increase?

12. A steel rod of length l_0 at 0° C will increase in length to l_t at t° C according to the formula:

$$l_t = l_0(1 + 10.7 \times 10^{-6} t)$$

(a) Solve this formula for t.

(b) At what temperature, to the nearest tenth of a degree, will the steel rod increase in length by 1%?

Hint: Let $l_t = 1.01\ l_0$.

13. An old model cellular phone is selling for $325, which is 35% less than the original price. What was the original price?

14. A high-volume discount store makes only 5% profit on sales. A shoplifter walks out with a $30 calculator. How much did the store pay for the calculator to the nearest cent?

15. Two trains 180 mi apart leave at the same time and travel toward each other. If the average rate of one train is 40 mi/h and the average rate of the other is 50 mi/h, how long will it take for the trains to meet?

16. The grades on three class tests are 70, 73, and 84. What do you need on the fourth test to raise your average to 80?

17. From Einstein's theory of relativity, the relationship between the rest mass m_0, and the mass m_v of a body at a high velocity, v, close to the speed of light, is given by:

$$\left(\frac{m_0}{m_v}\right)^2 = 1 - \frac{v_2}{c_2}$$

where the speed of light $c = 3.0 \times 10^8$ m/s. Find v in scientific notation to two significant digits when the mass $m_v = 2m_0$.

18. Given the temperature conversion formula:

$$C = \frac{5(F - 32)}{9}$$

(a) Solve this formula for degrees fahrenheit F.

(b) What temperature in degrees Fahrenheit is 40° C?

(c) At what temperature is degrees Fahrenheit equal to degrees Celsius?

Applications to Electronics

19. Given the formula for the battery voltage in a series circuit:

$$V_1 = V_2 - (IR_1 + IR_2)$$

(a) Solve this formula for I.

(b) Find I in amps and milliamps when V_1 = 60V, V_2 = 80 V, R_1 = 10 Ω, and R_2 = 15 Ω.

20. Using the power formula $P = VI$ and Ohm's law $I = \frac{V}{R}$, derive the formula:

$$V = \sqrt{PR}$$

21. The formula for the total resistance of a series-parallel circuit is:

$$R_T = R_1 + \frac{R_2 R_3}{R_2 + R_3}$$

(a) Given $R_2 = R_3$, solve this formula for R_2 in terms of R_T and R_1.

(b) Find R_2 when R_1 = 50 Ω and R_T = 80 Ω.

22. In problem 21, given that R_1 = 50 Ω, R_3 = 50 Ω, and R_T = 75 Ω, show that $R_2 = R_1$.

23. The average of three voltmeter readings is 39 V. If two of the readings are 40 V and 35 V, what is the third reading?

24. A new electric heater consumes only one-fourth of the power of an old model. If the total power consumption of both heaters is 3 kW, what is the power consumption of each in watts?

25. A copper wire 10 ft long is to be cut into three pieces. The resistance of the shortest piece is to be one-half the resistance of the longest piece. The resistance of the middle piece is to be 20% more than the resistance of the shortest piece. Given that electrical resistance is directly proportional to length, how long should each piece be to the nearest tenth of a foot?

26. The total resistance of a series circuit containing three resistors is 300 Ω. If R_1 is 100 Ω less than R_2, R_2 is 50% greater than R_3, and the total resistance equals the sum of the three resistances, find the three resistances in ohms.

27. Figure 10-11 shows a variable resistor R_V in parallel with a fixed resistor R_0. The variable resistor is adjusted so that its current remains constant at $I_V = 0.30$ A when the voltage V changes. The total current in the circuit is given by:

$$I_T = 0.30 + \frac{V}{R_0}$$

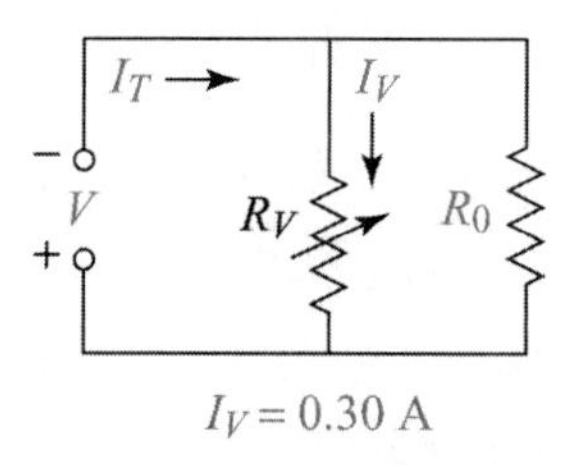

$I_V = 0.30$ A

FIGURE 10-11 Constant current in a parallel circuit for problems 27 and 28.

(a) If I_T changes from I_1 to I_2 when V changes from V_1 to V_2, find the formula for $\Delta I_T = I_2 - I_1$ in terms of $\Delta V = V_2 - V_1$.

(b) If $R_0 = 100$ Ω, find ΔI_T in milliamps when $\Delta V = 3$ V.

28. In problem 27, find the formula for ΔI_T if the voltage increases by one-third of its value from V_1 to $V_2 = V_1 + \frac{1}{3}V_1$.

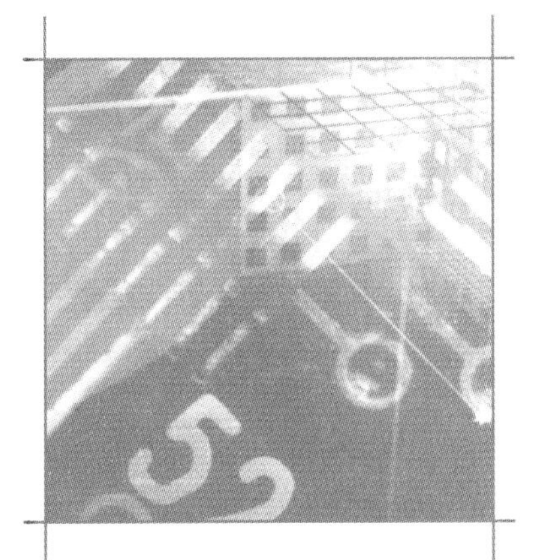

CHAPTER 11
Series and Parallel DC Circuits

CHAPTER OBJECTIVES

In this chapter, you will learn:

- The definition and properties of a series circuit.
- How to solve for unknown electrical values in a series circuit.
- The meaning of series-aiding and series-opposing voltages.
- The definition and properties of a parallel circuit.
- How to solve for unknown electrical values in a parallel circuit.
- The definition of conductance.
- How to analyze a series-parallel circuit and find its total resistance.

This chapter continues with the study of dc circuits presented in Chapter 7 and explores the two basic kinds of circuits: series circuits and parallel circuits. These two circuits form the basis for the analysis of more complex networks and have similar but opposite properties as follows: In a *series circuit* the *current is constant* and the *voltage divides;* in a *parallel circuit* the *voltage is constant* and the *current divides.* When Ohm's law and the power formula are applied to these circuits, many formulas result that provide solutions for voltage, current, resistance, and power. Many dc networks can be analyzed by combining the concepts of these two circuits.

• • •

11.1 Series Circuits

Figure 11-1 shows a series circuit containing n resistors $R_1, R_2, R_3, \ldots, R_n$ and defined as follows:

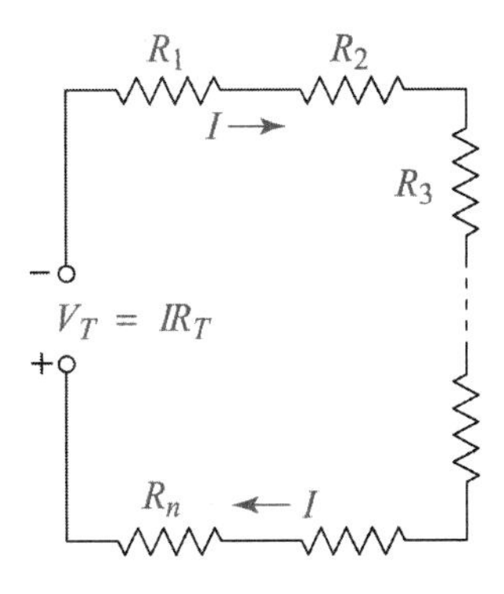

$$R_T = R_1 + R_2 + R_3 + \ldots + R_n$$

FIGURE 11-1 Series circuit.

Definition

Series Circuit

A series circuit is one in which all the components are connected in one path so that the same current flows through each component.

In a series circuit the sum of the resistances equals the total resistance:

Formula **Total Series Resistance**

$$R_T = R_1 + R_2 + R_3 + \cdots + R_n \quad (11.1)$$

The voltage drop or potential difference across each resistor is the voltage that exists between the two ends of the resistor. By Ohm's law, the voltage drop across each resistor is directly proportional to the resistance, and the sum of the voltage drops equals the applied voltage V_T:

Formula **Total Series Voltage**

$$V_T = V_1 + V_2 + V_3 + \cdots + V_n \quad (11.2)$$

EXAMPLE 11.1

Given the series circuit in Figure 11-2 where $R_1 = 10\ \Omega$, $R_2 = 30\ \Omega$, and the applied voltage $V_T = 3.0$ V.

(a) Find the current in the circuit in milliamps.

(b) Find the voltage drop in volts across each resistor.

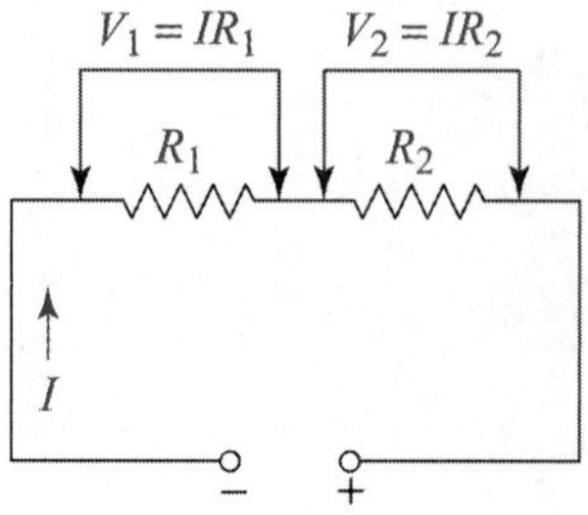

$V_1 = IR_1$ $V_2 = IR_2$

R_1 R_2

I

− +

$V_T = 3.0$ V

$R_T = R_1 + R_2$

$I = \frac{V_T}{R_T}$

FIGURE 11-2 Series circuit and voltage drops for Example 11.1.

Solution

(a) To find the current in the circuit, first find the total resistance R_T in the circuit. Apply formula (11.1):

$$R_T = R_1 + R_2 = 10\ \Omega + 30\ \Omega = 40\ \Omega$$

Then apply Ohm's law for $V_T = 3.0$ V and $R_T = 40\ \Omega$:

$$I = \frac{V_T}{R_T} = \frac{3.0\ \text{V}}{40\ \Omega} = 0.075\ \text{A} = 75\ \text{mA}$$

(b) To find the voltage drops V_1 and V_2, apply Ohm's law for each resistor:

$$V_1 = IR_1 = (0.075\ \text{A})(10\ \Omega) = 0.75\ \text{V}$$

$$V_2 = IR_2 = (0.075\ \text{A})(30\ \Omega) = 2.25\ \text{V}$$

Observe that the sum of the voltage drops equals the total voltage:

$$V_T = V_1 + V_2 = 0.75\ \text{V} + 2.25\ \text{V} = 3.0\ \text{V}$$

Note also that $R_2 = 30\ \Omega$ is *three times* $R_1 = 10\ \Omega$, and the voltage drop $V_2 = 2.25$ V is *three times* the voltage drop $V_1 = 0.75$ V.

The next example illustrates how a resistance is used to reduce the voltage across a load in a series circuit.

EXAMPLE 11.2

Figure 11-3 shows a series circuit containing a load whose resistance R_L is in series with a voltage dropping resistor R_D. The load operates at a voltage $V_L = 10$ V and uses 250 mW of power.

(a) If the total battery voltage $V_T = 14$ V, what should be the resistance of R_D?

(b) Find the total resistance and the resistance of the load.

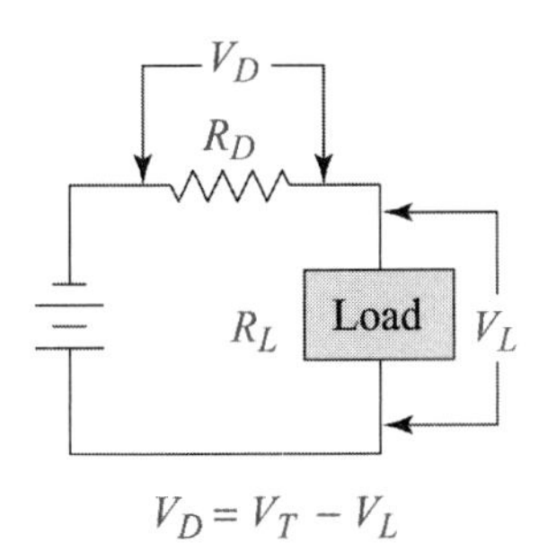

FIGURE 11-3 Voltage-dropping resistor for Example 11.2.

Solution

(a) Find the voltage drop across R_D by applying formula (11.2) for the sum of the voltages:

$$V_T = V_D + V_L$$

Solve for V_D:

$$V_D = V_T - V_L = 14\text{ V} - 10\text{ V} = 4.0\text{ V}$$

Find the current in the load by applying the power formula to the power in the load:

$$I = \frac{P}{V_L} = \frac{0.25\text{ W}}{10\text{ V}} = 0.025\text{ A} = 25\text{ mA}$$

The current through R_D is also 25 mA since the current is the same throughout a series circuit. Apply Ohm's law to find R_D:

$$R_D = \frac{V_D}{I} = \frac{4.0\text{ V}}{0.025\text{ A}} = 160\ \Omega$$

(b) Apply Ohm's law to find R_T:

$$R_T = \frac{V_T}{I} = \frac{14\text{ V}}{0.025\text{ A}} = 560\ \Omega$$

Apply formula (11.1) to find R_L:

$$R_L = R_T - R_D = 560\ \Omega - 160\ \Omega = 400\ \Omega$$

or apply Ohm's law to find R_L:

$$R_L = \frac{V_L}{I} = \frac{10\text{ V}}{0.025\text{ A}} = 400\ \Omega$$

Aiding and Opposing Voltages

When two voltages V_1 and V_2 are connected in series so that the positive terminal of one is connected to the negative terminal of the other (as shown in Figure 11-4(a)) the total voltage is equal to the sum of the voltages:

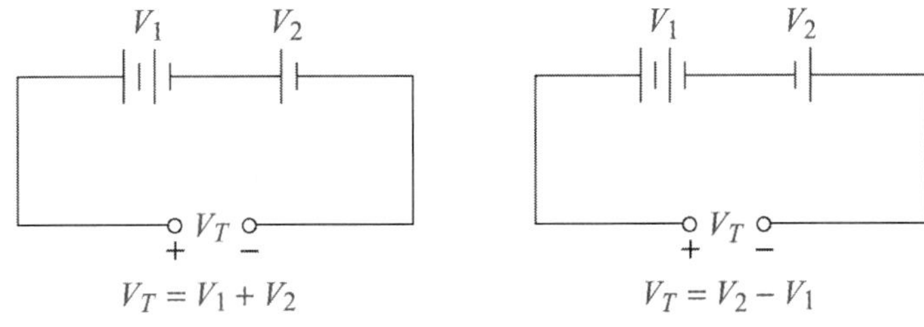

FIGURE 11-4 Aiding and opposing voltages.

Formula

Series-Aiding Voltages

$$V_T = V_1 + V_2 \tag{11.3}$$

The two voltages are said to be series-aiding because each voltage causes the current to flow in the same direction.

When two voltages V_1 and V_2 are connected in series so that the negative terminal of one is connected to the negative terminal of the other (as shown in Figure 11-4(b)) the total voltage is equal to the difference of the voltages:

Formula

Series-Opposing Voltages

$$V_T = V_2 - V_1 \quad (11.4)$$

The two voltages are said to be series-opposing because each voltage tends to cause the current to flow in the opposite direction. The larger voltage determines the resultant current flow.

EXAMPLE 11.3

Figure 11-5 shows two series-aiding voltages connected to a resistor R. Given $R = 75\ \Omega$, $V_2 = 2V_1$, and the ammeter in the circuit reads 100 mA:

(a) Find V_T, V_1 and V_2 in volts.

(b) Find the power in milliwatts dissipated in the resistor.

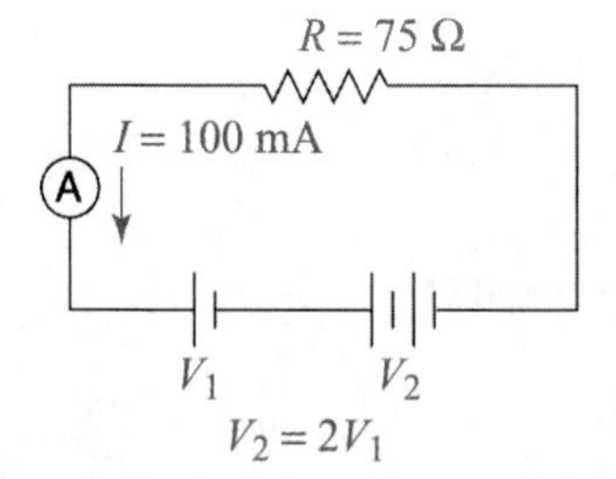

FIGURE 11-5 Series-aiding voltages for Example 11.3.

Solution

(a) Find V_T using Ohm's law:

$$V_T = IR = (0.10\text{ A})(75\ \Omega) = 7.5\text{ V}$$

To find V_1 and V_2, apply formula (11.3):

$$V_T = V_1 + V_2$$

Let $V_2 = 2V_1$:

$$V_T = V_1 + 2V_1 = 3V_1$$

Then:

$$V_T = 3V_1 = 7.5\text{ V}$$

$$V_1 = \frac{7.5\text{ V}}{3} = 2.5\text{ V}$$

and

$$V_2 = 2V_1 = 2(2.5\text{ V}) = 5.0\text{ V}$$

(b) Apply the power formula to find the power dissipated in the resistor:

$$P = V_T I = (7.5\text{ V})(0.10\text{ A}) = 0.75\text{ W} = 750\text{ mW}$$

Similar to voltage and resistance, the total power in a series circuit is equal to the sum of the powers dissipated:

Formula

Total Power in a Series Circuit

$$P_T = P_1 + P_2 + P_3 + \cdots + P_n \quad (11.5)$$

EXAMPLE 11.4

Given a series circuit with three resistors $R_1 = 1.3\text{ k}\Omega$, $R_2 = 1.0\text{ k}\Omega$, and $R_3 = 910\ \Omega$ and the applied voltage $V_T = 220$ V.

(a) Find the total resistance in kilohms and the current in milliamps.

(b) Find the voltage drop across each resistor in volts.

(c) Find the power dissipated in each resistor and the total power in watts.

Round answers to two significant digits.

Solution

(a) Using consistent units, the total resistance and the current are:

$$R_T = 1.3\ \text{k}\Omega + 1.0\ \text{k}\Omega + 0.91\ \text{k}\Omega = 3.21\ \text{k}\Omega \approx 3.2\ \text{k}\Omega$$

$$I = \frac{V_T}{R_T} = \frac{220\ \text{V}}{3210\ \Omega} \approx 0.0685\ \text{A} \approx 69\ \text{mA}$$

(b) The voltage drops across each resistor are:

$$\begin{aligned} V_1 &= IR_1 = (0.0685)(1300) \approx 89\ \text{V} \\ V_2 &= IR_2 = (0.0685)(1000) \approx 69\ \text{V} \\ V_3 &= IR_3 = (0.0685)(910) \approx 62\ \text{V} \end{aligned}$$

Note that the total voltage equals the sum of the voltage drops:

$$V_T = V_1 + V_2 + V_3 = 89\ \text{V} + 69\ \text{V} + 62\ \text{V} = 220\ \text{V}$$

(c) The power dissipated in each resistor and the total power are:

$$\begin{aligned} P_1 &= I^2R_1 = (0.0685)^2(1300) \approx 6.1\ \text{W} \\ P_2 &= I^2R_2 = (0.0685)^2(1000) \approx 4.7\ \text{W} \\ P_3 &= I^2R_3 = (0.0685)^2(910) \approx 4.3\ \text{W} \\ P_T &= I^2R_T = (0.0685)^2(3210) \approx 15.1\ \text{W} \approx 15\ \text{W} \end{aligned}$$

Note that the total power equals the sum of the powers:

$$P_T = P_1 + P_2 + P_3 = 6.1\ \text{W} + 4.7\ \text{W} + 4.3\ \text{W} = 15.1\ \text{W}$$

EXERCISE 11.1

For exercises 1 through 10, refer to the series circuit in Figure 11-6.

Express all answers in engineering notation (as numbers between 1 and 1000) with the appropriate units to two significant digits.

1. Given $R_1 = 100\ \Omega$, $R_2 = 200\ \Omega$, $V_T = 24$ V, find R_T, I, V_1, V_2.
2. Given $R_1 = 10\ \Omega$, $R_2 = 30\ \Omega$, $V_T = 12$ V, find R_T, I, V_1, V_2.
3. Given $R_1 = 1.5\ \text{k}\Omega$, $R_2 = 3.0\ \text{k}\Omega$, $V_T = 150$ V, find R_T, I, V_1, V_2.
4. Given $R_1 = 510\ \Omega$, $R_2 = 430\ \Omega$, $V_T = 60$ V, find R_T, I, V_1, V_2.
5. Given $I = 2.0$ A, $R_1 = 2R_2$, $V_T = 120$ V, find R_T, R_1, R_2, V_1, V_2.
6. Given $I = 1.5$ A, $R_2 = 3R_1$, $V_T = 36$ V, find R_T, R_1, R_2, V_1, V_2.
7. Given $I = 320$ mA, $V_1 = V_2$, $R_T = 3.0\ \text{k}\Omega$, find V_T, V_1, V_2, R_1, R_2.
8. Given $I = 750\ \mu\text{A}$, $V_1 = 2V_2$, $R_T = 5.5\ \text{k}\Omega$, find V_T, V_1, V_2, R_1, R_2.
9. Given R_2 is 50% greater than R_1, $V_T = 30$ V, $R_T = 300\ \Omega$, find I, R_1, R_2, V_1, V_2.
10. Given R_1 is 50% less than R_2, $V_T = 30$ V, $R_T = 300\ \Omega$, find I, R_1, R_2, V_1, V_2.

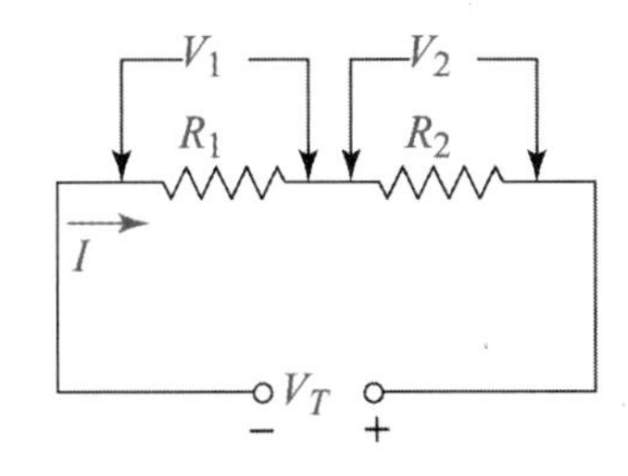

FIGURE 11-6 Series circuit for exercises 1 through 10.

For exercises 11 through 20, refer to the series circuit in Figure 11-7.

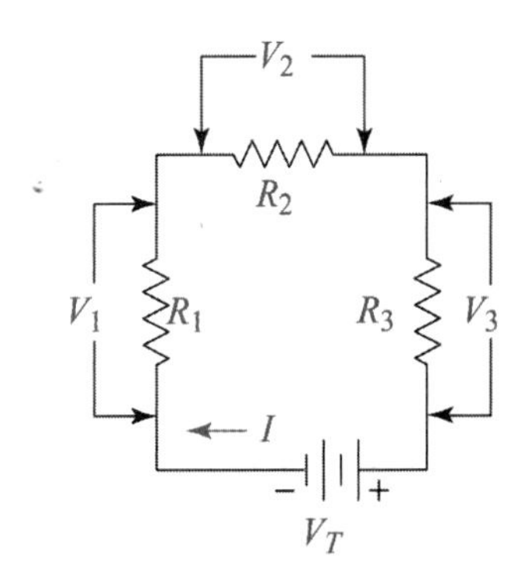

FIGURE 11-7 Series circuit for exercises 11 through 20.

11. Given $V_T = 10$ V, $R_1 = 15\ \Omega$, $R_2 = 22\ \Omega$, $R_3 = 33\ \Omega$, find R_T, I, V_1, V_2, V_3.
12. Given $V_T = 140$ V, $R_1 = 2.0$ kΩ, $R_2 = 1.2$ kΩ, $R_3 = 1.0$ kΩ, find R_T, I, V_1, V_2, V_3.
13. Given $V_T = 8.0$ V, $R_1 = 20\ \Omega$, $R_2 = 24\ \Omega$, $R_3 = 36\ \Omega$, find I, P_1, P_2, P_3, P_T.
14. Given $V_T = 110$ V, $R_1 = 15\ \Omega$, $R_2 = 18\ \Omega$, $R_3 = 22\ \Omega$, find I, P_1, P_2, P_3, P_T.
15. Given $I = 900$ μA, $R_1 = 2.0$ kΩ, $R_2 = 3.0$ kΩ, $R_3 = 3.0$ kΩ, find V_T, P_1, P_2, P_3, P_T.
16. Given $I = 50$ mA, $R_1 = 120\ \Omega$, $R_2 = 130\ \Omega$, $R_3 = 150\ \Omega$, find V_T, P_1, P_2, P_3, P_T.
17. Given $R_1 = R_2$, $R_3 = 2R_1$, $V_1 = 4.0$ V, $P_T = 600$ mW, find V_2, V_3, V_T, R_T.
18. Given $R_1 = R_3$, $R_2 = 2R_3$, $I = 1.0$ A, $P_T = 24$ W, find V_1, V_2, V_3, R_T.
19. Given $R_1 = R_2 = R_3$, $R_T = 30\ \Omega$, $P_T = 60$ W, find P_1, P_2, P_3, I.
20. Given $R_1 = 2R_2$, $R_2 = R_3$, $R_T = 120\ \Omega$, $P_T = 300$ W, find P_1, P_2, P_3, V_T.
21. Figure 11-3 shows a series circuit containing a load resistance R_L, a voltage dropping resistor R_D and a battery voltage = 14 V. If the load operates at 8.0 V and uses 200 mW of power, find R_D.
22. In problem 21, if the load operates at 12 V and uses 500 mW of power, find R_D.
23. A voltage-dropping resistor is in series with a load whose resistance $R_L = 330\ \Omega$. The total voltage drop across the load and the resistor is 100 V. If the load uses 14 W of power, what should be the value of the dropping resistor?
24. In problem 23, if $R_L = 91\ \Omega$, the total voltage drop across the load and the resistor is 80 V, and the load uses 30 W of power, what should be the value of the dropping resistor?
25. Figure 11-5 shows two series-aiding voltages connected to a resistor $R = 75\ \Omega$. If the ammeter reads 600 mA, find V_1 and V_2, and the power dissipated in the resistor.
26. Two voltages V_1 and V_2 are connected in series aiding to a resistor $R = 160\ \Omega$. An ammeter in the circuit reads 1.2 A. If V_1 is 50 V less than V_2, find V_1 and V_2.
27. Two voltages V_1 and V_2 are connected in series opposing to a resistor $R = 15\ \Omega$. An ammeter in the circuit reads 1.0 A. If $V_1 = 30$ V, what are the two possible values for V_2?
28. Two voltages V_1 and V_2 are connected in series opposing to a resistor $R = 16\ \Omega$. An ammeter in the circuit reads 750 mA. If $V_1 = 22$ V, what are the two possible values for V_2?
29. The total resistance of a circuit $R_T = 2.7$ kΩ. Assuming the voltage is constant, how much resistance must be added in series to reduce the current to one-half of its value?
30. Given the circuit in problem 29, how much resistance must be added in series to reduce the current to 75% of its value?

11.2 Parallel Circuits

A parallel circuit is defined as follows:

Definition

Parallel Circuit

A parallel circuit is one in which each component is connected directly to the voltage source so that each component has the same voltage drop as the voltage source.

The circuit in Figure 11-8 shows a parallel circuit containing n resistors $R_1, R_2, R_3, \ldots, R_n$ in parallel. Each resistor is connected so that the voltage across each resistor is equal to the applied voltage V. The total current I_T equals the sum of the currents through all the resistors:

Formula

Total Parallel Current

$$I_T = I_1 + I_2 + I_3 + \cdots + I_n \tag{11.6}$$

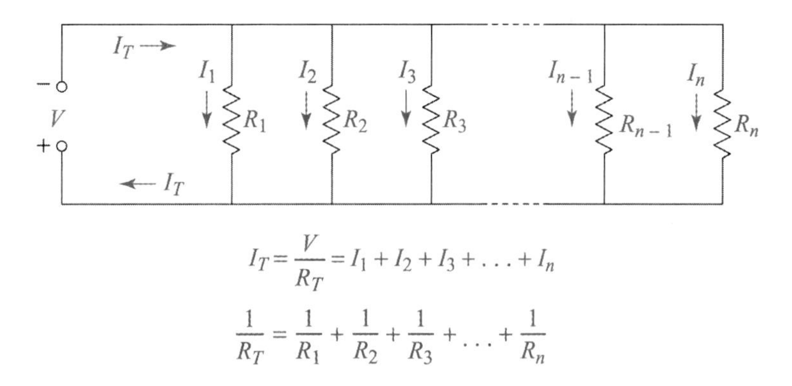

FIGURE 11-8 Parallel circuit.

By Ohm's law, it can be shown that the reciprocal of the total or equivalent resistance of n resistors in parallel equals the sum of the reciprocals of the resistances:

Formula **Total Parallel Resistance**

$$\frac{1}{R_T} = \frac{1}{R_1} + \frac{1}{R_2} + \frac{1}{R_3} + \cdots + \frac{1}{R_n} \tag{11.7}$$

Because parallel circuits provide additional current paths the following is true: The total or equivalent resistance R_T *is always less than the smallest of the parallel resistances.* A special case of formula (11.7) results when all the parallel resistances are equal. If there are N parallel resistances, each equal to R, then

$$\frac{1}{R_T} = \frac{1}{R} + \frac{1}{R} + \frac{1}{R} + \cdots + \frac{1}{R} \quad (N \text{ fractions})$$

Combining the N fractions,

$$\frac{1}{R_T} = \frac{1 + 1 + 1 + \cdots + 1}{R} = \frac{N}{R}$$

When two fractions are equal, their reciprocals are equal. Therefore you have the formula:

Formula ***N* Equal Parallel Resistances**

$$R_T = \frac{R}{N} \tag{11.8}$$

Formula (11.8) tells you that if there are two equal resistances in parallel, the equivalent resistance is one-half of one of the resistances. If there are three equal resistances in parallel the equivalent resistance is one-third of one of the resistances, and so on. For example, two 100-Ω resistances in parallel are equivalent to one resistance of 50 Ω. Three 30-Ω resistances in parallel are equivalent to one resistance of 10 Ω.

The following formula for two resistances in parallel follows from formula (11.7):

Formula **Two Parallel Resistances**

$$R_T = \frac{R_1 R_2}{R_1 + R_2} \tag{11.9}$$

Formula (11.9) is a convenient formula to calculate the equivalent resistance of two parallel resistances. One way to remember it is: *The product divided by the sum.* The following example shows the derivation of formula (11.9).

EXAMPLE 11.5

Show that the equivalent resistance of two resistances in parallel can be expressed as:

$$R_T = \frac{R_1 R_2}{R_1 + R_2}$$

Solution Apply formula (11.7) for two parallel resistances:

$$\frac{1}{R_T} = \frac{1}{R_1} + \frac{1}{R_2}$$

Combine the two fractions on the right side over the LCD = $R_1 R_2$:

$$\frac{1}{R_T} = \frac{1(R_2)}{R_1(R_2)} + \frac{(R_1)1}{(R_1)R_2}$$

$$\frac{1}{R_T} = \frac{R_2 + R_1}{R_1 R_2}$$

If two fractions are equal, their reciprocals are equal. Therefore, invert the fractions on both sides of the equation to produce the result:

$$R_T = \frac{R_1 R_2}{R_1 + R_2}$$

Example 11.5 can also be done by multiplying all the terms in the equation by the LCD = $R_T R_1 R_2$. This solution is left as problem 47 in Exercise 11.2.

EXAMPLE 11.6

Figure 11-9 shows a circuit containing two parallel resistors $R_1 = 150\ \Omega$ and $R_2 = 200\ \Omega$, connected to a voltage source $V = 20$ V.

(a) Find R_T in ohms two ways, using formula (11.9) and formula (11.7).

(b) Find I_T, I_1, and I_2 in milliamps.

Round answers to two significant digits.

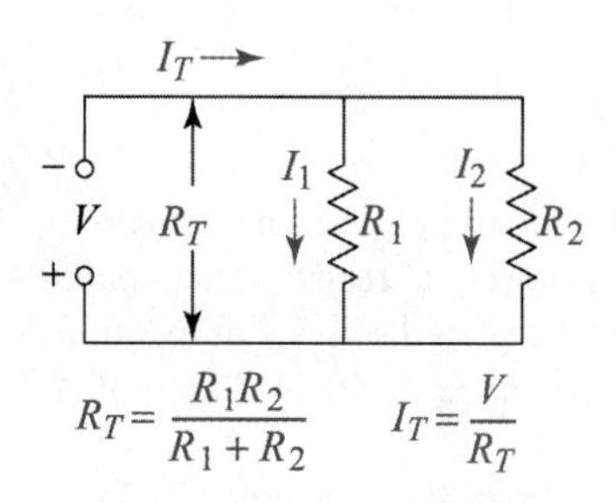

FIGURE 11-9 Two parallel resistors for Examples 11.6 and 11.7.

Solution

(a) The equivalent resistance R_T, using formula (11.9) and substituting the given values, is:

$$R_T = \frac{R_1 R_2}{R_1 + R_2} = \frac{(150)(200)}{150 + 200} \approx 86\ \Omega$$

The calculator solution is:

150 [×] 200 [÷] [(] 150 [+] 200 [)] [=] → 86 Ω

Note that *R_T is less than the smaller resistance.* The second method, using formula (11.7) for two resistances, gives you:

$$\frac{1}{R_T} = \frac{1}{R_1} + \frac{1}{R_2} = \frac{1}{150} + \frac{1}{200}$$

This calculation for R_T is readily done on the calculator in one series of steps using the reciprocal key $\boxed{x^{-1}}$ three times:

$$150\ \boxed{x^{-1}}\ \boxed{+}\ 200\ \boxed{x^{-1}}\ \boxed{=}\ \boxed{x^{-1}}\ \boxed{=} \rightarrow 86\ \Omega$$

(b) To find I_T, I_1, and I_2 apply Ohm's law for each resistance:

$$I_T = \frac{V}{R_T} = \frac{20\text{ V}}{86\ \Omega} \approx 230\text{ mA}$$

$$I_1 = \frac{V}{R_1} = \frac{20\text{ V}}{150\ \Omega} \approx 130\text{ mA}$$

$$I_2 = \frac{V}{R_2} = \frac{20\text{ V}}{200\ \Omega} = 100\text{ mA}$$

Note that the voltage across each resistor is equal to the voltage source 20 V, and $I_1 + I_2 = I_T$.

Given two parallel resistors, if you know the value of one resistance R_2 and the equivalent resistance R_T, you can find the value of the other resistance R_1 two ways. One is by using formula (11.9) solved for R_1:

Formula

Parallel Resistances Solved for R_1

$$R_1 = \frac{R_2 R_T}{R_2 - R_T} \quad \textbf{(11.10)}$$

The derivation of formula (11.10) is left as problem 48 in Exercise 11.2. The other way to find R_1 is by using formula (11.7) for two parallel resistances. Both ways are illustrated in the next example.

EXAMPLE 11.7

In Figure 11-9, if $R_2 = 75\ \Omega$, find R_1 so that $R_T = 50\ \Omega$ using:

(a) Formula (11.10).
(b) Formula (11.7).

Solution

(a) To use formula (11.10), substitute the values for $R_2 = 75\ \Omega$ and $R_T = 50\ \Omega$:

$$R_1 = \frac{(75)(50)}{75 - 50} = \frac{3750}{25} = 150\ \Omega$$

The calculator solution is:

$$75\ \boxed{\times}\ 50\ \boxed{\div}\ \boxed{(}\ 75\ \boxed{-}\ 50\ \boxed{)}\ \boxed{=} \rightarrow 150\ \Omega$$

(b) To use formula (11.7), solve the formula for $\frac{1}{R_1}$:

$$\frac{1}{R_1} = \frac{1}{R_T} - \frac{1}{R_2}$$

Find R_1 on the calculator using the reciprocal key three times:

$$50\ \boxed{x^{-1}}\ \boxed{-}\ 75\ \boxed{x^{-1}}\ \boxed{=}\ \boxed{x^{-1}}\ \boxed{=} \rightarrow 150\ \Omega$$

EXAMPLE 11.8 Given three parallel resistors $R_1 = 15\ \Omega$, $R_2 = 20\ \Omega$, and $R_3 = 20\ \Omega$, find R_T.

Solution First simplify this circuit by noting that $R_2 = R_3$, and you can apply formula (11.8). Two equal parallel resistances are equivalent to one-half the value of one resistance. Therefore, the equivalent of R_2 and R_3, is:

$$R_{2,3} = \frac{20}{2} = 10\ \Omega$$

The original parallel circuit then reduces to two parallel resistances: $R_1 = 15\ \Omega$ and $R_{2,3} = 10\ \Omega$.

Apply formula (11.9) to R_1 and $R_{2,3}$:

$$R_T = \frac{R_1 R_{2,3}}{R_1 + R_{2,3}} = \frac{(15)(10)}{15+10} = \frac{150}{25} = 6.0\ \Omega$$

You can also apply formula (11.7) to R_1 and $R_{2,3}$:

15 [x^{-1}] [+] 10 [x^{-1}] [=] [x^{-1}] [=] → 6.0 Ω

The conductance G of a resistance R is defined as its reciprocal:

Formula **Conductance (Siemens)**

$$G = \frac{1}{R} \tag{11.11}$$

Conductance is measured in siemens (S). Applying formula (11.11) to the parallel circuit in Figure 11-8 gives you:

$$G_T = \frac{1}{R_T},\ G_1 = \frac{1}{R_1},\ G_2 = \frac{1}{R_2},\ G_3 = \frac{1}{R_3},\ \cdots,\ G_n = \frac{1}{R_n}$$

Substituting into formula (11.7) gives you a more direct way to express this formula in terms of total conductance G_T:

Formula **Total Parallel Conductance**

$$G_T = G_1 + G_2 + G_3 + \cdots + G_n \tag{11.12}$$

EXAMPLE 11.9 Figure 11-10 shows a *parallel bank* of three resistors. Two or more resistors in parallel are called a parallel bank. If $R_1 = 20\ \Omega$, $R_2 = 30\ \Omega$, and $R_3 = 40\ \Omega$, find the total conductance G_T in millisiemens and the equivalent resistance R_T, to two significant digits.

Solution Two methods are shown. The first method, using formula (11.12) for the total conductance and combining fractions over the LCD = 120, gives you:

$$G_T = \frac{1}{R_T} = \frac{1}{20} + \frac{1}{30} + \frac{1}{40} = \frac{6+4+3}{120}$$

$$G_T = \frac{13}{120} \approx 0.11\ \text{S} = 110\ \text{mS}$$

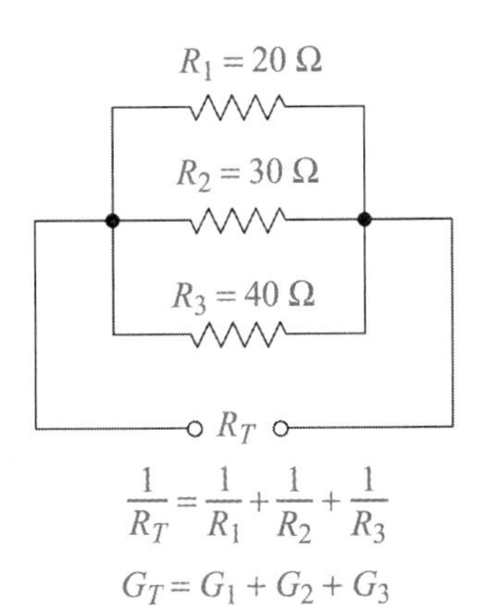

FIGURE 11-10 Total conductance of a parallel bank for Example 11.9.

To find R_T, invert the value for G_T:

$$R_T = \frac{1}{G_T} = \frac{120}{13} \approx 9.2\ \Omega$$

Observe that R_T is less than the smallest resistance of 20 Ω.

To apply the second method, use the calculator to compute G_T first:

20 [x⁻¹] [+] 30 [x⁻¹] [+] 40 [x⁻¹] [=] → 0.110 S

Then calculate R_T using the reciprocal key:

0.110 S [x⁻¹] [=] → 9.2 Ω

You can also find the equivalent resistance of three resistors in parallel by applying formula (11.9) twice. First find the equivalent resistance of R_1 and R_2, which is $R_{1,2}$. Then find the equivalent resistance of $R_{1,2}$ and R_3, which is equal to R_T.

EXAMPLE 11.10

Given R_1, R_2 and R_3 connected in parallel. If $R_2 = 75\ \Omega$ and $R_3 = 100\ \Omega$, find the value of R_1 that will make $R_T = 30\ \Omega$.

Solution This problem can be directly solved on the calculator. Solving formula (11.12) for G_1 gives you:

$$G_1 = G_T - G_2 - G_3$$

Find G_1 on the calculator and then take the reciprocal to find R_1:

30 [x⁻¹] [−] 75 [x⁻¹] [−] 100 [x⁻¹] [=] → 0.01 S [x⁻¹] [=] → 100 Ω

EXERCISE 11.2

Express all answers in engineering notation (as numbers between 1 and 1000) with the appropriate units to two significant digits.

In exercises 1 through 10, find R_T for the two resistors R_1 and R_2 connected in parallel.

1. $R_1 = 10\ \Omega, R_2 = 30\ \Omega$
2. $R_1 = 10\ \Omega, R_2 = 40\ \Omega$
3. $R_1 = 510\ \Omega, R_2 = 430\ \Omega$
4. $R_1 = 2.2\ k\Omega, R_2 = 3.3\ k\Omega$
5. $R_1 = 56\ k\Omega, R_2 = 56\ k\Omega$
6. $R_1 = 8.2\ \Omega, R_2 = 8.2\ \Omega$
7. $R_1 = 18\ \Omega, R_2 = 36\ \Omega$
8. $R_1 = 3.0\ \Omega, R_2 = 15\ \Omega$
9. $R_1 = 750\ \Omega, R_2 = 1.2\ k\Omega$
10. $R_1 = 1.0\ M\Omega, R_2 = 820\ k\Omega$

In exercises 11 through 20, R_1 and R_2 are connected in parallel. Find the value of R_1 that will make R_T equal to the given value. See Example 11.7.

11. $R_2 = 150\ \Omega, R_T = 75\ \Omega$
12. $R_2 = 20\ \Omega, R_T = 10\ \Omega$
13. $R_2 = 15\ \Omega, R_T = 10\ \Omega$
14. $R_2 = 300\ \Omega, R_T = 100\ \Omega$
15. $R_2 = 470\ \Omega, R_T = 100\ \Omega$
16. $R_2 = 330\ \Omega, R_T = 200\ \Omega$
17. $R_2 = 3.3\ \Omega, R_T = 1.2\ \Omega$
18. $R_2 = 27\ k\Omega, R_T = 7.5\ k\Omega$
19. $R_2 = 1.5\ k\Omega, R_T = 670\ \Omega$
20. $R_2 = 2.0\ k\Omega, R_T = 800\ \Omega$

In exercises 21 through 30, find G_T and R_T for the three resistors R_1, R_2, and R_3 connected in parallel. See Example 11.9.

21. $R_1 = 15\ \Omega, R_2 = 30\ \Omega, R_3 = 30\ \Omega$
22. $R_1 = 10\ \Omega, R_2 = 10\ \Omega, R_3 = 20\ \Omega$
23. $R_1 = 330\ \Omega, R_2 = 330\ \Omega, R_3 = 330\ \Omega$
24. $R_1 = 5.1\ k\Omega, R_2 = 5.1\ k\Omega, R_3 = 5.1\ k\Omega$
25. $R_1 = 18\ \Omega, R_2 = 12\ \Omega, R_3 = 36\ \Omega$
26. $R_1 = 100\ k\Omega, R_2 = 150\ k\Omega, R_3 = 200\ k\Omega$
27. $R_1 = 220\ \Omega, R_2 = 300\ \Omega, R_3 = 360\ \Omega$
28. $R_1 = 47\ \Omega, R_2 = 62\ \Omega, R_3 = 24\ \Omega$

29. $R_1 = 1.3\ \text{k}\Omega$, $R_2 = 820\ \Omega$, $R_3 = 1.8\ \text{k}\Omega$

30. $R_1 = 1.6\ \text{M}\Omega$, $R_2 = 910\ \text{k}\Omega$, $R_3 = 750\ \text{k}\Omega$

In exercises 31 through 40, R_1, R_2, and R_3 are connected in parallel. Find the value of R_1 that will make R_T equal to the given value.

31. $R_2 = 300\ \Omega$, $R_3 = 300\ \Omega$, $R_T = 100\ \Omega$

32. $R_2 = 24\ \text{k}\Omega$, $R_3 = 24\ \text{k}\Omega$, $R_T = 8.0\ \text{k}\Omega$

33. $R_2 = 2.0\ \text{k}\Omega$, $R_3 = 2.0\ \text{k}\Omega$, $R_T = 500\ \Omega$

34. $R_2 = 36\ \Omega$, $R_3 = 36\ \Omega$, $R_T = 9.0\ \Omega$

35. $R_2 = 11\ \Omega$, $R_3 = 22\ \Omega$, $R_T = 4.4\ \Omega$

36. $R_2 = 51\ \Omega$, $R_3 = 75\ \Omega$, $R_T = 19\ \Omega$

37. $R_2 = 30\ \Omega$, $R_3 = 20\ \Omega$, $R_T = 7.5\ \Omega$

38. $R_2 = 25\ \Omega$, $R_3 = 200\ \Omega$, $R_T = 20\ \Omega$

39. $R_2 = 100\ \Omega$, $R_3 = 300\ \Omega$, $R_T = 50\ \Omega$

40. $R_2 = 33\ \Omega$, $R_3 = 47\ \Omega$, $R_T = 13\ \Omega$

41. In Figure 11-9 on page 228, given $R_1 = 24\ \Omega$, $R_2 = 30\ \Omega$, and $V = 12$ V, find I_1, I_2, I_T, and R_T.

42. In Figure 11-9 on page 228, given $R_1 = 220\ \Omega$, $R_2 = 300\ \Omega$, and $V = 36$ V, find I_1, I_2, I_T, and R_T.

43. In Figure 11-9 on page 228, the power dissipated in R_1 is $P_1 = 20$ W, and the total power dissipated in the circuit $P_T = 30$ W. If $I_2 = 1.0$ A, find V, R_1, R_2, R_T, I_1, and I_T.
Note: $P_T = P_1 + P_2$.

44. In Figure 11-9 on page 228, the power dissipated in R_1 is $P_1 = 5$ W, and the power dissipated in R_2 is $P_2 = 10$ W. If $V = 20$ V, find R_1, R_2, R_T, I_1, I_2, and I_T.

45. Figure 11-11 shows three resistors in parallel connected to a battery whose voltage is V_B. If $V_B = 50$ V, $R_1 = 150\ \Omega$, $R_2 = 200\ \Omega$, and $R_3 = 180\ \Omega$, find I_1, I_2, I_3, I_T, G_T, and R_T.

46. Given the parallel circuit in Figure 11-11 with $V_B = 15$ V, $R_1 = 20\ \Omega$, $R_2 = 24\ \Omega$, and $R_3 = 30\ \Omega$. Find I_1, I_2, I_3, I_T, G_T, and R_T.

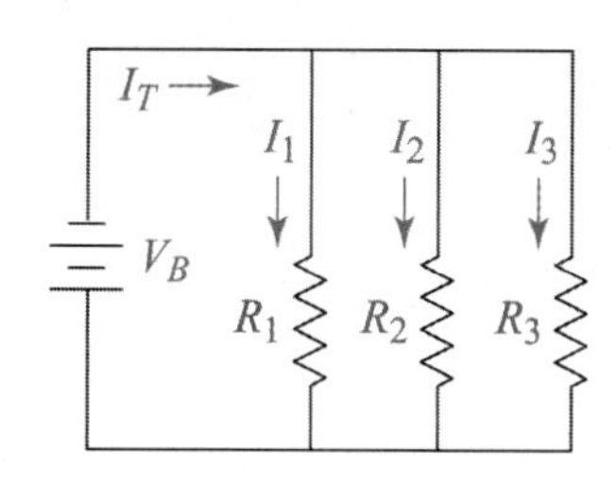

FIGURE 11-11 Parallel circuit for problems 45 and 46.

47. Starting with formula (11.7) for two parallel resistors:

$$\frac{1}{R_T} = \frac{1}{R_1} + \frac{1}{R_2}$$

Multiply all the terms by the LCD = $R_T R_1 R_2$ and derive formula (11.9):

$$R_T = \frac{R_1 R_2}{R_1 + R_2}$$

48. Starting with formula (11.9) in problem 47, derive formula (11.10):

$$R_1 = \frac{R_2 R_T}{R_2 - R_T}$$

49. Given two parallel resistors $R_1 = 100\ \Omega$ and R_2 is 50% more than R_1. Show that the equivalent resistance is 60% of the smaller resistance and 40% of the larger resistance.

50. Given a circuit with two resistors in parallel where $R_2 = 2R_1$. Use formula (11.9) and show that the equivalent resistance is:

$$R_{eq} = \frac{2R_1}{3}$$

11.3 Series-Parallel Circuits

Series circuits and parallel circuits are combined in various ways to create different types of series-parallel circuits. Some of the basic combinations are considered here. Two or more resistors in series are called a *series branch.* Two or more resistors in parallel are called a *parallel bank.*

To analyze a series-parallel circuit, it is necessary to consider the series branches and the parallel banks separately. You reduce each series branch to an equivalent resistance, and you reduce each parallel bank to an equivalent resistance. The equivalent resistances are then combined to obtain the total resistance of the circuit. Study the following examples that illustrate the techniques used to solve various series-parallel circuits.

EXAMPLE 11.11 Figure 11-12(a) shows a series-parallel circuit containing a resistor R_1 in series with a parallel bank of two resistors R_2 and R_3.

(a) Find the formula for the total resistance of the circuit R_T.

(b) Find R_T when $R_1 = 15\ \Omega$, $R_2 = 20\ \Omega$, and $R_3 = 10\ \Omega$ to two significant digits.

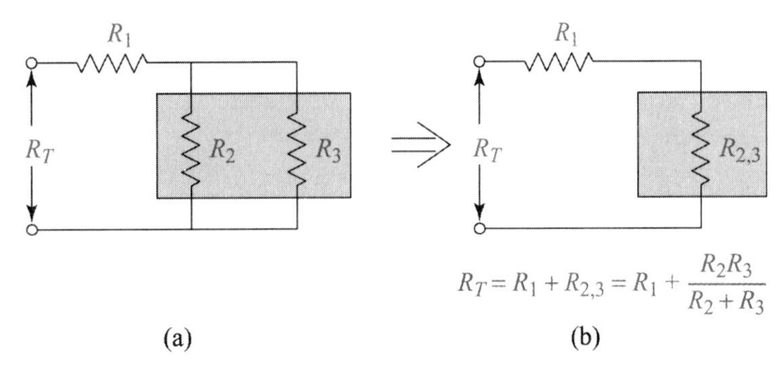

FIGURE 11-12 Series-parallel circuit for Example 11.11.

Solution

(a) First find the equivalent resistance $R_{2,3}$ of the parallel bank of R_2 and R_3. Apply formula (11.9) to R_2 and R_3:

$$R_{2,3} = \frac{R_2 R_3}{R_2 + R_3}$$

This reduces the circuit to two resistances in series, R_1 and $R_{2,3}$, as shown in Figure 11-12(b). The formula for the total resistance is then:

$$R_T = R_1 + R_{2,3} = R_1 + \frac{R_2 R_3}{R_2 + R_3}$$

(b) Substitute in the formula to find the value of R_T:

$$R_T = 15 + \frac{(20)(10)}{20 + 10} = 15 + \frac{200}{30} \approx 22\ \Omega$$

R_T can be found directly on the calculator using the reciprocal formula for $R_{2,3}$:

20 [x^{-1}] [+] 10 [x^{-1}] [=] [x^{-1}] [+] 15 [=] $\rightarrow 22\ \Omega$

EXAMPLE 11.12 Figure 11-13(a) on page 234 shows a series-parallel circuit that contains a series branch of two resistors, R_1 and R_2, connected to a parallel bank of two resistors, R_3 and R_4.

(a) Find the formula for the total resistance R_T.

(b) Find R_T given $R_1 = 200\ \Omega$, $R_2 = 240\ \Omega$, $R_3 = 150\ \Omega$, and $R_4 = 300\ \Omega$.

Solution

(a) Find the equivalent resistance of the series branch and the equivalent resistance of the parallel bank separately. The equivalent resistance of the series branch is:

$$R_{1,2} = R_1 + R_2$$

Apply formula (11.9) to R_3 and R_4 to find the equivalent resistance of the parallel bank:

$$R_{3,4} = \frac{R_3 R_4}{R_3 + R_4}$$

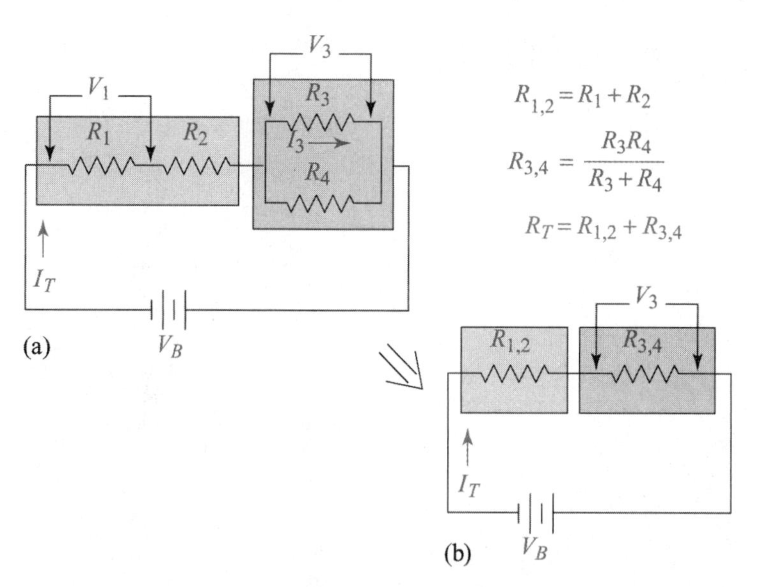

FIGURE 11-13 Series-parallel circuit for Examples 11.12 and 11.13.

The circuit can now be reduced to a series circuit containing $R_{1,2}$ and $R_{3,4}$, as shown in Figure 11-13(b). The formula for the total resistance R_T is then:

$$R_T = R_{1,2} + R_{3,4} = R_1 + R_2 + \frac{R_3 R_4}{R_3 + R_4}$$

(b) To find R_T, substitute the values $R_1 = 200\ \Omega$, $R_2 = 240\ \Omega$, $R_3 = 150\ \Omega$, and $R_4 = 300\ \Omega$ into the formula for R_T:

$$R_T = 200 + 240 + \frac{(150)(300)}{150 + 300} = 440 + 100 = 540\ \Omega$$

R_T can be found directly on the calculator using the reciprocal formula for $R_{3,4}$:

150 [x^{-1}] [+] 300 [x^{-1}] [=] [x^{-1}] [+] 200 [+] 240 [=] → 540 Ω

ERROR BOX

A common error to watch out for is not clearly identifying when resistors are in series or when they are in parallel.

For two resistors to be in series, the *same current* must travel through both resistors. As you follow the current through the two resistors, the end of one resistor must be connected to the beginning of the other resistor with no branch points in between.

For two resistors to be in parallel, the *same voltage drop* must be across each resistor. The ends of one resistor must be connected to the ends of the other resistor with no branch points in between. The same *total* current must enter and leave both resistors.

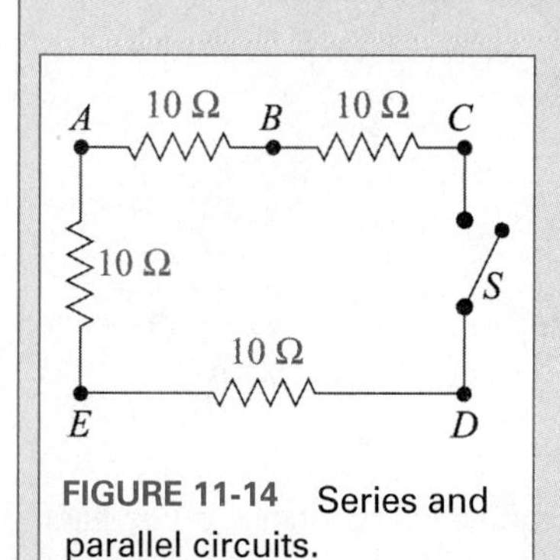

FIGURE 11-14 Series and parallel circuits.

Practice Problems: In the circuit in Figure 11-14, find the resistance between the two points indicated. *This is equivalent to the total resistance of the circuit when a voltage is applied across the two points.*

With switch S open, find: **1.** R_{AC} **2.** R_{CE} **3.** R_{AD} **4.** R_{CD} **5.** R_{AE}

With switch S closed, find: **6.** R_{AC} **7.** R_{CE} **8.** R_{AD} **9.** R_{CD} **10.** R_{AE}

The next example applies the results of Example 11.12 to find voltages and currents in the series-parallel circuit.

EXAMPLE 11.13

In Example 11.12, if the battery voltage $V_B = 24$ V, find the currents I_T and I_3 and the voltage drops V_1 and V_3 to two significant digits.

Solution From Example 11.12, the total resistance $R_T = 540\ \Omega$. Apply Ohm's law to find the total current:

$$I_T = \frac{V_T}{R_T} = \frac{24\text{ V}}{540\ \Omega} \approx 0.0444\text{ A} \approx 44\text{ mA}$$

To find the voltage drop V_1 across R_1, note that $I_T = I_1$. Therefore, by Ohm's law:

$$V_1 = I_T R_1 = (0.0444\text{ A})(200\ \Omega) \approx 8.9\text{ V}$$

Note that the calculation is done with three significant figures and then rounded off to two figures for greater accuracy. To find the voltage drop V_3 across R_3, note that V_3 is the same as the voltage drop across the equivalent parallel resistance $R_{3,4}$ in the series circuit of Figure 11-13(b). From Example 11.12:

$$R_{3,4} = \frac{(150)(300)}{150 + 300} = 100\ \Omega$$

or

150 Y + 300 Y = Y = → 100 Ω

The current through $R_{3,4}$ is the total current I_T. Apply Ohm's law to find the voltage drop across $R_{3,4}$:

$$V_3 = I_T R_{3,4} = (0.0444\text{ A})(100\ \Omega) = 4.44\text{ V} \approx 4.4\text{ V}$$

To find I_3, apply Ohm's law again to V_3 and R_3:

$$I_3 = \frac{V_3}{R_3} = \frac{4.44\text{ V}}{150\ \Omega} \approx 30\text{ mA}$$

EXAMPLE 11.14

Figure 11-15(a) on page 236 shows a series-parallel circuit that contains a resistance R_1 in series with a parallel bank. The parallel bank consists of a series branch of two resistors, R_2 and R_3, in parallel with R_4.

(a) Find the formula for the total resistance of the circuit.

(b) Calculate R_T given $R_1 = 330\ \Omega$, $R_2 = 510\ \Omega$, $R_3 = 620\ \Omega$, and $R_4 = 1.0\text{ k}\Omega$ to two significant digits.

Solution

(a) You must first reduce the series branch before you can reduce the parallel bank. The total series resistance of R_2 and R_3 is:

$$R_{2,3} = R_2 + R_3$$

This reduces the circuit to Figure 11-15(b). Then apply formula (11.9) to the parallel bank of $R_{2,3}$ and R_4:

$$R_{2,3,4} = \frac{R_{2,3}R_4}{R_{2,3} + R_4}$$

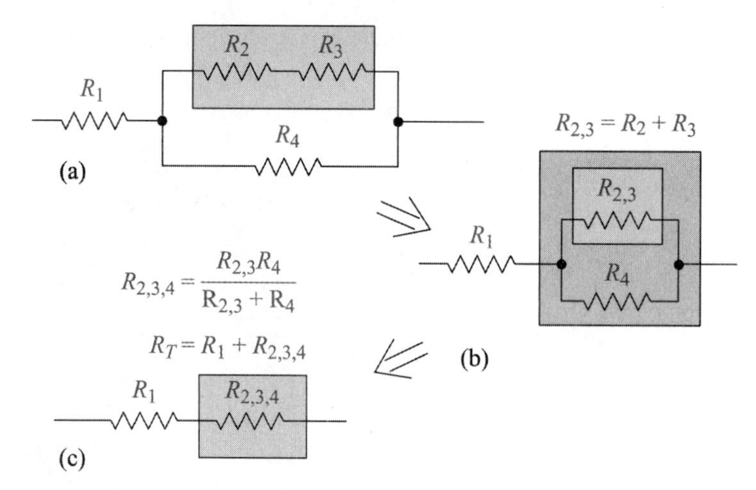

FIGURE 11-15 Series-parallel circuit for Example 11.14.

Substitute $R_2 + R_3$ for $R_{2,3}$ to obtain:

$$R_{2,3,4} = \frac{(R_2 + R_3)R_4}{(R_2 + R_3) + R_4}$$

This reduces the circuit to the series circuit in Figure 11-15(c). The total resistance of the circuit is then:

$$R_T = R_1 + R_{2,3,4} = R_1 + \frac{(R_2 + R_3)R_4}{(R_2 + R_3) + R_4}$$

(b) To calculate R_T, substitute the given values $R_1 = 330\ \Omega$, $R_2 = 510\ \Omega$, $R_3 = 620\ \Omega$, and $R_4 = 1.0\ \text{k}\Omega$ into the formula above:

$$R_T = 330 + \frac{(510 + 620)(1000)}{(510 + 620) + 1000} = 330 + \frac{(1130)(1000)}{2130} \approx 860\ \Omega$$

One calculator solution is:

[( ] 510 [+] 620 [ )] [x^{-1}] [+] 1000 [x^{-1}] [=] [x^{-1}] [+] 330 [=] → 860 Ω

Example 11.14 illustrates a basic procedure for a series-parallel circuit: You should always reduce any series branch that is part of a parallel bank before reducing the parallel bank to an equivalent resistance.

EXERCISE 11.3

Express all answers in engineering notation (as numbers between 1 and 1000) with the appropriate units to two significant digits.

1. In the series-parallel circuit of Figure 11-12(a), find R_T when $R_1 = 20\ \Omega$, $R_2 = 30\ \Omega$, and $R_3 = 30\ \Omega$.
2. In the series-parallel circuit of Figure 11-12(a), find R_T when $R_1 = 4.3\ \text{k}\Omega$, $R_2 = 6.8\ \text{k}\Omega$, and $R_3 = 6.8\ \text{k}\Omega$.
3. In the series-parallel circuit of Figure 11-12(a), find R_T when $R_1 = 2.2\ \text{k}\Omega$, $R_2 = 5.0\ \text{k}\Omega$, and $R_3 = 7.5\ \text{k}\Omega$.
4. In the series-parallel circuit of Figure 11-12(a), find R_T when $R_1 = 100\ \Omega$, $R_2 = 200\ \Omega$, and $R_3 = 100\ \Omega$.
5. In the series-parallel circuit of Figure 11-13(a), find R_T if $R_1 = 120\ \Omega$, $R_2 = 130\ \Omega$, $R_3 = 300\ \Omega$, and $R_4 = 200\ \Omega$.
6. In the series-parallel circuit of Figure 11-13(a), find R_T if $R_1 = 36\ \Omega$, $R_2 = 27\ \Omega$, $R_3 = 56\ \Omega$, and $R_4 = 56\ \Omega$.

Answers to Error Box Problems, page 234:

1. 20 Ω **2.** 30 Ω **3.** 20 Ω **4.** 40 Ω **5.** 10 Ω **6.** 10 Ω **7.** 7.5 Ω **8.** 10 Ω **9.** 0 Ω **10.** 7.5 Ω

7. In the series-parallel circuit of Figure 11-12(a), $R_1 = 12\ \Omega$ and $R_2 = 15\ \Omega$. If R_T is to be 22 Ω, what should be the value of R_3? Substitute in the formula for R_T and solve for R_3.
8. In the series-parallel circuit of Figure 11-12(a), $R_2 = 150\ \Omega$ and $R_3 = 100\ \Omega$. If R_T is to be 90 Ω, what should be the value of R_1? Substitute in the formula for R_T and solve for R_1.
9. In the series-parallel circuit of Figure 11-13(a), $R_2 = 10\ \Omega$, $R_3 = 33\ \Omega$, and $R_4 = 68\ \Omega$. If R_T is to be 45 Ω, what should be the value of R_1? Substitute in the formula for R_T and solve for R_1.
10. In the series-parallel circuit of Figure 11-13(a), $R_1 = 20\ \Omega$, $R_2 = 30\ \Omega$, and $R_3 = 40\ \Omega$. If R_T is to be 58 Ω, what should be the value of R_4? Substitute in the formula for R_T and solve for R_4.
11. In the circuit of Figure 11-16:
    (a) Find the formula for R_T.
    (b) Calculate R_T given $R_1 = 120\ \Omega$, $R_2 = 240\ \Omega$, and $R_3 = 100\ \Omega$.
12. In the circuit of Figure 11-17:
    (a) Find the formula for R_T.
    (b) Calculate R_T given $R_1 = 30\ \Omega$, $R_2 = 10\ \Omega$, $R_3 = 20\ \Omega$, and $R_4 = 75\ \Omega$.
13. In Example 11.12, if the battery voltage $V_B = 14$ V, find I_T, V_2, V_4, and I_4.
14. In the circuit of Figure 11-13(a), $R_1 = 18\ \Omega$, $R_2 = 22\ \Omega$, $R_3 = 75\ \Omega$, and $R_4 = 150\ \Omega$. If $V_B = 9.0$ V, find R_T, I_T, V_2, V_3, and I_3.
15. In the circuit of Figure 11-15(a), find R_T if $R_1 = 120\ \Omega$, $R_2 = 180\ \Omega$, $R_3 = 120\ \Omega$, and $R_4 = 200\ \Omega$.
16. In the circuit of Figure 11-15(a), $R_1 = 10\ \Omega$, $R_2 = 5\ \Omega$, and $R_3 = 5\ \Omega$. If R_T is to be 15 Ω, what should be the value of R_4?
17. In the circuit of Figure 11-16, find R_1 if $R_2 = 12\ \Omega$, $R_3 = 7.5\ \Omega$, and $R_T = 15\ \Omega$.
18. In the circuit of Figure 11-17, find R_1 if $R_2 = 1.5$ kΩ, $R_3 = 1.5$ kΩ, $R_4 = 1.2$ kΩ, and $R_T = 1.0$ kΩ.
19. In the circuit of Figure 11-18, find R_T if $R_1 = 20\ \Omega$, $R_2 = 10\ \Omega$, $R_3 = 20\ \Omega$, $R_4 = 30\ \Omega$, and $R_5 = 15\ \Omega$.
20. In the circuit of Figure 11-19, find R_T if $R_1 = 120\ \Omega$, $R_2 = 100\ \Omega$, $R_3 = 200\ \Omega$, and $R_4 = 100\ \Omega$.
21. In the circuit of Figure 11-18, if $R_1 = 100\ \Omega$, $R_2 = 50\ \Omega$, $R_3 = 50\ \Omega$, $R_4 = 150\ \Omega$, $R_5 = 100\ \Omega$, and $V_T = 12$ V, find the total current I_T in the circuit and the voltage drops V_1 and V_2.
22. In the circuit of Figure 11-19, if $R_1 = 43\ \Omega$, $R_2 = 120\ \Omega$, $R_3 = 120\ \Omega$, $R_4 = 120\ \Omega$, and $V_T = 24$ V, find the total current I_T and the voltage drops V_1 and V_2.
23. In the circuit of Figure 11-20, find the total power P_T in the circuit if $R_1 = 30\ \Omega$, $R_2 = 15\ \Omega$, $R_3 = 15\ \Omega$, $R_4 = 10\ \Omega$, and $V_T = 6.0$ V.

    Note: $P_T = \dfrac{V_T^2}{R_T}$
24. In the circuit of Figure 11-20, find the total voltage V_T if $I_T = 1.2$ A, $R_1 = 10\ \Omega$, $R_2 = 30\ \Omega$, $R_3 = 20\ \Omega$, and $R_4 = 20\ \Omega$.

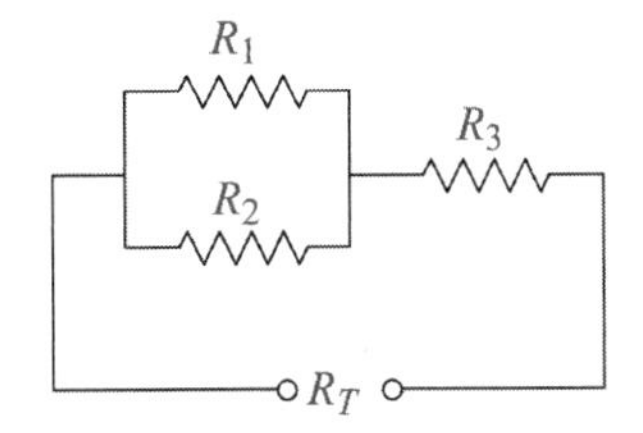

FIGURE 11-16 Series-parallel circuit for problems 11 and 17.

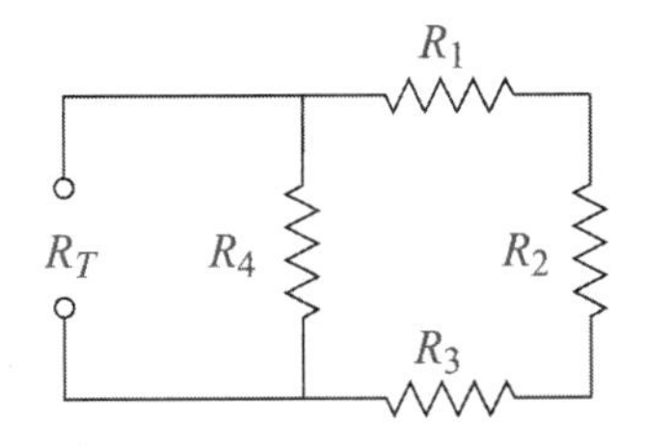

FIGURE 11-17 Series-parallel circuit for problems 12 and 18.

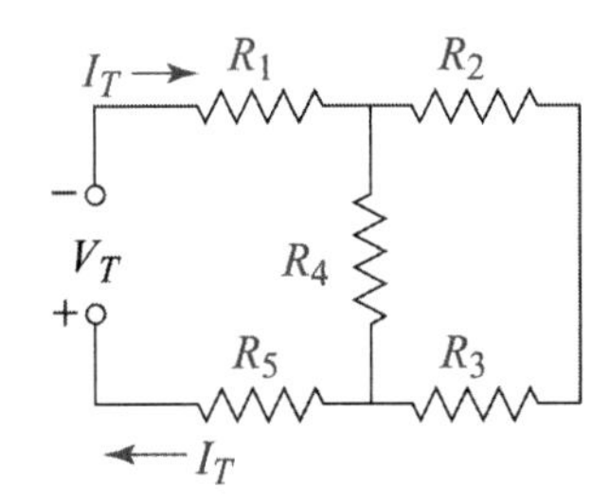

FIGURE 11-18 Series-parallel circuit for problems 19 and 21.

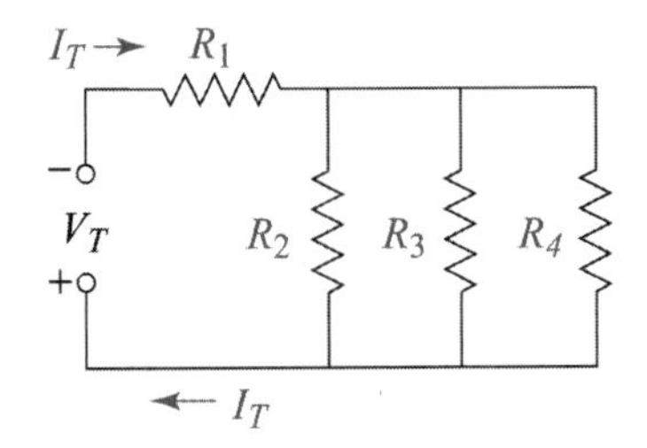

FIGURE 11-19 Series-parallel circuit for problems 20 and 22.

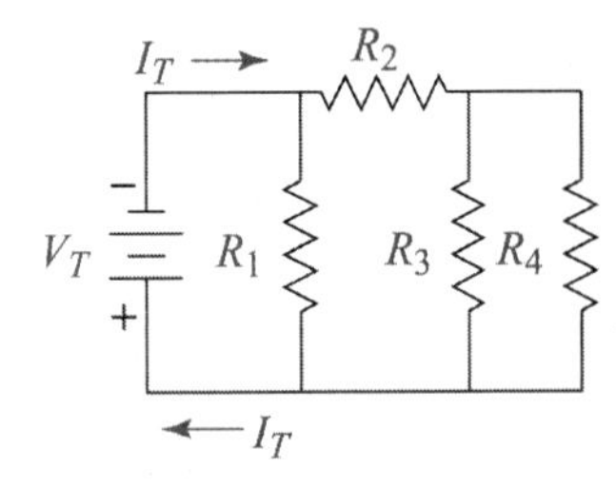

FIGURE 11-20 Series-parallel circuit for problems 23 and 24.

CHAPTER HIGHLIGHTS

11.1 SERIES CIRCUITS

Series Circuit

A series circuit is one in which all the components are connected in one path so that the same current flows through each component.

• • •

Total Series Resistance

$$R_T = R_1 + R_2 + R_3 + \cdots + R_n \tag{11.1}$$

• • •

Total Series Voltage

$$V_T = V_1 + V_2 + V_3 + \cdots + V_n \tag{11.2}$$

Key Example: In a series circuit with three resistors $R_1 = 50\ \Omega$, $R_2 = 100\ \Omega$, and $R_3 = 300\ \Omega$ and the applied voltage $V_T = 90$ V, the total resistance and the current are:

$$R_T = 50 + 100 + 300 = 450\ \Omega$$

$$I = \frac{V_T}{R_T} = \frac{90}{450} = 200 \text{ mA}$$

The voltage drops add up to $V_T = 90$ V:

$$\begin{aligned} V_1 &= IR_1 = (0.2)(50) = 10 \text{ V} \\ V_2 &= IR_2 = (0.2)(100) = 20 \text{ V} \\ V_3 &= IR_3 = (0.2)(300) = 60 \text{ V} \end{aligned}$$

The voltage drops are proportional to the resistances: V_3 is three times V_2, and V_2 is twice V_1.

Series-aiding voltages V_1 and V_2 are connected with the positive terminal of one to the negative terminal of the other:

Series-Aiding Voltages

$$V_T = V_1 + V_2 \tag{11.3}$$

Series-opposing voltages V_1 and V_2 are connected with the negative terminal of one to the negative terminal of the other:

Series-Opposing Voltages

$$V_T = V_2 - V_1 \tag{11.4}$$

• • •

Total Power in a Series Circuit

$$P_T = P_1 + P_2 + P_3 + \cdots + P_n \tag{11.5}$$

Key Example: Two series-opposing voltages $V_1 = 12$ V and $V_2 = 6.0$ V are connected to two series resistors $R_1 = 51\ \Omega$ and $R_2 = 43\ \Omega$. The total voltage and current are:

$$V_T = 12 - 6 = 6.0 \text{ V}$$

$$I = \frac{V_T}{R_T} = \frac{6.0}{51 + 43} \approx 64 \text{ mA}$$

The powers dissipated add up to the total power:

$$\begin{aligned} P_1 &= I^2R_1 = (0.064)^2(51) \approx 209 \text{ mW} \\ P_2 &= I^2R_2 = (0.064)^2(43) \approx 176 \text{ mW} \\ P_T &= I^2R_T = (0.064)^2(94) \approx 385 \text{ mW} \end{aligned}$$

11.2 PARALLEL CIRCUITS

Parallel Circuit

A parallel circuit is one in which each component is connected directly to the voltage source so that each component has the same voltage drop as the voltage source.

• • •

Total Parallel Current

$$I_T = I_1 + I_2 + I_3 + \cdots + I_n \tag{11.6}$$

• • •

Total Parallel Resistance

$$\frac{1}{R_T} = \frac{1}{R_1} + \frac{1}{R_2} + \frac{1}{R_3} + \cdots + \frac{1}{R_n} \tag{11.7}$$

The total or equivalent resistance R_T *is always less than the smallest of the parallel resistances.*

N Equal Parallel Resistances

$$R_T = \frac{R}{N} \tag{11.8}$$

A useful formula for the equivalent resistance R_T of two resistors R_1 and R_2 in parallel is given by *the product divided by the sum:*

Two Parallel Resistances

$$R_T = \frac{R_1 R_2}{R_1 + R_2} \quad (11.9)$$

A special case of formula (11.7) that lends itself to the calculator is:

$$\frac{1}{R_T} = \frac{1}{R_1} + \frac{1}{R_2}$$

Key Example: Given three parallel resistors $R_1 = 12\ \Omega$, $R_2 = 12\ \Omega$ and $R_3 = 16\ \Omega$. R_1 and R_2 are equivalent to one resistor $R_{1,2}$:

$$R_{1,2} = \frac{12}{2} = 6.0\ \Omega$$

Then:

$$R_T = \frac{R_{1,2} R_3}{R_{1,2} + R_3} = \frac{(6)(16)}{6 + 16} \approx 4.4\ \Omega$$

Using the calculator and formula (11.7):

12 [x⁻¹] [+] 12 [x⁻¹] [+] 16 [x⁻¹] [=] [x⁻¹] [=] → 4.4 Ω

Parallel Resistances Solved for R_1

$$R_1 = \frac{R_2 R_T}{R_2 - R_T} \quad (11.10)$$

Instead of formula (11.10), you can use the following formula, which lends itself to the calculator:

$$\frac{1}{R_1} = \frac{1}{R_T} - \frac{1}{R_2}$$

Key Example: Given two parallel resistors where $R_2 = 4.7\ \text{k}\Omega$ and $R_T = 3.0\ \text{k}\Omega$, R_1 is:

$$R_1 = \frac{(4.7)(3.0)}{4.7 - 3.0} = \frac{14.1}{1.7} \approx 8.3\ \text{k}\Omega$$

or

3 [x⁻¹] [−] 4.7 [x⁻¹] [=] [x⁻¹] [=] → 8.3 kΩ

Conductance (Siemens)

$$G = \frac{1}{R} \quad (11.11)$$

• • •

Total Parallel Conductance

$$G_T = G_1 + G_2 + G_3 + \cdots + G_n \quad (11.12)$$

Key Example: Given a parallel bank of three resistors, $R_1 = 75\ \Omega$, $R_2 = 150\ \Omega$, and $R_3 = 100\ \Omega$, the total conductance G_T and total resistance R_T are:

$$G_T = \frac{1}{R_T} = \frac{1}{75} + \frac{1}{150} + \frac{1}{100} = \frac{4 + 2 + 3}{300}$$

$$= 0.03\ \text{S} = 30\ \text{mS}$$

$$R_T = \frac{1}{G_T} = \frac{1}{30} \approx 33\ \Omega$$

Using the calculator:

75 [x⁻¹] [+] 150 [x⁻¹] [+] 100 [x⁻¹]
[=] → 0.03 S [x⁻¹] → 33 Ω

Remember: In a series circuit the *current is constant* and the *voltage divides.* In a parallel circuit the *voltage is constant* and the *current divides.*

11.3 SERIES-PARALLEL CIRCUITS

To solve a series-parallel circuit, reduce each series branch and each parallel bank, separately, to an equivalent resistance. When a series branch is part of a parallel bank, reduce the series branch first. Then combine the equivalent resistances to obtain the total resistance of the circuit.

Key Example: In Figure 11-15(a) on page 236, given $R_1 = 15\ \Omega$, $R_2 = 10\ \Omega$, $R_3 = 20\ \Omega$, and $R_4 = 30\ \Omega$, the total resistance is:

$$R_T = R_1 + \frac{(R_2 + R_3)R_4}{(R_2 + R_3) + R_4} = 15 + \frac{(10 + 20)(30)}{(10 + 20) + 30}$$

$$= 15 + \frac{(30)(30)}{60} = 30\ \Omega$$

or

30 [x⁻¹] [+] [(] 10 [+] 20 [)] [x⁻¹] [=] [x⁻¹] [+] 15
[=] → 30 Ω

Series-parallel circuits can have various configurations. Study Examples 11.12 through 11.14 to review the different types of circuits. Do the practice problems in the Error Box in Section 11.3 if you have not done them.

REVIEW EXERCISES

Express all answers in engineering notation (as numbers between 1 and 1000) with the appropriate units to two significant digits.

1. Given the series circuit in Figure 11-21 with $R_1 = 33\ \Omega$, $R_2 = 18\ \Omega$, and $V_T = 24$ V. Find R_T, I, V_1, and V_2.
2. Given the series circuit in Figure 11-21 with $R_1 = 1.5\ R_2$, $I = 500$ mA and $V_T = 10$ V. Find R_T, R_1, R_2, V_1, V_2, P_1, P_2, and P_T.

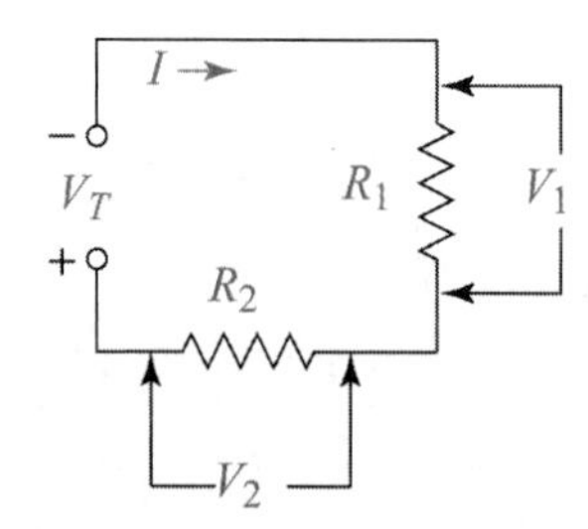

FIGURE 11-21 Series circuit for problems 1, 2, 3, and 4.

3. Given the series circuit in Figure 11-21 with $V_1 = 2V_2$, $I = 40$ mA and $V_T = 6.0$ V. Find R_T, R_1, R_2, V_1, V_2, P_1, P_2, and P_T.
4. Given the series circuit in Figure 11-21 with $R_1 = 10$ kΩ, $I = 1.5$ mA and $V_T = 36$ V. Find R_2, R_T, V_1, V_2, P_1, P_2, and P_T.
5. Three resistors R_1, R_2, and R_3 are connected in series to a voltage source. Given $I = 2.5$ A, $R_1 = 110\ \Omega$, $R_2 = 160\ \Omega$, and $R_3 = 180\ \Omega$, find V_T, P_1, P_2, P_3, P_T.
6. Three resistors R_1, R_2, and R_3 are connected in series to a voltage source. Given $V_T = 60$ V, $R_1 = 12\ \Omega$, $R_2 = 16\ \Omega$, and $I = 1.2$ A, find R_3, R_T, V_1, V_2, V_3, P_1, P_2, P_3, P_T.
7. Three resistors R_1, R_2, and R_3 are connected in series to a battery. If $R_T = 91\ \Omega$, $R_2 = 2R_1$ and $R_3 = 2R_2$, find each resistance.
8. Three resistors R_1, R_2, and R_3 are connected in series to a battery. The total resistance of the circuit is 1.1 kΩ. If $V_2 = V_1$ and $V_3 = 50\%$ of V_2, find each resistance.
9. Figure 11-22 shows a voltage dropping variable resistor in series with a bulb and a 9-V battery.
   (a) If the bulb uses 5.0 W of power and draws 800 mA, what should be the value of R_D?
   (b) What should be the value of R_D to decrease the voltage across the bulb by one-half of the voltage in part (a)?
10. Two voltages V_1 and V_2 are connected in series aiding to a resistor $R = 11$ kΩ. An ammeter in the circuit reads 20 mA. If $V_1 = V_2 + 20$ V, find V_1 and V_2.

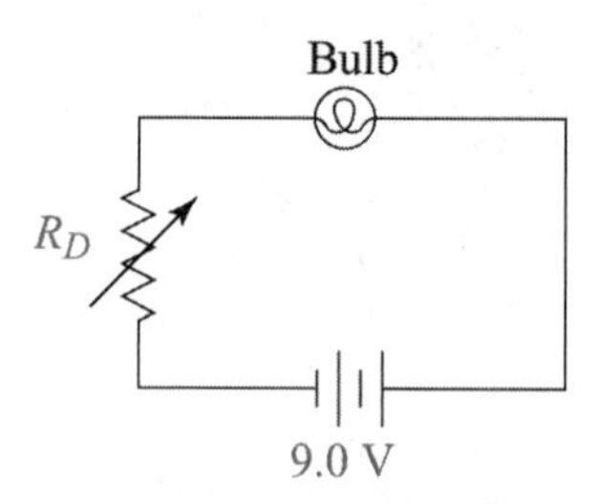

FIGURE 11-22 Voltage-dropping resistor for problem 9.

In exercises 11 through 16, given a parallel bank of two resistors R_1 and R_2, find the indicated values.

11. $R_1 = 22\ \Omega$, $R_2 = 43\ \Omega$; G_T and R_T.
12. $R_1 = 6.8$ kΩ, $R_2 = 5.6$ kΩ; G_T and R_T.
13. $R_1 = 120\ \Omega$, $R_2 = 130\ \Omega$; G_T and R_T.
14. $R_2 = 12\ \Omega$, $R_T = 3.0\ \Omega$; R_1.
15. $R_2 = 360\ \Omega$, $R_T = 160\ \Omega$; R_1.
16. $R_1 = 2.2$ kΩ, $R_T = 910$ kΩ; R_2.

In exercises 17 through 22, given a parallel bank of three resistors R_1, R_2 and R_3, find the indicated values.

17. $R_1 = 36\ \Omega$, $R_2 = 36\ \Omega$, $R_3 = 18\ \Omega$; G_T and R_T.
18. $R_1 = 1.6$ kΩ, $R_2 = 680\ \Omega$, $R_3 = 1.0$ kΩ; G_T and R_T.
19. $R_1 = 100\ \Omega$, $R_3 = 200\ \Omega$, $R_3 = 300\ \Omega$; G_T and R_T.
20. $R_1 = 240\ \Omega$, $R_3 = 120\ \Omega$, $R_T = 46\ \Omega$; R_2.
21. $R_1 = 820\ \Omega$, $R_2 = 750\ \Omega$, $R_T = 210\ \Omega$; R_3.
22. $R_2 = 1.1$ kΩ, $R_3 = 1.2$ kΩ, $R_T = 330\ \Omega$; R_1.
23. Two resistors $R_1 = 180\ \Omega$ and $R_2 = 270\ \Omega$ are connected in a parallel bank. How much resistance should be added in parallel to reduce the resistance of the bank by one half?
24. Two resistors R_1 and R_2 are connected in parallel to a battery whose voltage is V_B. If $I_T = 800$ mA, $R_1 = 9.1\ \Omega$, and $R_2 = 10\ \Omega$, find R_T, I_1, I_2, V_B, P_1, P_2, and P_T.
25. Three resistors R_1, R_2, and R_3 are connected in parallel across a voltage of 40 V. If $R_1 = 33\ \Omega$, $R_2 = 68\ \Omega$, and $R_3 = 100\ \Omega$, find I_1, I_2, I_3, I_T, R_T, and G_T for the circuit.
26. Three resistors R_1, R_2, and R_3 are connected in parallel across a voltage of 100 V. If $P_1 = 10$ W, $P_2 = 20$ W, and $P_T = 70$ W, find P_3, R_1, R_2, R_3, R_T, G_T, I_T, I_1, I_2, and I_3.
27. For the series-parallel circuit in Figure 11-23:
   (a) Find the formula for R_T.
   (b) Calculate R_T given $R_1 = 120\ \Omega$, $R_2 = 130\ \Omega$, and $R_3 = 150\ \Omega$.

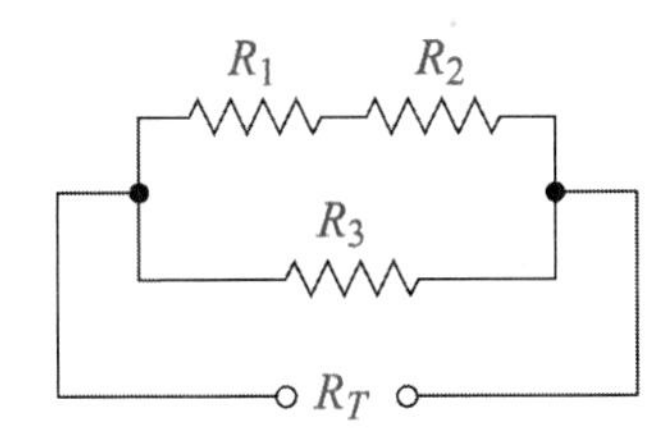

FIGURE 11-23 Series-parallel circuit for problems 27 and 28.

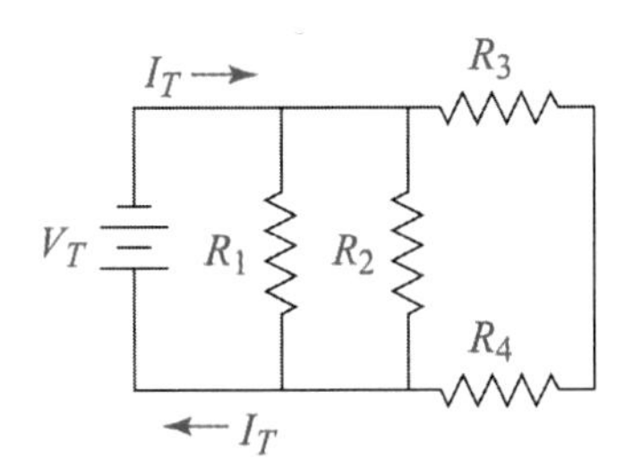

FIGURE 11-24 Series-parallel circuit for problems 29 and 30.

28. For the series-parallel circuit in Figure 11-23, find R_3 if $R_1 = 3.0\ \text{k}\Omega$, $R_2 = 2.0\ \text{k}\Omega$, and $R_T = 3.0\ \text{k}\Omega$.
29. For the series-parallel circuit in Figure 11-24, given $R_1 = 100\ \Omega$, $R_2 = 200\ \Omega$, $R_3 = 110\ \Omega$, $R_4 = 90\ \Omega$, and $I_T = 500$ mA, find R_T, V_1, V_3, and the total power dissipated.
30. For the series-parallel circuit in Figure 11-24, given $R_1 = 15\ \Omega$, $R_2 = 50\ \Omega$, $R_3 = 50\ \Omega$, $R_4 = 75\ \Omega$, and $V_T = 6.5$ V, find I_1, I_2, V_3, and the total power dissipated.
31. For the series-parallel circuit in Figure 11-25.
    (a) Find the formula for R_T.
    (b) Given $R_1 = 240\ \Omega$, $R_2 = 1.1\ \text{k}\Omega$, $R_3 = 1.6\ \text{k}\Omega$, $R_4 = 1.3\ \text{k}\Omega$, and $I_T = 35$ mA. Find R_T, V_T, and V_1.
32. For the series-parallel circuit in Figure 11-25, given $R_1 = 100\ \Omega$, $R_2 = 100\ \Omega$, $R_3 = 200\ \Omega$, $R_4 = 200\ \Omega$, and $V_T = 50$ V. Find R_T, I_T, and I_1.
33. Given a circuit with three resistors in parallel, R_1, R_2, and R_3. If $R_2 = R_1$ and $R_3 = 2R_1$, show that the equivalent resistance is:

$$R_T = \frac{2R_1}{5}$$

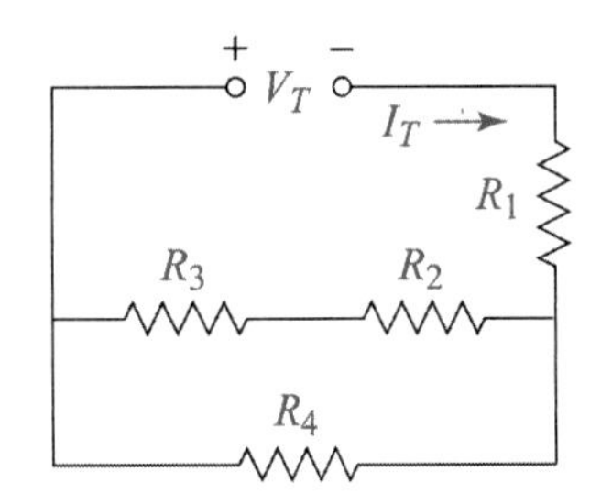

FIGURE 11-25 Series-parallel circuit for problems 31 and 32.

34. Given a circuit with three resistors in parallel, R_1, R_2, and R_3. If $R_3 = R_1$, show that the equivalent resistance is:

$$R_T = \frac{R_1 R_2}{R_1 + 2R_2}$$

CHAPTER 12

Graphs

CHAPTER OBJECTIVES

In this chapter, you will learn:

- How to plot points on a rectangular coordinate system.
- How to graph an equation in two variables.
- The definition of a mathematical function.
- The concept of a linear function and how to apply it to electrical formulas.
- The meaning of the slope of a graph.
- Some important nonlinear functions.
- How to graph experimental data.

Graphs are used throughout electricity and electronics to better illustrate ideas and to help visualize concepts. You need to be familiar with the techniques of graphing and how to analyze a graph. For example, in Chapter 7 Ohm's law tells us that the current is directly proportional to the voltage when resistance is constant. This can be seen very clearly from a straight line graph of current versus voltage for a circuit containing a constant resistance. A first-degree formula or linear function graphs as a straight line, which is the most basic and important graph because it is the key to understanding other types of graphs. Many of the formulas and relationships that have been introduced in previous chapters can be graphed as straight lines or simple curves and are studied in this chapter.

• • •

12.1 Rectangular Coordinate System

A graph is a line connecting a series of points where each point is plotted using two values, one measured in the horizontal direction and the other measured in the vertical direction. In a rectangular coordinate system, x is generally used for the horizontal axis and y for the vertical axis. See Figure 12-1. When plotting electrical or electronic data, the variables used would not be x and y but would represent the quantities being graphed, such as V for voltage, I for current, P for power, and so on.

The two values associated with each point are called coordinates. The first coordinate is the distance in the horizontal direction, and the second coordinate is the distance in the vertical direction. To the right and up is positive, to the left and down is negative. The following example shows the procedure for plotting points on a rectangular coordinate system and is important in understanding all types of graphs.

EXAMPLE 12.1

Draw a rectangular coordinate system and plot the following points: A (6,5), B (−3,2), C (0,4), D (4,0), E (−5,−6), F (−5,0), G (2,−3), H (0,−5).

Solution Use a ruler to draw the x and y axes on graph paper. The *origin* is where the axes cross. The axes divide the graph into four quadrants I, II, III, and IV, starting in the upper right and moving counterclockwise, as shown in Figure 12-1 on page 244. The scales are usually shown on each axis but they can be omitted if the boxes are all equal to one unit. Locate each point as follows: Start from the origin and measure the distance of the first coordinate along the horizontal or x axis. Move to the right for (+) and to the left for (−). Measure the distance of the second coordinate in the vertical or y direction. Move up for (+) and down for (−). For example, to plot point B(−3,2), move three boxes in the negative x direction and then two boxes in the positive y direction. To plot point H(0,−5),

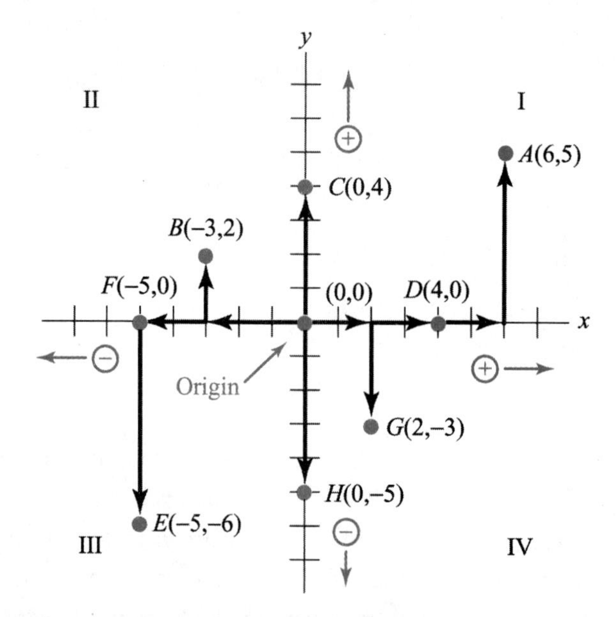

FIGURE 12-1 Rectangular coordinate system for Example 12.1.

move zero boxes in the x direction, which means that you are on the y axis, and then move five points down on the y axis. Figure 12-1 shows how each point is plotted.

Observe the following for certain points on the coordinate system:

- The origin where the axes intersect has the coordinates (0,0).
- Every point on the x axis has a y coordinate of zero.
- Every point on the y axis has an x coordinate of zero.
- In quadrant I, x and y are both positive.
- In quadrant II, x is negative and y is positive.
- In quadrant III, x and y are both negative
- In quadrant IV, x is positive and y is negative.

Any formula or literal equation in two variables expresses a relationship between the two variables that can be represented by a graph. The graph shows how one variable changes with respect to the other. A first-degree or linear equation is one that contains no exponents greater than one and no negative exponents. A linear equation in two variables produces the most basic and important graph, the straight line.

EXAMPLE 12.2 Draw the graph of the linear equation: $2x - y = -1$.

Solution You need to find corresponding values of x and y, which represent points that satisfy the equation. At least two points are necessary to graph a straight line. However, if one point is not correct, the entire line is wrong. Therefore, you should plot three or more points. You can assign values to either x or y. The values $x = 0$ and $y = 0$ are easy to work with and give the points where the line intercepts each axis as follows:

y intercept (Let $x = 0$)	x intercept (Let $y = 0$)
$2(0) - y = -1$	$2x - (0) = -1$
$y = 1$	$x = -0.5$

The point (0,1) is where the line crosses the y axis and is the y intercept. See Figure 12-2. The point (−0.5,0) is where the line crosses the x axis and is the x intercept. Now choose at least one more value for x as a check point and solve for y. Using $x = 1$ as the check point:

$$2(1) - y = -1$$
$$-y = -3$$
$$y = 3$$

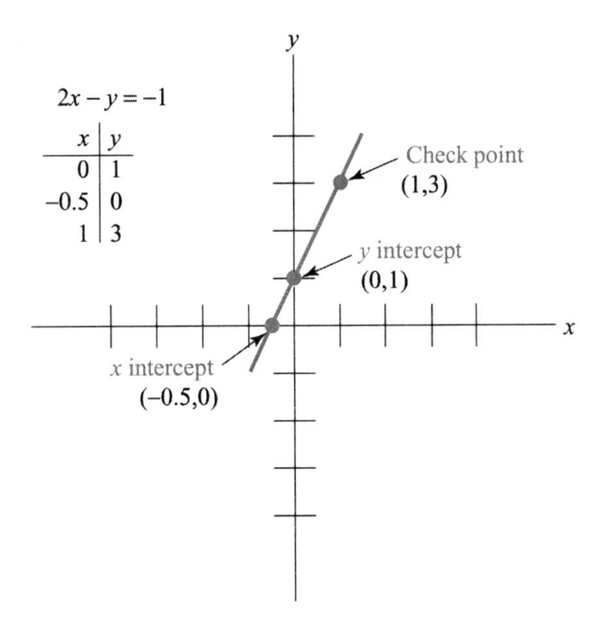

FIGURE 12-2 Intercept method for Example 12.2.

The table for the three points (0,1), (−0.5,0), and (1,3) and the line $2x - y = -1$ are shown in Figure 12-2. Since the x intercept falls halfway between 0 and −1 on the graph, it is necessary to estimate its position halfway between the two units.

ERROR BOX

To avoid confusing the x intercept with the y intercept, remember the opposite is true: At the x intercept, $y = 0$. At the y intercept, $x = 0$. The only exception to this is when a line passes through the origin (0,0). In this case there is only one intercept, and both x and y are zero. See if you can correctly identify the x and y intercepts for the lines in the practice problems.

Practice Problems: Find the x and y intercepts for each line:

1. $x - y = -1$ **2.** $2x - 3y = 6$ **3.** $0.5x + y = 0$ **4.** $x + 4y = 10$

5. $x = 1.5y$ **6.** $4x + 2y = 4$ **7.** $2x + 3y = 9$ **8.** $10x - y = 5$

9. $9x + y = -3$ **10.** $y = 7 + x$

EXAMPLE 12.3

Close the Circuit

A dc circuit contains a fixed resistor $R = 15\ \Omega$. Plot the current I in amps versus the voltage drop V across the resistor, using six points from $V = 0.0$ V to $V = 75$ V. Use the horizontal axis for V.

Solution Use Ohm's law to find the value of I for each value of V:

$$I = \frac{V}{R}$$

Let $R = 15\ \Omega$ as given: $I = \dfrac{V}{15\ \Omega}$

This is a linear equation in two variables, and the graph will therefore be a straight line. Only positive values of V apply. Substitute values of V every 15 V starting with $V = 0.0$ V into the equation to give six points:

V (V)	0.0	15	30	45	60	75
I (A)	0.0	1.0	2.0	3.0	4.0	5.0

The graph lies entirely in the first quadrant since both V and I assume positive values only. It is necessary to let each box be more than one unit on the V or horizontal axis to accomodate the range of values. Each box is set equal to 15 units, and the graph is shown in Figure 12-3.

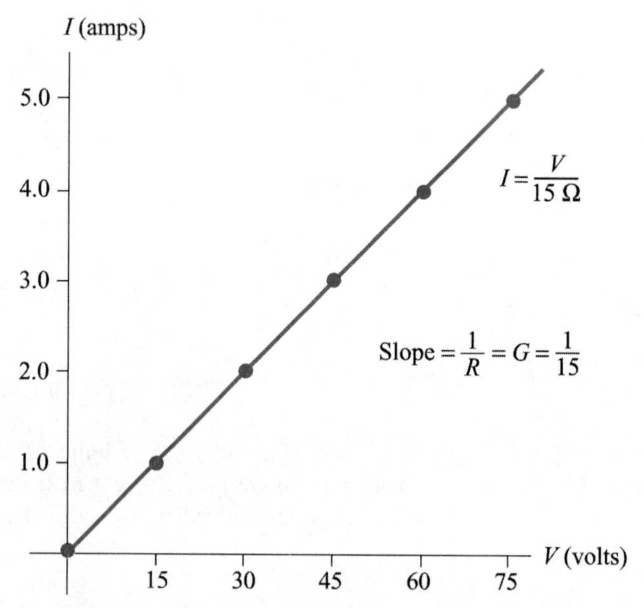

FIGURE 12-3 Ohm's law, current versus voltage, for Example 12.3.

EXERCISE 12.1

In exercises 1 through 16, plot each point on a rectangular coordinate system.

1. (3,2)
2. (1,4)
3. (0,5)
4. (–5,0)
5. (–2, 3)
6. (–4,5)
7. (–3,–2)
8. (–6,–4)
9. (5,–5)
10. (4,–4)
11. (0,0)
12. (0,–6)
13. (3,0.5)
14. (0.5,–2)
15. (–3.5,2.5)
16. (–1.5,–2.5)

In exercises 17 through 30, draw the graph of each linear equation using the intercept method. See Example 12.2.

17. $x - y = 3$
18. $x + y = -5$
19. $x - 2y = 4$
20. $2x + y = -2$
21. $x + y = 0$
22. $x - 2y = 0$
23. $2x - 3y = 6$
24. $5x - y = 10$

Answers to Error Box Problems, page 245:
1. (–1,0), (0,1) **2.** (3,0), (0,–2) **3.** (0,0) **4.** (10,0), (0,2.5) **5.** (0,0) **6.** (1,0), (0,2) **7.** (4.5,0), (0,3) **8.** (0.5,0), (0,–5) **9.** (–1/3,0), (0,–3) **10.** (–7,0), (0,7)

25. $3x + y = 6$
26. $x - 3y = 3$
27. $x + 2y = 3$
28. $3x - 2y = 3$
29. $y = 2x - 2$
30. $y = x + 4$

In exercises 31 through 40, draw the graph of each linear equation using the intercept method. Plot the variable in parentheses on the horizontal axis. Use appropriate scales. See Example 12.3.

31. $2a + b = 4$ (a)
32. $3a - 4b = 12$ (a)
33. $5r - 2t = 10$ (r)
34. $7r + t = 7$ (r)
35. $I = \frac{V}{10}$ (V)
36. $I = \frac{V}{150}$ (V)
37. $P - 20I = 0$ (I)
38. $P = 5I$ (I)
39. $2V_1 + V_2 = 5$ (V_1)
40. $3I_1 - 4I_2 = 6$ (I_1)

Applications to Electronics

In problems 41 through 46, draw the graph of each linear equation plotting at least five points. Use appropriate scales to accommodate the range of values. See Example 12.3.

41. A dc circuit contains a fixed resistor $R = 12\ \Omega$. Plot the current I in amps versus the voltage drop V across the resistor from $V = 0.0$ V to $V = 60$ V. Use the horizontal axis for V.
42. A dc circuit contains a fixed resistor $R = 1.5\ \text{k}\Omega$. Plot the current I in milliamps versus the voltage drop V across the resistor from $V = 0.0$ V to $V = 15$ V. Use the horizontal axis for V.
43. The applied voltage in a circuit is constant at $V = 120$ V. Plot the power P in the circuit in watts versus the current I from $I = 0.0$ A to $I = 15$ A. Use the horizontal axis for I.
44. The applied voltage in a circuit is constant at $V = 20$ V. Plot the power P in the circuit in watts versus the current I from $I = 0.0$ mA to $I = 500$ mA. Use the horizontal axis for I.
45. The current in a circuit is constant at $I = 200$ mA. Plot the power P in the circuit in milliwatts versus the voltage V from $V = 0.0$ V to $V = 5.0$ V. Use the horizontal axis for V.
46. The current in a circuit is constant at $I = 1.5$ A. Plot the power P in the circuit in watts versus the voltage V from $V = 0.0$ V to $V = 50$ V. Use the horizontal axis for V.

12.2 Linear Function

The concept of a function is a basic idea in mathematics. There are many examples of different types of functions throughout the text. In the examples of linear equations shown in Section 12.1, each value of x determines exactly one value of y. When that relationship exists between two variables, the second variable (y) is said to be a function of the first variable (x):

Definition **Function**

y is a function of x if, for each value of x, there corresponds *one and only one value* of y. (12.1)

The most basic function is the *linear function*. Consider the linear equation $2x - y = -1$ from Example 12.2. When this equation is solved for y, it becomes:

$$y = 2x + 1$$

The equation in this form expresses y as a linear function of x.

Formula **Linear Function**

$$y = mx + b \qquad (m, b \text{ constants}) \tag{12.2}$$
$$m = \text{slope} \qquad b = y \text{ intercept}$$

In the linear function (12.2), x is called the *independent* variable and y is called the *dependent* variable. The linear function always graphs as a straight line. The constant m is called the *slope* and is a measure of the steepness of the line:

Formula

Slope

$$m = \frac{\text{Change in } y}{\text{Change in } x} \tag{12.3}$$

The slope formula (12.3) tells you the following: If you move to the right on the line one unit in the x direction, y changes by m units. The constant b tells you that the line intercepts the y axis at the point $(0,b)$. For example, for the line $y = 3x - 1$, the slope $m = 3$. If you move to the right one unit in the x direction on this line, y increases 3 units. Also, the y intercept $b = -1$ so the line passes through the point $(0,-1)$ on the y axis. Examples 12.4, 12.5, and 12.6 further illustrate these concepts.

EXAMPLE 12.4 Draw the graph of the linear equation $2x - y = -1$ by first putting it in the form of a linear function. Plot at least three points and give the slope and y intercept.

Solution This is the same equation as Example 12.2. To put it in the form of a linear function, solve the equation for y:

$$2x - y = -1$$

Transpose $2x$ and isolate y:

$$-y = -2x - 1$$

Multiply by -1:

$$y = 2x + 1$$

The equation is now in the form of a linear function (12.2). The slope of the line is $m = 2$, which means that y increases 2 units for every increase of 1 unit in x. The y intercept of the line is $b = 1$, and it can be seen in Figure 12-4 that the line crosses the y axis at $(0,1)$.

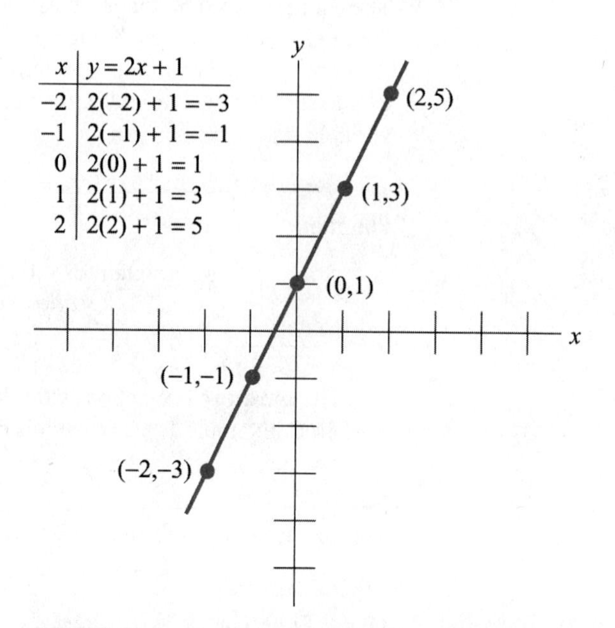

x	$y = 2x + 1$
-2	$2(-2) + 1 = -3$
-1	$2(-1) + 1 = -1$
0	$2(0) + 1 = 1$
1	$2(1) + 1 = 3$
2	$2(2) + 1 = 5$

FIGURE 12-4 Linear function for Example 12.4.

To graph this function, you can assign any value to x and calculate the corresponding value of y. You should choose convenient positive, zero, and negative values of x, such as: $x = -2, -1, 0, 1$, and 2. Make up a table listing the values of x and substitute each value of

x into the equation to calculate the corresponding value of y. Figure 12-4 shows the table of values of x with the calculated values of y for five plotted points, and the graph of the line $y = 2x + 1$.

When $b = 0$ in formula (12.2), the linear function becomes $y = mx$. This states that y is directly proportional to x, and m is the constant of proportionality. Study the next example, which illustrates different applications of linear functions where the variables are directly proportional.

EXAMPLE 12.5 List some applications of linear functions and direct proportion.

Solution The first five applications are linear functions where the constant $b = 0$. This means they are relations of direct proportion where the slope is the constant of proportionality.

1. $I = \frac{V}{R} = \left(\frac{1}{R}\right)V$. Ohm's law states that when the resistance R of a circuit is constant, the current I is a linear function of the voltage V. I is directly proportional to V and the constant of proportionality is the slope $m = \frac{1}{R}$, which is the conductance G. See Figure 12-3 on page 246 where the straight line shows the proportionality between V and I.
2. $C = 2\pi r$. The circumference C of a circle is a linear function of the radius r, and C is directly proportional to r. The constant of proportionality is $2\pi \approx 6.28$.
3. $P = VI$. When the voltage V of a circuit is constant and the load changes, the power P is a linear function of the current I, and P is directly proportional to I. The constant of proportionality is V.
4. $F = ma$. Newton's second law of motion states that when the mass m of a body is constant, the force F exerted on the body is a linear function of the acceleration a. The force F is directly proportional to the acceleration, and the constant of proportionality is the mass m.
5. $R = \left(\frac{\rho}{A}\right)l$. The resistance R of a wire is a linear function of its length l. ρ is a constant called the resistivity, which depends on the type of wire, and A is the cross-sectional area of the wire. The resistance R is directly proportional to the length l and the constant of proportionality is $\frac{\rho}{A}$.

The next three applications are linear functions where the constant b is not zero. In this case it is the *change in the variables* that is proportional rather than the variables themselves.

6. $F = 1.8C + 32$. This is the temperature conversion formula for changing Celsius to Fahrenheit. It states that F is a linear function of C with slope $m = 1.8$ and $b = 32$. See Example 12.7.
7. $R_t = (R_0\alpha)t + R_0$. Assuming α is constant, the resistance R_t of a material is a linear function of the temperature t in degrees Celsius where the slope $m = R_0\alpha$ and $b = R_0$. R_0 is the resistance of the material at 0°C, and α is the temperature coefficient of resistance at 0°C. See Example 12.8.
8. $v = v_0 + at$. The velocity v of a body with constant acceleration a is a linear function of the time t. The slope $m = a$ and $b = v_0$, where v_0 is the velocity at $t = 0$.

When the slope m is positive, y increases when x increases, and the line slopes up to the right. The greater the slope, the steeper the line. The line $y = 3x + 1$ ($m = 3$) rises at a faster rate than the line $y = 2x + 1$ ($m = 2$). The line $y = x + 1$ ($m = 1$) rises at a slower rate than the line $y = 2x + 1$.

When the slope m is negative, it means that y decreases when x increases, and the line slopes down to the right. For the line $y = -2x + 1$ ($m = -2$), y decreases 2 units when x increases 1 unit. The line $y = -3x + 1$ ($m = -3$) falls at a faster rate than the line $y = -2x + 1$. The line $y = -x + 1$ ($m = -1$) falls at a slower rate than the line $y = -2x + 1$.

EXAMPLE 12.6 Graph the linear equation $3x + y = 4$ and give the slope and y intercept.

Solution Put the equation in the form $y = mx + b$ by solving for y:

$$3x + y = 4$$

Transpose $3x$:

$$y = -3x + 4$$

The slope is $m = -3$, which means y decreases 3 units when x increases 1 unit. The line therefore slopes down to the right. The y intercept $b = 4$, so the line crosses the y axis at (0,4). The values chosen for x are -1, 0, 1, 2, and 3. The graph and the table of values for x and y are shown in Figure 12-5.

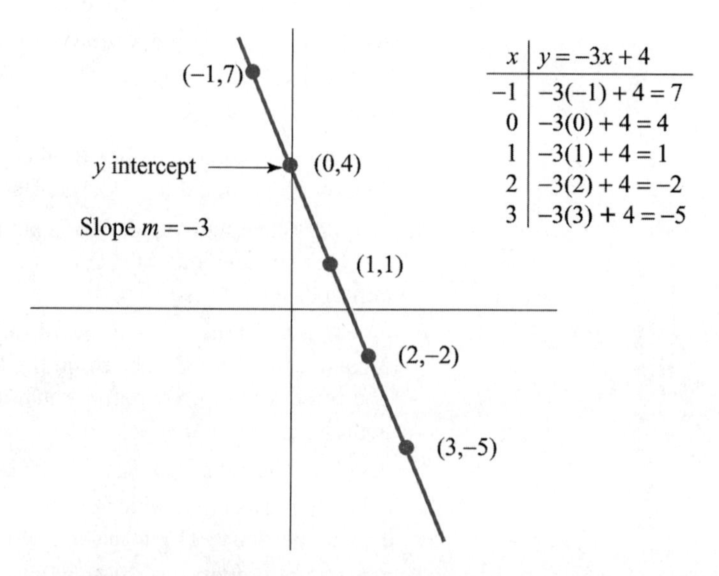

FIGURE 12-5 Linear function, negative slope, for Example 12.6.

When $m = 0$ in the linear function ($y = mx + b$) it becomes $y = b$ and the graph is a horizontal line that passes through the point $(0,b)$. Similarly, the equation $x = a$, which is not considered a function, is a vertical line that passes the point $(a,0)$.

EXAMPLE 12.7 Given the linear function for temperature conversion from Example 12.5, case 6:

$$F = 1.8C + 32$$

Construct a table of values for the independent variable C from −30°C to 40°C using values every 10°, and draw the graph of the function. This range of temperatures is what people experience in temperate climates around the world during the course of a year.

Solution The formula is solved for F, which means that F depends on values of C. C is then the independent variable, and F is the dependent variable. The slope $m = 1.8$, and the F intercept $b = 32$. Choose values of C every 10°, starting with −30°, and calculate the corresponding values of F. For example, when $C = -10°$:

$$F = 1.8(-10) + 32 = -18 + 32 = 14°$$

The table of values you obtain is:

C (°C)	−30	−20	−10	0	10	20	30	40
F (°F)	−22	−4	14	32	50	68	86	104

The table enables you to compare certain temperatures in degrees Celsius with those in degrees Fahrenheit. However, when you draw the graph, it provides you with a picture of how the temperature scales compare for all values of C and F. See Figure 12-6. The horizontal axis is used for the independent variable C, and the vertical axis is used for the dependent variable F.

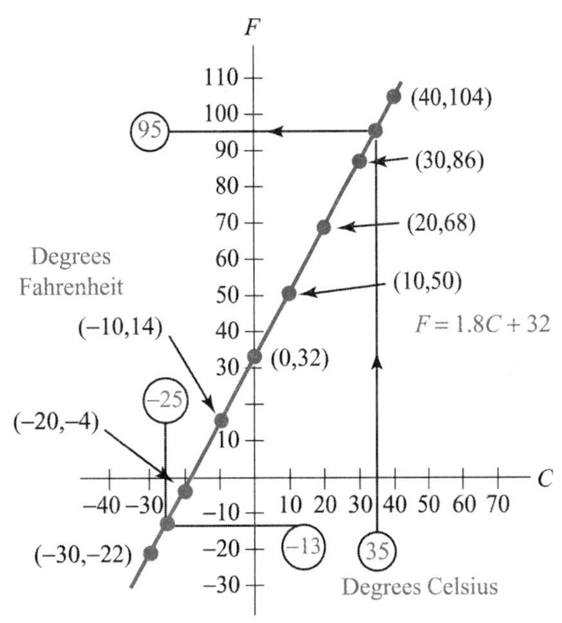

FIGURE 12-6 Fahrenheit versus Celsius for Example 12.7.

It is necessary to choose a scale of units to accommodate the range of values. Each box on the graph is set equal to 10 units, which uses 11 boxes on the positive F axis. The values from the table are then plotted on the graph.

From the line you can determine any value of F for a given value of C, or vice versa. For example, when $C = 35°$, the arrows on the graph show that $F = 95°$. Similarly, when $F = -13°$, the arrows show that $C = -25°$. Notice that the intercept on the F axis is (0,32). To find the intercept on the C axis, let $F = 0$ and solve for C:

$$0 = 1.8C + 32$$

$$C = \frac{-32}{1.8} \approx -17.8°$$

The C intercept (−17.8,0) can be seen approximately in Figure 12-6.

EXAMPLE 12.8

Close the Circuit

As shown in Example 12.5, case 7, the resistance R_t of a material is a linear function of the temperature t given by:

$$R_t = (R_0\alpha)t + R_0$$

where R_t = resistance at temperature t in °C, R_0 = resistance at 0°C, and α = temperature coefficient of resistance at 0°C. Given an aluminum wire where $\alpha = 4.00 \times 10^{-3}$ per °C and $R_0 = 6.00\ \Omega$, graph R_t versus t from $t = -10$°C to $t = 25$°C. Note that α is assumed constant, and the units for α are 1/°C.

Solution Substitute the given values in the above linear function to obtain the formula for the resistance of the aluminum wire in ohms:

$$R_t = 6.00(0.00400)t + 6.00 = 0.0240t + 6.00$$

The slope $m = 0.0240$, and the R_t intercept $b = 6.00$. Choose values of t every 5° to provide enough points for the graph. Substitute each value of t in the formula to obtain the corresponding value of R_t. For example, when $t = 15°$:

$$R_t = 0.0240(15) + 6.00 = 0.36 + 6.00 = 6.36\ \Omega$$

The table of values is then:

t (°C)	−10	−5	0	5	10	15	20	25
R_t (Ω)	5.76	5.88	6.00	6.12	6.24	6.36	6.48	6.60

For the graph of this function, you must choose an appropiate scale for each axis. Each box is set equal to 5 units on the t (horizontal) axis. For the vertical axis, the difference between the largest and smallest values of R_t to be graphed is $6.60 - 5.76 = 0.84$. Divide this by 10 to obtain $0.084 \approx 0.1$. Use this value to set each box on the R_t axis equal to 0.1 Ω. See the graph in Figure 12-7.

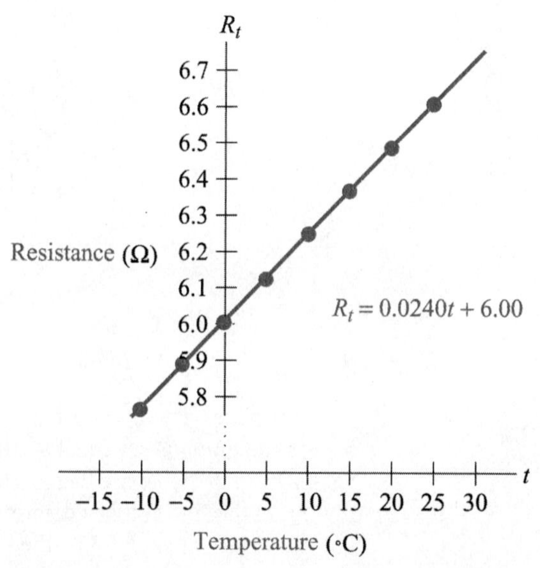

FIGURE 12-7 Resistance versus temperature for an aluminum wire for Example 12.8.

Observe how only the portion of the vertical axis that is needed is shown. The vertical axis is broken from below the graph to zero. The graph shows that R_t is a linear function for the given range of values. Over a larger range of temperature values, α may not be constant, which would mean that R_t might not be a linear function of t.

EXERCISE 12.2

In exercises 1 through 24, put each linear function in the form $y = mx + b$ if not already in this form. Draw the graph of the line showing at least three points and give the slope and y intercept.

1. $y = x$
2. $y = 3x$
3. $y = -3x$
4. $y = -2x$
5. $y = 2x - 2$
6. $y = x - 3$
7. $y = -2x - 2$
8. $y = -x + 1$
9. $x - y = 3$
10. $y - x = 2$
11. $x + y + 4 = 0$
12. $x + y - 5 = 0$
13. $4x + y + 4 = 0$
14. $3x + y - 3 = 0$

15. $3x - y = 4$
16. $3x - y = -1$
17. $3x - 2y = 0$
18. $3x + 2y = 0$
19. $x + 2y = 0$
20. $x - 2y = 0$
21. $2x - 2y + 3 = 0$
22. $2x + 2y = 5$
23. $3x - 2y = 6$
24. $2x + 3y = 6$

In exercises 25 through 36, draw the graph of each linear function showing at least three points. Use the horizontal axis for the independent variable. Choose appropriate scales and give the slope and y intercept.

25. $P = 3.5I$
26. $P = 2.5I$
27. $I = 0.9V$
28. $I = 0.10V$
29. $V = 3.2R + 2.3$
30. $V = 1.6R + 0.8$
31. $V = 4.0Z - 3.0$
32. $Z = 5.5X - 1.5$
33. $R = -1.6t + 5.0$
34. $Q = -0.4t - 1.0$
35. $I_2 = 0.5I_1 + 1$
36. $E_2 = 20E_1 - 10$

Applied Problems

In problems 37 through 46, draw the graph of each linear function plotting at least four points for each line. Use appropriate scales and the horizontal axis for the independent variable.

37. Given the linear function $v = v_0 + at$ from Example 12.5, case 8. If $v_0 = 100$ cm/s and $a = 50$ cm/s², graph v versus t from $t = 0$ sec to $t = 5$ sec.

38. The velocity of sound in air in meters per second is a linear function of the temperature t in degrees Celsius given by:

$$v = 0.607t + 332$$

Graph v versus t from $t = -20°$ to $t = 40°$. Use different scales for v and t. Show the v axis starting from $v = 310$ m/s. Use Figure 12-7 as a guide.

39. The cost C in dollars for the use of a powerful computer system is a linear function of the time t in minutes given by:

$$C = 150t + 1200$$

(a) Graph this function from $t = 5$ minutes to $t = 60$ minutes.
(b) What is the minimum cost for the use of the computer?

40. The cost C in dollars for the rental of a car is a linear function of the time t in days given by:

$$C = 60t + 120$$

(a) Graph this function from $t = 1$ day to $t = 7$ days.
(b) What is the minimum cost for the car rental?

Applications to Electronics

41. Figure 12-8 shows a circuit containing a resistor $R_0 = 150\ \Omega$. Assuming that the resistance is constant, graph the current I in milliamps versus the voltage drop V_0 for $V = 0.0$ V to $V = 5.0$ V. Give the slope of the line.

42. In problem 41, show that the power P in the circuit is directly proportional to the square of the current I by graphing P in milliwatts versus I^2 in (mA)² for $I = 0$ mA to $I = 30$ mA.

43. Figure 12-9 shows a dc circuit containing a series voltage-dropping resistor R_S and a load R_L. The voltage V_L across R_L is a linear function of the voltage drop V_S across R_S given by:

$$V_L = V_B - V_S$$

where V_B = battery voltage. If $V_B = 9.0$ V, graph V_L versus V_S from $V_S = 1.0$ V to $V_S = 6.0$ V.

44. Figure 12-10 shows a circuit containing two series-opposing voltages V_X and V_0. The positive terminals of V_X and V_0 are connected so that the voltages oppose each other. If V_0 is a constant voltage and V_X is a variable voltage, the total voltage V_T is a linear function of V_X given by:

$$V_T = V_0 - V_X$$

If $V_0 = 12$ V, graph V_T versus V_X from $V_X = 0.0$ V to $V_X = 3.0$ V.

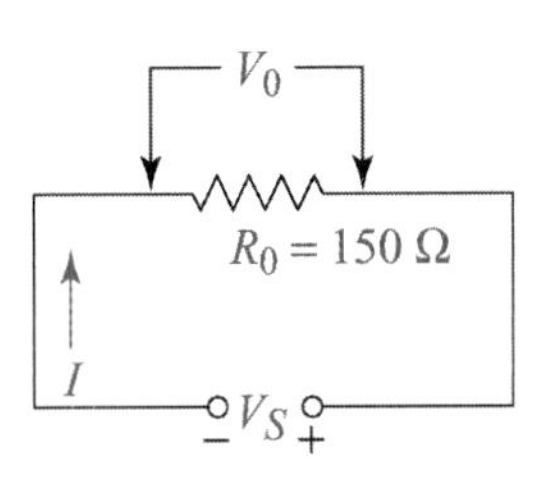

FIGURE 12-8 Voltage drop versus current in a resistance for problems 41 and 42.

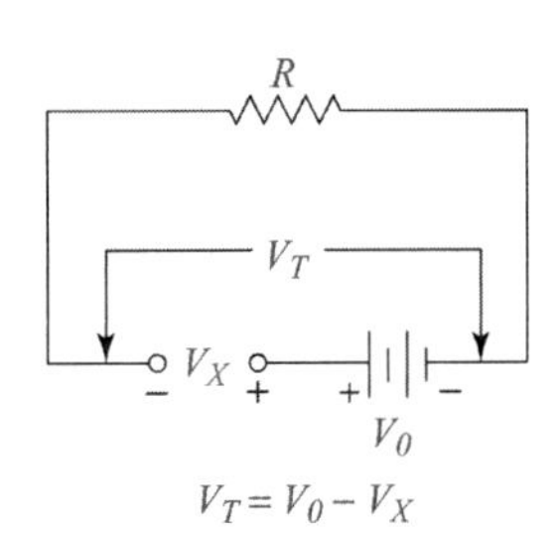

FIGURE 12-10 Series-opposing voltages for problem 44.

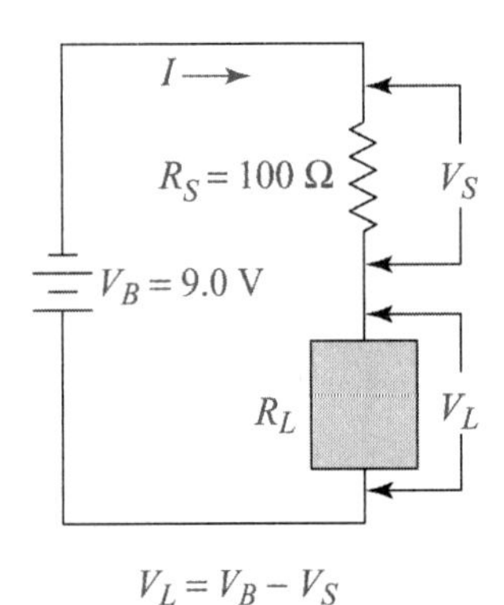

FIGURE 12-9 Series voltage-dropping resistor for problem 43.

45. A brass wire has a temperature coefficient of resistance $\alpha = 2.00 \times 10^{-3}$ per °C. If the resistance at 0°C, $R_0 = 15.0\ \Omega$, graph resistance versus temperature from $t = -10$°C to $t = 20$°C. See Example 12.8 for the linear function.

46. A copper wire has a resistivity $\rho = 10$ CM · Ω/ft (CM = circular mils) and a cross-sectional area $A = 1000$ CM. Graph the resistance in milohms versus length from $l = 1.0$ ft to $l = 6.0$ ft. See Example 12.5, case 5, for the linear function.

12.3 Nonlinear Graphs

Nonlinear graphs include a large number of different curves and other shapes. This section studies three types: a basic second-degree curve, a curve of inverse proportion, and a curve drawn from experimental data. Other types of nonlinear graphs appear in the chapters on trigonometry and logarithms. A second-degree or quadratic function contains the square of the independent variable and produces a curve called a parabola, as shown in Figure 12-11:

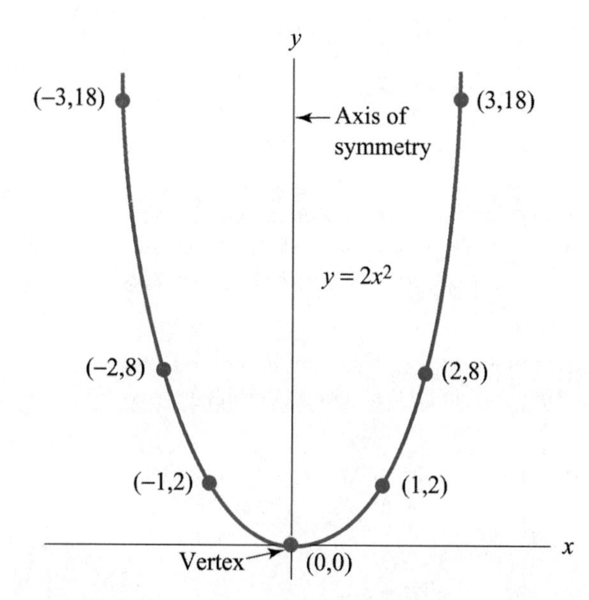

FIGURE 12-11 Parabola, second-degree function, for Example 12.9.

Formula **Basic Second-Degree Function**

$$y = ax^2 \tag{12.4}$$

The turning point or *vertex* of the parabola is at the origin. When a is positive, the parabola opens up, and vice versa.

EXAMPLE 12.9 Graph the second-degree function $y = 2x^2$, showing seven points.

Solution Make a table of values, choosing positive and negative values of x with zero in the middle, and substitute each to find the corresponding value of y:

x	−3	−2	−1	0	1	2	3
$y = 2x^2$	18	8	2	0	2	8	18

For example when $x = -3$:

$$y = 2(-3)^2 = 2(9) = 18$$

Figure 12-11 shows the points plotted on a graph and a smooth curve drawn through them. Notice the vertex and the symmetry on both sides of the y axis, which is the axis of symmetry of the parabola.

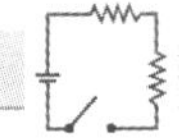

EXAMPLE 12.10

Close the Circuit

Given a resistor $R = 15\ \Omega$ in a dc circuit, graph the power P dissipated in the resistor versus the current I from $I = 0.0$ A to $I = 5.0$ A. Plot six points.

Solution Apply the power law and substitute $R = 15\ \Omega$ to set up the second-degree function:

$$P = I^2R = I^2\,(15\ \Omega) = 15I^2$$

Construct a table of values using values of I every 1 A:

I (A)	0.0	1.0	2.0	3.0	4.0	5.0
$P = 15I^2$ (W)	0.0	15	60	135	240	375

Since the values of I and P are all positive, the entire graph is in the first quadrant. This is often the case as many electrical applications use only positive values. See Figure 12-12.

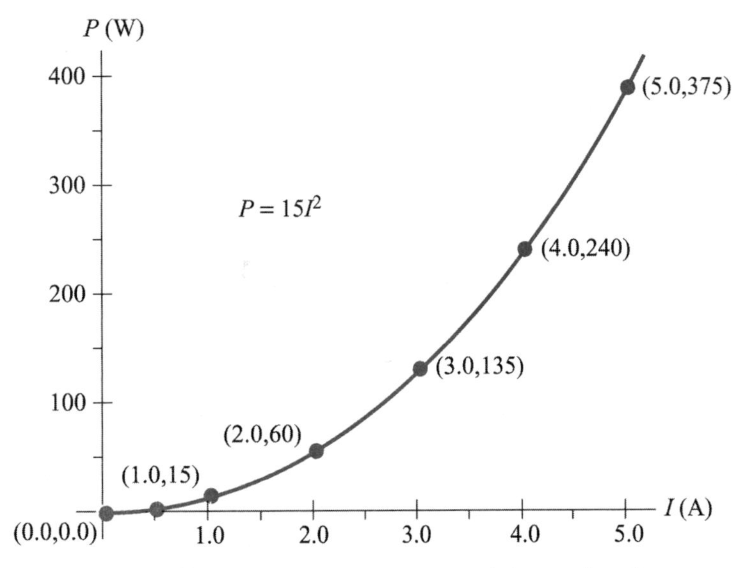

FIGURE 12-12 Power versus current, second-degree function, for Example 12.10.

Each box on the I axis is set equal to 0.5 A, and each box on the P axis is set equal to 50 W to clearly show the points. The curve is one-half of a parabola with the vertex at the origin. The slope, which is a measure of the steepness of the graph, increases as you move to the right, which shows that power increases at a greater and greater rate as the current increases.

Another important nonlinear function is a curve of inverse proportion, where y is inversely proportional to x:

Formula

Inverse Proportion

$$y = \frac{k}{x} \quad (k = \text{constant of proportionality}) \tag{12.5}$$

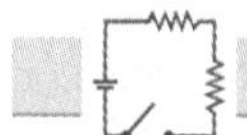

EXAMPLE 12.11

Close the Circuit

The voltage drop across a variable resistor is constant at $V = 120$ V. Graph the current I versus the resistance R from $R = 5.0\ \Omega$ to $R = 30\ \Omega$. Plot six points.

Solution Apply Ohm's law to set up the function:

$$I = \frac{V}{R} = \frac{120 \text{ V}}{R}$$

This is the inverse proportion function (12.5), where V is the constant of proportionality k and I is inversely proportional to R. This relationship is shown in Chapter 7, DC Circuits. Choose values of R every 5 Ω and construct a table:

R (Ω)	5.0	10	15	20	25	30
$I = \frac{120}{R}$ (A)	24	12	8	6	4.8	4

The graph is shown in Figure 12-13 in the first quadrant.

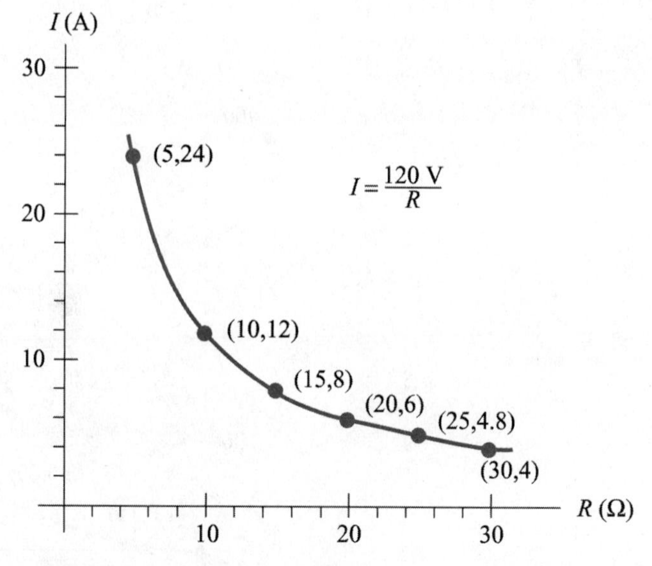

FIGURE12-13 Ohm's law, current versus resistance, inverse proportion for Example 12.11.

Each box on the R axis is set equal to 2 Ω and each box on the I axis is set equal to 2 A to clearly show the points. Notice that as R increases, I decreases by the reciprocal ratio. When R doubles from 10 Ω to 20 Ω, I decreases by one-half from 12 A to 6 A. The reverse is also true. As you extend the curve, it gets closer and closer to the two axes. This curve is called a hyperbola.

Experimental data is subject to many factors: accuracy of instruments, changes in electrical components, human limitations, etc., so results generally do not fit theory exactly. Ohm's law, for example, says that the current is proportional to the voltage when the resistance is constant. However, as current increases, a resistance increases in temperature. This can change the value of the resistance, resulting in a nonlinear graph for current versus voltage over a large enough range of values. The next example illustrates this situation.

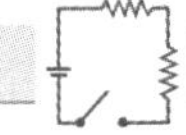

EXAMPLE 12.12

Close the Circuit

Given the following experimental data for a resistance in a dc circuit:

Voltage (V)	0	2	5	8	10	12	14
Current (mA)	0	45	110	160	180	190	200

Plot a graph of current versus voltage using voltage as the independent variable.

Solution Choose an appropriate scale for voltage and current. For voltage, each box = 1 V, and for current each box = 20 mA for the graph in Figure 12-14.

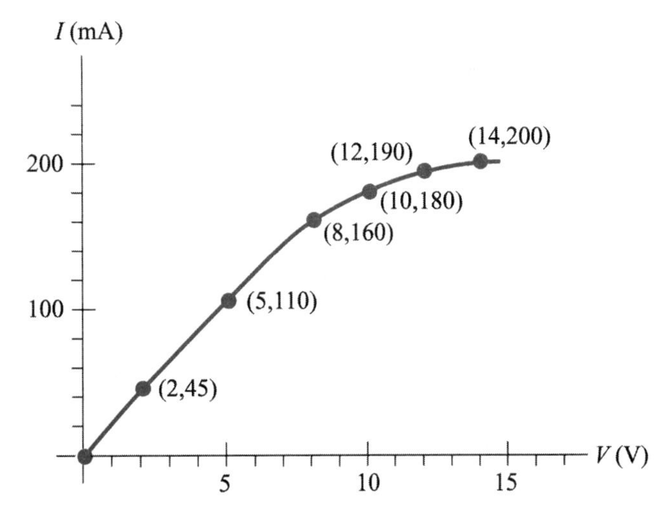

FIGURE 12-14 Nonlinear data, current versus voltage, for Example 12.12.

Observe that for low values of the voltage the graph is close to a straight line, but it becomes more nonlinear as the voltage increases. This is because the resistance is not constant and increases as the temperature of the resistance increases with the current. This is shown by applying Ohm's law to the data and calculating the values of the resistance when $V = 2$ V and when $V = 14$ V:

$$R = \frac{2.0\ \text{V}}{45\ \text{mA}} = 44\ \Omega$$

$$R = \frac{14\ \text{V}}{200\ \text{mA}} = 70\ \Omega$$

EXERCISE 12.3

In exercises 1 through 12, graph each function showing at least seven points. Use appropriate scales.

1. $y = x^2$
2. $y = 3x^2$
3. $y = 0.5x^2$
4. $y = 0.8x^2$
5. $y = 1.2x^2$
6. $y = 2.5x^2$
7. $y = 1.5x^2$
8. $y = 1.2x^2$
9. $y = -2x^2$
10. $y = -x^2$
11. $y = -0.6x^2$
12. $y = -1.5x^2$

In exercises 13 through 28, graph each function in the first quadrant showing at least four points. Use appropriate scales.

13. $P = 10I^2$

14. $P = 20I^2$

15. $P = 75I^2$

16. $P = 330I^2$

17. $P = 0.1V^2$

18. $P = 0.02V^2$

19. $P = 0.05V^2$

20. $P = 0.2V^2$

21. $I = \frac{12}{R}$

22. $I = \frac{36}{R}$

23. $I = \frac{220}{R}$

24. $I = \frac{60}{R}$

25. $I = \frac{9.2}{R}$

26. $I = \frac{12.6}{R}$

27. $I = \frac{5.5}{V}$

28. $I = \frac{40}{V}$

Applied Problems

In problems 29 through 38, graph each function showing at least four points. Use appropriate scales.

29. The distance d in feet an object falls is a second-degree function of the time t elapsed given by:

$$d = 16t^2$$

Graph d versus t for $t = 0.0$ sec to $t = 3.0$ sec.

30. A baseball is thrown horizontally. The vertical distance h in feet that it falls is a second-degree function of the horizontal distance x it travels given by:

$$h = 0.001x^2$$

Graph h versus x for $x = 0.0$ ft to $x = 60$ ft.

Applications to Electronics

31. A resistor in a dc circuit $R = 22\ \Omega$. The power dissipated P is a second-degree function of the current I. Graph P versus I from $I = 0.0$ A to $I = 3.0$ A.

32. A resistor in a dc circuit $R = 680\ \Omega$. The power dissipated P is a second-degree function of the current I. Graph P versus I from $I = 0.0$ mA to $I = 600$ mA.

33. A resistor in a dc circuit $R = 200\ \Omega$. The power dissipated P is a second-degree function of the voltage V. Graph P versus V from $V = 0.0$ V to $V = 120$ V.

34. A resistor in a dc circuit $R = 1.2\ \Omega$. The power dissipated P is a second-degree function of the voltage V. Graph P versus V from $V = 0.0$ V to $V = 3.0$ V.

35. The voltage drop across a variable resistor is constant at $V = 9.0$ V. The current I is inversely proportional to the resistance R. Graph I in milliamps versus R from $R = 20\ \Omega$ to $R = 100\ \Omega$.

36. The voltage drop across a variable resistor is constant at $V = 220$ V. The current I is inversely proportional to the resistance R. Graph I in amps versus R from $R = 50\ \Omega$ to $R = 250\ \Omega$.

37. The voltage drop across a variable resistor is constant at $V = 80$ V. The power dissipated in the resistor P is inversely proportional to the resistance R. Graph P versus R from $R = 40\ \Omega$ to $R = 160\ \Omega$.

38. The voltage drop across a variable resistor is constant at $V = 2.5$ V. The power dissipated in the resistor P is inversely proportional to the resistance R. Graph P versus R from $R = 1.6\ \Omega$ to $R = 5.1\ \Omega$.

In problems 39 through 44, given the experimental data for a resistance in a dc circuit:

(a) Plot a graph of current versus voltage, using voltage as the independent variable.

(b) Calculate the dc value of the resistance for each point, except zero, to two significant digits.

39.

Voltage (V)	0.0	8.0	16	26	37	48
Current (mA)	0.0	100	200	300	400	500

40.

Voltage (V)	0.0	0.5	1.0	1.5	2.0	2.5
Current (mA)	0.0	12	23	31	36	42

41.

Voltage (V)	0.0	3.0	6.0	9.0	12	15
Current (mA)	0.0	52	100	140	170	195

42.

Voltage (V)	0.0	1.2	2.5	4.1	6.0	8.3
Current (mA)	0.0	5.0	10	15	20	25

43. Given the following experimental data for a resistance and its temperature in a dc circuit:

t (°C)	−5	0	5	10	15	20
R_t (Ω)	11.4	12.0	12.7	13.5	14.5	15.6

Graph R_t versus t using t as the independent variable. Show only a part of the vertical axis. See Figure 12-7.

44. Given the following experimental data for a resistance and its temperature in a dc circuit:

t (°C)	10	20	30	40	50	60
R_t (kΩ)	3.3	3.3	3.4	3.6	3.9	4.2

Graph R_t versus t using t as the independent variable. Show only a part of the vertical axis. See Figure 12-7.

CHAPTER HIGHLIGHTS

12.1 RECTANGULAR COORDINATE SYSTEM

Every point on the x axis has a y coordinate of zero, and every point on the y axis has an x coordinate of zero. The origin, where the axes intersect, has coordinates (0,0). The graph of a first-degree or linear equation is always a straight line.

Key Example: To graph $3x - y = 6$ using the x and y intercepts, set $x = 0$ to find the y intercept, and vice versa. Choose a third value for x as a check point. See Figure 12-15.

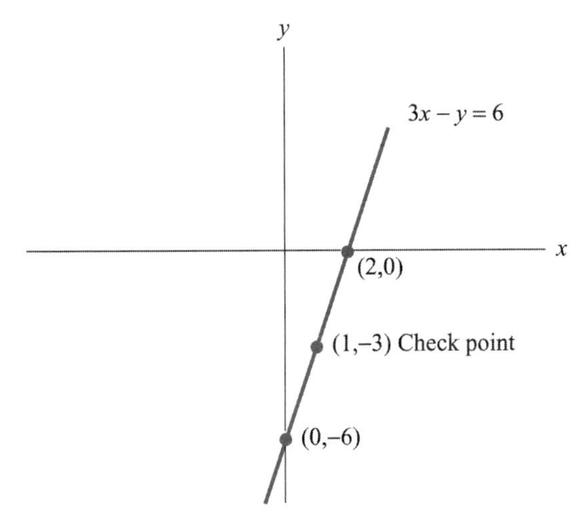

FIGURE 12-15 Key Example: Graph of a line, intercept method.

12.2 LINEAR FUNCTION

Function

y is a function of x if, for each value of x, there corresponds one and only one value of y. (12.1)

• • •

Linear Function

$$y = mx + b \quad (m,b \text{ constants}) \tag{12.2}$$
$$m = \text{slope} \quad b = y \text{ intercept}$$

The *independent* variable is x, and the *dependent* variable is y. When $b = 0$, $y = mx$ states that y is directly proportional to x. The slope m is the constant of proportionality. In Ohm's law, $V = IR$, when R is constant, V is proportional to I. In the power law, $P = VI$, when V is constant, P is proportional to I.

Slope Formula

$$m = \frac{\text{Change in } y}{\text{Change in } x} \tag{12.3}$$

If you move to the right on the line one unit in the x direction, y changes by m units. When m is positive, the line slopes up, and vice versa.

Key Example: To graph $2x - y - 3 = 0$ as a linear function, first solve for y:

$$y = 2x - 3$$

The slope $m = 2$, and the y intercept $= -3$. Plot the y intercept (0,–3) and choose at least two more values for x. See Figure 12-16.

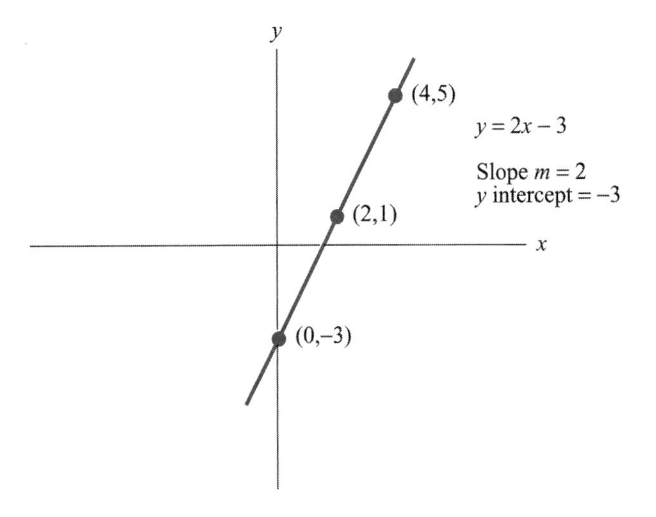

FIGURE 12-16 Key Example: Linear function.

12.3 NONLINEAR GRAPHS

Basic Second-Degree Function

$$y = ax^2 \tag{12.4}$$

The basic second-degree function graphs as a parabola with the vertex at the origin.

Key Example: A resistor in a dc circuit $R = 30\ \Omega$. Graph the power P dissipated in the resistance versus the voltage V from $V = 0.0$ V to $V = 50$ V.

Write the second-degree function:

$$P = \frac{V^2}{R} = \frac{V^2}{30}$$

Construct a table of values. Use convenient values of V every 10 V:

V (V)	0.0	10	20	30	40	50
P (W)	0.0	3.3	13	30	53	83

Choose appropiate scales, graph the points, and connect them with a smooth curve. See Figure 12-17.

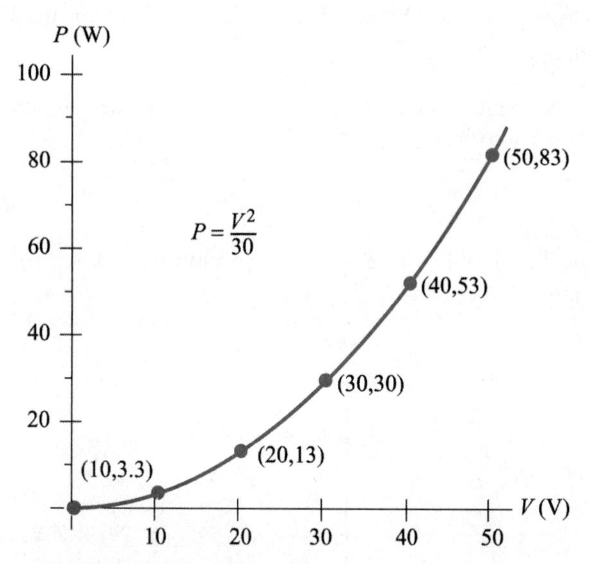

FIGURE 12-17 Key Example: Second-degree function, power versus voltage.

Inverse Proportion

$$y = \frac{k}{x} \quad (k = \text{constant of proportionality}) \qquad (12.5)$$

Key Example: See Figure 12-13 for an example of inverse proportion, current versus resistance.

A third type of nonlinear function is one that results from experimental data and may not produce exact theoretical results.

Key Example: Graph current versus voltage given the experimental data for a resistor in a dc circuit:

Voltage (V)	0.0	2.0	4.0	6.0	8.0	10
Current (mA)	0.0	16	32	47	59	62

Choose appropriate scales for the axes, using voltage as the independent variable. See Figure 12-18.

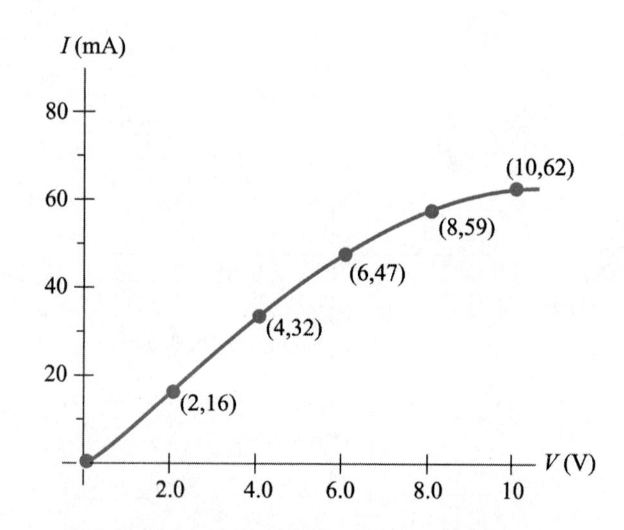

FIGURE 12-18 Key Example: Nonlinear data, current versus voltage.

REVIEW EXERCISES

In exercises 1 through 10, draw the graph of each linear function. Show at least three points, including the x and y intercepts, and give the slope.

1. $2x + y = 2$
2. $3x - y = 6$
3. $2x - 5y = 5$
4. $4x + y = 8$
5. $y = 2x - 5$
6. $y = 1.5x + 5$
7. $y = -x + 2$
8. $y = -3x + 4$
9. $5x + 2y = 0$
10. $2x - 2y = 3$

In exercises 11 through 20, draw the graph of each linear function adjusting scales if necessary. Plot the variable in parentheses on the horizontal axis, showing at least three points.

11. $I_1 + I_2 = 5\ (I_1)$
12. $V_1 - 2V_2 = 4\ (V_1)$
13. $P - 10V = 0\ (V)$
14. $5I - V = 0\ (V)$
15. $I = 0.04V\ (V)$
16. $P_T = 9.4I\ (I)$
17. $Z = 3.5X - 0.5\ (X)$
18. $E = 120 - 100d\ (d)$

19. $V = 1.6R + 3.2$ (R)

20. $R = 0.023t + 0.05$ (t)

In exercises 21 through 26, graph each function, showing at least seven points. Use appropriate scales.

21. $y = 1.5x^2$

22. $y = 4x^2$

23. $y = 2.2x^2$

24. $y = 0.8x^2$

25. $y = -3x^2$

26. $y = -0.5x^2$

In exercises 27 through 34, graph each function in the first quadrant, showing at least four points. Use appropriate scales.

27. $P = 12I^2$

28. $P = 0.5V^2$

29. $P = 1.4V^2$

30. $P = 0.6I^2$

31. $I = \dfrac{28}{R}$

32. $I = \dfrac{6.4}{R}$

33. $I = \dfrac{60}{V}$

34. $R = \dfrac{3.2}{I}$

Applied Problems

In problems 35 through 46, draw the graph of each function, plotting at least four points. Use appropriate scales and the horizontal axis for the independent variable.

35. Hooke's law says that the force F exerted on a spring is a linear function of the distance x the spring stretches given by $F = kx$. The number k is a constant of proportionality that depends on the spring. F is directly proportional to x. If $k = 1.5$ lb/ft for a given spring, graph F versus x from $x = 0$ ft to $x = 6$ ft.

36. The "normal" weight of a person is approximately a linear function of their height given by the formula:

$$W = 0.97H - 100$$

where W is in kilograms and H is in centimeters. Graph W versus H from $H = 150$ cm (4 ft 11 in) to $H = 200$ cm (6 ft 7 in).

37. Given the formula for the circumference of a circle $C = 2\pi r$, graph C versus r from $r = 0.0$ in to $r = 12$ in.

38. Refer to Figure 12-6.

(a) Extend the graph for values of $C < -30$ and find the point where $F = C$.

(b) Calculate this point algebraically by letting $F = C$ in the formula and solving for C.

Applications to Electronics

39. A dc circuit contains a constant resistor $R = 3.6$ kΩ. The current I is a linear function of the voltage drop V across the resistor. Graph I in milliamps versus V from $V = 0.0$ V to $V = 36$ V.

40. In an amplifier circuit, the dc collector voltage V_C is a linear function of the voltage V_L across the load resistor given by:

$$V_C = V_{CC} - V_L$$

where V_{CC} is the fixed voltage supply for the collector current. Given $V_{CC} = 16$ V, graph V_C versus V_L from $V_L = 2.5$ V to $V_L = 12.5$ V.

41. The thermoelectric power of copper with respect to lead is a linear function of the temperature t given by:

$$Q = 2.76 + 1.22t$$

where Q = power in μV/°C. Graph Q versus t from $t = -10$°C to $t = 10$°C.

42. Figure 12-19 shows a variable resistor R_V in parallel with a fixed resistor $R_0 = 40$ Ω. If the applied voltage is constant at 60 V, the total power P_T is a linear function of the total current I_T. Graph P_T versus I_T when R_V varies from 20 Ω to 60 Ω.

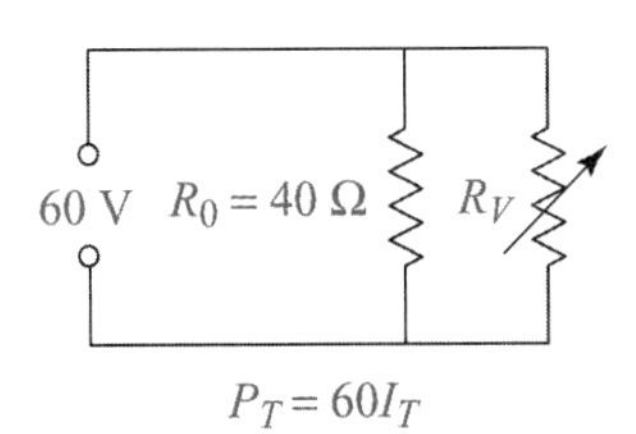

FIGURE 12-19 Power versus current in a parallel circuit for problem 42.

43. A resistor in a dc circuit $R = 120$ Ω. The power dissipated in the resistance P is a second-degree function of the current I. Graph P versus I from $I = 0.0$ A to $I = 2.0$ A.

44. The voltage drop across a variable resistor is constant at $V = 12$ V. The current I is inversely proportional to the resistance R. Graph I in milliamps versus R from $R = 30$ Ω to $R = 150$ Ω.

45. Given the following experimental data for a resistor in a dc circuit:

Voltage (V)	0.0	2.0	4.0	6.0	8.0	10
Current (mA)	0.0	10	19	26	33	40

(a) Plot a graph of current versus voltage, using voltage as the independent variable.

(b) Calculate the dc value of the resistance for each point.

46. Given the following experimental data for a resistor and its temperature in a dc circuit:

t (°C)	−10	0	10	20	30	40
R_t (Ω)	19.4	20.0	20.8	21.4	22.1	23.8

Plot a graph of resistance versus temperature using temperature as the independent variable. Show only a part of the vertical axis. See Figure 12-7.

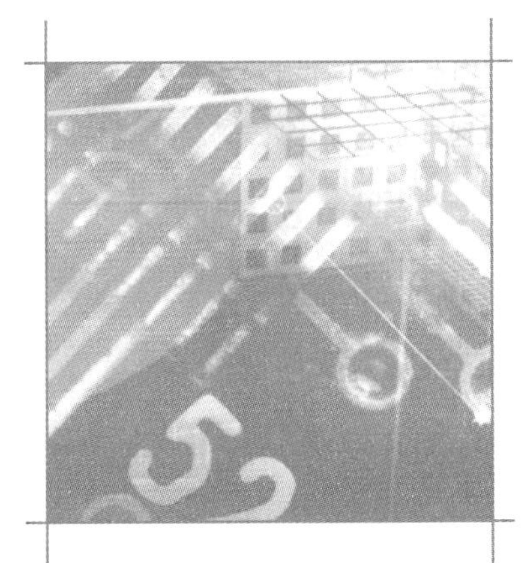

CHAPTER 13

Linear Systems

CHAPTER OBJECTIVES

In this chapter, you will learn:

- How to solve two equations with two unknowns graphically.
- How to solve two equations with two unknowns algebraically.
- The meaning of a matrix and a determinant.
- How to expand a 2×2 determinant.
- How to solve a linear system of two equations using determinants.
- How to expand a 3×3 determinant.
- How to solve a linear system of three equations using determinants.

When electrical circuits and networks become more complex and there are several currents or voltages to determine, it leads to linear equations containing two or more unknowns. These equations are called linear systems and must be solved simultaneously, that is, at the same time. There are several methods of solution that can be used. Graphical methods of solution provide a good picture of the problem and the interrelationships between the variables. Algebraic methods of solution provide a direct way to find precise values. More recent methods of solution, employing matrices and determinants, use a direct formula approach, which can be readily done on a calculator or computer. Matrices were first introduced in 1858 and have now become an important tool in many applications in electricity and electronics.

• • •

13.1 Graphical Solution of Two Linear Equations

Two linear equations in two unknowns usually have a unique solution for both variables. Each equation can be graphed as a straight line, and the point where the two lines intersect satisfies both equations simultaneously.

EXAMPLE 13.1

Solve the linear system of two equations graphically:

$$2x - y = -5$$
$$3x + 2y = 3$$

Solution Apply the graphing techniques from Chapter 12. For each equation, construct a table of values for at least three points. The most direct way is to calculate the x and y intercepts and a check point for each line, as shown in the table in Figure 13-1.

Let each box equal one unit, and carefully draw each line on the same graph with a ruler. Figure 13-1 on page 264 shows the graph for each line using the intercepts and a check point. You can also put each equation in the linear function form $y = mx + b$ and calculate the points as shown in Example 12.5. The point of intersection $(-1,3)$ represents the values of x and y that satisfy both equations simultaneously; this is the graphical solution. To check the solution, show that the values satisfy each equation:

$$2(-1) - (3) = -2 - 3 = -5 \checkmark$$
$$3(-1) + 2(3) = -3 + 6 = 3 \checkmark$$

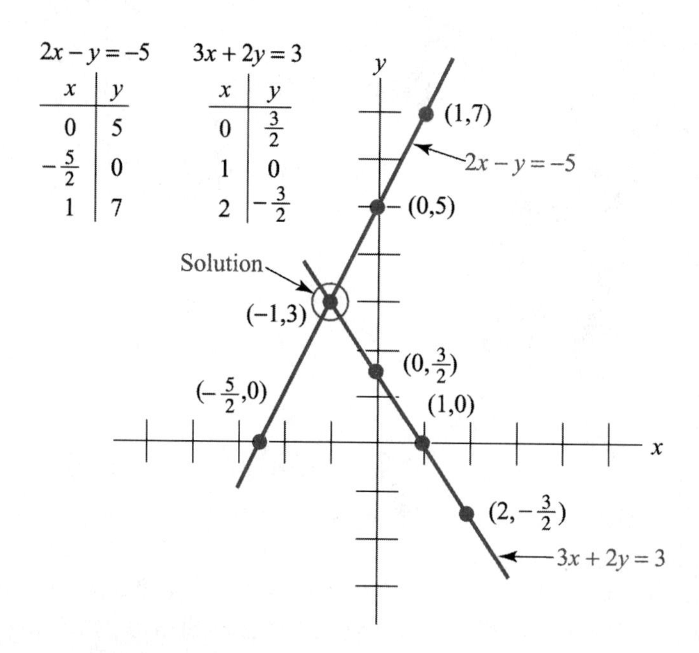

FIGURE 13-1 Graphical solution of simultaneous equations for Example 13.1.

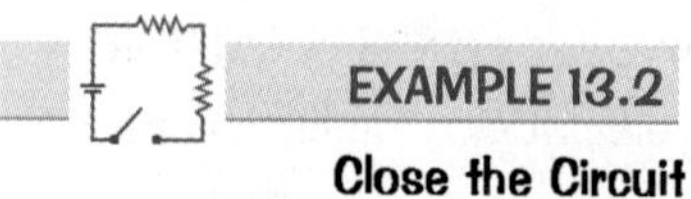

EXAMPLE 13.2

Close the Circuit

Two unknown voltages V_1 and V_2 are connected in series to a resistor $R = 12\ \Omega$. When the voltages are series aiding with the positive terminal of one connected to the negative terminal of the other, the ammeter in the circuit reads 1.5 A. See Figure 13-2(a). When the voltages are series opposing with the negative terminal of one connected to the negative terminal of the other, the ammeter reads 500 mA for the current in the same direction. See Figure 13-2(b). Find the two voltages using a graphical method of solution.

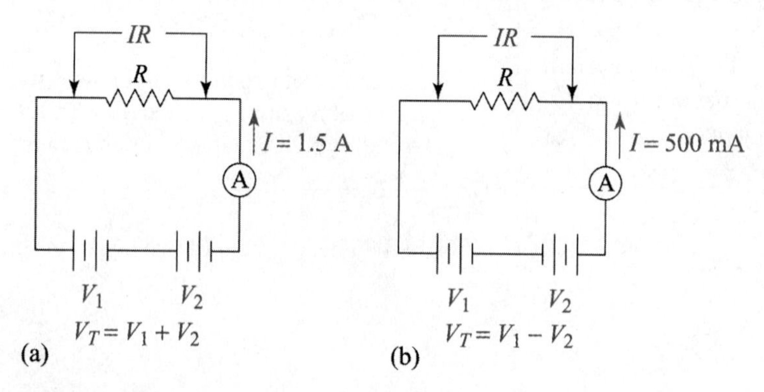

FIGURE 13-2 Series-aiding and series-opposing voltages for Example 13.2.

Solution Apply the problem-solving techniques from Section 10.2. Let V_1 = the larger voltage and V_2 = the smaller voltage. When two voltages V_1 and V_2 are series aiding, their sum equals the total voltage V_T:

$$V_1 + V_2 = V_T$$

When two voltages are series opposing, V_1 is greater than V_2, and their difference equals the total voltage:

$$V_1 - V_2 = V_T$$

Since the IR voltage drop across the resistance equals the total voltage V_T for each circuit, you can write two simultaneous linear equations substituting the values for I and R:

$$V_1 + V_2 = IR = (1.5\text{ A})(12\ \Omega) = 18\text{ V}$$
$$V_1 - V_2 = IR = (0.50\text{ A})(12\ \Omega) = 6.0\text{ V}$$

Each equation graphs as a straight line.The intersection of the two lines represents the values of V_1 and V_2 that satisfy both equations simultaneously and is the solution of the system. See Figure 13-3.

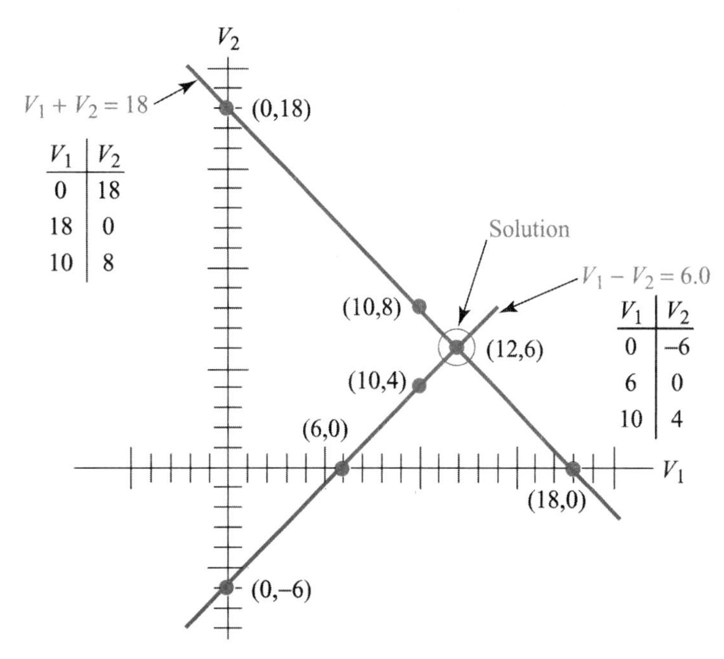

FIGURE 13-3 Graphical solution of simultaneous equations for Example 13.2.

For each equation, construct a table of values for at least three points, and carefully draw each line on the same graph. Figure 13-3 shows the table of values and the graph for each line where each box equals one unit. V_1 is plotted on the horizontal axis. Each table shows the intercepts and a check point for each line. The point of intersection (12,6) is the graphical solution. Therefore $V_1 = 12$ V and $V_2 = 6.0$ V, which check in the original equations:

$$12\ V + 6\text{ V} = 18\text{ V} \checkmark$$
$$12\ V - 6\text{ V} = 6\text{ V} \checkmark$$

Graphs help to further the understanding and solution of a problem and serve as a check on other methods. Graphical methods adapt more readily to programming on a computer and at times can provide a more direct solution to a technical problem.

EXAMPLE 13.3

Solve the linear system graphically to the nearest tenth:

$$5R + 2V = 4$$
$$3R + 5V = 6$$

Solution Plot three points for each line using the x axis for R and estimate decimal values. The graph in Figure 13-4 uses two boxes for each unit, so each box = 0.5 units, for better accuracy. It shows the intercepts and one check point for each equation.

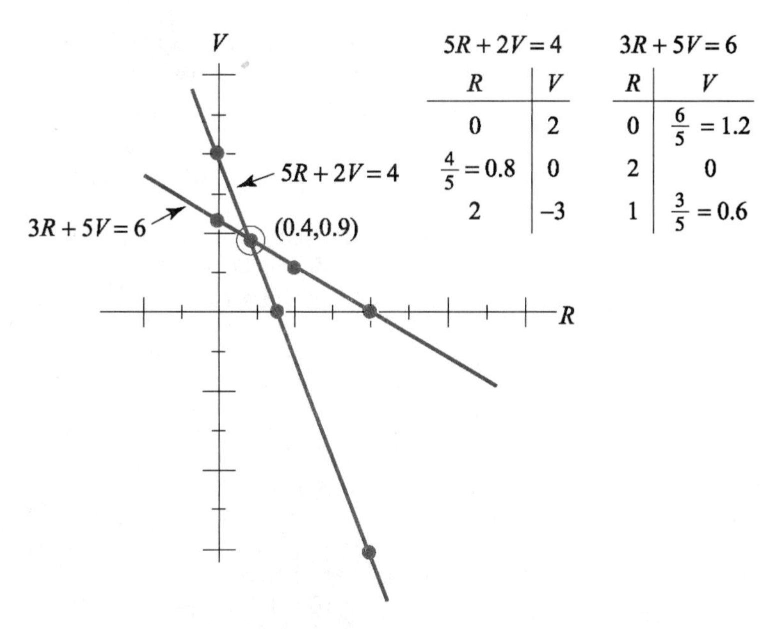

$5R + 2V = 4$	
R	V
0	2
$\frac{4}{5} = 0.8$	0
2	−3

$3R + 5V = 6$	
R	V
0	$\frac{6}{5} = 1.2$
2	0
1	$\frac{3}{5} = 0.6$

FIGURE 13-4 Graphical solution of simultaneous equations for Example 13.3.

The coordinates of the point of intersection (0.4,0.9) are estimated on the graph to the nearest tenth. Checking the answers in the original equations gives close approximate results:

$$5(0.4) + 2(0.9) = 2.0 + 1.8 = 3.8 \approx 4$$
$$3(0.4) + 5(0.9) = 1.2 + 4.5 = 5.7 \approx 6$$

You can always make the graphical solution as accurate as you like by using more boxes for each unit and choosing more accurate points.

EXERCISE 13.1

In exercises 1 through 22, solve each linear system graphically and check solutions by substitution. Use appropriate scales for each graph and estimate decimal solutions to the nearest tenth.

1. $x + y = 3$
$x - 2y = 6$

2. $x - y = 2$
$2x + y = 1$

3. $2x - y = 4$
$x + y = 5$

4. $3x + y = 6$
$x - 2y = 2$

5. $x - 2y = 4$
$3x + 2y = 4$

6. $x + 2y = 5$
$2x - y = 0$

7. $y = 2x + 1$
$y = 3x - 1$

8. $y = 3x - 1$
$y = 2x - 2$

9. $x = y + 4$
$y = 2x - 7$

10. $y = x + 6$
$x = 2 - y$

11. $a + 2b = 4$
$2a + b = 5$

12. $3a - b = 7$
$a - 2b = 9$

13. $3A - B = 1$
$A + 2B = 12$

14. $X - 2Y = 7$
$2X + 3Y = 0$

15. $2R + 3V = -1$
$2R - 3V = 3$

16. $5I - 2V = 20$
$I + 4V = 15$

17. $2V_1 - 3V_2 - 3 = 0$
$3V_1 - 2V_2 + 3 = 0$

18. $I_1 + 2I_2 - 3 = 0$
$I_1 + I_2 - 1 = 0$

19. $3P_1 = 4 - 8P_2$
$P_1 = 3P_2$

20. $R_1 = 1 - R_2$
$R_2 = 2R_1 + 5$

21. $r - t = 1.5$
$r + t = 2.5$

22. $0.5d + n = 1$
$1.5d + 2n = 3$

Applied Problems

23. The total number of students in an electronics class is 12. If there are 6 more men than women, how many of each are there? If m = number of men and w = number of women, solve the resulting linear equations graphically:

$$m + w = 12$$
$$m - w = 6$$

24. Julia and Jordan pull in the same direction on an object with a total force of 100 lb. When they pull in opposite directions, the result is a force of 20 lb in Julia's direction. Find the force exerted by each. If f_1 = Julia's force and f_2 = Jordan's force, solve the resulting linear equations graphically:

$$f_1 + f_2 = 100$$
$$f_1 - f_2 = 20$$

Applications to Electronics

25. The following equations come from a problem involving the currents in a circuit:

$$2I_1 + I_2 = 12$$
$$2I_1 - I_2 = 0$$

Solve this linear system graphically for the currents I_1 and I_2 in amps.

26. The following equations come from a problem involving two parallel resistors in a circuit:

$$2R_x + R_y = 20$$
$$2R_x - R_y = 0$$

Solve this linear system graphically for R_x and R_y in ohms.

27. Two unknown voltages are connected in series to a resistor R. When the voltages are series aiding, as shown in Figure 13-5(a), the ammeter reads 2.0 A. When the voltages are series opposing, as shown in Figure 13-5(b), the ammeter reads 1.0 A in the same direction. If $R = 10\ \Omega$, find the two voltages by setting up two linear equations and solving them graphically. See Example 13.2.

28. In Figure 13-5, when the voltages are series aiding, $I = 800$ mA. When the voltages are series opposing, $I = 200$ mA. If $R = 10\ \Omega$, find the two voltages by setting up two linear equations and solving them graphically. See Example 13.2.

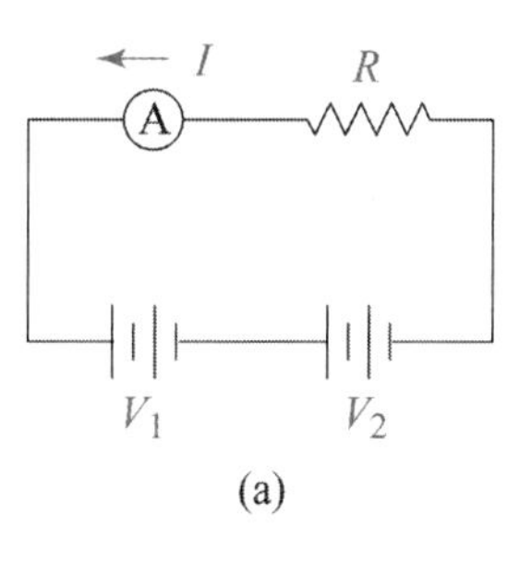

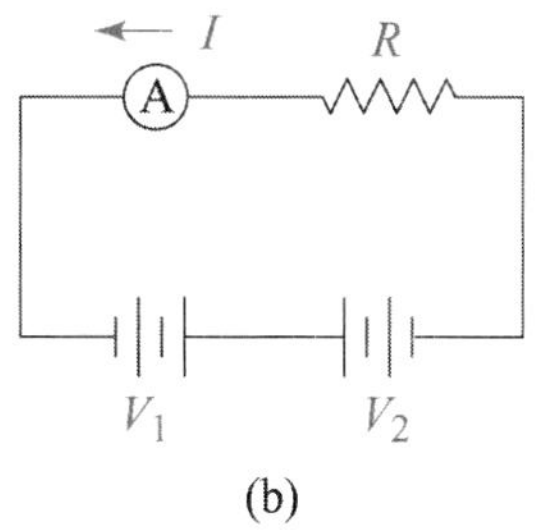

FIGURE 13-5 Series-aiding and series-opposing voltages for problems 27 and 28.

13.2 Algebraic Solution of Two Linear Equations

Two algebraic methods that are used for solving a system of linear equations are the addition method and the substitution method. The first two examples illustrate both methods.

EXAMPLE 13.4 Solve the linear system of Example 13.1 algebraically:

$$2x - y = -5$$
$$3x + 2y = 3$$

Solution

Addition Method. To solve by the addition method, you multiply one or both equations so that the coefficients of x, or y, are the same, but opposite in sign. Then you add the equations and eliminate x or y. To eliminate x in this example, you need to multiply both equations, the first by 3 and the second by −2. However, it is easier to eliminate y by just multiplying the first equation by 2 and adding equations. Therefore, y is eliminated as follows:

Multiply each term by 2: $2(2x - y = -5) \rightarrow$

$$\begin{aligned} 4x - 2y &= -10 \\ 3x + 2y &= 3 \\ \hline \end{aligned}$$

Add the equations: $7x + 0 = -7$

Divide by 7: $x = -1$

Find y by substituting $x = -1$ into either of the original equations. Using the first equation:

$$2x - y = -5$$
$$2(-1) - y = -5$$
$$-y = -3$$
$$y = 3$$

The solution is therefore the point (−1,3) and agrees with the graphical solution shown in Figure 13-1.

Substitution Method. To solve by substitution, first solve either of the equations for one of the variables. Then substitute this expression into the other equation. The first equation, $2x - y = -5$, can be readily solved for y and substituted into the second equation:

$$2x - y = -5$$

Transpose $2x$: $-y = -2x - 5$

Multiply by −1: $y = 2x + 5$

Now substitute $(2x + 5)$ for y in the second equation: $3x + 2y = 3$

$$3x + 2(2x + 5) = 3$$

Clear parentheses: $3x + 4x + 10 = 3$

Combine terms and transpose 10: $7x = -7$

Divide by 7: $x = -1$

Now find y by substituting −1 for x in the equation solved for y:

$$y = 2x + 5 = 2(-1) + 5 = 3$$

The substitution method is useful when variables cannot be easily eliminated by the addition method.

EXAMPLE 13.5 Solve the linear system of Example 13.2 algebraically:

$$V_1 + V_2 = 18$$
$$V_1 - V_2 = 6.0$$

Solution

Addition Method. Adding the two equations as given eliminates V_2:

$$V_1 + V_2 = 18$$
$$V_1 - V_2 = 6.0$$
$$2V_1 + 0 = 24$$

Divide by 2: $V_1 = 12\text{ V}$

Find V_2 by substituting $V_1 = 12$ into either of the original equations. Using the first equation:

$$12 + V_2 = 18$$
$$V_2 = 6.0\text{ V}$$

Substitution Method. Solve the first equation for V_1:

$$V_1 + V_2 = 18$$

Transpose V_2: $V_1 = 18 - V_2$

Substitute $(18 - V_2)$ for V_1 in the second equation and solve for V_2:

$$V_1 - V_2 = 6.0$$

$$(18 - V_2) - V_2 = 6.0$$

Clear parentheses: $18 - 2V_2 = 6.0$

Transpose 18: $-2V_2 = -12$

Divide by –2: $V_2 = 6.0 \text{ V}$

Substitute back to find V_1: $V_1 = 18 - V_2 = 18 - 6.0 = 12 \text{ V}$

The solution agrees with the graphical solution shown in Figure 13-3.

ERROR BOX

When using the addition method to solve a system of equations, one of the variables is usually easier to eliminate than the other one. If one of the variables has the same coefficients or a coefficient of one, that variable is easier to eliminate. If the variables have different coefficients and neither is one, the variable whose coefficients have the smaller least common multiple (LCM) is usually easier to eliminate. The LCM is the lowest number that is a multiple of each number and is similar to the lowest common denominator. For example, consider the two equations:

$$3x + 2y = 4$$
$$4x - 3y = 11$$

The LCM of the x coefficients is 12, and the LCM of the y coefficients is 6. It is easier to eliminate y because it has the smaller LCM, and the resulting numbers will be smaller. Multiply the first equation by 3 and the second equation by 2 and add:

$$3(3x + 2y = 4) \rightarrow 9x + 6y = 12$$
$$2(4x - 3y = 11) \rightarrow 8x - 6y = 22$$
$$17x + 0 = 34$$
$$x = 2$$

Substituting back: $3(2) + 2y = 4 \rightarrow y = -1$

Practice Problems: Eliminate the easier variable and solve the linear equation:

1. $2x - 2y = 6$, $3x + y = 5$ **2.** $x - 4y = 4$, $2x + 2y = 3$ **3.** $4x + 3y = 3$, $2x + 2y = 1$ **4.** $3x - 6y = 6$, $5x + 2y = 4$

5. $4x - 3y = 4$, $3x + 2y = 3$ **6.** $3x - 5y = 1$, $-4x + 2y = 1$ **7.** $6x + 5y = 10$, $3x - 10y = 10$ **8.** $3x + 5y = 3$, $5x + 6y = -2$

EXAMPLE 13.6 Solve the linear system of Example 13.3 algebraically to two significant digits:

$$5R + 2V = 4$$
$$3R + 5V = 6$$

Solution The addition method is more direct and is shown here. V is easier to eliminate because the lowest common multiple (LCM) of the coefficients is 10 and is smaller than the LCM of the coefficients of R, which is 15.

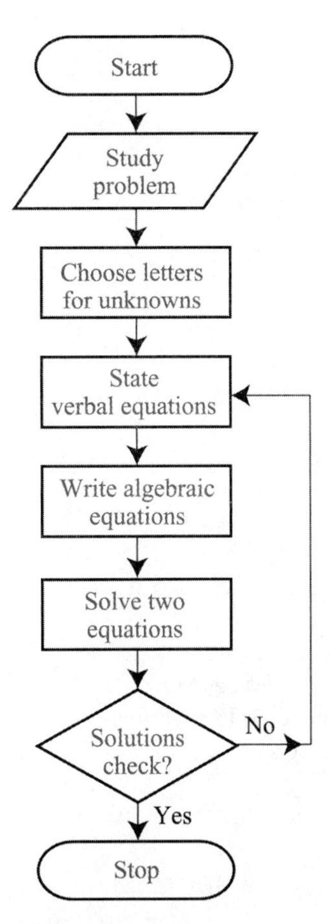

FIGURE 13-6 Algorithm for solving verbal problems with two unknowns.

Multiply the first equation by –5 and the second equation by 2 (you could also use 5 and –2) and add the equations:

$$\begin{aligned} (-5)(5R + 2V = 4) &\rightarrow & -25R - 10V &= -20 \\ (2)(3R + 5V = 6) &\rightarrow & 6R + 10V &= 12 \\ \hline & & -19R + 0 &= -8 \end{aligned}$$

Divide by –19:

$$R = \frac{-8}{-19} \approx 0.42$$

Substitute into the first equation to find V:

$$5(0.42) + 2V = 4$$
$$2V = 4 - 2.1 = 1.9$$
$$V = \frac{1.9}{2} \approx 0.95$$

The solution is therefore (0.42,0.95). These answers are more accurate than the graphical solution in Figure 13-4 where they are rounded off to the nearest tenth (0.4,0.9). You can check them by substituting into the original equations, which gives you very close results:

$$5(0.42) + 2(0.95) = 4 \ \checkmark$$
$$3(0.42) + 5(0.95) = 6.01 \approx 6 \ \checkmark$$

The next two examples show applications of linear systems to a geometric problem and an electrical problem. The solutions of these verbal problems apply the problem-solving techniques discussed in Section 10.2. See the flowchart in Figure 13-6 that illustrates the procedure.

EXAMPLE 13.7

Adam bought 320 ft of fencing to enclose the front and two sides of his house within a rectangular fence. See Figure 13-7. He decides that the front section will be three times as long as each side. What should be the dimensions of the rectangle in feet and meters?

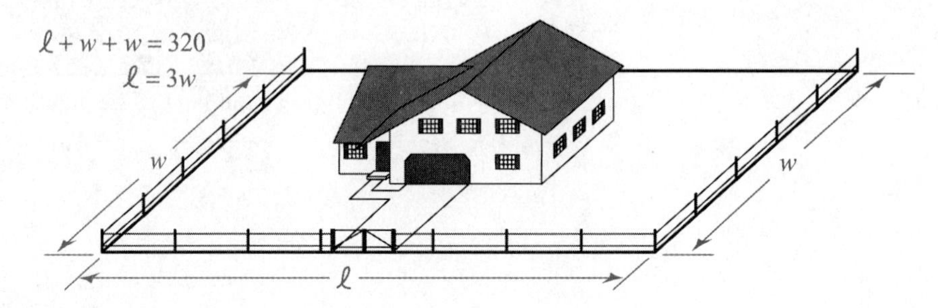

FIGURE 13-7 House fence for Example 13.7.

Solution You can solve this problem using two unknowns and two equations:

$$\text{Let } l = \text{length} \quad \text{and} \quad w = \text{width}$$

From the first sentence you have the verbal equation:

$$(\text{length}) + (\text{width}) + (\text{width}) = \text{total fencing}$$

Answers to Error Box Problems in the form (x,y), page 269:
1. (2,–1) **2.** (2,–0.5) **3.** (1.5,–1) **4.** (1,–0.5) **5.** (1,0) **6.** (–0.5,–0.5) **7.** (2,–0.4) **8.** (–4,3)

This gives the first algebraic equation:

$$l + w + w = 320$$

The second sentence gives the verbal equation:

$$\text{Length} = 3 \times (\text{Width})$$

This gives the second algebraic equation:

$$l = 3w$$

These two equations can be solved by the addition method or the substitution method. Since the second equation is already solved for l, the substitution method lends itself to the solution. Substitute $3w$ for l in the first equation and solve for w:

$$l = 3w$$

$$\begin{aligned} l + w + w &= 320 \\ 3w + w + w &= 320 \\ 5w &= 320 \\ w &= 64 \text{ ft} \end{aligned}$$

Substitute back and find l: $l = 3(64) = 192$ ft

Check by substituting the values for w and l into the first equation:

$$192 + 2(64) = 192 + 128 = 320 \checkmark$$

Using the conversion 1 m = 3.281 ft from Section 4.3, the answers in meters are:

$$w = 64 \text{ ft}\left(\frac{1 \text{ m}}{3.281 \text{ ft}}\right) \approx 19.5 \text{ m}$$

$$l = 192 \text{ ft}\left(\frac{1 \text{ m}}{3.281 \text{ ft}}\right) \approx 58.5 \text{ m}$$

EXAMPLE 13.8

Close the Circuit

Steven has two types of unknown resistors, an ammeter and a 12-V battery supply. To find the value of the resistors, he first connects two resistors of the first type and one resistor of the second type in series with the 12-V battery. See Figure 13-8(a). The ammeter in the circuit reads 250 mA. Then he connects two resistors of the second type and one resistor of the first type in series, and the ammeter reads 200 mA. See Figure 13-8(b). Find the values of the two resistors.

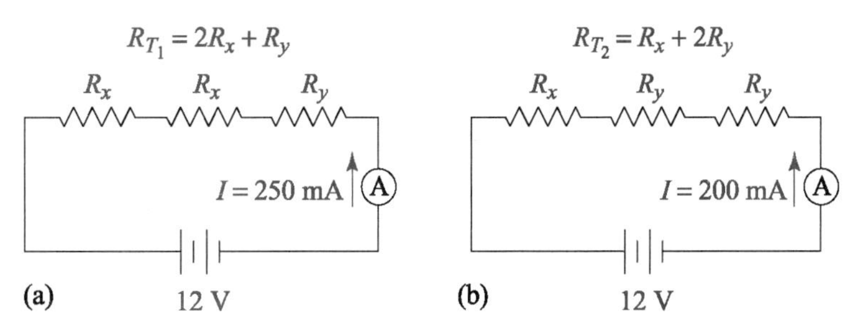

FIGURE 13-8 Resistors in series for Example 13.8.

Solution Let R_x = first type resistance and R_y = second type resistance. Because the resistances are connected in series, the verbal equation that applies to both circuits is:

$$\text{Sum of the resistances} = \text{Total resistance}$$

The algebraic equations are then:

$$2R_x + R_y = R_{T_1}$$
$$R_x + 2R_y = R_{T_2}$$

where R_{T_1} is the first total resistance and R_{T_2} is the second total resistance. Before you can solve these equations, it is necessary to find the two total resistances using Ohm's law:

$$R_{T_1} = \frac{12 \text{ V}}{0.25 \text{ A}} = 48\ \Omega$$
$$R_{T_2} = \frac{12 \text{ V}}{0.20 \text{ A}} = 60\ \Omega$$

The equations for each circuit are then:

$$2R_x + R_y = 48\ \Omega$$
$$R_x + 2R_y = 60\ \Omega$$

The addition method is shown for the solution of the two equations. Multiply the first equation by −2 and add the equations eliminating R_y:

$$-2(2R_x + R_y = 48) \rightarrow \quad -4R_x - 2R_y = -96\ \Omega$$
$$R_x + 2R_y = 60\ \Omega$$
$$-3R_x + 0 = -36\ \Omega$$

Divide by −3: $R_x = 12\ \Omega$

Substitute into the first equation to find R_y: $2(12\ \Omega) + R_y = 48\ \Omega$

$$R_y = 48\ \Omega - 2(12\ \Omega) = 24\ \Omega$$

EXERCISE 13.2

In exercises 1 through 14, solve each linear system algebraically using two methods: addition and substitution, or as directed by the instructor. Check that the answers satisfy the equations. Round decimal answers to two significant digits.

1. $x + y = 3$
$x - 2y = 6$

2. $x - y = 2$
$2x + y = 1$

3. $2x - y = 4$
$x + y = 5$

4. $3x + y = 6$
$x - 2y = 2$

5. $x - 2y = 4$
$3x + 2y = 4$

6. $x + 2y = 5$
$2x - y = 0$

7. $y = 2x + 1$
$y = 3x - 1$

8. $y = 3x - 1$
$y = 2x - 2$

9. $x = y + 4$
$y = 2x - 7$

10. $y = x + 6$
$x = 2 - y$

11. $a + 2b = 4$
$2a + b = 5$

12. $3a - b = 7$
$a - 2b = 9$

13. $3A - B = 1$
$A + 2B = 12$

14. $X - 2Y = 7$
$2X + 3Y = 0$

In exercises 15 through 30, solve each system by either method or as directed by the instructor. Round answers to two significant digits.

15. $3x - 2y = 6$
$2x - 3y = 4$

16. $5x + y = 10$
$x - 5y = 2$

17. $3R + 2X = -1$
$3R - 2X = -2$

18. $5V - 2I = 20$
$V + 4I = 15$

19. $2V_1 - 3V_2 - 3 = 0$
$3V_1 - 2V_2 + 3 = 0$

20. $I_1 + 2I_2 - 3 = 0$
$3I_1 + 4I_2 - 5 = 0$

21. $3P_1 = 4 + 8P_2$
$P_1 = 3P_2$

22. $R_1 = 1 - R_2$
$R_1 - 4 = 2R_1$

23. $r - t = 2.1$
$r + 2t = 1.5$

24. $3p + q = -0.30$
$2p - q = 2.3$

25. $\frac{e}{3} - f = 1$
$e + 2f = 8$

26. $\frac{d}{2} + n = 3$
$2d + n = 2$

27. $2f - 3g = 4$
$3f - 2g = -2$

28. $6a - 5b = -6$
$4a + 4b = 7$

29. $0.5V_x + V_y = 4$
$V_x - 0.5V_y = -5$

30. $1.5E_x - 2.0E_y = -1.0$
$3.0E_x + 2.5E_y = 3.2$

Applied Problems

In problems 31 through 48, solve each problem algebraically by setting up and solving two simultaneous equations. Round answers to two significant digits.

31. Fran has 108 m of fencing to enclose her house on all four sides in the shape of a rectangle. If the width is to be one-half the length, what should be the dimensions of the rectangle in meters and feet? See Example 13.7.

32. Betty bought 105 ft of fencing to enclose a vegetable garden in her backyard. The garden is to be a rectangle, twice as long as it is wide, with a fence across the middle parallel to the width. What should be the dimensions of the garden in feet and meters? See Example 13.7.

33. Mr. Chang and his wife together had an unpaid balance of \$500 on their credit cards for one month. The bank that issued Mr. Chang's credit card charges an interest rate of $1\frac{1}{4}\%$ a month on the unpaid balance; his wife's bank charges $1\frac{1}{2}\%$ a month. If the total finance charge was \$6.75, how much was each unpaid balance? Write one equation for the balances and one for the finance charge.

Note: Finance charge = (balance)(interest rate).

34. Helen invests a total of \$1000 in two CDs (certificates of deposit). The total interest after one year is \$50.50. If the interest rate is $5\frac{1}{2}\%$ on one CD and $4\frac{1}{2}\%$ on the other, how much was invested in each CD? Write one equation for the amounts invested and one for the finance charge.

35. Erin takes a total of 5 hours in her boat to travel a distance upriver and return downriver the same distance. If the time traveling upriver is twice the time traveling downriver, how many hours does each trip take?

36. Monica lives 16 mi from her job. One morning she decides to get some exercise by jogging part of the way and taking the bus the rest of the way. She estimates she can average 4 mi/h jogging and 20 mi/h on the bus. If she allows 1 hour to get to work, how many minutes should she jog and how many should she ride? Write one equation for time and one for distance.

Note: Distance = (Rate)(Time).

Applications to Electronics

37. In Figure 13-8 on page 271, find R_x and R_y in ohms for the following results with the 12-V battery supply: With two R_x resistors and one R_y resistor connected in series, the ammeter reads 30 mA. With one R_x resistor and two R_y resistors connected in series, the ammeter reads 24 mA.

38. The following equations come from a circuit problem involving two currents:

$$0.75I_1 + 0.25I_2 = 5.0$$
$$0.25I_1 + 0.75I_2 = 4.0$$

Solve this linear system for the currents in amps.

39. Two unknown voltages V_1 and V_2 are connected in series to a resistance R. When the voltages are series aiding, $I = 1.7$ A. When the voltages are series opposing, $I = 500$ mA in the same direction. If $R = 20\ \Omega$, what are the two voltages in volts? See Example 13.2.

40. Two voltages V_1 and V_2 are connected in series to a resistance R. When the voltages are series aiding, $I = 750$ mA. When the voltages are series opposing, $I = 150$ mA. If $V_1 = 9$ V, how much is V_2 in volts and R in ohms?

41. The following equations result when solving for the currents I_1 and I_2 in the series-parallel circuit in Figure 13-9:

$$3.5I_1 + 1.5I_2 = 0.50$$
$$1.5I_1 + 4.5I_2 = 0.50$$

Solve this linear system for the currents I_1 and I_2 in amps, and then express the answers in milliamps.

42. The following equations result when solving for the mesh currents I_A and I_B in the series-parallel circuit in Figure 13-9:

$$3.5I_A - 2I_B = 0.5$$
$$5I_B - 2I_A = 0$$

(a) Solve this system for I_A and I_B in amps, and then express the answers in milliamps.

(b) Find I_1 and I_2 in problem 35 using the two relationships: $I_2 = I_B$ and $I_1 = I_A - I_B$.

43. In Figure 13-10(a), the total conductance $G_1 + G_2$ of two resistors in parallel equals 90 mS. In Figure 13-10(b), when another resistor whose conductance $G_3 = 2G_2$ is added in parallel, the total conductance equals 190 mS.

(a) Find G_1 and G_2.

(b) Given that resistance is the reciprocal of conductance, find R_1 and R_2.

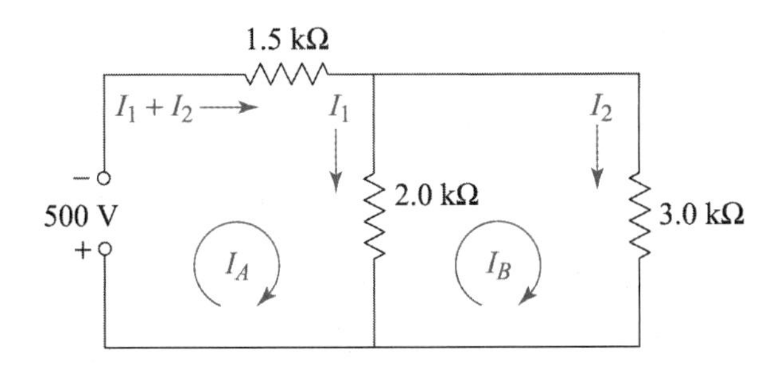

FIGURE 13-9 Kirchhoff's laws applied to a series-parallel circuit for problems 41 and 42.

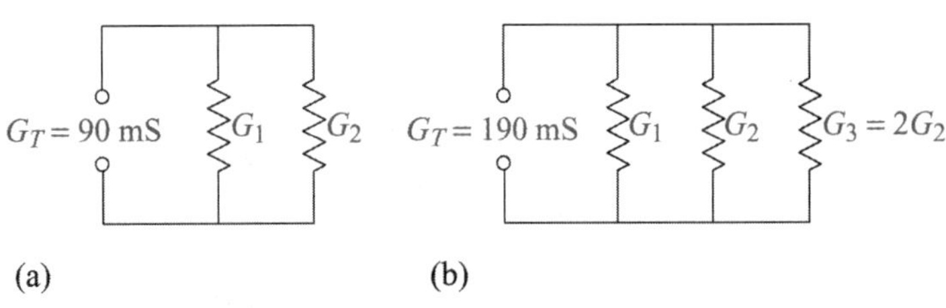

FIGURE 13-10 Total conductance of parallel resistors for problem 43.

44. When two inductances are connected in series so that they are series aiding, as shown in Figure 13-11(a), the total inductance is given by:

$$L_T = L_1 + L_2 + L_M$$

where L_M is the mutual inductance. When the same two inductances are connected so that they are series opposing, as shown in Figure 13-11(b), the total inductance is given by:

$$L_T = L_1 + L_2 - L_M$$

If $L_1 = 400\ \mu\text{H}$ and $L_T = 1.20$ mH when they are series aiding, and $L_T = 500\ \mu\text{H}$ when they are series opposing, find L_2 and L_M in microhenries.

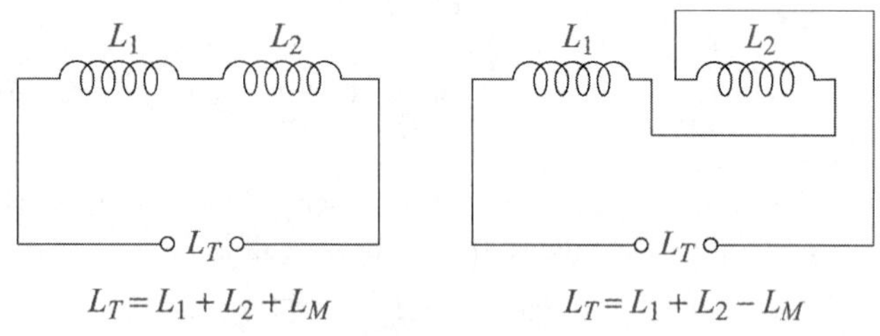

FIGURE 13-11 Series-aiding and series-opposing inductances for problem 44.

45. A powerful computer takes 60 ns (nanoseconds) to perform 5 operations of one type and 9 operations of a second type. The same computer takes 81 ns to perform 7 operations of the first type and 12 operations of the second type. How many nanoseconds are required for each operation?

46. When an unknown resistor R_x is connected in series with a resistance of 10 Ω to a battery, an ammeter in the circuit reads 250 mA, as shown in Figure 13-12(a). When R_x is connected in series with 20 Ω to the same battery, the ammeter reads 200 mA as shown in Figure 13-12(b). What is the value of the resistor and the voltage of the battery?

47. When two unknown resistors R_x and R_y are connected in series to a 12-volt battery, an ammeter in the circuit reads 100 mA. See Figure 13-13(a). When a third resistor, equal to R_x, is added in series to the circuit, the ammeter reads 60 mA. See Figure 13-13(b). What are the values of the two resistors? See Example 13.8.

48. In problem 47, if the voltage of the battery is 20 V and the current readings are first 80 mA and then 50 mA, what are the values of the two resistors?

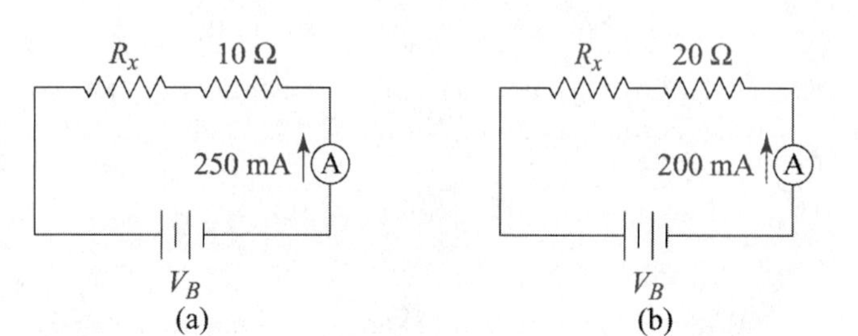

FIGURE 13-12 Unknown resistors in series for problem 46.

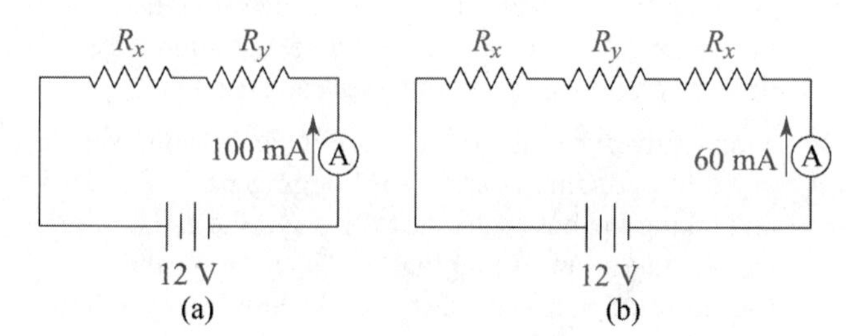

FIGURE 13-13 Unknown resistors in series for problem 47.

13.3 Solution by Determinants

Linear systems increase in difficulty when the coefficients are not simple numbers and there are more than two variables. Matrices and determinants provide methods for solving linear systems that adapt to use on calculators and computers. Matrices were first introduced in 1858 and proved to be an important tool in many applications in electricity, electronics, and other technical areas.

Matrices

A *matrix* is a rectangular array of numbers, called elements, arranged in rows and columns. A matrix is enclosed in parentheses or brackets, and its size is given by: (number of rows) × (number of columns). Some examples of matrices and their sizes are as follows:

$$\begin{pmatrix} 3 & -4 & 5 & -1 \\ 7 & 5 & 8 & 3 \end{pmatrix} 2 \times 4 \qquad \begin{pmatrix} 9 \\ 0 \\ -6 \end{pmatrix} 3 \times 1 \qquad \begin{pmatrix} 4.3 & -1.2 \\ -6.8 & 3.3 \\ 5.5 & 7.1 \end{pmatrix} 3 \times 2$$

The following matrices have the same number of rows as columns and are called *square matrices:*

$$\begin{pmatrix} 2 & 4 \\ -2 & 3 \end{pmatrix}_{2 \times 2} \qquad \begin{pmatrix} 4 & -2 & 1 \\ 1 & -4 & 3 \\ 5 & 7 & -5 \end{pmatrix}_{3 \times 3}$$

Every square matrix has a number associated with it called its *determinant,* which is defined as follows for a 2×2 matrix:

Definition

Determinant of 2×2 Matrix

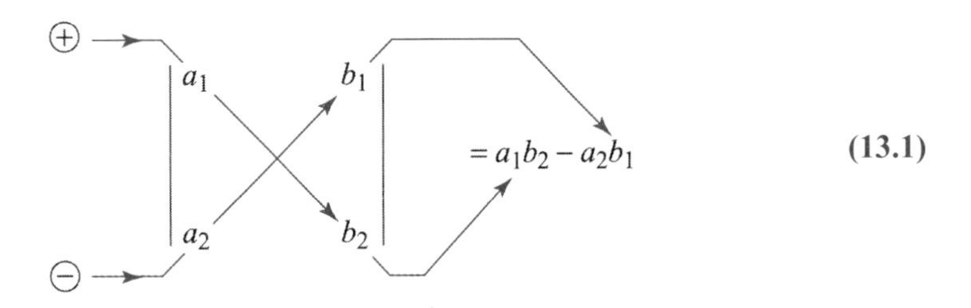

(13.1)

To calculate the determinant of a 2×2 matrix, multiply the elements in the upper left diagonal and subtract the product of the elements in the lower left diagonal. The determinant of the 2×2 matrix shown above is:

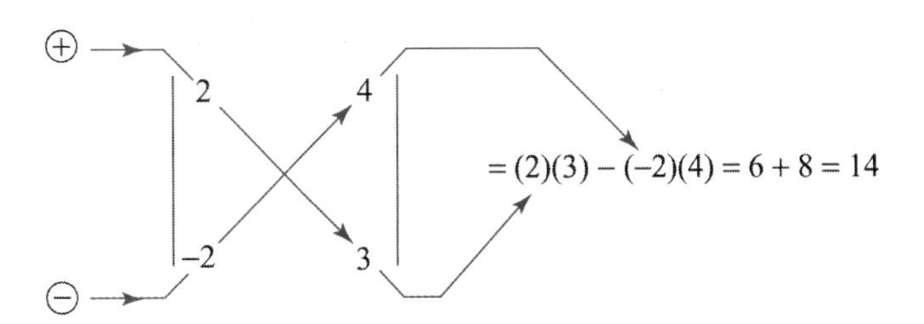

EXAMPLE 13.9 Evaluate each 2×2 determinant:

(a) $\begin{vmatrix} 1/4 & -3/4 \\ -1/2 & 1/2 \end{vmatrix}$ **(b)** $\begin{vmatrix} 0.75 & -0.60 \\ 0.25 & -0.44 \end{vmatrix}$

Solution Multiply the upper left diagonal and subtract the lower left diagonal product:

(a) $\begin{vmatrix} 1/4 & -3/4 \\ -1/2 & 1/2 \end{vmatrix} = (1/4)(1/2) - (-3/4)(-1/2) = 1/8 - 3/8 = -2/8 = -1/4$

(b) $\begin{vmatrix} 0.75 & -0.60 \\ 0.25 & -0.44 \end{vmatrix} = (0.75)(-0.44) - (0.25)(-0.60) = -0.33 + 0.15 = -0.18$

Solution of Two Linear Equations

The procedure for solving two equations using determinants is expressed by *Cramer's Rule* (no relation to the author) as follows. Given a linear system of two equations in *standard form,*

$$\begin{aligned} a_1x + b_1y &= k_1 \qquad (k_1, k_2 \text{ constants}) \\ a_2x + b_2y &= k_2 \end{aligned}$$

the solutions for x and y are:

Rule **Cramer's Rule for Two Equations**

$$x = \frac{\begin{vmatrix} k_1 & b_1 \\ k_2 & b_2 \end{vmatrix}}{\begin{vmatrix} a_1 & b_1 \\ a_2 & b_2 \end{vmatrix}} \qquad y = \frac{\begin{vmatrix} a_1 & k_1 \\ a_2 & k_2 \end{vmatrix}}{\begin{vmatrix} a_1 & b_1 \\ a_2 & b_2 \end{vmatrix}} \tag{13.2}$$

Study the next example, which shows how to apply Cramer's rule.

EXAMPLE 13.10 Solve the linear system of Example 13.4 using determinants:

$$2x - y = -5$$
$$3x + 2y = 3$$

Solution In Example 13.4, the solution of $(-1,3)$ is first found by the addition method. The addition method of solution is equivalent to the method shown here using determinants. It follows a logical pattern. *The equations must first be in the standard form,* shown above. The solutions for x and y are each a fraction whose numerator and denominator are determinants. The denominator of each fraction has the same determinant. It contains the four coefficients of x and y arranged exactly as they appear in the equations above and is called the *coefficient matrix:*

Coefficient Matrix: $\begin{vmatrix} 2 & -1 \\ 3 & 2 \end{vmatrix} = (2)(2) - (3)(-1) = 4 + 3 = 7$

The numerator for x is obtained from the coefficient matrix by *replacing the column of* x *coefficients (first column) by the constants* as they appear on the right side of the equations:

Constants

$$\begin{vmatrix} -5 & -1 \\ 3 & 2 \end{vmatrix} = (-5)(2) - (3)(-1) = -10 + 3 = -7$$

Putting the two determinants together, the complete solution for x is:

Constants

$$x = \frac{\begin{vmatrix} -5 & -1 \\ 3 & 2 \end{vmatrix}}{\begin{vmatrix} 2 & -1 \\ 3 & 2 \end{vmatrix}} = \frac{(-5)(2) - (3)(-1)}{(2)(2) - (3)(-1)} = \frac{-10 + 3}{4 + 3} = \frac{-7}{7} = -1$$

Coefficient Matrix

The solution for y is similar. The denominator for y is the same coefficient matrix as for x. The numerator for y is obtained from the coefficient matrix by *replacing the column of* y *coefficients (second column) by the constants:*

Constants

$$y = \frac{\begin{vmatrix} 2 & -5 \\ 3 & 3 \end{vmatrix}}{\begin{vmatrix} 2 & -1 \\ 3 & 2 \end{vmatrix}} = \frac{(2)(3) - (3)(-5)}{(2)(2) - (3)(-1)} = \frac{6 + 15}{4 + 3} = \frac{21}{7} = 3$$

Coefficient Matrix

Observe how the constants $k_1 = -5$ and $k_2 = 3$ replace the coefficients in each numerator. Remember that *the equations must be in standard form* before applying Cramer's rule.

EXAMPLE 13.11

Close the Circuit

The resistance of 1000 ft of no. 14 gauge copper wire at 20°C is measured to be 2.6 Ω. At 75°C it is measured to be 3.1 Ω. Find the resistance at 0°C and the temperature coefficient of resistance at 0°C for the wire to two significant digits.

Solution From Example 12.8, the resistance of a material is a linear function of the temperature t given by:

$$R_t = (R_0\alpha)t + R_0$$

where: R_t = resistance at t°C

R_0 = resistance at 0°C

α = temperature coefficient of resistance at 0°C

This function can be written as:

$$R_t = at + R_0$$

where $a = R_0\alpha$. In this form the formula more closely resembles the linear function $y = mx + b$, where R_t replaces y, t replaces x, the slope $m = a$, and the y intercept $b = R_0$. Substituting the two sets of values of R_t and t, a linear system of two equations results where the unknowns are a and R_0:

$$2.6 = a(20) + R_0$$
$$3.1 = a(75) + R_0$$

Write the equations in standard form for a and R_0:

$$20a + R_0 = 2.6$$
$$75a + R_0 = 3.1$$

Then apply Cramer's rule (13.2):

$$a = \frac{\begin{vmatrix} 2.6 & 1 \\ 3.1 & 1 \end{vmatrix}}{\begin{vmatrix} 20 & 1 \\ 75 & 1 \end{vmatrix}} = \frac{(2.6)(1) - (3.1)(1)}{(20)(1) - (75)(1)} = \frac{-0.50}{-55} \approx 0.0091$$

The entire calculation for a can be done on the calculator:

[(] 2.6 [−] 3.1 [)] [÷] [(] 20 [−] 75 [)] [=] → 0.0091

It is not necessary to calculate the denominator again for R_0:

$$R_0 = \frac{\begin{vmatrix} 20 & 2.6 \\ 75 & 3.1 \end{vmatrix}}{-55} = \frac{(20)(3.1) - (2.6)(75)}{-55} = \frac{-133}{-55} \approx 2.4\ \Omega$$

Then, since $a = R_0\alpha$:

$$\alpha = \frac{a}{R_0} = \frac{0.0091}{2.4} \approx 0.0038 \text{ per degree Celsius}$$

EXERCISE 13.3

In exercises 1 through 6, evaluate each determinant.

1. $\begin{vmatrix} 2 & 3 \\ 1 & 4 \end{vmatrix}$

2. $\begin{vmatrix} -1 & 5 \\ -3 & 1 \end{vmatrix}$

3. $\begin{vmatrix} 0 & 7 \\ -3 & 5 \end{vmatrix}$

4. $\begin{vmatrix} \frac{1}{2} & \frac{1}{4} \\ -\frac{1}{2} & \frac{3}{4} \end{vmatrix}$

5. $\begin{vmatrix} 3.2 & 2.5 \\ 1.2 & -1.5 \end{vmatrix}$

6. $\begin{vmatrix} 0.45 & -0.20 \\ 0.75 & 0.40 \end{vmatrix}$

In exercises 7 through 34, solve each linear system using determinants. Round answers to two significant digits.

7. $3x - y = 1$; $x + 2y = 12$

8. $x - 2y = 7$; $2x + 3y = 0$

9. $10x + 2y = -3$; $5x - 4y = -4$

10. $5x - 2y = 20$; $x + 4y = 15$

11. $3X - Y = 8$; $2X + 3Y = 9$

12. $a + 3b = 6$; $4a + b = 2$

13. $2R + 3V = 0$; $4R - 3V = -3$

14. $8C - 2L = 3$; $C + 4L = 11$

15. $2f - 3g = 4$; $3f - 2g = -2$

16. $6c - 5d = -6$; $4c + 4d = 7$

17. $2x - 3y - 3 = 0$; $3x - 2y + 3 = 0$

18. $x + 2y - 3 = 0$; $3x + 4y - 5 = 0$

19. $3R = 4 - 8t$; $R = 3t$

20. $R = 1 - a$; $R = 2a + 5$

21. $3I_1 - 2I_2 - 2 = 0$; $2I_1 - 3I_2 - 8 = 0$

22. $V_1 + 2V_2 - 1 = 0$; $4V_1 + 3V_2 + 6 = 0$

23. $2R_x = 8R_y - 2$; $R_y = 4R_x - 2$

24. $X_1 = 1 - X_2$; $X_2 = 10 - 6X_1$

25. $2A - B = 1.5$; $3A + 2B = 6.1$

26. $4G + 3H = 13$; $2G - 2H = 11$

27. $3a - \frac{x}{5} = 10$; $\frac{a}{10} + 3x = 5$

28. $\frac{P}{2} + Q = 4$; $2P - Q = 0$

29. $I_A - 0.5I_B = 0$; $0.5I_A + I_B + 5.5 = 0$

30. $2.5e_1 - 2.0e_2 + 1.1 = 0$; $3.0e_1 + 2.5e_2 - 2.6 = 0$

31. $0.5V_x + V_y = 4$; $V_x - 0.5V_y = -5$

32. $1.5E_x - 2.0E_y = -1.0$; $3.0E_x + 2.5E_y = 3.2$

33. $r - t = 2.1$; $r + 2t = 1.4$

34. $\frac{d}{2} + n = 3$; $2d + n = 2$

Applied Problems

In problems 35 through 44, solve each applied problem by setting up and solving a linear system using determinants. Round answers to two significant digits.

35. An electronics firm produces two types of integrated circuits using robots. A problem involving the most effective programming of the robots for production leads to the following linear system:

$$1.8N_1 + 2.4N_2 = 5400$$
$$2.8N_1 - 1.2N_2 = 1000$$

where N_1 = the number of one type produced and N_2 = the number of the other type produced. Solve this system for N_1 and N_2.

36. A shipment of 650 computer parts contains two types of parts. The first type costs $1.50 each, and the second part costs $1.70 each. If the total cost for the shipment is $1045.00, how many parts of each type are in the shipment? Write two equations, one for the parts and one for the total cost.

37. Nuria takes a total of 4 hours to paddle her kayak downriver a certain distance and return back upriver the same distance. If the time traveling downriver is one-half the time traveling upriver, how long does each trip take in hours and minutes?

38. Elizabeth drives her car over a mountain pass 10 mi long. She drives uphill for 21 min and downhill for 6 min. On the return trip, she drives uphill for 9 min and downhill for 14 min. If the speeds uphill are the same each way, and the speeds downhill are the same each way, find the two speeds in *miles per hour*. Write two equations for the distances, observing the units.

Applications to Electronics

39. In a circuit containing two voltage sources the following equations result for the currents in amps:

$$70I_1 - 50I_2 = 8$$
$$30I_1 + 80I_2 = 14$$

Solve this linear system for I_1 and I_2 in milliamps.

40. Two magnetic forces are found to satisfy the following equations:

$$6.4F_1 - 4.6F_2 = 36 \times 10^{-6}$$
$$3.2F_1 + 2.0F_2 = 74 \times 10^{-6}$$

Solve this linear system for F_1 and F_2 in newtons (N).

41. Two unknown voltages V_1 and V_2 are connected in series to a resistance of 50 Ω. When the voltages are series aiding, the current in the circuit is 1.12 A. When the voltages are series opposing, the current is 160 mA. Find the two voltages in volts.

42. The total current in two circuits is 4.5 A. If the difference of the currents in the two circuits equals one-half of the larger current, find the current in each circuit in amps.

43. The resistance of 1000 ft of no. 21 aluminum wire at 50°C is measured to be 23.3 Ω. At 100°C it is measured to be 27.1 Ω. Find the resistance at 0°C and the temperature coefficient of resistance at 0°C. See Example 13.11.

44. The resistance of a carbon rod at 20°C is measured to be 90 Ω. At 80°C it is found to decrease to 84 Ω. Find the resistance at 0°C and the temperature coefficient of resistance at 0°C, which will be negative. See Example 13.11.

13.4 Solution of Three Linear Equations

A system of three linear equations can be solved using determinants of 3 × 3 matrices similar to the procedure for solving two linear equations. A convenient method for evaluating the determinant of a 3 × 3 matrix uses six diagonals as follows. Write the 3 × 3 determinant repeating the first two columns on the right:

$$\begin{vmatrix} a_1 & b_1 & c_1 \\ a_2 & b_2 & c_2 \\ a_3 & b_3 & c_3 \end{vmatrix} \begin{matrix} a_1 & b_1 \\ a_2 & b_2 \\ a_3 & b_3 \end{matrix}$$

Draw three diagonals from the *upper* left. These are the *positive* products. Draw three diagonals from the *lower* left. These are the *negative* products. The sum of the three positive products and the three negative products is the value of the determinant:

Definition

Determinant of 3 × 3 Matrix

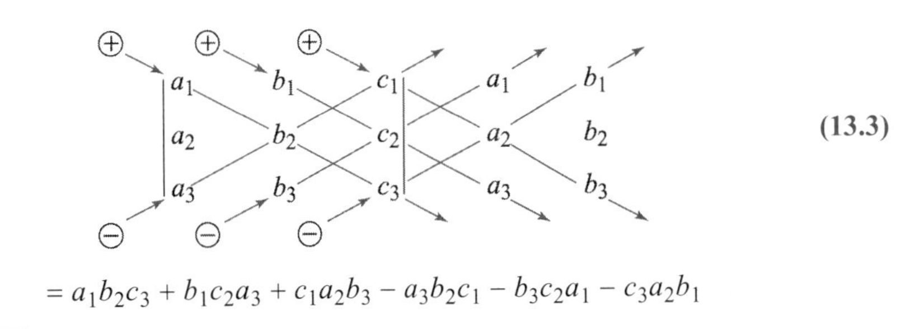

$$= a_1b_2c_3 + b_1c_2a_3 + c_1a_2b_3 - a_3b_2c_1 - b_3c_2a_1 - c_3a_2b_1 \quad (13.3)$$

Study the next example, which shows how to apply formula (13.3).

EXAMPLE 13.12 Evaluate the following 3 × 3 determinant:

$$\begin{vmatrix} 2 & -1 & 3 \\ 3 & 2 & -2 \\ 1 & -3 & 4 \end{vmatrix}$$

Solution Repeat the first two columns and expand the determinant applying formula (13.3):

$$\begin{vmatrix} 2 & -1 & 3 \\ 3 & 2 & -2 \\ 1 & -3 & 4 \end{vmatrix} \begin{matrix} 2 & -1 \\ 3 & 2 \\ 1 & -3 \end{matrix}$$

$$= (2)(2)(4) + (-1)(-2)(1) + (3)(3)(-3)$$
$$- (1)(2)(3) - (-3)(-2)(2) - (4)(3)(-1)$$
$$= 16 + 2 - 27 - 6 - 12 + 12 = 30 - 45 = -15$$

This can be done in one series of steps on the calculator:

2 [×] 2 [×] 4 [+] [(−)] 1 [×] [(−)] 2 [+] 3 [×] 3 [×] [(−)] 3 [−] 2

[×] 3 [−] [(−)] 3 [×] [(−)] 2 [×] 2 [−] 4 [×] 3 [×] [(−)] 1 [=] →−15

Cramer's rule (13.2) can be extended to the solution of a linear system of three equations. The expressions for x, y, and z are similar to those for x and y in formula (13.2) for two equations. Given a linear system of three equations in *standard form*,

$$\begin{aligned} a_1x + b_1y + c_1z &= k_1 \\ a_2x + b_2y + c_2z &= k_2 \quad (k_1, k_2, k_3 \text{ constants}) \\ a_3x + b_3y + c_3z &= k_3 \end{aligned}$$

the solutions for x, y, and z are:

Cramer's Rule for Three Equations

$$x = \frac{\begin{vmatrix} k_1 & b_1 & c_1 \\ k_2 & b_2 & c_2 \\ k_3 & b_3 & c_3 \end{vmatrix}}{\begin{vmatrix} a_1 & b_1 & c_1 \\ a_2 & b_2 & c_2 \\ a_3 & b_3 & c_3 \end{vmatrix}} \qquad y = \frac{\begin{vmatrix} a_1 & k_1 & c_1 \\ a_2 & k_2 & c_2 \\ a_3 & k_3 & c_3 \end{vmatrix}}{\begin{vmatrix} a_1 & b_1 & c_1 \\ a_2 & b_2 & c_2 \\ a_3 & b_3 & c_3 \end{vmatrix}} \qquad z = \frac{\begin{vmatrix} a_1 & b_1 & k_1 \\ a_2 & b_2 & k_2 \\ a_3 & b_3 & k_3 \end{vmatrix}}{\begin{vmatrix} a_1 & b_1 & c_1 \\ a_2 & b_2 & c_2 \\ a_3 & b_3 & c_3 \end{vmatrix}} \quad (13.4)$$

Look at Cramer's rule (13.4) closely. The denominator for x, y, and z is the coefficient matrix of a's, b's, and c's as they appear in the standard equations. Each numerator matrix has two columns, which are the same as the denominator matrix. The third column, which corresponds to the variable you are solving for, contains the constants, k_1, k_2, and k_3. For example, in the formula for x the first column in the numerator contains the constants.

EXAMPLE 13.13 Solve the linear system for x, y, and z:

$$\begin{aligned} 2x + 3y + z &= 5 \\ 3x - 2y + 4z &= -3 \\ 5x + y + 2z &= -1 \end{aligned}$$

Solution Apply Cramer's rule (13.4) for x, and evaluate the numerator and denominator using formula (13.3):

$$x = \frac{\begin{vmatrix} 5 & 3 & 1 \\ -3 & -2 & 4 \\ -1 & 1 & 2 \end{vmatrix} \begin{matrix} 5 & 3 \\ -3 & -2 \\ -1 & 1 \end{matrix}}{\begin{vmatrix} 2 & 3 & 1 \\ 3 & -2 & 4 \\ 5 & 1 & 2 \end{vmatrix} \begin{matrix} 2 & 3 \\ 3 & -2 \\ 5 & 1 \end{matrix}}$$

$$x = \frac{5(-2)(2) + 3(4)(-1) + 1(-3)(1) - (-1)(-2)(1) - 1(4)(5) - 2(-3)(3)}{2(-2)(2) + 3(4)(5) + 1(3)(1) - (5)(-2)(1) - 1(4)(2) - 2(3)(3)}$$

$$x = \frac{-20 - 12 - 3 - 2 - 20 + 18}{-8 + 60 + 3 + 10 - 8 - 18} = \frac{-39}{39} = -1$$

Once you have computed the value of the denominator, you do not have to do it again for y and z:

$$y = \frac{\begin{vmatrix} 2 & 5 & 1 \\ 3 & -3 & 4 \\ 5 & -1 & 2 \end{vmatrix}}{39} = \frac{78}{39} = 2 \qquad z = \frac{\begin{vmatrix} 2 & 3 & 5 \\ 3 & -2 & -3 \\ 5 & 1 & -1 \end{vmatrix}}{39} = \frac{39}{39} = 1$$

You should evaluate these determinants and check that you get the values shown. You can then check the results by substituting the values for x, y, and z in each of the original equations. The set of values must satisfy all three equations. As a check for the first equation, you have:

$$2(-1) + 3(2) + (1) = 5 \ ?$$
$$-2 + 6 + 1 = 5 \ \checkmark$$

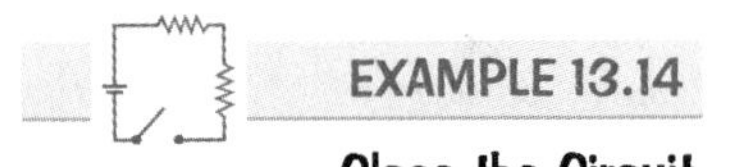

EXAMPLE 13.14

Close the Circuit

Figure 13-14 shows a series-parallel circuit containing two voltage sources. The following three equations result from applying Kirchhoff's laws to the circuit:

$$\begin{aligned} I_1 + I_2 - I_3 &= 0 \\ 4I_1 + 3I_3 - 9 &= 0 \\ 6I_2 + 3I_3 - 5 &= 0 \end{aligned}$$

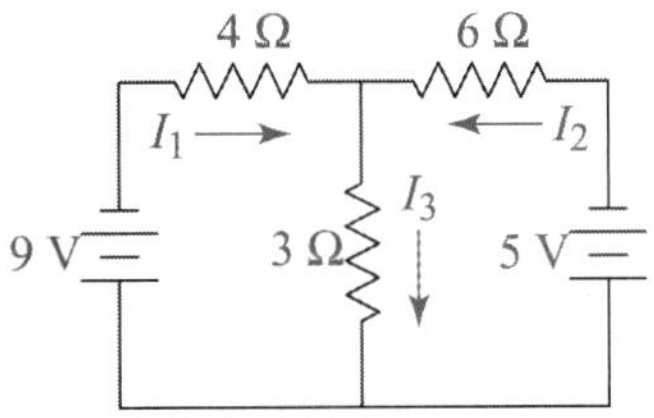

Kirchhoff's laws are studied in Chapter 14. Solve this linear system for the three currents I_1, I_2, and I_3.

FIGURE 13-14 Kirchhoff's laws for Example 13.14.

Solution First put the three equations into standard form lining up the terms:

$$\begin{aligned} I_1 + I_2 - I_3 &= 0 \\ 4I_1 \qquad + 3I_3 &= 9 \\ 6I_2 + 3I_3 &= 5 \end{aligned}$$

Apply Cramer's rule (13.4) for I_1 and evaluate the determinants. A missing term has a coefficient of zero. The solution for I_1 to two significant digits is:

$$I_1 = \frac{\begin{vmatrix} 0 & 1 & -1 \\ 9 & 0 & 3 \\ 5 & 6 & 3 \end{vmatrix} \begin{matrix} 0 & 1 \\ 9 & 0 \\ 5 & 6 \end{matrix}}{\begin{vmatrix} 1 & 1 & -1 \\ 4 & 0 & 3 \\ 0 & 6 & 3 \end{vmatrix} \begin{matrix} 1 & 1 \\ 4 & 0 \\ 0 & 6 \end{matrix}} = \frac{0 + (1)(3)(5) + (-1)(9)(6) - (0) - (0) - (3)(9)(1)}{0 + 0 + (-1)(4)(6) - 0 - (6)(3)(1) - (3)(4)(1)}$$

$$= \frac{-66}{-54} = 1.2 \text{ A}$$

Then compute I_2 and I_3:

$$I_2 = \frac{\begin{vmatrix} 1 & 0 & -1 \\ 4 & 9 & 3 \\ 0 & 5 & 3 \end{vmatrix} \begin{matrix} 1 & 0 \\ 4 & 9 \\ 0 & 5 \end{matrix}}{-54} = \frac{(1)(9)(3) + 0 + (-1)(4)(5) - 0 - (5)(3)(1) - 0}{-54}$$

$$= \frac{-8}{-54} = 0.15 \text{ A} = 150 \text{ mA}$$

$$I_3 = \frac{\begin{vmatrix} 1 & 1 & 0 \\ 4 & 0 & 9 \\ 0 & 6 & 5 \end{vmatrix} \begin{matrix} 1 & 1 \\ 4 & 0 \\ 0 & 6 \end{matrix}}{-54} = \frac{0 + 0 + 0 - 0 - (6)(9)(1) - (5)(4)(1)}{-54}$$

$$= \frac{-74}{-54} = 1.4 \text{ A}$$

The solution must satisfy all three equations. As a check for the first equation, you have:

$$1.2 + 0.15 - 1.4 = -0.05 \approx 0 \checkmark$$

Since the answers are rounded off, the check will not necessarily satisfy the equation exactly.

EXERCISE 13.4

In exercises 1 through 6, evaluate each determinant. See Example 13.12.

1. $\begin{vmatrix} 2 & 1 & -2 \\ 3 & 2 & -1 \\ 1 & 1 & 3 \end{vmatrix}$

2. $\begin{vmatrix} 1 & -2 & 4 \\ 3 & 2 & -1 \\ 4 & -1 & -2 \end{vmatrix}$

3. $\begin{vmatrix} -2 & 1 & 0 \\ 0 & 8 & -3 \\ 4 & -6 & 5 \end{vmatrix}$

4. $\begin{vmatrix} 4 & 0 & -3 \\ -2 & 1 & 0 \\ 5 & -3 & 1 \end{vmatrix}$

5. $\begin{vmatrix} 1.2 & 2.0 & 0.80 \\ 0.0 & -1.0 & 3.0 \\ 2.0 & 0.0 & 0.50 \end{vmatrix}$

6. $\begin{vmatrix} 2.6 & 1.5 & 4.0 \\ 0.0 & 3.5 & 5.0 \\ 1.2 & 4.4 & 0.0 \end{vmatrix}$

In exercises 7 through 20, solve each linear system using determinants. Round answers to two significant digits.

7. $x + y + z = 6$
   $x - y + 2z = 5$
   $2x - y - z = -3$

8. $3x + y - z = 2$
   $x - 2y - 2z = 6$
   $4x + y + 2z = 7$

9. $2x + y - 3z = -2$
   $x + 3y - 2z = -5$
   $3x + 2y - z = 7$

10. $2x + 3y + z = 4$
    $x + 5y - 2z = -1$
    $3x - 2y + 4z = 3$

11. $a + b + 2c = 9$
    $2a + b - c = 2$
    $a + b - 3c = -6$

12. $a + 2b - c = 2$
    $3a + 7b - 5c = 6$
    $a + 2b = 3$

13. $8a + b + c = 1$
    $7a - 2b + 9c = -3$
    $4a - 6b + 8c = -5$

14. $2a + 3b + 6c = 3$
    $3a + 4b - 5c = 2$
    $4a - 2b - 2c = 1$

15. $2x_1 + x_2 + 3x_3 = -2$
    $5x_1 + 2x_2 - 5 = 0$
    $2x_2 - 3x_3 + 7 = 0$

16. $2.3x_1 = 1.3x_2 + 1.7$
    $2.5x_2 = 2.5x_3 + 5.0$
    $0.6x_3 = 1.2x_1 - 2.4$

17. $2.0V_1 - 3.0V_2 = -1.0$
    $3.0V_1 + 5.0V_3 = 4.5$
    $1.0V_1 + 6.0V_2 - 10V_3 = -1.5$

18. $0.2R_1 + 0.1R_2 + 0.2R_3 = 0.8$
    $0.4R_1 + 0.2R_2 - 0.4R_3 = 0.4$
    $0.4R_1 - 0.3R_2 + 0.6R_3 = 1.4$

19. $3I_1 + 0.5I_2 + I_3 = 1.5$
    $0.5I_1 + 3I_2 + 2.5I_3 = 2$
    $4I_1 + 1.5I_2 - 2I_3 = 4.5$

20. $3I_1 - 2I_2 + 3I_3 = 0$
    $2I_1 - I_2 = 0$
    $I_1 + I_2 + I_3 = 1.1$

Applied Problems

In problems 21 through 28, solve each applied problem by setting up and solving a linear system using determinants. Round answers to two significant digits.

21. Roman wants to determine the volume of three oddly shaped bottles. All he has is a gallon container filled with water. After much experimenting, he discovers the following facts. The gallon container fills the first and the second container. The first and the third container exactly fill the second container. Three times the volume of the first container fills the second. What is the volume of each container?

22. Three pumps can fill a swimming pool that holds 6,000 gallons in 10 hours. When only the first and second pumps are operating, it takes 20 hours to fill the pool. When only the first and third pumps are operating, it takes 15 hours to fill the pool. What is the pumping rate in gallons per hour for each pump?

Applications to Electronics

23. Analysis of the voltages in an amplifier circuit produces the following three equations:

$$0.2V_A + 0.1V_B = 30$$
$$0.2V_B - 0.1V_C = 25$$
$$0.1V_A + 0.2V_C = 35$$

Solve this linear system for V_A, V_B, and V_C in volts.

24. A circuit analysis produces the following three equations:

$$2a_1 - a_2 + 4a_3 = 1.9$$
$$a_1 + 5a_2 - a_3 = -1.6$$
$$a_1 - 2a_2 - a_3 = 0.50$$

Solve this linear system for a_1, a_2, and a_3.

25. Applying Kirchhoff's laws to a circuit like that shown in Figure 13-14 with different resistances and voltages

produces the following equations for the currents in amps:

$$I_1 - I_2 + I_3 = 0$$
$$10I_1 + 10I_2 - 16 = 0$$
$$10I_2 + 20I_3 - 12 = 0$$

Solve this system for I_1, I_2, and I_3 and then express the answers in milliamps.

26. Applying Kirchhoff's voltage law to a circuit like that shown in Figure 13-14 with different resistances and voltages produces the following equations for the currents in amps:

$$I_1 - I_2 + I_3 = 0$$
$$10I_1 - 20I_3 - 0.07 = 0$$
$$10I_2 + 20I_3 - 0.23 = 0$$

Solve this system for I_1, I_2, and I_3 and express the answers in milliamps.

27. In Figure 13-15, $R_3 = 4\ \Omega$ and $R_4 = 5\ \Omega$. When Kirchhoff's laws are applied to the circuit, the following three equations result:

$$5I_1 + 5I_2 - 9I_3 = 0$$
$$3I_2 + 4I_3 - 10 = 0$$
$$3I_1 + 4I_3 - 12 = 0$$

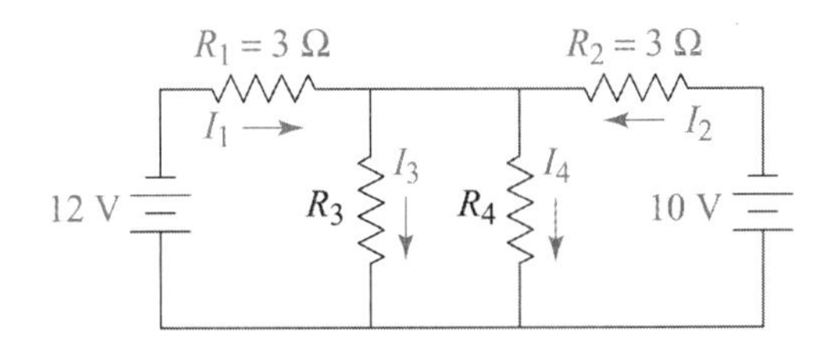

FIGURE 13-15 Kirchhoff's laws for problems 27 and 28.

Solve this linear system for the three currents I_1, I_2, and I_3 in amps.

28. In Figure 13-15, if $R_3 = R_4 = 4\ \Omega$, Kirchhoff's laws produce the following four equations:

$$I_1 + I_2 - I_3 - I_4 = 0$$
$$3I_1 + 4I_3 - 12 = 0$$
$$3I_2 + 4I_4 - 10 = 0$$
$$4I_3 - 4I_4 = 0$$

Solve this linear system for the currents in amps. First reduce it to three equations by solving the last equation for I_4 and substituting the expression in the first and third equations.

CHAPTER HIGHLIGHTS

13.1 GRAPHICAL SOLUTION OF TWO LINEAR EQUATIONS

To solve two linear equations graphically, draw the line for each equation on the same graph, and determine the point of intersection.

Key Example: See Figure 13-16.

13.2 ALGEBRAIC SOLUTION OF TWO LINEAR EQUATIONS

Addition Method. Multiply the equations so the coefficients for one unknown cancel when the equations are added.

Key Example:

$$\begin{aligned} 2x - y = -5 \rightarrow 2(2x - y = -5) &\rightarrow 4x - 2y = -10 \\ 3x + 2y = 3 \rightarrow 3x + 2y = 3 &\rightarrow 3x + 2y = 3 \\ &\quad\ \ 7x + 0 = -7 \\ &\quad\ \ x = -1 \end{aligned}$$

Substitute back the value of x to find y:

$$2(-1) - y = -5 \rightarrow y = 3$$

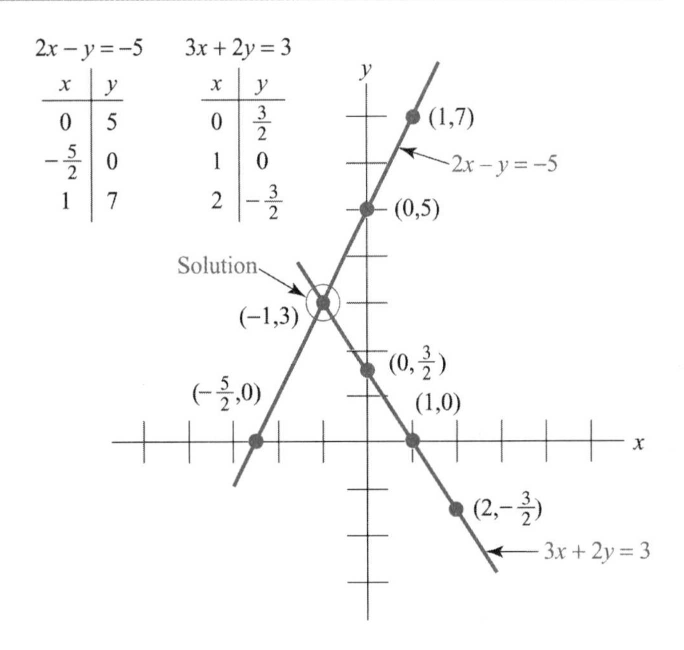

FIGURE 13-16 Key Example: Graphical solution of simultaneous equations.

Substitution Method. Solve one equation for one of the unknowns and substitute this expression into the other equation.

Key Example Above:

$$2x - y = -5 \rightarrow y = 2x + 5$$

$$3x + 2y = 3 \rightarrow 3x + 2(2x + 5) = 3$$

$$7x + 10 = 3 \rightarrow x = -1$$

13.3 SOLUTION BY DETERMINANTS

Determinant of 2 × 2 Matrix

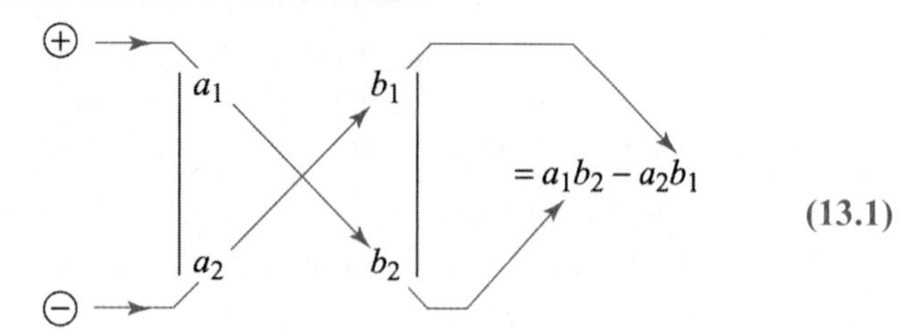

$$= a_1b_2 - a_2b_1 \qquad (13.1)$$

For a linear system of two equations in *standard form:*

$$a_1x + b_1y = k_1 \quad (k_1, k_2 \text{ constants})$$
$$a_2x + b_2y = k_2$$

The solution is:

Cramer's Rule for Two Equations

$$x = \frac{\begin{vmatrix} k_1 & b_1 \\ k_2 & b_2 \end{vmatrix}}{\begin{vmatrix} a_1 & b_1 \\ a_2 & b_2 \end{vmatrix}} \qquad y = \frac{\begin{vmatrix} a_1 & k_1 \\ a_2 & k_2 \end{vmatrix}}{\begin{vmatrix} a_1 & b_1 \\ a_2 & b_2 \end{vmatrix}} \qquad (13.2)$$

Key Example:

$$2x - y = -5$$
$$3x + 2y = 3$$

$$x = \frac{\begin{vmatrix} -5 & -1 \\ 3 & 2 \end{vmatrix}}{\begin{vmatrix} 2 & -1 \\ 3 & 2 \end{vmatrix}} = \frac{(-5)(2) - (3)(-1)}{(2)(2) - (3)(-1)} = -1$$

$$y = \frac{\begin{vmatrix} 2 & -5 \\ 3 & 3 \end{vmatrix}}{\begin{vmatrix} 2 & -1 \\ 3 & 2 \end{vmatrix}} = \frac{(2)(3) - (3)(-5)}{(2)(2) - (3)(-1)} = 3$$

13.4 SOLUTION OF THREE LINEAR EQUATIONS

Determinant of 3 × 3 Matrix

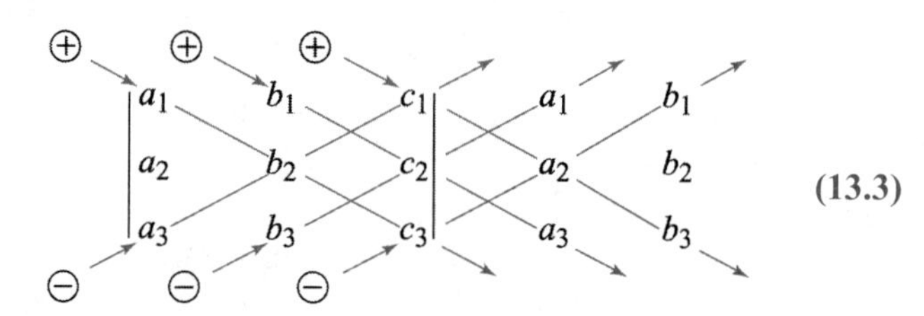

(13.3)

$$= a_1b_2c_3 + b_1c_2a_3 + c_1a_2b_3 - a_3b_2c_1 - b_3c_2a_1 - c_3a_2b_1$$

For a linear system of three equations in standard form:

$$a_1x + b_1y + c_1z = k_1$$
$$a_2x + b_2y + c_2z = k_2 \qquad (k_1, k_2, k_3 \text{ constants})$$
$$a_2x + b_3y + c_3z = k_3$$

The solution is:

Cramer's Rule for Three Equations (13.4)

$$x = \frac{\begin{vmatrix} k_1 & b_1 & c_1 \\ k_2 & b_2 & c_2 \\ k_3 & b_3 & c_3 \end{vmatrix}}{\begin{vmatrix} a_1 & b_1 & c_1 \\ a_2 & b_2 & c_2 \\ a_3 & b_3 & c_3 \end{vmatrix}} \qquad y = \frac{\begin{vmatrix} a_1 & k_1 & c_1 \\ a_2 & k_2 & c_2 \\ a_3 & k_3 & c_3 \end{vmatrix}}{\begin{vmatrix} a_1 & b_1 & c_1 \\ a_2 & b_2 & c_2 \\ a_3 & b_3 & c_3 \end{vmatrix}} \qquad z = \frac{\begin{vmatrix} a_1 & b_1 & k_1 \\ a_2 & b_2 & k_2 \\ a_3 & b_3 & k_3 \end{vmatrix}}{\begin{vmatrix} a_1 & b_1 & c_1 \\ a_2 & b_2 & c_2 \\ a_3 & b_3 & c_3 \end{vmatrix}}$$

Key Example:

$$2x + 3y + z = 5$$
$$3x - 2y + 4z = -3$$
$$5x + y + 2z = -1$$

$$x = \frac{\begin{vmatrix} 5 & 3 & 1 \\ -3 & -2 & 4 \\ -1 & 1 & 2 \end{vmatrix}}{\begin{vmatrix} 2 & 3 & 1 \\ 3 & -2 & 4 \\ 5 & 1 & 2 \end{vmatrix}}$$

$$= \frac{5(-2)(2) + 3(4)(-1) + 1(-3)(1) - (-1)(-2)(1) - 1(4)(5) - 2(-3)(3)}{2(-2)(2) + 3(4)(5) + 1(3)(1) - (5)(-2)(1) - 1(4)(2) - 2(3)(3)}$$

$$= \frac{-39}{39} = -1$$

$$y = \frac{\begin{vmatrix} 2 & 5 & 1 \\ 3 & -3 & 4 \\ 5 & 1 & 2 \end{vmatrix}}{39} = \frac{78}{39} = 2 \qquad z = \frac{\begin{vmatrix} 2 & 3 & 5 \\ 3 & -2 & -3 \\ 5 & 1 & -1 \end{vmatrix}}{39} = \frac{39}{39} = 1$$

REVIEW EXERCISES

In exercises 1 through 10, solve each linear system graphically and check by finding the algebraic solution.

1. $x + 2y = 6$
   $x - 2y = 8$
2. $x - 2y = 0$
   $x - y = 2$
3. $2x + y = 3$
   $3x + y = 5$
4. $3x - y = -5$
   $x + y = 1$
5. $4a - b = -1$
   $2a - b = -3$
6. $R - 2t = -1$
   $R - t = -3$
7. $I_1 + 2I_2 - 4 = 0$
   $3I_1 + 2I_2 - 6 = 0$
8. $3P_1 - P_2 + 4 = 0$
   $P_1 - P_2 + 1 = 0$
9. $X - 5Y = 30$
   $5X + Y = 20$
10. $X + 2Y = 0$
    $X - Y = 6$

In exercises 11 through 24, solve each linear system algebraically two ways: addition and substitution, or as directed by the instructor. Check that the answers satisfy the equations. Round answers to two significant digits.

11. $x + 2y = 6$
    $x - 2y = 2$
12. $x - 2y = 1$
    $x - y = 2$
13. $2x + 3y - 1 = 0$
    $3x + y - 5 = 0$
14. $3x - y = -5$
    $x + 4y = 7$
15. $a = 4b - 1$
    $a = 2b + 3$
16. $5A - 2B = 20$
    $A + 4B = 15$
17. $b = 1 - y$
    $y = 7b + 7$
18. $R = 2t - 1$
    $R = t + 3$
19. $I_1 + 2I_2 - 2 = 0$
    $3I_1 + 2I_2 - 6 = 0$
20. $3P_1 - P_2 + 4 = 0$
    $P_1 - P_2 + 2 = 0$
21. $X - 5Y = 20$
    $5X + Y = 22$
22. $X + 2Y = 2$
    $X - Y = 5$
23. $20I_1 - 10I_2 = -1$
    $4I_1 + 6I_2 = 7$
24. $V_1 = V_2 + 2$
    $V_2 = 1 - 3V_1$

In exercises 25 through 32, evaluate each determinant.

25. $\begin{vmatrix} 3 & 8 \\ 2 & -4 \end{vmatrix}$
26. $\begin{vmatrix} 11 & -18 \\ 10 & -15 \end{vmatrix}$
27. $\begin{vmatrix} -0.56 & 0.12 \\ -0.44 & 0.88 \end{vmatrix}$
28. $\begin{vmatrix} -1.6 & 3.3 \\ 2.3 & -1.1 \end{vmatrix}$
29. $\begin{vmatrix} 1 & 3 & -4 \\ 2 & -3 & 1 \\ 4 & 2 & -6 \end{vmatrix}$
30. $\begin{vmatrix} 0.2 & 0.5 & -0.2 \\ 1.0 & -0.5 & 1.0 \\ -2.0 & 2.0 & -1.0 \end{vmatrix}$
31. $\begin{vmatrix} 10 & 5 & -10 \\ 2 & -1 & 0 \\ 5 & -10 & -5 \end{vmatrix}$
32. $\begin{vmatrix} 1.2 & -3.5 & 2.0 \\ 7.0 & 0.0 & 6.1 \\ -3.3 & 4.1 & -4.2 \end{vmatrix}$

In exercises 33 through 52, solve each linear system using determinants. Round answers to two significant digits.

33. $x - 3y = 2$
    $2x - 3y = 5$
34. $4x + 3y = 2$
    $3x + 4y = -2$
35. $I_1 + 2I_2 = 3$
    $3I_1 - 2I_2 = 5$
36. $3P_1 - P_2 = -4$
    $2P_1 - 4P_2 = -1$
37. $20I_1 - 10I_2 = -1$
    $4I_1 + 6I_2 = 7$
38. $V_1 = V_2 + 2$
    $V_2 = 1 - 3V_1$
39. $1.6a - 1.4b + 0.9 = 0$
    $3.2a + 0.4b - 9.4 = 0$
40. $6.6A = 10B - 4.2$
    $12B = 8.2A + 4.9$
41. $\frac{X}{5} - \frac{Y}{2} = 6$
    $\frac{X}{2} - \frac{Y}{4} = 7$
42. $\frac{X}{4} + \frac{Y}{2} = -1$
    $\frac{X}{2} - \frac{Y}{4} = 3$
43. $0.2x - 0.3y = 1.2$
    $0.1x - 0.4y = 1.1$
44. $1.0p + 0.1v = 0.7$
    $2.0p + 0.7v = -3.1$
45. $x + 3y + z = 2$
    $2x - y - 2z = -1$
    $3x + 2y - z = 1$
46. $2x + 2y - z = 3$
    $3x + y - z = 6$
    $x + y - 4z = 5$
47. $3x + 2y + 3z = 8$
    $4x - 4y - 2z = 1$
    $x + y - 3z = 1$
48. $x - 2y + 3z = -4$
    $3x + 4y - z = 6$
    $3x - 2y + 2z = 0$
49. $x_1 - x_2 + 2x_3 = 6$
    $2x_1 + x_2 - x_3 = 4$
    $x_1 + 3x_2 + 3x_3 = 3$
50. $2a_1 + 2a_2 - 3a_3 = 5$
    $4a_1 - 3a_2 + 2a_3 = -4$
    $6a_1 + 5a_2 - 4a_3 = 13$
51. $I_1 + I_2 - I_3 = 0$
    $2I_1 + 3I_2 - 9 = 0$
    $3I_2 + 4I_3 - 20 = 0$
52. $4.5R = 3.0V + 1.5$
    $4.8V = 6.0I - 6.0$
    $1.2I = 1.1R + 1.4$

Applied Problems

53. The length of a rectangular field is 150% of its width. If the fencing around the perimeter is 155 m, what are the dimensions of the field in meters?
54. Rachel invests $1000 in two CDs (certificates of deposit). The total interest after one year is $54.00. If the interest rate is 5% on one CD and 6% on the other, how much did she invest in each CD?
55. Jack, a resourceful car salesman, offers some specials on a new Ford model. The standard model with a compact disc player is priced at $11,700. The standard model with a compact disc player and 4-wheel drive is priced at $12,800. If the value of the 4-wheel drive is five and

one-half times the value of the compact disc player, how much is the standard car alone worth?

56. Derek leaves a harbor at 6 A.M. and motors his sailboat under power with no wind. At 10 A.M., the wind picks up, and he shuts off the boat's engine and sails until 12 noon. At this time, Derek determines that the boat has traveled only 26 mi since departing the harbor, so he decides to use both power and sail to move faster. The sailboat arrives in port under power and sail at 6 P.M. after traveling a total of 74 mi. If the speed under power and sail equals the speed under power plus the speed under sail, how fast did the boat travel under power alone, and under sail alone?

Applications to Electronics

57. Two unknown voltages V_1 and V_2 are connected in series to a resistance $R = 10\ \Omega$. When the voltages are series aiding, $I = 2.5$ A. When the voltages are series opposing, $I = 1.5$ A in the same direction. Find the two voltages.

58. The total conductance of two resistors R_1 and R_2 in parallel is 100 mS. When the resistance of R_1 is doubled, the total conductance decreases to 60 mS. Find the two conductances and resistances.

59. The total conductance of two resistors R_1 and R_2 in parallel is 28 mS. When a third resistor, whose conductance is equal to R_1, is added in parallel, the total conductance is 38 mS. Find the conductance and the resistance of each resistor.

60. In order to accurately find the voltage V of a battery and its small internal resistance r, Mohammed connects a small resistance of 3.0 Ω to the battery and accurately measures the current with a digital ammeter to be 1.88 A. He then repeats the experiment with a resistance of 5.1 Ω and measures the current to be 1.15 A. What are the voltage of the battery in volts and its internal resistance in milohms to two significant digits?

61. A circuit analysis gives rise to the following linear system of three equations:

$$\begin{aligned} 2.2x_2 + 3.4x_3 &= 1.56 \\ 2.6x_1 + 5.2x_3 &= 2.60 \\ 3.8x_1 - 5.4x_2 &= 0.12 \end{aligned}$$

Solve this system for x_1, x_2, and x_3.

62. Three unknown resistances R_1, R_2, and R_3 are connected in series with a 6-V battery. The current in the circuit I = 10 mA. When only R_1 and R_3 are connected in series to the battery, $I = 20$ mA, and when only R_2 and R_3 are connected in series, $I = 15$ mA. Find the three resistances.

63. For the circuit in Figure 13-17, if $R_0 = 10\ \Omega$, application of Kirchhoff's laws produces the two equations for the currents in amps:

$$\begin{aligned} 40I_1 + 10I_2 &= 9 \\ 10I_1 + 40I_2 &= 6 \end{aligned}$$

Solve this linear system for I_1 and I_2 and express answers in milliamps.

64. For the circuit in Figure 13-17, if $R_0 = 20\ \Omega$, application of Kirchhoff's laws produces the three equations for the currents in amps:

$$\begin{aligned} I_1 + I_2 &= I_3 \\ 30I_1 + 20I_3 &= 9 \\ 30I_2 + 20I_3 &= 6 \end{aligned}$$

Solve this linear system for I_1, I_2, and I_3 and express the answers to the nearest milliamp.

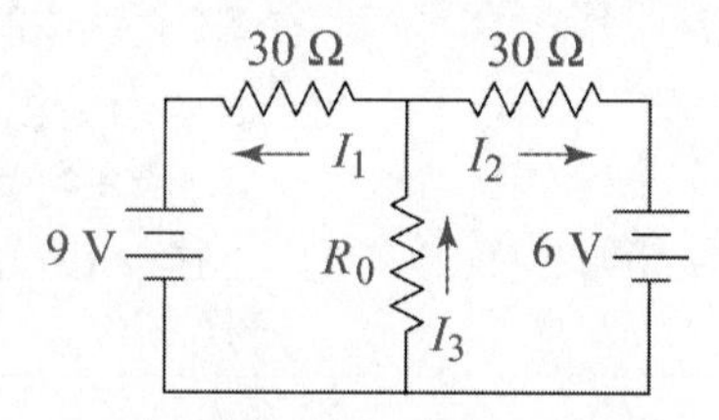

FIGURE 13-17 Kirchhoff's laws for problems 63 and 64.

CHAPTER 14
Network Analysis

CHAPTER OBJECTIVES

In this chapter, you will learn:

- The voltage division formula.
- Kirchhoff's current law.
- Kirchhoff's voltage law.
- The concept of polarity and how to apply it.
- How to solve dc circuits using Kirchhoff's laws.

As a circuit becomes more complex, it is necessary to have additional tools to analyze the circuit. In addition to Ohm's law and the power formulas, another useful formula, which comes directly from Ohm's law, is the voltage division formula for resistances in series. Two other basic electrical laws that apply to current and voltage were formulated in 1847 by the German physicist Gustav R. Kirchhoff. Kirchhoff's current law and voltage law are fundamental tools that can be used to analyze any network. When you apply Kirchhoff's laws to a dc circuit with one or more voltage sources, it leads to a system of linear equations for the currents, which can be solved by the methods shown in Chapter 13.

• • •

14.1 Voltage Division

In a series circuit or a series branch of a series-parallel circuit, if R_X = any resistance and V_X = voltage drop across R_X, the current through R_X is given by Ohm's law:

$$I_X = \frac{V_X}{R_X}$$

If R_T = total resistance of the series circuit or branch, and V_T = total voltage of the circuit or branch, the total current is also given by Ohm's law:

$$I_T = \frac{V_T}{R_T}$$

In a series circuit or series branch, the current through any resistance I_X equals the total current I_T. Therefore, the two expressions for the currents are equal:

$$\frac{V_X}{R_X} = \frac{V_T}{R_T}$$

This equation is a proportion, which says that the voltage drop across any resistance in a series circuit or series branch is proportional to the resistance. For example, if R_1 and R_2 are in series and R_2 is twice R_1, then $V_2 = IR_2$ will be twice $V_1 = IR_1$. This relationship is illustrated in Chapter 11, Section 11.1. Multiplying both sides of the above equation by R_X leads to the proportional voltage or voltage division formula:

Formula **Voltage Division Formula**

$$V_X = \frac{R_X}{R_T} V_T \tag{14.1}$$

Formula (14.1) provides a convenient way to find voltage drops in a series circuit without having to find the current, as the next two examples illustrate.

EXAMPLE 14.1

Given the series circuit in Figure 14-1 with $R_1 = 15\ \Omega$, $R_2 = 30\ \Omega$, $R_3 = 20\ \Omega$, and an applied voltage $V_T = 10$ V. Find V_1, V_2, and V_3.

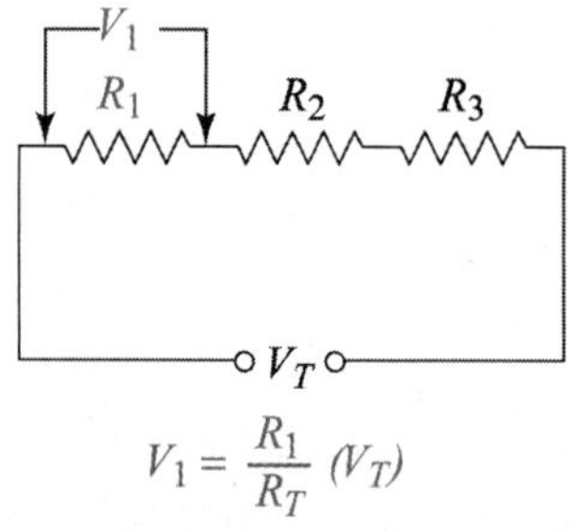

$$V_1 = \frac{R_1}{R_T}(V_T)$$

FIGURE 14-1 Voltage division in a series circuit for Example 14.1.

Solution It is not necessary to find the current. Apply formula (14.1) to find the voltage drops:

$$V_1 = \frac{R_1}{R_T}(V_T) = \frac{15}{15+30+20}(10\text{ V}) = \frac{15}{65}(10\text{ V}) = 2.3\text{ V}$$

$$V_2 = \frac{R_2}{R_T}(V_T) = \frac{30}{15+30+20}(10\text{ V}) = \frac{30}{65}(10\text{ V}) = 4.6\text{ V}$$

$$V_3 = \frac{R_3}{R_T}(V_T) = \frac{20}{15+30+20}(10\text{ V}) = \frac{20}{65}(10\text{ V}) = 3.1\text{ V}$$

Note the proportional relationship between the resistances and the voltage drops. R_2 is double R_1, and $V_2 = 4.6$ V is double $V_1 = 2.3$ V. Also R_2 is 1.5 times R_3, and $V_2 = 4.6$ V is 1.5 times $V_3 = 3.1$ V.

EXAMPLE 14.2

Given the series-parallel circuit in Figure 14-2(a) with an applied voltage $V_T = 12$ V. Find R_T in ohms, V_1, V_2, V_3, and V_4 in volts, and I_T, I_1, I_2, I_3, and I_4 in milliamps.

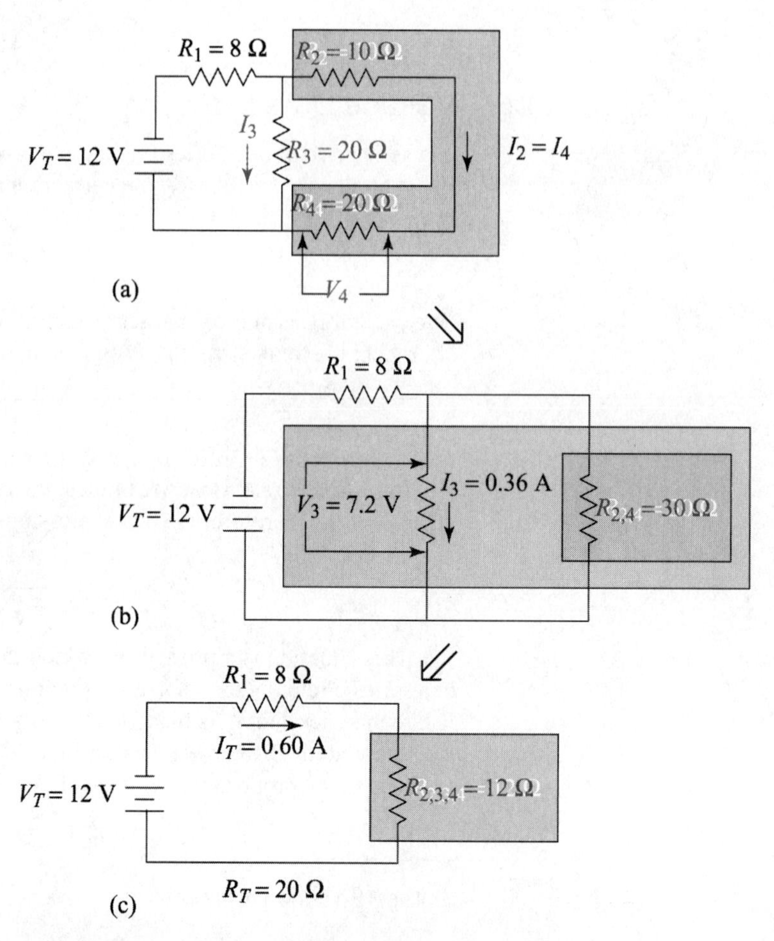

FIGURE 14-2 Current and voltage in a series-parallel circuit for Example 14.2.

Solution To find R_T, the series branch containing R_2 and R_4 can first be added and represented as one resistance $R_{2,4}$:

$$R_2 + R_4 = R_{2,4} = 10\ \Omega + 20\ \Omega = 30\ \Omega$$

The circuit can then be reduced to that in Figure 14-2(b).

Since $R_3 = 20\ \Omega$ is in parallel with $R_{2,4}$, the equivalent resistance for R_3 and $R_{2,4}$, written $R_{2,3,4}$, can be found using the reciprocal formula and the reciprocal key [x⁻¹] or [1/x] on the calculator:

$R_{2,3,4}$: 20 [x⁻¹] [+] 30 [x⁻¹] [=] [x⁻¹] [=] → 12 Ω

or applying the product sum formula (12.9):

$$R_{2,3,4} = \frac{(20)(30)}{20 + 30} = 12\ \Omega$$

The circuit can now be reduced to that in Figure 14-2(c), where R_1 is in series with $R_{2,3,4}$. The total resistance of the circuit is then:

$$R_T = 8\ \Omega + 12\ \Omega = 20\ \Omega$$

Using Ohm's law, the total current is:

$$I_T = \frac{V_T}{R_T} = \frac{12\ \text{V}}{20\ \Omega} = 0.60\ \text{A} = 600\ \text{mA}$$

Since $I_1 = I_T = 600$ mA, apply Ohm's law to find V_1:

$$V_1 = I_1 R_1 = (0.6)(8) = 4.8\ \text{V}$$

The voltage across $R_{2,3,4}$ is then:

$$V_{2,3,4} = 12\ \text{V} - 4.8\ \text{V} = 7.2\ \text{V}$$

To find V_2 and V_4, observe that $V_{2,3,4} = V_{2,4}$, the voltage across $R_{2,4}$. You can then apply the voltage division formula (14.1) to the series branch containing R_2 and R_4 by letting $R_T = R_{2,4} = 30\ \Omega$ and $V_T = V_{2,4} = 7.2$ V:

$$V_2 = \frac{R_2}{R_T}(V_T) = \frac{R_2}{R_{2,4}}(V_{2,4}) = \frac{10}{30}(7.2\ \text{V}) = 2.4\ \text{V}$$

$$V_4 = \frac{R_4}{R_T}(V_{2,4}) = \frac{R_2}{R_{2,4}}(V_{2,4}) = \frac{20}{30}(7.2\ \text{V}) = 4.8\ \text{V}$$

To find I_3, note that $V_{2,4} = V_3$, the voltage across R_3, and you can apply Ohm's law:

$$I_3 = \frac{V_3}{R_3} = \frac{7.2\ \text{V}}{20\ \Omega} = 0.36\ \text{A} = 360\ \text{mA}$$

To find I_2 and I_4, which are equal, apply Ohm's law again:

$$I_2 = \frac{V_2}{R_2} = \frac{2.4\ \text{V}}{10\ \Omega} = 0.24\ \text{A} = 240\ \text{mA},\ I_4 = \frac{V_4}{R_4} = \frac{4.8\ \text{V}}{20\ \Omega} = 240\ \text{mA}$$

There are other ways to solve this circuit. This example demonstrates one way, using Ohm's law and the voltage division formula.

EXERCISE 14.1

Express all answers in engineering notation (as numbers between 1 and 1000) with the appropiate units to two significant digits.

1. In Figure 14-1, given $V_T = 9$ V, $R_1 = 25\ \Omega$, $R_2 = 15\ \Omega$, and $R_3 = 30\ \Omega$, find V_1, V_2, and V_3.
2. In Figure 14-1, given $V_T = 32$ V, $R_1 = 300\ \Omega$, $R_2 = 750\ \Omega$, and $R_3 = 200\ \Omega$, find V_1, V_2, and V_3.
3. In Figure 14-2(a), given $V_T = 15$ V, $R_1 = 10\ \Omega$, $R_2 = 20\ \Omega$, $R_3 = 20\ \Omega$, and $R_4 = 20\ \Omega$, find R_T, V_1, V_2, V_3, V_4, in volts and I_T, I_1, I_2, I_3, and I_4 in milliamps.
4. In Figure 14-2(a), given $V_T = 50$ V, $R_1 = 100\ \Omega$, $R_2 = 150\ \Omega$, $R_3 = 300\ \Omega$, and $R_4 = 150\ \Omega$, find R_T, V_1, V_2, V_3, V_4, in volts and I_T, I_1, I_2, I_3, and I_4 in milliamps.
5. Given the series-parallel circuit in Figure 14-3 with $V_T = 12$ V, $R_1 = 300\ \Omega$, $R_2 = 400\ \Omega$, and $R_3 = 200\ \Omega$, find V_1, V_2, and V_3.
6. Given the series-parallel circuit in Figure 14-3 with $V_T = 200$ V, $R_1 = 10$ kΩ, $R_2 = 10$ kΩ, and $R_3 = 30$ kΩ, find V_1, V_2, and V_3.

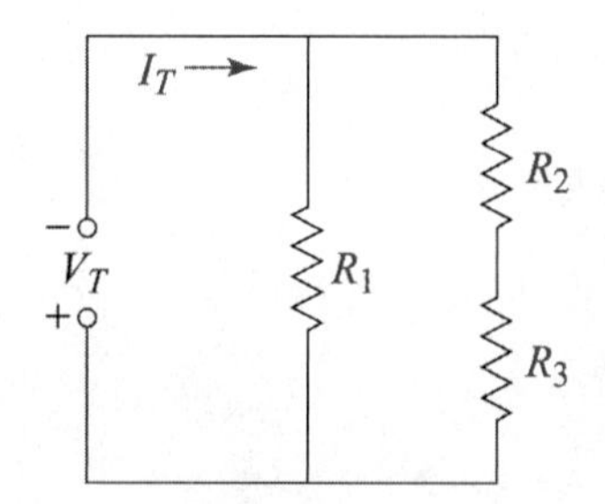

FIGURE 14-3 Voltage division for problems 5 and 6.

7. Given the series-parallel circuit in Figure 14-4 with $V_T = 24$ V, $R_1 = 10\ \Omega$, $R_2 = 7.5\ \Omega$, $R_3 = 7.5\ \Omega$, and $R_4 = 10\ \Omega$, find R_T, I_T, I_1, I_2, V_1, V_2, V_3, and V_4.

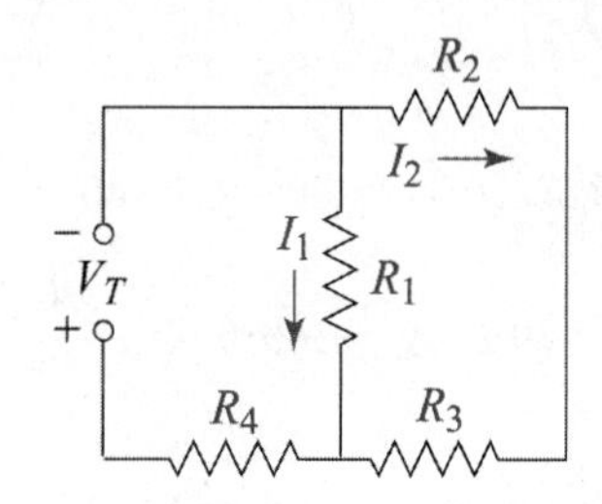

FIGURE 14-4 Voltages and currents for problems 7 and 8.

8. Given the series-parallel circuit in Figure 14-4 with $V_T = 60$ V, $R_1 = 30\ \Omega$, $R_2 = 10\ \Omega$, $R_3 = 20\ \Omega$, and $R_4 = 15\ \Omega$, find R_T, I_T, I_1, I_2, V_1, V_2, V_3, and V_4.
9. Given the series-parallel circuit in Figure 14-5 with $V_T = 200$ V, $R_1 = 10$ kΩ, $R_2 = 20$ kΩ, $R_3 = 2.0$ kΩ, and $R_4 = 3.0$ kΩ, find R_T, I_T, I_1, I_2, I_4, V_1, V_2, V_3, and V_4 in volts.
10. Given the series-parallel circuit in Figure 14-5 with $V_T = 1.5$ V, $R_1 = 12\ \Omega$, $R_2 = 15\ \Omega$, $R_3 = 20\ \Omega$, and $R_4 = 10\ \Omega$, find R_T, I_T, I_1, I_2, I_4, V_1, V_2, V_3, and V_4 in volts.

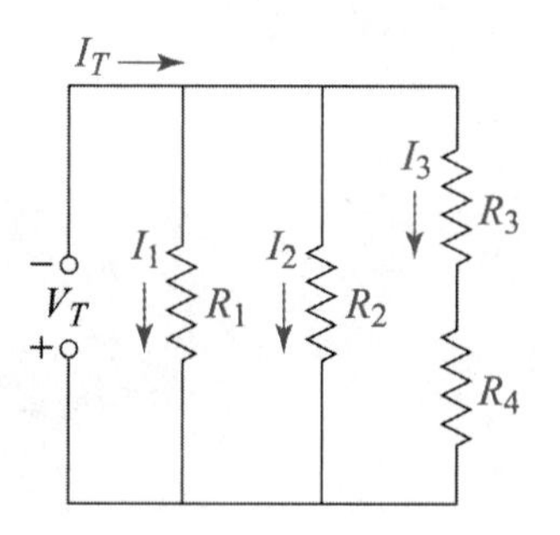

FIGURE 14-5 Voltages and currents for problems 9 and 10.

11. Two resistors R_1 and R_2 are in series. If $R_1 = 100\ \Omega$, what should be the value of R_2 so that V_2 equals 20% of the total voltage drop across both resistors?
12. Two resistors R_1 and R_2 are in series. If $R_1 = 120\ \Omega$, what should be the value of R_2 so that V_2 equals one-fourth of the total voltage drop across both resistors?
13. Given the series-parallel circuit in Figure 14-6 with $R_2 = 2R_1$ and $R_4 = 3R_3$, if $V_1 = 12$ V, find V_2, V_3, and V_4.
14. Given the series-parallel circuit in Figure 14-6 with $R_2 = 50\%(R_1)$ and $R_4 = 25\%(R_3)$, if $V_1 = 20$ V, find V_2, V_3, and V_4.

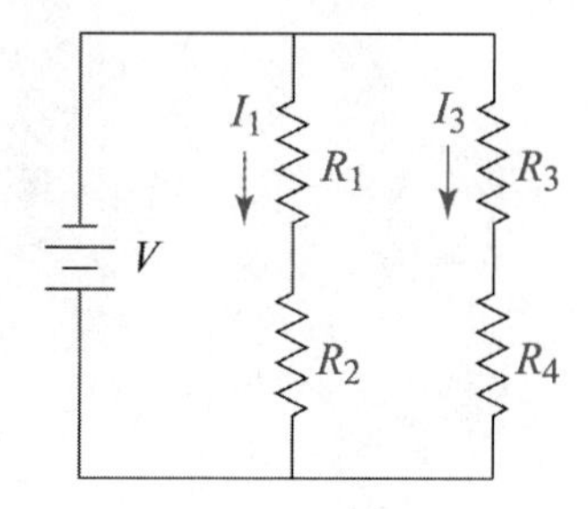

FIGURE 14-6 Voltage division for problems 13 and 14.

14.2 Kirchhoff's Laws

Current is the rate of flow of charge. There are two ways to consider current, as electron current or as conventional current. *Electron current* flows from (−) to (+) and is used throughout the text to analyze and solve dc circuits. Conventional current is an older model that flows from (+) to (−).

Kirchhoff's current law demonstrates that charge cannot accumulate at any point in a conductor:

Law

Kirchhoff's Current Law

The algebraic sum of all currents entering and leaving any branch point in a circuit equals zero. **(14.2)**

Charge that flows into a point in a circuit must also flow out, otherwise the charge would build indefinitely. To apply Kirchhoff's current law, assign a *positive* value to currents *entering* a branch point and a *negative* value to currents *leaving* a branch point.

EXAMPLE 14.3

Given the series-parallel circuit in Figure 14-7, containing one voltage source and three resistors, apply Kirchhoff's current law to the branch point **X.**

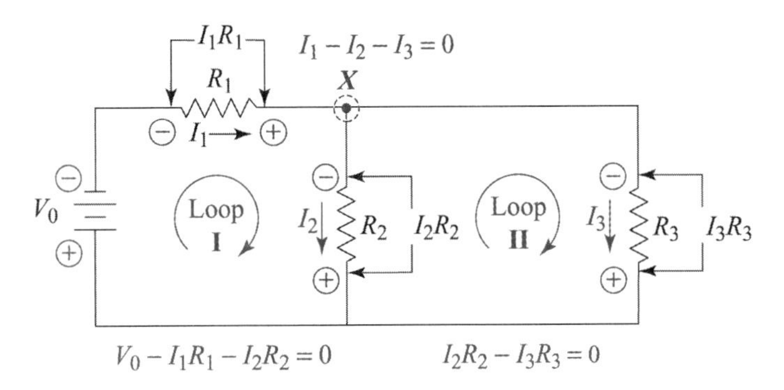

FIGURE 14-7 Kirchhoff's laws for Examples 14.3, 14.5, and 14.7.

Solution In Figure 14-7, the polarities are shown for the battery and the resistances. The end of resistance R_1 is connected to the negative terminal of the battery and is assigned a negative polarity. The electron current I_1 flows from (−) to (+) and splits into I_2 and I_3 at branch point **X.** The polarities for R_2 and R_3 are marked so I_2 and I_3 flow from (−) to (+). In this way, the three currents are consistent and illustrate electron current. Notice that the current flows through the battery from (+) to (−). At branch point **X**, I_1 is entering and is assigned a positive value while I_2 and I_3 are leaving and are assigned negative values. Kirchhoff's current law applied at branch point **X** is then:

$$I_1 - I_2 - I_3 = 0 \text{ or } I_1 = I_2 + I_3$$

Notice that the equation on the right shows that the current entering I_1 equals the sum of the currents leaving $I_2 + I_3$.

EXAMPLE 14.4

Given the series-parallel circuit in Figure 14-8 containing two voltage sources and three resistors, apply Kirchhoff's current law to the branch point **X.**

Solution This circuit has two voltage sources. If the values of the voltages are close, the currents will flow from (+) to (−) through both batteries. However, if one voltage is

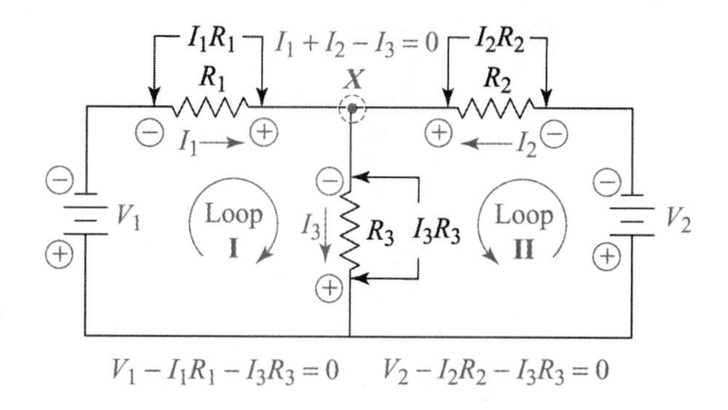

FIGURE 14-8 Kirchhoff's laws for Examples 14.4, 14.6, and 14.8.

sufficiently greater than the other, it is possible for the current to flow from (−) to (+) through the lower voltage battery. For purposes of assigning polarity for the current flow, it is not necessary to know the actual situation. *As long as the electron currents are consistently shown flowing from* (−) *to* (+) *with the assigned polarities, Kirchhoff's current law will be correctly applied.* The actual values of the currents in the equation can be positive *or* negative, always making the equation a true statement. In Figure 14-8, the ends of R_1 and R_2 are connected to the negative terminals of the batteries and are assigned negative polarities with the currents I_1 and I_2 shown accordingly. The polarity of R_3 is then assigned so that I_3 is logically shown leaving point **X.** Kirchhoff's current law at branch point **X** is then:

$$I_1 + I_2 - I_3 = 0 \text{ or } I_1 + I_2 = I_3$$

Note that the currents could all be shown flowing in the opposite directions, with all the polarities reversed, and you would obtain the same equation with all the signs reversed.

Kirchhoff's voltage law demonstrates that electrical potential is conserved throughout a closed path or loop:

Law

Kirchhoff's Voltage Law

The algebraic sum of the voltages around any closed path equals zero. (14.3)

Whatever potential you start at in a closed path, you must return to the same potential after traveling around the path. As a result, the voltage increases must equal the voltage drops. To apply Kirchhoff's voltage law as you move around a closed path or loop, you assign a *positive* sign to a voltage when *positive* polarity is encountered first and a *negative* sign to a voltage when *negative* polarity is encountered first.

EXAMPLE 14.5

Given the circuit in Figure 14-7, apply Kirchhoff's voltage law to loop **I** [V_0, R_1, R_2] and loop **II** [R_2, R_3].

Solution It is first necessary to assign polarities and show the current flow as done in Example 14.3 and Figure 14-7. Then, moving clockwise around loop **I** starting at the positive terminal of the battery, the voltage V_0 is assigned positive. The voltage across R_1 (I_1R_1) is assigned negative, and the voltage across R_2 (I_2R_2) is also assigned negative, because these signs are encountered first when moving clockwise. Kirchhoff's voltage law for loop **I** is then:

$$V_0 - I_1R_1 - I_2R_2 = 0 \text{ or } V_0 = I_1R_1 + I_2R_2$$

Notice that the equation on the right shows that the total voltage increase V_0 equals the sum of the voltage drops $I_1R_1 + I_2R_2$. Moving clockwise around loop **II** starting at the positive side of R_2, the voltage across R_2 (I_2R_2) is assigned positive, and the voltage across R_3 (I_3R_3) is assigned negative. Kirchhoff's voltage law for loop **II** is:

$$I_2R_2 - I_3R_3 = 0 \text{ or } I_2R_2 = I_3R_3$$

Observe that a loop does not necessarily include a voltage source. You can move in either direction around a loop, and you will get the same equation. If you move counterclockwise around loop **I** or loop **II**, all the signs in the equation will change, giving you the same result.

EXAMPLE 14.6

Given the circuit in Figure 14-8, apply Kirchhoff's voltage law to the loop **I** $[V_1, R_1, R_3]$ and loop **II** $[V_2, R_2, R_3]$.

Solution It is first necessary to assign polarities and show the current flow as done in Example 14.4 and Figure 14-8. Starting at the positive terminal of V_1 and traveling clockwise around loop **I**, Kirchhoff's voltage law gives the equation:

$$V_1 - I_1R_1 - I_3R_3 = 0 \text{ or } V_1 = I_1R_1 + I_3R_3$$

The equation on the right shows that the total voltage increase V_1 equals the sum of the voltage drops $I_1R_1 + I_3R_3$. You can also travel around the same loop *counterclockwise* and apply Kirchhoff's voltage law. This reverses the sign of each voltage but results in the same equation:

$$I_3R_3 + I_1R_1 - V_1 = 0$$

Kirchhoff's voltage law going counterclockwise around loop **II** is:

$$V_2 - I_2R_2 - I_3R_3 = 0 \text{ or } V_2 = I_2R_2 + I_3R_3$$

EXERCISE 14.2

1. State Kirchhoff's current law.
2. State Kirchhoff's voltage law.

In exercises 3 through 7, refer to the circuit in Figure 14-9, and write the equation that expresses each of the following:

3. Kirchhoff's current law at branch point **X**.
4. Kirchhoff's current law at branch point **Y**.
5. Kirchhoff's voltage law for the loop $[V_0, R_1, R_2]$.
6. Kirchhoff's voltage law for the loop $[V_0, R_1, R_3]$.
7. Kirchhoff's voltage law for the loop $[R_2, R_3]$.

In exercises 8 through 12, refer to the circuit in Figure 14-10 and write the equation that expresses each of the following:

8. Kirchhoff's current law at branch point **X**.
9. Kirchhoff's current law at branch point **Y**.

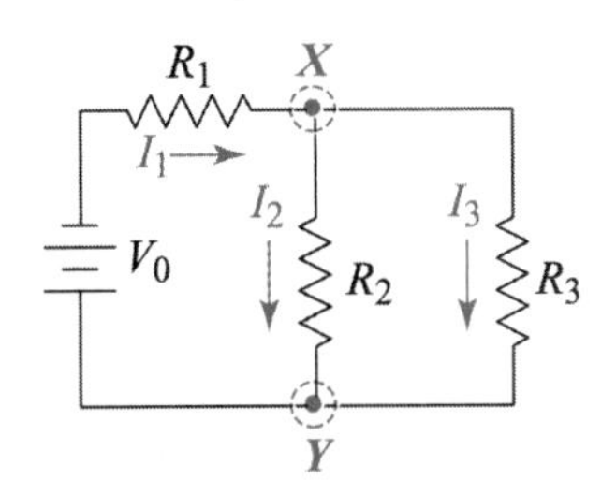

FIGURE 14-9 Kirchhoff's laws for exercises 3 through 7.

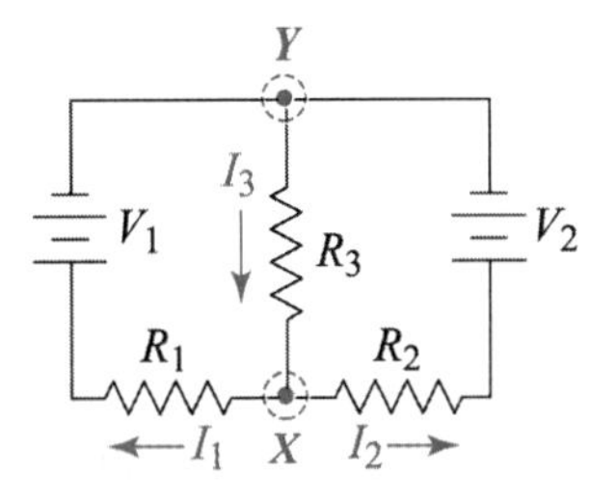

FIGURE 14-10 Kirchhoff's laws for exercises 8 through 12.

10. Kirchhoff's voltage law for the loop $[V_1, V_2, R_2, R_1]$.
11. Kirchhoff's voltage law for the loop $[V_1, R_3, R_1]$.
12. Kirchhoff's voltage law for the loop $[V_2, R_3, R_2]$.

In exercises 13 through 20, refer to the circuit in Figure 14-11 and write the equation that expresses each of the following:

13. Kirchhoff's current law at branch point **X**.
14. Kirchhoff's current law at branch point **Y**.
15. Kirchhoff's voltage law for the loop $[V_1, R_1, V_2, R_4]$.
16. Kirchhoff's voltage law for the loop $[V_1, R_1, R_2]$.
17. Kirchhoff's voltage law for the loop $[V_2, R_3, R_4]$.
18. Kirchhoff's voltage law for the loop $[R_2, R_3]$.
19. Kirchhoff's voltage law for the loop $[V_1, R_1, R_3]$.
20. Kirchhoff's voltage law for the loop $[V_2, R_2, R_4]$.

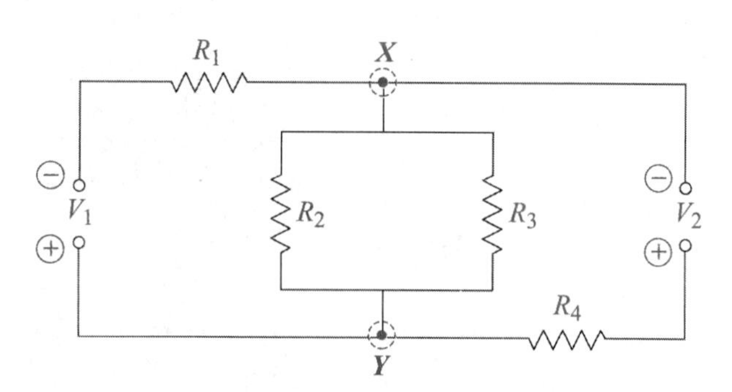

FIGURE 14-11 Kirchhoff's laws for exercises 13 through 20.

14.3 Solving Circuits

If you know the resistances and applied voltages in a circuit, you can apply Kirchhoff's laws and produce simultaneous linear equations with the currents as the unknowns. These equations can then be solved using the methods shown in Chapter 13.

EXAMPLE 14.7

Given the circuit in Figure 14-7 with $V_0 = 16$ V, $R_1 = 10\ \Omega$, $R_2 = 10\ \Omega$, and $R_3 = 15\ \Omega$, find the currents I_1, I_2, and I_3 by applying Kirchhoff's laws.

Solution Since there are three unknowns, I_1, I_2, and I_3, it is necessary to write three equations containing the currents to find their values. First assign polarities and show the current flow as done in Example 14.3 and Figure 14-7. Then apply Kirchhoff's current law at point **X** to give the current equation in Example 14.3:

$$I_1 - I_2 - I_3 = 0$$

Apply Kirchhoff's voltage law to loop **I**, $[V_0, R_1, R_2]$, to give the equation in Example 14.5:

$$V_0 - I_1R_1 - I_2R_2 = 0$$

Apply Kirchhoff's voltage law to loop **II**, $[R_2, R_3]$, to give the other equation in Example 14.5:

$$I_2R_2 - I_3R_3 = 0$$

Substitute the values for the resistances to give the three simultaneous equations for the currents:

$$\begin{aligned} I_1 - I_2 - I_3 &= 0 \\ 16 - 10I_1 - 10I_2 &= 0 \\ 10I_2 - 15I_3 &= 0 \end{aligned}$$

This linear system can be solved by different methods. The substitution method shown in Section 13.2 is used here. Reduce the system to two equations as follows. Use the first equation, which is simplest to work with, and solve it for I_1:

$$I_1 = I_2 + I_3$$

Then substitute the expression for $I_1 = (I_2 + I_3)$ in the second equation, and put it in the standard form:

$$16 - 10(I_2 + I_3) - 10I_2 = 0$$

Clear parentheses: $$16 - 10I_2 - 10I_3 - 10I_2 = 0$$

Transpose $20I_2$ and $10I_3$: $$20I_2 + 10I_3 = 16$$

This equation and the third equation now give you two simultaneous equations for I_2 and I_3:

$$20I_2 + 10I_3 = 16$$
$$10I_2 - 15I_3 = 0$$

Using the substitution method again for these two equations, solve the second equation for I_2 and substitute into the first equation:

$$I_2 = \left(\frac{15}{10}\right)I_3 = 1.5I_3 \rightarrow 20(1.5I_3) + 10I_3 = 16$$

Solve for I_3:

$$30I_3 + 10I_3 = 16 \rightarrow I_3 = \frac{16}{40} = 0.40 \text{ A} = 400 \text{ mA}$$

Substitute back to find I_2:

$$I_2 = 1.5I_3 = 1.5(400 \text{ mA}) = 600 \text{ mA}$$

Find I_1 by substituting back into the current equation solved for I_1:

$$I_1 = I_2 + I_3 = 600 \text{ mA} + 400 \text{ mA} = 1000 \text{ mA} = 1.0 \text{ A}$$

The currents in this circuit can also be found using the methods of series-parallel circuits shown in Section 11.3 since there is only one voltage source. However, these methods cannot be used in the next example because the circuit contains two voltage sources.

EXAMPLE 14.8

Given the circuit in Figure 14-8 with $V_1 = 14$ V, $V_2 = 12$ V, $R_1 = 6\ \Omega$, $R_2 = 5\ \Omega$, and $R_3 = 4\ \Omega$, find the currents I_1, I_2, and I_3.

Solution It is necessary to write three equations containing the three unknown currents to find their values. First assign polarities and show the current flow as done in Example 14.4 and Figure 14-8. Apply Kirchhoff's current law at point **X** to give the current equation in Example 14.4:

$$I_1 + I_2 - I_3 = 0$$

Apply Kirchhoff's voltage law to loop **I**, $[V_1, R_1, R_3]$, to give the equation in Example 14.6:

$$V_1 - I_1R_1 - I_3R_3 = 0$$

Apply Kirchhoff's voltage law to loop **II**, $[V_2, R_2, R_3]$, to give the other equation in Example 14.6:

$$V_2 - I_2R_2 - I_3R_3 = 0$$

Substitute the given values into the equations to give you the linear system of three equations for the currents:

$$I_1 + I_2 - I_3 = 0$$
$$14 - 6I_1 - 4I_3 = 0$$
$$12 - 5I_2 - 4I_3 = 0$$

You can solve this system by using different methods. The method of determinants shown in Section 13.4 is used here. First put the equations into standard form with the variables on one side and the constants on the other side:

$$\begin{aligned} I_1 + I_2 - I_3 &= 0 \\ 6I_1 + \quad\quad 4I_3 &= 14 \\ 5I_2 + 4I_3 &= 12 \end{aligned}$$

Apply Cramer's rule (13.4) and insert zeros for missing terms. The solutions to two significant digits are:

$$I_1 = \frac{\begin{vmatrix} 0 & 1 & -1 \\ 14 & 0 & 4 \\ 12 & 5 & 4 \end{vmatrix}}{\begin{vmatrix} 1 & 1 & -1 \\ 6 & 0 & 4 \\ 0 & 5 & 4 \end{vmatrix}} = \frac{0 + 48 - 70 - 0 - 0 - 56}{0 + 0 - 30 - 0 - 20 - 24}$$

$$= \frac{-78}{-74} \approx 1.1 \text{ A}$$

$$I_2 = \frac{\begin{vmatrix} 1 & 0 & -1 \\ 6 & 14 & 4 \\ 0 & 12 & 4 \end{vmatrix}}{-74} = \frac{56 + 0 - 72 - 0 - 48 - 0}{-74}$$

$$= \frac{-64}{-74} \approx 0.86 \text{ A} = 860 \text{ mA}$$

$$I_3 = \frac{\begin{vmatrix} 1 & 1 & 0 \\ 6 & 0 & 14 \\ 0 & 5 & 12 \end{vmatrix}}{-74} = \frac{0 + 0 + 0 - 0 - 70 - 72}{-74}$$

$$= \frac{-142}{-74} \approx 1.9 \text{ A}$$

The solutions for all three currents are positive. This indicates that the directions assumed in Figure 14-8 are correct. If the solution for a current is negative, this indicates that the direction assumed for the current is actually the reverse. This can happen in a circuit like Figure 14-8 that contains two voltage sources, if the voltages differ by a significant amount. See the following Error Box and problems 9 and 10 in Exercise 14.3.

ERROR BOX

A common error to watch out for is not being consistent when you assign polarities to resistances in a circuit. *As long as the polarities agree with the direction of the current, the Kirchhoff equations will be correct.* If a current actually flows opposite to your assumed direction, then the answer for that current will be negative. See if you can do the practice problem and discover the negative current. Refer to Figure 14-12.

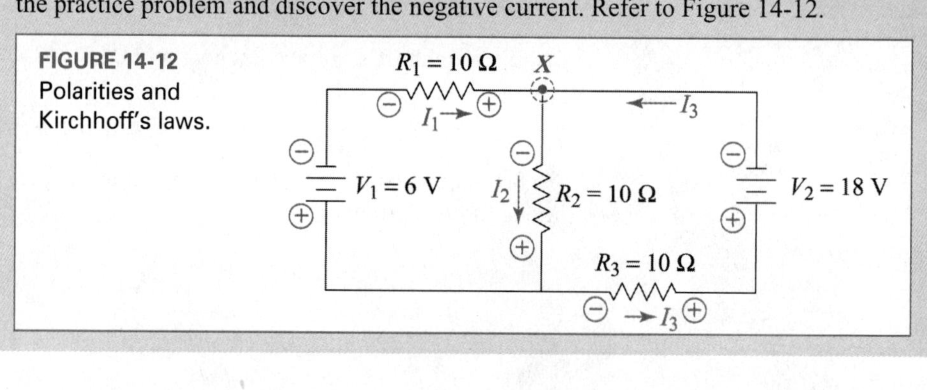

FIGURE 14-12 Polarities and Kirchhoff's laws.

Practice Problem: Given the circuit in Figure 14-12 with the polarities and currents shown, find I_1, I_2, and I_3 by solving the three equations, which result from Kirchhoff's laws:

$$I_1 + I_3 - I_2 = 0$$
$$6 - 10I_1 - 10I_2 = 0$$
$$18 - 10I_2 - 10I_3 = 0$$

EXERCISE 14.3

Solve each problem by applying Kirchhoff's laws and express all answers in engineering notation (as numbers between 1 and 1000) with the appropriate units to two significant digits.

1. Given the circuit in Figure 14-9, Exercise 14.2 with $V_0 = 8$ V, $R_1 = 3\ \Omega$, $R_2 = 10\ \Omega$, and $R_3 = 10\ \Omega$, find the currents I_1, I_2, and I_3.
2. Given the circuit in Figure 14-9, Exercise 14.2 with $V_0 = 15$ V, $R_1 = 18\ \Omega$, $R_2 = 20\ \Omega$, and $R_3 = 30\ \Omega$, find the currents I_1, I_2, and I_3.
3. Given the circuit in Figure 14-9, Exercise 14.2 with $V_0 = 40$ V, $R_1 = 75\ \Omega$, $R_2 = 100\ \Omega$, and $R_3 = 200\ \Omega$, find the currents I_1, I_2, and I_3.
4. Given the circuit in Figure 14-9, Exercise 14.2 with $V_0 = 100$ V, $R_1 = 1.0$ kΩ, $R_2 = 2.0$ kΩ, and $R_3 = 3.0$ kΩ, find the currents I_1, I_2, and I_3.
5. Given the circuit in Figure 14-10, Exercise 14.2, with $V_1 = 10$ V, $V_2 = 15$ V, $R_1 = 20\ \Omega$, $R_2 = 10\ \Omega$, and $R_3 = 10\ \Omega$, find the currents I_1, I_2, and I_3.
6. Given the circuit in Figure 14-10, Exercise 14.2, with $V_1 = 24$ V, $V_2 = 12$ V, $R_1 = 10\ \Omega$, $R_2 = 3\ \Omega$, and $R_3 = 5\ \Omega$, find the currents I_1, I_2, and I_3.
7. Given the circuit in Figure 14-10, Exercise 14.2, with $V_1 = 30$ V, $V_2 = 50$ V, $R_1 = 100\ \Omega$, $R_2 = 100\ \Omega$, and $R_3 = 50\ \Omega$, find the currents I_1, I_2, and I_3.
8. Given the circuit in Figure 14-10, Exercise 14.2, with $V_1 = 100$ V, $V_2 = 80$ V, $R_1 = 2.0$ kΩ, $R_2 = 3.0$ kΩ, and $R_3 = 2.0$ kΩ, find the currents I_1, I_2, and I_3.
9. Given the circuit in Figure 14-13 with $V_1 = 6$ V, $V_2 = 12$ V, $R_1 = 4\ \Omega$, $R_2 = 6\ \Omega$, and $R_3 = 3\ \Omega$, find the currents I_1, I_2, and I_3. Note the polarity of the voltage sources.
10. Given the circuit in Figure 14-13 with $V_1 = 32$ V, $V_2 = 6$ V, $R_1 = 20\ \Omega$, $R_2 = 10\ \Omega$, and $R_3 = 10\ \Omega$, find the currents I_1, I_2, and I_3. Note the polarity of the voltage sources.
11. An unknown voltage V_X is connected in series to an unknown resistance R_X and a resistor of 10 Ω. An ammeter in the circuit reads 200 mA. See Figure 14-14(a). When the 10-Ω resistor is replaced by a 30-Ω resistor, the ammeter reads 100 mA. See Figure 14-14(b). Use Kirchhoff's voltage law to find V_X and R_X.
12. A battery is connected to a resistance of 300 Ω. When a 20-V battery and a resistance of 1.0 kΩ are added in series to the circuit so that the voltages are series aiding, the current remains the same. See Figure 14-15. Use Kirchhoff's voltage law to find V_X and I_X.
13. Given the circuit in Figure 14-11, Exercise 14.2, with $V_1 = 12$ V, $V_2 = 18$ V, $R_1 = 10\ \Omega$, $R_2 = 20\ \Omega$,

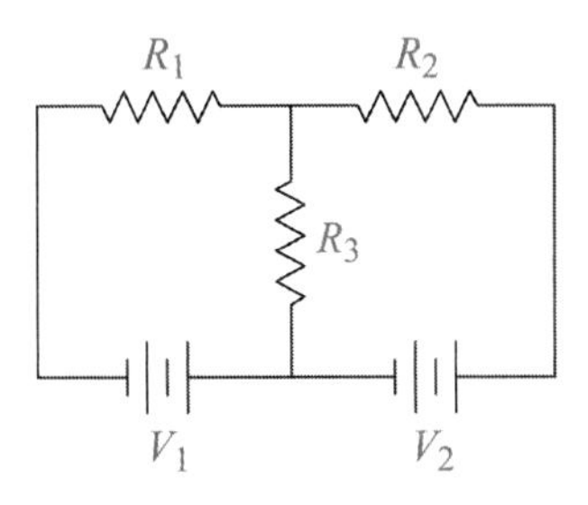

FIGURE 14-13 Kirchhoff's laws for problems 9 and 10.

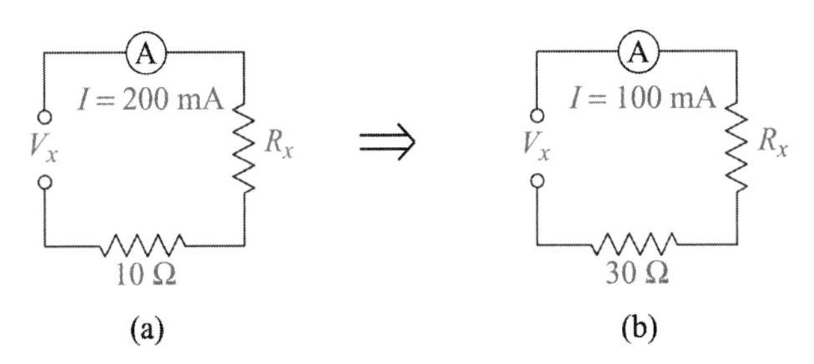

FIGURE 14-14 Unknown voltage and resistance for problem 11.

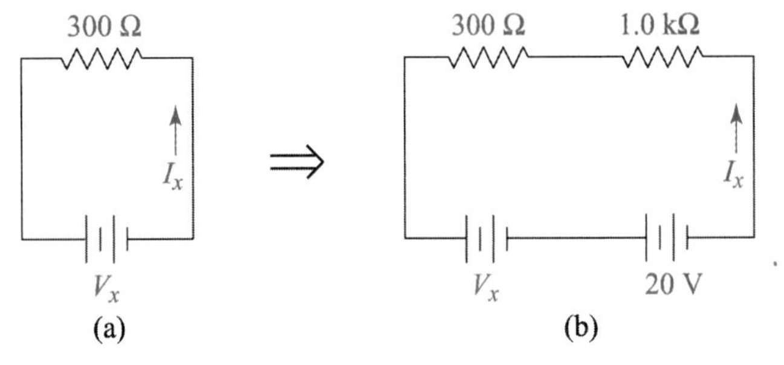

FIGURE 14-15 Unknown voltage and current for problem 12.

$R_3 = 20\ \Omega$, and $R_4 = 10\ \Omega$, find the current through each resistor.

Hint: Change R_2 and R_3 to one equivalent resistance.

14. Given the circuit in Figure 14-11, Exercise 14.2, with $V_1 = 9.0$ V, $V_2 = 12$ V, $R_1 = 1.0$ kΩ, $R_2 = 2.0$ kΩ, $R_3 = 2.0$ kΩ, and $R_4 = 1.0$ kΩ, find the current through each resistor.

Hint: Change R_2 and R_3 to one equivalent resistance.

15. Given the circuit in the Error Box, Figure 14-12, with $V_1 = 30$ V, $V_2 = 10$ V, $R_1 = 1.0$ kΩ, $R_2 = 2.0$ kΩ, and $R_3 = 1.0$ kΩ. Find the current through each resistor using Kirchhoff's laws.

16. Given the circuit in the Error Box, Figure 14-12, with $V_1 = 24$ V, $V_2 = 6.0$ V, $R_1 = 100\ \Omega$, $R_2 = 100\ \Omega$, and $R_3 = 200\ \Omega$. Find the current through each resistor using Kirchhoff's laws.

CHAPTER HIGHLIGHTS

14.1 VOLTAGE DIVISION

For a resistance R in a series circuit or a series branch where V_X = voltage across R_X and V_T = total voltage or branch voltage:

Voltage Division Formula

$$V_X = \frac{R_X}{R_T} V_T \tag{14.1}$$

Key Example: In Figure 14-16, $R_1 = 10\ \Omega$, $R_2 = 20\ \Omega$, $R_3 = 5\ \Omega$, and $R_4 = 20\ \Omega$:

If $V_T = 10$ V:

$$V_1 = 10 \text{ V}$$

$$V_2 = \frac{5}{5 + 20}(10 \text{ V}) = 2 \text{ V}$$

$$V_3 = \frac{20}{5 + 20}(10 \text{ V}) = 8 \text{ V}$$

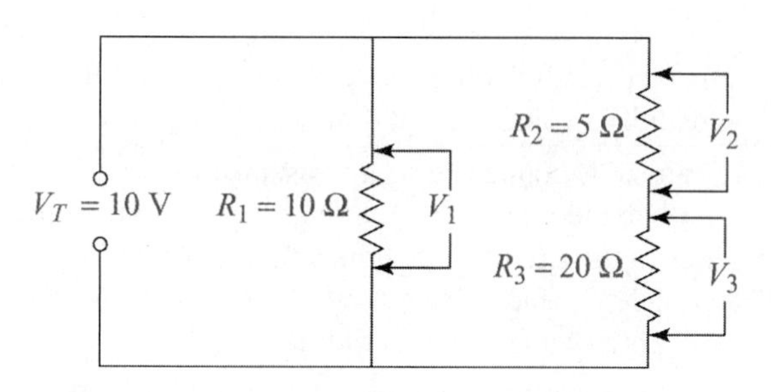

FIGURE 14-16 Key Example: Voltage division.

14.2 KIRCHHOFF'S LAWS

Kirchhoff's Current Law

The algebraic sum of all currents entering and leaving any branch point in a circuit equals zero. (14.2)

At a branch point assign a positive value to a current entering and a negative value to a current leaving.

Kirchhoff's Voltage Law

The algebraic sum of the voltages around any closed path equals zero. (14.3)

As you move around a closed path, assign a *positive sign* to a voltage when *positive* polarity is encountered first, and a *negative sign* to a voltage when *negative* polarity is encountered first.

Key Example: In Figure 14-17 at branch point **X**, Kirchhoff's current law gives:

$$I_1 + I_2 - I_3 = 0$$

For loops **I** and **II**, Kirchhoff's voltage law gives:

$$V_1 - I_1R_1 - I_3R_3 = 0 \quad \text{or} \quad I_1R_1 + I_3R_3 = V_1$$
$$V_2 - I_2R_2 - I_3R_3 = 0 \quad \text{or} \quad I_2R_2 + I_3R_3 = V_2$$

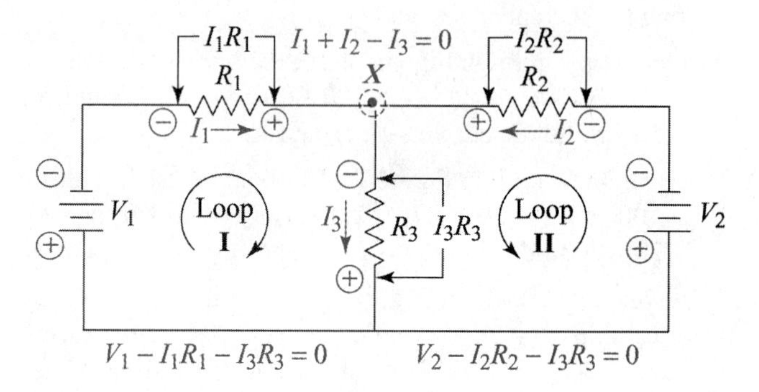

FIGURE 14-17 Key Example: Kirchhoff's laws.

14.3 SOLVING CIRCUITS

Key Example: Given the circuit in Figure 14-17 with $V_1 = 14$ V, $V_2 = 12$ V, $R_1 = 6\ \Omega$, $R_2 = 5\ \Omega$, and $R_3 = 4\ \Omega$, find the

Answers to Error Box Problem, page 297:
$I_1 = -200$ mA, $I_2 = 800$ mA, $I_3 = 1.0$ A

currents by substituting in the above equations and solving the system. The equations in standard form and the solution using determinants are:

$$\begin{aligned} I_1 + I_2 - I_3 &= 0 \\ 6I_1 + 4I_3 &= 14 \\ 5I_2 + 4I_3 &= 12 \end{aligned}$$

$$I_1 = \frac{\begin{vmatrix} 0 & 1 & -1 \\ 14 & 0 & 4 \\ 12 & 5 & 4 \end{vmatrix}}{\begin{vmatrix} 1 & 1 & -1 \\ 6 & 0 & 4 \\ 0 & 5 & 4 \end{vmatrix}} = \frac{-78}{-74} \approx 1.1 \text{ A}$$

$$I_2 = \frac{\begin{vmatrix} 1 & 0 & -1 \\ 6 & 14 & 4 \\ 0 & 12 & 4 \end{vmatrix}}{-74} \approx 0.86 \text{ A} = 860 \text{ mA}$$

$$I_3 = \frac{\begin{vmatrix} 1 & 1 & 0 \\ 6 & 0 & 14 \\ 0 & 5 & 12 \end{vmatrix}}{-74} \approx 1.9 \text{ A}$$

REVIEW EXERCISES

Express all answers in engineering notation (as numbers between 1 and 1000) with the appropriate units to two significant digits unless specified.

1. Given the series-parallel circuit in Figure 14-18 with V_T = 26 V, find V_1, V_2, V_3, and V_4.
2. Given the series-parallel circuit in Figure 14-19 with V_T = 100 V, find V_1, V_2, V_3, and V_4.
3. For the series-parallel circuit in Figure 14-18, given V_T = 13 V, find R_T, I_T, V_1, V_2, V_3, V_4, I_1, I_2, I_3, and I_4.
4. For the series-parallel circuit in Figure 14-19, given V_T = 60 V, find R_T, I_T, V_1, V_2, V_3, V_4, I_1, I_2, I_3, and I_4.
5. For the circuit in Figure 14-20, write Kirchhoff's current equations for branch points **X** and **Y** and show that they are equivalent.

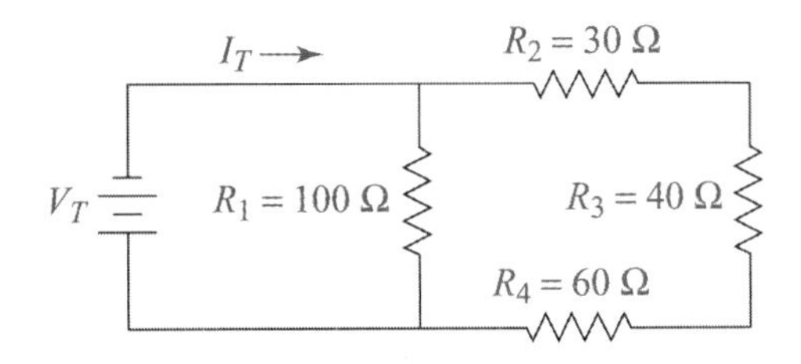

FIGURE 14-18 Series-parallel circuit for problems 1 and 3.

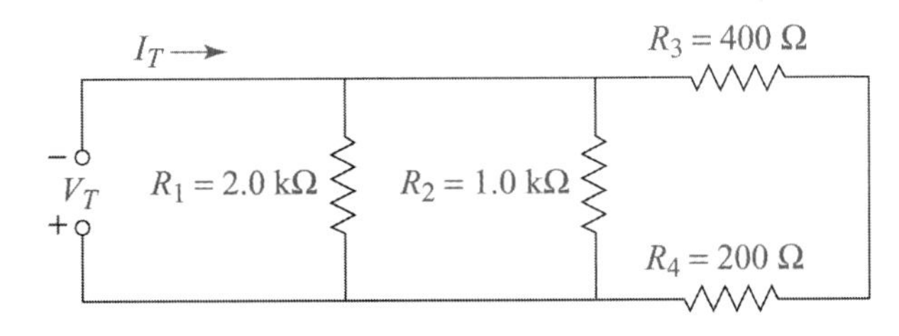

FIGURE 14-19 Series-parallel circuit for problems 2 and 4.

6. For the circuit in Figure 14-20, write Kirchhoff's voltage equations for the three loops: [V_1, R_1, R_2], [V_2, R_3, R_2], and [V_1, R_1, R_3, V_2].
7. Given the circuit in Figure 14-20, with V_1 = 6 V, V_2 = 6 V, $R_1 = 2\ \Omega$, $R_2 = 4\ \Omega$, and $R_3 = 2\ \Omega$, find the three currents I_1, I_2, and I_3, using Kirchhoff's laws.
8. Given the circuit in Figure 14-20, with V_1 = 12 V, V_2 = 9 V, $R_1 = 40\ \Omega$, $R_2 = 60\ \Omega$, and $R_3 = 40\ \Omega$, find the three currents I_1, I_2, and I_3, to the nearest milliamp using Kirchhoff's laws.
9. Given the circuit in Figure 14-21, find I_1, I_2, I_3, V_1, V_2, and V_3, using Kirchhoff's laws.

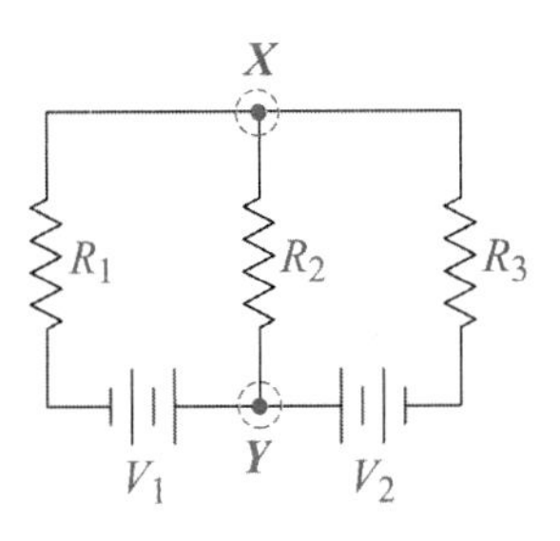

FIGURE 14-20 Kirchhoff's laws for problems 5 through 8.

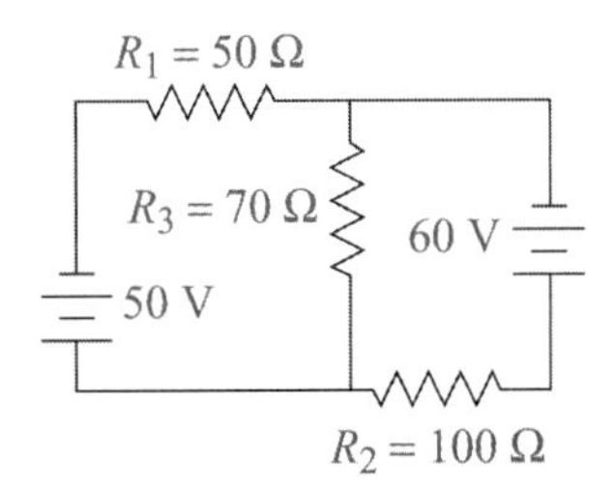

FIGURE 14-21 Kirchhoff's laws for problem 9.

10. Given the circuit in Figure 14-22, find I_1, I_2, I_3, V_1, V_2, and V_3, using Kirchhoff's laws.

11. Given the circuit in Figure 14-23, with $V_1 = 12$ V, $V_2 = 6$ V, $R_1 = 6\ \Omega$, $R_2 = 6\ \Omega$, $R_3 = 4\ \Omega$, and $R_4 = 4\ \Omega$, find the current through each resistor using Kirchhoff's laws.

    Hint: Change R_3 and R_4 to one equivalent resistance.

12. Given the circuit in Figure 14-23, with $V_1 = 120$ V, $V_2 = 120$ V, $R_1 = 100\ \Omega$, $R_2 = 100\ \Omega$, $R_3 = 200\ \Omega$, and $R_4 = 200\ \Omega$. Find the current through each resistor using Kirchhoff's laws.

    Hint: Change R_3 and R_4 to one equivalent resistance.

13. Given the circuit in Figure 14-13, Exercise 14.3, with $V_1 = 15$ V, $V_2 = 10$ V, $R_1 = 100\ \Omega$, $R_2 = 200\ \Omega$, and $R_3 = 200\ \Omega$. Find the current through each resistor using Kirchhoff's laws.

14. Given the circuit in Figure 14-13, Exercise 14.3, with $V_1 = 100$ V, $V_2 = 200$ V, $R_1 = 1.0$ kΩ, $R_2 = 2.0$ kΩ, and $R = 3.0$ kΩ. Find the current through each resistor using Kirchhoff's laws.

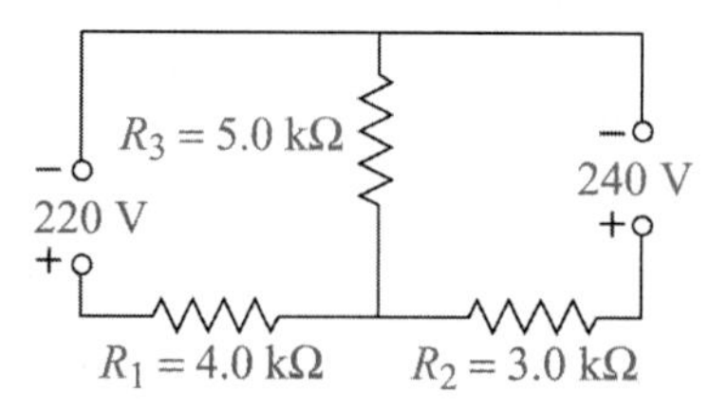

FIGURE 14-22 Kirchhoff's laws for problem 10.

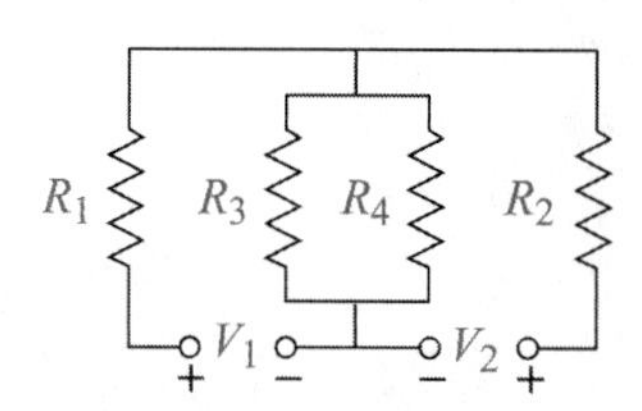

FIGURE 14-23 Kirchhoff's laws for problems 11 and 12.

CHAPTER 15

Network Theorems

CHAPTER OBJECTIVES

In this chapter, you will learn:

- How to apply the Superposition theorem to find currents and voltages in a circuit.
- The meaning of a Thevenin equivalent circuit.
- How to obtain the Thevenin voltage and Thevenin resistance for a load in a circuit.
- The meaning of a Norton equivalent circuit.
- How to obtain the Norton current and Norton resistance for a load in a circuit.

This chapter continues with the analysis of circuits studied in Chapter 14. Three important network theorems are studied: the Superposition theorem, Thevenin's theorem, and Norton's theorem. Each theorem analyzes a circuit in a similar way. They replace the original circuit with one or more simpler equivalent circuits that can be solved by applying basic concepts from previous chapters. The results of the simpler circuits are then combined to find currents and voltages in the original circuit.

• • •

15.1 Superposition Theorem

The Superposition theorem helps you analyze a circuit having *two voltage sources* by considering the effect of only one voltage source at a time. The procedure is:

1. Redraw the circuit as two separate circuits. Each circuit contains one voltage source, with the other voltage source shorted out.
2. Combine algebraically, or superimpose, the two results to produce the currents or voltages in the original circuit.

EXAMPLE 15.1

Given the series circuit in Figure 15-1(a) containing two voltage sources $V_1 = 12$ V and $V_2 = 18$ V. Using the Superposition theorem, find the total current I_T in the circuit and the voltage drop across R_1.

Solution To apply the Superposition theorem, redraw the circuit as two separate circuits. In each circuit, short out one of the voltage sources, leaving only the other voltage source. See circuit 1 and circuit 2 in Figure 15-1(b) on page 302.

Circuit 1: Apply Ohm's law to find the current I_{T_1} and the voltage drop across R_1:

$$I_{T_1} = \frac{V_1}{R_T} = \frac{12\text{ V}}{10\ \Omega + 5\ \Omega + 15\ \Omega} = \frac{12\text{ V}}{30\ \Omega} = 0.40\text{ A} = 400\text{ mA}$$

$$I_{T_1}R_1 = (0.40\text{ A})(15\ \Omega) = 6.0\text{ V}$$

Circuit 2: Apply Ohm's law to find the current I_{T_2} and the voltage drop across R_1:

$$I_{T_2} = \frac{V_2}{R_T} = \frac{18\text{ V}}{30\ \Omega} = 0.60\text{ A} = 600\text{ mA}$$

$$I_{T_2}R_1 = (0.6\text{ A})(15\ \Omega) = 9.0\text{ V}$$

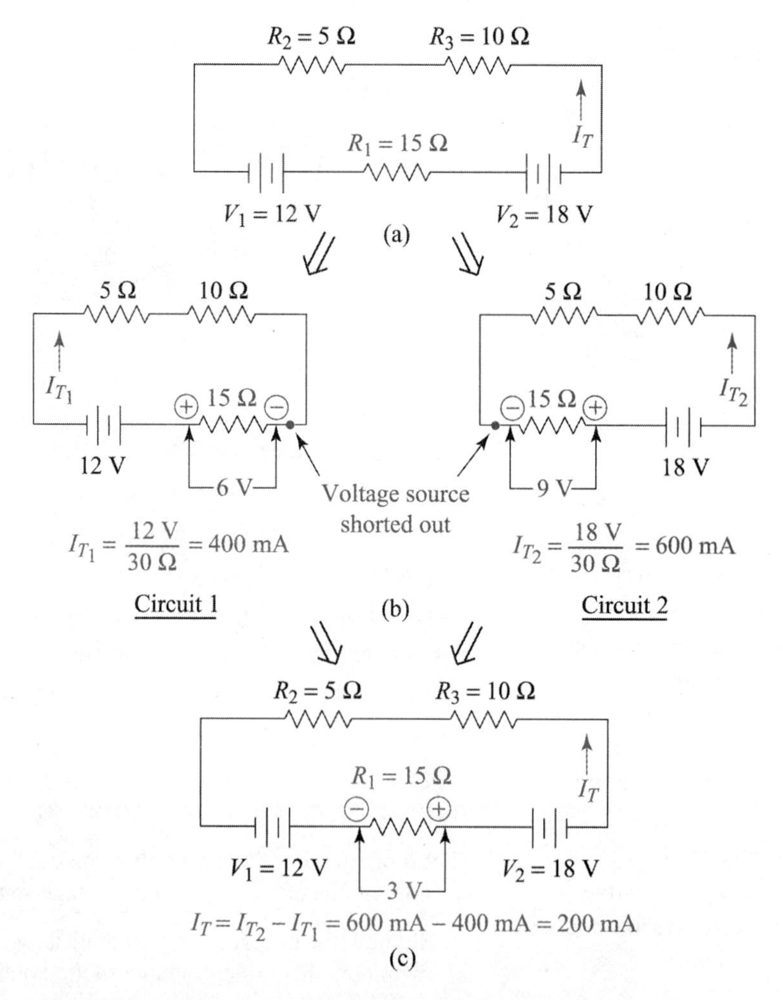

FIGURE 15-1 Superposition theorem for Example 15.1.

Now observe that the direction of I_{T_2} in circuit 2 is opposite to the direction of I_{T_1} in circuit 1. Combine the results algebraically as shown in Figure 15-1(c). Since I_{T_2} flows opposite to I_{T_1}, the total current I_T in the original circuit is the difference of the two currents:

$$I_T = I_{T_2} - I_{T_1} = 600\text{ mA} - 400\text{ mA} = 200\text{ mA}$$

Also, since I_{T_2} is greater than I_{T_1}, the direction of I_T is the same as the direction of I_{T_2}.

Similarly, the voltage drop across R_1 in the original circuit is the difference of the two voltage drops:

$$V_{R_1} = 9.0 - 6.0 = 3.0\text{ V}$$

The polarities across R_1 agree with those in circuit 2 since the voltage drop is greater in circuit 2.

To verify the results, the original circuit can be solved using Ohm's law and the concept of series-opposing voltages. The total voltage, V_T, is the difference of the voltages:

$$V_T = V_2 - V_1 = 18\text{ V} - 12\text{ V} = 6.0\text{ V}$$

The total current and the voltage drop across R_1 are then:

$$I_T = \frac{6.0\text{ V}}{30\ \Omega} = 0.20\text{ V} = 200\text{ mA}$$

$$I_T R_1 = (0.20\text{ A})(15\ \Omega) = 3.0\text{ V}$$

EXAMPLE 15.2 Given the circuit in Figure 15-2(a) with two voltage sources and three resistors. Find the currents I_1, I_2, and I_3 by applying the Superposition theorem.

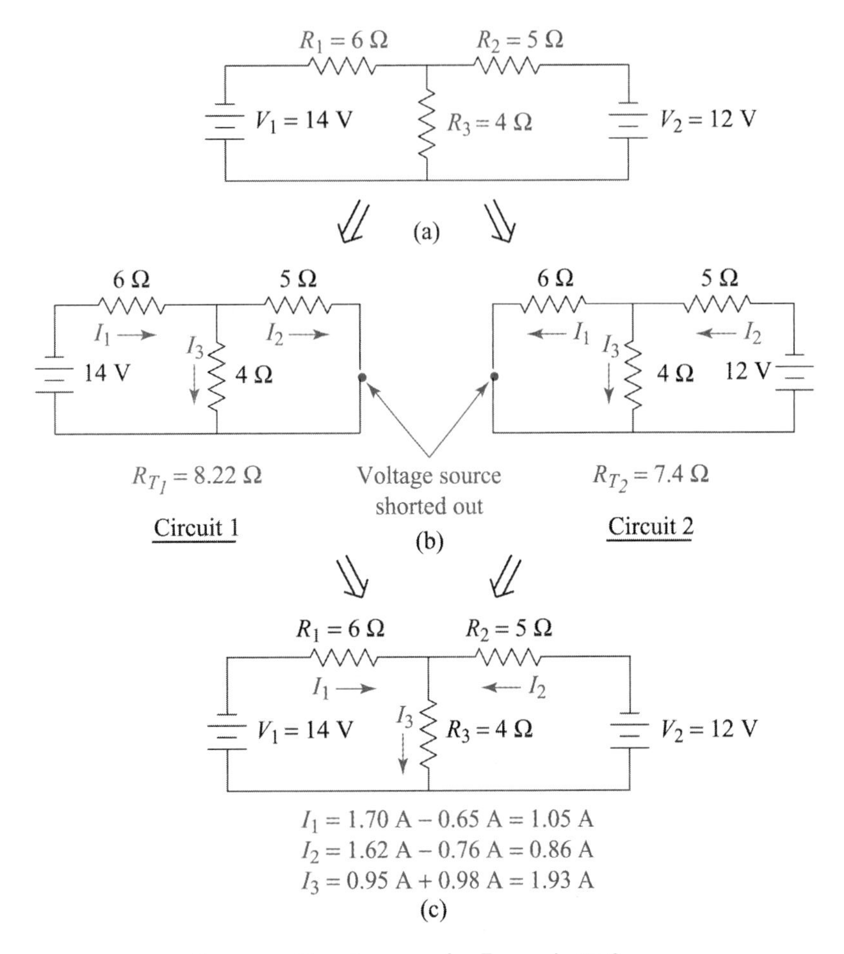

FIGURE 15-2 Superposition theorem for Example 15.2.

Solution This is the same circuit as Example 14.8 where the currents are found using Kirchhoff's laws and a system of three linear equations. The solution here uses the Superposition theorem, and the results will agree with those of Example 14.8. First redraw the circuit as two circuits, each with one voltage source. In each circuit, short out the other voltage source. See circuit 1 and circuit 2 in Figure 15-2(b). Show the current flow for the three currents in each circuit. Note that I_3 flows in the same direction in both circuits, but I_1 and I_2 flow in opposite directions. The calculations for each circuit are similar as follows:

Circuit 1: First find the total resistance of the circuit. Resistances 5 Ω and 4 Ω are in parallel, and this parallel branch is in series with 6 Ω. Calculate the total resistance R_{T_1} using the reciprocal formula on the calculator:

5 [x^{-1}] [+] 4 [x^{-1}] [=] [x^{-1}] [+] 6 [=] → 8.22 Ω

or the product sum formula:

$$R_{T_1} = \frac{(5)(4)}{5+4} + 6 = \frac{20}{9} + 6 \approx 8.22\ \Omega$$

Using Ohm's law, the total current I_1 through the battery is:

$$I_1 = \frac{V_1}{R_{T_1}} = \frac{14\ \text{V}}{8.22\ \Omega} \approx 1.70\ \text{A}$$

The voltage drop across R_1 is:

$$V_1 = I_1R_1 = (1.70\ \text{A})(6\ \Omega) = 10.2\ \text{V}$$

The voltage drops across R_2 and R_3 are then each 14 V – 10.2 V = 3.8 V, and the currents I_2 and I_3 are:

$$I_2 = \frac{3.8\ V}{5\ \Omega} = 0.76\ \text{A}$$

$$I_3 = \frac{3.8\ V}{4\ \Omega} = 0.95\ \text{A}$$

Circuit 2: Resistances 6 Ω and 4 Ω are in parallel, and this parallel branch is in series with 5 Ω. The total resistance R_{T_2} is then:

6 [x⁻¹] [+] 4 [x⁻¹] [=] [x⁻¹] [+] 5 [=] → 7.4 Ω

or

$$R_{T_2} = \frac{(6)(4)}{6+4} + 5 = 2.4\ \Omega + 5\ \Omega = 7.4\ \Omega$$

The total current through the battery, I_2, is:

$$I_2 = \frac{12\ \text{V}}{7.4\ \Omega} \approx 1.62\ \text{A}$$

The voltage drop across R_2 is:

$$V_2 = I_2R_2 = (1.62\ \text{A})(5\ \Omega) = 8.1\ \text{V}$$

The voltage drops across R_1 and R_3 are then each 12 V – 8.1 V = 3.9 V, and the currents I_1 and I_3 are:

$$I_1 = \frac{3.9\ V}{6\ \Omega} = 0.65\ \text{A}$$

$$I_3 = \frac{3.9\ V}{4\ \Omega} \approx 0.98\ \text{A}$$

Now combine the results of the two circuits algebraically as shown in Figure 15-2(c). Since the direction of I_1 in circuit 1 is *opposite* to its direction in circuit 2, its value in the original circuit is the *difference* of the values in the two circuits:

$$I_1 = 1.70\ \text{A} - 0.65\ \text{A} = 1.05\ \text{A}$$

Since I_1 in circuit 1 is *greater* than I_1 in circuit 2, its direction in the original circuit is the *same* as that in circuit 1. Similarly, I_2 is the difference of the values in the two circuits:

$$I_2 = 1.62\ \text{A} - 0.76\ \text{A} = 0.86\ \text{A}$$

Since I_2 in circuit 2 is *greater* than I_2 in circuit 1, the direction of I_2 in the original circuit is the same as that in circuit 2. Since I_3 flows in the same direction in both circuits, its value is the sum of the currents in the two circuits:

$$I_3 = 0.95\ \text{A} + 0.98\ \text{A} = 1.93\ \text{A}$$

The direction of I_3 in the original circuit is the same as in circuit 1 or circuit 2.

The Superposition theorem can be used for networks that contain components that are linear and bilateral, such as resistors, capacitors, and certain inductors. *Linear* means that the current is proportional to the voltage for each component. *Bilateral* means that the currents are the same when the polarities are reversed.

EXERCISE 15.1

In problems 1 through 10, use the Superposition theorem to find the currents and voltages. Express all answers in engineering notation (as numbers between 1 and 1000) with the appropriate units to two significant digits.

1. Given the series circuit in Figure 15-3, find the total current and the voltage across each resistor.
2. Given the parallel circuit in Figure 15-4, where the two voltage sources are connected to a negative ground, find the currents I_1, I_2, and the voltage drop across each resistor.

   Note: Voltages are series opposing.
3. In the series-parallel circuit in Figure 15-5, $V_1 = 20$ V, $V_2 = 24$ V, $R_1 = 200\ \Omega$, and $R_2 = 300\ \Omega$. Find I_1, I_2, and the voltage drop across each resistor.

   Note: A short across a resistance results in a resistance of zero.
4. In the series-parallel circuit in Figure 15-5, $V_1 = 20$ V, $V_2 = 24$ V, $R_1 = 100\ \Omega$, and $R_2 = 50\ \Omega$. Find I_1, I_2, and the voltage drop across each resistor.

   Note: A short across a resistance results in a resistance of zero.
5. In the series-parallel circuit in Figure 15-6, $V_1 = 12$ V, $V_2 = 18$ V, $R_1 = 10\ \Omega$, $R_2 = 20\ \Omega$, and $R_3 = 5\ \Omega$. Find I_1, I_2, I_3, and the voltage drop across each resistor.
6. In the series-parallel circuit in Figure 15-6, $V_1 = 15$ V, $V_2 = 30$ V, $R_1 = 10\ \Omega$, $R_2 = 20\ \Omega$, and $R_3 = 10\ \Omega$. Find I_1, I_2, I_3, and the voltage drop across each resistor.
7. In the series-parallel circuit in Figure 15-6, $V_1 = 9$ V, $V_2 = 3$ V, $R_1 = 10\ \Omega$, $R_2 = 10\ \Omega$, and $R_3 = 10\ \Omega$. Find I_1, I_2, I_3, and the voltage drop across each resistor.
8. In the series-parallel circuit in Figure 15-6, $V_1 = 20$ V, $V_2 = 12$ V, $R_1 = 30\ \Omega$, $R_2 = 30\ \Omega$, and $R_3 = 10\ \Omega$. Find I_1, I_2, I_3, and the voltage drop across each resistor.
9. In the series-parallel circuit in Figure 15-7, $V_1 = 10$ V, $V_2 = 12$ V, $R_1 = 2.0$ kΩ, $R_2 = 3.0$ kΩ, and $R_3 = 2.0$ kΩ. Find I_1, I_2, I_3, and the voltage drop across each resistor. Note the polarity of the voltages.
10. In the series-parallel circuit in Figure 15-7, $V_1 = 9$ V, $V_2 = 6$ V, $R_1 = 15\ \Omega$, $R_2 = 15\ \Omega$, and $R_3 = 10\ \Omega$. Find I_1, I_2, I_3, and the voltage drop across each resistor. Note the polarity of the voltages.

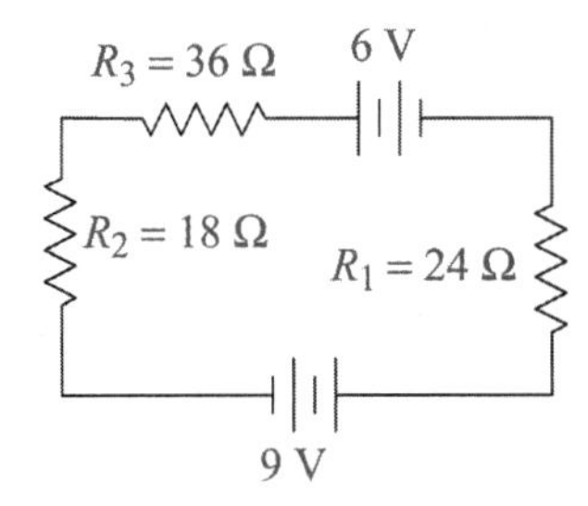

FIGURE 15-3 Series circuit with two voltage sources for problem 1.

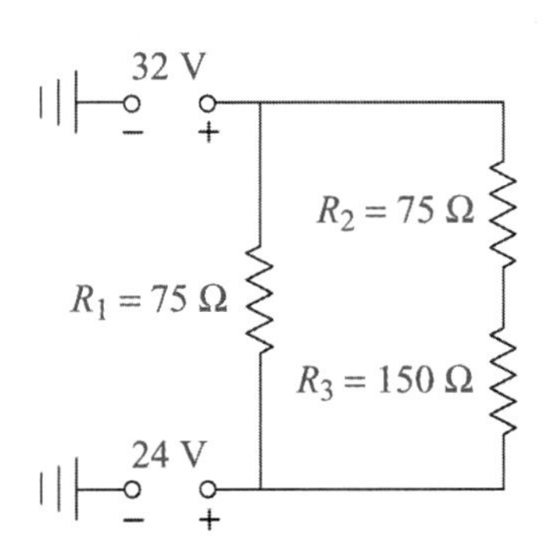

FIGURE 15-4 Parallel circuit with two voltage sources for problem 2.

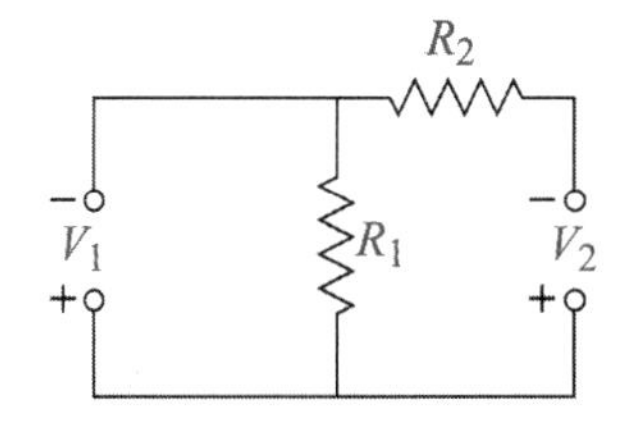

FIGURE 15-5 Series-parallel circuit for problems 3 and 4.

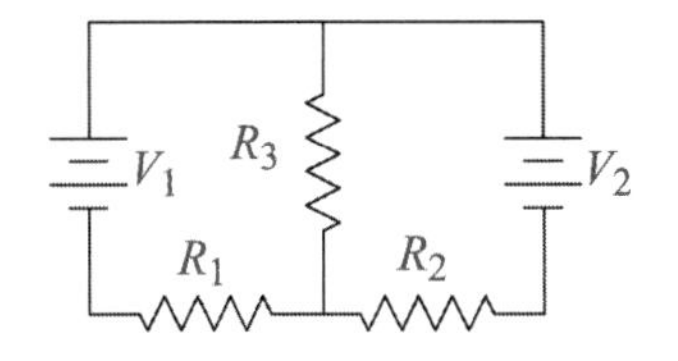

FIGURE 15-6 Series-parallel circuit for problems 5, 6, 7, and 8.

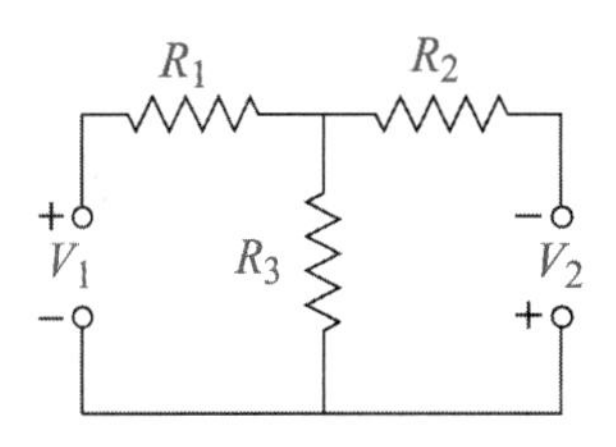

FIGURE 15-7 Series-parallel circuit for problems 9 and 10.

15.2 Thevenin's Theorem

M.L. Thevenin, a French engineer, discovered one of the most useful theorems in network analysis. It is a simple but powerful idea that allows you to reduce a complex circuit to the simplest series circuit: a voltage source and a resistance. Consider a load resistance R_L in some complex circuit represented by a black box as shown in Figure 15-8(a). No matter how complex the circuit, what happens in R_L is determined by two factors: the voltage experienced by the load and the equivalent resistance of the rest of the circuit. The voltage experienced by the load is called the Thevenin voltage V_{TH}. The equivalent resistance is called the Thevenin resistance R_{TH}. Thevenin's theorem says that the entire circuit, without the load, can be represented by a series circuit having a source voltage V_{TH} and a series resistance R_{TH}.

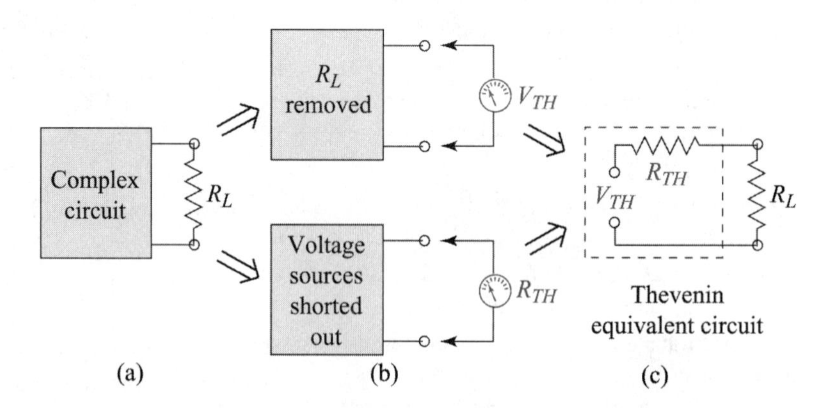

FIGURE 15-8 Thevenizing a circuit.

Given a circuit with a load resistance R_L, the steps to *Thevenizing a circuit* are:

1. Find the Thevenin voltage V_{TH}: Remove R_L and calculate the open circuit voltage across the terminals of R_L. See Figure 15-8(b).
2. Find the Thevenin resistance R_{TH}: Short out the voltage sources and calculate the equivalent resistance across the open terminals of R_L. See Figure 15-8(b).
3. Draw the Thevenin equivalent circuit and find I_L (the current through R_L) and V_L (the voltage across R_L). See Figure 15-8(c).

In a laboratory, with an actual circuit, you can measure V_{TH} and R_{TH} directly with a multimeter.

EXAMPLE 15.3 Thevenize the circuit of Figure 15-9(a). Find the Thevenin voltage and the Thevenin resistance for the load R_L. Then draw the Thevenin equivalent circuit and find I_L and V_L.

Solution First remove R_L as shown in Figure 15-9(b). Now observe that since V_{TH} is an open circuit voltage, it is the same as V_3, the voltage across R_3, because there is no current through R_2. Calculate V_{TH} as V_3, using the voltage division formula (14.1):

$$V_{TH} = V_3 = \frac{R_3}{R_1 + R_3}(V_T) = \frac{6\ \Omega}{4\ \Omega + 6\ \Omega}(12\ \text{V}) = 7.2\ \text{V}$$

To find R_{TH}, short out the voltage source and compute the total resistance of the series-parallel circuit shown in Figure 15-9(b), using the reciprocal formula on the calculator:

4 [x^{-1}] [+] 6 [x^{-1}] [=] [x^{-1}] [+] 4 [=] $\rightarrow 6.4\ \Omega$

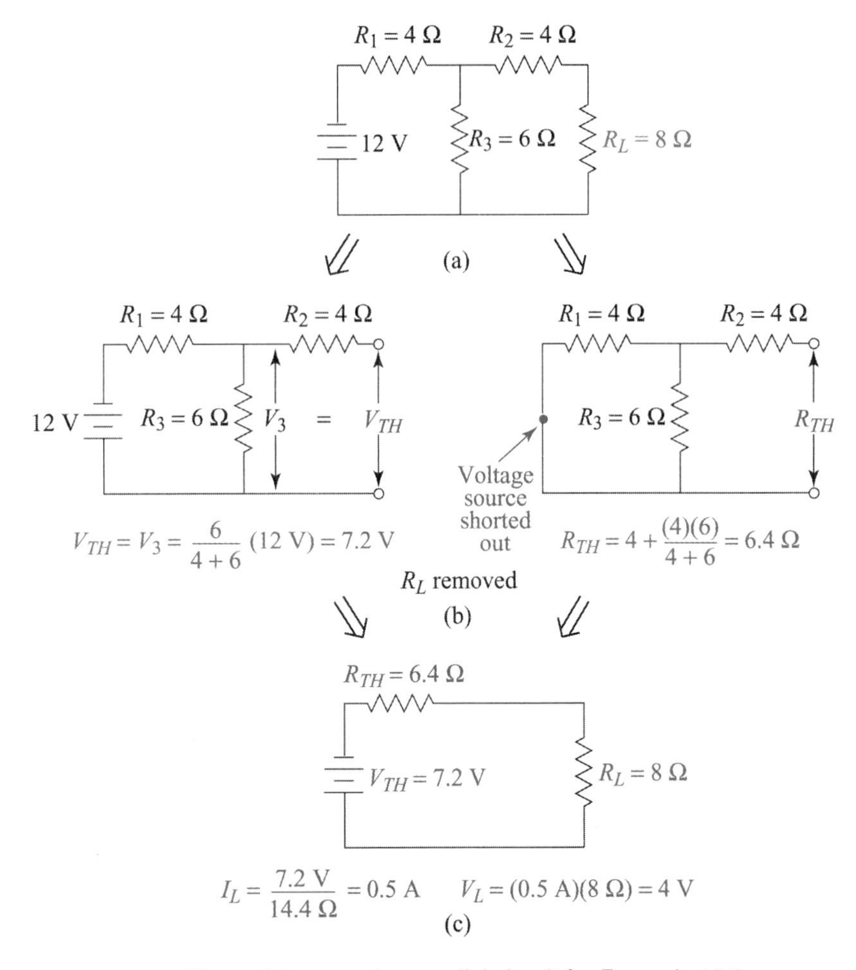

FIGURE 15-9 Thevenizing a series-parallel circuit for Example 15.3.

or the product sum formula:

$$R_{TH} = 4 + \frac{(4)(6)}{4 + 6} = 4\ \Omega + 2.4\ \Omega = 6.4\ \Omega$$

Now draw the Thevenin equivalent circuit as shown in Figure 15-9(c). Find I_L and V_L by applying Ohm's law to this series circuit:

$$I_L = \frac{7.2\ V}{(8 + 6.4)\ \Omega} = 0.5\ A = 500\ mA$$

$$V_L = (0.5\ A)(8\ \Omega) = 4\ V$$

ERROR BOX

A common error to watch out for when applying network theorems is not analyzing an open circuit voltage and a short circuit current correctly.

Consider first the open circuit voltage. When there is no current flowing in a resistor, there is *no* voltage drop across that resistor. The voltage at *each* end of the resistor is the *same* with respect to any other circuit point.

In the case of a short circuit current, if the two ends of a resistor are connected by a short wire, there will be *no* current through the resistor. The current will flow through the short circuit. The resistor can be removed from the circuit, and it will not change the circuit.

See if you can correctly find the voltage or current in each of the practice problems.

Practice Problems: In Figure 15-10, find the open circuit voltage V in circuits (a), (b), and (c). Find the short circuit current I in circuits (d), (e), and (f).

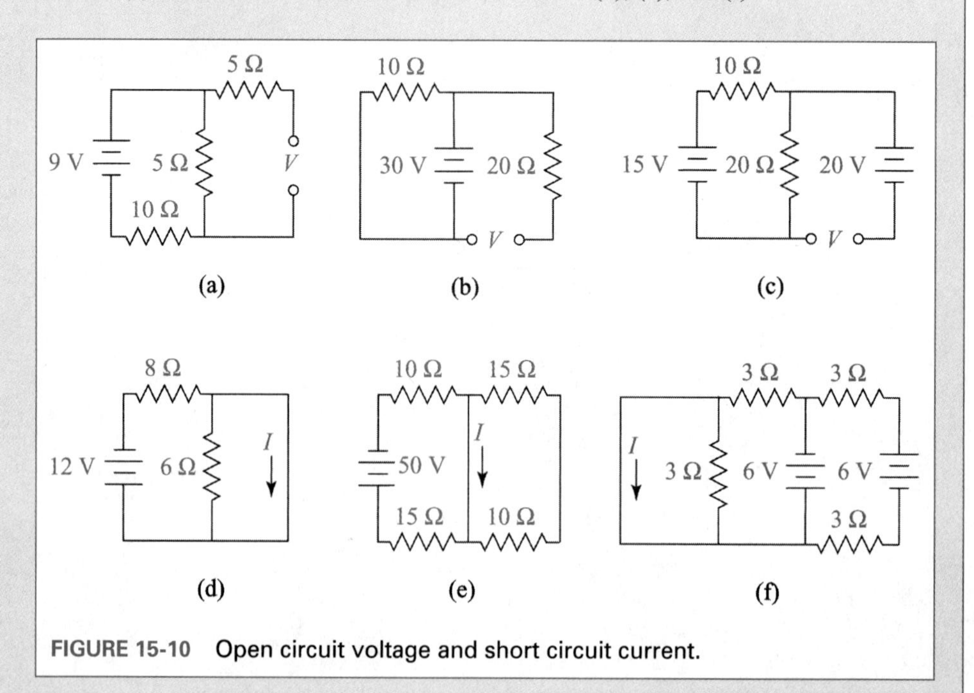

FIGURE 15-10 Open circuit voltage and short circuit current.

The next example shows how to Thevenize a circuit with two voltage sources.

EXAMPLE 15.4 Thevenize the circuit in Figure 15-11(a) and find I_L and V_L.

Solution To find V_{TH}, remove R_L as shown in Figure 15-11(b). The circuit is now a series circuit with two series-opposing voltages. The total voltage across R_1 and R_2 is then:

$$V_T = 30\text{ V} - 24\text{ V} = 6\text{ V}$$

The open circuit voltage V_{TH} is equal to the voltage across R_2 combined with the voltage of the source V_2, as shown in Figure 15-11(c). Observe that the polarity across R_2 is such that its voltage *adds* to V_2 to produce V_{TH}. Using the voltage division formula, V_{TH} is:

$$V_{TH} = \frac{50}{100 + 50}(6\text{ V}) + 24\text{ V} = 2\text{ V} + 24\text{ V} = 26\text{ V}$$

V_{TH} can also be found by *subtracting* the voltage across R_1 from 30 V.

To find R_{TH}, remove R_L and short out both voltages as shown in Figure 15-11(b). Redraw the circuit to show that R_{TH} is equivalent to R_1 and R_2 in parallel as shown in Figure 15-11(c).

Using the reciprocal formula on the calculator, R_{TH} is:

50 [x^{-1}] [+] 100 [x^{-1}] [=] [x^{-1}] [=] → 33 Ω

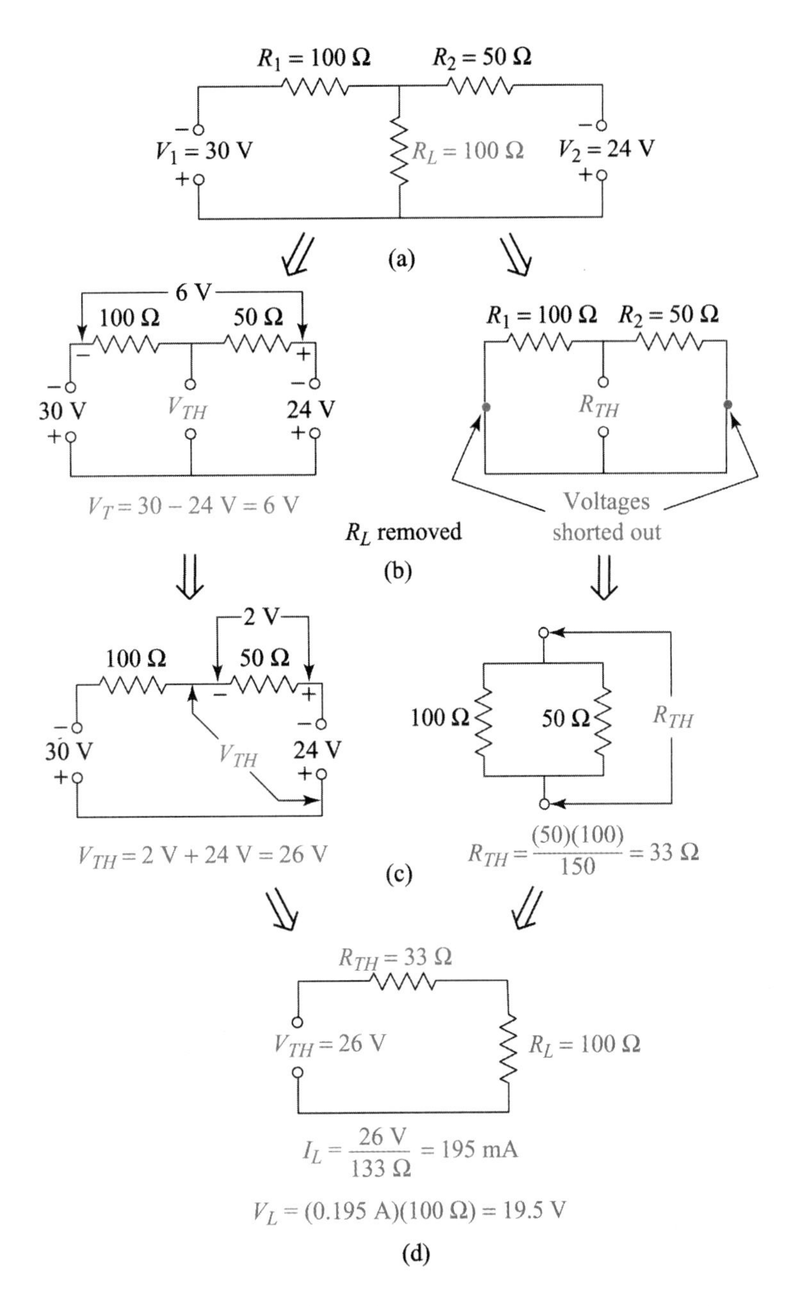

FIGURE 15-11 Thevenizing a circuit with two voltage sources for Example 15.4.

or the product sum formula:

$$R_{TH} = \frac{(50)(100)}{50 + 100} \approx 33\ \Omega$$

The Thevenin equivalent circuit is shown in Figure 15-11(d). I_L and V_L are then as follows:

$$I_L = \frac{26\text{ V}}{100\ \Omega + 33\ \Omega} \approx 0.195\text{ A} = 195\text{ mA}$$

$$V_L = (0.195\text{ A})(100\ \Omega) = 19.5\text{ V}$$

EXERCISE 15.2

Express all answers in engineering notation (as numbers between 1 and 1000) with the appropriate units to two significant digits.

1. Given the series-parallel circuit in Figure 15-12 with $V_T = 9$ V, $R_1 = 5\ \Omega$, $R_2 = 3\ \Omega$, $R_3 = 6\ \Omega$, and $R_L = 9\ \Omega$. Thevenize this circuit for R_L by first finding V_{TH} and R_{TH}, and then finding I_L and V_L. See Example 15.3.

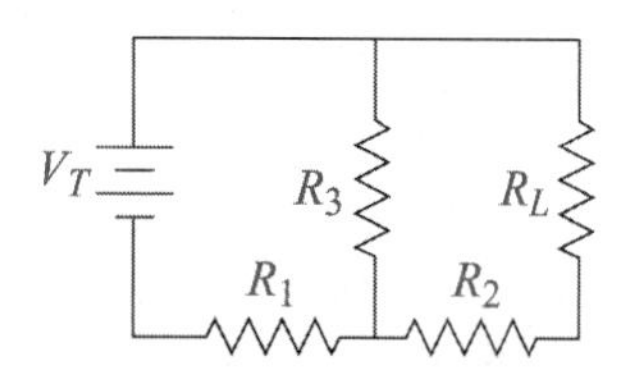

FIGURE 15-12 Series-parallel circuit for problems 1 and 2.

2. Given the series-parallel circuit in Figure 15-12 with $V_T = 60$ V, $R_1 = 2.0$ kΩ, $R_2 = 1.0$ kΩ, $R_3 = 2.0$ kΩ, and $R_L = 1.0$ kΩ. Thevenize this circuit for R_L by first finding V_{TH} and R_{TH}, and then finding I_L and V_L. See Example 15.3.

3. Given the series-parallel circuit in Figure 15-13 with $V_T = 120$ V, $R_1 = 1.0$ kΩ, $R_2 = 2.0$ kΩ, and $R_L = 3.0$ kΩ. Find I_L and V_L two ways:
   (a) Use Thevenin's theorem and first find V_{TH} and R_{TH}.
   (b) Find R_T and I_T directly. Check that the values for I_L and V_L agree with (a).

4. Given the series-parallel circuit in Figure 15-13 with $V_T = 36$ V, $R_1 = 100\ \Omega$, $R_2 = 100\ \Omega$, and $R_L = 50\ \Omega$. Find I_L and V_L two ways:
   (a) Use Thevenin's theorem and first find V_{TH} and R_{TH}.
   (b) Find R_T and I_T directly. Check that the values for I_L and V_L agree with (a).

5. Given the circuit in Figure 15-14 with $V_1 = 12$ V and $V_2 = 15$ V. Thevenize this circuit when $R_L = 6\ \Omega$, $R_1 = 4\ \Omega$, and $R_2 = 5\ \Omega$. Find V_{TH}, R_{TH}, I_L, and V_L. See Example 15.4.

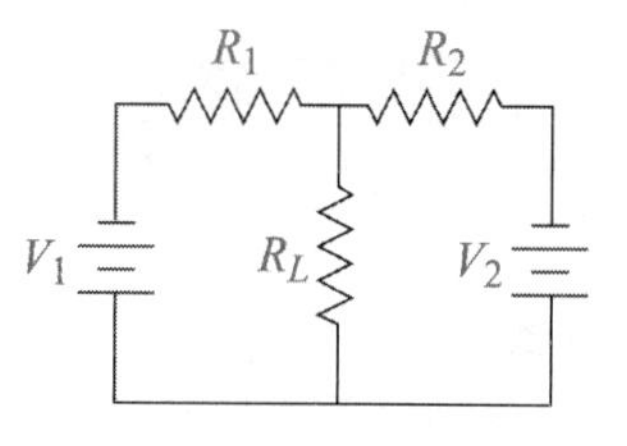

FIGURE 15-14 Circuit with two voltage sources for problems 5, 6, 7, and 8.

6. Given the circuit in Figure 15-14 with $V_1 = 80$ V and $V_2 = 65$ V. Thevenize this circuit when $R_L = 100\ \Omega$, $R_1 = 100\ \Omega$, and $R_2 = 200\ \Omega$. Find V_{TH}, R_{TH}, I_L, and V_L. See Example 15.4.

7. Given the circuit in Figure 15-14 with $V_1 = 50$ V and $V_2 = 70$ V. Theveninize this circuit when $R_L = 200\ \Omega$, $R_1 = 100\ \Omega$, and $R_2 = 100\ \Omega$. Find V_{TH}, R_{TH}, I_L, and V_L. See Example 15.4.

8. Given the circuit in Figure 15-14 with $V_1 = 12$ V and $V_2 = 18$ V. Thevenize this circuit when $R_L = 30\ \Omega$, $R_1 = 10\ \Omega$, and $R_2 = 20\ \Omega$. Find V_{TH}, R_{TH}, I_L, and V_L. See Example 15.4.

9. Given the circuit in Figure 15-15 with $V_1 = 22$ V and $V_2 = 24$ V. Thevenize this circuit when $R_L = 1.5$ kΩ, $R_1 = 3.0$ kΩ, and $R_2 = 2.0$ kΩ. Find V_{TH}, R_{TH}, I_L, and V_L. Note the polarity of the voltage sources and the location of R_L. See Example 15.4.

10. Given the circuit in Figure 15-15 with $V_1 = 12$ V, and $V_2 = 9.0$ V. Thevenize this circuit when $R_L = 50\ \Omega$, $R_1 = 200\ \Omega$, and $R_2 = 100\ \Omega$. Find V_{TH}, R_{TH}, I_L, and V_L. Note the polarity of the voltage sources and the location of R_L. See Example 15.4.

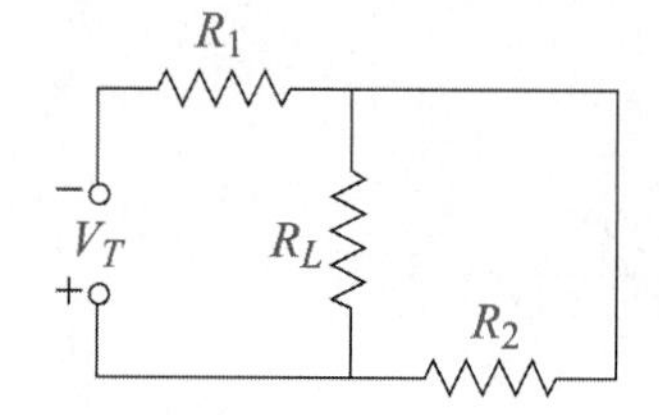

FIGURE 15-13 Series-parallel circuit for problems 3 and 4.

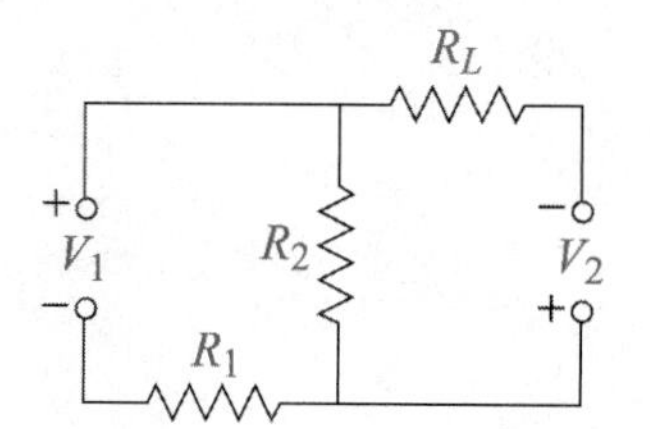

FIGURE 15-15 Circuit with two voltage sources for problems 9 and 10.

Answers to Error Box Problems, page 307:
(a) 3 V **(b)** 30 V **(c)** 5 V **(d)** 1.5 A **(e)** 2.0 A **(f)** 2.0 A

11. Given the circuit in Figure 15-16 with $V_1 = 6.0$ V, $V_2 = 8.0$ V, $R_1 = 20\ \Omega$, $R_2 = 30\ \Omega$, and $R_L = 25\ \Omega$. Find I_L and V_L two ways:
  (a) Apply Thevenin's theorem.
  (b) Apply the Superposition theorem. Check that the two sets of results agree.

12. Given the circuit in Figure 15-16 with $V_1 = 28$ V, $V_2 = 20$ V, $R_1 = 60\ \Omega$, $R_2 = 40\ \Omega$, and $R_L = 100\ \Omega$. Find I_L and V_L two ways:
  (a) Apply Thevenin's theorem.
  (b) Apply the Superposition theorem. Check that the two sets of results agree.

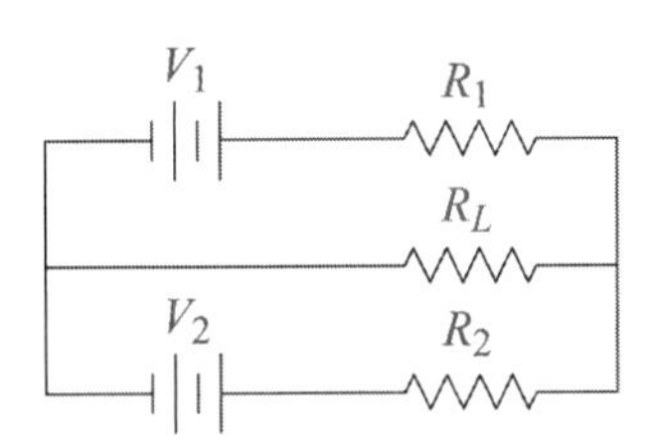

FIGURE 15-16 Thevenin's theorem and superposition theorem for problems 11 and 12.

15.3 Norton's Theorem

E.L. Norton, an American scientist, devised another way to simplify a circuit that is similar to Thevenin's idea, but it uses a current source instead of a voltage source. An *ideal current source* is an electrical component that supplies a constant current, no matter what the load, in the same way that an ideal voltage source supplies a constant voltage. It is represented in a schematic as a circle with a current arrow, as shown in the Norton circuit in Figure 15-17.

Voltage and Current Sources

A Thevenin circuit consisting of a voltage source and a series resistance can be shown to be equivalent to a Norton circuit containing a constant current source I_N and a *parallel* resistance R_N, as the next example shows.

EXAMPLE 15.5 Given the Thevenin circuit in Figure 15-17(a) with $V_{TH} = 12$ V and $R_{TH} = 6\ \Omega$. Find the Norton circuit that is equivalent to this Thevenin circuit.

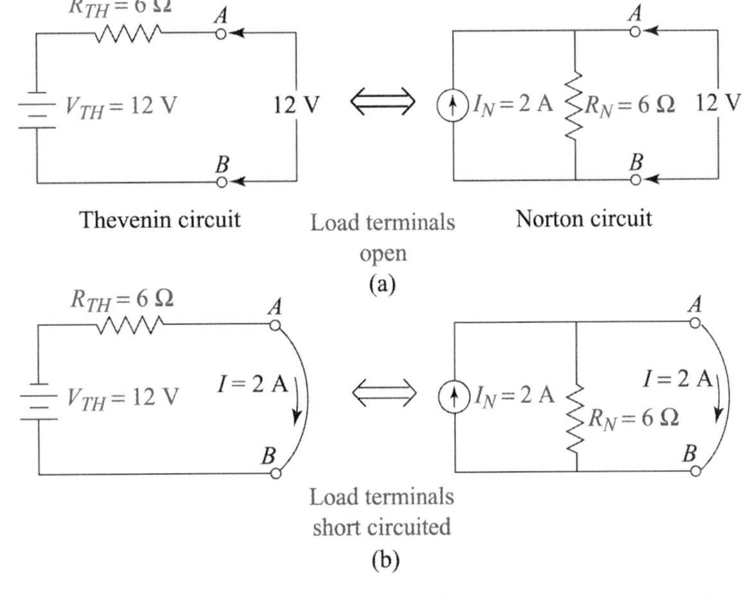

FIGURE 15-17 Equivalence of Thevenin circuit and Norton circuit for Example 15.5.

Solution The current source I_N in the equivalent Norton circuit is equal to the current in the Thevenin circuit with the load terminals short circuited, Figure 15-17(b). Apply Ohm's law to the Thevenin circuit to find I_N:

$$I_N = \frac{V_{TH}}{R_{TH}} = \frac{12 \text{ V}}{6 \ \Omega} = 2 \text{ A}$$

In the equivalent Norton circuit, the parallel resistance R_N is the same as the Thevenin resistance R_{TH}:

$$R_N = R_{TH} = 6 \ \Omega$$

The Norton equivalent circuit is shown in Figure 15-17(a). To demonstrate that these circuits are equivalent, consider when the load terminals *A* and *B* are open for each. The voltage across these terminals will be the same for both circuits, equal to 12 V. When the load terminals are short circuited, as shown in Figure 15-17(b), the current through these terminals will be the same in both circuits, equal to 2 A. Since the open voltages and short-circuit currents are the same in the Norton and Thevenin circuits, each circuit will have the same effect on a load resistance connected across *AB* and are therefore equivalent circuits.

Nortonizing a Circuit

Given a circuit with a load resistance R_L, the steps to *Nortonizing a circuit* are:

1. Find the Norton current I_N: Remove the load resistance and short circuit the terminals of R_L. Calculate the short-circuit current through these terminals.
2. Find the Norton resistance R_N in the same way as R_{TH}: Short out the voltage sources and calculate the equivalent resistance across the open terminals of R_L.
3. Draw the Norton equivalent parallel circuit with current source I_N and parallel R_N and R_L. Find I_L and V_L.

EXAMPLE 15.6

Nortonize the circuit of Figure 15-18(a): Find the Norton current and the Norton resistance for the load R_L. Then draw the Norton equivalent circuit and find I_L and V_L.

Solution To find I_N, remove R_L as shown in Figure 15-18(b). Short circuit the load terminals and find the current through these terminals. In this circuit, 3 Ω is in series with the parallel bank of 4 Ω and 6 Ω. The total resistance is then:

4 [x⁻¹] [+] 6 [x⁻¹] [=] [x⁻¹] [+] 3 [=] → 5.4 Ω

or

$$R_T = 3 + \frac{(4)(6)}{4+6} = 3 \ \Omega + 2.4 \ \Omega = 5.4 \ \Omega$$

The total current is: $$I_T = \frac{12 \text{ V}}{5.4 \ \Omega} \approx 2.22 \text{ A}$$

The voltage drop across R_1 is: $$V_1 = (2.22 \text{ A})(3 \ \Omega) = 6.66 \text{ V}$$

The voltage across R_3 and R_2 is then 12 V – 6.66 V = 5.34 V, and I_N is:

$$I_N = \frac{5.34 \text{ V}}{4 \ \Omega} \approx 1.34 \text{ A}$$

To find R_N, short out the voltage source and compute the equivalent resistance across the open terminals of R_L as shown in Figure 15-18(b). In this circuit, 4 Ω is in series with the parallel bank of 3 Ω and 6 Ω. R_N is then:

3 [x⁻¹] [+] 6 [x⁻¹] [=] [x⁻¹] [+] 4 [=] → 6 Ω

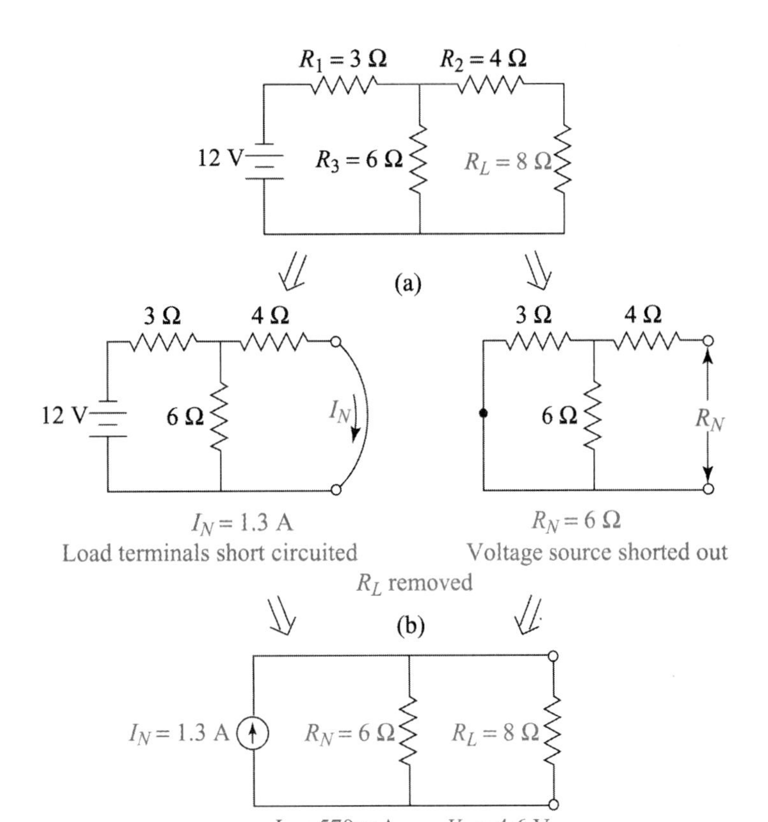

FIGURE 15-18 Nortonizing a circuit for Example 15.6.

or

$$R_N = 4 + \frac{(3)(6)}{3+6} = 4\ \Omega + 2\ \Omega = 6\ \Omega$$

The Norton equivalent circuit is shown in Figure 15-18(c) with a current source $I_N =$ 1.3 A and a resistance $R_N = 6\ \Omega$ parallel to the load. The equivalent resistance $R_{N,L}$ of this circuit and the voltage across R_N are:

$$R_{N,L} = \frac{(6)(8)}{6+8} \approx 3.43\ \Omega$$

$$V_N = (1.34\ \text{A})(3.43\ \Omega) \approx 4.6\ \text{V}$$

I_L and V_L are then:

$$I_L = \frac{4.60\ \text{V}}{8\ \Omega} \approx 0.57\ \text{A} = 570\ \text{mA}$$

$$V_L = (0.57\ \text{A})(8\ \Omega) \approx 4.6\ \text{V}$$

Note that $V_N = V_L$ in the Norton equivalent circuit.

EXERCISE 15.3

Express all answers in engineering notation (as numbers between 1 and 1000) with the appropriate units to two significant digits.

1. Given the series-parallel circuit in Figure 15-19 with $V_T = 14$ V, $R_1 = 6\ \Omega$, $R_2 = 8\ \Omega$, and $R_3 = 10\ \Omega$.
   (a) Nortonize this circuit for R_3 as the load resistance by first finding I_N and R_N, and then finding I_L and V_L.

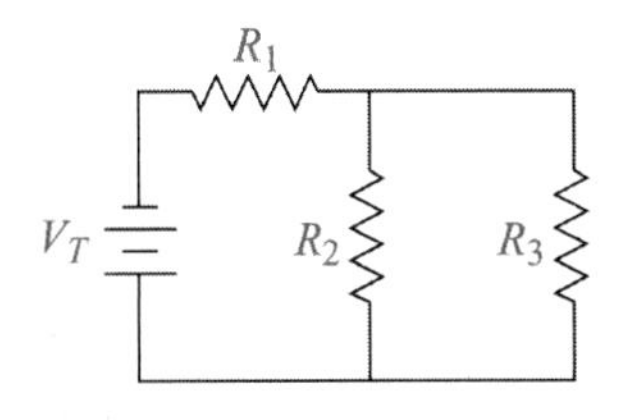

FIGURE 15-19 Series-parallel circuit for problems 1 and 2.

(b) Check your results by finding I_3 and V_3 by another method.

2. Given the series-parallel circuit in Figure 15-19 with $V_T = 24$ V, $R_1 = 50\ \Omega$, $R_2 = 100\ \Omega$, and $R_3 = 300\ \Omega$.
(a) Nortonize this circuit for R_1 as the load resistance by first finding I_N and R_N, and then finding I_L and V_L.
(b) Check your results by finding I_1 and V_1 by another method.

3. Given the circuit in Figure 15-20 with $V_T = 50$ V, $R_1 = 200\ \Omega$, $R_2 = 100\ \Omega$, $R_3 = 100\ \Omega$, and $R_L = 300\ \Omega$.
(a) Find I_L and V_L by applying Norton's theorem.
(b) Show that you get the same results for I_L and V_L by applying Thevenin's theorem.

4. Given the circuit in Figure 15-20 with $V_T = 12$ V, $R_1 = 5\ \Omega$, $R_2 = 3\ \Omega$, $R_3 = 7\ \Omega$, and $R_L = 5\ \Omega$,
(a) Find I_L and V_L by applying Norton's theorem.
(b) Show that you get the same results for I_L and V_L by applying Thevenin's theorem.

5. Given the circuit in Figure 15-14, Exercise 15.2 with $V_1 = 10$ V, $V_2 = 20$ V, $R_1 = 10\ \Omega$, $R_2 = 10\ \Omega$, and $R_L = 20\ \Omega$.
(a) Find I_L and V_L by applying Norton's theorem.
(b) Show that you get the same results for I_L and V_L by applying Thevenin's theorem.

6. Given the circuit in Figure 15-14, Exercise 15.2 with $V_1 = 12$ V, $V_2 = 9.0$ V, $R_1 = 10\ \Omega$, $R_2 = 20\ \Omega$, and $R_L = 12\ \Omega$.
(a) Find I_L and V_L by applying Norton's theorem.
(b) Show that you get the same results for I_L and V_L by applying Thevenin's theorem.

7. Given the Thevenin circuit and the equivalent Norton circuit in Figure 15-21 where $V_{TH} = 10$ V, $R_{TH} = R_N = 20\ \Omega$, $R_L = 5\ \Omega$, and:

$$I_N = \frac{V_{TH}}{R_{TH}} = \frac{10\ V}{20\ \Omega} = 0.5\ \text{A} = 500\ \text{mA}$$

Show that I_L and V_L are the same for each circuit.

8. Do the same as problem 7 when $V_{TH} = 150$ V, $R_{TH} = R_N = 100\ \Omega$, $R_L = 1.0\ \text{k}\Omega$, and:

$$I_N = \frac{V_{TH}}{R_{TH}} = \frac{150\ V}{100\ \Omega} = 1.5\ \text{A}$$

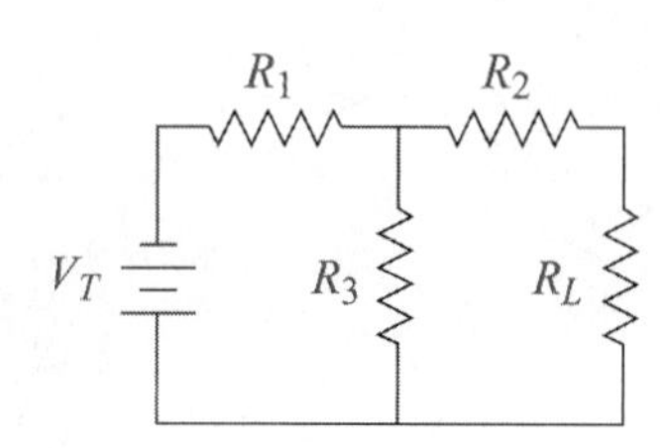

FIGURE 15-20 Series-parallel circuit for problems 3 and 4.

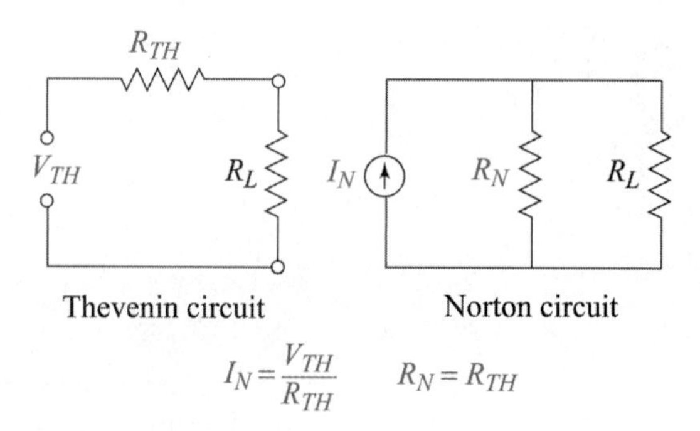

FIGURE 15-21 Equivalent Thevenin and Norton circuits for problems 7 and 8.

CHAPTER HIGHLIGHTS

15.1 SUPERPOSITION THEOREM

1. Redraw the circuit as two separate circuits. Each circuit contains one voltage source, with the other voltage source shorted out.

2. Combine the two results algebraically to produce the currents or voltages of the original circuit. See Figure 15-22.

15.2 THEVENIN'S THEOREM

Given a circuit with a load resistance R_L, the steps to *Thevenizing a circuit* are:

1. Find the Thevenin voltage V_{TH}: Remove R_L and calculate the open circuit voltage across the terminals of R_L.

2. Find the Thevenin resistance R_{TH}: Short out the voltage sources and calculate the equivalent resistance across the open terminals of R_L.

3. Draw the Thevenin equivalent circuit with V_{TH}, R_{TH}, and R_L in series. Find I_L and V_L. See Figure 15-23. Also, study the Error Box in Section 15.2.

15.3 NORTON'S THEOREM

Given a circuit with a load resistance R_L, the steps to *Nortonizing a circuit* are:

1. Find the Norton current I_N: Remove the load resistance and short circuit the terminals of R_L. Calculate the short-circuit current through these terminals.

2. Find the Norton resistance R_N in the same way as R_{TH}: Short out the voltage sources and calculate the equivalent resistance across the open terminals of R_L.
3. Draw the Norton equivalent parallel circuit with current source I_N and parallel resistances R_N and R_L. Find I_L and V_L. See Figure 15-24.

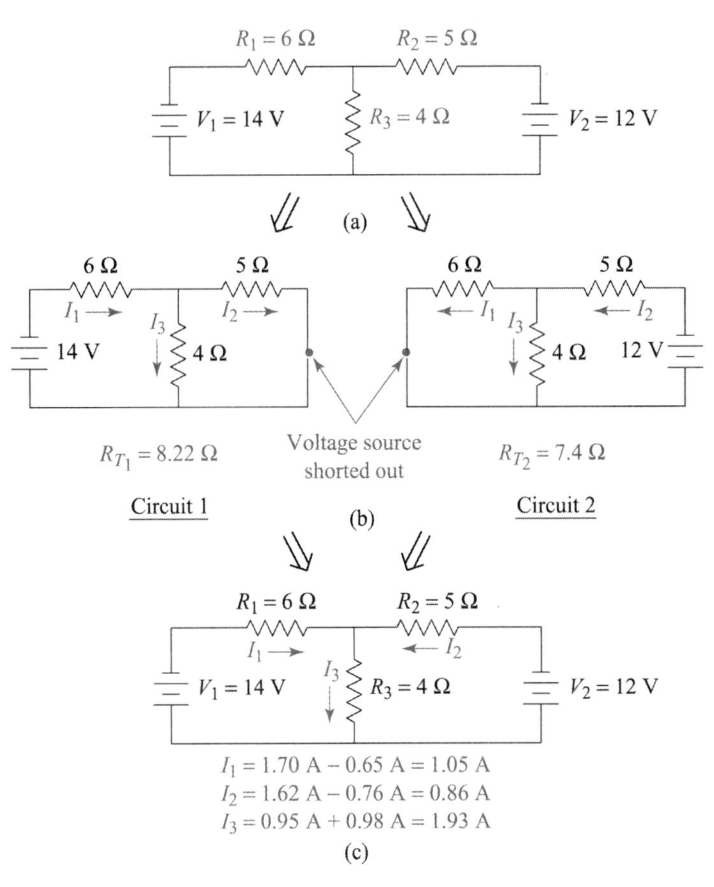

FIGURE 15-22 Key Example: Superposition theorem.

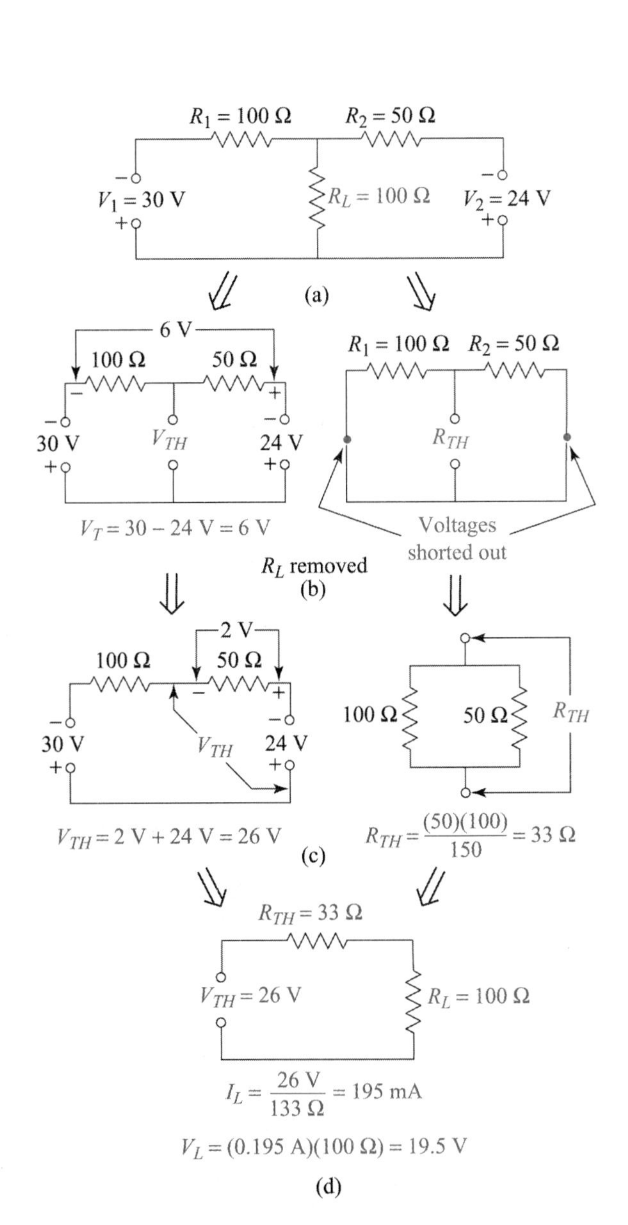

FIGURE 15-23 Key Example: Thevenin's theorem.

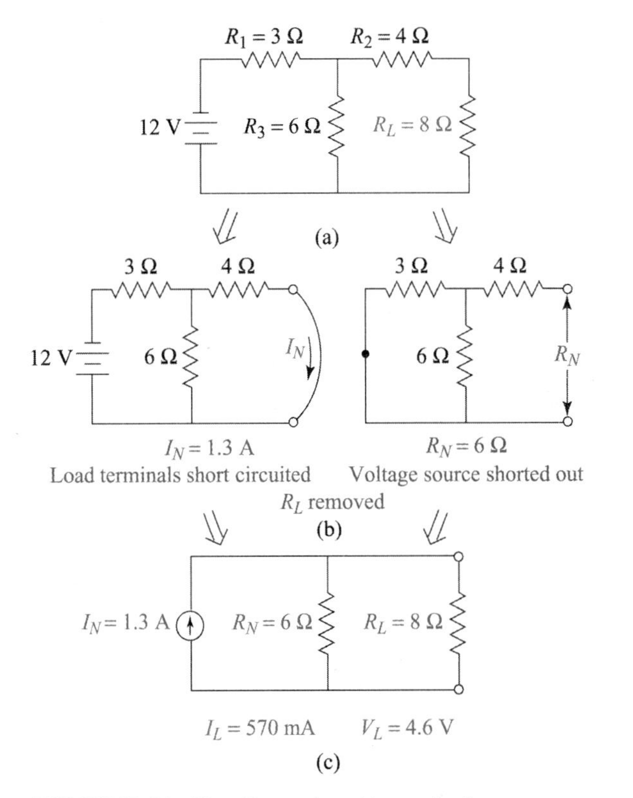

FIGURE 15-24 Key Example: Norton's theorem.

REVIEW EXERCISES

Express all answers in engineering notation (as numbers between 1 and 1000) with the appropriate units to two significant digits.

1. In the circuit in Figure 15-25:
   (a) Find the total current and the voltage across each resistance using the Superposition theorem.
   (b) Solve this circuit directly using Ohm's law and check that the results agree with (a).

2. In the circuit in Figure 15-26, find I_1, I_2, I_3, and the voltage across each resistance using the Superposition theorem.

3. In the circuit in Figure 15-27, find I_1, I_2, I_3, and the voltage across each resistance using the Superposition theorem.

4. In the circuit in Figure 15-28, find I_1, I_2, I_3, and the voltage across each resistance using the Superposition theorem.

5. Given the circuit in Figure 15-26 with R_3 the load resistance, find the equivalent Thevenin voltage V_{TH} and Thevenin resistance R_{TH}. Then find I_3 and V_3. Check that I_3 is the same value as that found in problem 2.

6. Given the circuit in Figure 15-27 with R_3 the load resistance, find the equivalent Thevenin voltage V_{TH} and Thevenin resistance R_{TH}. Then find I_3 and V_3. Check that I_3 is the same value as that found in problem 3.

7. Given the circuit in Figure 15-26 with R_2 the load resistance, find the equivalent Thevenin voltage V_{TH} and Thevenin resistance R_{TH}. Then find I_2 and V_2. Check that I_2 is the same value as that found in problem 2.

8. Given the circuit in Figure 15-27 with R_2 the load resistance, find the equivalent Thevenin voltage V_{TH} and Thevenin resistance R_{TH}. Then find I_2 and V_2. Check that I_2 is the same value as that found in problem 3.

9. Given the series-parallel circuit in Figure 15-29:
   (a) Thevenize this circuit for the load resistance R_L. Find V_{TH}, R_{TH}, I_L, and V_L.
   (b) Solve this circuit directly by first finding R_T and I_T. Check that the results agree with (a).

10. Given the circuit in Figure 15-28 with R_3 the load resistance, find the equivalent Thevenin voltage V_{TH} and Thevenin resistance R_{TH}. Then find I_3 and V_3. Check that I_3 is the same value as that found in problem 4.

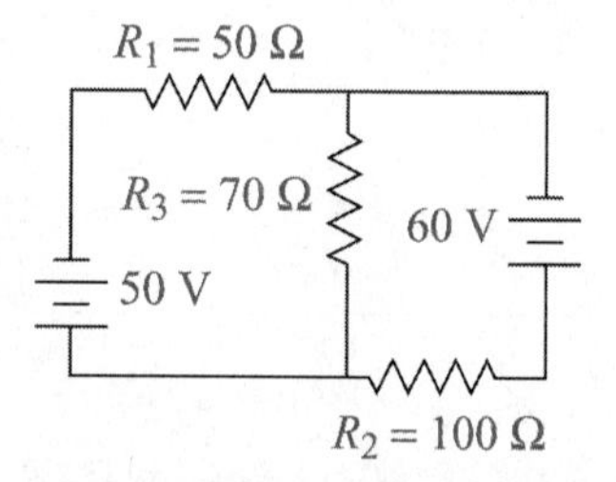

FIGURE 15-27 Superposition theorem for problem 3 and Thevenin's theorem for problems 6 and 8.

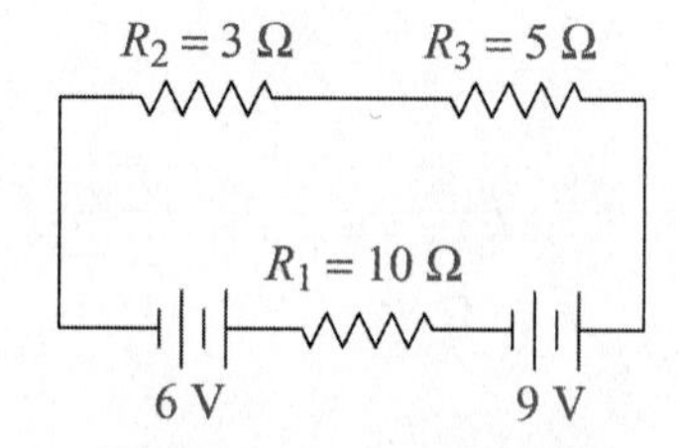

FIGURE 15-25 Superposition theorem for problem 1.

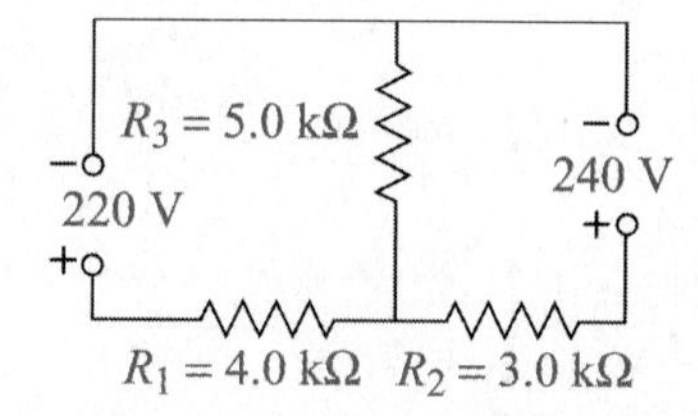

FIGURE 15-28 Superposition theorem for problem 4 and Thevenin's theorem for problem 10.

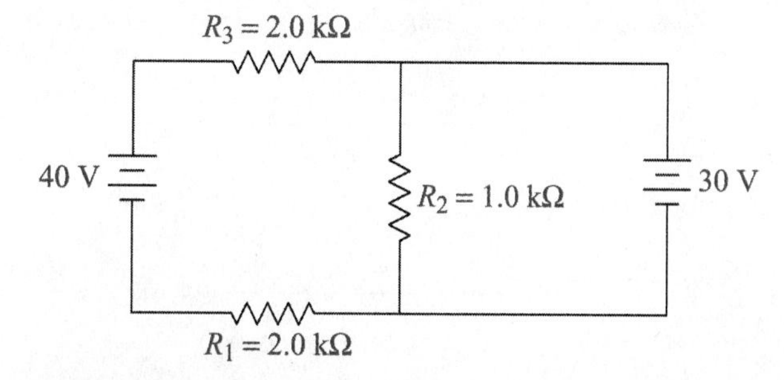

FIGURE 15-26 Network theorems for problems 2, 5, 7, and 13.

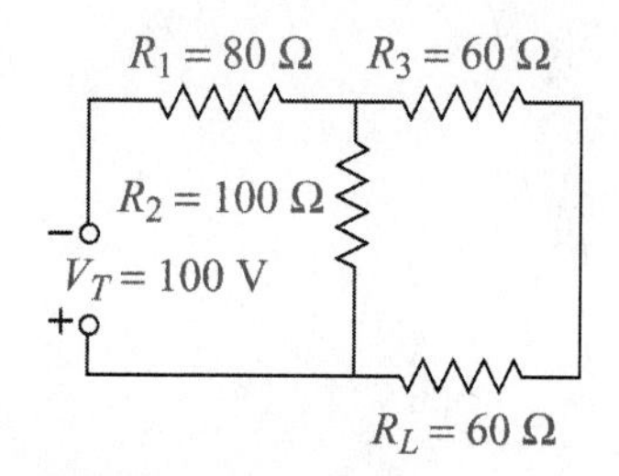

FIGURE 15-29 Thevenin's theorem for problem 7 and Norton's theorem for problem 14.

11. Given the series-parallel circuit in Figure 15-30 with $V_T = 100$ V, $R_1 = 1.2$ kΩ, $R_2 = 3.3$ kΩ, and $R_3 = 1.5$ kΩ.

(a) Nortonize this circuit for R_3 as the load resistance. Find I_N and R_N, and then find I_L and V_L.

(b) Check your results by finding I_3 and V_3 by another method.

12. Given the circuit in Figure 15-31 with $V_T = 20$ V, $R_1 = 10\ \Omega$, $R_2 = 5\ \Omega$, $R_3 = 5\ \Omega$, and $R_L = 10\ \Omega$:

(a) Find I_L and V_L by applying Norton's theorem.

(b) Show that you get the same results for I_L and V_L by applying Thevenin's theorem.

13. Nortonize the circuit in Figure 15-26 for the load resistance R_3. Find I_N, R_N, I_L, and V_L. Check that the results agree with those in problem 5.

14. the circuit in Figure 15-29 for the load resistance R_L. Find I_N, R_N, I_L, and V_L. Check that the results agree with those in problem 7.

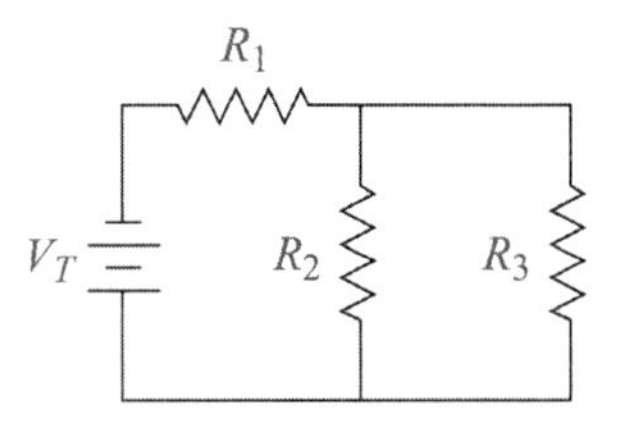

FIGURE 15-30 Norton's theorem for problem 11.

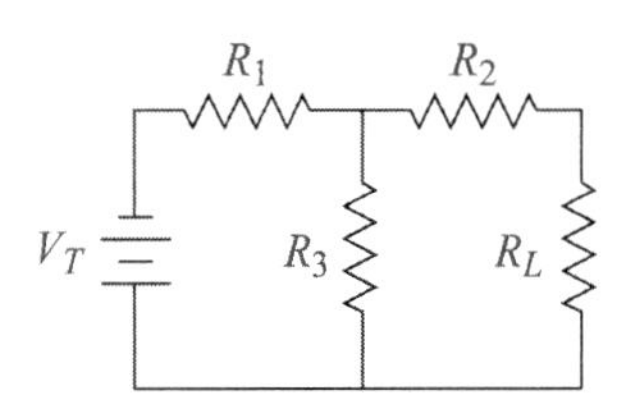

FIGURE 15-31 Norton's theorem for problem 12.

CHAPTER 16
Trigonometry of the Right Triangle

CHAPTER OBJECTIVES

In this chapter, you will learn:

- The basic geometric concepts of angles.
- The meaning of degree and radian measure of an angle.
- How to change from degrees to radians and vice versa.
- The definition of similar triangles.
- How to find the sides of similar triangles using proportions.
- The Pythagorean theorem and how to apply it.
- The trigonometric ratios of the right triangle: sine, cosine, and tangent.
- How to solve triangles using the trignometric ratios.

Trigonometry is the study of angles and triangles, which are the basic building blocks for much of geometry. The most important triangle is the right triangle. It has many technical applications especially in alternating-current electricity and ac circuits. One of the most useful geometric relationships concerning the sides of a right triangle, and a major theorem in geometry, is the Pythagorean theorem. This theorem is at least 500 years older than Christianity. Also associated with the right triangle are the trigonometric ratios: sine, cosine, and tangent. These ratios, combined with the Pythagorean theorem, enable you to solve right triangles that occur in a wide variety of electronic problems.

• • •

16.1 Angles and Triangles

Angles and Radian Measure

An *angle is a measure of rotation* between two radii in a circle. The center of the circle where the radii meet is called the *vertex* of the angle. See Figure 16-1. An angle of one complete rotation is defined as 360° (degrees). Half of a rotation is 180° and one-quarter of a rotation is 90°. Another way of measuring angles or rotation is radian measure. One complete rotation is defined as 2π radians (rad) where $\pi \approx 3.14$ and $2\pi \text{ rad} \approx 2(3.14) = 6.28$ rad. It follows that:

$$2\pi \text{ rad} = 360°$$

Dividing both sides of this equation by 2 gives you the basic formula:

Formula **Radian Measure**

$$\pi \text{ rad} = 180° \tag{16.1}$$

The following radian equivalents for common angles follow from formula (16.1):

$$\frac{\pi}{2} \text{ rad} = 90° \qquad \frac{\pi}{3} \text{ rad} = 60° \qquad \frac{\pi}{4} \text{ rad} = 45°$$

$$\frac{\pi}{6} \text{ rad} = 30° \qquad \frac{3\pi}{2} \text{ rad} = 270° \qquad 2\pi \text{ rad} = 360°$$

See Figure 16-1, which shows these common angles in a circle. Rotation is counterclockwise starting from the right at 0°. Dividing both sides of formula (16.1) by π tells us that:

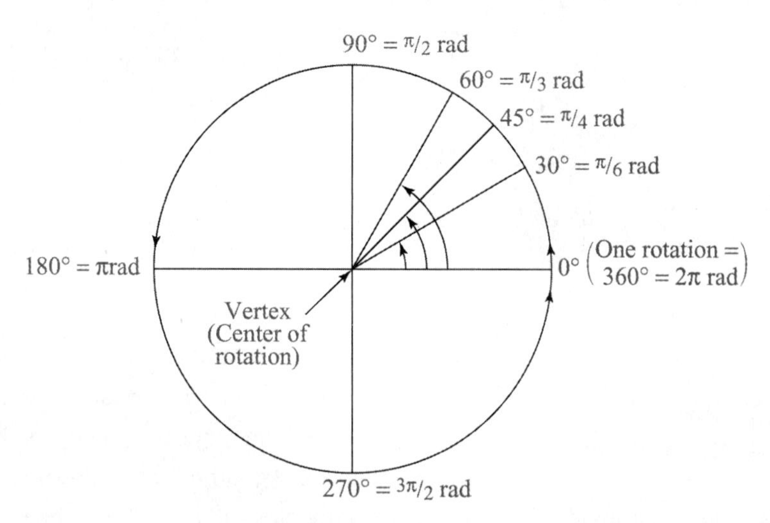

FIGURE 16-1 Basic angles.

Definition

Definition of Radian

$$1 \text{ rad} = \frac{180°}{\pi} \approx \frac{180°}{3.142} \approx 57.3° \quad \textbf{(16.2)}$$

An angle of one radian ($\approx 57.3°$) cuts off an arc on the circle equal to one *radius*, which is why the name "radian" is used. Formula 16.2 provides the following conversion factors:

Procedure

Conversion Factors

- To change from radians to degrees, multiply by $\frac{180°}{\pi}$ or 57.3°.
- To change from degrees to radians, multiply by $\frac{\pi}{180°}$ or divide by 57.3°.

EXAMPLE 16.1 Change to degrees:

(a) $\frac{\pi}{9}$ rad **(b)** 1.7 rad

Solution

(a) When π is used to express the angle in radians, you can multiply by $\frac{180°}{\pi}$ to divide out the π:

$$\frac{\pi}{9} \text{ rad} = \frac{\cancel{\pi}}{\cancel{9}}\left(\frac{\overset{20}{\cancel{180°}}}{\cancel{\pi}}\right) = 20°$$

(b) When π is not used to express the angle in radians, multiply by $\frac{180°}{\pi}$ or 57.3°:

$$1.7 \text{ rad}\left(\frac{180°}{\pi}\right) \approx 97° \text{ or } 1.7 \text{ rad} = 1.7(57.3°) \approx 97°$$

Use the [π] key on the calculator when converting:

1.7 [×] 180 [÷] [π] [=] → 97

EXAMPLE 16.2 Change to radians:

(a) 135° **(b)** 44°

Solution

(a) When the angle is a multiple of one of the common angles 30°, 45°, or 60°, you can multiply by $\frac{\pi}{180°}$ to express the angle conveniently in terms of π by dividing out the common angle:

$$135° = (135°)\left(\frac{\pi}{180°}\right) = 3(\cancel{45})\left(\frac{\pi}{4(\cancel{45})}\right) = \frac{3\pi}{4} \text{ rad}$$

(b) When the angle is not a multiple of one of the angles 30°, 45°, or 60°, multiply by $\frac{\pi}{180°}$ or divide by 57.3°:

$$44° = (44°)\left(\frac{\pi}{180°}\right) \approx 0.77 \text{ rad or } \frac{44°}{57.3°} \approx 0.77 \text{ rad}$$

The calculations in Examples 16.1 and 16.2 can easily be done on the calculator. Most calculators have two function keys [DRG] and [DRG>]. The letters stand for Degrees, Radians, and Gradient, which are three ways to measure angles. Gradient measure is used in civil engineering and is defined as 400 grads = 360°. The first key [DRG] is used for *setting the calculator* for degrees, radians, or gradient. The second key [DRG>] is used for *converting between radians, degrees, and gradient.* For example, to convert 44° to radians you first set your calculator for degrees by pressing [DRG] until you see "D" or "DEG" on the display and then enter:

44 [DRG>] [=] → 0.77

Angles are denoted in different ways. See Figure 16-2 for the following angles:

- An *acute angle* is greater than zero and less than 90°. Angle A is an acute angle.
- A *right angle* is 90°. Angle C is a right angle.
- An *obtuse angle* is greater than a right angle and less than 180°. Angle E is an obtuse angle.
- A *straight angle* is 180°. Angle $(A + E)$ is a straight angle.
- *Perpendicular lines* meet at right angles denoted by a square at the vertex.
- *Complementary angles* are two angles that add up to 90°. Angles A and B are complementary.

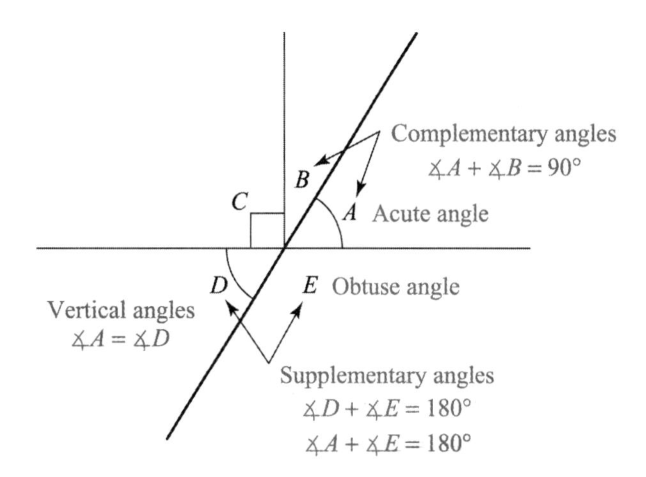

FIGURE 16-2 Types of angles for Example 16.3.

- *Supplementary angles* are two angles that add up to 180°. Angles D and E are supplementary.
- *Vertical angles* are opposite angles formed by two intersecting lines and are equal. Angles A and D are vertical angles and $\measuredangle A = \measuredangle D$.

EXAMPLE 16.3 In Figure 16-2, given $\measuredangle E = 110°$, find $\measuredangle A$, $\measuredangle B$, and $\measuredangle D$.

Solution Observe $\measuredangle D$ is the supplement of $\measuredangle E$ and $\measuredangle A = \measuredangle D$ because they are vertical angles. Therefore,

$$\measuredangle A = \measuredangle D = 180° - 110° = 70°$$

Note $\measuredangle B$ is the complement of $\measuredangle A$. Therefore,

$$\measuredangle B = 90° - 70° = 20°$$

An important theorem for the internal angles of any triangle is the following:

Theorem

Angles of a Triangle

The sum of the internal angles of any triangle equals 180°.

Triangles are classified as follows:

- A *right triangle* is a triangle containing a right angle.
- An *acute triangle* is a triangle containing three acute angles.
- An *obtuse triangle* is a triangle containing an obtuse angle.

EXAMPLE 16.4 Given the acute triangle divided into two right triangles in Figure 16-3, find the lettered angles $\measuredangle A$, $\measuredangle B$, $\measuredangle C$, and $\measuredangle D$.

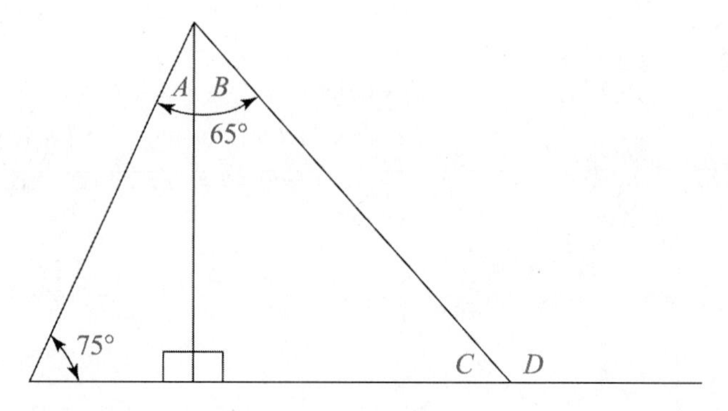

FIGURE 16-3 Sum of the angles of a triangle for Example 16.4.

Solution Since the sum of the angles of a triangle add up to 180°, find angle A in the small right triangle and $\measuredangle C$ in the acute triangle by subtracting the other angles in each triangle from 180°:

$$A = 180° – 90° – 75° = 15°$$
$$C = 180° – 75° – 65° = 40°$$

Then:

$$B = 65° – 15° = 50°$$
$$D = 180° – 40° = 140°$$

EXERCISE 16.1

In exercises 1 through 16, change each angle to degrees.

1. $\frac{\pi}{6}$ rad

2. $\frac{\pi}{2}$ rad

3. $\frac{3\pi}{2}$ rad

4. $\frac{\pi}{4}$ rad

5. $\frac{3\pi}{4}$ rad

6. $\frac{4\pi}{3}$ rad

7. $\frac{\pi}{10}$ rad

8. $\frac{2\pi}{5}$ rad

9. 1.1 rad

10. 3.1 rad

11. 0.50 rad

12. 0.75 rad

13. 1.6 rad

14. 2.1 rad

15. 0.091 rad

16. 0.026 rad

In exercises 17 through 24, change each angle to radians in terms of π or as directed by the instructor.

17. 60°

18. 45°

19. 360°

20. 180°

21. 120°

22. 150°

23. 10°

24. 20°

In exercises 25 through 34, change each angle to radians.

25. 55°

26. 27°

27. 98°

28. 130°

29. 8.0°

30. 17°

31. 170°

32. 110°

33. 1.0°

34. 2.5°

In exercises 35 through 44, find all the lettered angles in degrees or radians.

35.

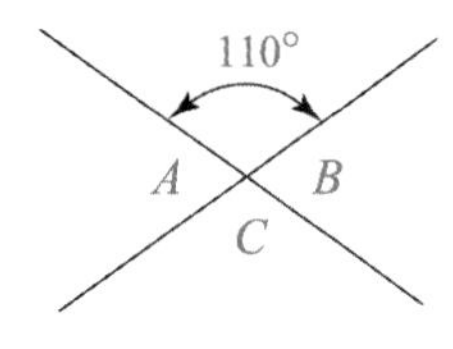

FIGURE 16-4

36.

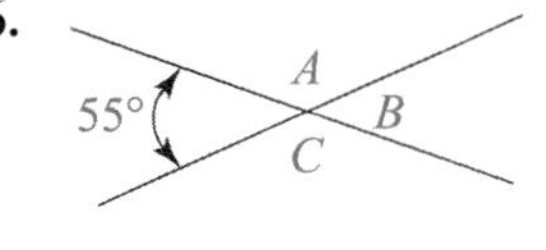

FIGURE 16-5

37.

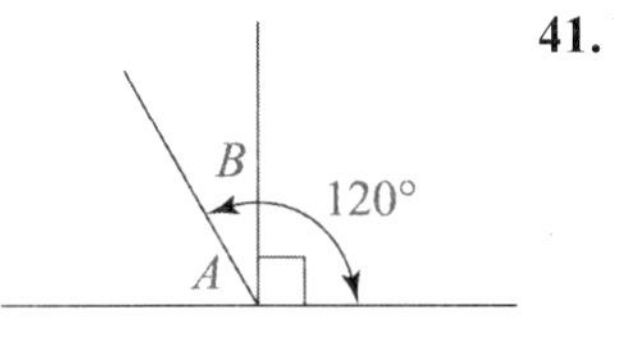

FIGURE 16-6

38.

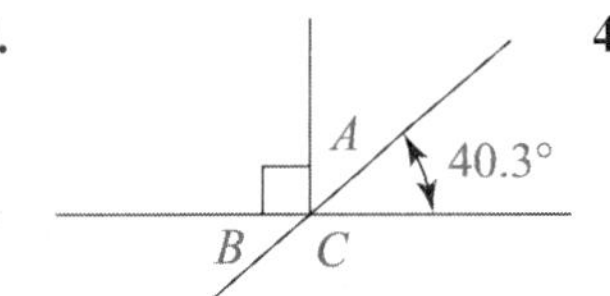

FIGURE 16-7

39.

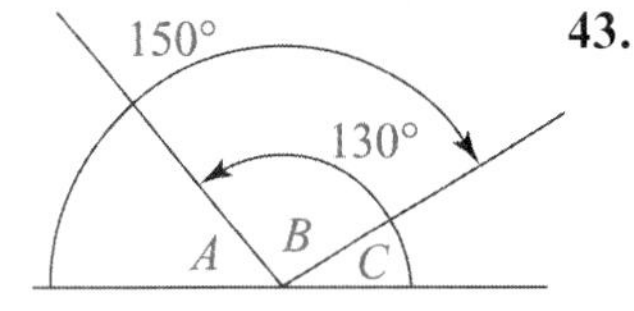

FIGURE 16-8

40.

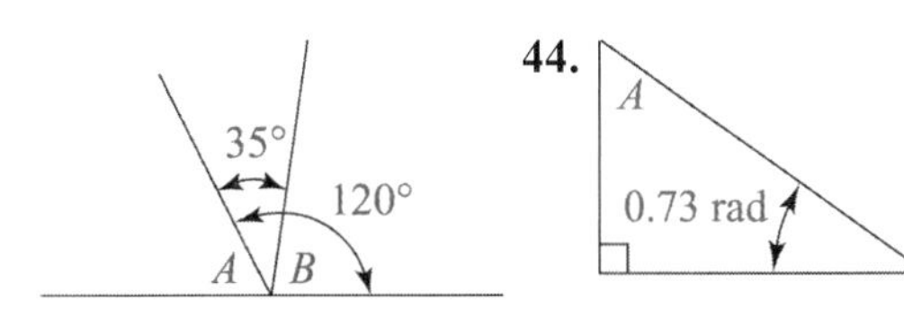

FIGURE 16-9

41.

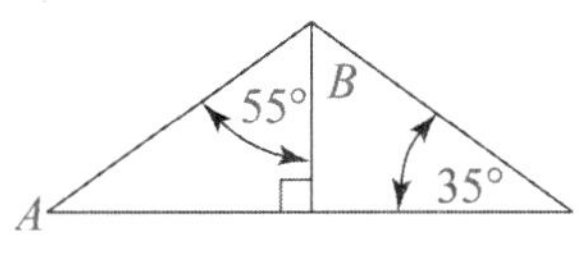

FIGURE 16-10

42.

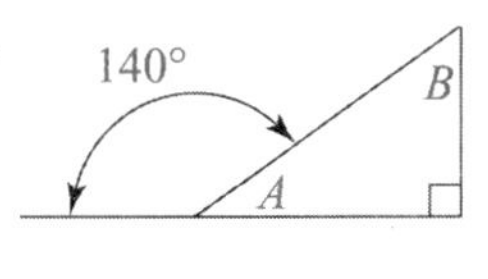

FIGURE 16-11

43.

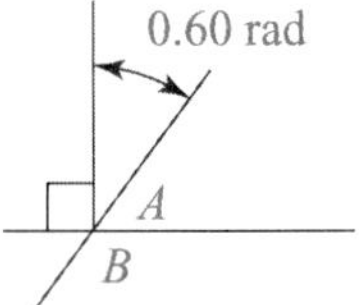

FIGURE 16-12

44.

FIGURE 16-13

Applied Problems

45. If it takes you 5 min to do this problem, how many degrees and radians does the minute hand on an analog watch move in that time?

46. Longitude is measured from 0° (London) traveling west parallel to the equator. There are 24 one-hour time zones. How many degrees and radians are in each time zone?

Applications to Electronics

47. **(a)** The phase angle of the impedance of an ac circuit is 0.63 rad. How many degrees is this phase angle?

(b) The phase angle of the voltage of an ac circuit is 67°. How many radians is this phase angle?

48. The angular velocity ω (Greek omega) of ordinary house current is $\omega = 120\pi$ rad/s (radians per second). What is the angular velocity in degrees per second?

16.2 Pythagorean Theorem

A *polygon* is a plane figure bounded by straight lines. Common polygons are a triangle (three-sided polygon), a rectangle (four-sided polygon), a pentagon (five-sided polygon), and a hexagon (six-sided polygon). Every polygon can be divided into triangles, and any triangle can be divided into two right triangles. The right triangle is therefore a basic building block for many figures. Figure 16-14 shows a typical right triangle. The longest side of the right triangle, which is opposite the right angle, is called the *hypotenuse* and labeled c. The other two sides (also called the legs) are labeled a and b. The angles opposite these sides are labeled with corresponding capital letters. Angle A is opposite side a, angle B is opposite side b, and the right angle C is opposite the hypotenuse c.

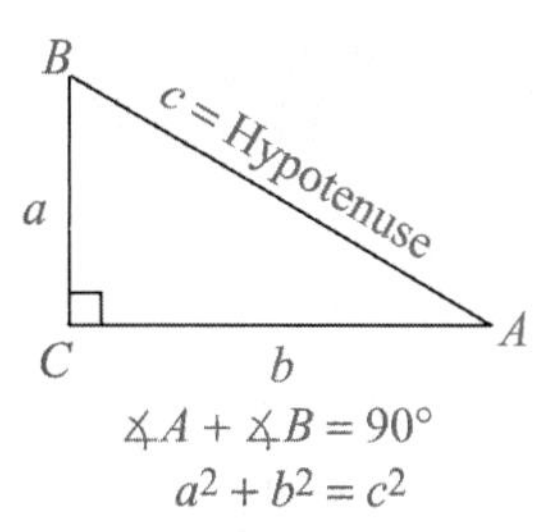

FIGURE 16-14 Right triangle and Pythagorean theorem for Example 16.5.

The sum of the angles of any triangle add up to 180°, therefore angles A and B in a right triangle add up to 90° and are complementary:

$$\measuredangle A + \measuredangle B = 90°$$

One of the most important and useful theorems in all of geometry, and mathematics, is the Pythagorean theorem. Although Babylonian and Egyptian surveyors used the theorem more than 3000 years ago, the earliest record of its formal proof was left by the Greek Pythagoras around 520 B.C. The Pythagorean theorem applied to the right triangle in Figure 16-14 is:

Theorem

Pythagorean Theorem

In a right triangle, the sum of the squares of the sides equals the square of the hypotenuse:

$$a^2 + b^2 = c^2. \tag{16.3}$$

EXAMPLE 16.5

Given the right triangle ABC in Figure 16-14:

(a) If $a = 15$ and $b = 20$, find side c:

(b) If $a = 5.5$ and $c = 9.2$, find side b:

Solution

(a) To find side c apply the Pythagorean theorem (16.3):

$$a^2 + b^2 = c^2$$
$$15^2 + 20^2 = c^2$$

Solve for c^2 and take the positive square root (a negative answer does not apply):

$$c^2 = 225 + 400 = 625$$
$$c = \sqrt{625} = 25$$

(b) To find side b, apply the Pythagorean theorem and solve for b:

$$5.5^2 + b^2 = 9.2^2$$
$$30.25 + b^2 = 84.64$$
$$b^2 = 84.64 - 30.25 = 54.39$$
$$b = \sqrt{54.39} \approx 7.4$$

EXAMPLE 16.6

A boat travels 5 mi due south and then turns and travels due east for 10 mi. How far is the boat from its starting point?

Solution Figure 16-15 shows the route that the boat travels. North is straight up and east is to the right, perpendicular to north, and south is straight down, opposite of

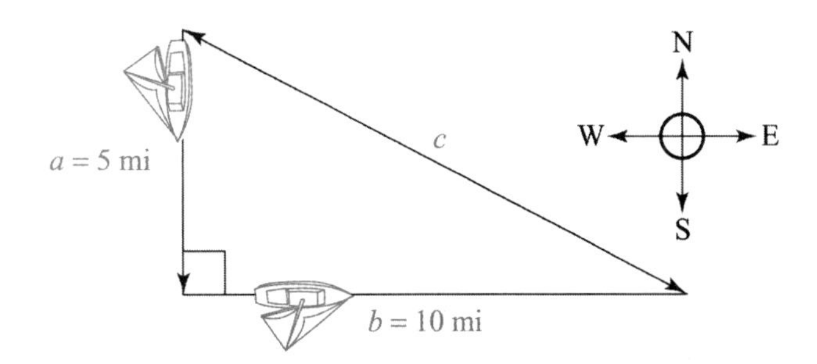

FIGURE 16-15 Boat trip for Example 16.6.

north. Therefore, the distance of the boat from its starting point is the hypotenuse of a right triangle. You are given $a = 5$ mi and $b = 10$ mi, and you need to find c. Apply the Pythagorean theorem:

$$a^2 + b^2 = c^2$$
$$(5)^2 + (10)^2 = c^2$$
$$25 + 100 = c^2$$
$$c^2 = 125$$
$$c = \sqrt{125} \approx 11.2 \text{ mi}$$

There are more than 400 proofs of the Pythagorean theorem, including one by a president of the United States, James Garfield. Study the next example that closes the circuit and shows an important application of the Pythagorean theorem to a problem in electronics.

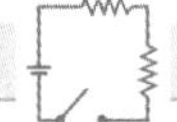

EXAMPLE 16.7

Close the Circuit

Figure 16-16(a) shows an ac circuit containing a resistance R and a capacitance C in series (RC series circuit). The total impedance Z is related to the resistance and the capacitive reactance X_C by the impedance triangle shown in Figure 16-16(b). If $R = 4.3$ kΩ and $Z = 8.2$ kΩ, find X_C in kilohms.

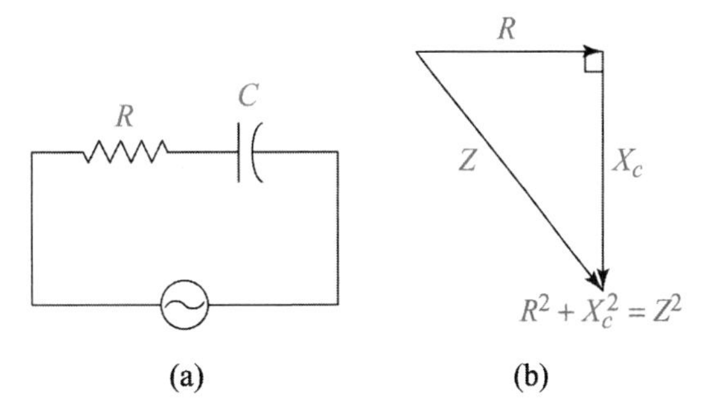

FIGURE 16-16 *RC* series circuit for Example 16.7.

Solution The impedance triangle is a right triangle with sides R and X_C, and hypotenuse Z. The three component values are therefore related by the Pythagorean theorem:

$$R^2 + X_C^2 = Z^2$$

Substitute the values in this Pythagorean formula using $R = 4.3$ kΩ and $Z = 8.2$ kΩ and solve for X_C:

$$(4.3 \text{ k}\Omega)^2 + X_C^2 = (8.2 \text{ k}\Omega)^2$$
$$X_C^2 = (8.2 \text{ k}\Omega)^2 - (4.3 \text{ k}\Omega)^2 = 48.75$$

Take the square root of both sides. The answer to two significant digits is:

$$X_C = \sqrt{48.75} \approx 7.0 \text{ k}\Omega$$

The answer can be found in one series of steps on the calculator:

DAL: [√] [(] 8.2 [x^2] [−] 4.3 [x^2] [)] [=] → 7.0

Not DAL: 8.2 [x^2] [−] 4.3 [x^2] [=] [√] → 7.0

Similar Polygons and Triangles

An important relationship between triangles is that of similarity:

Definition

Similar Triangles

Two triangles are similar when all the angles of one triangle are equal to the corresponding angles of the other triangle.

The angles of a triangle determine the shape of the triangle. If the triangles are similar, then they will have the same shape *but not necessarily the same size.* Figure 16-17 shows two similar right triangles *ABC* and *DEF.* The corresponding angles are equal(indicated by the same slash marks):

$$A = D$$
$$B = E$$
$$C = F$$

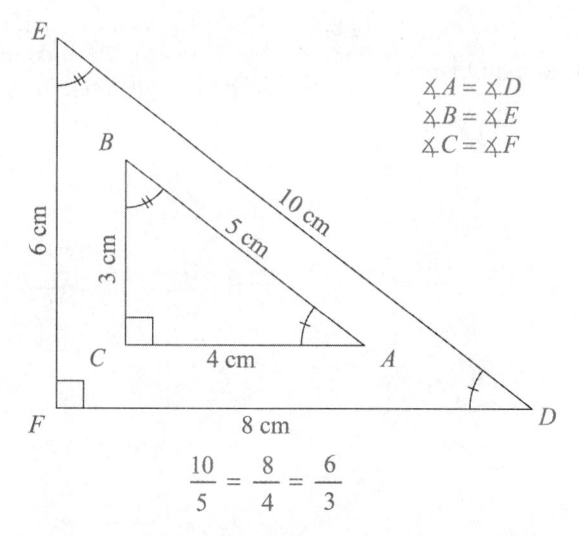

$$\frac{10}{5} = \frac{8}{4} = \frac{6}{3}$$

FIGURE 16-17 Similar triangles and proportions.

Similar triangles have their *corresponding sides in proportion.* A proportion is an equality between ratios. Consider the two right triangles in Figure 16-17. The sides of the large triangle are *twice* the sides of the small triangle. Therefore, you can write the following proportion for the three ratios of the sides of the large triangle to the corresponding sides of the small triangle:

$$\frac{\text{Large Triangle}}{\text{Small Triangle}}: \frac{10}{5} = \frac{8}{4} = \frac{6}{3}$$

These ratios can also be written from the small triangle to the large triangle:

$$\frac{\text{Small Triangle}}{\text{Large Triangle}}: \frac{5}{10} = \frac{4}{8} = \frac{3}{6}$$

EXAMPLE 16.8

Given two right triangles: ABC and BEF in Figure 16-18 with $CE = 5$, $BE = 10$, and $FE = 6$.

(a) Find side AC.

(b) Find the hypotenuse AB to two significant digits.

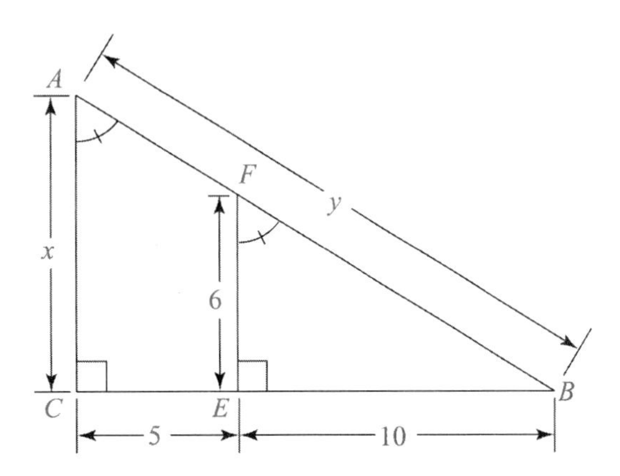

FIGURE 16-18 Similar right triangles for Example 16.8.

Solution

(a) The two right triangles are similar for the following reasons: The right angles are equal, and angle B is the same in both. When two angles of a triangle are equal to two corresponding angles in another triangle, the third angles must also be equal because the three angles of any triangle add up to 180°. Therefore $\measuredangle A = \measuredangle BFE$, and the triangles are similar. This is indicated by slash marks on $\measuredangle A$ and $\measuredangle BFE$ in Figure 16-18.

Identify the corresponding sides. *Corresponding sides are opposite equal angles.* Side AC corresponds to side FE, and side BC corresponds to side BE. Therefore, you can write the following proportion going from the large triangle to the small triangle:

$$\frac{\text{Large Triangle}}{\text{Small Triangle}}: \frac{AC}{FE} = \frac{BC}{BE}$$

Let x = side AC, and note that side $BC = 10 + 5 = 15$. Then, substituting for the values in the proportion gives you:

$$\frac{x}{6} = \frac{15}{10}$$

You can solve a proportion by cross multiplying, which means multiplying the numerator of one fraction by the denominator of the other fraction and vice versa:

$$\frac{x}{6} \times \frac{15}{10}$$

$$(x)(10) = (15)(6)$$

$$10x = 90$$

$$x = 9$$

Cross multiplication is equivalent to multiplying both sides of this equation by the product of the denominators = (6)(10) = 60. You can also solve the proportion by multiplying both sides by the LCD = 30.

(b) To find side AB, let $y = AB$ and apply the Pythagorean theorem:

$$9^2 + 15^2 = y^2$$
$$y^2 = 225 + 81 = 306$$
$$y = \sqrt{306} \approx 17$$

Similar polygons are two polygons that have the same angles and are encountered in many technical problems. For example, the computer design model of a microcircuit is constructed similar to that of the actual circuit. The corresponding angles are equal, and the corresponding sides are in proportion. The scale of the model gives the ratio of the model distances to the actual distances. A scale of 100:1 means that the distances in the computer model are 100 times the distances in the actual circuit. Maps, photographs, or video images of any object are all examples of similarity. The next example closes the circuit and illustrates an application of similar polygons in electronics.

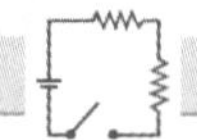

EXAMPLE 16.9

Close the Circuit

A rectangular microprocessor chip has width = 3.3 mm and length = 9.4 mm. If the image of the chip on a computer monitor has a width equal to 10 cm:

(a) What is the length of the image in centimeters and meters?

(b) What is the scale of the image?

Solution

(a) The computer image is similar to the actual chip but is much larger. The sides of the image and the chip are in proportion. Let l = length of the image. To find l, you can write the following proportion going from the image to the object using consistent units:

$$\frac{\text{Image}}{\text{Object}}: \frac{l}{9.4 \text{ mm}} = \frac{100 \text{ mm}}{3.3 \text{ mm}}$$

Note that 10 cm = 100 mm. Cross multiply to obtain the solution:

$$\frac{l}{9.4 \text{ mm}} \times \frac{100 \text{ mm}}{3.3 \text{ mm}}$$

$$(l)(3.3 \text{ mm}) = (100 \text{ mm})(9.4 \text{ mm})$$

Divide both sides by 3.3 mm:

$$l = \frac{(100 \text{ mm})(9.4 \text{ mm})}{3.3 \text{ mm}} \approx 285 \text{ mm} = 28.5 \text{ cm}$$

(b) To find the scale of the image, divide one of the sides of the image by the corresponding side of the actual chip:

$$\text{Scale} = \frac{\text{Image}}{\text{Object}} = \frac{100 \text{ mm}}{3.3 \text{ mm}} \approx 30.3{:}1$$

This means that the image is 30.3 times larger than the actual chip.

EXERCISE 16.2

Round all answers in the exercise to two significant digits.

In exercises 1 through 12, find the missing side of each right triangle (c = hypotenuse).

1. $a = 5, b = 12$
2. $a = 8, b = 15$
3. $a = 12, b = 19$
4. $a = 3.6, b = 6.7$
5. $b = 2.1, c = 3.8$
6. $b = 150, c = 330$
7. $a = 7.5, c = 9.1$
8. $a = 2.7, c = 4.7$
9. $a = 0.51, b = 0.68$
10. $a = 0.033, b = 0.044$
11. $b = 560, c = 820$
12. $a = 690, b = 960$

In exercises 13 through 20, given that the triangles are similar, find x and y. See Example 16.8.

13.

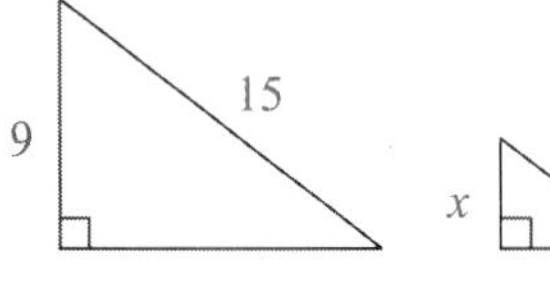

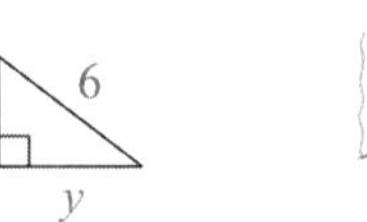

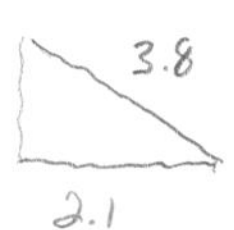

FIGURE 16-19

14.

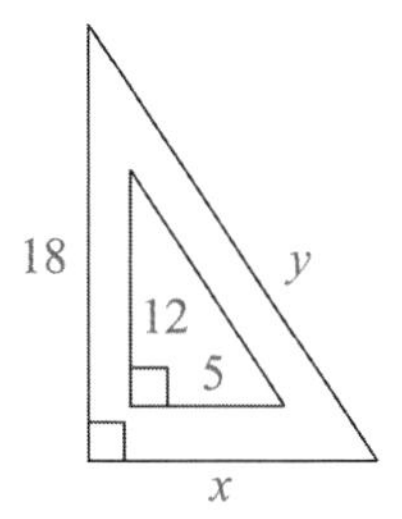

FIGURE 16-20

15.

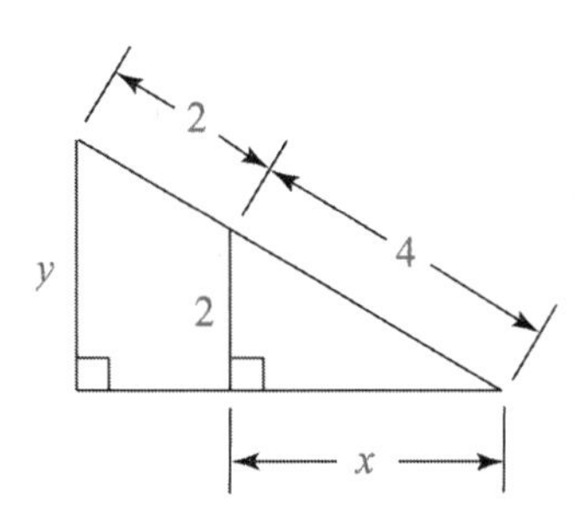

FIGURE 16-21

16.

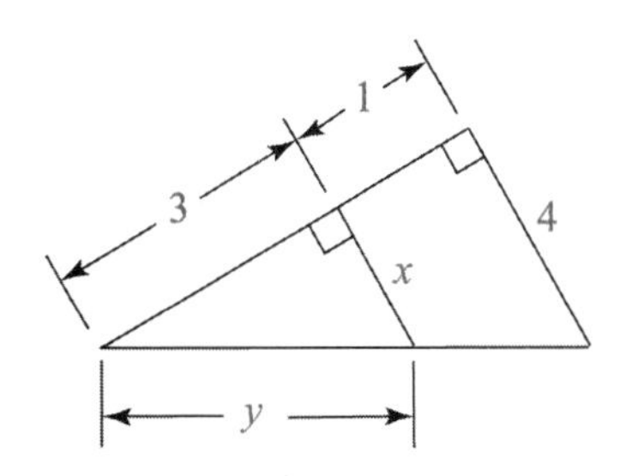

FIGURE 16-22

17.

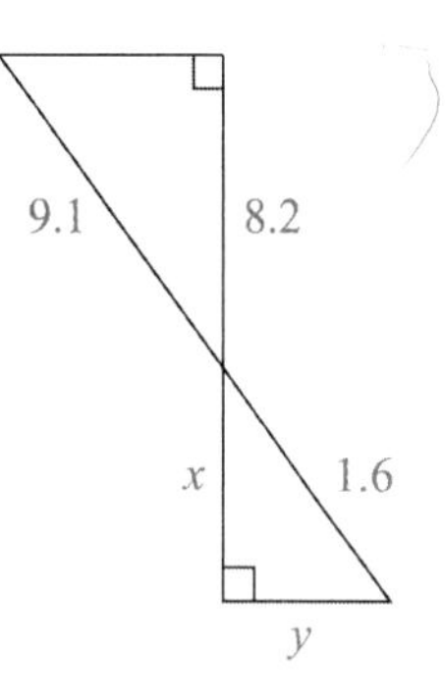

FIGURE 16-23

18.

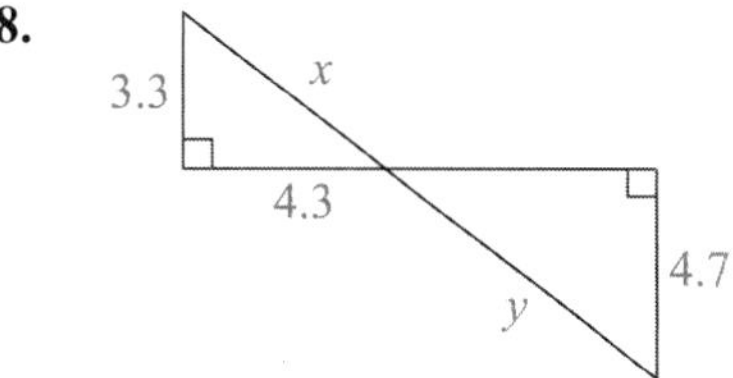

FIGURE 16-24

19.

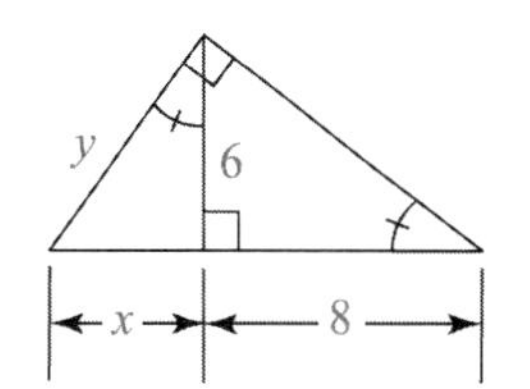

FIGURE 16-25 (Note equal angles)

20.

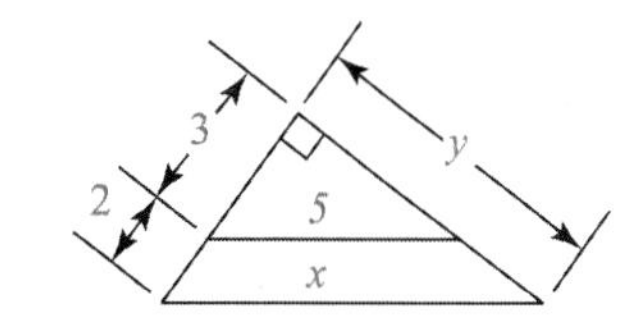

FIGURE 16-26

Applied Problems

21. A ladder 8.0 ft long is placed 4.8 ft from a wall. How far up the wall does the ladder reach?

22. Two high-speed commuter trains leave a station at exactly 6 P.M. One train travels due north averaging 80 mi/h, and the other travels due west averaging 70 mi/h. How far apart are they at 6:30 P.M.?

Hint: First find the distances traveled where Distance = (Rate)(Time).

23. Roxanne cycles 8 km south, 9 km east, and then 4 km further south. How far is she from her starting point?

Hint: Draw one large right triangle.

24. A rectangular doorway is 1.4 m wide by 2.9 m high. Can a circular tabletop 3.3 m in diameter fit through the doorway? Find the diagonal of the rectangle.

25. The shadow of a building is 50 ft long. At the same time, a tree 36 ft tall casts a shadow 20 ft long. How tall is the building?

    Note: Triangles are similar.

26. The shadow of a woman standing 10 m from a streetlight is 1.5 m long. If the woman is 1.8 m tall, how high is the streetlight?

    Note: Triangles are similar.

Applications to Electronics

27. In Figure 16.16, find X_C in kilohms when $R = 3.9$ kΩ and $Z = 7.5$ kΩ.

28. In Figure 16.16, find Z in kilohms when $X_C = 680$ kΩ and $R = 1.2$ kΩ.

29. Figure 16-27(a) shows an ac circuit containing a resistance R and an inductance L in series (RL series circuit). The impedance Z is related to the resistance R and the inductive reactance X_L by the impedance triangle shown in Figure 16-27(b):

$$R^2 + X_L^2 = Z^2$$

    Find Z in kilohms when $X_L = 25$ kΩ and $R = 15$ kΩ. See Example 16.7.

30. In problem 29, find R in kilohms when $X_L = 68$ kΩ and $Z = 90$ kΩ. See Example 16.7.

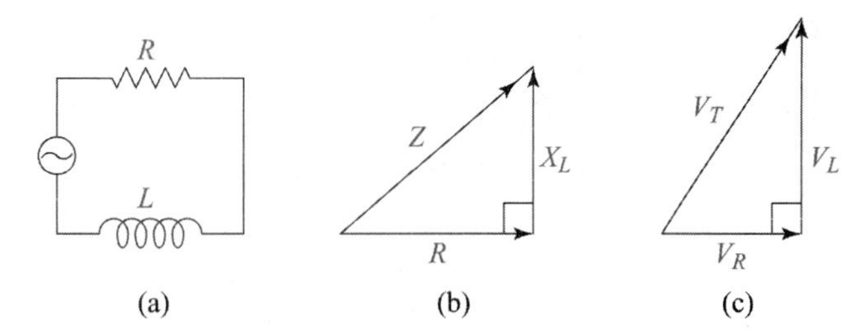

FIGURE 16-27 *RL* series circuit for problems 29 through 32.

31. In the RL series circuit in Figure 16-27, the total voltage V_T is related to the voltage across the inductance V_L and the voltage across the resistance V_R by the voltage triangle shown in Figure 16-27(c):

$$V_R^2 + V_L^2 = V_T^2$$

    Find V_L in volts when $V_T = 20$ V and $V_R = 12$ V.

32. In problem 31, find V_T in volts when $V_L = 80$ V and $V_R = 110$ V.

33. The dimensions of a rectangular computer screen are 27 cm × 20 cm. The screen size is to be scaled down proportionally for a notebook model. The width of the notebook screen is to be 125 mm. What should be the length of the notebook screen in millimeters? See Example 16.9.

34. The scale of a computer circuit image is 35:1. If the length of a diode in the image is 2.5 cm, how long is the actual diode in millimeters? See Example 16.9.

16.3 Trigonometric Ratios

Right triangles that have the same angles are similar. Therefore, the sides are in proportion, and they have the same ratios for any two corresponding sides. These trigonometric ratios or trigonometric functions are very useful for finding unknown sides and angles of right triangles when you know some of the sides or angles. For the right triangle in Figure 16-28, the three basic trigonometric ratios for $\angle A$ are:

Definitions **Trigonometric Ratios**

$$\sin A \text{ (sine)} = \frac{\text{Opposite side}}{\text{Hypotenuse}} = \frac{a}{c}$$

$$\cos A \text{ (cosine)} = \frac{\text{Adjacent side}}{\text{Hypotenuse}} = \frac{b}{c} \qquad \textbf{(16.4)}$$

$$\tan A \text{ (tangent)} = \frac{\text{Opposite side}}{\text{Adjacent side}} = \frac{a}{b}$$

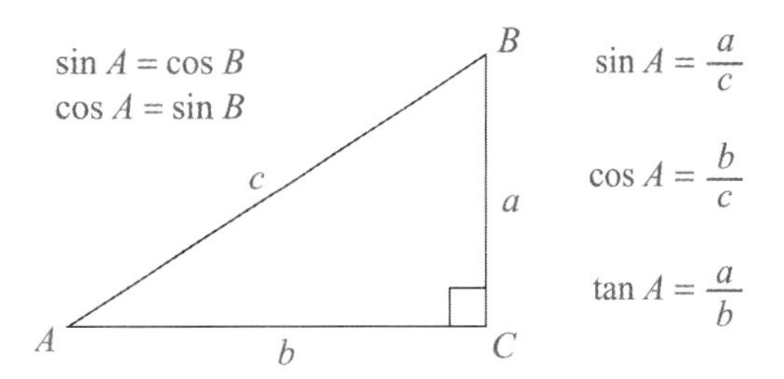

FIGURE 16-28 Trigonometric ratios for Example 16.10.

Following are three mnemonics for remembering the trigonometric ratios (**B**old **L**etters):

"**S**–**OH**, **C**–**AH**, **T**–**OA**" (**S**in–**O**pp/**H**yp, **C**os–**A**dj/**H**yp, **T**an–**O**pp/**A**dj)

"**S**ome **O**ld **H**orses **C**an **A**lways **H**ear **T**heir **O**wners **A**pproaching"

"**O**scar **H**ad **A** **H**eap **O**f **A**pples"

Study the next example, which shows how to apply the trigonometric ratios (16.4).

EXAMPLE 16.10

Given the triangle in Figure 16-28 with $a = 9$ and $b = 12$, find the values of the sine, cosine, and tangent of $\measuredangle A$ and $\measuredangle B$.

Solution First calculate the hypotenuse c using the Pythagorean theorem:

$$c^2 = 9^2 + 12^2 = 81 + 144 = 225$$
$$c = \sqrt{225} = 15$$

Then apply the definitions of the trigonometric ratios (16.4):

$$\sin A = \frac{a}{c} = \frac{9}{15} = 0.60 \qquad \sin B = \frac{b}{c} = \frac{12}{15} = 0.80$$

$$\cos A = \frac{b}{c} = \frac{12}{15} = 0.80 \qquad \cos B = \frac{a}{c} = \frac{9}{15} = 0.60$$

$$\tan A = \frac{a}{b} = \frac{9}{12} = 0.75 \qquad \tan B = \frac{b}{a} = \frac{12}{9} = 1.33$$

(EQUAL: $\sin A$ and $\cos B$; EQUAL: $\cos A$ and $\sin B$)

Note that for angle A, the opposite side is a, and the adjacent side is b. For angle B, side b is opposite, and side a is adjacent. Observe that $\sin A = \cos B$ and $\cos A = \sin B$. This is always true when $\measuredangle A$ and $\measuredangle B$ are complementary, since the side opposite $\measuredangle A$ is the same side that is adjacent to $\measuredangle B$. It follows that: *The sine of an angle equals the cosine of its complement.* For example, $\sin 60° = \cos 30°$, $\sin 40° = \cos 50°$, $\sin 10.5° = \cos 79.5°$, and so on. Because of this relationship, sine and cosine are called cofunctions.

EXAMPLE 16.11

Find the value of each trigonometric ratio using the calculator.

(a) $\sin 50°$ **(b)** $\tan 1.2$ rad **(c)** $\cos \frac{\pi}{5}$ rad

Solution

(a) To find the sin 50°, set the calculator for degrees by pressing the [DRG] or [MODE] key. The display should show "DEG" or "D." Some calculators are in the degree mode when you turn them on. Press the trigonometric function key and then enter the angle, or vice versa, depending on your calculator:

DAL: [sin] 50 [=] → 0.766

Not DAL: 50 [sin] → 0.766

(b) To find the tan 1.2 rad, first set the calculator for radians using the [DRG] or [MODE] key. The display should show "RAD" or "R." Then proceed as in (a):

DAL: [DRG] [tan] 1.2 [=] → 2.57

Not DAL: [DRG] 1.2 [tan] → 2.57

Remember to set the calculator back to degrees after you are finished working with radians.

(c) To find the cos $\frac{\pi}{5}$ rad, first set the calculator for radians as in (b):

DAL: [DRG] [cos] [(] [π] [÷] 5 [)] [=] → 0.809

Not DAL: [DRG] [π] [÷] 5 [=] cos → 0.809

You can also first change the angle to degrees and then find the cosine:

[π] [÷] 5 [×] 180 [÷] [π] [=] → 36°

DAL: [cos] 36 [=] → 0.809

Not DAL: 36 [cos] → 0.809

If any two sides of a right triangle are known, or if one side and one acute angle are known, any other side or angle can be found using the trigonometric ratios as shown in the following examples.

EXAMPLE 16.12

Given a right triangle with $\measuredangle A = 36°$ and hypotenuse $c = 8$, find the missing angles and sides of the triangle.

Solution You need to find $\measuredangle B$ and the two legs a and b of the triangle. Since $\measuredangle B$ is complementary to $\measuredangle A$:

$$\measuredangle B = 90° - 36° = 54°$$

When you know the hypotenuse, you can use sine or cosine to find a and b. Using $\sin A$, set up an equation to find side a as follows:

$$\sin A = \frac{a}{c}$$

Then substitute the values for $\measuredangle A$ and c and solve for a:

$$\sin 36° = \frac{a}{8}$$

Multiply both sides by 8: $$a = 8(\sin 36°) \approx 8(0.588) \approx 4.70$$

Using $\cos A$, solve for b as follows:

$$\cos A = \frac{b}{c}$$

$$\cos 36° = \frac{b}{8}$$

Multiply both sides by 8: $$b = 8(\cos 36°) \approx 8(0.809) \approx 6.47$$

You can also use the Pythagorean theorem to find b once you have the value for a.

ERROR BOX

A common error to watch out for is choosing the wrong trigonometric ratio to find the side of a triangle. If you know one acute angle and one side of a right triangle, remember this rule to help you determine which trigonometric ratio to use to find another side:

If you know or are looking for the hypotenuse, use sine or cosine; otherwise use tangent. For example, suppose you are given $\measuredangle A = 40°$ and the hypotenuse $c = 5$. If you want to find side b, use cosine:

$$\cos A = \frac{b}{c} \rightarrow \cos 40° = \frac{b}{5}$$

You can also find side b using the sine with the complementary angle $90° - 40° = 50°$:

$$\sin B = \frac{b}{c} \rightarrow \sin 50° = \frac{b}{5}$$

If you are given $\measuredangle B = 25°$ and side $a = 7$ and want to find side b, use tangent:

$$\tan B = \frac{b}{a} \rightarrow \tan 25° = \frac{b}{7}$$

See if you can do the practice problems correctly.

Practice Problems: For each given angle and side, find the side in parentheses to three significant digits:

1. $A = 50°, c = 5; (a)$ **2.** $B = 20°, c = 10; (b)$ **3.** $B = 30°, c = 4; (a)$
4. $A = 55°, c = 24; (b)$ **5.** $A = 70°, b = 2; (a)$ **6.** $B = 9°, a = 7.5; (b)$
7. $B = 15°, b = 6; (a)$ **8.** $A = 26°, a = 0.45; (b)$ **9.** $A = 20°, a = 3; (c)$
10. $B = 60°, a = 1.2; (c)$

EXAMPLE 16.13

Given a right triangle with $a = 6.2$ and $c = 8.5$. Find $\measuredangle A$.

Solution In this example, you are given the sides of the triangle and need to find $\measuredangle A$. You have to work the trigonometric ratios in reverse as follows. First set up an equation to find sin A by applying the definition (16.4):

$$\sin A = \frac{a}{c} = \frac{6.2}{8.5} \approx 0.729$$

Then, to find an angle knowing the value of a trigonometric function of the angle, you use the *inverse* trigonometric function: $\sin^{-1}$(or $\cos^{-1}$ or $\tan^{-1}$):

$$\text{Angle } A = \sin^{-1}(0.729)$$

On the calculator, inverse sine is [sin⁻¹], [INV] [sin], or [2nd] [sin]. Press the inverse function key and enter the value (or vice versa):

DAL: [sin⁻¹] 0.729 [=] $\rightarrow 46.8°$
Not DAL: 0.729 [sin⁻¹] $\rightarrow 46.8°$

EXAMPLE 16.14

Close the Circuit

Figure 16-29(a) on page 334 shows an ac circuit containing a resistance R and a inductance L in series (RL series circuit). The impedance Z is related to the resistance R and the inductive reactance X_L by the impedance triangle shown in Figure 16-29(b). If $X_L = 10$ kΩ and $R = 5.6$ kΩ, find the angle θ (Greek theta) between Z and R and the value of the impedance Z. Round answers to two significant digits.

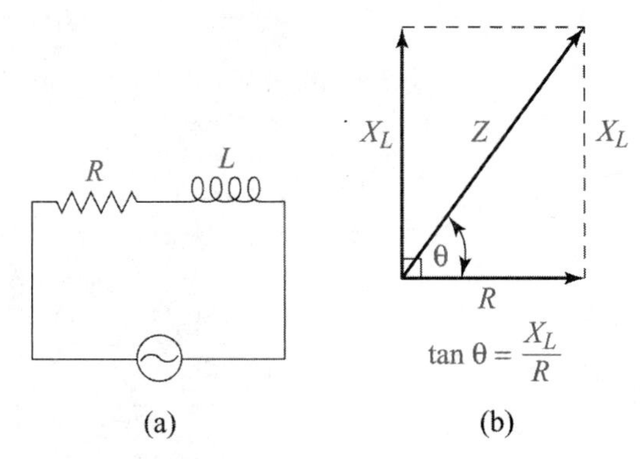

FIGURE 16-29 *RL* series circuit and impedance triangle for Example 16.14.

Solution The trigonometric ratio of the given values X_L and R is the tangent, since X_L is the opposite side and R is the adjacent side. Therefore, use the tangent to find θ:

$$\tan\theta = \frac{\text{Opposite}}{\text{Adjacent}} = \frac{X_L}{R}$$

Substitute the given values and calculate the value of the tangent:

$$\tan\theta = \frac{10 \text{ k}\Omega}{5.6 \text{ k}\Omega} \approx 1.79$$

Then find angle θ using the inverse tangent [tan⁻¹] on the calculator:

DAL: [tan⁻¹] 1.79 [=] → 61°

Not DAL: 1.79 [tan⁻¹] → 61°

To find Z, use the Pythagorean theorem:

$$X_L^2 + R^2 = Z^2$$

$$10^2 + 5.6^2 = Z^2$$

$$Z = \sqrt{10^2 + 5.6^2} = \sqrt{131.36} \approx 11 \text{ k}\Omega$$

The next example shows how a surveyor uses trigonometry to find the height of a cliff by measuring an angle.

EXAMPLE 16.15 To find the height of a cliff, Sergio, a surveyor, chooses a point 100 m from the base of a cliff and sets up his transit (a telescope that can measure horizontal and vertical angles). He measures the *angle of elevation* (angle above the horizontal) of the top of the cliff to be 36.4°. See Figure 16-30. Assuming the angle of elevation is measured from the ground, how high is the cliff?

Solution Set up an equation using the side of the triangle you know, the side you are trying to find, and the trigonometric function of the angle that is the ratio of these two sides. The known side (100 m) is adjacent to the angle of 36.4°. The unknown side (the cliff height) is opposite to 36.4°. Therefore, choose the tangent function, which

---

Answers to Error Box Problems, page 333:
1. 3.83 **2.** 3.42 **3.** 3.46 **4.** 14.3 **5.** 5.49 **6.** 1.19 **7.** 22.4
8. 0.923 **9.** 8.77 **10.** 2.40

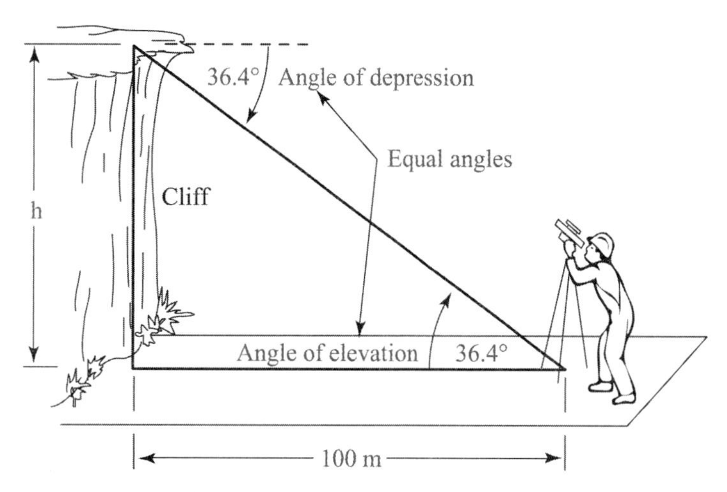

FIGURE 16-30 Surveyor and cliff height for Example 16.15.

is opposite over adjacent. Letting h = cliff height, the equation and solution to two significant digits is:

$$\tan 36.4° = \frac{h}{100}$$
$$h = 100(\tan 36.4°)$$
$$h = 100(0.737) \approx 74 \text{ m}$$

DAL: 100 [×] [tan] 36.4 [=] → 74 m

Not DAL: 36.4 [tan] [×] 100 [=] → 74 m

For an observer on top of the cliff looking down at the surveyor, the *angle of depression* is the angle measured below the horizontal and is equal to the angle of elevation.

EXERCISE 16.3

In exercises 1 through 54, round answers to three significant digits.

In exercises 1 through 20, apply the definitions and find the sine, cosine, and tangent of ∡A and ∡B for each right triangle.

1. $a = 8, b = 15, c = 17$
2. $a = 6, b = 8, c = 10$
3. $a = 6, b = 4.5, c = 7.5$
4. $a = 2.5, b = 6, c = 6.5$
5. $a = 5, b = 7, c = 8.6$
6. $a = 0.34, b = 0.12, c = 0.36$
7. $a = 10, b = 12$
8. $a = 2.3, b = 3.2$
9. $b = 4.5, c = 8.2$
10. $b = 20, c = 25$
11. $a = 0.30, c = 0.90$
12. $a = 6.6, c = 12.2$
13. $a = 1.0, b = 1.0$
14. $a = 9.1, b = 9$
15. $b = 5.0, c = 8.66$
16. $b = 1.0, c = 2.0$
17. $a = 120, c = 160$
18. $a = 500, c = 800$
19. $a = x, c = 2x$
20. $a = 3x, b = 4x$

In exercises 21 through 36, find the value of each trigonometric ratio.

21. $\sin 30°$
22. $\cos 45°$
23. $\tan 10°$
24. $\tan 60°$
25. $\cos 80.6°$
26. $\sin 39.3°$
27. $\tan 3.5°$
28. $\tan 7.2°$
29. $\tan 1.1$ rad
30. ttan 1.3 rad
31. $\cos 0.55$ rad
32. $\sin 0.21$ rad
33. $\cos \frac{\pi}{6}$ rad
34. $\sin \frac{\pi}{3}$ rad
35. $\tan \frac{\pi}{4}$ rad
36. $\tan \frac{\pi}{10}$ rad

In exercises 37 through 50, using the inverse function, find the value of the given angle between 0° and 90°. See Example 16.13.

37. $\cos A = 0.50$
38. $\cos A = 0.40$
39. $\sin B = 0.33$
40. $\sin B = 0.65$
41. $\tan \theta = 1.00$
42. $\tan \theta = 1.73$
43. $\tan \theta = 2.14$
44. $\tan \theta = 0.577$
45. $\tan \theta = 0.233$
46. $\tan \theta = 5.61$
47. $\tan \phi = 0.416$ (ϕ = phi)
48. an $\phi = 0.0989$
49. $\sin x = 0.866$
50. $\cos x = 0.707$

In exercises 51 through 62, find all the missing sides and acute angles for each right triangle. In exercises 59 and 60, find the angles in radians. See Example 16.12.

51. $A = 50°, c = 3.0$

52. $B = 17°, c = 8.2$

53. $B = 25.5°, a = 10.3$

54. $A = 53.2°, a = 0.45$

55. $a = 5.0, c = 8.0$

56. $b = 330, c = 550$

57. $a = 2.5, b = 6.2$

58. $a = 35, b = 25$

59. $B = 0.73$ rad, $a = 4.5$

60. $A = 1.1$ rad, $a = 3.3$

61. $\tan A = 1.0, a = 2.0$

62. $\sin B = 0.80, c = 10$

Applied Problems

In problems 63 through 72, round answers to two significant digits.

63. The angle of elevation of a building is 56° at a point 60 m from its base. How high is the building? See Example 16.15.

64. The angle of elevation of a cliff is 23° at a point 100 ft from its base. How high is the cliff? See Example 16.15.

65. The angle of elevation of a lighthouse from a sailboat is 11°. If the top of the lighthouse is 150 ft above sea level, how far away is the sailboat from the base of the lighthouse?

66. The safe angle θ for the bank of a highway curve is given by:

$$\tan\theta = \frac{v^2}{rg}$$

where v = velocity, r = radius of curve, and gravitational acceleration $g = 9.81$. If $v = 25$ and $r = 290$, find the value of the safe angle. Substitute the values, find $\tan\theta$, and then find θ.

Applications to Electronics

67. In Figure 16.29, given $X_L = 2.6$ kΩ and $R = 3.7$ kΩ, find the angle θ and the impedance Z.

68. For the RL series circuit in Example 16.14, the voltage across the resistance V_R and the voltage across the inductance V_L are related to the total voltage V_T by the voltage triangle shown in Figure 16-31. If $V_R = 80$ V and $V_L = 60$ V, find V_T and the angle θ.

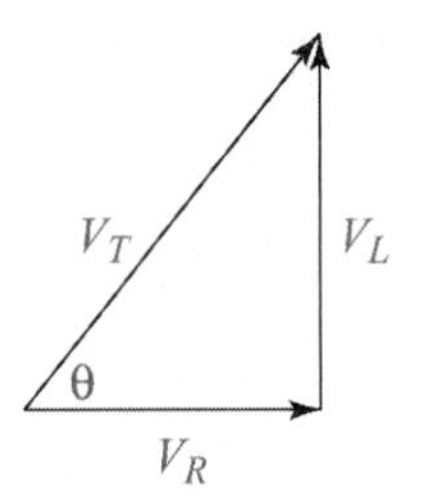

FIGURE 16-31 Voltage triangle in an *RL* circuit for problem 68.

69. In an ac circuit, the instantaneous voltage v is given by:

$$v = V_P \sin\theta$$

where V_P = peak voltage and θ = angle of rotation in radians.

Find v when $V_P = 300$ V and $\theta = 0.785$ rad.

70. In problem 69, find the angle θ between 0 and π/2 radians when $V_P = 300$ V and $v = 150$ V.

71. The true or active power of an ac circuit due to the resistance is given by:

$$P = VI\cos\phi$$

where ϕ (Greek phi) is called the phase angle. Find ϕ in degrees and radians when $V = 120$ V, $I = 1.5$ A and $P = 100$ W.

72. The reactive power of an ac circuit due to the inductance and/or the capacitance is given by

$$P_x = VI\sin\phi$$

where ϕ is called the phase angle. Find ϕ in degrees and radians when $V = 240$ V, $I = 500$ mA, and $P_x = 100$ W.

CHAPTER HIGHLIGHTS

16.1 ANGLES AND TRIANGLES

Radian Measure

$$\pi \text{ rad} = 180° \quad (16.1)$$

$$\frac{\pi}{2}\text{ rad} = 90° \qquad \frac{\pi}{3}\text{ rad} = 60° \qquad \frac{\pi}{4}\text{ rad} = 45°$$

$$\frac{\pi}{6}\text{ rad} = 30° \qquad \frac{3\pi}{2}\text{ rad} = 270° \qquad 2\pi\text{ rad} = 360°$$

Definition of Radian

$$1\text{ rad} = \frac{180°}{\pi} \approx \frac{180°}{3.142} \approx 57.3° \quad (16.2)$$

• • •

Conversion Factors

- To change from radians to degrees, multiply by $\frac{180°}{\pi}$ or 57.3°.
- To change from degrees to radians, multiply by $\frac{\pi}{180°}$ or divide by 57.3°.

Key Example:

$$\frac{\pi}{10}\text{ rad} = \left(\frac{\pi}{10}\right)\left(\frac{180}{\pi}\right) = 18°$$

$$2.1\text{ rad} = 2.1\text{ rad}\,(57.3°) \approx 120°$$

$$15° = (15°)\left(\frac{\pi}{180°}\right) = \frac{\pi}{12}\text{ rad} \approx 0.26\text{ rad}$$

Angles of a Triangle

The sum of the internal angles of any triangle equals 180°.

16.2 PYTHAGOREAN THEOREM

Pythagorean Theorem

In a right triangle, the sum of the squares of the sides equals the square of the hypotenuse:

$$a^2 + b^2 = c^2 \qquad \textbf{(16.3)}$$

Key Example: In a right triangle:

If $a = 5$ and $b = 7$

then $5^2 + 7^2 = c^2 \rightarrow c = \sqrt{5^2 + 7^2} \approx 8.6$

If $b = 6.8$ and $c = 8.2$

then $a^2 + 6.8^2 = 8.2^2 \rightarrow a = \sqrt{8.2^2 - 6.8^2} \approx 4.6$

Similar Triangles

Two triangles are similar when all the angles of one triangle are equal to the corresponding angles of the other triangle.

Similar triangles have corresponding sides in proportion.

Key Example: Given two similar right triangles *ABC* and *DEF* in Figure 16-32. The proportion for the three ratios of the sides of the large triangle to the corresponding sides of the small triangle is:

$$\frac{\text{Large Triangle}}{\text{Small Triangle}} : \frac{10}{5} = \frac{x}{4} = \frac{6}{3}$$

To find x, solve the proportion:

$$\frac{x}{4} \bowtie \frac{6}{3} \rightarrow 3x = 24 \rightarrow x = 8$$

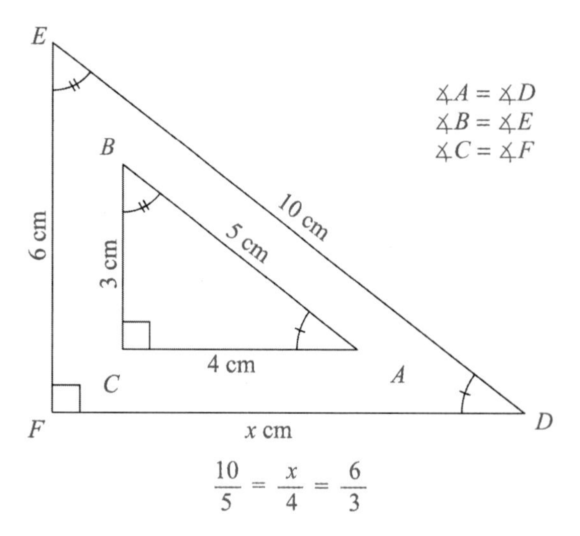

FIGURE 16-32 Key Example: Similar triangles and proportions.

16.3 TRIGONOMETRIC RATIOS

Trigonometric Ratios

$$\sin A\,(\text{sine}) = \frac{\text{Opposite side}}{\text{Hypotenuse}} = \frac{a}{c} \qquad \textbf{(16.4)}$$

$$\cos A\,(\text{cosine}) = \frac{\text{Adjacent side}}{\text{Hypotenuse}} = \frac{b}{c}$$

$$\tan A\,(\text{tangent}) = \frac{\text{Opposite side}}{\text{Adjacent side}} = \frac{a}{b}$$

Remember the mnemonic:

"**S**–OH, **C**–AH, **T**–OA"
(**S**in–**O**pp/**H**yp, **C**os–Adj/**H**yp, **T**an–**O**pp/**A**dj)

Key Example:

If $c = 4.6$ and $\measuredangle B = 40°$, side a is:

$$\cos B = \cos 40° = \frac{a}{c} = \frac{a}{4.6} \rightarrow a = 4.6(\cos 40°) \approx 3.5$$

If $a = 1.2$ and $b = 1.8$, $\measuredangle A$ is:

$$\tan A = \frac{a}{b} = \frac{1.2}{1.8} \approx 0.667 \rightarrow A = \tan^{-1}(0.667) \approx 33.7°$$

REVIEW EXERCISES

Round answers to three significant digits unless specified.

In exercises 1 through 16, change from radians to degrees or vice versa.

1. $\frac{\pi}{3}$ rad
2. $\frac{5\pi}{6}$ rad
3. $\frac{\pi}{15}$ rad
4. $\frac{3\pi}{8}$ rad
5. 1.5 rad
6. 0.34 rad
7. 0.014 rad
8. 2.3 rad
9. 30°
10. 135°
11. 20°
12. 180°
13. 300°
14. 36°
15. 140°
16. 46°

In exercises 17 and 18, find angles A, B, and C.

17.

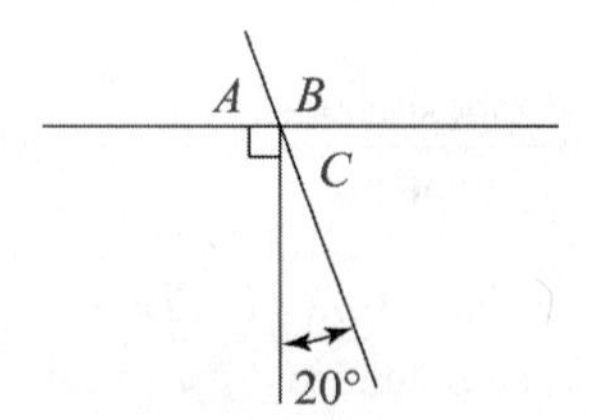

FIGURE 16-33

18.

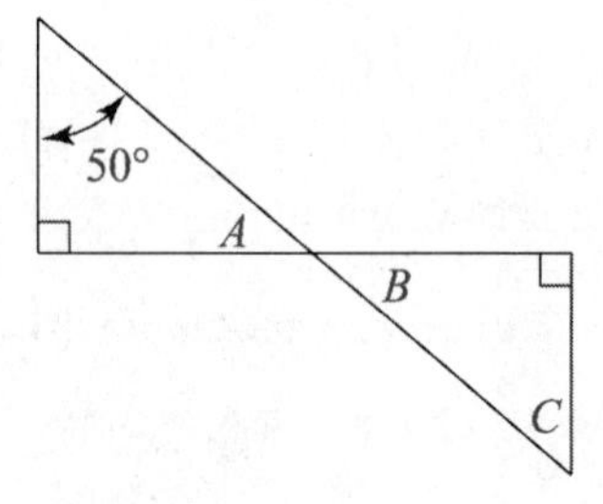

FIGURE 16-34

In exercises 19 through 26, find the missing side of each right triangle.

19. $a = 1.5, b = 2.0$
20. $a = 24, b = 10$
21. $b = 504, c = 80$
22. $a = 4.7, c = 5.6$
23. $a = 150, c = 250$
24. $b = 7.5, c = 15$
25. $a = 13, b = 13$
26. $a = 90, b = 75$

In exercises 27 and 28, given that the triangles are similar, find x and y.

27.

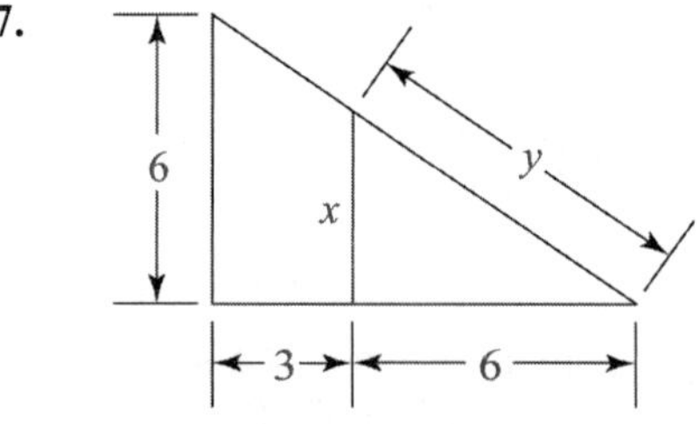

FIGURE 16-35

28.

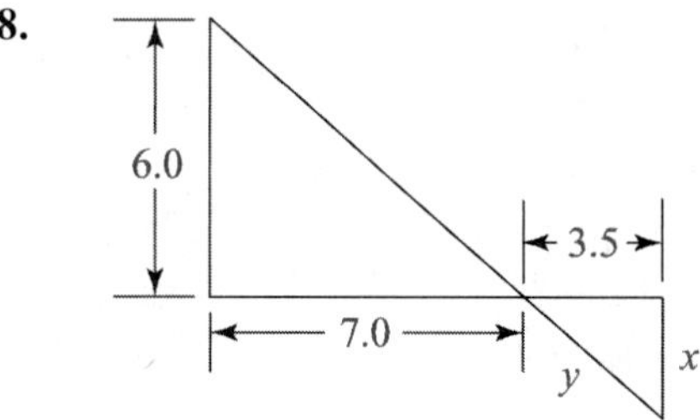

FIGURE 16-36

In exercises 29 through 38, find the sine, cosine, and tangent of ∡A and ∡B for each right triangle.

29. $a = 9.6, b = 7.2, c = 12$
30. $a = 16, b = 30, c = 34$
31. $a = 10, b = 24$
32. $a = 1.5, b = 2.5$
33. $a = 51, c = 75$
34. $a = 0.12, c = 0.16$
35. $b = 120, c = 150$
36. $b = 560, c = 680$
37. $a = 40, b = 60$
38. $a = 0.055, b = 0.034$

In exercises 39 through 48, find the value of each trigonometric function.

39. sin 60°
40. cos 45°
41. tan 56.7°
42. tan 70°
43. cos 89°
44. sin 2.5°
45. cos 0.77 rad
46. sin 2.2 rad
47. $\tan \frac{2\pi}{9}$ rad
48. $\tan \frac{\pi}{6}$ rad

In exercises 49 through 56, using the inverse function, find the value of the given angle between 0° and 90°.

49. $\sin A = 0.200$
50. $\cos A = 0.350$
51. $\tan B = 0.577$
52. $\tan B = 5.67$

53. $\cos B = 0.955$

54. $\sin B = 0.055$

55. $\tan \theta = 1.60$

56. $\tan \theta = 0.977$

In exercises 57 through 64, find all the missing sides and acute angles for each right triangle.

57. $A = 30°, b = 5$

58. $B = 25°, b = 8.8$

59. $B = 56°, c = 33$

60. $A = 28°, c = 1.3$

61. $a = 1.5, c = 2.0$

62. $b = 2.5, c = 6.5$

63. $a = 10, b = 15$

64. $a = 120, b = 300$

Applied Problems

Round all answers to two significant digits.

65. Etta sails her boat 20 mi south and then 30 mi east. How far is she from her starting point?

66. The shadow of a building is 100 ft long. At the same time, a telephone pole 30 ft tall casts a shadow 18 ft long. How tall is the building?

67. The angle of elevation of the top of a wind generator is 75° at a point 20 m from the base of the tower. How high is the generator?

68. A rocket travels 500 m at a constant angle of ascent (angle above the horizontal) while its vertical increase in altitude is 200 m. Find the angle of ascent.

Applications to Electronics

69. Figure 16-37 shows the impedance triangle for an *RL* series circuit. If $R = 8.2$ kΩ and $X_L = 2.4$ kΩ:
(a) Find the impedance Z in kilohms.
(b) Find the phase angle θ.

70. In problem 69, given $R = 820\ \Omega$ and $Z = 1.1$ kΩ:
(a) Find the inductive reactance X_L in ohms.
(b) Find the phase angle θ.

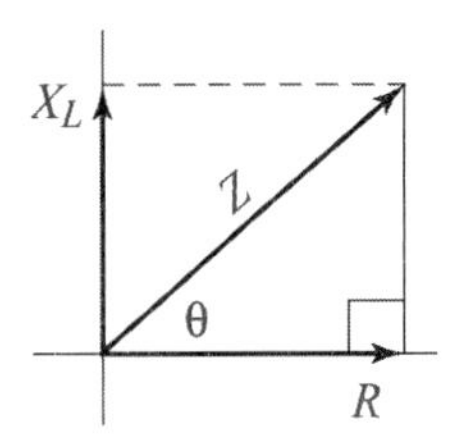

FIGURE 16-37 *RL* series circuit impedance triangle for problem 69.

71. An integrated circuit schematic on a computer monitor is shown at a scale of 10:3. If the dimensions of a rectangular component in the schematic are 1.2 cm by 2.4 cm, what are the actual dimensions of the component in millimeters? Note the units.

72. A rectangular circuit chip measures 1.3 cm by 2.8 cm. What are the angles that the diagonal makes with the sides of the rectangle?

73. The angle α in degrees of a laser beam is expressed by:

$$\alpha = \tan^{-1}\left(\frac{w}{2d}\right)$$

where w = width of the beam and d = distance from the source. Find α in degrees when:
(a) $w = 1.0$ m and $d = 1000$ m.
(b) $w = 50$ cm and $d = 100$ m.

Note: Units must be the same.

74. The following equation occurs when computing the rotational velocity of an electrical generator:

$$\omega = (10^2)(\sin\theta)\left(1 + \frac{\cos\theta}{3}\right)$$

Find ω (Greek omega) in rad/s when θ equals:
(a) $\frac{\pi}{4}$
(b) 1.19 rad
(c) 60° (change to radians first)

CHAPTER 17

Trigonometry of the Circle

CHAPTER OBJECTIVES

In this chapter, you will learn:

- The definitions of the trigonometric functions for any angle in a circle.
- How to work with reference angles and reference triangles.
- How to find the trigonometric functions for special angles.
- How to use the inverse trigonometric functions to find angles in a circle.

An angle is a measure of rotation, or circular motion, and can be any positive or negative value, depending on the direction and amount of rotation. The trigonometric functions of an angle can be defined for any positive or negative angle. This extends their definition for an angle in a right triangle and leads to important applications in ac circuits where rotational angles of any value are encountered. As a result, the inverse trigonometric functions $\sin^{-1}$ and $\tan^{-1}$ can assume values that are negative angles, and $\cos^{-1}$ can assume values that are more than 90°. These inverse functions are used to find angles in electronic circuit problems. This chapter is therefore a basic building block for Chapters 18 through 22 on vectors, phasors, and ac circuits.

• • •

17.1 Trigonometric Functions

Study the circle in Figure 17-1 whose center is at the origin of a rectangular coordinate system. The angle θ in the circle determines the direction of the radius r. It is a positive angle measured from the positive x axis (initial side), to the radius (terminal side) in a counterclockwise direction. Counterclockwise rotation is therefore positive, and clockwise rotation is negative. The circle in Figure 17-1 contains four quadrants: Quadrant I: 0° to 90°, Quadrant II: 90° to 180°, Quadrant III: 180° to 270°, and Quadrant IV: 270° to 360°.

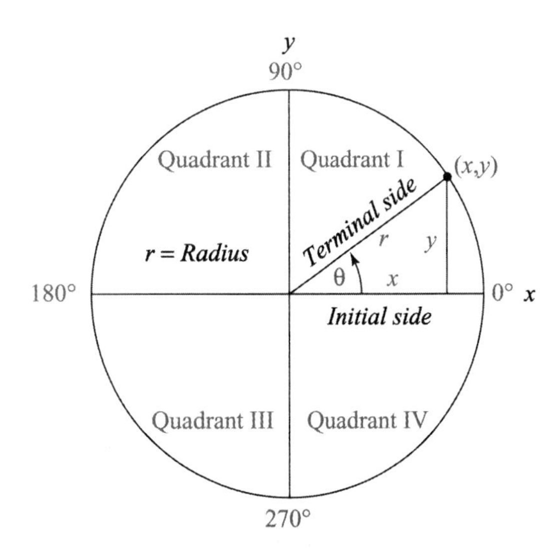

FIGURE 17-1 Trigonometry of the circle.

The three trigonometric functions of an angle θ whose terminal side passes through a point (x,y) on a circle of radius r are:

Definitions

Trigonometric Functions

$$\sin\theta = \frac{y}{r} \qquad \cos\theta = \frac{x}{r} \qquad \tan\theta = \frac{y}{x} \tag{17.1}$$

It is important to note that definitions (17.1) apply to any angle, positive or negative, such as 10°, 150°, 300°, 450°, –80° –360°, etc. Definitions (17.1) extend the definitions of the trigonometric ratios for a right triangle (16.4) given by the mnemonic: **S**-OH **C**-AH **T**-OA. When the angle is in the first quadrant, the right triangle definitions apply where side x = adjacent side, side y = opposite side, and the radius r = hypotenuse. When the angle is in quadrant II, III, or IV, the definitions are applied using the coordinates (x,y) of the point on the circle and the radius r, which is always considered positive. Carefully study the following example, which shows how to apply these ideas.

EXAMPLE 17.1 Draw each angle whose terminal side passes through the given point, and find the three trigonometric functions of each angle.

(a) $\theta_1 : (4,3)$ **(b)** $\theta_2 : (-4,3)$ **(c)** $\theta_3 : (-4,-3)$ **(d)** $\theta_4 : (4,-3)$

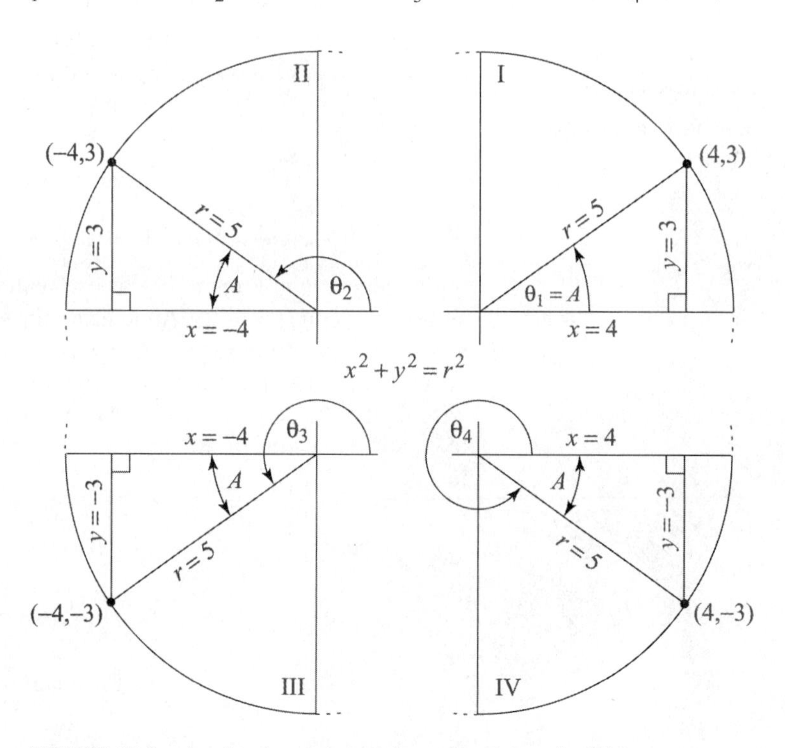

FIGURE 17-2 Angles in each quadrant for Example 17.1.

Solution The four angles, one in each quadrant, are drawn in Figure 17-2. Study the diagram closely. For each case, a perpendicular is drawn from the point on the circle to the x axis forming a right triangle called the *reference triangle.* Angle A in the triangle is called the *reference angle* and is an acute angle equal to angle θ_1 in quadrant I. The *perpendicular is always drawn to the x axis* so that side y is always opposite the reference angle A, and side x is always adjacent to angle A. This allows you to refer to angle A and the definitions of the trigonometric ratios for a right triangle when working with the angles θ_2, θ_3, or θ_4. The sides of each reference triangle are the same length in each quadrant, but x and y are not always positive. They have different signs in each quadrant. For all angles, the Pythagorean theorem applies to the right triangle and can be solved for the radius r, which is always considered positive:

$$x^2 + y^2 = r^2 \rightarrow r = +\sqrt{x^2 + y^2}$$

For $\theta_1, \theta_2, \theta_3$, or θ_4, r is the same value. If $x = +4$ or -4, and $y = +3$ or -3 it follows that $x^2 = 9$, $y^2 = 16$ and:

$$r = +\sqrt{(\pm 4)^2 + (\pm 3)^2} = \sqrt{16 + 9} = \sqrt{25} = 5$$

Apply definitions (17.1) to obtain the three trigonometric functions for each angle, observing the different signs for x and y:

(a) Quadrant I: $x = 4, y = 3, r = 5$

$$\sin \theta_1 = \frac{3}{5} = 0.60$$

$$\cos \theta_1 = \frac{4}{5} = 0.80$$

$$\tan \theta_1 = \frac{3}{4} = 0.75$$

(b) Quadrant II: $x = -4, y = 3, r = 5$

$$\oplus \Rightarrow \sin \theta_2 = \frac{+3}{+5} = \frac{3}{5} = 0.60$$

$$\cos \theta_2 = \frac{-4}{+5} = -\frac{4}{5} = -0.80$$

$$\tan \theta_2 = \frac{+3}{-4} = -\frac{3}{4} = -0.75$$

(c) Quadrant III: $x = -4, y = -3, r = 5$

$$\sin \theta_3 = \frac{-3}{+5} = -\frac{3}{5} = -0.60$$

$$\cos \theta_3 = \frac{-4}{+5} = -\frac{4}{5} = -0.80$$

$$\oplus \Rightarrow \tan \theta_3 = \frac{-3}{-4} = \frac{3}{4} = 0.75$$

(d) Quadrant IV: $x = 4, y = -3, r = 5$

$$\sin \theta_4 = \frac{-3}{+5} = -\frac{3}{5} = -0.60$$

$$\oplus \Rightarrow \cos \theta_4 = \frac{+4}{+5} = \frac{4}{5} = 0.80$$

$$\tan \theta_4 = \frac{-3}{+4} = -\frac{3}{4} = -0.75$$

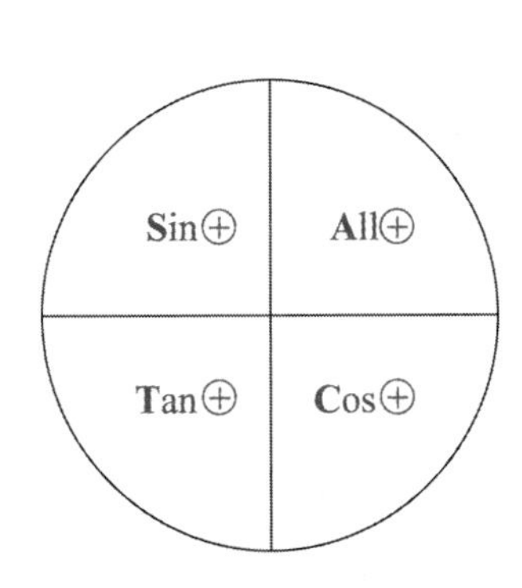

FIGURE 17-3 Positive trigonometric functions: "All Star Trig Class."

Observe that the absolute values of the functions without the signs are the same in each quadrant and are equal to the functions of the reference angle A. The actual values differ only in sign. The signs are determined by the point (x,y) on the terminal side of the angle. All the functions are positive in the first quadrant, only sine is positive in quadrant II, only tangent is positive in quadrant III, and only cosine is positive in quadrant IV. See the arrows in (b), (c), and (d) above. Study Figure 17-3, which shows a way to remember the positive functions with the phrase: "All Star Trig Class." The first letters **A, S, T, C** stand for **A**ll, **S**in, **T**an, **C**os, starting from the first quadrant and reading counterclockwise.

EXAMPLE 17.2 Given that $\sin \theta = -\frac{1}{2}$ and $\cos \theta$ is positive,

(a) Draw θ, showing the values for x, y, and r.

(b) Find the values of $\cos \theta$ and $\tan \theta$ applying definitions (17.1).

Solution

(a) First determine which quadrant the angle θ is in. Figure 17-3 shows that $\cos \theta$ is positive in the first and fourth quadrant. However, since it is also given that $\sin \theta$ is negative, θ must be in the fourth quadrant. Draw the sketch of the angle as shown in Figure 17-4.

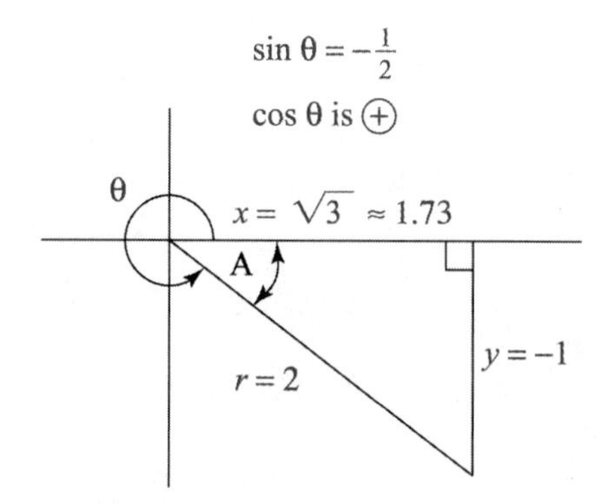

FIGURE 17-4 Fourth quadrant angle and reference triangle for Example 17.2.

You can choose any negative value for y, and any positive value for r, as long as:

$$\sin \theta = \frac{y}{r} = -\frac{1}{2}$$

The simplest set of values is $y = -1$ and $r = 2$. Then since x is positive in quadrant IV:

$$x = +\sqrt{r^2 - y^2} = \sqrt{(2)^2 - (-1)^2} = \sqrt{3} \approx 1.73$$

In Figure 17-4, the triangle is called the reference triangle. It is always drawn to the x axis so that y is opposite the reference angle A. Reference triangles and reference angles are studied further in the next section.

(b) Find the values of $\cos \theta$ and $\tan \theta$ by using the values of x, y, and r from the reference triangle in Figure 17-4:

$$\cos \theta = \frac{x}{r} = \frac{\sqrt{3}}{2} \approx 0.866$$

$$\tan \theta = \frac{y}{x} = \frac{-1}{\sqrt{3}} \approx -0.577$$

You can check these answers with the calculator by first finding the reference angle A and then angle θ. Since A is a first quadrant angle, $\sin A$ equals the absolute value of $\sin \theta$:

$$\sin A = |\sin \theta| = \left|-\frac{1}{2}\right| = \frac{1}{2} = 0.5 \text{ and therefore angle } A = \sin^{-1} 0.5$$

Then, using the calculator you have for the reference angle A:

DAL: [sin-1] 0.5 [=] → 30°

Not DAL: 0.5 [sin-1] → 30°

Since θ is in the fourth quadrant: θ = 360° – 30° = 330° and

DAL: [cos] 330 [=] → –0.866
DAL: [tan] 330 [=] → –0.577
Not DAL: 330 [cos] → –0.866
Not DAL: 330 [tan] → –0.577

These values agree with the above answers and provide a check for the problem.

EXAMPLE 17.3

Close the Circuit

Figure 17-5(a) shows an ac circuit containing a resistance R and a capacitance C in series. The voltage across the resistance V_R and the voltage across the capacitance V_C are related to the total voltage V_T as shown by the voltage triangle in Figure 17-5(b). Note that angle θ goes clockwise and is a *negative angle in the fourth quadrant.* If V_R = 12 V and V_C = 15 V:

(a) Find V_T in volts.

(b) Find sin θ, cos θ, and tan θ to three significant digits.

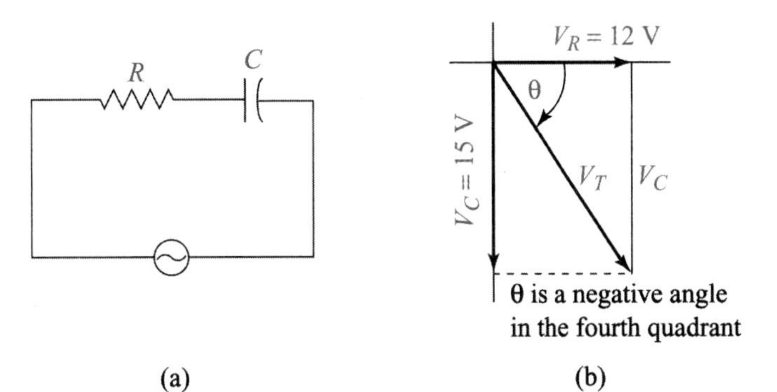

FIGURE 17-5 *RC* series circuit and voltage triangle for Example 17.3.

Solution

(a) Apply the Pythagorean theorem to find V_T:

$$V_T = \sqrt{V_R^2 + V_C^2} = \sqrt{(12)^2 + (15)^2} = \sqrt{144 + 225} = \sqrt{369} \approx 19.2 \text{ V}$$

(b) To find sin θ, cos θ, and tan θ, note that θ is in the fourth quadrant where y is negative. You can apply definition (17.1) letting $x = V_R$, $y = -V_C$, and $r = V_T$:

$$\sin\theta = \frac{y}{r} = \frac{-V_C}{V_T} = \frac{-15}{19.2} \approx -0.781$$

$$\cos\theta = \frac{x}{r} = \frac{V_R}{V_T} = \frac{12}{19.2} \approx 0.625$$

$$\tan\theta = \frac{y}{x} = \frac{-V_C}{V_R} = \frac{-15}{12} = -1.25$$

EXERCISE 17.1

Round all answers in the exercise to three significant digits.

In exercises 1 through 20, draw the angle whose terminal side passes through the given point and find the three trigonometric functions of the angle applying formulas (17.1).

1. (3,4)
2. (8,6)
3. (–8,15)
4. (–5,12)
5. (–12,–5)
6. (–9,–12)
7. (8,–6)
8. (15,–8)

9. (1.1,–1.1)
10. (2.7,–3.9)
11. (4.3,5.6)
12. (2.1,2.8)
13. (–1.6,2.2)
14. (–7.5,6.2)
15. (–15,–12)
16. (–13,–10)
17. (3.9,–8.2)
18. (9.1,–5.6)
19. (69,92)
20. (288,120)

In exercises 21 through 34, draw the angle and find the values of the other two trigonometric functions applying definitions (17.1).

21. $\sin\theta = \frac{3}{5}$, tan θ positive.

22. $\sin\theta = \frac{12}{13}$, cos θ positive.

23. $\tan\theta = -\frac{12}{5}$, sin θ negative

24. $\tan\theta = -\frac{3}{4}$, cos θ positive

25. $\cos\theta = -\frac{1}{3}$, sin θ negative

26. $\cos\theta = -\frac{1}{2}$, tan θ negative

27. $\tan\theta = \frac{3}{2}$, cos θ positive

28. tan θ = 3.00, sin θ positive

29. cos θ = 0.700, θ in quadrant IV

30. $\sin\theta = \frac{2}{5}$, θ in quadrant I

31. tan θ = –1.00, θ in quadrant II

32. $\tan\theta = \frac{4}{3}$, θ in quadrant III

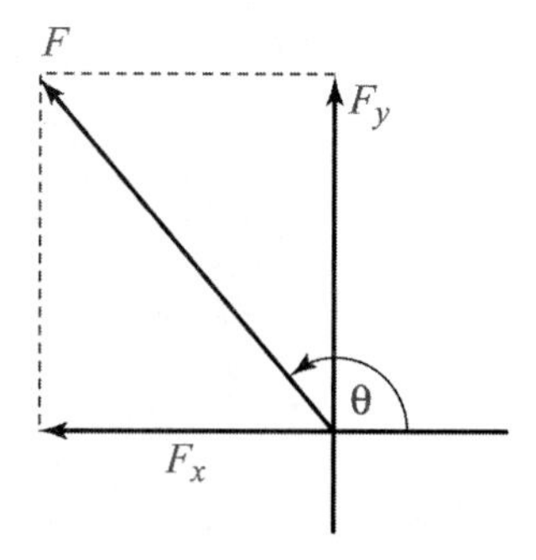

FIGURE 17-6 Components of force vector for problem 35.

33. sin θ = –0.60, θ in quadrant III

34. cos θ = 0.80, θ in quadrant IV

Applied Problems

35. A force vector F in the second quadrant has components $F_x = -7.50$ N (N = newtons) and $F_y = 3.50$ N, as shown in Figure 17-6. Find sin θ, cos θ, and tan θ.

36. Similar to problem 35, find sin θ, cos θ, and tan θ for a force vector F in the *fourth quadrant* when $F_x = 5.3$ N and $F_y = -6.1$ N.

Applications to Electronics

37. In Example 17.3, given $V_R = 14$ V and $V_C = 12$ V, find: (a) V_T (b) sin θ, cos θ, and tan θ.

38. In Example 17.3, given $V_R = 28$ V and $V_T = 40$ V, find: (a) V_C (b) sin θ, cos θ, and tan θ.

39. In an ac circuit, $\tan\phi_Z = \frac{X}{R}$ (ϕ = phi), where ϕ_Z = phase angle of the impedance, and $\tan\phi_Y = -\frac{X}{R}$, where ϕ_Y = phase angle of the admittance. If $X^2 + R^2 = Z^2$, find $\tan\phi_Z$ and $\tan\phi_Y$ when $X = 4.5$ kΩ and $Z = 5.5$ kΩ. First find R.

40. In problem 39, find $\tan\phi_Z$ and $\tan\phi_Y$ when $R = 5.1$ kΩ and $Z = 6.5$ kΩ.

17.2 Reference Angles and Special Angles

Look back at Figure 17-2. The triangle in each quadrant is called the reference triangle. *It is always drawn to the x axis so that side y is always opposite the reference angle A in the triangle.* The following rule applies to the reference angle:

Rule

Reference Angle

The absolute, or positive, value of the trigonometric function of an angle in any quadrant is equal to the trigonometric function of its reference angle.

The above rule tells you that the value of the trigonometric function of an angle will either be the same as the trigonometric function of its reference angle (positive) or will be opposite in sign (negative), depending on the quadrant. In Quadrant I the reference angle A is the same as a given angle θ For the other quadrants study Figure 17-7 carefully. It shows how the reference angle A is related to the angle θ in quadrants II, III, and IV. In quadrants II and III, angle A is equal to the difference between θ and 180°. In quadrant IV, angle A is the difference between θ and 360°. *You never calculate with 90° or 270° when working with reference angles.*

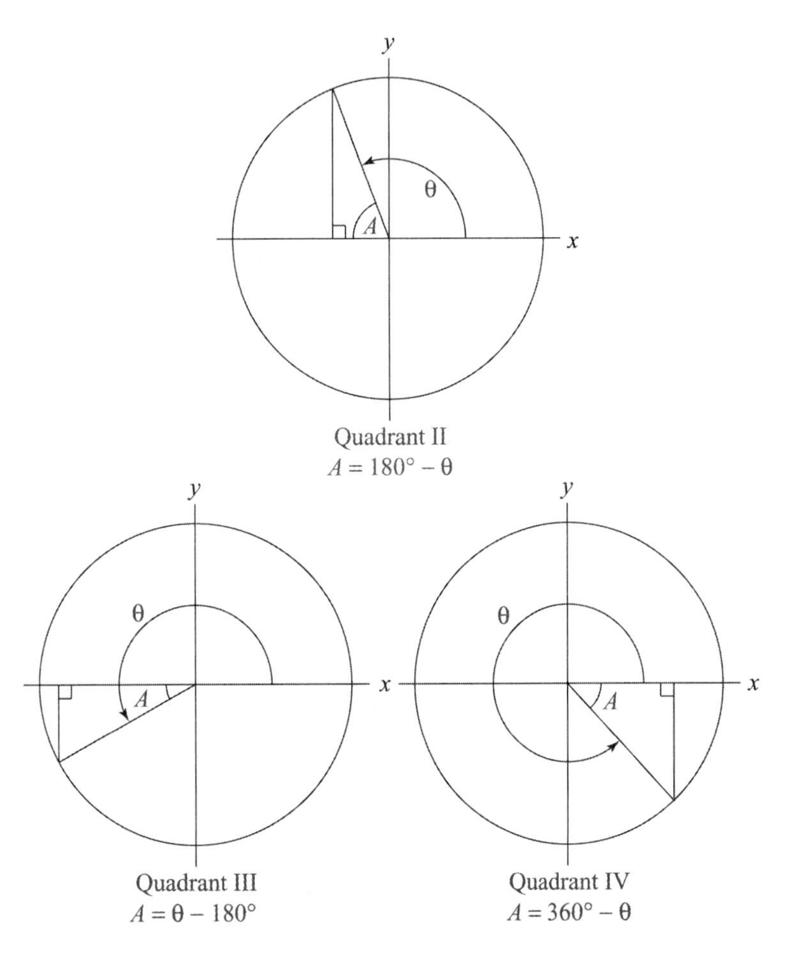

FIGURE 17-7 Reference angles and reference triangles.

EXAMPLE 17.4 Express each trigonometric function as a function of its reference angle in degrees and find the value of this function. Check by finding the value of the given angle on the calculator.

(a) sin 95° **(b)** cos 225° **(c)** tan 290° **(d)** tan 5.8 rad

Solution Study Figure 17-7 for each solution explained below.

(a) To find sin 95°, which is an angle in quadrant II, subtract 95° from 180° to find the reference angle. Then, since the sine is positive in quadrant II, the positive sine of the reference angle is equal to sin 95°:

$$\sin 95° = +\sin(180° - 95°) = +\sin 85° \approx 0.996$$

Check:

DAL: [sin] 95 [=] → 0.996

Not DAL: 95 [sin] → 0.996

(b) To find cos 225°, which is an angle in quadrant III, subtract 180° from 225° to find the reference angle. Since cosine is negative in quadrant III, the negative of the cosine of the reference angle is equal to cos 225°:

$$\cos 225° = -\cos(225° - 180°) = -\cos 45° \approx -0.707$$

Check:

DAL: [cos] 225 → [=] −0.707

Not DAL: 225 [cos] → − 0.707

(c) To find tan 290°, which is an angle in quadrant IV, subtract 290° from 360° to find the reference angle. Since tangent is negative in quadrant IV, the negative tangent of the reference angle is equal to tan 290°:

$$\tan 290° = -\tan(360° - 290°) = -\tan 70° \approx -2.75$$

Check:

DAL: [tan] 290 [=] → −2.75

Not DAL: 290 [tan] → −2.75

(d) To find tan 5.8 rad, first change the angle to degrees:

$$(5.8 \text{ rad})(57.3°) \approx 332.3°$$

Then find the negative of the tangent of the reference angle as in (c):

$$\tan 332.3° = -\tan(360° - 332.3°) = -\tan 27.7° \approx -0.525$$

You can check the original example by first setting your calculator for radians:

DAL: [DRG] [tan] 5.8 → −0.525

Not DAL: [DRG] 5.8 [tan] → −0.525

Although the value of the trigonometric function of any angle can be found directly on the calculator, it is necessary to know how to use reference angles because they must be applied when working with inverse trigonometric functions and solving equations with trigonometric functions. This is shown in the next section.

Quadrant Angles

The quadrant angles are 0°, 90°, 180°, and 270°. It is not possible to draw a reference triangle for these angles. The values of their sine, cosine, and tangent can be found directly with the calculator or by using values of x, y, and r for a point that is on the terminal side of each angle. For 0°and 180°, the x axis is the terminal side: for 90° and 270°, the y axis is the terminal side. The exceptions are tan 90° and tan 270°, which are infinitely large and have no defined value. The calculator indicates an error if you try to find these values.

Negative Angles and Angles Greater Than 360°

To find the reference angle for negative angles and angles greater than 360°, first find the *coterminal angle* between 0° and 360°. The coterminal angle is the angle having the same terminal side as the given angle and, therefore, the same reference angle and reference triangle. Study the next example.

EXAMPLE 17.5

Find the value of each trigonometric function by using a reference angle or quadrant angle:

(a) cos 400° **(b)** sin 450° **(c)** sin(−47°) **(d)** tan(−1.05 rad)

Solution Each of the given angles and their reference angles are shown in Figure 17-8.

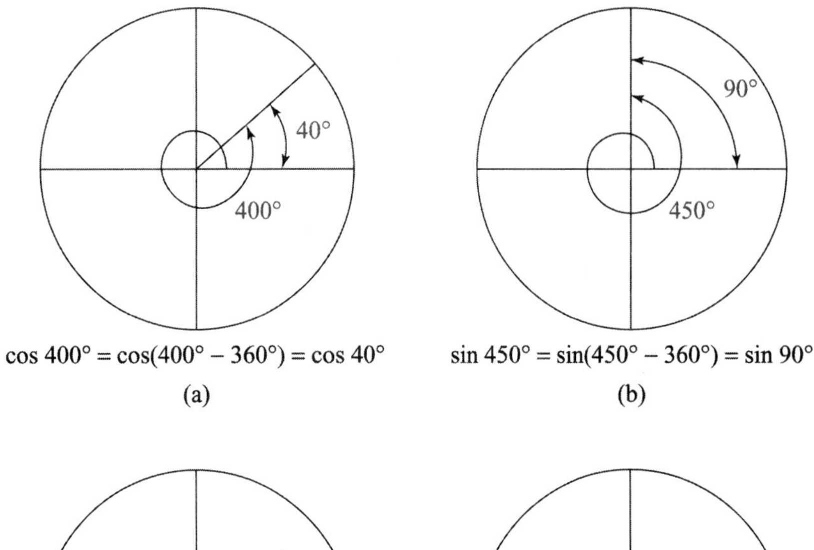

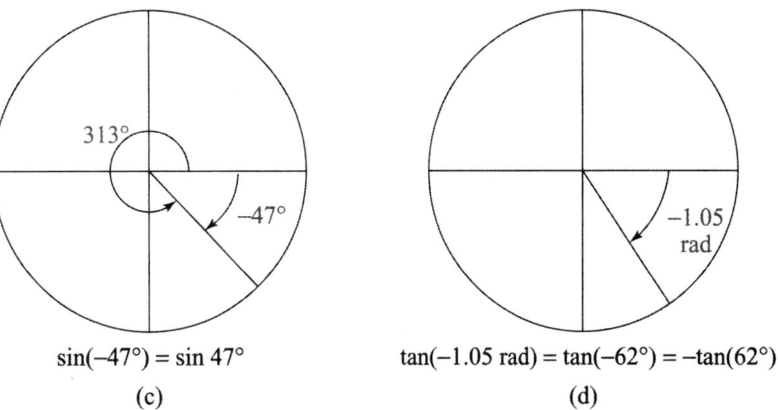

FIGURE 17-8 Coterminal angles and reference angles for Example 17.5.

(a) To find cos 400°, find the coterminal angle by subtracting 360°:

$$\cos 400^\circ = \cos(400^\circ - 360^\circ) = \cos 40^\circ \approx 0.766$$

Check:

DAL: [cos] 400° [=] $\rightarrow$0.766

Not DAL: 400° [cos] $\rightarrow$0.766

(b) To find sin 450°, find the coterminal angle by subtracting 360°:

$$\sin 450^\circ = \sin(450^\circ - 360^\circ) = \sin 90^\circ = 1.00$$

Check:

DAL: [sin] 450 [=] $\rightarrow$1.00

Not DAL: 450 [sin] $\rightarrow$1.00

(c) To find sin (–47°), note that for a negative angle in the fourth quadrant, the reference angle is equal to the positive value of the angle as follows: The angle – 47° has the same terminal side as 360° – 47° = 313°. The reference angle is then 360° – 313° = 47°. Since sine is negative in the fourth quadrant:

$$\sin(-47^\circ) = -\sin 47^\circ \approx -0.731$$

Check:

DAL: [sin] [(–)] 47 [=] $\rightarrow$–0.731

Not DAL: 47 [+/-] [sin] $\rightarrow$–0.731

(d) To find tan (−1.05 rad), first change the angle to degrees:

$$(-1.05 \text{ rad})(57.3°) \approx -60.2°$$

Then, as in (c) above, for a negative angle in the fourth quadrant, the reference angle is equal to the positive value of the angle. Since tangent is negative in the fourth quadrant:

$$\tan(-60.2°) = -\tan 60.2° \approx -1.75$$

Check:

DAL: DRG tan (−) 1.05 = → −1.75

Not DAL: DRG 1.05 +/− tan → −1.75

EXERCISE 17.2

Round all answers in the exercise to three significant digits.

In exercises 1 through 32, express each function as a function of its reference angle or quadrant angle in degrees and find the value of the function.

1. sin 130°
2. sin 120°
3. tan 150°
4. tan 110°
5. cos 345°
6. cos 300°
7. tan 280°
8. tan 315°
9. sin 330°
10. sin 200°
11. cos 225°
12. cos 106°
13. tan (5.4 rad)
14. tan (2.3 rad)
15. tan (1.1 rad)
16. tan (4.3 rad)
17. cos 450°
18. cos 540°
19. tan 540°
20. tan 360°
21. sin 630°
22. sin 480°
23. tan(−53°)
24. tan(−6.7°)
25. sin(−100°)
26. cos(−150°)
27. cos(−720°)
28. sin (−450°)
29. tan(−0.785 rad)
30. tan(−0.524 rad)
31. tan(−0.393 rad)
32. tan(−0.262 rad)

Applied Problems

33. Find the resultant velocity v of a ship in knots (nautical mph) given by:

$$v = \sqrt{67 - 28 \cos 120°}$$

34. The range of a projectile fired with velocity v_0 at an angle of elevation ϕ is given by:

$$R = \frac{v_0^2 \sin 2\phi}{g}$$

where the acceleration due to gravity $g = 9.81$ m/s^2. Find R in meters when:
(a) $v_0 = 100$ m/s and $\phi = 45°$
(b) $v_0 = 100$ m/s and $\phi = 75°$

Applications to Electronics

35. In an RC series circuit, the resistance R and the capacitive reactance X_C are related to the impedance Z by the impedance triangle shown in Figure 17-9. If $\theta = -23°$ and the resistance $R = 4.6$ kΩ, find (a) tan θ, and (b) Z and X_C as positive values in kΩ.

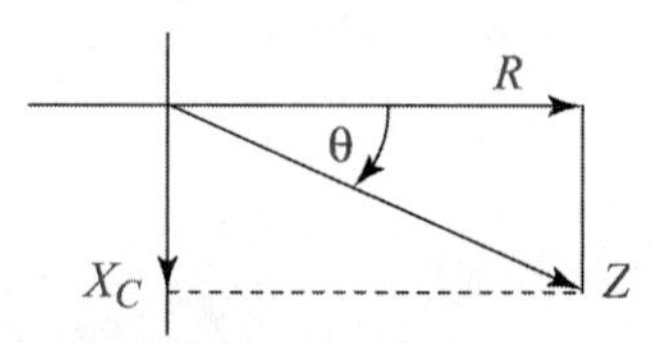

FIGURE 17-9 *RC* series circuit impedance triangle for problems 35 and 36.

36. In problem 35, if $\theta = -56°$ and $Z = 3.80$ kΩ, find (a) tan θ, and (b) R and X_C as positive values in kΩ.

37. In an ac circuit, the instantaneous current $i = I_P \sin\theta$ where I_P = peak or maximum current and θ = angle of rotation of a generator. Find i in mA when $I_P = 850$ mA and θ is:
(a) 90°
(b) 120°
(c) 540°
(d) 3.93 rad

38. In an ac circuit, the instantaneous voltage $v = V_P \cos\theta$, where V_P = peak or maximum voltage. Find v in volts when $V_P = 10$ V and θ is:
(a) 60°
(b) 360°
(c) 460°
(d) 3.5 rad

39. In an ac circuit, the instantaneous voltage v is given by:

$$v = V_P \sin(2\pi f t)$$

where V_P = peak or maximum voltage, f = frequency in hertz, t = time in *seconds*, and $(2\pi f t)$ is the angle *in radians*. Find v in volts when $V_P = 120$ V, $f = 60$ Hz and:
(a) $t = 5$ ms
(b) $t = 10$ ms
(c) $t = 25$ ms

40. In problem 39, find v in volts when $V_P = 50$ V, $f = 60$ Hz and:
(a) $t = 2$ ms
(b) $t = 12$ ms
(c) $t = 50$ ms

17.3 Inverse Functions

The inverse trigonometric functions $\sin^{-1}x$, $\cos^{-1}x$, and $\tan^{-1}x$ are introduced in Chapter 16 for positive values of x corresponding to acute angles in a right triangle. However, they are also defined for negative values of x, which can be entered in the calculator. It is necessary to clearly understand how inverse trigonometric functions work for positive and negative values and how these values relate to reference angles. The following rule applies:

Rule

Inverse Function Rule

The inverse function of the *absolute or positive* value of a trigonometric function is always the reference angle.

For example, suppose $\tan\theta = -1$ and θ is in quadrant II. Then the reference angle = $\tan^{-1}(+1) = 45°$ and $\theta = 180° - 45° = 135°$. Study the following examples.

EXAMPLE 17.6 Given $\cos\theta = -0.250$, find angle θ such that $0° \le \theta < 360°$ (θ greater than or equal to 0° and less than 360°).

Solution Since cosine is negative in the second and third quadrants, there are two values for the angle θ that satisfy $\cos\theta = -0.250$. First apply the Inverse Function Rule and find the reference angle by entering the *positive* value, + 0.250 with the inverse cosine function in the calculator:

DAL: Reference angle = [cos⁻¹] 0.25 [=] → 75.5°

Not DAL: Reference angle = 0.25 [cos⁻¹] → 75.5°

The reference angle for θ is then 75.5°. This is expressed mathematically as follows:

$$\cos^{-1}(0.250) = 75.5°$$

and means "The angle whose cosine is 0.250 = 75.5°." See Figure 17-10.

Next find the two values of θ whose reference angle is 75.5° in the second and third quadrants. To find the angle in the second quadrant, subtract the reference angle from 180°:

$$\text{Quadrant II:} \quad \theta = 180° - 75.5° = 104.5°$$

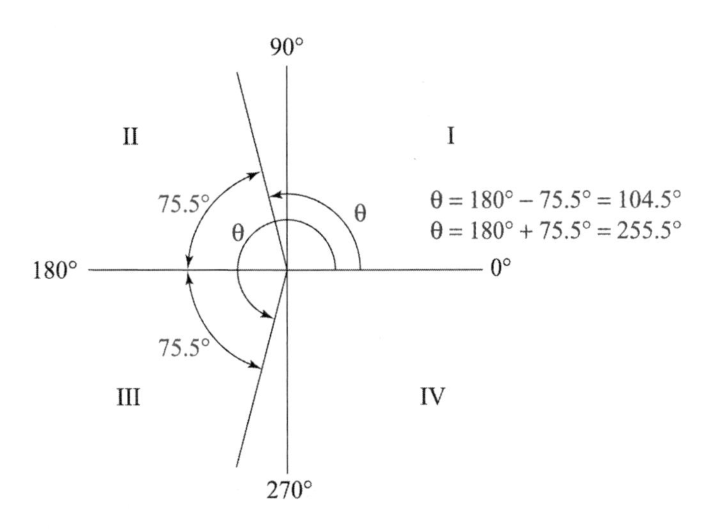

FIGURE 17-10 Inverse functions for Example 17.6.

To find the angle in the third quadrant, add the reference angle to 180°:

$$\text{Quadrant III:} \quad \theta = 180° + 75.5° = 255.5°$$

Note in this example, if you needed to find an angle in the fourth quadrant, you would subtract the reference angle from 360°. You always work with 180° or 360° (not 90° or 270°).

For each value of an inverse trigonometric function you enter, the calculator is programmed to give you only one defined value for the angle called the *principal value.* That value is defined within the following specific range of angles for each function:

Definitions

Inverse Functions (Principal Values)

$$-90° \le \sin^{-1}x \le 90°$$

$$0° \le \cos^{-1}x \le 180° \qquad (17.2)$$

$$-90° < \tan^{-1}x < 90°$$

The inverse functions (17.2) on the calculator always give you a reference angle in the first quadrant when you enter a positive value of x. When you enter a negative value of x, [sin⁻¹] and [tan⁻¹] each give you a negative angle in the fourth quadrant, while [cos⁻¹] gives you a positive angle in the second quadrant. The reason for these definitions is to give only one value for the angle within a *continuous* range of angles that includes all possible values. For example, the principal value of $\tan^{-1}(-1.67)$ on the calculator is a negative angle in the fourth quadrant:

DAL: [tan⁻¹] [(-)] 1.67 [=] → −59.1°

Not DAL: 1.67 [+/-] [tan⁻¹] → −59.1°

Note that the principal value of $\sin^{-1}(-0.5) = -30°$ is also a negative angle in the fourth quadrant, whereas the principal value of $\cos^{-1}(-0.5) = 120°$ is a positive angle in the fourth quadrant.

EXAMPLE 17.7 Given $3 \tan \theta + 1 = 0$. Find θ in degrees and radians for $-90° < \theta < 90°$.

Solution This is a basic trigonometric equation. First solve for $\tan \theta$ and then find θ:

$$3 \tan \theta + 1 = 0$$

$$3 \tan \theta = -1$$

$$\tan \theta = -\frac{1}{3} \approx -0.333$$

The given range of values for θ ($-90 < \theta < 90$) are the principal values for $\tan^{-1}x$ defined in (17.2). Therefore, there is only one value for θ, which is a negative angle in the fourth quadrant:

$$\theta = \tan^{-1}(-0.333) \approx -18.4^0 = \frac{-18.4^0}{57.3^0} \text{ rad} \approx -0.321 \text{ rad}$$

ERROR BOX

You can avoid an error when working with inverse functions by applying the following rules. Suppose you enter a negative value instead of a positive value with an inverse trigonometric function. Here's how to determine the reference angle from the result:

1. When a negative value is entered with [sin⁻¹] and [tan⁻¹], the result is the negative of the reference angle. Change it to a positive value, and you have the reference angle.
2. When a negative value is entered with [cos⁻¹], the result is a positive angle in the second quadrant. Subtract this angle from 180°, and you have the reference angle.

Practice Problems: Find the reference angle for θ by entering the negative value.

1. $\sin\theta = -0.45$ **2.** $\cos\theta = -0.76$ **3.** $\tan\theta = -1.50$

4. $\cos\theta = -0.12$ **5.** $\tan\theta = -0.015$ **6.** $\sin\theta = -0.80$

7. $\tan\theta = -0.75$ **8.** $\cos\theta = -0.90$ **9.** $\sin\theta = -0.04$

10. $\tan\theta = -3.3$

The next example shows how radians are used in an application to electricity. Most technical applications use radians instead of degrees because they are based on the formula for the circumference of a circle and lend to a better interpretation of the ideas. Computers are programmed to work with angles in radian measure.

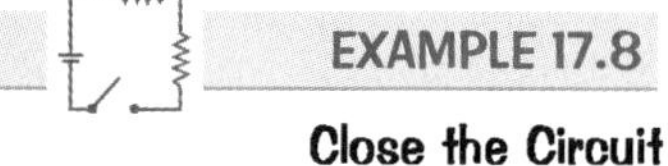

EXAMPLE 17.8

Close the Circuit

The instantaneous voltage of ordinary house current is given by $v = 170 \sin(2\pi ft)$, where the frequency $f = 60$ Hz, t = time in seconds, and the angle $2\pi ft$ *is in radians.* Note that for the time t to be in seconds the angle must be expressed in radians.

(a) The angle $2\pi ft$ in degrees and radians for $0 \le 2\pi ft < 2\pi$.

(b) The time t.

Solution

(a) To find the angle $2\pi ft$, substitute 120 for v and solve for $\sin(2\pi ft)$ by dividing by 170:

$$120 = 170 \sin(2\pi ft)$$

$$\sin 2\pi ft = \frac{120}{170} \approx 0.7059$$

Then find the angle $2\pi ft$. The reference angle in degrees and radians is:

$$\text{Reference angle} = \sin^{-1}(0.7059) = 44.9° \approx \frac{44.9°}{57.3°} \text{ rad} \approx 0.784 \text{ rad}$$

Sine is positive in the first and second quadrants, so the two solutions are:

Quadrant I: $2\pi ft = 44.9° = 0.784$ rad

Quadrant II: $2\pi ft = 180° - 44.9° = 135.1° = \dfrac{135.1°}{57.3°} = 2.36$ rad

(b) To find the time t, let $f = 60$ Hz, and solve for t using only the above solutions for the angle *in radians:*

Quadrant I: $2\pi ft = 2\pi(60)t = 0.784$ rad

$120\pi t = 0.784$

Divide by 120π: $t = \dfrac{0.784}{120\pi} = 0.00208 \text{ s} = 2.08 \text{ ms}$

Quadrant II: $2\pi(60)t = 2.36$ rad

Divide by 120π: $t = \dfrac{2.36}{120\pi} = 0.00626 \text{ s} = 6.26 \text{ ms}$

EXERCISE 17.3

In exercises 1 through 12, find the principal value of each inverse trigonometric function to the nearest 0.1°.

1. $\sin^{-1}(0.500)$
2. $\cos^{-1}(0.866)$
3. $\tan^{-1}(1.19)$
4. $\tan^{-1}(2.46)$
5. $\tan^{-1}(3.22)$
6. $\tan^{-1}(0.226)$
7. $\cos^{-1}(-0.787)$
8. $\sin^{-1}(-0.0997)$
9. $\tan^{-1}(-0.897)$
10. $\tan^{-1}(-0.543)$
11. $\tan^{-1}(-1.73)$
12. $\tan^{-1}(-1.00)$

In exercises 13 through 22, find θ to the nearest 0.1° for 0° ≤ θ < 360°.

13. $\cos\theta = 0.906$
14. $\sin\theta = 0.629$
15. $\tan\theta = 0.845$
16. $\tan\theta = 1.25$
17. $\sin\theta = -0.556$
18. $\cos\theta = -0.133$
19. $\tan\theta = -0.503$
20. $\tan\theta = -2.00$
21. $\tan\theta = -6.34$
22. $\tan\theta = -0.0567$

In exercises 23 through 28, find θ to the nearest 0.1° for −90° < θ < 90°.

23. $\tan\theta = 1.56$
24. $\tan\theta = 0.225$
25. $\sin\theta = -0.556$
26. $\sin\theta = 0.133$
27. $\tan\theta = -4.48$
28. $\tan\theta = -0.346$

In exercises 29 through 38, solve each equation for θ in radians between 0 ≤ θ < 2π to three significant digits. Note that there are four solutions for 37 and 38.

29. $\cos\theta - 0.556 = 0$
30. $\sin\theta + 0.631 = 0$
31. $\tan\theta + 1.0 = 0$
32. $2\tan\theta = 3.0$
33. $2\sin\theta + 1.0 = 0$
34. $3\cos\theta - 2.0 = 0$
35. $\tan\theta - 2.5 = 0$
36. $\tan\theta + 3.3 = 0$
37. $\sin^2\theta = \frac{1}{2}$
38. $\tan^2\theta = 3$

In exercises 39 through 42, find θ in radians to three significant digits for −π/2 < θ < π/2.

39. $\tan\theta = 1.65$
40. $\tan\theta = 0.925$
41. $\tan\theta = -0.848$
42. $\tan\theta = -3.34$

Applied Problems

In problems 43 through 52, solve each applied problem to three significant digits.

43. The components F_x and F_y of a resultant force F are shown in Figure 17-6, Exercise 17.1. If $F_x = -6.20$ N (newtons) and $F_y = 8.70$ N, find $\tan\theta$ and the angle θ in degrees.

44. The components F_x and F_y of a resultant force F are shown in Figure 17-6, Exercise 17.1. If $F_x = -1.5$ lb and $F_y = 1.70$ lb, find $\tan\theta$ and the angle θ in degrees.

45. The original mathematical terms for $\sin^{-1}$, $\cos^{-1}$, and $\tan^{-1}$ are arcsin, arccos, and arctan. Many computers and some calculators still use this notation. Find the difference D between the true heading and the compass heading of an airplane, given by the following expression:

$$D = 60° - \arctan(1.4\sin 65°)$$

46. As in problem 45, evaluate the following expression which comes from a problem involving the propagation of a sound wave:

$$\arcsin(0.45) + \arctan(0.25\tan 60°)$$

Applications to Electronics

47. Figure 17-11(a) on page 355 shows an ac circuit containing a resistor R and an inductor L in parallel (RL parallel circuit). The current through the resistor I_R and the current through the inductor I_L are related to the total current I_T by the current diagram shown in Figure 17-11(b). If $I_R = 20$ mA and $I_L = 30$ mA, find $\tan\theta$ and θ, which is a negative angle in the fourth quadrant, in degrees.

48. In problem 47, find $\tan\theta$ and θ in degrees if $I_T = 380$ μA when $I_L = 150$ μA.

49. An electric field vector has components $E_x = 3.99$ N/C (newtons/coulomb) and $E_y = -2.03$ N/C as shown in Figure 17-12. Find the positive angle α and the negative angle θ in degrees.

50. In a problem involving an electric field, the following expression arises:

$$\alpha = \tan^{-1} x + \tan^{-1}(1/x).$$

Answers to Error Box Problems, page 353:
1. 26.7° **2.** 40.5° **3.** 56.3° **4.** 83.1° **5.** 0.86°
6. 53.1° **7.** 36.9° **8.** 25.8° **9.** 2.3° **10.** 73.1°

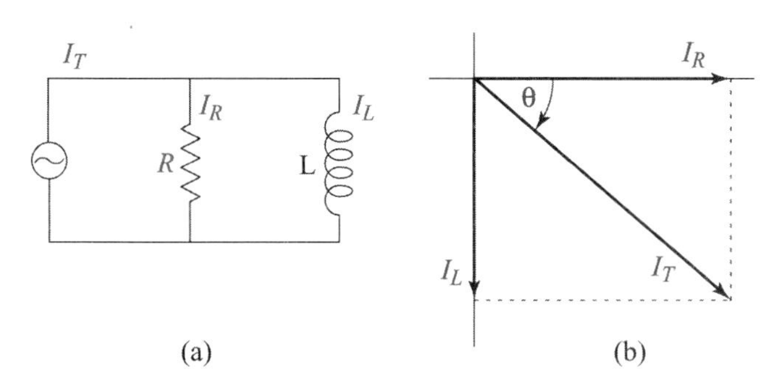

FIGURE 17-11 *RL* parallel circuit for problems 47 and 48.

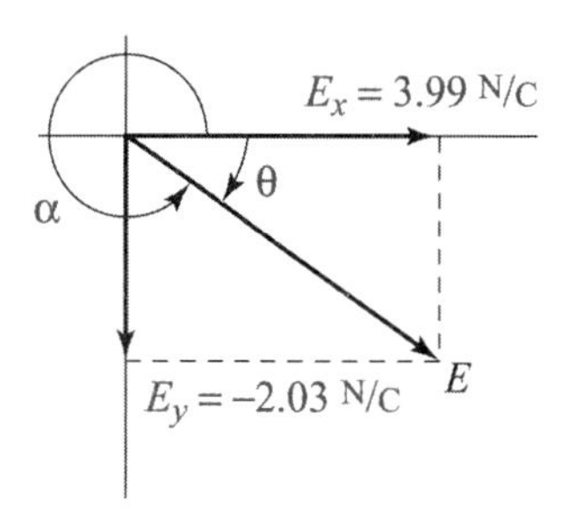

FIGURE 17-12 Electric field vectors for problem 49.

Find α in radians when x is:

(a) 0.50 **(b)** 1.5 **(c)** 2.5

(d) Can you show that $\alpha = \pi/2$ rad (90°) for any value of x?

Hint: Draw a diagram.

51. In Example 17.8, given that $v = 100$ V, find the two values of:

(a) The angle $2\pi ft$ in radians for $0 \le 2\pi ft < 2\pi$.

(b) The time t.

52. In an ac circuit, the instantaneous current in milliamps is given by $i = 2.5 \cos(\omega t)$ where ω = angular velocity in rad/s and t = time in seconds. If $i = 1.5$ mA, find the two values of:

(a) The angle ωt in radians for $0 \le \omega t < 2\pi$.

(b) The time t if $\omega = 400$ rad/s.

CHAPTER HIGHLIGHTS

17.1 TRIGONOMETRIC FUNCTIONS

Trigonometric Functions

$$\sin\theta = \frac{y}{r} \qquad \cos\theta = \frac{x}{r} \qquad \tan\theta = \frac{y}{x} \tag{17.1}$$

See Figure 17-13 for the signs of the trigonometric functions.

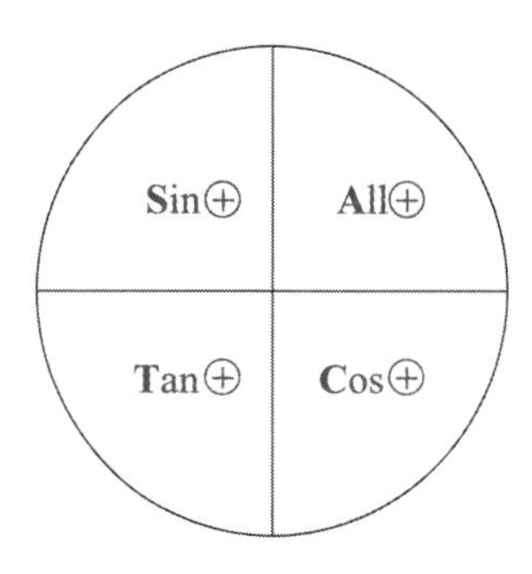

FIGURE 17-13 Positive trigonometric functions: "All Star Trig Class."

Key Example: Given the terminal side of angle θ passes through (−5,12). Then $r = \sqrt{(-5)^2 + (12)^2} = 13$ and:

$$\sin\theta = \frac{12}{13}$$
$$\cos\theta = \frac{-5}{13}$$
$$\tan\theta = \frac{12}{-5}$$

17.2 REFERENCE ANGLES AND SPECIAL ANGLES

Reference triangles are always drawn to the x axis with the reference angle A between the terminal side of the angle and the x axis. See Figure 17-14 on page 356.

Reference Angle

The absolute, or positive, value of the trigonometric function of an angle in any quadrant is equal to the trigonometric function of its reference angle.

Quadrant angles 0°, 90°, 180°, and 270° have no reference angles, and angles greater than 360° have the same trigonometric functions as coterminal angles less than 360°.

Key Example:

$\sin 95° = +\sin(180° - 95°) = \sin 85° \approx 0.996$
$\tan 332.3° = -\tan(360° - 332.3°) = -\tan 27.7° \approx -0.525$
$\cos 540° = \cos(540° - 360°) = \cos 180° = -1.0$
$\tan(-47°) = -\tan 47° \approx -1.07$

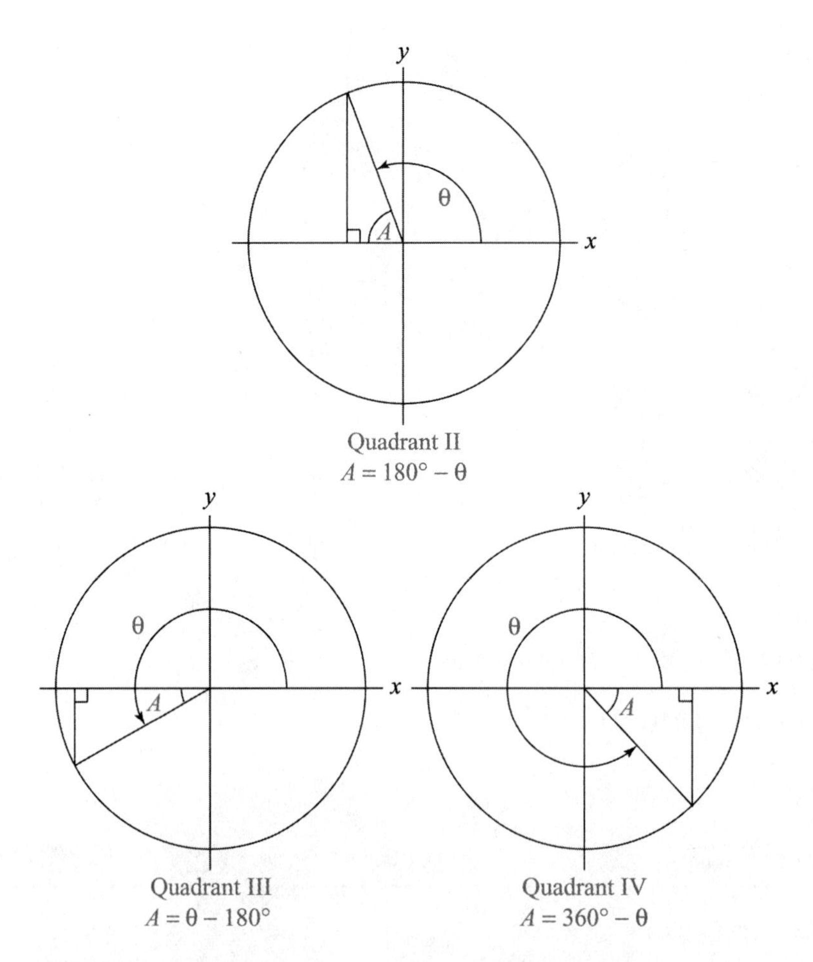

FIGURE 17-14 Reference angles and reference triangles.

17.3 INVERSE FUNCTIONS

Inverse Function Rule

The inverse function of the *absolute* or *positive* value of a trigonometric function is always the reference angle.

Key Example: Given $\cos\theta = -0.250$, the reference angle and values of θ such that $0° \le \theta < 360°$ are:

Reference angle $= \cos^{-1}(0.250) \approx 75.5°$

Quadrant II: $\theta = 180° - 75.5° = 104.5°$

Quadrant III: $\theta = 180° + 75.5° = 255.5°$

Inverse Functions (Principal Values)

$$\begin{aligned} -90° &\le \sin^{-1}x \le 90° \\ 0° &\le \cos^{-1}x \le 180° \\ -90° &< \tan^{-1}x < 90° \end{aligned} \tag{17.2}$$

Key Example: Given $3\tan\theta + 2.5 = 0$. For $-90 < \theta < 90$, the angle in degrees and radians is:

$$3\tan\theta = -2.5$$

$$\tan\theta = -\frac{2.5}{3} \approx -0.833$$

$$\theta = \tan^{-1}(-0.833) \approx -39.8° = \frac{-39.8°}{57.3°} \approx -0.695 \text{ rad}$$

REVIEW EXERCISES

Round all answers in the exercise to three significant digits.

In exercises 1 through 8, draw the angle θ and find sin θ, cos θ, and tan θ.

1. The terminal side of θ passes through (2.5,6).
2. The terminal side of θ passes through (−0.8,−0.6).
3. The terminal side of θ passes through (−8,15).
4. The terminal side of θ passes through (12,−3.5).
5. $\sin\theta = -0.300$, $\cos\theta$ positive
6. $\tan\theta = 2.00$, θ in quadrant III

7. $\cos \theta = 0.671$, θ in quadrant I
8. $\tan \theta = -1.17$, $\sin \theta$ positive

In exercises 9 through 16, express each function as a function of the reference angle or quadrant ngle in degrees and find the value of the function.

9. $\sin 230°$
10. $\cos 276°$
11. $\tan 245°$
12. $\tan (-56°)$
13. $\cos 400°$
14. $\sin (-630°)$
15. $\tan 1.60$ rad
16. $\tan (-77°$ rad)

In exercises 17 through 30, find θ in degrees and radians for $0° \leq \theta < 360°$.

17. $\theta = \tan^{-1}(4.52)$
18. $\theta = \tan^{-1}(-0.586)$
19. $\theta = \sin^{-1}(0.910)$
20. $\theta = \cos^{-1}(0.876)$
21. $\cos \theta = 0.500$
22. $\sin \theta = -0.696$
23. $\tan \theta = -0.966$
24. $\tan \theta = 2.30$
25. $\sin \theta = 0.663$
26. $\cos \theta = -0.901$
27. $5 \sin \theta = 3$
28. $3 \cos \theta + 1 = 0$
29. $\tan \theta + 3 = 0$
30. $2 \tan \theta = 4.4$

In exercises 31 through 36, find θ in degrees and radians for $-90° < \theta < 90°$.

31. $\tan \theta = 2.12$
32. $\tan \theta = -0.876$
33. $\sin \theta = 0.600$
34. $\sin \theta = -0.202$
35. $\tan \theta = -0.588$
36. $\tan \theta = 4.22$

Applied Problems

37. The formula for the area A of a parallelogram with sides a and b and included angle θ is:

$$A = ab \sin \theta$$

Find the area of a structure in the shape of a parallelogram if $a = 5.30$ cm, $b = 6.20$ cm, and the included angle $\theta = 145°$.

38. A velocity vector V has components V_x and V_y as shown in Figure 17-15. If $V_x = 15$ mi/h and $V_y = 12$ mi/h, find:
    (a) $\sin \theta$, $\cos \theta$, and $\tan \theta$
    (b) The magnitude of V and the negative angle θ in degrees and radians.

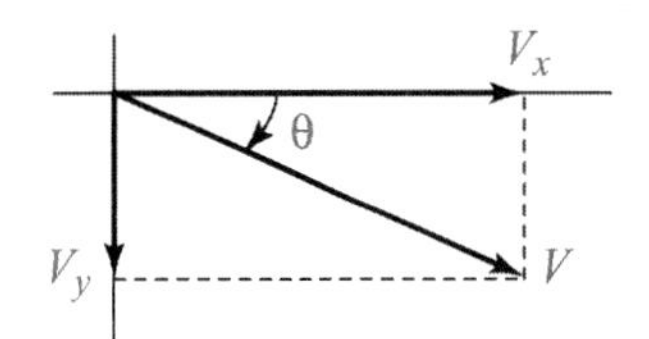

FIGURE 17-15 Components of velocity vector for problem 38.

Applications to Electronics

39. In an ac circuit, the current i at any time t is given by:

$$i = I_P \cos (2\pi ft)$$

where I_P = peak current, f = frequency in hertz, and t = time in seconds. Find i in amps when $I_P = 4.0$ A, $f = 60$ Hz, and t is:
    (a) 25 ms
    (b) 22 ms
    (c) 4.4 ms

40. In problem 39, find the angle $(2\pi ft)$ in radians for $0 \leq 2\pi ft < 2\pi$ and the time t when:
    (a) $i = 3.0$ A
    (b) $i = -2.0$ A

41. In Figure 17-11(b), Exercise 17.3, find I_T and the angle θ in degrees when $I_R = 34$ mA and $I_L = 22$ mA.

42. In Figure 17-12, Exercise 17.3, find E, angle α, and angle θ in degrees when $E_x = 5.67$ N/C and $E_y = -8.92$ N/C.

CHAPTER 18
Alternating Current

CHAPTER OBJECTIVES

In this chapter, you will learn:

- How to graph a sine and cosine curve.
- How to graph alternating-current and voltage waves and find instantaneous values.
- How to find peak and root mean square, or effective, values of voltage and current.
- How to find the period and frequency of an ac wave.
- How to determine the phase angle of an ac wave.

Alternating current electricity is produced in a coil of wire rotating in a magnetic field. This circular movement generates an ac current wave and an ac voltage wave in the wire, which behave like sine curves. The theory of alternating current is therefore linked to the mathematics of the sine function. The sine function is a periodic function that repeats its values after a certain number of degrees known as its period or cycle. Hence, ac current and ac voltage are periodic in nature. The fundamentals of ac electricity, including peak values, effective or root mean square values, frequency, period, and phase angles are all studied in this chapter.

• • •

18.1 Alternating Current Waves

Sine Curve

The graph of the sine function is generated by the movement of a point around a circle. Example 18.1 shows the basic sine curve using angles in degrees and radians. Radians, which are introduced in Section 16.1, are often used instead of degrees in electronic applications. Study Example 18.1 well. It is the basis for many ideas in this and subsequent chapters.

EXAMPLE 18.1 Graph $y = \sin\theta$ from $\theta = 0°$ to $\theta = 360°$.

Solution To graph the basic sine function, you need to construct a table of values of θ and $y = \sin\theta$ using a convenient interval. The angle 30° ($\pi/6$ rad) is used as the interval because it is a common angle and provides enough points to sketch the graph. Using a calculator, find the values of sin θ for each angle shown in the table below from 0° to 360°. For example, to calculate the sin 210°:

DAL: [sin] 210 [=] → −0.50

Not DAL: 210 [sin] → −0.50

In the following table, angles are shown in both degrees and radians in terms of π to help you become familiar with the radian values.

θ (deg)	0	30	60	90	120	150	180	210	240	270	300	330	360
θ (rad)	0	$\frac{\pi}{6}$	$\frac{\pi}{3}$	$\frac{\pi}{2}$	$\frac{2\pi}{3}$	$\frac{5\pi}{6}$	π	$\frac{7\pi}{6}$	$\frac{4\pi}{3}$	$\frac{3\pi}{2}$	$\frac{5\pi}{3}$	$\frac{11\pi}{6}$	2π
y = sin θ	0	0.50	0.87	1.0	0.87	0.50	0	−0.50	−0.87	−1.0	−0.87	−0.50	0

The graph of the sine function is shown in Figure 18-1 with points plotted every 30°. It illustrates how the sine curve is generated by the movement of a point in a circle. The radius of the circle $r = 1$, so $\sin\theta = y/r = y/1$ or simply $y = \sin\theta$. Therefore, *the height of the point on the circle for each angle θ equals the height of the sine curve.* By projecting these heights to the graph on the right as the radius revolves in a circle, you can see how the movement of the point around the circle produces the sine curve.

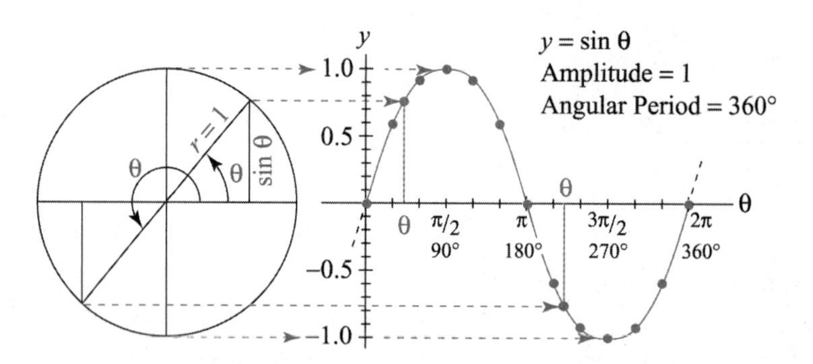

FIGURE 18-1 Sine curve showing projections of points on the unit circle for Example 18.1.

Observe that the sine curve begins at zero, reaches a maximum of 1 at 90° (π/2 rad), and returns to zero at 180° (π rad). The maximum value is called the *amplitude* of the curve. The sine becomes negative after 180°, reaches a minimum of −1 at 270° (3π/2 rad), and returns to zero at 360° (2π rad). The curve starts the same cycle again at 360° and repeats itself every 360°. The number of degrees in one cycle is called the *angular period* of the curve. The amplitude of the sine curve is therefore 1, and its angular period is 360°.

Cosine Curve

The cosine curve is essentially the same as the sine curve except for its position. Instead of beginning at zero, it begins and ends its cycle at the maximum value 1.

EXAMPLE 18.2 Graph $y = \cos\theta$ from $\theta = 0$ to $\theta = 360°$ for values of θ every 30°.

Solution The table of values for θ and cos θ every 30° is:

θ (deg)	0	30	60	90	120	150	180	210	240	270	300	330	360
θ (rad)	0	$\frac{\pi}{6}$	$\frac{\pi}{3}$	$\frac{\pi}{2}$	$\frac{2\pi}{3}$	$\frac{5\pi}{6}$	π	$\frac{7\pi}{6}$	$\frac{4\pi}{3}$	$\frac{3\pi}{2}$	$\frac{5\pi}{3}$	$\frac{11\pi}{6}$	2π
y = cos θ	1.0	0.87	0.50	0	−0.50	−0.87	−1.0	−0.87	−0.50	0	0.50	0.87	1.0

Figure 18-2 shows the cosine curve with the sine curve superimposed in dashed lines for comparison.

Both curves have an amplitude = 1 and an angular period = 360°. The cosine curve is extended back to −90° ($-\frac{\pi}{2}$ rad) where $\cos\theta = 0$. This point is equivalent to $\theta = 0°$ for the sine curve. If you move sin θ to the left 90°, both curves will match exactly. The difference is that: *cos θ is 90° out of phase with sin θ.* Since cosine starts 90° before sine: *cos θ leads sin θ by 90°*. Because the sine and cosine curves are essentially the same, both are called *sinusoidal* curves.

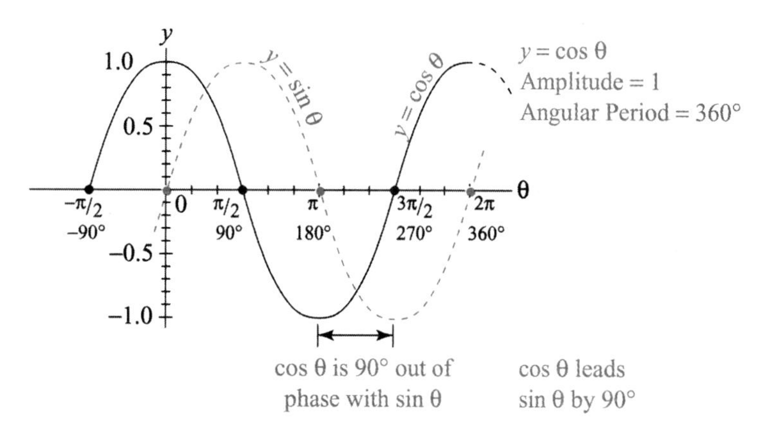

FIGURE 18-2 Cosine curve with sine curve compared for Example 18.2.

You can change the amplitude of the sine or cosine curve by multiplying by a constant.

Definition

Amplitude of Sine or Cosine

The amplitude of $y = a \sin\theta$ or $y = a \cos\theta$ is $|a|$. (18.1)

$|a|$ = absolute or positive value.

EXAMPLE 18.3

Graph $y = 3 \sin\theta$ from $\theta = 0$ to $\theta = 2\pi$.

Solution The values for $y = 3 \sin\theta$ are three times the values for $y = \sin\theta$ in Example 18.1. The curve looks the same except for its amplitude. See Figure 18-3. $y = 3 \sin\theta$ begins at 0, reaches a maximum of 3 at 90°, returns to 0 at 180°, reaches a minimum of −3 at 270°, and returns to 0 at 360°. The amplitude = $|3| = 3$ and the angular period = 360°.

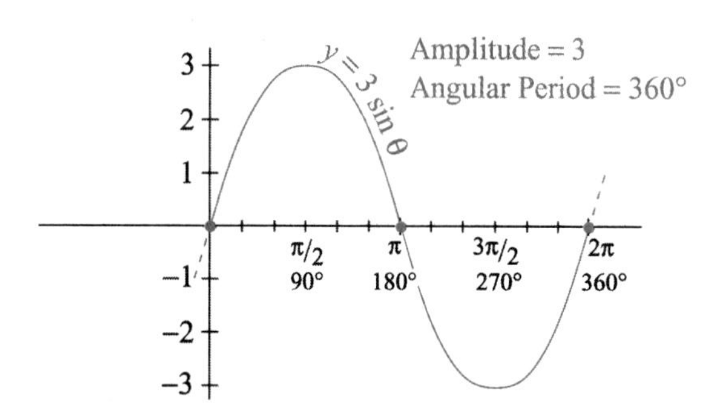

FIGURE 18-3 Graph of $y = 3 \sin\theta$ for Example 18.3.

Graphs of the sine and cosine curves can be done on a graphing calculator such as Texas Instruments calculators TI-83 to TI-86. For example, to draw the graph of $y = 3 \sin\theta$ on the TI-83, first press the MODE key and then the down arrow to set the calculator for degrees. One set of keystrokes for the TI-83 to graph $y = 3 \sin\theta$ is:

ZOOM 7 Y= CLEAR 3 SIN X,T,Θ GRAPH

The keys ZOOM 7 set the scale on the θ axis from $\theta = -360°$ to $\theta = 360°$ and the scale on the y axis from −4 to 4. You can set your own scale for x and y by pressing WINDOW. Appendix A shows some basic graphing functions on the TI-83.

Alternating Current

Alternating current behaves in the same way as a sinusoidal curve. It is produced in a coil of wire that rotates in a circle in a magnetic field. See Figure 18-4.

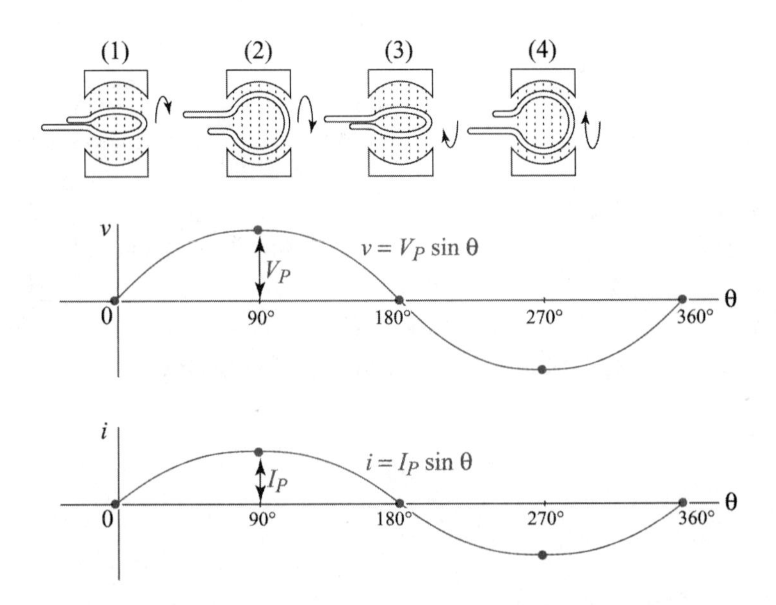

FIGURE 18-4 Induced ac voltage and ac current in a coil of wire.

When the coil is in position (1) so it is perpendicular to the magnetic lines of force, there is no induced voltage and no current flow in the wire. As the coil turns and cuts the lines of force, there is an induced voltage that increases, and the current begins to flow in one direction. The voltage and current reach a maximum in position (2) when the coil has turned 90°. As the coil turns more, the induced voltage and current decrease. The voltage and current reach zero at position (3) after turning 180°. As the coil continues turning, there is an induced voltage of opposite polarity, and the current begins to move in the other direction. The voltage and current reach a maximum in the other direction at position (4) after turning 270°. As the coil continues to turn, the induced voltage again decreases, and the current decreases to zero at 360°, which is back to position (1). The cycle then repeats itself.

The graph of the induced voltage and the induced current are both sine waves expressed by the following equations:

Formulas

AC Voltage and Current Waves

$$v = V_P \sin\theta \qquad i = I_P \sin\theta \tag{18.2}$$

In formulas (18.2), v and i represent instantaneous values of the voltage and current. The amplitude or maximum value is called the *peak value* and is equal to V_P for the voltage and I_P for the current. The peak to peak value, V_{PP} or I_{PP}, is the difference between the maximum and minimum values and is twice the peak value. The number of degrees in one cycle or angular period is 360°.

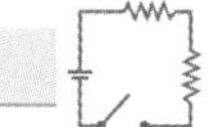

EXAMPLE 18.4

Close the Circuit

Graph the ac voltage wave $v = 170 \sin \theta$ for one cycle. Give the peak value, the peak-to-peak value, and the angular period.

Solution The peak value $V_P = 170$ V and the angular period = 360°. Construct the following table of values from 0° to 360° for angles every 30°. For example, when $\theta = 240°$ the value of v is found on the calculator as follows:

DAL: 170 [×] [sin] 240 [=] → −147 V

Not DAL: 240 [sin] [×] 170 [=] → −147 V

The table of values is:

θ (deg)	0	30	60	90	120	150	180	210	240	270	300	330	360
θ (rad)	0	$\frac{\pi}{6}$	$\frac{\pi}{3}$	$\frac{\pi}{2}$	$\frac{2\pi}{3}$	$\frac{5\pi}{6}$	π	$\frac{7\pi}{6}$	$\frac{4\pi}{3}$	$\frac{3\pi}{2}$	$\frac{5\pi}{3}$	$\frac{11\pi}{6}$	2π
$v = 170 \sin \theta$ (V)	0	85	147	170	147	85	0	−85	−147	−170	−147	−85	0

Carefully set up a scale and plot the points to show the sine wave voltage as in Figure 18-5. The peak-to-peak value V_{PP} is the difference between the maximum and minimum values and is therefore equal to:

$$V_{PP} = 170 \text{ V} - (-170 \text{ V}) = 340 \text{ V}$$

or

$$V_{PP} = 2V_P = 2(170 \text{ V}) = 340 \text{ V}$$

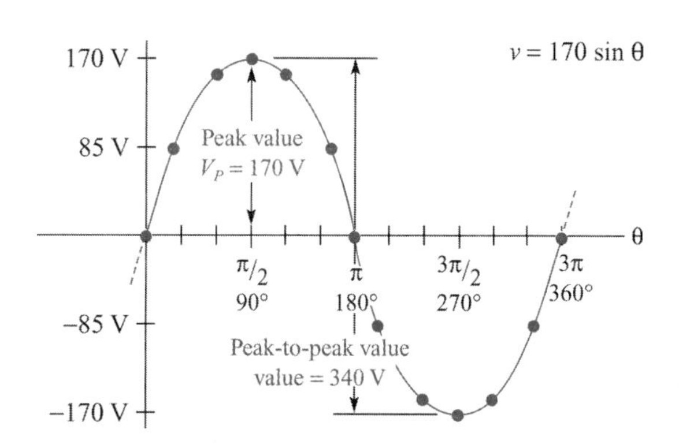

FIGURE 18-5 Sine wave voltage for Example 18.4.

To do Example 18.4 on the TI-83 graphing calculator, first press [WINDOW] to enter the range of values for θ (x) and y. Set x from 0° to 360° and y from −180 to 180. Then enter the following:

[Y=] [CLEAR] 170 [SIN] [X, T, Θ] [GRAPH]

In an ac circuit containing only a resistance, the sine wave current is *in phase* with the sine wave voltage, as shown in Figure 18-4. Both waves start at 0 and end at 360°. The sine wave current is graphed in the same way as the sine wave voltage. There are many other technical applications of sinusoidal curves such as sound waves, light waves, radio waves, water waves, and the mechanical vibration of a wire, spring, and beam.

EXERCISE 18.1

In exercises 1 through 12, graph each sinusoidal curve from θ = 0 to θ = 360° for θ every 30°. Give the amplitude and the angular period.

1. $y = 2 \sin \theta$
2. $y = 5 \sin \theta$
3. $y = 50 \sin \theta$
4. $y = 120 \sin \theta$
5. $y = 10 \sin \theta$
6. $y = 15 \sin \theta$

7. $y = 3 \cos \theta$
8. $y = 4 \cos \theta$
9. $y = 2.5 \cos \theta$
10. $y = 3.5 \cos \theta$
11. $y = 100 \cos \theta$
12. $y = 50 \cos \theta$

Applications to Electronics

In exercises 13 through 24, graph each ac voltage or current wave from $\theta = 0$ to $\theta = 360°$ for θ every 30°. Give the peak value, the peak-to-peak value, and the angular period.

13. $v = 100 \sin \theta$
14. $v = 60 \sin \theta$
15. $i = 20 \sin \theta$
16. $i = 30 \sin \theta$
17. $v = 300 \sin \theta$
18. $v = 150 \sin \theta$
19. $i = 1.5 \cos \theta$
20. $i = 1.2 \cos \theta$
21. $v = 12 \cos \theta$
22. $v = 3.6 \cos \theta$
23. $i = 6.0 \cos \theta$
24. $i = 0.50 \cos \theta$

In problems 25 through 28, solve each applied problem to two significant digits.

25. The current in an ac circuit containing an inductance is given by $i = 2.0 \sin \theta$, while the voltage is given by $v = 110 \cos \theta$. Sketch these two ac waves on the same graph to show that the current and the voltage are 90° out of phase. Use different scales for current and voltage.
26. The voltage in an ac circuit containing a capacitance is given by $v = 200 \sin \theta$, while the current is given by $i = 4.0 \cos \theta$. Sketch these two ac waves on the same graph to show that the voltage and the current are 90° out of phase. Use different scales for voltage and current.
27. The voltage in volts in an ac circuit containing a resistance is given by $v = 100 \sin \omega t$ where ω = angular velocity in rad/s, t = time in seconds, and ωt = angle in radians. If the angular velocity $\omega = 400$ rad/s,
    (a) Find v in volts when $t = 1.0$ ms.
    (b) Find the angle ωt in radians between 0 and 2π rad when $v = 50$ V.
28. The current in amps in an ac circuit containing a resistance is given by $i = 1.5 \sin \omega t$ where ω = angular velocity, t = time in seconds, and ωt = angle in radians. If the angular velocity $\omega = 300$ rad/s
    (a) Find i in milliamps when $t = 2.0$ ms.
    (b) Find the angle ωt in radians between 0 and 2π rad when $i = 500$ mA.

18.2 Root Mean Square Values

The current and voltage in an ac circuit are continuously changing and assume different values every instant. However, we assign only one value to ac voltage or current, such as 120 V for ordinary household voltage, or 15 A for the current rating of an ac fuse. These values represent a type of mathematical "average" calculated over one cycle of the ac sine wave and are called the *root mean square* or *effective* values. The root mean square () value of an ac wave is equal to the value of a dc voltage or current that produces the same electrical energy as the ac voltage or current. For example, an ac voltage wave whose peak value is 170 V has an rms value of 120 V. When this ac source is connected to a certain resistance, it will dissipate the same amount of electrical energy as a dc voltage of 120 V connected to the same resistance. The name "root mean square" is used because of the way it is calculated. First, the instantaneous values of the voltage or current are squared. Second, the average or mean of these squared values is calculated. Third, the square root of this mean is computed. When this is done precisely, using the methods of calculus applied to a sine wave, it yields the following formulas:

Formulas **Root Mean Square Values**

$$V_{rms} = \left(\frac{1}{\sqrt{2}}\right) V_P \approx (0.707)\ V_P$$

$$I_{rms} = \left(\frac{1}{\sqrt{2}}\right) I_P \approx (0.707)\ I_P \qquad (18.3)$$

EXAMPLE 18.5 Given $V_P = 170$ V and $I_P = 4.4$ A, find V_{PP}, I_{PP}, V_{rms} and I_{rms}.

Solution The peak-to-peak values are equal to twice the peak values:

$$V_{PP} = 2V_P = 2(170\text{ V}) = 340\text{ V}$$
$$I_{PP} = 2I_P = 2(4.4\text{ A}) = 8.8\text{ A}$$

To find the rms values, apply formulas (18.3):

$$V_{rms} = 0.707(170\text{ V}) \approx 120\text{ V}$$
$$I_{rms} = 0.707(4.4\text{ A}) \approx 3.11\text{ A}$$

Table 18-1 shows how an approximate calculation for the rms value of the voltage $v = 170 \sin \theta$ is done. Values of v and v^2 are calculated for θ every 15° from $\theta = 15°$ to $\theta = 180°$. The average of the v^2 values or mean square voltage is found from the total shown:

$$\text{Mean Square Voltage} = \frac{173{,}400}{12} = 14{,}450$$

TABLE 18-1 ROOT MEAN SQUARE VALUE

Angle	Sine	v	v^2
θ	**Sin θ**	**170 sin θ**	**(170 sin $\theta)^2$**
15°	0.2588	44.0	1,936
30°	0.5000	85.0	7,225
45°	0.7071	120.2	14,450
60°	0.8660	147.2	21,675
75°	0.9659	164.2	26,964
90°	1.000	170.0	28,900
105°	0.9659	164.2	26,964
120°	0.8660	147.2	21,675
135°	0.7071	120.2	14,450
150°	0.5000	85.0	7,225
165°	0.2588	44.0	1,936
180°	0.0000	0.0	0
		Total	173,400

The approximate rms voltage is then the square root of the mean square voltage:

$$V_{rms} \approx \sqrt{14{,}450} = 120.2\text{ V}$$

Observe that this value agrees to four significant digits with that obtained using formula (18.3):

$$V_{rms} = 0.707(170\text{ V}) = 120.2\text{ V}$$

To find the peak values from the rms values, use formulas (18.3) solved for V_P and I_P:

Formulas

Peak Values

$$V_P = \left(\sqrt{2}\right) V_{rms} \approx (1.41)\ V_{rms}$$
$$I_P = \left(\sqrt{2}\right) I_{rms} \approx (1.41)\ I_{rms} \quad (18.4)$$

EXAMPLE 18.6

Given $V_{rms} = 50$ V and $I_{rms} = 1.5$ A, find V_P, I_P, V_{PP}, and I_{PP}.

Solution Apply formulas (18.4) to find the peak values:

$$V_P = 1.41\ (50\ \text{V}) \approx 70.5\ \text{V}$$
$$I_P = 1.41\ (1.5\ \text{A}) \approx 2.12\ \text{A}$$

Double the peak values to find the peak-to-peak values:

$$V_{PP} = 2(70.5\ \text{V}) = 141\ \text{V}$$
$$I_{PP} = 2(2.12\ \text{A}) = 4.24\ \text{A}$$

Whenever voltage and current values are given for an ac circuit, they are always rms values unless stated otherwise. Since the rms value of an ac source provides the same power to a resistance as a dc source equal to the rms value, Ohm's law and the power formula can be applied to rms values in an ac circuit containing only resistances:

Formulas

Ohm's Law and Power Formula (Resistive circuit only)

$$I_{rms} = \frac{V_{rms}}{R}$$
$$P = (V_{rms})(I_{rms}) \quad (18.5)$$

EXAMPLE 18.7

Given an ac series circuit with resistors $R_1 = 100\ \Omega$, $R_2 = 150\ \Omega$, and a source voltage $V = 120$ V.

(a) Find I_{rms}, I_P, V_1 = the voltage drop across R_1, and V_2 = the voltage drop across R_2.

(b) Find P_1 = the power dissipated in R_1, P_2 = the power dissipated in R_2, and P_T = total power.

Solution

(a) Since it is not indicated otherwise, the ac voltage given is the rms value: V_{rms} = 120 V. The total resistance $R_T = 100\ \Omega + 150\ \Omega = 250\ \Omega$. To find I_{rms}, apply Ohm's law (18.5):

$$I_{rms} = \frac{V_{rms}}{R_T} = \frac{120\ \text{V}}{250\ \Omega} = 0.48\ \text{A} = 480\ \text{mA}$$

Apply (18.4) to find I_P:

$$I_P = 1.41(480\ \text{mA}) \approx 680\ \text{mA}$$

To find the voltage drops, apply Ohm's law:

$$V_1 = I_{rms}R_1 = (0.48 \text{ A})(100 \ \Omega) = 48 \text{ V}$$
$$V_2 = I_{rms}R_2 = (0.48 \text{ A})(150 \ \Omega) = 72 \text{ V}$$

(b) To find P_1, P_2, and P_T, apply the power formula (18.5) using the rms values:

$$P_1 = V_1 I_{rms} = (48 \text{ V})(0.48 \text{ A}) \approx 23 \text{ W}$$
$$P_2 = V_2 I_{rms} = (72 \text{ V})(0.48 \text{ A}) \approx 35 \text{ W}$$
$$P_T = V_{rms} I_{rms} = (120 \text{ V})(0.48 \text{ A}) \approx 58 \text{ W}$$

EXERCISE 18.2

In all exercises, round answers to two significant digits.

In exercises 1 through 12, find the peak-to-peak and rms values of the ac voltage or current.

1. $V_P = 300$ V
2. $V_P = 160$ V
3. $I_P = 2.2$ A
4. $I_P = 5.6$ A
5. $V_P = 140$ V
6. $V_P = 100$ V
7. $I_P = 150$ mA
8. $I_P = 380$ mA
9. $I_P = 5.5$ A
10. $I_P = 38$ mA
11. $V_P = 200$ V
12. $V_P = 60$ V

In exercises 13 through 24, find the peak and peak-to-peak values of the ac voltage or current.

13. $V_{rms} = 110$ V
14. $V_{rms} = 230$ V
15. $I_{rms} = 1.2$ A
16. $I_{rms} = 3.3$ A
17. $I_{rms} = 600$ mA
18. $I_{rms} = 25$ mA
19. $V_{rms} = 75$ V
20. $V_{rms} = 36$ V
21. $I_{rms} = 80$ mA
22. $I_{rms} = 120$ mA
23. $V_{rms} = 12$ V
24. $V_{rms} = 9.0$ V
25. An ac circuit contains a resistor $R = 50 \ \Omega$ connected to a voltage source $V = 110$ V. Find V_{rms}, V_P, I_{rms}, I_P, and the power P dissipated in the resistor.
26. A circuit contains a resistor $R = 5.1$ kΩ connected to an ac voltage source. If $I = 22$ mA, find V_{rms}, V_P, I_{rms}, I_P, and the power P dissipated in the resistor.
27. Given the ac series circuit in Figure 18-6 with $R_1 = 200 \ \Omega$, $R_2 = 300 \ \Omega$, $R_3 = 100 \ \Omega$, and the current $I = 55$ mA. Find:
    (a) The source voltage V and the voltage drops V_1, V_2, and V_3.
    (b) The power dissipated by each resistor P_1, P_2, P_3, and the total power P_T.
28. Given the ac series circuit in Figure 18-6 with $R_1 = 750 \ \Omega$, $R_2 = 1.2$ kΩ, $R_3 = 1.6$ kΩ, and the source voltage $V = 100$ V. Find:

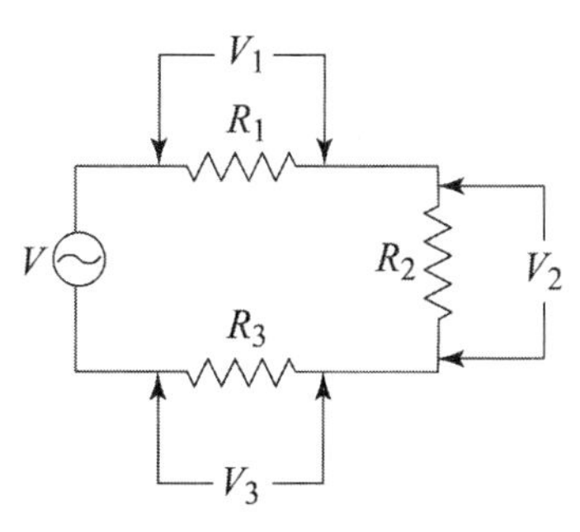

FIGURE 18-6 AC series circuit for problems 27 and 28

    (a) The current I and the voltage drops V_1, V_2, and V_3.
    (b) The power dissipated by each resistor P_1, P_2, P_3, and the total power P_T.
29. Given the ac parallel circuit in Figure 18-7 with $R_1 = 1.2$ kΩ, $R_2 = 1.5$ kΩ, and $V = 80$ V. Find:
    (a) I_1 and I_2, and the total current I_T.
    (b) The power dissipated in each resistance P_1 and P_2, and the total power P_T.

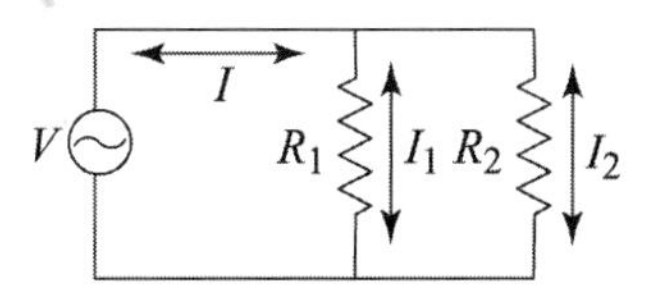

FIGURE 18-7 AC parallel circuit for problems 29 and 30.

30. Given the ac parallel circuit in Figure 18-7 with $R_1 = 620 \ \Omega$, $R_2 = 510 \ \Omega$, and $I_1 = 350$ mA. Find:
    (a) The source voltage V, I_2 and I_T.
    (b) The power dissipated in each resistance P_1 and P_2, and the total power P_T.

18.3 Frequency and Period

When alternating current is generated by revolving a coil of wire in a magnetic field, the speed of rotation is an important factor in the sine wave voltage and current. The speed of rotation ω (omega) is called the *angular velocity*. It is the rate of change of the angle θ with respect to time t and is measured in radians per second (rad/s):

Formula **Angular Velocity (rad/s)**

$$\omega = \frac{\theta}{t} \quad \text{or} \quad \theta = \omega t \tag{18.6}$$

EXAMPLE 18.8 Given the angular velocity of an ac wave $\omega = 120\pi$ rad/s, find the cycles per second of rotation.

Solution Using 2π rad = 1 cycle, divide ω by 2π:

$$\frac{\omega}{2\pi} = \frac{120\pi}{2\pi} = 60 \text{ cycles per second}$$

Cycles per second is called the *frequency* f and the units for f are hertz (Hz). The frequency of the ac wave is therefore $f = 60$ Hz.

From Example 18.8 the relationship between frequency and angular velocity is:

Formula **Frequency (Hz)**

$$f = \frac{\omega}{2\pi} \quad \text{or} \quad \omega = 2\pi f \tag{18.7}$$

Substituting $2\pi f$ for ω in (18.6) gives you the relationship between the angle θ and the frequency:

Formula **Angle (θ)**

$$\theta = \omega t = 2\pi f t \tag{18.8}$$

The equations for the sine wave voltage and current can then be written in terms of angular velocity ω or frequency f:

Formula **AC Voltage and Current Waves**

$$\begin{aligned} v &= V_P \sin \omega t \quad \text{or} \quad v = V_P \sin 2\pi f t \\ i &= I_P \sin \omega t \quad \text{or} \quad i = I_P \sin 2\pi f t \end{aligned} \tag{18.9}$$

The *period* T of an ac wave in seconds is the time for one cycle and is the reciprocal of the frequency:

Formula **Period (seconds)**

$$T = \frac{1}{f} = \frac{2\pi}{\omega} \tag{18.10}$$

EXAMPLE 18.9

(a) Given $f = 60$ Hz for an ac wave, find the angular velocity ω and the period T.

(b) Given $f = 880$ kHz for an ac wave, find the angular velocity ω and the period T.

Solution

(a) Apply formulas (18.7) and (18.10):

$$\omega = 2\pi f = 2\pi(60) = 120\pi \text{ rad/s}$$

$$T \approx \frac{1}{f} = \frac{1}{60 \text{ Hz}} \approx 0.017 \text{ s} = 17 \text{ ms}$$

(b) Apply formulas (18.7) and (18.10) and change 880 kHz to Hz:

$$\omega = 2\pi f = 2\pi(880 \times 10^3) = 1.76\pi \times 10^6 \text{ rad/s}$$

$$T \approx \frac{1}{f} = \frac{1}{880 \times 10^3 \text{ Hz}} \approx 1.1 \times 10^{-6} \text{ s} = 1.1 \text{ μs}$$

Note that the period and the frequency are inversely proportional. As one increases the other decreases. Higher frequencies and smaller periods are encountered in circuits that transmit and receive radio waves.

EXAMPLE 18.10

Given the ac current wave $i = 3.0 \sin 500\pi t$:

(a) Find the angular velocity, frequency, and the period of the wave.

(b) Sketch the graph of current versus time for one cycle from $t = 0$.

Solution

(a) Apply formula (18.9) to find the angular velocity, which is the coefficient of t:

$$\omega = 500\pi \text{ rad/s}$$

Apply formula (18.7) to find the frequency:

$$f = \frac{\omega}{2\pi} = \frac{500\pi}{2\pi} = 250 \text{ Hz}$$

From formula (18.10), the period is the reciprocal of the frequency:

$$T = \frac{1}{250 \text{ Hz}} \text{ s} = 0.0040 \text{ s} = 4.0 \text{ ms}$$

(b) Since the period of the wave is given in units of time, the values of the current are plotted against time rather than radians. To sketch the sine wave you need to plot the maximum, minimum, and zero values. These values correspond to $\theta = 0$, $\pi/2(90°)$, $\pi(180°)$, $3\pi/2(270°)$, and 2π radians (360°), and are enough to sketch the curve, since you know the shape is sinusoidal. For each of these values, first calculate t in milliseconds using formula (18.6) solved for t:

$$t = \frac{\theta}{\omega} = \frac{\theta}{500\pi}$$

For example, when $\theta = 3\pi/2$ rad:

$$t = \frac{3\pi}{2} \div 500\pi = \left(\frac{3\cancel{\pi}}{2}\right)\left(\frac{1}{500\cancel{\pi}}\right) = 0.003 \text{ s} = 3 \text{ ms}$$

The values of t that correspond to the five values of θ are then:

θ (rad)	0	$\frac{\pi}{2}$	π	$\frac{3\pi}{2}$	2π
t (ms)	0	1.0	2.0	3.0	4.0

Using the formula $i = 3.0 \sin 500\pi t$ and the values of t above, construct the table of values for current versus time:

t (ms)	0	1.0	2.0	3.0	4.0
i (A)	0	3.0	0.0	–3.0	0.0

Plot the values of i versus t and sketch the curve shown in Figure 18-8. See the Error Box for more information on ac waves.

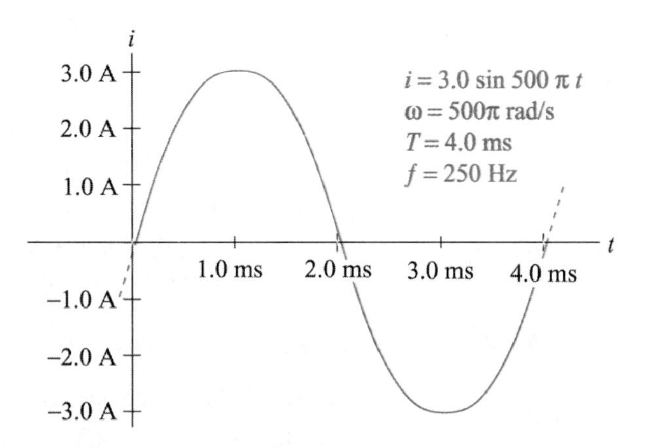

FIGURE 18-8 Frequency and period of ac sine wave for Example 18.10.

ERROR BOX

A common error to watch out for when working with ac waves is not to confuse frequency, the angular period, which is number of *radians* in one cycle, and the time period T, which is the number of *seconds* in one cycle. For an ac wave $V_P \sin 2\pi ft$ or $I_P \sin 2\pi ft$, the angular period is 2π rad, but the time *period* T is not equal to 2π rad. It is equal to the value of t that makes the angle $2\pi ft$ equal to 2π. This is when $ft = 1$, which gives rise to the formula $T = \frac{1}{f}$. Using this information, see if you can do the practice problems correctly.

Practice Problems: For each ac wave, find the angular period, the frequency f, and the period T.

1. $170 \sin 120\pi t$ **2.** $170 \sin 377t$ **3.** $3 \sin 100\pi t$ **4.** $3 \sin 314t$

5. $80 \sin 150\pi t$ **6.** $80 \sin 471t$ **7.** $50 \sin 200\pi t$ **8.** $50 \sin 628t$

9. $1.5 \sin 50\pi t$ **10.** $1.5 \sin 157t$

EXAMPLE 18.11 Given the ac voltage wave $v = 160 \sin 100\pi t$. Find to two significant digits:

(a) The value of v when $t = 5$ ms and $t = 13$ ms.

(b) The two values of t in the first cycle when $v = 50$ V.

Solution

(a) To find the value of v when $t = 5$ ms and 13 ms, substitute into the equation of the curve:

$$v = 160 \sin 100\pi(5 \times 10^{-3}) = 160 \sin (0.5\pi \text{ rad}) = 160 \text{ V}$$
$$v = 160 \sin 100\pi(13 \times 10^{-3}) = 160 \sin (1.3\pi \text{ rad}) \approx -130 \text{ V}$$

Note that 0.5π rad = 90° and that 1.3π rad = 234°, which is in the third quadrant. This gives a negative value for the voltage. You need to set your calculator for radians or convert to degrees by letting π = 180°:

$$v = 160 \sin 100(180°)(5 \times 10^{-3}) = 160 \sin (90°) = 160 \text{ V}$$

$$v = 160 \sin 100(180°)(13 \times 10^{-3}) = 160 \sin 234° \approx -130 \text{ V}$$

(b) To find the values of t, you must use the $\sin^{-1}$ with radian measure. Substitute 50 V for v, solve for the sine of the angle, and then solve for the angle $100\pi t$:

$$50 = 160 \sin 100\pi t$$

$$\sin 100\pi t = \frac{50}{160} = 0.3125$$

The reference angle in radians is:

$$\sin^{-1}(0.3125) \approx 0.318 \text{ rad}$$

Since the value of the sine is positive, there are two solutions for the angle: one in the first quadrant, which is the reference angle:

$$\text{Quadrant I: } 100\pi t \approx 0.318 \text{ rad}$$

and one in the second quadrant, which is the difference from $\pi(180°)$ and the reference angle:

$$\text{Quadrant II: } 100\pi t = \pi - 0.318 \text{ rad} \approx 2.82 \text{ rad}$$

Each value of the angle yields a value of t:

$$100\pi t = 0.318 \text{ rad} \Rightarrow t = \frac{0.318}{100\pi} \approx 1.0 \text{ ms}$$

$$100\pi t = 2.82 \text{ rad} \Rightarrow t = \frac{2.82}{100\pi} \approx 9.0 \text{ ms}$$

These solutions represent two points on the voltage wave that have the same value of v = 50 V, one when the curve is increasing and one when the curve is decreasing.

EXERCISE 18.3

Round all answers in the exercise to two significant digits.

In exercises 1 through 14, find the frequency f *and the period* T *for each value of the angular velocity* ϖ *for an ac wave.*

1. 110π rad/s

2. 130π rad/s

3. 140π rad/s

4. 100π rad/s

5. 380 rad/s

6. 370 rad/s

7. 300 rad/s

8. 320 rad/s

9. 50π rad/s

10. 200π rad/s

11. 5.3×10^6 rad/s

12. 600×10^3 rad/s

13. 75×10^3 rad/s

14. 8.8×10^6 rad/s

In exercises 15 through 26, find the period T *and the angular velocity* ϖ *for each value of the frequency* f *for an ac wave.*

15. 60 Hz

16. 50 Hz

17. 45 Hz

18. 75 Hz

19. 500 Hz

20. 100 Hz

21. 200 Hz

22. 150 Hz

23. 910 kHz

24. 60 kHz

25. 70 MHz

26. 100 Mhz

In exercises 27 through 38, give the peak value, angular velocity ϖ*, frequency* f*, and the period* T *for each ac wave.*

27. $i = 4.0 \sin 120\pi t$

28. $i = 1.0 \sin 100\pi t$

29. $v = 110 \sin 140\pi t$

30. $v = 90 \sin 150\pi t$

31. $v = 310 \sin 300t$

32. $v = 170 \sin 400t$

33. $i = 0.45 \sin 500t$

34. $i = 0.65 \sin 800t$

35. $i = 1.2 \sin 200\pi t$

36. $i = 2.5 \sin 50\pi t$

37. $v = 72 \sin (3.0 \times 10^6)t$

38. $i = 6.4 \sin (200 \times 10^6)t$

39. Find the angular velocity, frequency, and period for the ac voltage wave in Figure 18-9.

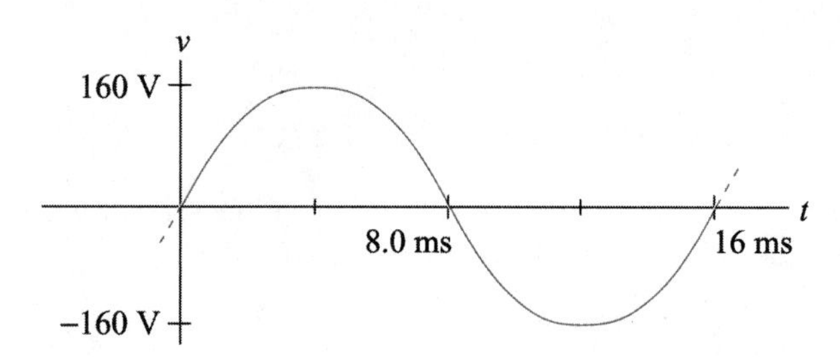

FIGURE 18-9 AC voltage wave for problem 39.

40. Find the angular velocity, frequency, and period for the ac current wave in Figure 18-10.

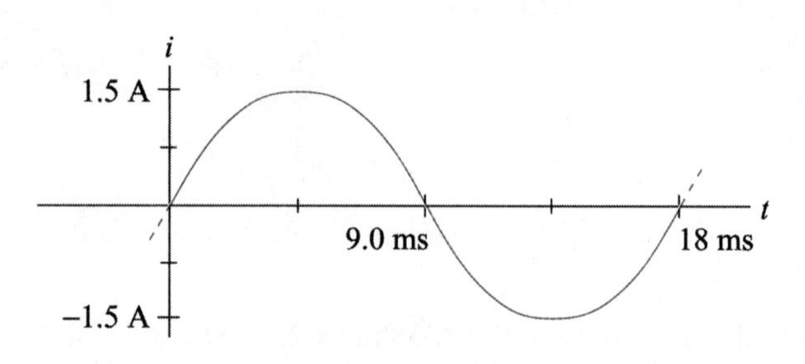

FIGURE 18-10 AC current wave for problem 40.

In exercises 41 through 48 , write the equation of the ac sine wave having the given conditions.

41. $V_P = 180$ V, $f = 60$ Hz

42. $V_P = 90$ V, $f = 50$ Hz

43. $I_P = 2.5$ A, $f = 50$ Hz

44. $I_P = 330$ mA, $f = 60$ Hz

45. $I_P = 450$ mA, $T = 15$ ms

46. $I_P = 4.0$ A, $T = 20$ ms

47. $V_P = 160$ V, $T = 22$ ms

48. $V_P = 130$ V, $T = 18$ ms

In exercises 49 through 56, sketch each ac wave for one cycle plotting the maximum, minimum, and zero values. Give the angular velocity ω*, the frequency* f*, and the period* T.

49. $v = 340 \sin 100\pi t$

50. $v = 170 \sin 120\pi t$

51. $v = 170 \sin 377t$

52. $v = 340 \sin 314t$

53. $i = 2.5 \sin 120\pi t$

54. $i = 1.0 \sin 200\pi t$

55. $i = 12 \sin 1570t$

56. $i = 15 \sin 377t$

For problems 57 through 60, see Example 18.11.

57. Given the ac voltage wave $v = 160 \sin 120\pi t$, find:
(a) The values of v when $t = 7.0$ ms and $t = 1.5$ ms.
(b) The two values of t in the first cycle when $v = 100$ V.

58. Given the ac voltage wave $v = 300 \sin 100\pi t$, find:
(a) The values of v when $t = 17$ ms and $t = 5.5$ ms.
(b) The two values of t in the first cycle when $v = -100$ V.

59. Given the ac current wave $i = 2.0 \sin 100\pi t$, find:
(a) The values of i when $t = 8.0$ ms and $t = 1.5$ ms.
(b) The two values of t in the first cycle when $I = -1.0$ A.

60. Given the ac current wave $i = 3.0 \sin 120\pi t$, find:
(a) The values of i when $t = 15$ ms and $t = 1.2$ ms.
(b) The two values of t in the first cycle when $i = 2.0$ A.

18.4 Phase Angle

Figure 18-2 compares the cosine curve with the sine curve. It shows that cos θ is 90° out of phase with sin θ and that cos θ *leads* sin θ by 90° (or sin θ *lags* cos θ by 90°). This is because cos θ is zero at −90° and therefore "starts" 90° before sin θ. Cos θ is already at the peak value when sin θ is just starting at 0°. If you move sin θ to the left 90°, it will match cos θ exactly. This movement to the left is done by adding a *phase angle* of π/2 to sin θ:

$$\sin\left(\theta + \frac{\pi}{2}\right) = \cos\theta$$

When you compute values of $\sin(\theta + \frac{\pi}{2})$, you will get exactly the same values as cos θ. For example, when θ = 135°:

$$\sin(135° + 90°) = \sin(225°) = -0.707$$
$$\text{and } \cos 135° = -0.707$$

Answers to Error Box Problems, page 370:

1. 2π, 60 Hz, 17 ms **2.** 2π, 60 Hz, 17 ms **3.** 2π, 50 Hz, 20 ms **4.** 2π, 50 Hz, 20 ms
5. 2π, 75 Hz, 13 ms **6.** 2π, 75 Hz, 13 ms **7.** 2π, 100 Hz, 10 ms
8. 2π, 100 Hz, 10 ms **9.** 2π, 25 Hz, 40 ms **10.** 2π, 25 Hz, 40 ms

The concept of lead or lag for ac waves is important in the theory of alternating current. The equations of the ac voltage wave and ac current wave with a phase angle ϕ (phi) take the form:

Formulas

AC Voltage and Current Waves

$$v = V_P \sin(\varpi t + \phi) \qquad i = I_P \sin(\varpi t + \phi) \tag{18.11}$$

Equations (18.11) use only the sine function since the cosine function is essentially the same function. The phase angle ϕ tells you how many degrees the wave leads a basic sine wave with the same frequency where the phase angle $\phi = 0$. Phase angle differences can only apply to waves of the same frequency. Study the following examples, which illustrate the concept of phase angle in an ac circuit.

EXAMPLE 18.12

Given the ac wave $v = 180 \sin(130\pi t - \frac{\pi}{2})$. Find the peak value, angular velocity, phase angle, frequency, and period.

Solution Compare the equation of the curve to formula (18.11):

$$\text{Peak value} = V_P = 180 \text{ V}$$

$$\text{Angular velocity} = \omega = 130\pi$$

$$\text{Phase angle} = \phi = -\frac{\pi}{2}$$

Use formulas (18.7) and (18.10) to find the frequency and period:

$$f = \frac{\omega}{2\pi} = \frac{130\pi}{2\pi} = 65 \text{ Hz}$$

$$T = \frac{1}{f} = \frac{1}{65 \text{ Hz}} \approx 15 \text{ ms}$$

EXAMPLE 18.13

In an ac circuit containing an inductance L, the current is given by $i = I_P \sin \omega t$ and the voltage across the inductance by:

$$v = V_P \sin\left(\omega t + \frac{\pi}{2}\right)$$

where the phase angle $\phi = \frac{\pi}{2}$. Given $f = 60$ Hz, $I_P = 2.0$ A, and $V_P = 100$ V:

(a) Write the equations for v and i and sketch each on the same graph for one cycle from $t = 0$.

(b) Determine the time difference between the two waves.

Solution

(a) Use formula (18.7) to express ω in terms of f:

$$\omega = 2\pi f = 2\pi(60) = 120\pi$$

Write the current and voltage equations substituting the values for V_P, I_P, and ω:

$$i = 2.0 \sin 120\pi t$$

$$v = 100 \sin\left(120\pi t + \frac{\pi}{2}\right)$$

Find the period for both curves applying formula (18.10):

$$T = \frac{1}{f} = \frac{1}{60}\text{ s} \approx 16.7\text{ ms}$$

The current wave has a phase angle $\phi = 0$ rad and therefore starts at $t = 0$. The voltage wave has a phase angle $\phi = \pi/2$ rad and therefore *leads* the current wave by $\pi/2$ radians. It starts at $-\pi/2$ radians or a quarter of a period sooner, and reaches peak value at $t = 0$. Note that when the phase angle is *positive it shifts the curve to the left and vice versa.* The sketch of both sine waves is shown in Figure 18-11 with separate scales for i and v.

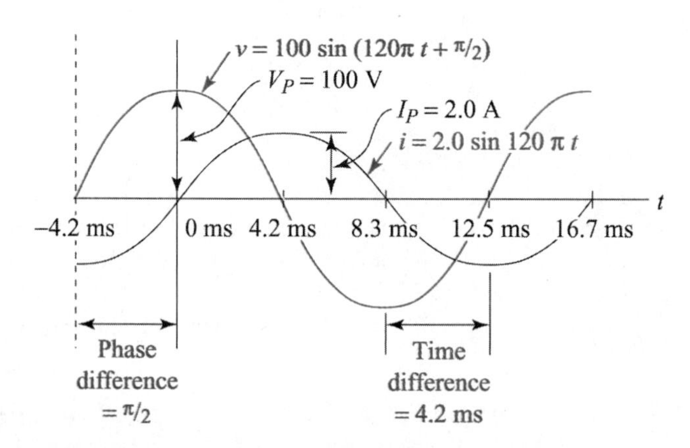

FIGURE 18-11 Voltage leads current in an ac inductive circuit for Example 18.13.

(b) To find the time difference between the two waves, you must convert the phase angle difference into time. Since $\theta = \omega t$, you have:

$$t = \frac{\theta}{\omega}$$

Therefore a phase angle difference of $\pi/2$ represents a time difference of:

$$t = \frac{(\pi/2)}{120\pi} = \frac{1}{240}\text{ s} \approx 4.2\text{ ms}$$

Observe that the phase angle difference is one-fourth of a period, which is $\frac{16.7}{4} \approx 4.2$ ms.

EXERCISE 18.4

Round all answers in the exercise to two significant digits.

In exercises 1 through 10, give the peak value, frequency, period, and phase angle for each ac wave.

1. $i = 1.5\sin\left(100\pi t + \frac{\pi}{2}\right)$

2. $i = 2.3\sin\left(120\pi t + \frac{\pi}{2}\right)$

3. $v = 300\sin\left(200\pi t + \frac{\pi}{2}\right)$

4. $v = 170\sin\left(110\pi t - \frac{\pi}{2}\right)$

5. $v = 60\sin\left(377t - \frac{\pi}{2}\right)$

6. $v = 80\sin\left(314t + \frac{\pi}{2}\right)$

7. $i = 18\sin\left(314t + \frac{\pi}{2}\right)$

8. $i = 16\sin\left(377t + \frac{\pi}{2}\right)$

9. $i = 5.5\sin\left(628t - \frac{\pi}{2}\right)$

10. $i = 3.5\sin\left(377t - \frac{\pi}{2}\right)$

In exercises 11 through 22, write the equation of the ac sine wave having the given conditions.

11. $I_P = 3.3$ A, $\omega = 120\pi$, $\phi = \pi/2$ rad

12. $I_P = 7.5$ A, $\omega = 120\pi$, $\phi = 0$ rad

13. $V_P = 400$ V, $\omega = 100\pi$, $\phi = 0$ rad

14. $V_P = 380$ V, $\omega = 100\pi$, $\phi = \pi/2$ rad

15. $V_P = 350$ V, $f = 50$ Hz, $\phi = \pi/2$ rad

16. $V_P = 150$ V, $f = 60$ Hz, $\phi = -\pi/2$ rad

17. $I_P = 12$ A, $f = 60$ Hz, $\phi = 0$ rad

18. $I_P = 10$ A, $f = 50$ Hz, $\phi = 0$ rad

19. $V_P = 170$ V, $T = 18$ ms, $\phi = \pi/2$ rad

20. $V_P = 160$ V, $T = 16$ ms, $\phi = \pi$ rad

21. $I_P = 500$ mA, $T = 20$ ms, $\phi = -\pi/2$ rad

22. $I_P = 800$ mA, $T = 25$ ms, $\phi = \pi/2$ rad

23. In an ac circuit containing a capacitance C, the current leads the voltage by $\pi/2$. The voltage across the capacitance and the current in the circuit are given by:

$$v = 200 \sin 120\pi t$$

$$i = 1.0 \sin\left(120\pi t + \frac{\pi}{2}\right)$$

(a) Sketch each wave on the same graph for one cycle from $t = 0$.

(b) Determine the time difference between the two waves. See Example 18.13.

24. In an ac circuit containing a resistance and an inductance in series (*RL* circuit), the voltage across the inductance, v_L, leads the voltage across the resistance, v_R, by $\pi/2$. The voltages are given by:

$$v_R = 100 \sin 100\pi t$$

$$v_L = 100 \sin\left(100\pi t + \frac{\pi}{2}\right)$$

(a) Sketch each wave on the same graph for one cycle from $t = 0$.

(b) Determine the time difference between the two waves. See Example 18.13.

25. For the ac waves in Figure 18-12 find:

(a) The frequency, period, and phase angle for each wave.

(b) The time difference. Tell whether v leads or lags i and the phase angle difference.

26. For the ac waves in Figure 18-13 find:

(a) The frequency, period, and phase angle for each wave.

(b) The time difference. Tell whether v leads or lags i and the phase angle difference.

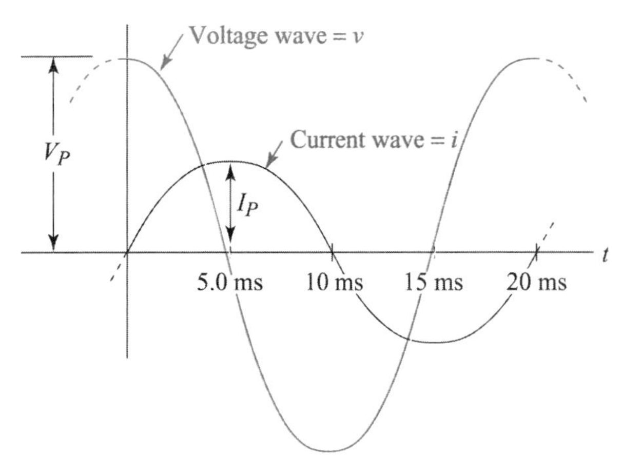

FIGURE 18-12 AC voltage and current waves for problem 25.

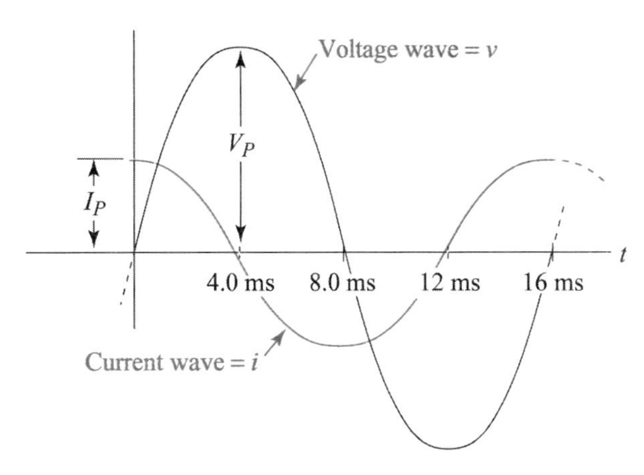

FIGURE 18-13 AC voltage and current waves for problem 26.

27. For the following current wave:

$$i = 3.1 \sin\left(120\pi t + \frac{\pi}{2}\right)$$

(a) Find i when $t = 6.0$ ms and $t = 14$ ms.

(b) Find the two positive values of t in the first cycle when $i = 0.0$ A.

Hint: Sketch the curve.

28. For the following voltage wave:

$$v = 200 \sin\left(100\pi t - \frac{\pi}{2}\right)$$

(a) Find v when $t = 5.0$ ms and $t = 7.5$ ms.

(b) Find the one positive value of t in the first cycle when $v = 200$ V.

Hint: Sketch the curve.

CHAPTER HIGHLIGHTS

18.1 ALTERNATING CURRENT WAVES

The cosine curve leads the sine curve by 90°($\pi/2$ rad). See Figure 18-14. The number of degrees in one cycle or *angular period* = 360°.

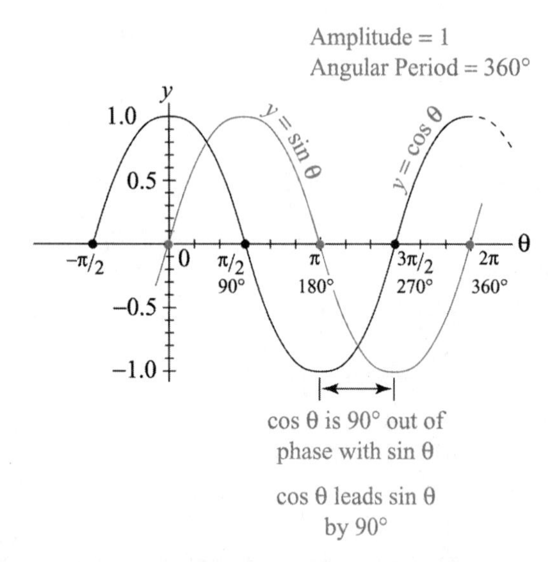

FIGURE 18-14 Key Example: Sine curve and cosine curve.

Amplitude of Sine or Cosine

The amplitude of $y = a \sin\theta$ or $y = a \cos\theta$ is $|a|$. (18.1)

$|a|$ = absolute or positive value.

• • •

AC Voltage and Current Waves

$$v = V_P \sin\theta \qquad i = I_P \sin\theta \tag{18.2}$$

Key Example: Figure 18-15 shows the graph of the ac voltage wave $v = 170 \sin\theta$ for one cycle. The peak value V_P = 170 V and the peak-to-peak value $V_{PP} = 2V_P = 2(170 \text{ V}) = 340$ V.

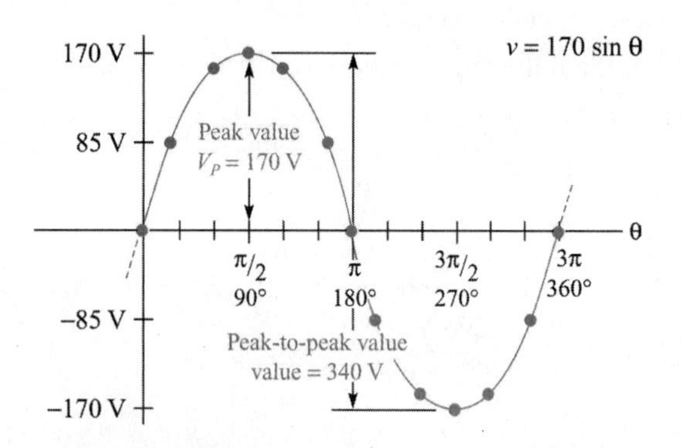

FIGURE 18-15 Key Example: Sine wave voltage.

18.2 ROOT MEAN SQUARE VALUES

The root mean square or effective value of an ac voltage wave or ac current wave is the dc value that produces the same electrical energy or heat dissipation in a resistance.

Root Mean Square Values

$$V_{\text{rms}} = \left(\frac{1}{\sqrt{2}}\right)V_P \approx (0.707)\,V_P$$

$$I_{\text{rms}} = \left(\frac{1}{\sqrt{2}}\right)I_P \approx (0.707)\,I_P \tag{18.3}$$

• • •

Peak Values

$$V_P = (\sqrt{2})V_{\text{rms}} \approx (1.41)\,V_{\text{rms}}$$
$$I_P = (\sqrt{2})I_{\text{rms}} \approx (1.41)\,I_{\text{rms}} \tag{18.4}$$

The rms values are always understood to be the ones given for an ac circuit unless stated otherwise.

Ohm's Law and Power Formula (Resistive circuit only)

$$I_{rms} = \frac{V_{rms}}{R}$$

$$P = (V_{\text{rms}})(I_{\text{rms}}) \tag{18.5}$$

Key Example: In a series ac circuit with resistors R_1 = 100 Ω, R_2 = 150 Ω, and source voltage V = 120 V:

$$I_{\text{rms}} = \frac{V_{\text{rms}}}{R_T} = \frac{120\text{ V}}{250\ \Omega} = 0.48\text{ A} = 480\text{ mA}$$
$$I_P = 1.41(480\text{ mA}) = 677\text{ mA} \approx 680\text{ mA}$$
$$V_1 = I_{\text{rms}}R_1 = (0.48\text{ A})(100\ \Omega) = 48\text{ V}$$
$$V_2 = I_{\text{rms}}R_2 = (0.48\text{ A})(150\ \Omega) = 72\text{ V}$$
$$P_1 = V_1 I_{\text{rms}} = (48\text{ V})(0.48\text{ A}) \approx 23\text{ W}$$
$$P_2 = V_2 I_{\text{rms}} = (72\text{ V})(0.48\text{ A}) \approx 35\text{ W}$$
$$P_T = V_{\text{rms}} I_{\text{rms}} = (120\text{ V})(0.48\text{ A}) \approx 58\text{ W}$$

18.3 FREQUENCY AND PERIOD

Angular Velocity (rad/s)

$$\omega = \frac{\theta}{t} \text{ or } \theta = \omega t \tag{18.6}$$

• • •

Frequency (Hz)

$$f = \frac{\omega}{2\pi} \text{ or } \omega = 2\pi f \quad (18.7)$$

• • •

Angle (θ)

$$\theta = \omega t = 2\pi f t \quad (18.8)$$

• • •

AC Voltage and Current Waves

$$v = V_P \sin \omega t \text{ or } v = V_P \sin 2\pi f t$$
$$i = I_P \sin \omega t \text{ or } i = I_P \sin 2\pi f t \quad (18.9)$$

• • •

Period (seconds)

$$T = \frac{1}{f} = \frac{2\pi}{\omega} \quad (18.10)$$

Key Example: For the ac current wave $i = 3.0 \sin 500\pi t$:

$$\omega = 500\pi$$
$$f = \frac{\omega}{2\pi} = \frac{500\pi}{2\pi} = 250 \text{ Hz}$$
$$T = \frac{1}{f} = \frac{1}{250 \text{ Hz}} = 0.004 \text{ s} = 4.0 \text{ ms}$$

18.4 PHASE ANGLE

AC Voltage and Current Waves

$$v = V_P \sin(\omega t + \phi) \quad i = I_P \sin(\omega t + \phi) \quad (18.11)$$

The phase angle ϕ tells you how many degrees the wave leads a sine wave with the same frequency whose phase angle is 0.

Key Example: For the ac wave:

$$v = 100 \sin(120\pi t + \tfrac{\pi}{2}):$$

$$\phi = \frac{\pi}{2}$$
$$\omega = 120\pi$$
$$f = \frac{\omega}{2\pi} = \frac{120\pi}{2\pi} = 60 \text{ Hz}$$
$$T = \frac{1}{f} = \frac{1}{60 \text{ Hz}} \approx 0.167 \text{ s} = 16.7 \text{ ms}$$

REVIEW EXERCISES

Round all answers in the exercise to two significant digits.

In exercises 1 through 4, graph each ac wave from θ = 0 to θ = 360° for θ every 30°. Give the peak value, the peak-to-peak value, and the angular period.

1. $v = 120 \sin \theta$
2. $v = 200 \sin \theta$
3. $i = 15 \sin \theta$
4. $i = 2.1 \sin \theta$

In exercises 5 through 16, give the peak value, rms value, angular velocity, frequency, period, and phase angle for each ac wave.

5. $v = 160 \sin 120\pi t$
6. $v = 100 \sin 150\pi t$
7. $i = 3.0 \sin 200\pi t$
8. $i = 11 \sin 100\pi t$
9. $v = 170 \sin 350t$
10. $v = 300 \sin 314t$
11. $i = 1.6 \sin 400t$
12. $i = 2.4 \sin 380t$
13. $i = 1.2 \sin\left(140\pi t + \frac{\pi}{2}\right)$
14. $i = 5.6 \sin\left(314t + \frac{\pi}{2}\right)$
15. $v = 320 \sin\left(377t + \frac{\pi}{2}\right)$
16. $v = 50 \sin\left(120\pi t - \frac{\pi}{2}\right)$

In exercises 17 through 28, write the equation of the ac sine wave having the given conditions.

17. $I_P = 2.4$ A, $f = 60$ Hz, $\phi = 0$ rad
18. $I_P = 5.5$ A, $\omega = 120\pi$, $\phi = 0$ rad
19. $V_P = 110$ V, $\omega = 100\pi$, $\phi = 0$ rad
20. $V_P = 50$ V, $f = 50$ Hz, $\phi = 0$ rad
21. $V_P = 120$ V, $T = 17$ ms, $\phi = 0$ rad
22. $I_P = 14$ A, $T = 19$ ms, $\phi = 0$ rad
23. $I_P = 1.3$ A, $T = 10$ ms, $\phi = 0$ rad
24. $V_P = 220$ V, $T = 20$ ms, $\phi = 0$ rad
25. $I_P = 14$ A, $\omega = 100\pi$, $\phi = \pi/2$ rad
26. $V_P = 300$ V, $f = 50$ Hz, $\phi = \pi/2$ rad
27. $V_P = 220$ V, $f = 60$ Hz, $\phi = \pi/2$ rad

28. $I_P = 1.5$ A, $T = 40$ ms, $\phi = -\pi/2$ rad

29. An ac circuit contains a resistor $R = 1.5$ kΩ connected to a voltage source. If the current $I = 50$ mA, find I_{rms}, I_P, V_{rms}, V_P, and the power dissipated by the resistor.

30. A series circuit contains two parallel resistors connected to an ac power supply of 120 V. If $R_1 = 30$ Ω and $R_2 = 20$ Ω, find:

(a) The total current I, I_P, V_P, and the current through each resistor, I_1 and I_2.

(b) The power dissipated by each resistor P_1, P_2, and the total power P_T.

31 **(a)** An ac voltage wave takes 3 ms to go from 0 to peak value. What are the frequency, period, and angular velocity of the wave?

(b) If the peak value of the voltage wave is 120 V and the phase angle = 0 rad, write the equation of the wave.

32. In Figure 18-16 find:

(a) The frequency, period, and angular velocity of each ac wave.

(b) The phase angle difference and time difference of the two ac waves.

33. Sketch the voltage wave $v = 90 \sin 150\pi t$ for one cycle from t = 0. Give the peak value, frequency, and period of the wave.

34. An ac circuit contains a resistor and a capacitor in series (*RC* circuit). The voltage across the capacitor is given by $v_C = 200 \sin 120\pi t$. The voltage across the resistor v_R leads the voltage across the capacitor by 90° and has the same peak value.

(a) Write the equation of the voltage wave v_R.

(b) Sketch the two waves on the same graph for one cycle from $t = 0$, and find the time difference between the two waves.

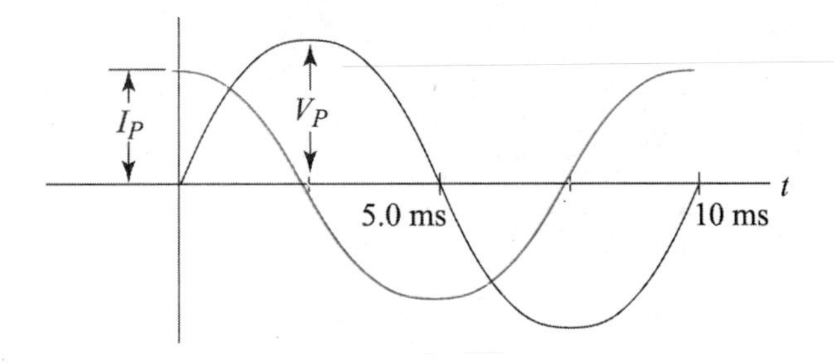

FIGURE 18-16 AC waves for problem 32.

35. Given the current wave $i = 0.50 \sin 150\pi t$:

(a) Find i when $t = 2.0$ ms and 10 ms.

(b) Find the values of t in the first cycle when $i = 250$ mA.

(c) Find the values of t in the first cycle when $i = 0$ mA.

36. Given the voltage wave $v = 400 \sin (200\pi t + \frac{\pi}{2})$:

(a) Find v when $t = 2.0$ ms. and 5.5 ms.

(b) At what value of t in the first cycle is $v = 400$ V?

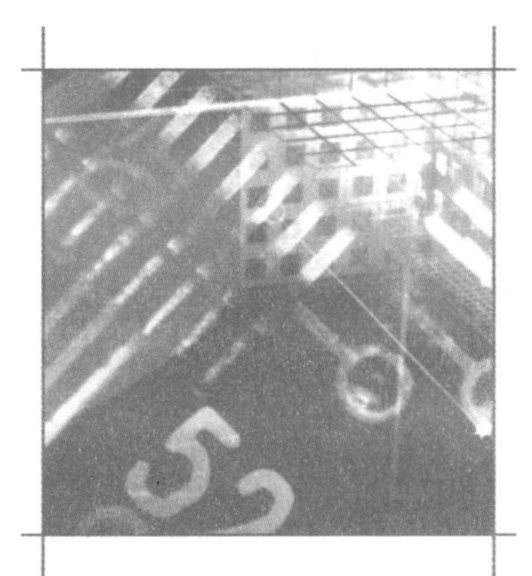

CHAPTER 19
Trigonometry of Vectors

CHAPTER OBJECTIVES

In this chapter, you will learn:

- The definition of a vector and a phasor.
- How to solve problems with vectors and phasors.
- The law of sines and the law of cosines.
- How to use the law of sines and law of cosines to solve problems with vectors and phasors.

A vector or a phasor is a quantity that can be described by two numbers which represent its magnitude and direction. It can therefore be drawn as a directed line segment or arrow. The difference between a phasor and a vector is in the meaning of the direction. A vector's direction is its orientation in space, such as a force vector. A phasor's direction represents *a time difference* between two quantities in an ac circuit, such as the voltage across a resistor and the voltage across an inductor. Vectors and phasors are both important in the study of ac circuits and use the trigonometric concepts in Chapters 16 and 17. Two other trigonometric laws, the law of sines and the law of cosines, are also useful in solving problems with vectors and phasors.

• • •

19.1 Vectors and Phasors

Figure 19-1 shows a vector or phasor as an arrow on a graph starting at the origin and ending at a point. The *magnitude* r is the length of the arrow, and the *direction* θ is the angle that the arrow makes with the positive x axis. When a vector or phasor is described in terms of r and θ, it is called *polar form*. The length and direction of the arrow is also determined by the rectangular coordinates of the point (x,y) at the head of the arrow. When a vector or phasor is described in terms of x and y, it is called *rectangular form*. Vectors and phasors are treated the same way mathematically.

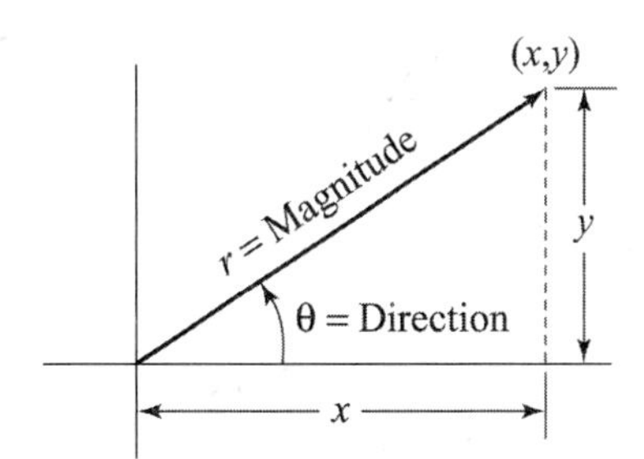

FIGURE 19-1 Vector or phasor.

The following example illustrates some important vector and phasor quantities that you may be familiar with.

EXAMPLE 19.1

List some examples of vectors and phasors.

Solution Many quantities can be described with one number. They are called *scalars*. For example, height, weight, time, and distance are scalars. The following quantities need two numbers to describe them and are vectors or phasors.

(a) **Radius Vector:** The head of a radius vector is at the point (3,4). This is rectangular form. This same radius vector in polar form is 5 units long and makes an angle of 53° with the x axis. This is written in polar form as $5\ \angle 53°$.

(b) **Velocity Vector:** The velocity of the wind is 20 mi/h *from* the north and therefore blowing in a southerly direction. This can be written in polar form as $20\ \angle 180°$ mi/h, where 0° corresponds to north, or in rectangular form as (0,–20).

(c) **Force vector:** The force acting on a body is 13 N (newtons) at an angle $\theta = 45^0$ with the horizontal. This can be written in polar form as $13\ \angle 45°$ N..

(d) **Voltage Phasor:** The total voltage in an *RL* series circuit is 14 V with a phase angle $\theta = 47°$. This is written in polar form as $14\ \angle 47°$ V.

(e) **Impedance Phasor:** The impedance in an *RLC* series circuit is 3.2 kΩ, and the phase angle is −34°. This is written in polar form as $3.2\ \angle{-34°}$ kΩ.

(f) **Current Phasor:** The total current in an *RL* parallel circuit is 50 mA with a phase angle of $\theta = -15°$. This is written in polar form as $50\ \angle{-15°}$ mA.

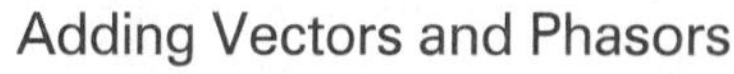

Adding Vectors and Phasors

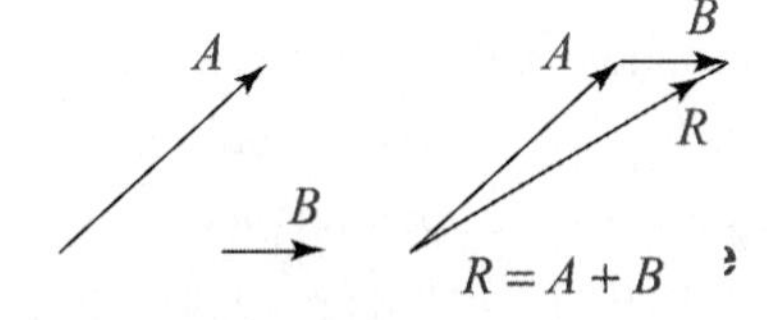

FIGURE 19-2 Addition of vectors or phasors.

Figure 19-2 shows you how to add two vectors or phasors graphically.

The procedure is as follows:

Adding Vectors and Phasors

To add two vectors or phasors A and B, place the tail of B at the head of A without changing its direction. The resultant $R = A + B$ goes from the tail of A to the head of B.

The next example shows you how to apply the above rule for adding two vectors or phasors mathematically.

EXAMPLE 19.2

Figure 19-3 shows a body acted on by a horizontal force $F_x = 18$ N (newtons) at an angle $\theta_x = 180°$ and a vertical force $F_y = 10$ N at an angle $\theta_y = 90°$. Find the resultant force F acting on the body.

Solution Add F_y to F_x by moving the vector (F_y) parallel to itself so that its tail is on the head of F_x. The resultant F is the hypotenuse of the right triangle whose sides are F_x and F_y. Since force is a vector you need to find both its magnitude *and* its direction. To find the magnitude of F, apply the Pythagorean theorem to the right triangle:

$$F_x^2 + F_y^2 = F^2$$

$$F = \sqrt{F_x^2 + F_y^2} = \sqrt{18^2 + 10^2} = \sqrt{424} \approx 20.6 \text{ N}$$

To find the direction, you need to find the angle θ in the second quadrant. First find the reference angle and then subtract from 180° to find θ:

$$\text{Reference angle} = \tan^{-1}\left|\frac{F_y}{F_x}\right| = \tan^{-1}\left(\frac{10}{18}\right) \approx 29°$$

$$\theta = 180° - 29° = 151°$$

The resultant force can then be written in polar form as $F = 20.6\ \angle 151°$ N.

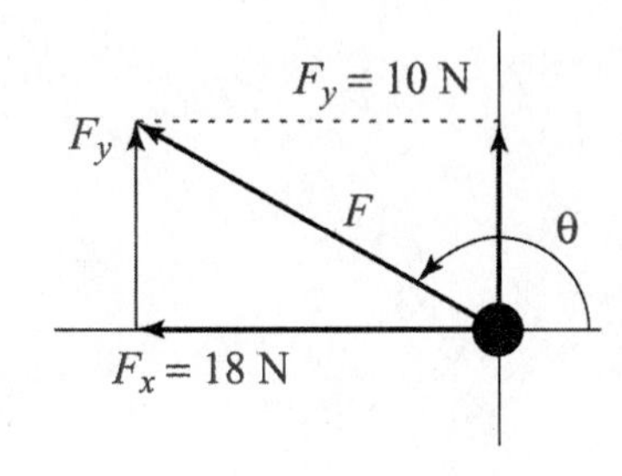

$$F = \sqrt{F_x^2 + F_y^2}$$

$$\text{Reference Angle} = \tan^{-1}\left|\frac{F_y}{F_x}\right|$$

FIGURE 19-3 Force vectors for Example 19.2.

The vectors F_x and F_y in the above example can be considered the horizontal and vertical *components* of the force F. The formulas for combining vector components are then as follows:

Formulas

Adding Vector Components F_x and F_y

$$\text{Magnitude of the Resultant } F = \sqrt{F_x^2 + F_y^2} \tag{19.1}$$

$$\text{Reference Angle} = \tan^{-1}\left|\frac{F_y}{F_x}\right| \qquad 0° \le \theta < 360°$$

Components are further explained below.

Study the next example that closes the circuit and shows an application of phasors in an ac circuit.

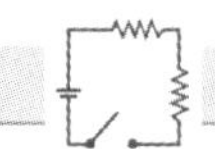

EXAMPLE 19.3

Close the Circuit

Figure 19-4 shows the phasor diagram for the voltages in a series RC circuit. The voltage across the resistance $V_R = 4.7 \angle 0°$ V, and the voltage across the capacitance $V_C = 3.3 \angle{-90°}$ V. The negative phase angle indicates that V_C *lags* V_R by 90°. Find the total voltage V_T, which is the sum of the two phasors.

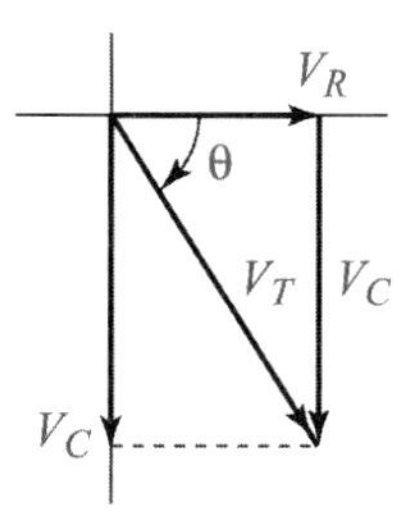

$V_T = \sqrt{V_r^2 + V_C^2}$

$\theta = \tan^{-1}\left(\frac{-V_C}{V_R}\right)$

FIGURE 19-4 Voltages in an RC series circuit for Example 19.3.

Solution For a phasor, you must find the magnitude *and* the angle. Apply the Pythagorean theorem to find the magnitude of V_T:

$$V_T = \sqrt{V_R^2 + V_C^2} = \sqrt{4.7^2 + 3.3^2} = \sqrt{32.98} \approx 5.7 \text{ V}$$

The phase angle θ represents a time difference and is *a negative angle in the fourth quadrant*. It can be found directly on the calculator by using the inverse tangent function and making V_C negative:

$$\theta = \tan^{-1}\left(\frac{-V_C}{V_R}\right) = \tan^{-1}\left(\frac{-3.3}{4.7}\right) \approx -35°$$

The solution in polar form is then $V_T = 5.7 \angle{-35°}$ V.

The formulas for combining phasor components are as follows:

Formulas

Adding Phasor Components Vx and Vy

$$\text{Magnitude of the Resultant } V = \sqrt{V_x^2 + V_y^2} \tag{19.2}$$

$$\text{Angle } \theta = \tan^{-1}\left(\frac{V_y}{V_x}\right) \qquad -90° < \theta < 90°$$

The error box on page 382 explains more about finding the angle of a vector phasor.

When two vectors or phasors directly oppose each other and are 180° apart, the smaller one directly cancels part of the larger. The resultant magnitude is the *difference* in the magnitudes, and the resultant direction is that of the vector or phasor with the greater magnitude.

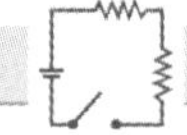

EXAMPLE 19.4

Close the Circuit

Figure 19-5(a) on page 382 shows the phasor diagram for an ac circuit containing a resistance, an inductance, and a capacitance in series (RLC series circuit). Given the resistance $R = 2.0 \angle 0°$ kΩ, the capacitive reactance $X_C = 3.0 \angle{-90°}$ kΩ, and the inductive reactance $X_L = 4.0 \angle 90°$ kΩ. Find the circuit impedance Z, which is the sum of these three phasors.

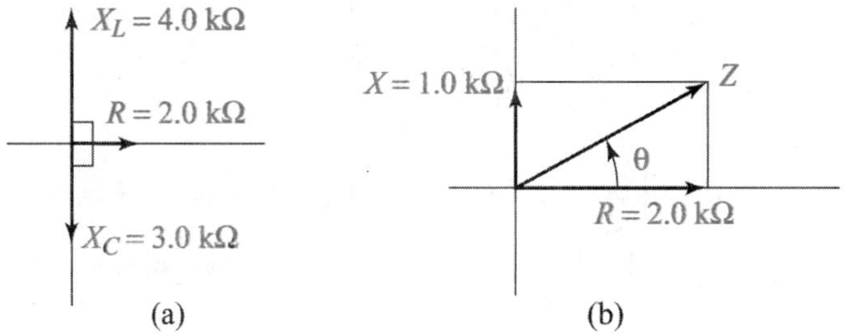

FIGURE 19-5 *RLC* series circuit impedance phasors for Example 19.4.

Solution The phasors X_C and X_L oppose each other in this ac circuit. Add them together first to obtain the magnitude of the net reactance X, which is their difference:

$$X = X_L - X_C = 4.0 - 3.0 = 1.0 \text{ k}\Omega$$

The phase angle of X is the same as that of X_L, 90°, since X_L is larger than X_C. Therefore, $X = 1.0 \angle 90° \text{ k}\Omega$.

The phasor diagram can then be reduced to two phasors, as shown in Figure 19-5(b). You can now add these two perpendicular phasors to find the circuit impedance Z. The magnitude of Z, which is in the first quadrant, is:

$$Z = \sqrt{1.0^2 + 2.0^2} = \sqrt{5.0} \approx 2.2 \text{ k}\Omega$$

The phase angle of Z, which is in the first quadrant, is:

$$\theta = \tan^{-1}\left(\frac{X}{R}\right) = \tan^{-1}\left(\frac{1.0}{2.0}\right) \approx 27°$$

The circuit impedance in polar form is then $Z = 2.2 \angle 27° \text{ k}\Omega$.

ERROR BOX

A common error when working with vectors and phasors is not getting the correct angle. For a *vector* in the *fourth* quadrant we usually use a *positive* angle between 270° and 360°; for a *phasor* in the fourth quadrant, we use a *negative* angle between 0° and −90°. The calculator will give you a negative angle for [tan⁻¹] when you enter a negative number. *You can therefore use a negative value with inverse tangent when working with phasors.*

For example: [tan⁻¹] [(−)] 2.14 [=] → −65°

This is the same direction as the positive angle 360° − 65° = 295°. Apply this idea and see if you can find the correct angle for each vector or phasor sum.

Practice Problems: For each pair of vectors or phasors, find the *angle* of the resultant to the nearest degree.

1. Force vectors:	$F_x = 0.25$ N, $\theta_x = 0°$;	$F_y = 0.15$ N, $\theta_y = 270°$
2. Velocity vectors:	$v_x = 3.6$ m/s, $\theta_x = 0°$;	$v_y = 8.4$ m/s, $\theta_y = 270°$
3. Voltage phasors:	$V_R = 60 \angle 0°$ V;	$V_C = 20 \angle -90°$ V
4. Current phasors:	$I_R = 22 \angle 0°$ mA;	$I_L = 33 \angle -90°$ mA
5. Acceleration vectors:	$a_x = 9.2$ ft/s/s $\theta_x = 0°$	$a_y = 5.5$ ft/s/s $\theta_y = 270°$
6. Reactance phasors:	$X_L = 5.4 \angle 90°$ kΩ;	$X_C = 6.2 \angle -90°$ kΩ
7. Radius vectors:	$x = 12$ in $\angle 0°$	$y = 15$ in $\angle 270°$
8. Impedance phasors:	$Z_x = 2.2 \angle 0°$ kΩ;	$Z_y = 1.4 \angle -90°$ kΩ

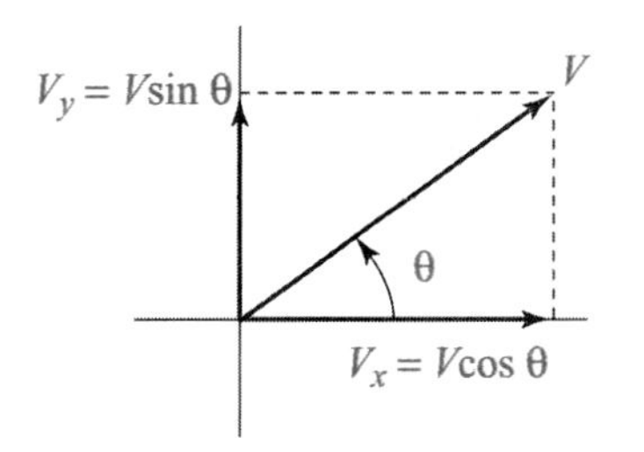

FIGURE 19-6 Vector or phasor components.

Components

In the preceding examples, two perpendicular vectors or phasors are combined into one resultant. The reverse of this process is also very useful when adding vectors or phasors that are not perpendicular. When a vector or phasor is placed on a rectangular coordinate system with its tail at the origin, it can be resolved into components along the x and y axes. In Figure 19-3, F_x is the x component of F, and F_y is the y component of F. In Figure 19-4, V_R is the x component of V_T, and V_C is the y component. The relationship between a vector or phasor V and its components V_x and V_y is shown in Figure 19-6. The definitions of sine and cosine yield the formulas for the components:

Formulas

Vector or Phasor Components

$$\cos\theta = \frac{V_x}{V} \rightarrow V_x = V\cos\theta \ (x \text{ component})$$
$$\sin\theta = \frac{V_y}{V} \rightarrow V_y = V\sin\theta \ (y \text{ component}) \quad (19.3)$$

These formulas apply for any angle θ in any quadrant. When you need to add two or more vectors or phasors that are not perpendicular, proceed as follows:

1. Resolve each vector or phasor into x and y components using formulas (19.3).
2. Algebraically add the x components and the y components separately to give the x and y components of the resultant.
3. Add the x and y components of the resultant applying formulas (19.1) for a vector or (19.2) for a phasor.

Study the next example, which closes the circuit and illustrates this process in an application with impedance phasors.

EXAMPLE 19.5 Given the two impedance phasors $Z_1 = 7.3 \angle 25°$ kΩ and $Z_2 = 5.5 \angle -55°$ kΩ. Find the resultant phasor Z to two significant digits.

Solution Draw the phasors on a coordinate graph as shown in Figure 19-7(a). Set up a table and calculate the components of each phasor by applying formulas (19.3). Then add the components to produce the x and y components of the resultant:

Phasors	x components	y components
Z_1	7.3 cos 25° kΩ = 6.62 kΩ	7.3 sin 25° kΩ = 3.09 kΩ
Z_2	5.5 cos(−55°) kΩ = 3.15 kΩ	5.5 sin(−55°) kΩ = −4.51 kΩ
Resultant Z	$Z_x = 9.77$ kΩ	$Z_y = -1.42$ kΩ

Note that formulas (19.3) produce the correct signs for the components, corresponding to the quadrant the angle is in.. Finally add the components of the resultant phasor applying formulas (19.2). Figure 19-7(b) on page 384 shows the components of the resultant and how they are added to produce the resultant Z. The magnitude of Z is found by applying the Pythagorean theorem:

$$Z = \sqrt{9.77^2 + (-1.42)^2} = \sqrt{97.5} \approx 9.9 \text{ k}\Omega$$

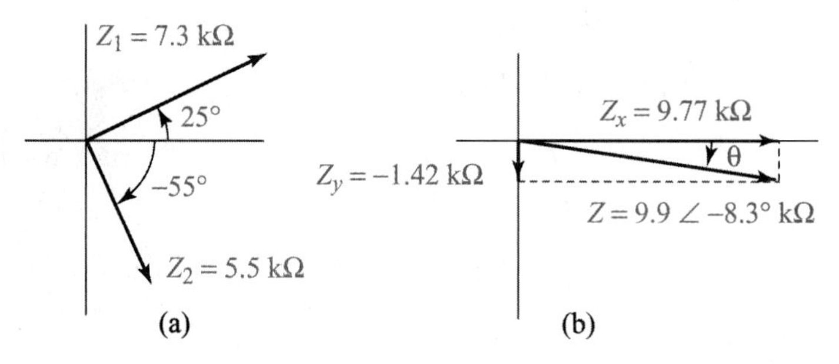

FIGURE 19-7 Adding impedance phasors for Example 19.5.

The negative phase angle of Z is found by applying the inverse tangent to Z_y and Z_x:

$$\theta = \tan^{-1}\left(\frac{Z_y}{Z_x}\right) = \tan^{-1}\left(\frac{-1.42}{9.77}\right) \approx -8.3°$$

Observe that using the negative component Z_y from the table with the inverse tangent produces the correct negative angle. The resultant phasor is then $Z = 9.9 \angle -8.3°$ kΩ.

EXERCISE 19.1

Round angles to the nearest degree and all other answers to two significant digits.

In exercises 1 through 8, find the resultant of each pair of vectors with an angle between 0° and 360°.

1. $F_x = 10$ N, $\theta_x = 0°$; $F_y = 24$ N, $\theta_y = 90°$
2. $F_x = 44$ N, $\theta_x = 180°$; $F_y = 26$ N, $\theta_y = 90°$
3. $v_x = 53$ ft/s, $\theta_x = 180°$; $v_y = 26$ ft/s, $\theta_y = 90°$
4. $x = 12$ m, $\theta_x = 0°$; $y = 9.8$ m, $\theta_y = 270°$
5. $F_1 = 4 \times 10^{-6}$ N, $\theta_1 = 180°$; $F_2 = 5 \times 10^{-6}$ N, $\theta_2 = 270°$
6. $E_1 = 400$ N/C, $\theta_1 = 180°$; $E_2 = 300$ N/C, $\theta_2 = 270°$
7. $a_x = 25$ m/s/s, $\theta_x = 0°$; $a_y = 42$ m/s/s, $\theta_y = 90°$
8. $v_1 = 2.3$ m/s, $\theta_1 = 0°$; $v_2 = 6.2$ m/s, $\theta_2 = 270°$

In exercises 9 through 22, find the resultant of each pair of phasors with an angle between −90° and 90°.

9. $R = 33 \angle 0°$ kΩ; $X_L = 22 \angle 90°$ kΩ
10. $R = 1.2 \angle 0°$ kΩ; $X_L = 1.5 \angle 90°$ kΩ
11. $R = 2.4 \angle 0°$ kΩ; $X_C = 4.7 \angle -90°$ kΩ
12. $R = 18 \angle 0°$ kΩ; $X_C = 8.2 \angle -90°$ kΩ
13. $V_L = 24 \angle 90°$ V; $V_C = 12 \angle -90°$ V
14. $I_C = 340 \angle 90°$ mA; $I_L = 770 \angle -90°$ mA
15. $R = 1.0 \angle 0°$ kΩ; $X = 820 \angle -90°$ Ω
16. $R = 910 \angle 0°$ Ω; $X = 1.6 \angle 90°$ kΩ
17. $I_R = 160 \angle 0°$ μA; $I_C = 250 \angle 90°$ μA
18. $V_R = 28 \angle 0°$ V; $V_C = 18 \angle -90°$ V
19. $Z_1 = 3.1 \angle 65°$ kΩ; $Z_2 = 6.1 \angle -15°$ kΩ
20. $Z_1 = 4.3 \angle 30°$ kΩ; $Z_2 = 5.2 \angle -60°$ kΩ
21. $Z_1 = 12 \angle 26°$ kΩ; $Z_2 = 15 \angle -56°$ kΩ
22. $Z_1 = 22 \angle 74°$ kΩ; $Z_2 = 38 \angle -46°$ kΩ

Applied Problems

23. A horizontal force of 23 N acts to the right on a body, and a vertical force of 40 N acts directly down on the body. Find the magnitude and direction of the resultant force.
24. A taxi driver traveling at 15 mi/h throws a cigarette out of the window perpendicular to his direction and at a speed of 5 mi/h. What is the velocity of the cigarette as it leaves his hand? Find the magnitude and the angle with respect to the taxi's direction. Let the taxi's direction be x and the cigarette's direction be y.
25. Tom sails his boat 5.5 mi due east and then sails 2.0 mi at an angle of 40° north of east (counterclockwise from east). What is the distance and direction from his starting point? Let east be 0° and add components as in Example 19.5.
26. In Figure 19-8, Doug is flying a light airplane due north at 130 kn (knots = nautical miles per hour). The plane is experiencing a 40-kn headwind from the northwest (45° west of north). What is the resultant velocity of Doug's plane? Find the magnitude and the angle measured from due north. (Place the tail of the wind vector at the origin and add components as in Example 19.5.)

Answers to Error Box Problems, page 382:
1. 329° **2.** 293° **3.** −18° **4.** −56° **5.** 329° **6.** −90° **7.** 309° **8.** −32.5°

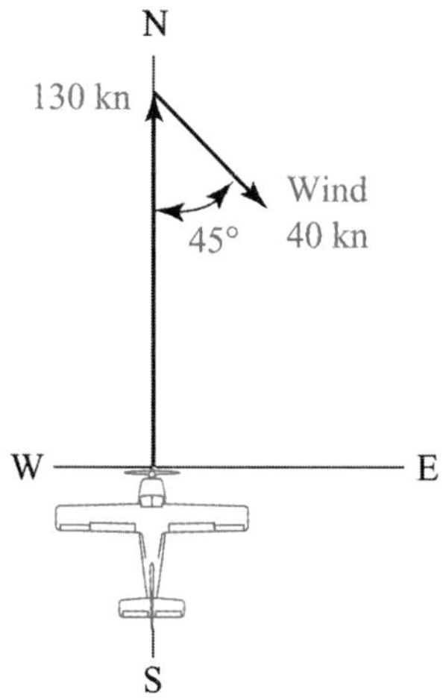

N
130 kn
Wind
40 kn
45°
W E
S

FIGURE 19-8 Airplane velocity for problem 26.

Applications to Electronics

27. In Figure 19-4, find V_T if the magnitude of $V_R = 33$ V and the magnitude of $V_C = 63$ V.
28. In Figure 19-4, find V_T if the magnitude of $V_R = 4.5$ V and the magnitude of $V_C = 2.6$ V.
29. In a parallel RC circuit, the current through the capacitance I_C leads the current through the resistance I_R by 90°. The total current I_T is the phasor sum of I_C and I_R. Find I_T when $I_R = 50 \angle 0°$ mA and $I_C = 70 \angle 90°$ mA.
30. In problem 29, find I_T when $I_R = 90 \angle 0°$ μA and $I_C = 200 \angle 90°$ μA.
31. In Figure 19-5, find the circuit impedance Z when $R = 15 \angle 0°$ kΩ, $X_L = 30 \angle 90°$ kΩ, and $X_C = 20 \angle -90°$ kΩ.
32. In Figure 19-5, find the circuit impedance Z when $R = 4.3 \angle 0°$ kΩ, $X_L = 3.6 \angle 90°$ kΩ, and $X_C = 7.5 \angle -90°$ kΩ.

    Note: X_C is larger than X_L.
33. Given the two impedance phasors $Z_1 = 44 \angle -56°$ kΩ and $Z_2 = 22 \angle 86°$ kΩ. Find the resultant phasor. See Example 19.5.
34. Given the two impedance phasors $Z_1 = 5.2 \angle 12°$ kΩ and $Z_2 = 6.7 \angle -15°$ kΩ. Find the resultant phasor. See Example 19.5.
35. At a point in an electric field, two electric charges produce the intensity vectors: $E_1 = (1.9 \times 10^3) \angle 130°$ N/C (newtons/coulomb) and $E_2 = 2.3 \times 10^3 \angle 200°$ N/C. Find the resultant vector. See Example 19.5.
36. At a point in a magnetic field, the magnetic flux density B_1 due to current in one wire is $4.0 \times 10^{-6} \angle 0°$ T (tesla), and the flux density B_2 due to current in another wire is $6.0 \times 10^{-6} \angle 45°$ T. Find the resultant flux density. See Example 19.5.

19.2 Law of Sines

The formulas for sin, cos, and tan apply to right triangles. There are two important laws, however, that can be used to find the sides or angles *of any triangle* using sine and cosine. For any triangle (Figure 19-9), the law of sines states:

Law

Law of Sines

$$\frac{a}{\sin A} = \frac{b}{\sin B} = \frac{c}{\sin C} \qquad (19.4)$$

Formula (19.4) represents three different formulas. Each one is between two different sides and their opposite angles in a triangle. The law of sines says that the side of a triangle is proportional to the sine of the opposite angle. The value of the sine increases as the angle increases from 0° to 90°, which means that *the largest side is opposite the largest angle and the smallest side is opposite the smallest angle.*

The law of sines can be used to find the missing parts of a triangle when you are given:

1. Two angles and a side opposite one of the angles, or
2. Two sides and an angle opposite one of the sides.

EXAMPLE 19.6

In Figure 19-9, given $\measuredangle A = 40°$, $\measuredangle B = 60°$, and side $c = 5.0$, find the three missing parts of triangle ABC to two significant digits.

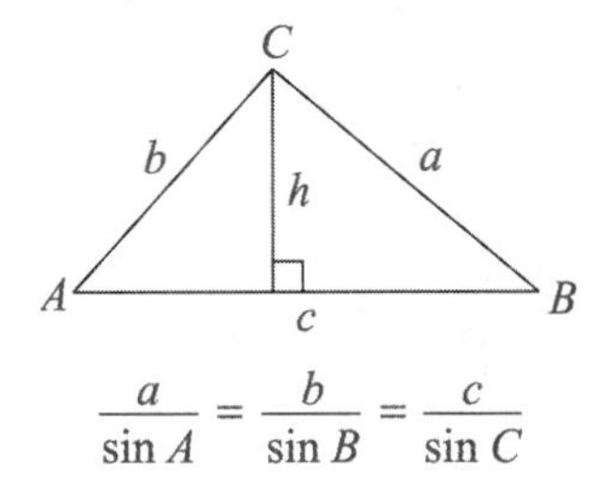

$$\frac{a}{\sin A} = \frac{b}{\sin B} = \frac{c}{\sin C}$$

FIGURE 19-9 Law of sines for Example 19.6.

Solution You know two angles and side c but you need $\measuredangle C$ to use the law of sines. Subtract the two angles from 180° to find $\measuredangle C$:

$$\measuredangle C = 180° - \measuredangle A - \measuredangle B = 180° - 40° - 60° = 80°$$

You now have side c and $\measuredangle C$ and can use the law of sines to find sides a and b. To find side a, use formula (19.4) with a and c:

$$\frac{a}{\sin A} = \frac{c}{\sin C}$$

Multiply both sides by sin A *to solve for* a and substitute the given values:

$$(\cancel{\sin A})\frac{a}{\cancel{\sin A}} = \frac{c}{\sin C}(\sin A)$$

$$a = \frac{c(\sin A)}{\sin C} = \frac{5.0(\sin 40°)}{\sin 80°} = \frac{5.0(0.985)}{(0.643)} \approx 3.3$$

To find side b, use formula (19.4) with b and c:

$$\frac{b}{\sin B} = \frac{c}{\sin C}$$

Multiply both sides by sin B to solve for b and substitute the given values:

$$b = \frac{c(\sin B)}{\sin C} = \frac{5.0(\sin 60°)}{\sin 80°} = \frac{5.0(0.866)}{(0.985)} \approx 4.4$$

Proof of the Law of Sines

The following proof of the law of sines applies to acute triangle ABC in Figure 19-9. The proof for an obtuse or right triangle is similar. In triangle ABC,

$$\frac{h}{b} = \sin A \rightarrow h = b(\sin A)$$

$$\frac{h}{a} = \sin B \rightarrow h = a(\sin B)$$

Equate the two expressions for h:

$$a(\sin B) = b(\sin A)$$

Divide by (sin A)(sin B) to obtain the first part of the law of sines:

$$\frac{a(\cancel{\sin B})}{(\sin A)(\cancel{\sin B})} = \frac{b(\cancel{\sin A})}{(\sin B)(\cancel{\sin A})}$$

Similarly, it can be shown that

$$\frac{b}{\sin B} = \frac{c}{\sin C}$$

Therefore, any two of the three ratios in formula (19.4) are equal to each other.

Sometimes the law of sines provides two possible solutions, but only one of them applies. Study the next example that closes the circuit and illustrates this situation in an electrical problem with vectors.

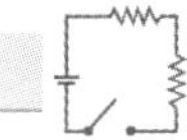

EXAMPLE 19.7

Close the Circuit

Two electrical forces F_1 and F_2 act on a particle in an electrical field at an angle of 130° when added head to tail. If $F_1 = 8.0 \times 10^{-6}$ N (Newtons) and the resultant R is measured to be 10×10^{-6} N, find the magnitude of F_2 to two significant digits. See Figure 19-10, which shows the vector diagram with F_2 added to F_1.

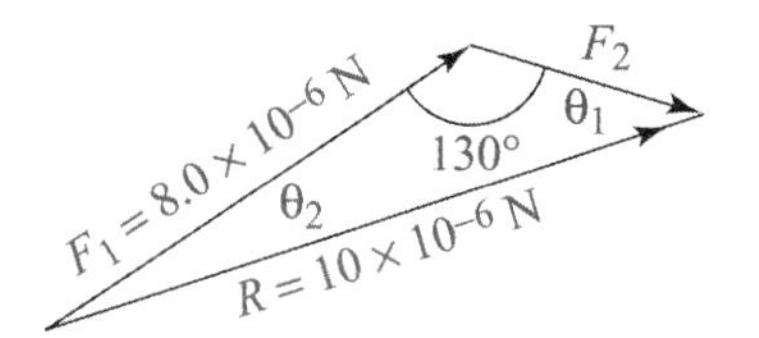

FIGURE 19-10 Electrical forces for Example 19.7.

Solution You know two sides of the triangle and the angle opposite one of them. In order to find F_2, you need to first find the angle θ_1 between R and F_2 using the law of sines. You can then find θ_2 and apply the law of sines again to find F_2. Apply the law of sines to sides F_1 and R and solve for $\sin\theta_1$:

$$\frac{F_1}{\sin\theta_1} = \frac{R}{\sin 130°}$$

Cross multiply:

$$F_1 \sin 130° = R \sin\theta_1$$

Divide by R and substitute values:

$$\sin\theta_1 = \frac{F_1(\sin 130°)}{R} = \frac{(8 \times 10^{-6})(0.766)}{10 \times 10^{-6}} \approx 0.6128$$

At this point, you need to find the inverse sine of 0.6128. However, there are two possible angles that could apply, one in the first quadrant and one in the second quadrant. The first quadrant angle is:

$$\theta_1 = \sin^{-1}(0.6128) \approx 37.8°$$

The second quadrant angle is:

$$\theta_1 = 180° - 37.8° = 142.2°$$

However, 142.2° cannot be a solution to the problem because the triangle of forces can have only one obtuse angle. Therefore,

$$\theta_1 = 37.8° \text{ and } \theta_2 = 180° - 130° - 37.8° = 12.2°$$

Now find F_2 by using the law of sines between F_2 and R:

$$F_2 = \frac{R(\sin 12.2°)}{\sin 130°} = \frac{(10 \times 10^{-6})(0.211)}{0.766} \approx 2.8 \times 10^{-6} \text{ N}$$

EXERCISE 19.2

Round angles to the nearest degree and all other answers to two significant digits.

In exercises 1 through 24, find the missing parts of each triangle by applying the law of sines.

1. $A = 65°, B = 75°, b = 8.0$

2. $A = 35°, B = 80°, b = 6.0$

3. $A = 29°, B = 68°, a = 15$

4. $A = 72°, B = 42°, a = 22$

5. $A = 100°, C = 35°, a = 9.0$

6. $A = 120°, C = 20°, a = 10$

7. $A = 57°, C = 48°, c = 30$

8. $A = 29°, C = 81°, c = 50$

9. $B = 98°, C = 40°, a = 14$

10. $B = 101°, C = 55°, a = 25$

11. $A = 57°, B = 48°, c = 2.5$

12. $A = 38°, B = 56°, c = 7.5$

13. $A = 35°, C = 82°, b = 6.5$

14. $A = 65°, C = 50°, b = 4.6$

15. $B = 120°, C = 40°, a = 27$

16. $B = 12°, C = 95°, a = 80$

17. $A = 67°, a = 5.0, b = 4.0$

18. $A = 79°, a = 12, b = 8$

19. $B = 87°, b = 860, c = 620$

20. $B = 62°, b = 80, c = 50$

21. $C = 105°, a = 0.46, c = 0.57$

22. $C = 125°, a = 1.7, c = 2.6$

23. $B = 151°, a = 11, b = 16$

24. $B = 104°, a = 140, b = 190$

Applied Problems

25. Two power lines are to be strung across a river from tower A to towers B and C, which are 840 ft apart, on the other side of the river. Fernando, a surveyor, measures $\measuredangle ABC$ to be 51° and $\measuredangle BCA$ to be 96°. What are the distances AB and AC?

26. Two fire lookout towers A and B are 17 km apart. Tower A spots a fire at C and measures $\measuredangle CAB$ to be 45°, whereas tower B measures $\measuredangle ABC$ to be 38°. Which tower is closer to the fire, and by how many kilometers?

Applications to Electronics

27. Two electrical forces F_1 and F_2 act on a particle in an electrical field at an angle θ as shown in Figure 19-11. If $F_1 = 9.0 \times 10^{-6}$ N, $\theta = 60°$, and $R = 13 \times 10^{-6}$ N, find the magnitude of F_2. See Example 19.7.

28. In Figure 19-11, find the magnitude of F_1, if $F_2 = 1.5 \times 10^{-6}$ N, $\theta = 100°$, and $R = 2.5 \times 10^{-6}$ N. See Example 19.7.

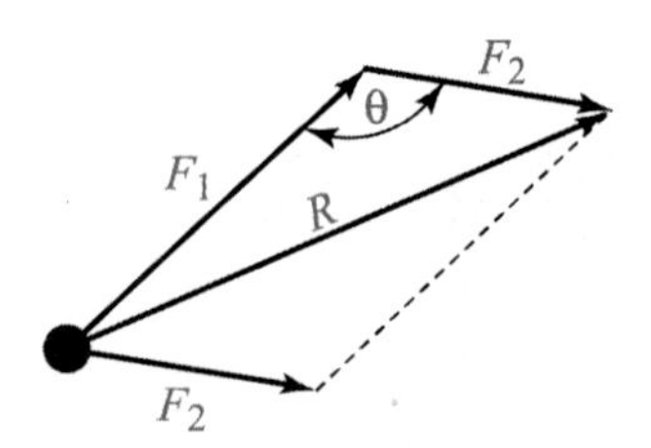

FIGURE 19-11 Electrical forces for problems 27 and 28.

19.3 Law of Cosines

Given any triangle ABC, as shown in Figure 19-12, the law of cosines states:

Law

Law of Cosines

$$c^2 = a^2 + b^2 - 2ab(\cos C) \quad (19.5)$$

The law of cosines is a generalization of the Pythagorean theorem. When $\measuredangle C = 90°$, $\cos 90° = 0$, and formula (19.5) becomes $c^2 = a^2 + b^2$. The law of cosines is used to solve a triangle when the law of sines does not apply and you are given:

1. Two sides and the included angle, or
2. Three sides.

EXAMPLE 19.8 Given triangle ABC in Figure 19-12 with $\measuredangle C = 55°$, $a = 12$, and $b = 15$, find the three missing parts of triangle ABC to two significant digits.

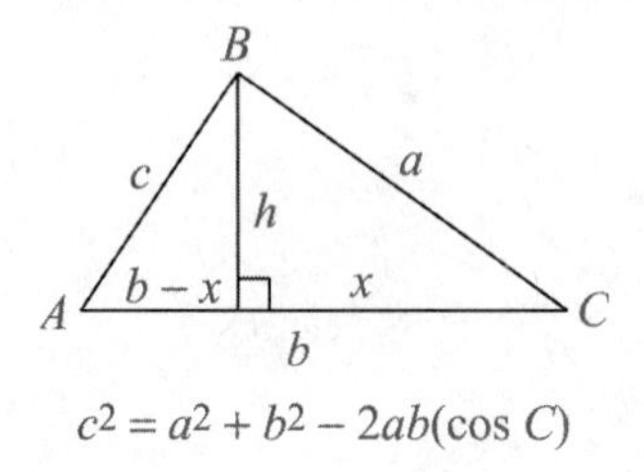

FIGURE 19-12 Law of cosines for Example 19.8.

Solution Since you are given two sides and the included angle, use the law of cosines and find side c first:

$$c^2 = a^2 + b^2 - 2ab(\cos C)$$
$$c^2 = (12)^2 + (15)^2 - 2(12)(15)(\cos 55°)$$
$$c^2 = 144 + 225 - 360(0.5736) = 162.5$$
$$c = \sqrt{162.5} = 12.75 \approx 13$$

Now you can use the law of sines to find A (or B):

$$\sin A = \frac{a(\sin C)}{c} = \frac{(12)(\sin 55°)}{12.75} = 0.7711$$

Then
$$A = \sin^{-1}(0.7711) \approx 50°$$
$$B = 180° - 55° - 50° = 75°$$

Note that 12.75 is used for side c and not the rounded result 13. You need to calculate with three or four figures and then round the result to two figures.

Proof of the Law of Cosines

The following proof of the law of cosines applies to triangle ABC in Figure 19-12. The proof for an obtuse triangle is similar. Apply the Pythagorean theorem to each of the right triangles in triangle ABC:

$$c^2 = (b - x)^2 + h^2$$
$$a^2 = x^2 + h^2$$

Subtract the two equations canceling out h^2:

$$c^2 - a^2 = (b - x)^2 - x^2$$

Expand $(b - x)^2$, simplify and solve for c^2:

$$c^2 - a^2 = b^2 - 2bx + x^2 - x^2$$
$$c^2 = a^2 + b^2 - 2bx$$

Now

$$\frac{x}{a} = \cos C \rightarrow x = a(\cos C)$$

Substitute $a(\cos C)$ for x to produce the law of cosines:

$$c^2 = a^2 + b^2 - 2ab(\cos C)$$

By switching the letters in the triangle in Figure 19-12, the law of cosines can be written two other ways:

Law

Law of Cosines

$$b^2 = a^2 + c^2 - 2ac(\cos B) \quad \textbf{(19.5a)}$$

$$a^2 = b^2 + c^2 - 2bc(\cos A) \quad \textbf{(19.5b)}$$

EXAMPLE 19.9 Given triangle ABC with $a = 4.5$, $b = 6.8$, and $c = 3.0$.

(a) Find $\measuredangle A$

(b) Find $\measuredangle B$

Solution

(a) To find $\measuredangle A$, use formula (19.5b). Substitute the given values, solve for $\cos A$ and then $\measuredangle A$:

$$(4.5)^2 = (6.8)^2 + (3.0)^2 - 2(6.8)(3.0)(\cos A)$$
$$20.25 = 46.24 + 9.0 - 40.8(\cos A)$$
$$\cos A = \frac{46.24 + 9.0 - 20.25}{40.8} = 0.8576$$
$$A = \cos^{-1}(0.8576) = 30.95° \approx 31°$$

(b) Use the law of sines to find $\measuredangle B$ using $\measuredangle A = 30.95°$:

$$\sin B = \frac{b(\sin A)}{a} = \frac{(6.8)(0.5143)}{4.5} \approx 0.7771$$

At this point, a word of caution is necessary. The reference angle = $\sin^{-1} 0.7771 \approx 51°$ and $\measuredangle B$ can be 51° or 129°. If you choose $\measuredangle B = 51°$, then $\measuredangle C = 180° - 51° - 31° = 98°$. This is not possible because *$\measuredangle B$ must be the largest angle since it is opposite the largest side.* Hence the only solution is: $\measuredangle B = 129°$.

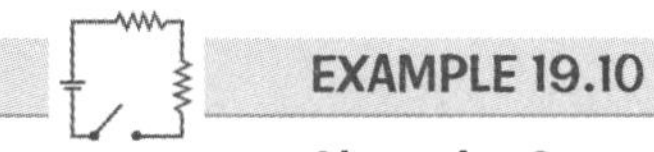

EXAMPLE 19.10

Close the Circuit

Figure 19-13 shows the phasor diagram for the currents in a three-phase alternator. If the phase current $I_P = 14\ A$, find the line current I_T to two significant digits.

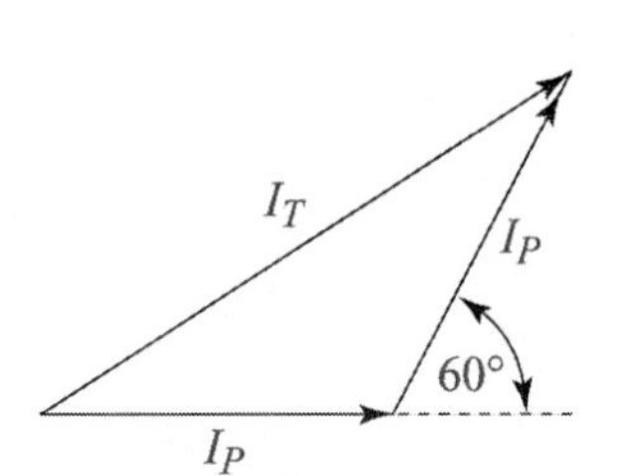

FIGURE 19-13 Three-phase alternator currents for Example 19.10.

Solution The phasor diagram is an isosceles triangle with two of the sides equal to I_P. Figure 19-13 shows the exterior angle at the vertex of the equal sides to be 60°. Therefore, the interior angle formed by the equal sides is 180° – 60° = 120°. The other two angles in the triangle are equal, and therefore each is 30°. You can use the law of sines or the law of cosines to find I_T. Using the law of cosines, let I_T = side c in formula (19.5). Then $\measuredangle C = 120°$, $a = b = I_P = 14$ A, and you have

$$c^2 = a^2 + b^2 - 2ab(\cos C)$$

$$I_T^2 = (14)^2 + (14)^2 - 2(14)(14)(\cos 120°)$$

$$I_T^2 = 196 + 196 - 392(-0.5) = 588$$

$$I_T = \sqrt{588} \approx 24 \text{ A}$$

EXERCISE 19.3

Round angles to the nearest degree and all other answers to two significant digits.

In exercises 1 through 20, find the missing parts of each triangle by applying the law of cosines.

1. $C = 70°, a = 5, b = 7$
2. $C = 55°, a = 12, b = 18$
3. $C = 130°, a = 35, b = 54$
4. $C = 110°, a = 67, b = 34$
5. $C = 18°, a = 3.5, b = 5.7$
6. $C = 22°, a = 9.3, b = 6.3$
7. $B = 145°, a = 0.13, c = 0.31$
8. $B = 27°, a = 11, c = 22$
9. $A = 16°, b = 1.3, c = 1.5$
10. $A = 110°, b = 180, c = 230$
11. $a = 2.0, b = 3.0, c = 4.0$
12. $a = 4.0, b = 7.0, c = 9.0$
13. $a = 37, b = 40, c = 29$
14. $a = 16, b = 12, c = 10$
15. $a = 19, b = 11, c = 10$
16. $a = 40, b = 21, c = 23$
17. $a = 14, b = 14, c = 11$
18. $a = 24, b = 42, c = 24$
19. $a = 3.1, b = 4.2, c = 4.9$
20. $a = 4.5, b = 5.5, c = 2.2$

Applied Problems

21. Each of the sides of the Pentagon building in Virginia is 280 m long and each interior angle is 108°. How far is it in meters from one corner of the building to a nonadjacent corner?
22. Jann hikes into the woods in a northeasterly direction (45° east of north) for 5 mi and then turns due east and hikes 6 mi. How far is she from her starting point?
23. A triangular bridge structure has sides equal to 23 ft, 34 ft, and 48 ft. What are the angles of the structure?
24. A tugboat is towing a barge, as shown in Figure 19-14. The tension in each cable is 1000 lb. What is the magnitude of the resultant force exerted by the tug? Add the two force vectors to create a triangle and find the third side.

Applications to Electronics

25. In Figure 19-13:
    (a) Find the line current I_T when the phase current I_P = 25 A.
    (b) Find the phase current I_P when the line current I_T = 40 A.
    (c) Find a formula for I_T in terms of I_P.
26. A triangular circuit board has dimensions 3.2 mm × 5.3 mm × 4.4 mm.
    (a) Find the largest angle of the board.
    (b) Find the area of the board using the formula $A = \frac{1}{2}ab(\sin C)$.

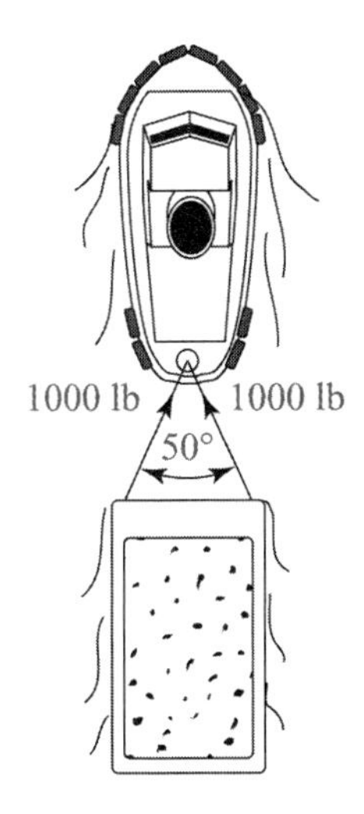

FIGURE 19-14 Tug pulling barge for problem 24.

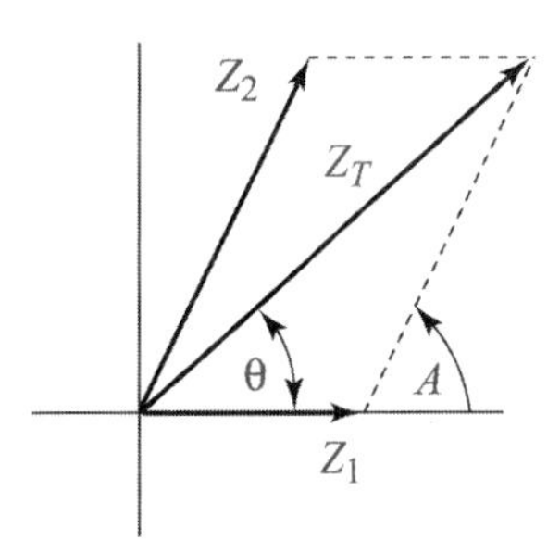

FIGURE 19-15 Addition of phasors for problems 27 and 28.

27. Two phasors when added together produce the parallelogram shown in Figure 19-15 where the resultant is the diagonal of the parallelogram. If the magnitude of Z_1 = 7.5 kΩ, the magnitude of Z_2 = 11 kΩ, and $\measuredangle A = 50°$, use the law of cosines to find the magnitude of Z_T and the angle θ.

28. In Figure 19-15, if the magnitude of Z_1 = 2.5 kΩ, the magnitude of Z_2 = 4.5 kΩ, and $\measuredangle A = 45°$, find the magnitude of Z_T and θ.

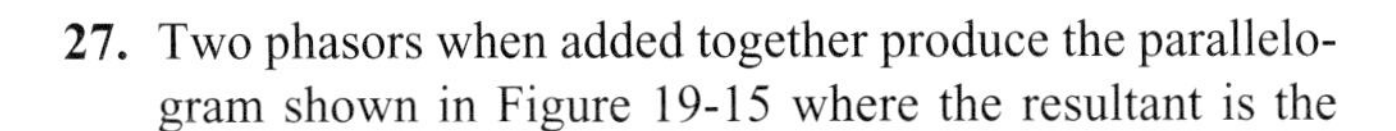

CHAPTER HIGHLIGHTS

19.1 VECTORS AND PHASORS

Adding Vectors and Phasors

To add two vectors or A and B, place the tail of B at the head of A without changing its direction. The resultant $R = A + B$ goes from the tail of A to the head of B.

Adding Vector Components F_x and F_y

$$\text{Magnitude of the Resultant } F = \sqrt{F_x^2 + F_y^2} \qquad (19.1)$$

$$\text{Reference Angle} = \tan^{-1}\left|\frac{F_y}{F_x}\right| \qquad 0° \le \theta < 360°$$

Key Example: To add force vectors $F_x = 20\text{ N}\angle 0°$ and $F_y = 10\text{ N}\angle 270°$

$$F = \sqrt{F_x^2 + F_y^2} = \sqrt{20^2 + 10^2}$$

$$= \sqrt{500} \approx 22.4\text{ N}$$

$$\text{Reference Angle} = \tan^{-1}\left|\frac{F_y}{F_x}\right| = \tan^{-1}\left(\frac{10}{20}\right) \approx 26.6°$$

$$\theta = 360° - 26.6° = 333.4°$$

Adding Phasor Components V_x and V_y

$$\text{Magnitude of the Resultant } V = \sqrt{V_x^2 + V_y^2}$$

$$\text{Angle } \theta = \tan^{-1}\left(\frac{V_y}{V_x}\right) \qquad -90° < \theta < 90° \qquad (19.2)$$

Key Example: To find V_T in Figure 19-16, given $V_R = 4.7$ V and $V_C = 3.3$ V, use the Pythagorean theorem to find the magnitude and the inverse tangent to find the negative angle θ:

$$V_T = \sqrt{V_R^2 + V_C^2} = \sqrt{4.7^2 + 3.3^2} = \sqrt{32.98} \approx 5.7\text{ V}$$

$$\theta = \tan^{-1}\left(\frac{-V_C}{V_R}\right) = \tan^{-1}\left(\frac{-3.3}{4.7}\right) \approx -35°$$

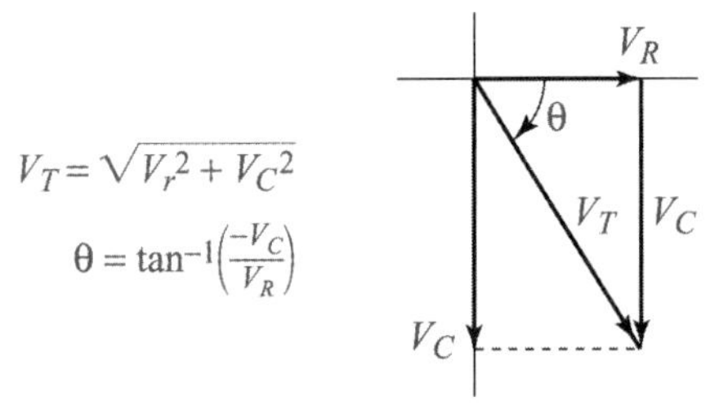

FIGURE 19-16 Key Example: Voltages in an *RC* series circuit.

Vector or Phasor Components

$$\cos\theta = \frac{V_x}{V} \rightarrow V_x = V\cos\theta \ (x \text{ component})$$
$$\sin\theta = \frac{V_y}{V} \rightarrow V_y = V\sin\theta \ (y \text{ component}) \qquad (19.3)$$

Key Example: To add the two impedance phasors in Figure 19-17, find the x and y components of each and combine to find the resultant:

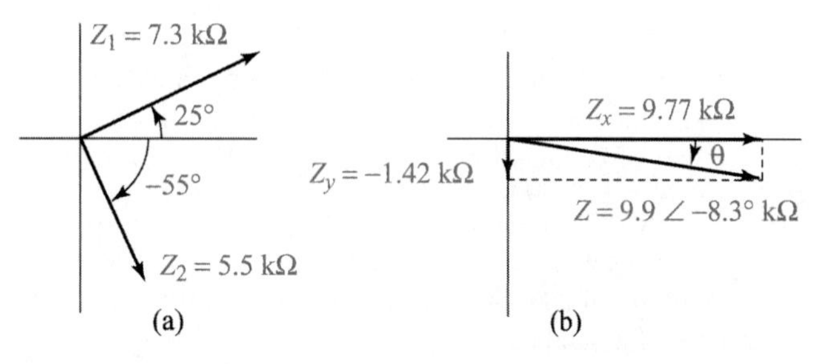

FIGURE 19-17 Key Example: Adding impedance phasors.

Phasors	x components	y components
Z_1	7.3 cos 25° kΩ = 6.62 kΩ	7.3 sin 25° kΩ = 3.09 kΩ
Z_2	5.5 cos(−55°) kΩ = 3.15 kΩ	5.5 sin(−55°) kΩ = −4.51 kΩ
Resultant Z	$Z_x = 9.77$ kΩ	$Z_y = -1.42$ kΩ

$$Z = \sqrt{Z_x^2 + Z_y^2} = \sqrt{9.77^2 + (-1.42)^2} = \sqrt{97.5} \approx 9.9 \text{ k}\Omega$$

$$\theta = \tan^{-1}\left(\frac{Z_y}{Z_x}\right) = \tan^{-1}\left(\frac{-1.42}{9.77}\right) \approx -8.3°$$

19.2 LAW OF SINES

Law of Sines

$$\frac{a}{\sin A} = \frac{b}{\sin B} = \frac{c}{\sin C} \qquad (19.4)$$

Use the law of sines to find the parts of any triangle when you are given:

1. Two angles and a side opposite one of the angles, or
2. Two sides and an angle opposite one of the sides.

Key Example: Given $\measuredangle A = 40°$, $\measuredangle B = 60°$ and $c = 5.0$:

$$C = 180° - \measuredangle A - \measuredangle B = 180° - 40° - 60° = 80°$$

$$a = \frac{c(\sin A)}{\sin C} = \frac{(5.0)(\sin 40°)}{\sin 80°} \approx 3.3$$

$$b = \frac{c(\sin B)}{\sin C} = \frac{5.0(\sin 60°)}{\sin 80°} \approx 4.4$$

19.3 LAW OF COSINES

Law of Cosines

$$c^2 = a^2 + b^2 - 2ab(\cos C) \qquad (19.5)$$
$$b^2 = a^2 + c^2 - 2ac(\cos B) \qquad (19.5a)$$
$$a^2 = b^2 + c^2 - 2bc(\cos A) \qquad (19.5b)$$

Use the law of cosines when the law of sines does not apply and you are given:

1. Two sides and the included angle, or
2. Three sides.

Key Example: Given $\measuredangle C = 55°$, $a = 12$, and $b = 15$:

$$c^2 = (12)^2 + (15)^2 - 2(12)(15)(\cos 55°) = 162.5$$

$$c = \sqrt{162.5} = 12.75 \approx 13$$

REVIEW EXERCISES

Round angles to the nearest degree and all other answers to two significant digits.

In exercises 1 through 6, find the resultant of each pair of vectors with a positive angle between 0° and 360°.

1. $F_x = 16$ N, $\theta_x = 0°$; $F_y = 12$ N, $\theta_y = 90°$
2. $F_x = 54$ N, $\theta_x = 90°$; $F_y = 77$ N, $\theta_y = 180°$
3. $v_1 = 3.2$ m/s, $\theta_1 = 180°$; $v_2 = 1.1$ m/s, $\theta_2 = 270°$
4. $E_1 = 43$ N/C, $\theta_1 = 0°$; $E_2 = 62$ N/C, $\theta_2 = 270°$
5. $a_x = 6.7$ ft/s/s, $\theta_x = 0°$; $a_y = 8.3$ ft/s/s, $\theta_y = 270°$
6. $x = 27$ m, $\theta_x = 180°$; $y = 36$ m, $\theta_y = 270°$

In exercises 7 through 12, find the resultant of each pair of phasors with an angle between −90° and 90°.

7. $R = 10 \angle 0°$ kΩ; $X_C = 7.5 \angle -90°$ kΩ

8. $R = 4.3 \angle 0° \text{ k}\Omega$; $X_L = 8.2 \angle 90° \text{ k}\Omega$
9. $I_C = 750 \angle 90° \text{ mA}$; $I_L = 900 \angle -90° \text{ mA}$
10. $V_L = 24 \angle 90° \text{ V}$; $V_C = 12 \angle -90° \text{ V}$
11. $Z_1 = 6.8 \angle 20° \text{ k}\Omega$; $Z_2 = 9.1 \angle -50° \text{ k}\Omega$
12. $Z_1 = 19 \angle 62° \text{ k}\Omega$; $Z_2 = 235$

In exercises 13 through 28, find the missing parts of each triangle.

13. $A = 85°, B = 75°, b = 7.0$
14. $B = 108°, C = 42°, c = 4.1$
15. $A = 46°, B = 68°, c = 24$
16. $A = 43°, C = 51°, b = 410$
17. $A = 40°, a = 11, b = 13$
18. $C = 26°, a = 16, c = 27$
19. $B = 97°, b = 8.3, c = 6.8$
20. $B = 130°, a = 150, b = 220$
21. $C = 110°, a = 18, b = 12$
22. $C = 25°, a = 40, b = 60$
23. $A = 60°, b = 10, c = 15$
24. $B = 105°, a = 5.5, c = 8.2$
25. $a = 12, b = 16, c = 19$
26. $a = 1.0, b = 1.5, c = 2.1$
27. $a = 6.7, b = 2.4, c = 5.1$
28. $a = 130, b = 140, c = 85$

Applied Problems

29. A body is acted on by a horizontal force of 7.8 lb acting to the left and a vertical force of 5.1 lb acting down. Find the magnitude and direction of the resultant force.
30. Grace hikes 10 miles north and then 6 miles northeast (45° north of east). How far is she from her starting point, and what is her direction? Find the angle measured from due east.
31. Carol measures the three sides of her triangular garden to be 9.0 m, 10 m, and 13 m. She believes it to be a right triangle. Is she correct? Find the three angles of the triangle.
32. Mr. Chen wants to fly his plane southwest. See Figure 19-18. If his airspeed is 220 km/h and there is an 80 km/h wind from duc north, how many degrees south of west should he fly so that his resultant direction will be southwest? Find θ in Figure 19-18.

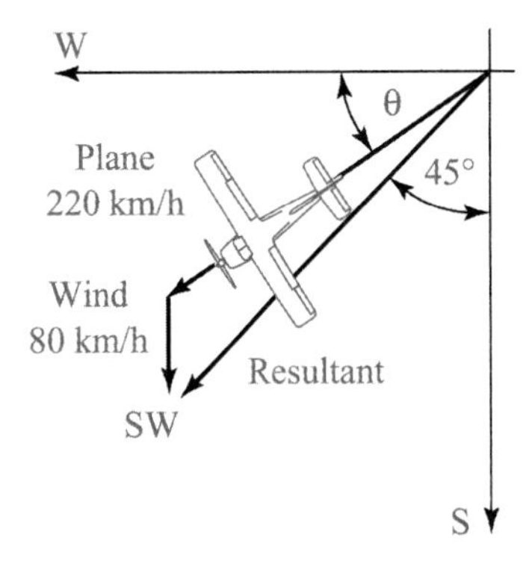

FIGURE 19-18 Flight direction for problem 32.

Applications to Electronics

33. Given each pair of phasors for an ac series circuit:
    (a) RL series circuit: $R = 16 \angle 0° \text{ k}\Omega, X_L = 20 \angle 90° \text{ k}\Omega$. Find the resultant Z.
    (b) RC series circuit: $V_R = 5 \angle 0° \text{ V}, V_C = 5 \angle -90° \text{ V}$. Find the resultant V_T.
    (c) RC parallel circuit: $I_R = 26 \angle 0° \text{ mA}, I_C = 13 \angle 90° \text{ mA}$. Find the resultant I_T.
34. Given the two impedance phasors $Z_1 = 22 \angle 20° \text{ k}\Omega$ and $Z_2 = 13 \angle -30° \text{ k}\Omega$, find the resultant phasor Z.
35. In Figure 19-13 find:
    (a) The line current I_T when the phase current $I_P = 5.5$ A.
    (b) The phase current I_P when the line current $I_T = 16$ A.
36. A triangular circuit board has two sides equal to 1.6 mm and 2.1 mm. If the angle opposite the larger side is 100°, find the other side and the two other angles.
37. Figure 19-19 shows two phasors added together with an angle of 25° between them. If $Z_1 = 1.0 \text{ k}\Omega$ and $Z_2 = 2.0 \text{ k}\Omega$, find the magnitude of the resultant Z and the angle θ it makes with Z_1.
38. In Figure 19-19, find θ and Z when $Z_1 = 850\ \Omega$ and $Z_2 = 1.5 \text{ k}\Omega$.

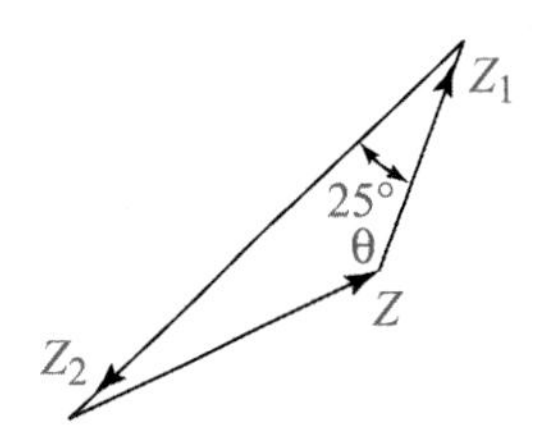

FIGURE 19-19 Impedance phasors for problems 37 and 38.

CHAPTER 20
Complex Numbers and Phasors

CHAPTER OBJECTIVES

In this chapter, you will learn:

- The *j* operator and its basic operations.
- How to add, multiply, and divide complex numbers.
- Complex phasors and how to graph them.
- How to change a complex phasor from rectangular form to polar form and vice versa.
- How to add, multiply, and divide complex phasors in polar form.

All the numbers that you have been using so far are called *real* numbers. For example 10, $-\frac{1}{2}$, 1.5, $\sqrt{2}$, and π are all real numbers. However, in ac circuits, square roots of negative numbers are encountered, such as $\sqrt{-4}$. This number is not a real number but is called an *imaginary number.* When square roots of negative numbers first appeared about 500 years ago, they were found to have little application; hence they were called imaginary. Today, however, they have a very important application in electronics and are as real as "real" numbers, but different rules apply to their operations.

When a real number is combined with an imaginary number, a *complex number* is formed. For example, $3 + \sqrt{-4}$ is a complex number. Complex numbers are the basis for the mathematics of ac networks. The real part represents resistance, and the imaginary part represents reactance. Complex numbers are not really "complex" as their name implies and are not difficult to work with.

• • •

20.1 The *j* Operator

The *imaginary unit* or *j operator* is defined as:

Definition

j Operator

$$j = \sqrt{-1} \text{ or } j^2 = (\sqrt{-1})^2 = -1 \tag{20.1}$$

It is important to understand that *j* is just a symbol for $\sqrt{-1}$ and is used because it is easier to work with than the radical. In mathematics, the letter *i* is used to represent $\sqrt{-1}$. However, since *i* is used for instantaneous current in electricity, *j* is used for the imaginary unit in electronics. Square roots of negative numbers are called *imaginary numbers.* The *j* operator allows you to simplify imaginary numbers by separating the *j* as follows.

EXAMPLE 20.1

Simplify:

(a) $\sqrt{-4}$ **(b)** $\sqrt{-25}$ **(c)** $\sqrt{-0.16}$ **(d)** $\sqrt{-2}$

Solution For each radical, separate the $\sqrt{-1}$, change it to *j*, and then simplify the positive radical:

(a) $\sqrt{-4} = \left(\sqrt{-1}\right)\left(\sqrt{4}\right) = j\sqrt{4} = j2$

Observe that the j is written first and then the coefficient. The negative sign in the radical simply becomes a j on the outside. Similarly, for the other radicals:

(b) $\sqrt{-25} = j\sqrt{25} = j5$

(c) $\sqrt{-0.16} = j\sqrt{0.16} = j0.4$

(d) $\sqrt{-2} = j\sqrt{2}$

The rule for simplifying the square root of a negative number is then:

Rule

Rule for Simplifying Radicals

$$\sqrt{-x} = j\sqrt{x} \qquad (x = \text{positive number}) \qquad (20.2)$$

EXAMPLE 20.2

Simplify:

(a) $4\sqrt{-0.09}$ **(b)** $-2\sqrt{-49}$

Solution Apply rule (20.2):

(a) $4\sqrt{-0.09} = 4j\sqrt{0.09} = 4j(0.3) = j1.2$

(b) $-2\sqrt{-49} = -2j\sqrt{49} = -2j(7) = -j14$

Note that the number $-j$ is a different number from j similar to the way that -1 is different from 1.

The multiplication and division rules for square roots of positive numbers do not apply to square roots of negative numbers. To multiply and divide imaginary numbers you must *first separate the j operator* as shown in the following examples:

EXAMPLE 20.3

Simplify:

$$\left(\sqrt{-9}\right)\left(\sqrt{-16}\right)$$

Solution First separate the j operator:

$$\left(\sqrt{-9}\right)\left(\sqrt{-16}\right) = \left(j\sqrt{9}\right)\left(j\sqrt{16}\right)$$

Then simplify the radicals and multiply the j's:

$$\left(j\sqrt{9}\right)\left(j\sqrt{16}\right) = (j3)(j4) = j^2(3)(4)$$

Now apply (20.1) and *whenever j^2 appears, replace it with -1*:

$$j^2(3)(4) = (-1)(3)(4) = -12$$

The result is a real number that contains no j operator. Observe that if you tried to multiply under the radicals before separating the j's, you would get $\sqrt{(-9)(-16)} = \sqrt{144} = 12$, which is not the correct answer.

ERROR BOX

A common error when working with the j operator is confusing j, $-j$, -1, and 1. Each of these is a different number and none of them can be changed to any one of the others. The j operator is just a *symbol* for $\sqrt{-1}$. The letter j is written *instead of* the radical symbol because it is easier to work with. Since j^2 is equal to -1, you should *always change* j^2 to -1 when it occurs. See if you can do the practice problems correctly.

Practice Problems: Simplify each of the following.

1. $(\sqrt{4})(\sqrt{-4})$ **2.** $(-\sqrt{4})(\sqrt{-4})$ **3.** $(\sqrt{-4})(\sqrt{-4})$ **4.** $(-\sqrt{-4})(\sqrt{-4})$
5. $(-\sqrt{-4})(-\sqrt{-4})$ **6.** $(-\sqrt{4})(-\sqrt{-4})$ **7.** $(\sqrt{-4})^2$ **8.** $(-\sqrt{-4})^2$
9. $(\sqrt{-4})(\sqrt{4})^2$ **10.** $(\sqrt{4})(\sqrt{-4})^2$

EXAMPLE 20.4 Simplify:

$$\left(5\sqrt{-0.8}\right)\left(-3\sqrt{0.2}\right)$$

Solution First separate the j operator from the first radical:

$$\left(5\sqrt{-0.8}\right)\left(-3\sqrt{0.2}\right) = \left(j5\sqrt{0.8}\right)\left(-3\sqrt{0.2}\right)$$

You can now multiply under the radicals, since the numbers under the radicals are both positive:

$$\begin{aligned}\left(j5\sqrt{0.8}\right)\left(-3\sqrt{0.2}\right) &= (j5)(-3)\sqrt{(0.8)(0.2)}\\ &= -j15\sqrt{0.16} = -j15(0.4) = -j6\end{aligned}$$

Since j is a radical, multiplying the numerator and denominator of a fraction by j is the way that you divide imaginary numbers, as the next example shows.

EXAMPLE 20.5 Divide:

$$\frac{\sqrt{25}}{10\sqrt{-4}}$$

Solution Separate the j first, then simplify the radicals and the fraction:

$$\frac{\sqrt{25}}{10\sqrt{-4}} = \frac{\sqrt{25}}{j10\sqrt{4}} = \frac{5}{j10(2)} = \frac{5}{j20} = \frac{1}{j4}$$

Now, to divide by j, multiply the numerator and denominator by j and simplify:

$$\frac{1}{j4} = \frac{(j)1}{(j)j4} = \frac{j}{(j^2)4} = \frac{j}{(-1)4} = -\frac{j}{4} \text{ or } -j0.25$$

The imaginary unit is called the j operator because of its special properties. When you repeatedly multiply by the j operator, a cycle occurs that repeats after four multiplications. Observe the pattern that results by applying the rules for exponents to powers of j:

$$j^0 = +1$$
$$j^1 = j$$
$$j^2 = -1$$
$$j^3 = j^2(j) = (-1)(j) = -j$$
$$j^4 = (j^2)(j^2) = (-1)(-1) = +1$$
$$j^5 = (j^4)(j) = (+1)(j) = j \quad \text{etc.}$$

This is represented graphically where the horizontal axis is the real axis containing the numbers 1 and −1 and the vertical axis is the imaginary axis containing j and $-j$. See Figure 20-1.

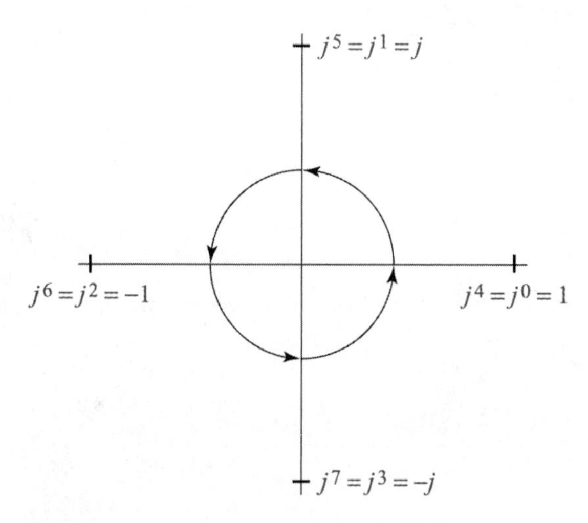

FIGURE 20-1 90° rotation of j operator.

The j operator performs the "operation" of 90° rotation *counterclockwise* each time you multiply by j. This is important when working with phasors in ac networks and is studied further in Section 20.3.

EXERCISE 20.1

In exercises 1 through 22, simplify and express each in terms of the j operator.

1. $\sqrt{-16}$
2. $\sqrt{-9}$
3. $\sqrt{-36}$
4. $\sqrt{-100}$
5. $\sqrt{-0.04}$
6. $\sqrt{-0.25}$
7. $\sqrt{-1.44}$
8. $\sqrt{-1.21}$
9. $\sqrt{-0.0049}$
10. $\sqrt{-0.0081}$
11. $2\sqrt{-1}$
12. $-2\sqrt{-1}$
13. $-3\sqrt{-0.01}$
14. $-5\sqrt{-0.09}$
15. $6\sqrt{-\frac{1}{9}}$
16. $-8\sqrt{-\frac{1}{4}}$
17. $\sqrt{-3}$
18. $\sqrt{-5}$
19. $2\sqrt{-2}$
20. $2\sqrt{-6}$
21. $\sqrt{-4 \times 10^6}$
22. $\sqrt{-16 \times 10^{-6}}$

Answers to Error Box Problems, page 397:
1. $j4$ **2.** $-j4$ **3.** −4 **4.** 4 **5.** −4 **6.** $j4$ **7.** −4 **8.** 4 **9.** $j8$ **10.** −8

In exercises 23 through 40, multiply and simplify each product.

23. $(\sqrt{-36})(\sqrt{-4})$
24. $(\sqrt{-2})(\sqrt{-8})$
25. $(\sqrt{-1})(\sqrt{1})$
26. $(\sqrt{0.5})(\sqrt{-0.5})$
27. $(-\sqrt{-5})(\sqrt{-20})$
28. $(-\sqrt{-8})(\sqrt{2})$
29. $(-\sqrt{-2})^2$
30. $-(\sqrt{-3})^2$
31. $(\sqrt{0.4})(-\sqrt{-0.9})$
32. $(-\sqrt{-10})(-\sqrt{-2.5})$
33. $(4\sqrt{-10})(-10\sqrt{40})$
34. $(-5\sqrt{-3})(-5\sqrt{-27})$
35. $(\sqrt{2\times10^3})(\sqrt{-8\times10^3})$
36. $(\sqrt{-1.5\times10^{-3}})(\sqrt{-13.5\times10^{-3}})$
37. $(j2)(j6)(j4)$
38. $(j3)(-j7)(-j)(j)$
39. $(j)^2(j)^3(j)$
40. $(-j)^2(-j)^2(-j)$

In exercises 41 through 56, divide and simplify each quotient.

41. $\frac{5}{j}$
42. $\frac{-4}{j}$
43. $\frac{6}{j2}$
44. $\frac{10}{-j4}$
45. $\frac{4}{\sqrt{-64}}$
46. $\frac{\sqrt{2}}{\sqrt{-25}}$
47. $\frac{-5}{\sqrt{-100}}$
48. $\frac{-3}{\sqrt{-0.09}}$
49. $\frac{\sqrt{36}}{\sqrt{-4}}$
50. $\frac{\sqrt{100}}{\sqrt{-25}}$
51. $\frac{\sqrt{-4}}{\sqrt{-16}}$
52. $\frac{\sqrt{-81}}{\sqrt{-144}}$
53. $\frac{2\sqrt{0.01}}{\sqrt{-0.04}}$
54. $\frac{-5.5\sqrt{4}}{\sqrt{-0.25}}$
55. $\frac{-6\times10^3}{\sqrt{-16\times10^2}}$
56. $\frac{12\times10^3}{\sqrt{-9\times10^6}}$

Applications to Electronics

In problems 57 through 60, solve each applied problem to two significant digits.

57. The voltage in an ac circuit is given by Ohm's law $V = IZ$ where I = current in amps and Z = impedance in ohms. Find V in volts when $I = j10$ mA and $Z = -j3.5$ kΩ.
58. Find V in volts in problem 57 when $I = j620$ μA and $Z = j15$ kΩ.
59. The impedance in an ac circuit is given by Ohm's law $Z = \frac{V}{I}$ where V = voltage in volts and I = current in amps. Find Z in ohms when $V = 5.5$ V and $I = j25$ mA.
60. Find I in milliamps in problem 59 when $V = 12$ V and $Z = -j1.6$ kΩ.

20.2 Complex Numbers

When a real number is combined with an imaginary number, it is called a complex number. For example, $2 - j5$ is a complex number. The number 2 is the real part, and $-j5$ is the imaginary part. All complex numbers have the form $x + jy$, where x and y are both real numbers. The group of complex numbers contains all the mathematical numbers needed for technical calculations as shown in Figure 20-2 on page 400. When $x = 0$, you have a pure imaginary number jy. When $y = 0$, you simply have the real number x.

Rule **Addition of Complex Numbers**

Add the real and imaginary parts separately.

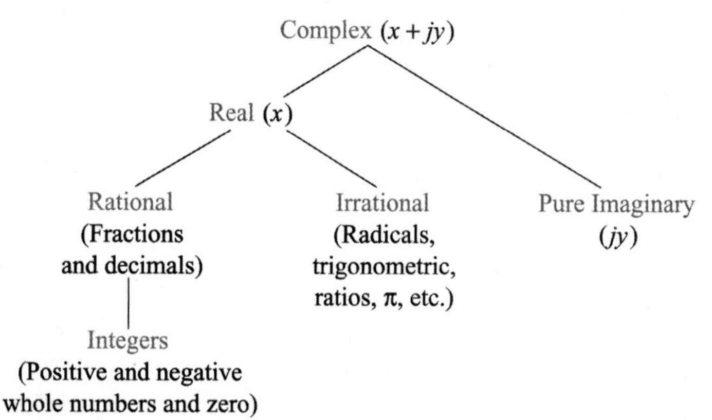

FIGURE 20-2 Mathematical number systems.

EXAMPLE 20.6 Add:

(a) $(2.1 - j5.3) + (3.4 + j4.2)$

(b) $(1 - j2) - (3 - j5) + (-2 + j3)$

Solution

(a) Add the real and imaginary parts separately, treating j as you would any variable in algebra:

$$(2.1 - j5.3) + (3.4 + j4.2) = (2.1 + 3.4) + (-j5.3 + j4.2) = 5.5 - j1.1$$

(b) Remove the parentheses and add all the real parts and all the imaginary parts separately:

$$(1 - j2) - (3 - j5) + (-2 + j3) = 1 - j2 - 3 + j5 - 2 + j3$$
$$= (1 - 3 - 2) + (-j2 + j5 + j3) = -4 + j6$$

Rule

Multiplication of Complex Numbers

Use the FOIL method to multiply the four products and replace j^2 with -1.

EXAMPLE 20.7 Multiply:

$$(3 - j4)(2 + j5)$$

Solution Multiply the binomials using the FOIL method: **F**irst, **O**uter, **I**nner, **L**ast:

$$(3 - j4)(2 + j5) = \overset{\textbf{F}}{(3)(2)} + \overset{\textbf{O}}{(3)(j5)} + \overset{\textbf{I}}{(-j4)(2)} + \overset{\textbf{L}}{(-j4)(j5)}$$

Simplify and replace j^2 with -1:

$$= 6 + j15 - j8 - (j^2)(20) = 6 + j7 - (-1)(20) = 26 + j7$$

Since j^2 is always replaced by -1, the product of two or more complex numbers will always be a complex number in the form $x + jy$. Study the next example which illustrates the product of three complex numbers.

EXAMPLE 20.8 Multiply:

$$(j2)(2 - j)(1 + j)$$

Solution Multiply the last two complex numbers first. Then simplify and multiply the result by $j2$:

$$(j2)(2 - j)(1 + j) = (j2)(2 + j2 - j - j^2) = (j2)(2 + j + 1)$$
$$= (j2)(3 + j) = j6 + j^2 2 = -2 + j6$$

Note that the real part of the complex number is always written first.

Definition

Complex Conjugate

The *complex conjugate* of $x + jy$ is $x - jy$.

For example, the conjugate of $2 - j3$ is $2 + j3$, and the conjugate of $-5 + j4$ is $-5 - j4$. The first sign remains the same, and the second sign is changed. When complex conjugates are multiplied together, the result is a real number since the outside and the inside products add up to zero:

$$(x + jy)(x - jy) = x^2 - \cancel{jxy} + \cancel{jxy} - j^2y^2 = x^2 - (-1)y^2 = x^2 + y^2$$

Applying this idea, division of complex numbers is done using the conjugate:

Rule

Division of Complex Numbers

Multiply the numerator and denominator by the conjugate of the denominator.

EXAMPLE 20.9 Divide:

$$\frac{2 + j3}{5 - j4}$$

Solution Multiply the numerator and denominator by the conjugate of the denominator, which is $5 + j4$:

$$\frac{(5 + j4)(2 + j3)}{(5 + j4)(5 - j4)} = \frac{10 + j15 + j8 + j^2 12}{25 - j^2 16}$$
$$= \frac{10 + j23 + (-1)12}{25 - (-1)16} = \frac{-2 + j23}{41}$$

The answer is usually expressed in the form $x + jy$ where x and y are decimals:

$$\frac{-2 + j23}{41} = \frac{-2}{41} + j\frac{23}{41} = -0.049 + j0.56$$

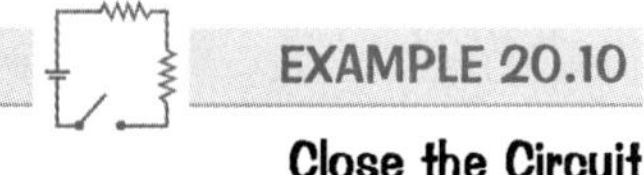

EXAMPLE 20.10

Close the Circuit

The total impedance of an ac circuit containing two impedances Z_1 and Z_2 in parallel is given by:

$$Z_T = \frac{Z_1 Z_2}{Z_1 + Z_2}$$

Find Z_T when $Z_1 = 1 + j$ kΩ and $Z_2 = 1 - j2$ kΩ

Solution Substitute into the formula to obtain:

$$Z_T = \frac{(1+j)(1-j2)}{(1+j)+(1-j2)}$$

Multiply the numbers in the numerator and add the numbers in the denominator:

$$Z_T = \frac{1 - j2 + j - j^2 2}{2 - j} = \frac{1 - j + 2}{2 - j} = \frac{3 - j}{2 - j}$$

Now divide the complex number by multiplying the numerator and the denominator by the conjugate of the denominator, which is $2 + j$:

$$Z_T = \frac{(2+j)}{(2+j)}\frac{(3-j)}{(2-j)} = \frac{6 - j2 + j3 - j^2}{4 - j^2}$$

$$= \frac{6 + j + 1}{4 - (-1)} = \frac{7 + j}{5}\ \text{k}\Omega$$

The answer in $x + jy$ form is:

$$Z_T = \frac{7}{5} + j\frac{1}{5}\ \text{k}\Omega = 1.4 + j0.20\ \text{k}\Omega$$

You can add, multiply, and divide complex numbers on your calculator if it has a [CPLX] or [a+bi] mode. You must first put it in the complex mode by pressing [MODE] and the key that corresponds to [CPLX] or [a+bi]. Then enter the real and imaginary parts separately, using the [i] key or an equivalent key for the j operator, and perform the operation. For example, one way to calculate Example 20.10 on a TI-83 calculator is:

[MODE] [a+bi] [(] 1 [+] [i] [)] [×] [(] 1 [−] 2 [i] [)] [÷] [(] [(] 1 [+] [i] [)] [+] [(] 1 [−] 2 [i] [)] [)] [)] [=] → 1.4 + .2i

Note that the coefficient is written in front of the imaginary unit. The notation for the imaginary number in mathematics is the reverse of that used in electronics.

EXERCISE 20.2

Round all answers to two significant digits.

In exercises 1 through 40, perform the operations and express the answers in the form $x + jy$.

1. $(1 - j2) + (3 + j4)$
2. $(5 + j) + (3 - j2)$
3. $(3 + j2) - (3 - j3)$
4. $(4 - j2) - (6 - j2)$
5. $(5.3 + j9.1) + (3.1 - j6.4)$
6. $(1.0 - j1.3) - (0.9 + j0.7)$
7. $(1 - j) - (1 + j) + (3 - j2)$
8. $(5 + j3) + 2(5 - j3)$
9. $(3 - j2)(5 + j)$
10. $(2 + j5)(3 - j4)$
11. $(-1 + j2)(1 - j2)$
12. $(3 - j6)(2 - j)$
13. $(2 + j2)^2$
14. $(1 - j)^2$
15. $(1.6 - j2.0)(3.5 + j1.5)$
16. $(-0.2 + j0.6)(2.5 - j0.5)$
17. $(5 + j6)(5 - j6)$
18. $(-7 + j2)(-7 - j2)$
19. $(3)(5 - j2)(1 + j4)$
20. $-(8 + j)(1 - j6)$
21. $j6(1 - j)(2 + j)$
22. $(j)(3 - j3)(3 + j4)$
23. $(j2)(1 + j)^2$
24. $(1 + j)(1 - j)^2$
25. $\dfrac{1}{2 + j4}$
26. $\dfrac{1}{5 - j5}$
27. $\dfrac{5 + j5}{1 - j2}$
28. $\dfrac{6 - j4}{1 + j}$
29. $\dfrac{j2}{-1 + j}$
30. $\dfrac{j10}{3 + j}$

31. $\dfrac{0.5}{0.1 + j0.2}$

32. $\dfrac{2.4}{1.2 - j1.6}$

33. $\dfrac{5 + j}{1 + j2}$

34. $\dfrac{4 - j3}{6 - j8}$

35. $\dfrac{0.1(50 - j40)}{1 - j}$

36. $\dfrac{30 - j20}{(j)(20 + j30)}$

37. $\dfrac{(1 + j)(2 - j)}{2 + j2}$

38. $\dfrac{3 - j}{(2 + j)(1 + j)}$

39. $\dfrac{(2 + j)(-2 + j)}{(2 + j) + (-2 + j)}$

40. $\dfrac{(4 - j)(1 + j)}{(4 - j) + (1 + j)}$

Applied Problems

41. Any quadratic equation can be factored if the factors can be expressed as irrational or complex numbers. Show that the factors of $x^2 + 4 = 0$ are $(x + j2)(x - j2) = 0$ by multiplying out the factors. Note that the roots are the negative of the factors: $x = -j2$ and $x = j2$.

42. Prove:

$$\sqrt{j} = \frac{1 + j}{\sqrt{2}}$$

by showing that:

$$\left(\frac{1 + j}{\sqrt{2}}\right)^2 = j$$

Applications to Electronics

43. The total impedance of an ac circuit containing two impedances Z_1 and Z_2 in series is given by: $Z_T = Z_1 + Z_2$. Find Z_T when $Z_1 = 5.5 + j6.7$ kΩ and $Z_2 = 8.2 - j3.5$ kΩ.

44. In problem 43, find Z_1 when $Z_T = 5.2 + j4.5$ kΩ and $Z_2 = 3.5 - j3.3$ kΩ.

45. In Example 20.10, find Z_T when $Z_1 = 1 - j$ kΩ and $Z_2 = 1 + j2$ kΩ.

46. In Example 20.10, find Z_T when $Z_1 = 10 + j10$ kΩ and $Z_2 = 10 - j20$ kΩ.

47. The voltage in an ac circuit is given by Ohm's law $V = IZ$ where I = current in amps and Z = circuit impedance in ohms. Find V in volts when $I = 20 + j10$ mA and $Z = 3 - j4$ kΩ.

48. In problem 47, find I in amps and milliamps when $V = 120$ V and $Z = 500 + j200$ Ω.

20.3 Complex Phasors

Complex numbers can be represented as with a magnitude and a direction angle. In phasor form they are very useful for applications to ac circuits. Given a complex number $x + jy$, the phasor is represented by the arrow drawn from the origin to the point (x,y) as shown in Figure 20-3. The x axis represents real numbers and the y axis represents imaginary numbers. Figure 20-3 is called the *complex plane*, or Gaussian plane after Karl Gauss (1777–1855), the foremost German mathematician of the nineteenth century.

EXAMPLE 20.11 Draw the complex phasors:

(a) $3 + j$ **(b)** $2 - j3$ **(c)** $-1 - j$ **(d)** $4 + j0$ **(e)** $0 + j2$

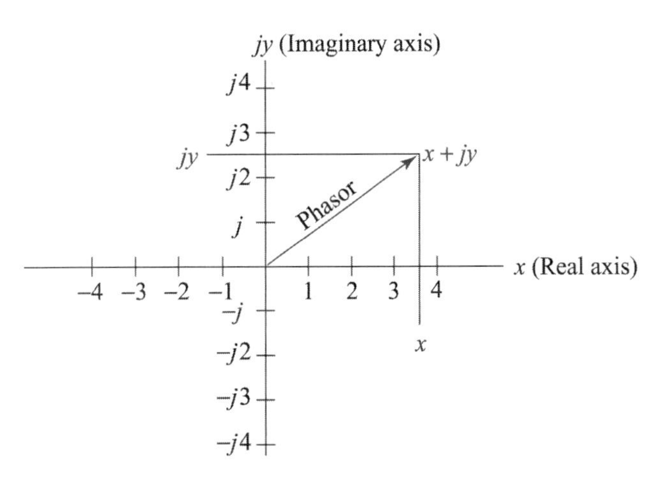

FIGURE 20-3 Complex plane or Gaussian plane.

Solution For each phasor, the real number corresponds to the x coordinate, and the coefficient of j corresponds to the y coordinate. Figure 20-4 shows the five complex phasors. Note that a real number $x + j0$ is a horizontal phasor, and a pure imaginary number $0 + jy$ is a vertical phasor.

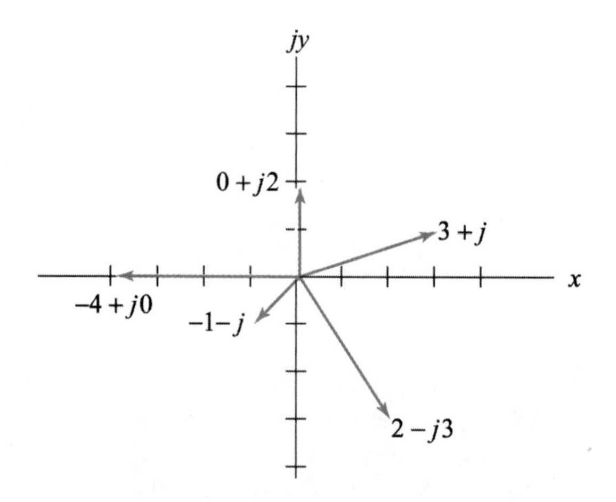

FIGURE 20-4 Complex phasors.

EXAMPLE 20.12 Add the complex phasors $2 + j$ and $1 + j3$ algebraically and graphically.

Solution The algebraic sum is done the same as adding complex numbers: $(2 + j) + (1 + j3) = 3 + j4$. The graphic sum is done by adding the x components $(2 + 1)$ and the y components $(j + j3)$ on the graph to find the components of the resultant. This is the same as adding vectors or phasors by placing the tail of one on the head of the other. The resultant $3 + j4$ is the diagonal of the parallelogram shown in Figure 20-5.

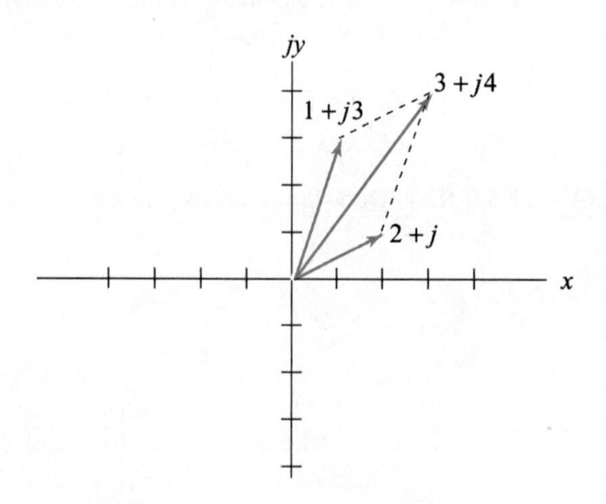

FIGURE 20-5 Adding complex phasors for Example 20.12.

Polar Form

The complex phasor $x + jy$ is in rectangular form. As with any vector or phasor it has magnitude, which is its length, and direction, which is the angle it makes with the positive x axis. Polar form expresses the phasor in terms of its length Z and angle θ as $Z \angle\theta$. The

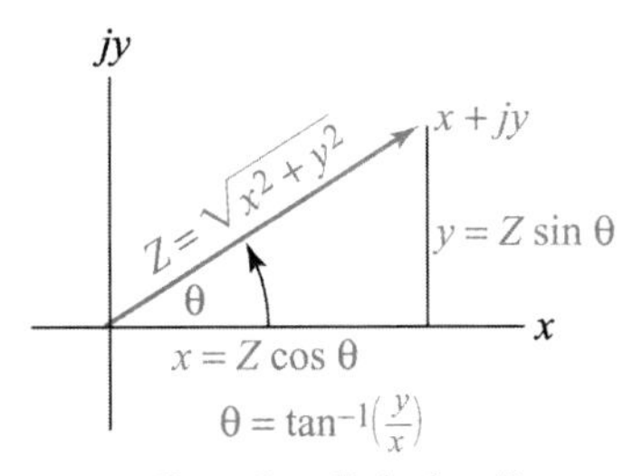

FIGURE 20-6 Polar form of complex phasor.

formulas to change from rectangular to polar form for a phasor in the first or fourth quadrant come from trigonometry as shown in Figure 20-6. They are the same as formulas (19.2) for adding phasor components:

$$Z = \sqrt{x^2 + y^2}$$

$$\theta = \tan^{-1}\left(\frac{y}{x}\right) \quad -90° < \theta < 90° \quad (x \neq 0)$$

Note that the angle θ is either a positive angle in the first quadrant or a negative angle in the fourth quadrant. These two formulas can be combined and written in rectangular and polar form as:

Formula

Rectangular to Polar Form

$$x + jy = \sqrt{x^2 + y^2} \angle \tan^{-1}\left(\frac{y}{x}\right) = Z \angle\theta \quad (x \neq 0) \tag{20.3}$$

EXAMPLE 20.13

Change the phasor $1 + j2$ to polar form.

Solution Apply (20.3) where $x = 1$ and $y = 2$:

$$1 + j2 = \sqrt{1^2 + 2^2} \angle \tan^{-1}\left(\frac{2}{1}\right) = 2.2 \angle 63°$$

EXAMPLE 20.14

Change $4 - j3$ to polar form.

Solution Apply (20.3) with $x = 4$ and $y = -3$:

$$4 - j3 = \sqrt{4^2 + (-3)^2} \angle \tan^{-1}\left(\frac{-3}{4}\right) = \sqrt{25} \angle \tan^{-1}(-0.75) = 5 \angle -37°$$

When you enter a negative number on the calculator and press [tan⁻¹], the calculator will correctly give you a negative angle in the fourth quadrant.

EXAMPLE 20.15

Change $0 - j25$ to polar form.

Solution To change $0 - j25$ to polar form, observe that $x = 0$ and formula (20.3) will not work when $x = 0$. If you draw the phasor, you see it lies along the negative y axis. The angle is therefore −90°. The magnitude is just the value of y = 25. The polar form is then 25 ∠−90°.

The formulas from trigonometry, $x = Z \cos\theta$ and $y = Z \sin\theta$, shown in Figure 20-6 are used to change from polar form to rectangular form. They are the same as formulas (19.3) for finding the components of a phasor:

Formula

Polar to Rectangular Form

$$Z \angle\theta = Z \cos\theta + j(Z \sin\theta) = x + jy \tag{20.4}$$

EXAMPLE 20.16 Change to rectangular form:

(a) $20 \angle 30°$ **(b)** $45 \angle -40°$

Solution

(a) Apply formula (20.4) where $Z = 20$ and $\theta = 30°$:

$$20 \angle 30° = 20 \cos 30° + j(20 \sin 30°) = 17 + j10$$

(b) Apply formula (20.4) with $Z = 45$ and $\theta = -40°$:

$$45 \angle -40° = 45 \cos(-40°) + j45 \sin(-40°) = 34 - j29$$

EXAMPLE 20.17 Change $3.5 \angle 0°$ to rectangular form.

Solution Observe that for an angle $\theta = 0°$, the phasor lies along the positive x axis. Therefore, $y = 0$ and $x = 3.5$. In rectangular form, the phasor is then $3.5 + j0$. You can also use formula (20.4).

You can use your calculator to change from rectangular to polar form or vice versa if you have the keys [R→P] and [P→R] or similar keys. Depending on your calculator, the key sequence may vary. For example, one way to change $4 - j3$ to polar form on some TI calculators is:

[(] 4 [,] [(−)] 3 [)] [2nd] [INV] [R→P] [=] → ($5 \angle -37°$)

One way to change $2.8 \angle 50°$ to rectangular form is:

[(] 2.8 [∠] 50 [)] [P→R] → (1.8,2.1)

EXERCISE 20.3

Round all answers to two significant digits.

In exercises 1 through 10, graph each complex phasor.

1. $1 + j2$
2. $4 + j3$
3. $2 - j4$
4. $5 - j6$
5. $0 + j$
6. $0 - j4$
7. $-4 + j0$
8. $5 + j0$
9. $-3.5 - j2.5$
10. $-5.5 + j7.5$

In exercises 11 through 20, add the complex phasors algebraically and graphically.

11. $(1 + j2) + (3 + j)$
12. $(3 + j4) + (2 + j)$
13. $(2 + j3) + (2 - j3)$
14. $(3 - j2) + (3 + j2)$
15. $(0 + j2) + (5 + j0)$
16. $(0 - j3) + (3 + j0)$
17. $(0 + j4) + (0 - j2)$
18. $(0 - j6) + (0 + j4)$
19. $(4.5 - j0.5) + (1.5 + j2.5)$
20. $(6.5 - j5.5) + (4.0 + j7.0)$

In exercises 21 through 38, change each complex phasor to polar form.

21. $2 + j2$
22. $1 + j$
23. $2 - j4$
24. $3 - j5$
25. $1 - j$
26. $7 - j3$
27. $4 + j2$
28. $3 + j2$
29. $0 - j55$
30. $0 - j40$
31. $75 + j0$
32. $-10 + j0$
33. $5 + j20$
34. $8 + j15$
35. $0.5 - j1.2$
36. $1.2 - j0.9$
37. $1.8 + j3.9$
38. $5.6 + j4.7$

In exercises 39 through 54, change each complex phasor to rectangular form.

39. $3 \angle 45°$
40. $5 \angle 60°$
41. $50 \angle -30°$
42. $45 \angle -20°$
43. $100 \angle 0°$
44. $60 \angle 0°$
45. $10 \angle -90°$
46. $75 \angle -90°$
47. $3.2 \angle -26°$
48. $3.9 \angle -10°$

49. $4.3 \angle 82°$

50. $5.1 \angle 13°$

51. $15 \angle 90°$

52. $9 \angle 90°$

53. $10 \angle -45°$

54. $20 \angle -45°$

Applications to Electronics

55. An impedance phasor in rectangular form is given by $Z = R + jX$ where R is the resistance and X is the reactance.

(a) If $R = 5.5$ kΩ and $X = 2.5$ kΩ, write Z in rectangular and polar form.

(b) If $R = 3.3$ kΩ and $X = -4.7$ kΩ, write Z in rectangular and polar form.

56. An impedance phasor in polar form is given by $Z = r \angle \theta$ where r is the magnitude and θ is the phase angle.

(a) If $r = 1.6$ kΩ and $\theta = -45°$, write Z in polar and rectangular form.

(b) If $r = 910$ Ω and $\theta = 60°$, write Z in polar and rectangular form.

20.4 Operations with Complex Phasors

To add complex phasors in polar form, *first change to rectangular form and add components.* Then change back to polar form. Study the next example.

EXAMPLE 20.18 Add the complex phasors:

$$40 \angle 25° + 50 \angle -45°$$

Solution To add complex phasors, they must be in rectangular form. Change each phasor to rectangular form by applying formula (20.4):

$$40 \angle 25° = 40 \cos 25° + j40 \sin 25° = 36.25 + j16.90$$
$$50 \angle -45° = 50 \cos(-45°) + j50 \sin(-45°) = 35.36 + -j35.36$$

Add components:

$$(36.25 + 35.36) + j(16.90 - 35.36) = 71.61 - j18.46$$

Change back to polar form by applying formula (20.3):

$$71.61 - j18.46 = \sqrt{71.61^2 + (-18.46)^2} \angle \tan^{-1}\left(\frac{-18.46}{71.61}\right) \approx 74 \angle -14°$$

Phasors can be directly multiplied and divided in polar form. To multiply two complex phasors in polar form, multiply the magnitudes and add the angles:

Formula **Multiplying Complex Phasors**

$$(Z_1 \angle \theta_1)(Z_2 \angle \theta_2) = Z_1 Z_2 \angle(\theta_1 + \theta_2) \quad \textbf{(20.5)}$$

To divide two complex phasors in polar form, divide the magnitudes and subtract the angles:

Formula **Dividing Complex Phasors in Polar Form**

$$\frac{Z_2 \angle \theta_2}{Z_1 \angle \theta_1} = \frac{Z_2}{Z_1} \angle(\theta_2 - \theta_1) \quad \textbf{(20.6)}$$

Study the next example, which illustrates formulas (20.5) and (20.6).

EXAMPLE 20.19

(a) Multiply in polar form:

$$(2 \angle 30°)(3 \angle 40°)$$

(b) Divide in polar form:

$$\frac{6 \angle 50°}{3 \angle 60°}$$

Solution

(a) Apply formula (20.5):

$$(2 \angle 30°)(3 \angle 40°) = (2)(3) \angle(30° + 40°) = 6 \angle 70°$$

(b) Apply formula (20.6):

$$\frac{6 \angle 50°}{3 \angle 60°} = \frac{6}{3} \angle(50° - 60°) = 2 \angle{-10°}$$

EXAMPLE 20.20 Perform the operations and express the answer in polar form:

$$\frac{(1 + j)(1 - j2)}{3 + j}$$

Solution Change each phasor to polar form:

$$1 + j = \sqrt{1^2 + 1^2} \angle \tan^{-1}\left(\frac{1}{1}\right) \approx 1.41 \angle 45°$$

$$1 - j2 = \sqrt{1^2 + (-2)^2} \angle \tan^{-1}\left(\frac{-2}{1}\right) \approx 2.24 \angle{-63.4°}$$

$$3 + j = \sqrt{3^2 + 1^2} \angle \tan^{-1}\left(\frac{1}{3}\right) \approx 3.16 \angle 18.4°$$

Express the example in polar form and multiply the numerator by applying formula (20.5):

$$\frac{(1 + j)(1 - j2)}{3 + j} = \frac{(1.41 \angle 45°)(2.24 \angle{-63.4°})}{3.16 \angle 18.4°}$$

$$= \frac{(1.41)(2.24) \angle(45° - 63.4°)}{3.16 \angle 18.4°} = \frac{3.16 \angle{-18.4°}}{3.16 \angle 18.4°}$$

Then divide by applying formula (20.6):

$$\frac{3.16 \angle{-18.4°}}{3.16 \angle 18.4°} = \frac{3.16}{3.16} \angle(-18.4° - 18.4°) = 1.0 \angle{-37°}$$

You can also do the operations in rectangular form and then change to polar form. However, if the phasors are given in polar form, it is easier to multiply and divide in that form.

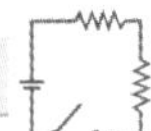

EXAMPLE 20.21

Close the Circuit

In an ac circuit Ohm's law $V = IZ$ applies when the voltage V, the current I, and the impedance Z are complex phasors. Given an ac circuit where $V = 20\ \angle 45°$ V and $Z = 2.5\ \angle{-45°}$ kΩ. Find the current I.

Solution To find I, apply Ohm's law in the form $I = \frac{V}{Z}$ and express Z in ohms:

$$I = \frac{V}{Z} = \frac{20\ \angle 45°\ \text{V}}{2.5\ \angle{-45°}\ \text{k}\Omega} = \frac{20\ \angle 45°\ \text{V}}{2500\ \angle{-45°}\ \Omega}$$

Apply formula (20.6) and divide in polar form:

$$I = \frac{20}{2500}\angle[45° - (-45°)] = 0.008\ \angle 90°\ \text{A} = 8.0\ \angle 90°\ \text{mA}$$

Complex phasors play an important part in the mathematics of ac circuits. Their importance is shown in the next three chapters on ac circuits.

EXERCISE 20.4

Round all answers in the exercise to two significant digits.

In exercises 1 through 14, add the complex phasors by first changing to rectangular form. Express answers in polar form. See Example 20.18.

1. $30\ \angle 30° + 60\ \angle 45°$
2. $25\ \angle 60° + 45\ \angle 30°$
3. $15\ \angle 75° + 18\ \angle 54°$
4. $23\ \angle 58° + 18\ \angle 42°$
5. $40\ \angle 10° + 60\ \angle{-20°}$
6. $3\ \angle 75° + 4\ \angle{-40°}$
7. $50\ \angle 90° + 60\ \angle 90°$
8. $2.2\ \angle{-90°} + 1.2\ \angle{-90°}$
9. $10\ \angle 90° + 20\ \angle{-90°}$
10. $220\ \angle 90° + 120\ \angle{-90°}$
11. $3.2\ \angle{-90°} + 5.6\ \angle 0°$
12. $6.8\ \angle 0° + 8.1\ \angle 90°$
13. $1.5\ \angle{-50°} + 4.4\ \angle{-85°}$
14. $7.0\ \angle{-36°} + 5.0\ \angle{-53°}$

In exercises 15 through 34, perform the operations and express the answer in polar form. Apply formulas (20.5) and (20.6).

15. $(2\ \angle 40°)(4\ \angle 20°)$
16. $(5\ \angle 50°)(10\ \angle 10°)$
17. $(22\ \angle 25°)(30\ \angle{-70°})$
18. $(15\ \angle{-50°})(30\ \angle 70°)$
19. $(10\ \angle{-30°})(20\ \angle{-20°})$
20. $(4\ \angle{-26°})(18\ \angle{-32°})$
21. $(30\ \angle 90°)(6\ \angle 0°)$
22. $(3.6\ \angle{-90°})(1.5\ \angle 0°)$
23. $(8.2\ \angle 90°)(2.2\ \angle{-90°})$
24. $(10\ \angle{-90°})(24\ \angle 90°)$
25. $\dfrac{60\ \angle 70°}{20\ \angle 20°}$
26. $\dfrac{15\ \angle 50°}{3\ \angle 30°}$
27. $\dfrac{11\ \angle{-12°}}{22\ \angle 56°}$
28. $\dfrac{10\ \angle{-45°}}{50\ \angle 15°}$
29. $\dfrac{8.0\ \angle 90°}{2.0\ \angle 0°}$
30. $\dfrac{120\ \angle{-90°}}{10\ \angle 0°}$
31. $\dfrac{3.5\ \angle{-25°}}{2.1\ \angle{-65°}}$
32. $\dfrac{5.9\ \angle{-16°}}{1.7\ \angle{-34°}}$
33. $\dfrac{12\ \angle 50°}{22\ \angle{-40°}}$
34. $\dfrac{5.2\ \angle 31°}{4.1\ \angle{-37°}}$

In exercises 35 through 44, perform the operations and express the answer in polar form.

35. $(1 + j)(2 - j2)$
36. $(4 - j3)(3 + j4)$
37. $(4 - j6)(3 - j2)$
38. $(1 - j5)(5 - j)$
39. $\dfrac{12 + j9}{6 - j8}$
40. $\dfrac{2 - j}{1 + j}$
41. $\dfrac{(1 + j2)(1 - j)}{1 + j}$
42. $\dfrac{(1 - j2)(1 + j)}{1 - j}$
43. $\dfrac{(1 + j)(2 + j)}{(1 + j) + (2 + j)}$
44. $\dfrac{(1 - j)(2 - j)}{(1 - j) + (2 - j)}$

Applications to Electronics

45. The phasor current in one branch of a parallel ac circuit is given by $I_1 = 2\ \angle 30°$ mA and in another branch by $I_2 = 3\ \angle 60°$ mA. Find the total current in both branches in polar form.

46. In an ac series circuit, one impedance is given by $Z_1 = 40\angle 0°$ kΩ and another by $Z_2 = 30\angle -45°$ kΩ. Find the total impedance $Z_T = Z_1 + Z_2$ in polar form.

47. In an ac circuit, the voltage $V = 12\angle 50°$ V and the impedance $Z = 3.3\angle -15°$ kΩ. Using Ohm's law find the current I in polar form. See Example 20.21.

48. In an ac circuit, the voltage $V = 30\angle -45°$ V and the current $I = 12\angle 30°$ mA. Using Ohm's law find the impedance Z in polar form. See Example 20.21.

49. The total impedance of an ac circuit containing two impedances Z_1 and Z_2 in parallel is given by:

$$Z_T = \frac{Z_1 Z_2}{Z_1 + Z_2}$$

Find Z_T in polar form when $Z_1 = 60\angle 30°$ kΩ and $Z_2 = 50\angle 90°$ kΩ.

50. In problem 49, find Z_T in polar form when $Z_1 = 100\angle -90°$ kΩ and $Z_2 = 100\angle 45°$ kΩ.

CHAPTER HIGHLIGHTS

20.1 THE *j* OPERATOR

j Operator

$$j = \sqrt{-1} \text{ or } j^2 = \left(\sqrt{-1}\right)^2 = -1 \qquad (20.1)$$

• • •

Rule for Simplifying Radicals

$$\sqrt{-x} = j\sqrt{x} \quad (x = \text{positive number}) \qquad (20.2)$$

To multiply or divide imaginary numbers, separate the j first. Treat j like an algebraic variable but replace j^2 by -1. To divide by j, multiply the numerator and denominator by j.

Key Example:

$$\left(\sqrt{-9}\right)\left(\sqrt{-16}\right) = \left(j\sqrt{9}\right)\left(j\sqrt{16}\right) = j^2(3)(4) = (-1)(12) = -12$$

$$\frac{\sqrt{18}}{12\sqrt{-2}} = \frac{\sqrt{18}}{j12\sqrt{2}} = \frac{\sqrt{9}}{j12} = \frac{3}{j12} = \frac{(j)3}{(j)j12} = \frac{j3}{(j^2)12} = \frac{j3}{-12} = -\frac{j}{4}$$

20.2 COMPLEX NUMBERS

Addition of Complex Numbers

Add the real and imaginary parts separately.

Key Example:

$$(2.1 - j5.3) + (3.4 + j4.2) = (2.1 + 3.4) + (-j5.3 + j4.2) = 5.5 - j1.1$$

Multiplication of Complex Numbers

Use the FOIL method to multiply the four products and replace j^2 with -1.

Key Example:

$$\begin{aligned}(3 - j4)(2 + j5) &= \overset{F}{(3)(2)} + \overset{O}{(3)(j5)} + \overset{I}{(-j4)(2)} + \overset{L}{(-j4)(j5)}\\ &= 6 + j15 - j8 - (j^2)(20)\\ &= 6 + j7 - (-1)(20) = 26 + j7\end{aligned}$$

Complex Conjugate

The *complex conjugate* of $x + jy$ is $x - jy$.

• • •

Division of Complex Numbers

Multiply the numerator and denominator by the conjugate of the denominator.

Key Example:

$$\begin{aligned}\frac{2 + j3}{5 - j4} &= \frac{(5 + j4)(2 + j3)}{(5 + j4)(5 - j4)} = \frac{10 + j15 + j8 + j^2 12}{25 - j^2 16}\\ &= \frac{10 + j23 + (-1)12}{25 - (-1)16} = \frac{-2 + j23}{41}\\ &= -0.049 + j0.56\end{aligned}$$

20.3 COMPLEX PHASORS

A complex phasor $x + jy$ is represented by the arrow drawn from the origin to the point (x,y).

Rectangular to Polar Form

$$x + jy = \sqrt{x^2 + y^2}\ \angle \tan^{-1}\left(\frac{y}{x}\right) = Z\angle\theta \quad (x \neq 0) \qquad (20.3)$$

Key Example:

$$1 + j2 = \sqrt{1^2 + 2^2} \angle \tan^{-1}\left(\frac{2}{1}\right) \approx 2.2 \angle 63°$$

$$0 - j4 = \sqrt{0^2 + 4^2} \angle -90° = 4 \angle -90°$$

Polar to Rectangular Form

$$Z \angle \theta = Z \cos \theta + j(Z \sin \theta) = x + jy \qquad (20.4)$$

Key Example:

$$45 \angle -40° = 45 \cos(-40°) + j45 \sin(-40°) \approx 34 - j29$$

20.4 OPERATIONS WITH COMPLEX PHASORS

To add complex phasors in polar form, *first change to rectangular form and add components.* Then change back to polar form.

Key Example:

$$\begin{aligned} 40 \angle 25° + 50 \angle -45° &= (36.25 + j16.90) + (35.36 - j35.36) \\ &= (36.25 + 35.36) + j(16.90 - 35.36) \\ &= 71.61 - j18.46 \approx 74 \angle -14° \end{aligned}$$

Multiplying Complex Phasors

$$(Z_1 \angle \theta_1)(Z_2 \angle \theta_2) = Z_1 Z_2 \angle(\theta_1 + \theta_2) \qquad (20.5)$$

Key Example:

$$(2 \angle 30°)(3 \angle 40°) = (2)(3) \angle(30° + 40°) = 6 \angle 70°$$

Dividing Complex Phasors

$$\frac{Z_2 \angle \theta_2}{Z_1 \angle \theta_1} = \frac{Z_2}{Z_1} \angle(\theta_2 - \theta_1) \qquad (20.6)$$

Key Example:

$$\frac{6 \angle 50°}{3 \angle 60°} = \frac{6}{3} \angle(50° - 60°) = 2 \angle -10°$$

REVIEW EXERCISES

Round all answers in the exercise to two significant digits.

In exercises 1 through 16, simplify and express as a real or imaginary number.

1. $\sqrt{-81}$

2. $\sqrt{-0.36}$

3. $-5\sqrt{-0.04}$

4. $-10\sqrt{-0.0001}$

5. $-8\sqrt{-\frac{1}{25}}$

6. $\sqrt{-64 \times 10^{-6}}$

7. $(\sqrt{-16})(\sqrt{-64})$

8. $(-\sqrt{-3})(\sqrt{-27})$

9. $(-8\sqrt{-2})(\sqrt{-72})$

10. $(3\sqrt{-12})(2\sqrt{-27})$

11. $(j6)^2(j2)$

12. $(-j)^3(-j2)$

13. $\frac{15}{6j}$

14. $\frac{5}{\sqrt{-100}}$

15. $\frac{\sqrt{-75}}{\sqrt{-3}}$

16. $\frac{1.5\sqrt{4}}{\sqrt{-0.25}}$

In exercises 17 through 28, perform the operations and express the result in rectangular form x + jy.

17. $(2 + j5) + (4 - j3)$

18. $2(1 - j) + (8 - j4)$

19. $(5 + j4)(2 - j3)$

20. $(1 + j2)^2$

21. $(1.2 - j1.5)(1.0 - j2.2)$

22. $j2(1 + j)(2 - j)$

23. $\frac{1}{1 - j2}$

24. $\frac{3 + j}{1 + j2}$

25. $\frac{6 - j2}{1 + j}$

26. $\frac{4 + j3}{4 - J3}$

27. $\frac{5 - j5}{(-j)(5 + j5)}$

28. $\frac{(2 + j)(1 + j)}{(2 + j) + (1 + j)}$

In exercises 29 through 38, change the complex phasor from rectangular to polar form or vice versa.

29. $1 + j3$

30. $3 - j2$

31. $0 - j45$

32. $0 + j7$

33. $5 + 0j$

34. $10 + 10j$

35. $100 \angle 30°$

36. $5 \angle 45$

37. $5.5 \angle 90°$

38. $6.8 \angle -90°$

In exercises 39 through 56, perform the operations and express the answer in polar form.

39. $5.0 \angle 30° + 4.0 \angle 50°$

40. $14 \angle 70° + 18 \angle 10°$

41. $20 \angle 45° + 40 \angle -60°$

42. $1.3 \angle -28° + 1.8 \angle -42°$

43. $7.5 \angle 90° + 4.5 \angle -90°$

44. $12 \angle 0° + 16 \angle -90°$

45. $(3.2 \angle 25°)(4.5 \angle 15°)$

46. $(48 \angle -65°)(1.5 \angle 35°)$

47. $(2.0 \angle -40°)(30 \angle -30°)$

48. $(2.4 \angle 90°)(1.5 \angle -90°)$

49. $(2.0 \angle 0°)(30 \angle 30°)$

50. $(2.4 \angle -40°)(1.5 \angle 0°)$

51. $\dfrac{30 \angle 50°}{10 \angle 30°}$

52. $\dfrac{12 \angle -35°}{48 \angle 20°}$

53. $\dfrac{5.1 \angle -62°}{1.7 \angle -54°}$

54. $\dfrac{42 \angle 90°}{20 \angle 0°}$

55. $\dfrac{(1 + j)(1 - j)}{(1 + j) + (1 - j)}$

56. $\dfrac{(10 \angle 45°)(10 \angle -90°)}{10 \angle 45° + 10 \angle -90°}$

Applications to Electronics

57. The total voltage in a series circuit is given by $V_T = V_1 + V_2$.

(a) Find V_T in polar form when $V_1 = 15 \angle 34°$ V and $V_2 = 10 \angle 90°$ V.

(b) Find V_T in polar form when $V_1 = 3.6 \angle -26°$ V and $V_2 = 2.9 \angle -45°$ V.

58. The admittance Y in siemens of an ac circuit is given by $Y = \dfrac{1}{Z}$ where Z = impedance in ohms.

(a) Find Y in polar form when $Z = 20 \angle -45°$ kΩ.

Note: $1 = 1 \angle 0°$.

(b) Find Y in polar form when $Z = 20 \angle 60°$ kΩ.

59. Given Ohm's law $V = IZ$ in an ac circuit:

(a) Find V in polar form when $I = 5 \angle 45°$ mA and $Z = 15 \angle 15°$ kΩ.

(b) Find V in polar form when $I = 810 \angle 20°$ μA and $Z = 1.3 \angle -48°$ kΩ.

60. Using Ohm's law $V = IZ$ in an ac circuit:

(a) Find I in polar form given $V = 120 \angle 45°$ V and the impedance $Z = 910 \angle 60°$ Ω.

(b) Find I in polar form given $V = 24 \angle 45°$ V and the impedance $Z = 5.6 \angle -24°$ kΩ.

61. A parallel ac circuit contains the two impedances $Z_1 = 30 \angle 0°$ kΩ and $Z_2 = 10 \angle 60°$ kΩ. Find the total impedance Z_T in kΩ, in polar form and rectangular form using the formula:

$$Z_T = \frac{Z_1 Z_2}{Z_1 + Z_2}$$

62. In problem 61, find Z_T in kΩ, in polar and rectangular form given $Z_1 = 2 - j$ and $Z_2 = 1 + j$.

CHAPTER 21

Series AC Circuits

CHAPTER OBJECTIVES

In this chapter, you will learn:

- How to calculate inductive reactance.
- How to apply Ohm's law to ac circuits.
- How to use complex phasors to solve an *RL* series circuit.
- How to calculate capacitive reactance.
- How to use complex phasors to solve an *RC* series circuit.
- How to use complex phasors to solve an *RLC* series circuit.
- How to find the resonant frequency of an *RLC* series circuit.

Alternating current electricity is introduced in Chapter 18, where ac circuits containing only resistors are studied. However, there are two other basic types of circuit components found in ac circuits: inductors and capacitors. Each of these tend to oppose the flow of current but in a different way than resistors. Their effect is called *reactance*. Resistance and reactance are two types of *impedances* found in ac circuits. Impedances can be combined in three basic ways to form a series ac circuit: a resistance-inductance or *RL* circuit, a resistance-capacitance or *RC* circuit, and a resistance-inductance-capacitance or *RLC* circuit. The mathematical relationships that apply to these circuits use complex phasors and the ideas studied in Chapter 20.

• • •

21.1 Inductive Reactance and *RL* Series Circuits

Inductive Reactance

An inductor is a coil of wire in an ac circuit. A voltage that opposes the applied voltage is induced in the coil by the ac current. The measure of the coil's capacity to produce voltage is called the inductance L, which is measured in henrys (H). The inductive reactance, X_L, is a measure of the inductor's effect on the applied voltage and current. X_L is measured in the same units as resistance, that is ohms, and is a function of the inductance L and the frequency f:

Formula

Inductive Reactance (Ω)

$$X_L = 2\pi f L \tag{21.1}$$

EXAMPLE 21.1

Given an inductance $L = 100$ mH in an ac circuit with a frequency $f = 2.0$ kHz. Find the inductive reactance.

Soluion Apply formula (21.1):

$$X_L = 2\pi(2.0 \text{ kHz})(100 \text{ mH}) \approx 1.3 \text{ k}\Omega$$

Formula (21.1) can also be used to find the frequency f, or the inductance L, given the other quantities, by solving the formula for f or L as follows.

EXAMPLE 21.2

Given an inductance $L = 70$ mH in an ac circuit whose inductive reactance $X_L = 500\ \Omega$. Find the frequency of the current.

Solution Solve formula (21.1) for the frequency by dividing both sides by 2πL and substitute the given values:

$$\frac{X_L}{2\pi L} = \frac{2\pi f L}{2\pi L}$$

$$f = \frac{X_L}{2\pi L} = \frac{500\ \Omega}{2\pi(70\ \text{mH})} \approx 1.1\ \text{kHz}$$

One of the reasons X_L is measured in ohms is because Ohm's law applies to the reactance X_L, the voltage across the inductance V_L, and the current through the inductance I:

Law

Ohm's Law with X_L

$$I = \frac{V_L}{X_L} \tag{21.2}$$

EXAMPLE 21.3

Figure 21-1 shows an ac source connected to an inductance. If $V_L = 10$ V and $X_L = 2.6$ kΩ, find the current in the circuit.

$$X_L = 2\pi f L$$
$$I = \frac{V_L}{X_L}$$

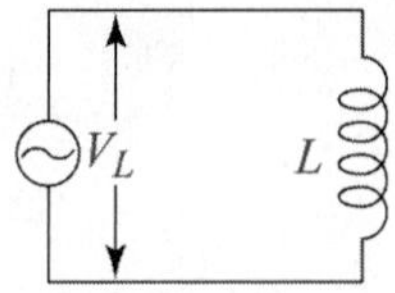

FIGURE 21-1 Inductive reactance for Example 21.3.

Solution Apply Ohm's law (21.2):

$$I = \frac{V_L}{X_L} = \frac{10\ \text{V}}{2.6\ \text{k}\Omega} \approx 3.8\ \text{mA}$$

EXAMPLE 21.4

Given the inductive circuit in Figure 21-1 with $V_L = 55$ V, $f = 30$ kHz, and $I = 4.5$ mA. Find X_L and L.

Solution Solve Ohm's law (21.2) for X_L and substitute the values:

$$X_L = \frac{V_L}{I} = \frac{55\ \text{V}}{4.5\ \text{mA}} = 12.22\ \text{k}\Omega \approx 12\ \text{k}\Omega$$

Then apply formula (21.1) solved for L:

$$L = \frac{X_L}{2\pi f} = \frac{12.22\ \text{k}\Omega}{2\pi(30\ \text{kHz})} \approx 65\ \text{mH}$$

RL Series Circuits

When a resistor and an inductor are connected in series to an ac source, the voltage across the inductor V_L is out of phase with both the current I and the voltage across the resistance V_R. The induced voltage in the inductor tends to oppose the current flow and the result is: *V_L leads I and V_R by 90°*. Figure 21-2(a) shows the *RL* series circuit and the voltages. Figure 21-2(b) shows the three ac sine waves. The current I and the voltage V_R are sine waves with a phase angle of 0°. The voltage V_L is a sine wave with a phase angle of 90°. Therefore V_L leads I and V_R by 90°.

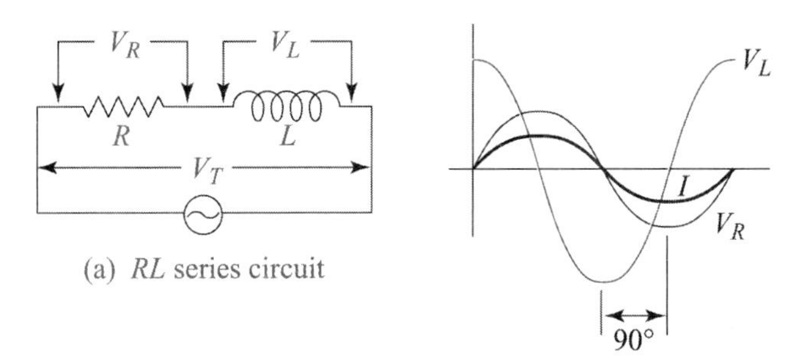

FIGURE 21-2 *RL* series circuit for Examples 21.6 and 21.7.

Figure 21-3(a) shows the phasor diagram of the voltages and the current in the *RL* series circuit. The voltage V_R is the real component, and the voltage V_L is the imaginary component of the resultant or total voltage V_T. Applying formula (20.3) for converting from rectangular to polar form, V_T is given by:

Formula

RL Series Voltage

$$V_T = V_R + jV_L = \left(\sqrt{V_R^2 + V_L^2}\right) \angle \tan^{-1}\left(\frac{V_L}{V_R}\right) \tag{21.3}$$

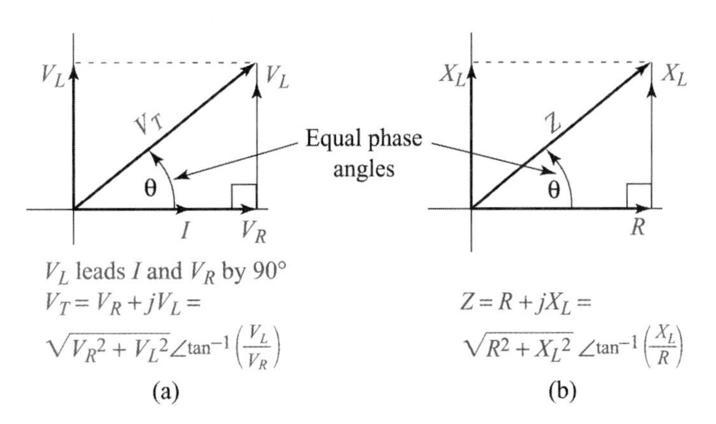

FIGURE 21-3 Phasor relationships in an *RL* series circuit.

The resistance R and the reactance X_L have the same phase relationship as the voltages V_R and V_L, respectively. R is the real component, and X_L is the imaginary component of the total impedance Z:

Formula **RL Series Impedance (Ω)**

$$Z = R + jX_L = \left(\sqrt{R^2 + X_L^2}\right) \angle \tan^{-1}\left(\frac{X_L}{R}\right) \qquad (21.4)$$

Figure 21-3(b) shows the impedance triangle. Note that Z has the same phase angle as V_T.

EXAMPLE 21.5 Given the RL series circuit in Figure 21-2(a) with $R = 800\ \Omega$ and $X_L = 1.2\ \text{k}\Omega$. Find the total impedance Z of the circuit in rectangular and polar form.

Solution Draw the impedance triangle with R as the real component and X_L as the imaginary component of the impedance Z. See Figure 21-4. Z in rectangular form is then:

$$Z = R + jX_L = 800 + j1200\ \Omega = 0.80 + j1.2\ \text{k}\Omega$$

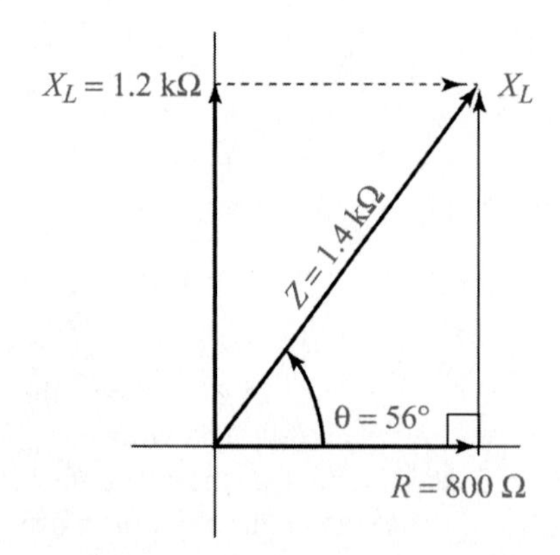

$$Z = 800 + j1200 = \sqrt{800^2 + 1200^2}\ \angle \tan^{-1}\left(\frac{1200}{800}\right) \approx 1.4\angle 56°\ \text{k}\Omega$$

FIGURE 21-4 Impedance triangle for Example 21.5.

Apply formula (21.4) to find Z in polar form:

$$Z = \left(\sqrt{R^2 + X_L^2}\right) \angle \tan^{-1}\left(\frac{X_L}{R}\right) = \left(\sqrt{0.80^2 + 1.2^2}\right) \angle \tan^{-1}\left(\frac{1.2}{0.80}\right)$$

$$Z = \sqrt{2.08}\ \angle \tan^{-1}(1.5) \approx 1.4\ \angle 56°\ \text{k}\Omega$$

The magnitude of Z is 1.4 kΩ, and the phase angle is 56°. The calculation from rectangular to polar form, and vice versa, can be done on the calculator if you have the keys [R→P] and [P→R] or similar keys. Depending on your calculator, the key sequence may vary. For example, one way to change $800 + j1200$ to polar form on some TI calculators is:

[(] 0.80 [,] 1.2 [)] [2nd] [INV] [P→R] [=] → (1.4 ∠56°) kΩ

In ac circuits, impedance is analogous to resistance in dc circuits, and Ohm's law applies for both scalar *and* phasor quantities:

Law **Ohm's Law with Z (Scalar and Phasor)**

$$I = \frac{V}{Z} \qquad (21.5)$$

EXAMPLE 21.6 Given the total voltage in the *RL* series circuit of Example 21.5, $V_T = 15\ \angle 56°$ V and the impedance $Z = 1.44\ \angle 56°$ kΩ, find the magnitude and the phase angle of the current in polar form.

Solution You can find I using scalars or phasors. Both methods are shown.

Scalars: Apply Ohm's law (21.5) to find the magnitude of I using the magnitudes of V_T and Z:

$$I = \frac{V_T}{Z} = \frac{15\text{ V}}{1.44\text{ k}\Omega} \approx 10\text{ mA}$$

The phase angle of I is the same as V_R, which is 0°. Therefore $I = 10\ \angle 0°$ mA.

Phasors: Apply Ohm's law (21.5) and divide the phasors in polar form:

$$I = \frac{V}{Z} = \frac{15\ \angle 56°\text{ V}}{1.44\ \angle 56°\text{ k}\Omega} = \frac{15}{1440}\angle(56° - 56°) \approx 0.010\ \angle 0°\text{ A} = 10\ \angle 0°\text{ mA}$$

Remember, when you divide complex phasors you divide the magnitudes and subtract the angles. Observe that the phase angles of V_T and Z are the same, 56°, which verifies that the phase angle of I is 0°.

EXAMPLE 21.7 Given the *RL* series circuit in Figure 21-2(a) with $R = 15$ kΩ, $X_L = 11$ kΩ, and $I = 800$ μA.

(a) Find V_R and V_L.

(b) Find the total voltage V_T and the impedance Z in polar form.

Solution

(a) Find the magnitudes of V_R and V_L by applying Ohm's law solved for V:

$$V_R = IR = (800\ \mu\text{A})(15\text{ k}\Omega) = 12\text{ V}$$
$$V_L = IX_L = (800\ \mu\text{A})(11\text{ k}\Omega) = 8.8\text{ V}$$

See Figure 21-5, which shows the voltage phasor diagram for the *RL* series circuit.

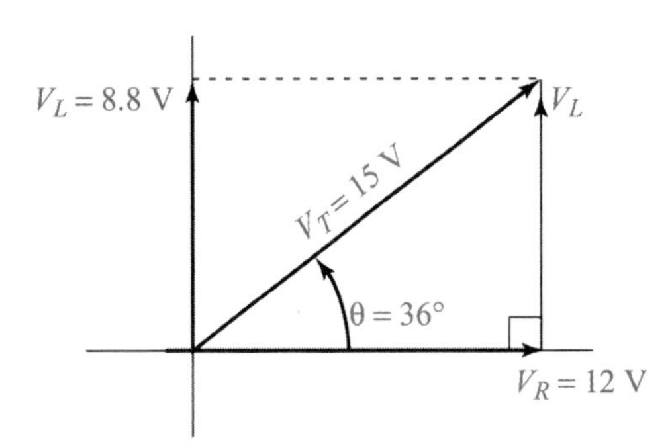

FIGURE 21-5 Voltage phasors in an *RL* series circuit for Example 21.7.

(b) Apply formula (21.3) to find V_T:

$$V_T = V_R + jV_L = 12 + j8.8 \text{ V} = \sqrt{(12)^2 + (8.8)^2} \angle \tan^{-1}\left(\frac{8.8}{12}\right)$$

$$= \sqrt{221} \angle \tan^{-1}(0.733) \approx 15 \angle 36° \text{ V}$$

The magnitude of V_T is then 15 V, and the phase angle $\theta = 36°$.

Use Ohm's law (21.5) to find the impedance Z from the voltage and the current. In a series circuit, the phase angle of Z is the same as the phase angle of V_T. See Figure 21-3. Therefore, it is only necessary to calculate the magnitude of Z:

$$Z = \frac{V_T}{I} = \frac{15 \text{ V}}{800 \text{ μA}} = 19 \text{ k}\Omega$$

The impedance in polar form is then $Z = 19 \angle 36° \text{ k}\Omega$.

You can also find the impedance using the given values of R and X_L and formula (21.4):

$$Z = R + jX_L = 15 + j11 \text{ k}\Omega = \sqrt{15^2 + 11^2} \angle \tan^{-1}\left(\frac{11}{15}\right)$$

$$= \sqrt{346} \angle \tan^{-1}(0.733) \approx 19 \angle 36° \text{ k}\Omega$$

EXAMPLE 21.8

Given an RL series circuit with $L = 500$ mH, $f = 400$ Hz, $V_L = 6.5$ V, and $Z = 3.0$ kΩ.

(a) Find X_L and I.
(b) Find R and V_R.
(c) Find V_T and Z in rectangular and polar form.

Solution

(a) Apply formula (21.1) to find X_L:

$$X_L = 2\pi f L = 2\pi(400 \text{ Hz})(500 \text{ mH}) \approx 1.26 \text{ k}\Omega$$

Apply Ohm's law with X_L to find I:

$$I = \frac{V_L}{X_L} = \frac{6.5 \text{ V}}{1.26 \text{ k}\Omega} \approx 5.2 \text{ mA}$$

(b) Apply the Pythagorean theorem to the impedance triangle to find R:

$$R = \sqrt{Z^2 - X_L^2} = \sqrt{3.0^2 - 1.26^2} \approx 2.7 \text{ k}\Omega$$

V_R can then be found using Ohm's law:

$$V_R = IR = (5.2 \text{ mA})(2.7 \text{ k}\Omega) \approx 14 \text{ V}$$

(c) Apply (21.3) to find V_T in polar form:

$$V_T = V_R + jV_L = 14 + j6.5 \text{ k}\Omega = \sqrt{14^2 + 6.5^2} \angle \tan^{-1}\left(\frac{6.5}{14}\right) \approx 15 \angle 25° \text{ V}$$

The phase angle of Z is the same as V_T, therefore $Z = 2.7 + j1.26 \text{ k}\Omega = 3.0 \angle 25° \text{ k}\Omega$.

EXERCISE 21.1

Round all answers in the exercise to two significant digits.

In exercises 1 through 14, using the given values for an inductance L, find the indicated quantity.

1. $L = 1.0$ H, $f = 600$ Hz; X_L
2. $L = 1.5$ H, $f = 500$ Hz; X_L
3. $L = 20$ mH, $f = 5.0$ kHz; X_L
4. $L = 100$ mH, $f = 6.0$ kHz; X_L
5. $L = 800$ mH, $f = 2.0$ kHz; X_L
6. $L = 700$ mH, $f = 1.0$ kHz; X_L
7. $X_L = 10$ kΩ, $f = 15$ kHz; L
8. $X_L = 15$ kΩ, $f = 20$ kHz; L
9. $X_L = 2.5$ kΩ, $f = 10$ kHz; L
10. $X_L = 2.7$ kΩ, $f = 15$ kHz; L
11. $X_L = 200$ Ω, $L = 50$ mH; f
12. $X_L = 330$ Ω, $L = 40$ mH; f
13. $X_L = 3.5$ kΩ, $L = 75$ mH; f
14. $X_L = 6.8$ kΩ, $L = 200$ mH; f

In exercises 15 through 22, using the given values for an ac circuit containing an inductance L, find the indicated quantity.

15. $V_L = 12$ V, $I = 15$ mA; X_L
16. $V_L = 4.5$ V, $I = 300$ μA; X_L
17. $V_L = 5.0$ V, $I = 3.5$ mA; X_L
18. $V_L = 18$ V, $I = 6.6$ mA; X_L
19. $X_L = 25$ kΩ, $I = 100$ μA; V_L
20. $X_L = 1.6$ kΩ, $I = 15$ mA; V_L
21. $X_L = 600$ Ω, $V_L = 1.5$ V; I
22. $X_L = 2.2$ kΩ, $V_L = 11$ V; I

In exercises 23 through 30, using the given values for an RL series circuit, find the impedance Z in rectangular and polar form.

23. $R = 750$ Ω, $X_L = 1.5$ kΩ
24. $R = 1.0$ kΩ, $X_L = 820$ Ω
25. $R = 3.3$ kΩ, $X_L = 4.5$ kΩ
26. $R = 6.8$ kΩ, $X_L = 3.9$ kΩ
27. $R = 2.0$ kΩ, $X_L = 1.2$ kΩ
28. $R = 1.5$ kΩ, $X_L = 1.5$ kΩ
29. $R = 910$ Ω, $X_L = 750$ Ω
30. $R = 620$ Ω, $X_L = 780$ Ω

In exercises 31 through 38, using the given values for an RL series circuit:

(a) Find V_R and V_L.
(b) Find V_T and Z in polar form.

31. $R = 620$ Ω, $X_L = 800$ Ω, $I = 20$ mA
32. $R = 680$ Ω, $X_L = 900$ Ω, $I = 50$ mA
33. $R = 1.0$ kΩ, $X_L = 1.5$ kΩ, $I = 5.0$ mA
34. $R = 5.6$ kΩ, $X_L = 3.3$ kΩ, $I = 3.6$ mA
35. $R = 22$ kΩ, $X_L = 18$ kΩ, $I = 750$ μA
36. $R = 16$ kΩ, $X_L = 16$ kΩ, $I = 820$ μA
37. $R = 2.7$ kΩ, $X_L = 2.2$ kΩ, $I = 15$ mA
38. $R = 2.7$ kΩ, $X_L = 880$ Ω, $I = 3.5$ mA

In exercises 39 through 46, using the given values for an RL series circuit:

(a) Find Z in polar form.
(b) Find I, V_R, and V_L.

39. $R = 620$ Ω, $X_L = 330$ Ω, $V_T = 15$ V
40. $R = 300$ Ω, $X_L = 500$ Ω, $V_T = 20$ V
41. $R = 3.0$ kΩ, $X_L = 1.0$ kΩ, $V_T = 12$ V
42. $R = 1.8$ kΩ, $X_L = 4.7$ kΩ, $V_T = 16$ V
43. $R = 4.3$ kΩ, $X_L = 2.4$ kΩ, $V_T = 60$ V
44. $R = 5.6$ kΩ, $X_L = 9.5$ kΩ, $V_T = 9.0$ V
45. $R = 11$ kΩ, $X_L = 11$ kΩ, $V_T = 50$ V
46. $R = 15$ kΩ, $X_L = 7.5$ kΩ, $V_T = 20$ V
47. Given an *RL* series circuit with $L = 150$ mH, $f = 1.5$ kHz, $V_L = 7.5$ V, and $Z = 2.3$ kΩ.
(a) Find X_L and I, R, and V_R. See Example 21.8.
(b) Find V_T in polar form.
48. Given an *RL* series circuit with $f = 800$ Hz, $R = 12$ kΩ, $V_R = 10$ V, and $X_L = 5.5$ kΩ.
(a) Find L, I, and V_L.
(b) Find V_T and Z in polar form.
49. Figure 21-6 shows a resistance in series with an inductance connected to a 120-V, 60-Hz generator. Given $R = 100$ Ω and $L = 750$ mH:
(a) Find X_L, I, V_R, and V_L
(b) Find Z in polar form.

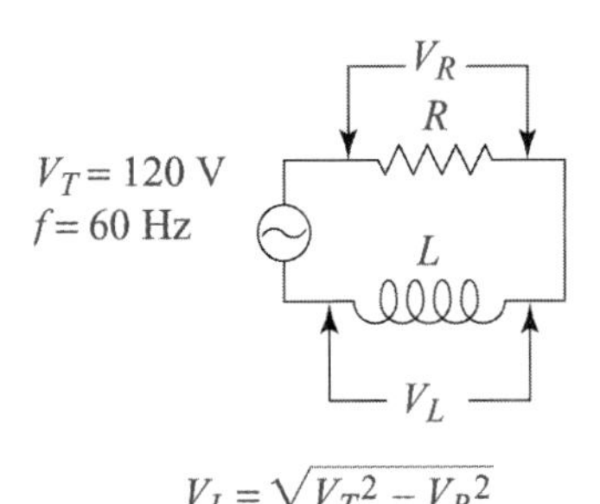

FIGURE 21-6 *RL* series circuit for problems 49 and 50.

50. In the circuit of Figure 21-6, the resistance $R = 150\ \Omega$ and V_R is measured to be 50 V.

(a) Find I, V_L, X_L, and L.

Note: $V_L = \sqrt{V_T^2 - V_R^2}$

(b) Find Z in polar form.

21.2 Capacitive Reactance and *RC* Series Circuits

Capacitive Reactance

A capacitor consists of two conductors separated by an insulator, or dielectric, and is capable of storing charge. See Figure 21-7 on page 421. The measure of the capacitor's ability to store charge is called capacitance C and is measured in farads (F). The capacitive reactance X_C is measured in ohms and depends on the capacitance C and the frequency f:

Formula

Capacitive Reactance (Ω)

$$X_C = \frac{1}{2\pi fC} \tag{21.6}$$

EXAMPLE 21.9

Given a capacitance $C = 10$ nF in an ac circuit with a frequency $f = 5.0$ kHz. Find the capacitive reactance.

Solution Apply formula (21.6):

$$X_C = \frac{1}{2\pi(5.0 \times 10^3\ \text{Hz})(10 \times 10^{-9}\ \text{F})} \approx 3.2\ \text{k}\Omega$$

This calculation can be done on the calculator using the reciprocal key:

DAL: 2 [×] [π] [×] 5 [EE] 3 [×] 10 [EE] [(-)] 9 [=] [x⁻¹] [=] → 3.2×10^3 Hz

Not DAL: 2 [×] [π] [×] 5 [EXP] 3 [×] 10 [EXP] 9 [+/-] [=] [1/x] [=] → 3.2×10^3 Hz

Formula (21.6) can also be used to find f or C, given the other quantities, by solving the formula for f or C as shown in the next examples.

EXAMPLE 21.10

Given a capacitance $C = 200$ nF in an ac circuit with a capacitive reactance $X_C = 750\ \Omega$. Find the frequency of the current.

Solution Solve formula (21.6) for f by multiplying both sides by f and dividing both sides by X_C. Then substitute the given values:

$$\left(\frac{f}{\cancel{X_C}}\right)\cancel{X_C} = \frac{1}{2\pi\cancel{f}C}\left(\frac{\cancel{f}}{X_C}\right)$$

$$f = \frac{1}{2\pi X_C C} = \frac{1}{2\pi(750\ \Omega)(200 \times 10^{-9}\ \text{F})} \approx 1.1\ \text{kHz}$$

Ohm's law also applies to X_C, the current I and the voltage V_C across the capacitor:

Law **Ohm's Law with X_C**

$$I = \frac{V_C}{X_C} \tag{21.7}$$

EXAMPLE 21.11

In Figure 21-7, if the voltage across the capacitor $V_C = 14$ V, $f = 2.0$ kHz, and $I = 15$ mA, find X_C and C.

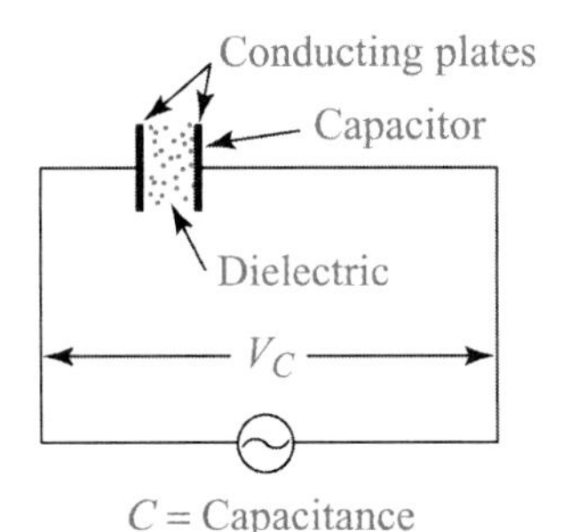

FIGURE 21-7 Capacitor and capacitive reactance for Example 21.11.

Solution To find X_C, solve Ohm's law (21.7) for X_C and substitute:

$$X_C = \frac{V_C}{I} = \frac{14 \text{ V}}{15 \text{ mA}} \approx 930\ \Omega$$

To find C, solve formula (21.6) for C by multiplying by C and dividing by X_C. Then substitute the given values:

$$\left(\frac{C}{\cancel{X_C}}\right)\cancel{X_C} = \frac{1}{2\pi f\cancel{C}}\left(\frac{\cancel{C}}{X_C}\right)$$

$$C = \frac{1}{2\pi f X_C} = \frac{1}{2\pi(2.0 \text{ kHz})(933\ \Omega)} \approx 85 \text{ nF}$$

RC Series Circuits

When a resistor and a capacitor are connected in series to an ac source, the voltage across the capacitor V_C is out of phase with both the current I and the voltage across the resistance V_R. The stored charge in the capacitor tends to oppose the applied voltage and the result is: *I and V_R lead V_C by 90°*. Figure 21-8(a) shows the *RC* circuit and the voltages. Figure 21-8(b) shows the three ac sine waves. The current I and the voltage V_R are sine waves with a phase angle of 0°. The voltage V_C is a sine wave with a phase angle of –90°. Therefore, I and V_R lead V_C by 90°.

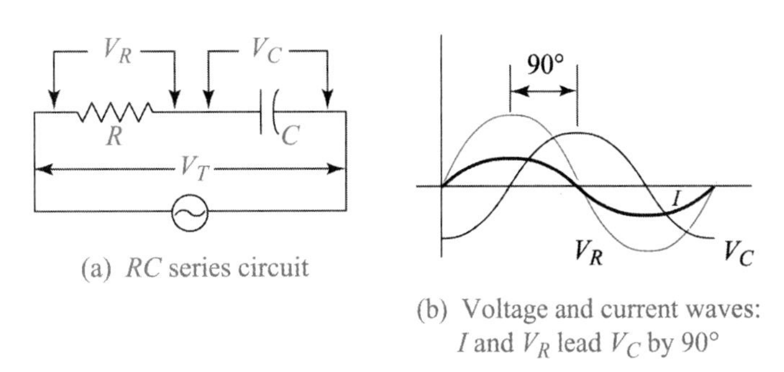

FIGURE 21-8 *RC* series circuit for Examples 21.12 and 21.13.

A good way to remember the relationship between the current and voltage in *RL* and *RC* circuits is by the following mnemonic where E (electromotive force) is used for voltage: "***ELI*** the ***ICE*** man" {Voltage (***E***) in an inductive circuit (***L***) leads current (***I***)} *and* {Current (***I***) in a capacitative circuit (***C***) leads Voltage (***E***)}. In a dc circuit, E is often used for voltage.

Figure 21-9(a) shows the phasor diagram of the voltages and the current in the *RC* series circuit. The phasor V_R is the real component, and $-V_C$ is the negative imaginary component of the resultant or total voltage V_T:

Formula

RC Series Voltage

$$V_T = V_R - jV_C = \left(\sqrt{V_R^2 + V_C^2}\right) \angle \tan^{-1}\left(\frac{-V_C}{V_R}\right) \tag{21.8}$$

Observe that the phasor V_C has a phase angle of –90°. As a result, the phase angle θ of V_T is a negative angle in the fourth quadrant.

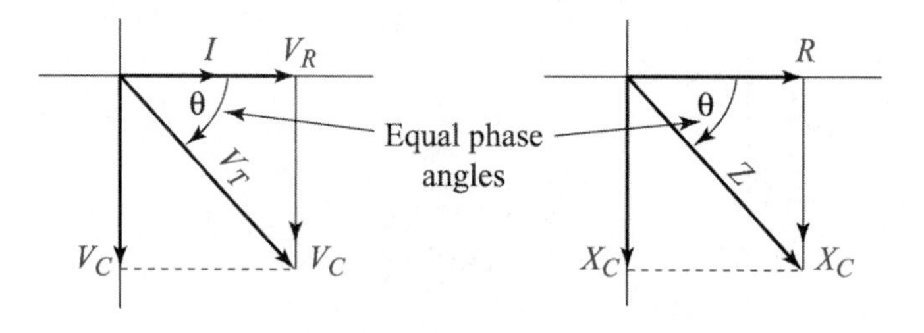

I and V_R lead V_C by 90°
$V_T = V_R - jV_C = \sqrt{V_R^2 + V_C^2} \angle \tan^{-1}\left(\frac{-V_C}{V_R}\right)$
(a)

$Z = R - jX_C = \sqrt{R^2 + X_C^2} \angle \tan^{-1}\left(\frac{-X_C}{R}\right)$
(b)

FIGURE 21-9 Phasor relationships in an *RC* series circuit.

The resistance and the reactance in an *RC* series circuit have the same phase relationship as V_R and V_C, respectively. The resistance *R* is the real component, and $-X_C$ is the negative imaginary component of the total impedance *Z*:

Formulas

RC Series Impedance

$$Z = R - jX_C = \left(\sqrt{R^2 + X_C^2}\right) \angle \tan^{-1}\left(\frac{-X_C}{R}\right) \tag{21.9}$$

Figure 21-9(b) shows the impedance triangle. Note that *Z* has the same phase angle as V_T.

EXAMPLE 21.12

Given the *RC* circuit in Figure 21-8(a) with $R = 3.0$ kΩ and $X_C = 1.0$ kΩ. Find the total impedance *Z* of the circuit in rectangular and polar form.

Solution Draw the impedance triangle with *R* as the real component and $-X_C$ as the negative imaginary component of *Z*. See Figure 21-10. *Z* in rectangular form is then:

$$Z = R - jX_C = 3.0 - j1.0 \text{ k}\Omega$$

Apply formula (21.9) to find *Z* in polar form:

$$Z = \left(\sqrt{R^2 + X_C^2}\right) \angle \tan^{-1}\left(\frac{-X_C}{R}\right) = \sqrt{3.0^2 + 1.0^2} \angle \tan^{-1}\left(\frac{-1.0}{3.0}\right)$$

$$Z = \sqrt{10} \angle \tan^{-1}(-0.333) \approx 3.2 \angle -18° \text{ k}\Omega$$

The magnitude of *Z* is then 3.2 kΩ, and the phase angle is –18°.

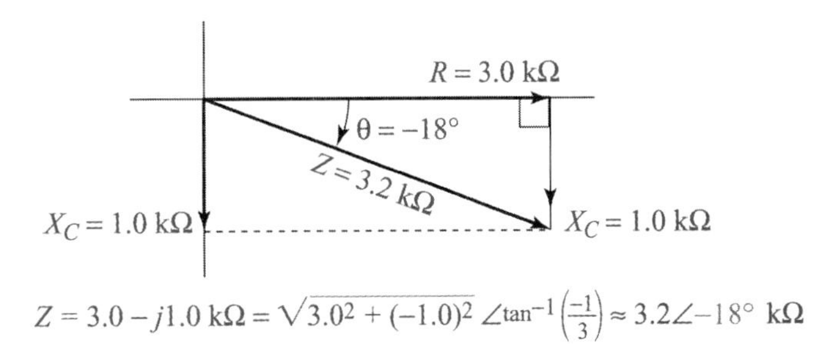

FIGURE 21-10 Impedance phasors for Example 21.12.

EXAMPLE 21.13

Given the *RC* circuit in Figure 21-8(a) with $R = 7.5\ \text{k}\Omega$, $C = 33$ nF, $f = 500$ Hz, and $V_T = 20$ V.

(a) Find X_C.

(b) Find Z in rectangular and polar form.

(b) Find I, V_C, and V_R.

Solution

(a) Apply formula (21.6) to find X_C:

$$X_C = \frac{1}{2\pi fC} = \frac{1}{2\pi(500\ \text{Hz})(33\ \text{nF})} \approx 9.65\ \text{k}\Omega$$

(b) Find Z by applying formula (21.9):

$$Z = 7.5 - j9.65\ \text{k}\Omega = \sqrt{(7.5)^2 + (9.65)^2}\ \tan^{-1}\left(\frac{-9.65}{7.5}\right) \approx 12.2\ \angle{-52^\circ}\ \text{k}\Omega$$

(c) Find the current I by applying Ohm's law with Z:

$$I = \frac{V_T}{Z} = \frac{20\ \text{V}}{12.2\ \text{k}\Omega} \approx 1.64\ \text{mA}$$

Now find V_C and V_R by applying Ohm's law using X_C and R:

$$V_C = IX_C = (1.64\ \text{mA})(9.65\ \text{k}\Omega) \approx 16\ \text{V}$$
$$V_R = IR = (1.64\ \text{mA})(7.5\ \text{k}\Omega) \approx 12\ \text{V}$$

Because of the many calculations in this example, results are calculated to three figures and then rounded to two figures for the voltages.

EXAMPLE 21.14

Given an *RC* series circuit with $R = 6.8\ \text{k}\Omega$, $I = 750\ \mu\text{A}$, and $V_C = 8.5$ V.

(a) Find X_C and V_R.

(b) Find V_T and Z in polar form.

Solution

(a) Use Ohm's law to find X_C and V_R:

$$X_C = \frac{V_C}{I} = \frac{8.5\ \text{V}}{750\ \mu\text{A}} \approx 11.3\ \text{k}\Omega$$

$$V_R = IR = (750\ \mu\text{A})(6.8\ \text{k}\Omega) = 5.1\ \text{V}$$

(b) You can now find V_T and Z by applying formulas (21.8) and (21.9):

$$V_T = \sqrt{5.1^2 + 8.5^2}\ \angle\tan^{-1}\left(\frac{-8.5}{5.1}\right) \approx 9.9\ \angle{-59°}\ \text{V}$$

$$Z = \sqrt{6.8^2 + 11.3^2}\ \angle\tan^{-1}\left(\frac{-11.3}{6.8}\right) \approx 13\ \angle{-59°}\ \text{k}\Omega$$

You can also find the magnitude of V_T and Z by finding the phase angle first using $\tan^{-1}$, and then using the reference angle θ (See Figure 21-11) with the sine and cosine functions.

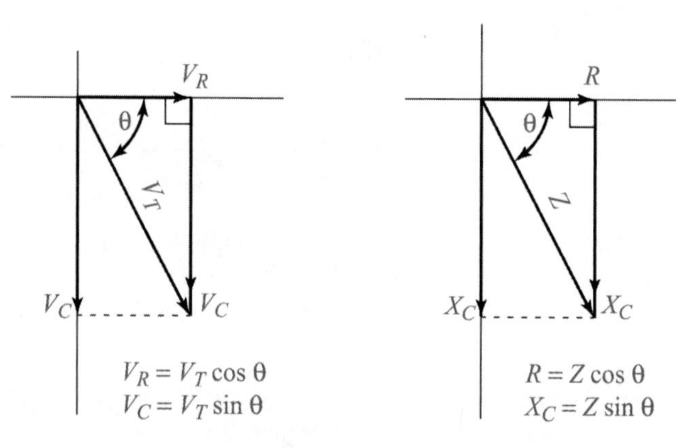

FIGURE 21-11 Phasor components for Example 21.14.

Using the components of V_T and Z given by:

$$V_C = V_T \sin\theta \text{ and } R = Z\cos\theta$$

It follows that:

$$V_T = \frac{V_C}{\sin\theta} = \frac{8.5\ \text{V}}{\sin 59°} \approx 9.9\ \text{V}$$

$$Z = \frac{R}{\cos\theta} = \frac{6.8\ \text{k}\Omega}{\cos 59°} \approx 13\ \text{k}\Omega$$

EXERCISE 21.2

Round answers to two significant digits.

In exercises 1 through 14, using the given values for a capacitance C, find the indicated quantity.

1. $C = 20$ nF, $f = 6.0$ kHz; X_C
2. $C = 50$ nF, $f = 5.0$ kHz; X_C
3. $C = 500$ pF, $f = 15$ kHz; X_C
4. $C = 400$ pF, $f = 20$ kHz; X_C
5. $C = 1.0$ μF, $f = 300$ Hz; X_C
6. $C = 2.0$ μF, $f = 200$ Hz; X_C
7. $X_C = 10$ kΩ, $f = 10$ kHz; C
8. $X_C = 16$ kΩ, $f = 12$ kHz; C
9. $X_C = 800$ Ω, $f = 2.0$ kHz; C
10. $X_C = 950$ Ω, $f = 3.0$ kHz; C
11. $X_C = 600$ Ω, $C = 200$ nF; f
12. $X_C = 500$ Ω, $C = 100$ nF; f
13. $X_C = 15$ kΩ, $C = 800$ pF; f
14. $X_C = 5.6$ kΩ, $C = 400$ pF; f

In exercises 15 through 22, using the given values for an ac circuit containing a capacitance C, find the indicated quantity.

15. $V_C = 20$ V, $I = 12$ mA; X_C
16. $V_C = 22$ V, $I = 34$ mA; X_C

17. $V_C = 12$ V, $I = 1.5$ mA; X_C
18. $V_C = 8.4$ V, $I = 850$ μA; X_C
19. $X_C = 6.2$ kΩ, $I = 780$ μA; V_C
20. $X_C = 750$ Ω, $I = 7.5$ mA; V_C
21. $X_C = 25$ kΩ, $V_C = 10$ V; I
22. $X_C = 630$ Ω, $V_C = 14$ V; I

In exercises 23 through 30, using the given values for an RC series circuit, find the impedance Z in rectangular and polar form.

23. $R = 1.6$ kΩ, $X_C = 1.2$ kΩ
24. $R = 8.2$ kΩ, $X_C = 16$ kΩ
25. $R = 680$ Ω, $X_C = 910$ Ω
26. $R = 3.3$ kΩ, $X_C = 2.2$ kΩ
27. $R = 550$ Ω, $X_C = 550$ Ω
28. $R = 1.5$ kΩ, $X_C = 750$ Ω
29. $R = 820$ Ω, $X_C = 1.0$ kΩ
30. $R = 710$ Ω, $X_C = 650$ Ω

In exercises 31 through 38, using the given values for an RC series circuit:

(a) Find V_R and V_C.
(b) Find V_T and Z in polar form.

31. $R = 7.5$ kΩ, $X_C = 4.0$ kΩ, $I = 3.2$ mA
32. $R = 12$ kΩ, $X_C = 15$ kΩ, $I = 3.2$ mA
33. $R = 10$ kΩ, $X_C = 25$ kΩ, $I = 500$ μA
34. $R = 24$ kΩ, $X_C = 10$ kΩ, $I = 650$ μA
35. $R = 500$ Ω, $X_C = 800$ Ω, $I = 4.0$ mA
36. $R = 620$ Ω, $X_C = 480$ Ω, $I = 4.0$ mA
37. $R = 4.3$ kΩ, $X_C = 2.7$ kΩ, $I = 6.5$ mA
38. $R = 1.2$ kΩ, $X_C = 2.4$ kΩ, $I = 25$ mA

In exercises 39 through 46, using the given values for an RC series circuit:

(a) Find Z in polar form.
(b) Find I, V_R, and V_C.

39. $R = 680$ Ω, $X_C = 620$ Ω, $V_T = 20$ V
40. $R = 470$ Ω, $X_C = 620$ Ω, $V_T = 8.4$ V
41. $R = 1.2$ kΩ, $X_C = 850$ Ω, $V_T = 24$ V
42. $R = 5.6$ kΩ, $X_C = 6.2$ kΩ, $V_T = 24$ V
43. $R = 8.2$ kΩ, $X_C = 13$ kΩ, $V_T = 12$ V
44. $R = 20$ kΩ, $X_C = 30$ kΩ, $V_T = 10$ V
45. $R = 750$ Ω, $X_C = 750$ Ω, $V_T = 9.7$ V
46. $R = 470$ Ω, $X_C = 1.1$ kΩ, $V_T = 12$ V
47. Given an *RC* series circuit with $R = 3.3$ kΩ, $C = 100$ nF, $f = 1.0$ kHz, and $V_T = 15$ V.
(a) Find X_C, I, V_C, and V_R.
(b) Find Z in polar form.
48. Given an *RC* series circuit with $R = 820$ Ω, $I = 9.5$ mA, and $V_C = 22$ V.
(a) Find X_C and V_R.
(b) Find V_T and Z in polar form.
49. Figure 21-12 shows a resistor in series with a capacitor connected to a 120-V, 60-Hz generator. Given R = 22 kΩ and $C = 150$ nF:
(a) Find X_C, V_R, V_C, and I.
(b) Find Z in polar form.

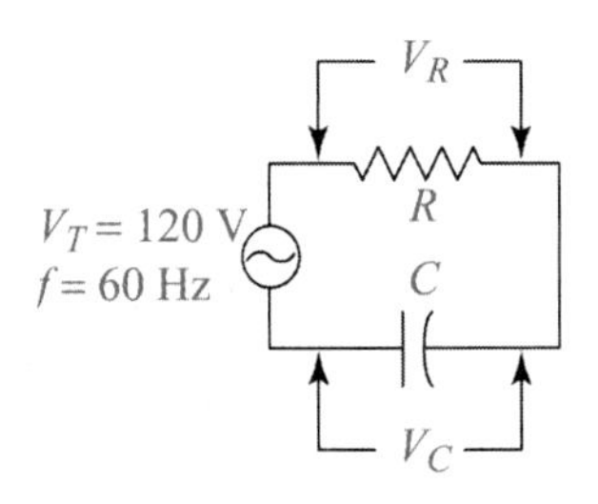

FIGURE 21-12 *RC* series circuit for problems 49 and 50.

50. In the circuit of Figure 21-12, given $X_C = 450$ Ω and $V_C = 100$ V.
(a) Find R, C, I, and V_R.

Note: $R = \sqrt{Z^2 - X_C^2}$

(b) Find Z in polar form.

21.3 *RLC* Series Circuits and Resonance

When an inductor and a capacitor are connected in series in an ac circuit, the reactances cancel each other because the phasors have opposite direction. Inductive reactance has aphase angle of 90°, while capacitive reactance has a phase angle of –90°. Consider the *RLC* series circuit in Figure 21-13 containing a resistor, an inductor, and a capacitor in series.

The total impedance of the *RLC* series circuit in rectangular form is:

Formula

RLC Series Impedance

$$Z = R + j(X_L - X_C) = R \pm jX \quad X = |X_L - X_C| \tag{21.10}$$

Inductive ($+jX$) Capacitive ($-jX$)

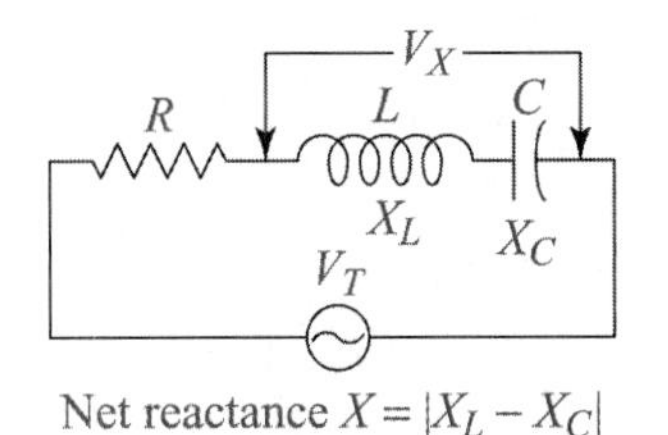

Net reactance $X = |X_L - X_C|$

FIGURE 21-13 *RLC* series circuit for Example 21.16.

where $X = |X_L - X_C|$ is the magnitude of the net reactance. When $X_L > X_C$, the net reactance X is inductive, and the imaginary term jX is positive. See Figure 21-14(a). When $X_L < X_C$, X is capacitive, and the imaginary term jX is negative. See Figure 21-14(b). To change Z to polar form, use formula (21.4) when X is inductive and formula (21.9) when X is capacitive.

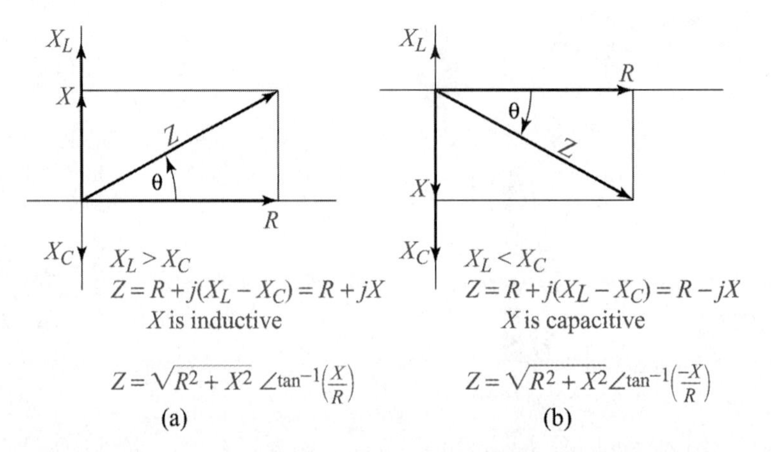

FIGURE 21-14 Impedance phasors in an *RLC* series circuit.

EXAMPLE 21.15 In the *RLC* series circuit in Figure 21-15(a), $R = 500\ \Omega$, $X_L = 600\ \Omega$, and $X_C = 1.0\ \text{k}\Omega$.

(a) Find the the net reactance X.

(b) Find the total impedance Z in rectangular and polar form.

Solution

(a) The magnitude of the net reactance is:

$$X = |X_L - X_C| = |600 - 1000| = 400\ \Omega$$

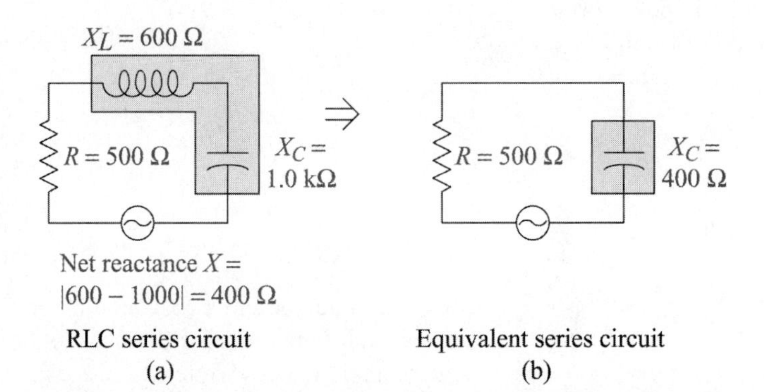

FIGURE 21-15 *RLC* series circuit and equivalent series circuit for Example 21.15.

The net reactance is capacitive since X_L is less than X_C. The circuit is equivalent to an *RC* series circuit where $R = 500\ \Omega$ and $X_C = X = 400\ \Omega$. See Figure 21-15(b). This is called the *equivalent series circuit.*

(b) Apply formula (21.10) to find the total impedance in rectangular form:

$$Z = 500 + j(600 - 1000)\ \Omega = 500 - j400\ \Omega$$

To find Z in polar form, apply formula (21.9) when X is capacitive. See Figure 21.14(b):

$$Z = 500 - j400\ \Omega = \left(\sqrt{500^2 + 400^2}\right) \angle \tan^{-1}\left(\frac{-400}{500}\right) \approx 640\ \angle{-39°}\ \Omega$$

The total voltage of an *RLC* series circuit in rectangular form is:

RLC Series Voltage

$$V_T = V_R + j(V_L - V_C) = V_R \pm jV_X \quad V_X = |V_L - V_C| \tag{21.11}$$

where $V_X = |V_L - V_C|$ is the magnitude of the net reactive voltage. To change V_T to polar form, use formula (21.3) when the net reactance X is inductive and formula (21.8) when X is capacitive. Study the next problem, which illustrates these concepts and others from Sections 21.1 and 21.2.

EXAMPLE 21.16

In the *RLC* series circuit in Figure 21-13, $R = 3.3\ \text{k}\Omega$, $L = 400$ mH, and $C = 10$ nF. If the applied voltage $V_T = 14$ V with a frequency $f = 3.0$ kHz:

(a) Find X_L and X_C. Find Z in rectangular and polar form.

(b) Find I, V_R, V_L, and V_C. Find V_T in rectangular and polar form..

Solution

(a) Find the reactances X_L and X_C using formulas (21.1) and (21.6):

$$X_L = 2\pi fL = 2\pi(3.0\ \text{kHz})(400\ \text{mH}) \approx 7.5\ \text{k}\Omega$$

$$X_C = \frac{1}{2\pi fC} = \frac{1}{2\pi(3.0\ \text{kHz})(10\ \text{nF})} \approx 5.3\ \text{k}\Omega$$

Apply formula (21.10) to find the total impedance in rectangular form:

$$Z = R + j(X_L - X_C) = 3.3 + j(7.5 - 5.3)\ \text{k}\Omega = 3.3 + j2.2\ \text{k}\Omega$$

Since $X_L > X_C$, the term jX is positive. The net reactance X is inductive, and the equivalent series circuit is an *RL* series circuit with $R = 3.3\ \text{k}\Omega$ and $X_L = X = 2.2\ \text{k}\Omega$. To change Z to polar form, apply formula (21.4):

$$Z = \left(\sqrt{R^2 + X_L^2}\right) \angle \tan^{-1}\left(\frac{X_L}{R}\right)$$

$$Z = \left(\sqrt{3.3^2 + 2.2^2}\right) \angle \tan^{-1}\left(\frac{2.2}{3.3}\right) \approx 4.0\ \angle 34°\ \text{k}\Omega$$

(b) To find I and the voltages across R, L and C, apply Ohm's law:

$$I = \frac{V}{Z} = \frac{14 \text{ V}}{4.0 \text{ kHz}} = 3.5 \text{ mA}$$
$$V_R = IR = (3.5 \text{ mA})(3.3 \text{ k}\Omega) \approx 11.6 \text{ V}$$
$$V_L = IX_L = (3.5 \text{ mA})(7.5 \text{ k}\Omega) \approx 26.3 \text{ V}$$
$$V_C = IX_C = (3.5 \text{ mA})(5.3 \text{ k}\Omega) \approx 18.6 \text{ V}$$

The voltages V_R, V_L, and V_C may appear to add up to more than the applied voltage 14 V. However, V_L tends to cancel V_C, and their phasor sum equals 14 V as follows. Apply (21.11) to find V_T in rectangular form:

$$V_T = 11.6 + j(26.3 - 18.6) \text{ V} = 11.6 + j7.7$$

The net reactive voltage $V_X = 7.7$ V. Since the circuit is inductive, apply formula (21.3) to change V_T to polar form where $V_L = V_X$:

$$V_T = \left(\sqrt{V_R^2 + V_L^2}\right) \angle \tan^{-1}\left(\frac{V_L}{V_R}\right)$$

$$V_T = \left(\sqrt{11.6^2 + 7.7^2}\right) \angle \tan^{-1}\left(\frac{7.7}{11.6}\right) \approx 14 \angle 34° \text{ V}$$

ERROR BOX

A common error when working with *RLC* circuits is getting the wrong sign for the phase angle θ. The net reactance X is considered a positive quantity, however you must know if it is inductive or capacitive and supply the correct sign for θ. If X_L is greater than X_C, θ is positive. If X_L is less than X_C, θ is negative. When using the calculator, if you enter a positive value and press [tan⁻¹], you will get a positive angle. If you enter the *same* value with a negative sign and press [tan⁻¹], you will get the *same* angle with a negative sign. Therefore, if X is inductive, enter a positive value for $\frac{X}{R}$, and if X is capacitive enter a negative value for $\frac{X}{R}$. See if you can get the correct angle in each of the practice problems.

Practice Problems: For each series *RLC* circuit, find the phase angle θ.

1. $R = 10\ \Omega, X_L = 20\ \Omega, X_C = 10\ \Omega$
2. $R = 10\ \Omega, X_L = 10\ \Omega, X_C = 20\ \Omega$
3. $R = 500\ \Omega, X_L = 600\ \Omega, X_C = 1.0\ \text{k}\Omega$
4. $R = 500\ \Omega, X_L = 2.7\ \text{k}\Omega, X_C = 2.4\ \text{k}\Omega$
5. $R = 100\ \Omega, X_L = 1.0\ \Omega, X_C = 1.0\ \text{k}\Omega$
6. $R = 100\ \Omega, X_L = 500\ \Omega, X_C = 2.0\ \text{k}\Omega$
7. $R = 1.0\ \text{k}\Omega, X_L = 1.1\ \text{k}\Omega, X_C = 750\ \Omega$
8. $R = 1.0\ \text{k}\Omega, X_L = 500\ \Omega, X_C = 2.4\ \text{k}\Omega$

When an ac series circuit contains more than one resistor, you can add the resistances, as in a dc circuit, to obtain the total resistance. When an ac series circuit contains more than one inductor, or more than one capacitor, you can add the similar reactances to obtain the total inductive reactance and the total capacitive reactance:

$$X_{L_T} = X_{L_1} + X_{L_2} + X_{L_3} + \cdots$$
$$X_{C_T} = X_{C_1} + X_{C_2} + X_{C_3} + \cdots$$

You can therefore always reduce the circuit to an *RLC* series circuit with one resistor, one inductor, and one capacitor. Then, as in Example 21.16, you can find the equivalent *RL* or *RC* series circuit.

Power in an AC Circuit

There are three types of power in an ac circuit:

Formulas

True or Real Power (W)

$$P = I^2R = VI\cos\theta \qquad (21.12a)$$

Reactive Power (VAR)

$$Q = VI\sin\theta \qquad (21.12b)$$

Apparent Power (VA)

$$S = VI \qquad (21.12c)$$

The true power is the power dissipated by the resistance and is measured in watts (W). The reactive power is the power expended by the reactance and is measured in voltampere-reactive (VAR). The apparent power is the phasor sum of the true power and the reactive power: $S = \sqrt{P^2 + Q^2}$. It is measured in voltamperes (VA). The ratio $P/S = \cos\theta$ is called the power factor.

EXAMPLE 21.17 Find the true power, reactive power, apparent power, and the power factor for the circuit of Example 21.16.

Solution Apply formulas (21.12) using $V = V_T = 14$ V, $I = 3.5$ mA and $\theta = 34°$:

$$\begin{aligned}\text{True power } P &= (14\text{ V})(3.5\text{ mA})\cos 34° \approx 41\text{ mW}\\ \text{Reactive power } Q &= (14\text{ V})(3.5\text{ mA})\sin 34° \approx 27\text{ mVAR}\\ \text{Apparent power } S &= (14\text{ V})(3.5\text{ mA}) = 49\text{ mVA}\\ \text{Power factor} = P/S &= \cos\theta = \cos 34° \approx 0.829\end{aligned}$$

Series Resonance

In an *RLC* circuit, the frequency at which the inductive reactance equals the capacitive reactance, that is, $X_L = X_C$, is called the *resonant frequency* f_r. At this frequency, the net reactance $X = 0$, and the circuit is purely resistive. That is, the total impedance $Z = R$, and the phase angle $\theta = 0°$. The true power is at a maximum because when $\theta = 0°$, $\cos\theta = 1$, and $P = VI$. Since $X_L = X_C$ at the resonant frequency f_r:

$$2\pi f_r L = \frac{1}{2\pi f_r C}$$

which leads to:

$$f_r^2 = \frac{1}{4\pi^2 LC}$$

Taking the square root of both sides, the formula for the resonant frequency f_r is:

Formula

Resonant Frequency (Hz)

$$f_r = \frac{1}{2\pi\sqrt{LC}} \qquad (21.13)$$

EXAMPLE 21.18 Given $L = 50$ mH and $C = 20$ nF, find the resonant frequency.

Solution Apply formula (21.13):

$$f_r = \frac{1}{2\pi\sqrt{(50\text{ mH})(20\text{ nF})}} \approx 5.0\text{ kHz}$$

This calculation can be done by multiplying all the factors in the denominator and then taking the reciprocal:

DAL: 2 [×] [π] [×] [√] [(] 50 [EE] [(−)] 3 [×] 20 [EE] [(−)] 9 [)] [=] [x^{-1}] [=] → 5.03×10^3

NOT DAL: 50 [EXP] 3 [+/-] [×] 20 [EXP] 9 [+/-] [=] [√] [×] 2 [×] [π] [=] [x^{-1}] → 5.03×10^3

Resonance is important in tuning radio frequency (RF) circuits because at the resonant frequency the impedance Z is minimized. The current is at a maximum, the reactive power is zero, and the true power is at a maximum, and equal to the apparent power.

EXAMPLE 21.19 At a frequency of 4.0 kHz, how much inductance is required in series with a capacitance of 10 nF to obtain resonance?

Solution You need to solve formula (21.13) for L. Square both sides:

$$f_r^2 = \frac{1}{4\pi^2 LC}$$

Multiply both sides by $\frac{L}{f_r^2}$:

$$\left(\frac{L}{\cancel{f_r^2}}\right)\cancel{f_r^2} = \frac{1}{4\pi^2 \cancel{L}C}\left(\frac{\cancel{L}}{f_r^2}\right)$$

$$L = \frac{1}{4\pi^2 f_r^2 C}$$

Substitute the given values to find L:

$$L = \frac{1}{4\pi^2(4.0\text{ kHz})^2(10\text{ nF})} \approx 160\text{ mH}$$

EXERCISE 21.3

Round all answers to two significant digits.

In exercises 1 through 10, using the given values for an RLC series circuit, find X and find Z in rectangular and polar form.

1. $R = 750\ \Omega, X_L = 800\ \Omega, X_C = 500\ \Omega$
2. $R = 750\ \Omega, X_L = 600\ \Omega, X_C = 900\ \Omega$
3. $R = 1.8\ \text{k}\Omega, X_L = 3.3\ \text{k}\Omega, X_C = 5.5\ \text{k}\Omega$
4. $R = 3.0\ \text{k}\Omega, X_L = 5.0\ \text{k}\Omega, X_C = 3.0\ \text{k}\Omega$
5. $R = 3.9\ \text{k}\Omega, X_L = 11\ \text{k}\Omega, X_C = 6.2\ \text{k}\Omega$
6. $R = 5.1\ \text{k}\Omega, X_L = 10\ \text{k}\Omega, X_C = 16\ \text{k}\Omega$
7. $R = 1.0\ \text{k}\Omega, X_L = 750\ \Omega, X_C = 1.2\ \text{k}\Omega$
8. $R = 680\ \Omega, X_L = 1.3\ \text{k}\Omega, X_C = 930\ \Omega$
9. $R = 200\ \Omega, X_L = 1.1\ \text{k}\Omega, X_C = 750\ \Omega$
10. $R = 200\ \Omega, X_L = 880\ \Omega, X_C = 1.5\ \text{k}\Omega$

In exercises 11 through 18, using the given values for an RLC series circuit, find I, V_R, V_L, V_C, and find V_T in polar form.

11. $R = 3.3\ \text{k}\Omega, X_L = 1.2\ \text{k}\Omega, X_C = 4.7\ \text{k}\Omega, V_T = 12$ V
12. $R = 3.0\ \text{k}\Omega, X_L = 1.5\ \text{k}\Omega, X_C = 1.2\ \text{k}\Omega, V_T = 24$ V

Answers to Error Box Problems, page 428:

1. 45° **2.** −45° **3.** −39° **4.** 31° **5.** 0° **6.** −86° **7.** 19° **8.** −62°

13. $R = 270\ \Omega, X_L = 510\ \Omega, X_C = 450\ \Omega, V_T = 20$ V

14. $R = 680\ \Omega, X_L = 660\ \Omega, X_C = 780\ \Omega, V_T = 15$ V

15. $R = 910\ \Omega, X_L = 4.3\ \text{k}\Omega, X_C = 1.5\ \text{k}\Omega, V_T = 14$ V

16. $R = 11\ \text{k}\Omega, X_L = 7.9\ \text{k}\Omega, X_C = 12\ \text{k}\Omega, V_T = 8.0$ V

17. $R = 600\ \Omega, X_L = 1.2\ \text{k}\Omega, X_C = 840\ \Omega, V_T = 30$ V

18. $R = 500\ \Omega, X_L = 920\ \Omega, X_C = 1.0\ \text{k}\Omega, V_T = 6.5$ V

In exercises 19 through 26, using the given values for series resonance, find the indicated value.

19. $L = 100$ mH, $C = 20$ nF; f_r

20. $L = 50$ mH, $C = 100$ nF; f_r

21. $L = 40$ mH, $C = 800$ pF; f_r

22. $L = 350$ mH, $C = 100$ pF; f_r

23. $L = 400\ \mu$H, $f_r = 6.5$ kHz; C

24. $L = 240$ mH, $f_r = 15$ kHz; C

25. $C = 1.0\ \mu$F, $f_r = 600$ Hz; L

26. $C = 60$ nF, $f_r = 3.2$ kHz; L

27. Given an *RLC* series circuit with $R = 330\ \Omega$, $L = 50$ mH, and $C = 100$ nF. If the applied voltage $V = 12$ V with a frequency $f = 2.6$ kHz:
(a) Find X_L and X_C. Find Z in rectangular and polar form.
(b) Find I, V_R, V_L, and V_C. Find V_T in rectangular and polar form.

28. Given an *RLC* series circuit with $R = 1.0\ \text{k}\Omega$, $L = 200$ mH, and $C = 40$ nF. If applied voltage $V = 24$ V with a frequency $f = 1.5$ kHz:
(a) Find X_L and X_C. Find Z in rectangular and polar form.
(b) Find I, V_R, V_L, and V_C. Find V_T in rectangular and polar form.

29. Given an *RLC* series circuit with $R = 2.2\ \text{k}\Omega$, $L = 500$ mH, and $C = 20$ nF. If the applied voltage $V = 20$ V with a frequency $f = 1.2$ kHz:
(a) Find X_L and X_C. Find Z in rectangular and polar form.
(b) Find I, V_R, V_L, and V_C. Find V_T in rectangular and polar form.

30. Given an *RLC* series circuit with $R = 500\ \Omega$, $L = 200$ mH, and $C = 100$ nF. If the applied voltage $V = 10$ V with a frequency $f = 1.6$ kHz:
(a) Find X_L and X_C. Find Z in rectangular and polar form.
(b) Find I, V_R, V_L, and V_C. Find V_T in rectangular and polar form.

31. Figure 21-16 shows an *RLC* series circuit containing two inductors, a capacitor, and a resistor connected to a voltage source V. If $V = 20$ V, $R = 600\ \Omega$, $X_{L_1} = 500\ \Omega$, $X_{L_2} = 750\ \Omega$, and $X_C = 800\ \Omega$:
(a) Find Z in rectangular and polar form.
Note: $X_L = X_{L_1} + X_{L_2}$.
(b) Find I, true power P, reactive power Q, apparent power S, and the power factor.

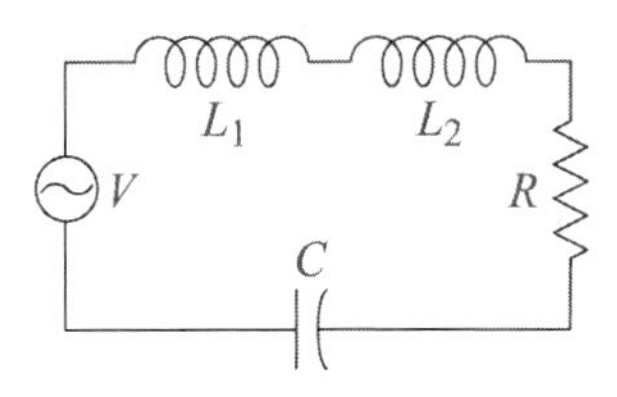

FIGURE 21-16 *RLC* series circuit for problems 31 and 32.

32. In the *RLC* series circuit in Figure 21-16, if $V = 10$ V, $R = 1.0\ \text{k}\Omega$, $X_{L_1} = 2.0\ \text{k}\Omega$, $X_{L_2} = 3.0\ \text{k}\Omega$, and $X_C = 7.5\ \text{k}\Omega$:
(a) Find Z in rectangular and polar form.
Note: $X_L = X_{L_1} + X_{L_2}$.
(b) Find I, true power P, reactive power Q, apparent power S, and the power factor.

33. In an *RLC* series circuit, $R = 1.8\ \text{k}\Omega$, $X_L = 3.6\ \text{k}\Omega$, $X_C = 1.6\ \text{k}\Omega$, and $I = 10$ mA:
(a) Find V_R, V_L, and V_C. Find V_T in rectangular and polar form.
(b) Find Z in rectangular and polar form.

34. In an *RLC* series circuit, $R = 510\ \Omega$, $X_L = 360\ \Omega$, $X_C = 750\ \Omega$, and $V_R = 5.0$ V:
(a) Find I, V_L, and V_C. Find V_T in rectangular and polar form.
(b) Find Z in rectangular and polar form.

35. Given a series circuit containing two resistances $R_1 = 2.7\ \text{k}\Omega$ and $R_2 = 1.2\ \text{k}\Omega$, an inductance $L = 300$ mH, and a capacitance $C = 50$ nF. If the applied voltage $V = 10$ V with $f = 1.0$ kHz:
(a) Find X_L and X_C. Find Z in rectangular and polar form.
(b) Find I, V_{R_1}, V_{R_2}, V_L, and V_C. Find V_T in rectangular and polar form.

36. In problem 35, given $R_1 = 330\ \Omega$, $R_2 = 180\ \Omega$, $L = 800\ \mu$H, and $C = 800$ pF. If the applied voltage $V = 20$ V with $f = 130$ kHz:
(a) Find X_L and X_C. Find Z in rectangular and polar form.
(b) Find I, V_R, V_L, and V_C. Find V_T in rectangular and polar form.

CHAPTER HIGHLIGHTS

21.1 INDUCTIVE REACTANCE AND *RL* SERIES CIRCUITS

Inductive Reactance (Ω)

$$X_L = 2\pi fL \quad (21.1)$$

• • •

Ohm's Law with X_L

$$I = \frac{V_L}{X_L} \quad (21.2)$$

In an RL series circuit, V_L leads I and V_R by 90°, and the phase angle θ is positive.

***RL* Series Voltage**

$$V_T = V_R + jV_L = \left(\sqrt{V_R^2 + V_L^2}\right) \angle \tan^{-1}\left(\frac{V_L}{V_R}\right) \quad (21.3)$$

• • •

***RL* Series Impedance (Ω)**

$$Z = R + jX_L = \left(\sqrt{R^2 + X_L^2}\right) \angle \tan^{-1}\left(\frac{X_L}{R}\right) \quad (21.4)$$

Key Example: For an RL series circuit with L = 500 mH, f = 400 Hz, V_L = 6.5 V, and Z = 3.0 kΩ:

$$X_L = 2\pi fL = 2\pi(400 \text{ Hz})(500 \text{ mH}) \approx 1.26 \text{ k}\Omega$$

$$I = \frac{V_L}{X_L} = \frac{6.5 \text{ V}}{1.26 \text{ k}\Omega} \approx 5.2 \text{ mA}$$

$$R = \sqrt{Z^2 - X_L^2} = \sqrt{3.0^2 - 1.26^2} \approx 2.7 \text{ k}\Omega$$

$$V_R = IR = (5.2 \text{ mA})(2.7 \text{ k}\Omega) \approx 14 \text{ V}$$

$$V_T = V_R + jV_L = 14 + j6.5$$

$$= \sqrt{14^2 + 6.5^2} \angle \tan^{-1}\left(\frac{6.5}{14}\right) \approx 15 \angle 25° \text{ V}$$

Ohm's Law with Z (Scalar and Phasor)

$$I = \frac{V}{Z} \quad (21.5)$$

21.2 CAPACITIVE REACTANCE AND *RC* SERIES CIRCUITS

Capacitive Reactance (Ω)

$$X_C = \frac{1}{2\pi fC} \quad (21.6)$$

• • •

Ohm's Law with X_C

$$I = \frac{V_C}{X_C} \quad (21.7)$$

In an RC series circuit, I and V_R lead V_C by 90°, and the phase angle θ is negative.

***RC* Series Voltage**

$$V_T = V_R - jV_C = \left(\sqrt{V_R^2 + V_C^2}\right) \angle \tan^{-1}\left(\frac{-V_C}{V_R}\right) \quad (21.8)$$

Remember: "***ELI*** the ***ICE*** man" {***E*** (voltage) in ***L*** (inductor) leads ***I***} *and* {***I*** in ***C*** (capacitor) leads ***E***}.

***RC* Series Impedance**

$$Z = R - jX_C = \left(\sqrt{R^2 + X_C^2}\right) \angle \tan^{-1}\left(\frac{-X_C}{R}\right) \quad (21.9)$$

Key Example: For an RC circuit with R = 7.5 kΩ, C = 33 nF, f = 500 Hz, and V_T = 20 V:

$$X_C = \frac{1}{2\pi fC} = \frac{1}{2\pi(500 \text{ Hz})(33 \text{ nF})} \approx 9.65 \text{ k}\Omega$$

$$Z = \sqrt{(7.5)^2 + (9.65)^2} \tan^{-1}\left(\frac{-9.65}{7.5}\right) \approx 12.2 \angle -52.1° \text{ k}\Omega$$

$$I = \frac{V_T}{Z} = \frac{20 \text{ V}}{12.2 \text{ k}\Omega} \approx 1.64 \text{ mA}$$

$$V_C = IX_C = (1.64 \text{ mA})(9.65 \text{ k}\Omega) \approx 16 \text{ V}$$

$$V_R = IR = (1.64 \text{ mA})(7.5 \text{ k}\Omega) \approx 12 \text{ V}$$

21.3 *RLC* SERIES CIRCUITS AND RESONANCE

***RLC* Series Impedance**

$$Z = R + j(X_L - X_C) = R \pm jX \quad (21.10)$$

$$X = |X_L - X_C|$$

Inductive (+jX) Capacitive (–jX)

When $X_L > X_C$, X is inductive, when $X_L < X_C$, X is capacitive. See Figure 21-14. To find Z in polar form, use formula (21.4) when X is inductive, and formula (21.9) when X is capacitive.

RLC Series Voltage

$$V_T = V_R + j(V_L - V_C) = V_R \pm jV_X \qquad (21.11)$$
$$V_X = |V_L - V_C|$$

To find V_T in polar form, use formula (21.3) when X is inductive, and formula (21.8) when X is capacitive.

Key Example: For an *RLC* series circuit with $R = 3.3$ kΩ, $L = 400$ mH, $C = 10$ nF, $V_T = 14$ V, and $f = 3.0$ kHz:

$$X_L = 2\pi fL = 2\pi(3.0 \text{ kHz})(400 \text{ mH}) \approx 7.5 \text{ k}\Omega$$
$$X_C = \frac{1}{2\pi fC} = \frac{1}{2\pi(3.0 \text{ kHz})(10 \text{ nF})} \approx 5.3 \text{ k}\Omega$$
$$Z = R + j(X_L - X_C) = 3.3 + j(7.5 - 5.3) \text{ k}\Omega$$
$$= 3.3 + j2.2 \text{ k}\Omega$$

$X_L > X_C$ and $X = 2.2$ kΩ is inductive. Use formula (21.4) for Z and formula (21.3) for V_T:

$$Z = \sqrt{3.3^2 + 2.2^2} \angle\tan^{-1}\left(\frac{2.2}{3.3}\right) \approx 4.0 \angle 34° \text{ k}\Omega$$
$$I = \frac{V}{Z} = \frac{14 \text{ V}}{4.0 \text{ kHz}} = 3.5 \text{ mA}$$

$$V_R = IR = (3.5 \text{ mA})(3.3 \text{ k}\Omega) \approx 11.6 \text{ V}$$
$$V_L = IX_L = (3.5 \text{ mA})(7.5 \text{ k}\Omega) \approx 26.3 \text{ V}$$
$$V_C = IX_C = (3.5 \text{ mA})(5.3 \text{ k}\Omega) \approx 18.6 \text{ V}$$
$$V_T = 11.6 + j(26.3 - 18.6) \text{ V} = 11.6 + j7.7$$
$$V_T = \sqrt{11.6^2 + 7.7^2} \angle\tan^{-1}\left(\frac{7.7}{11.6}\right) \approx 14 \angle 34° \text{ V}$$

True or Real Power (W)

$$P = I^2R = VI\cos\theta \qquad (21.12a)$$

Reactive Power (VAR)

$$Q = VI\sin\theta \qquad (21.12b)$$

Apparent Power (VA)

$$S = VI \qquad (21.12c)$$

Series resonance is when $X_L = X_C$, and the circuit is purely resistive with $Z = R$ and $\theta = 0°$. The current and true power are at a maximum.

Resonant Frequency (Hz)

$$f_r = \frac{1}{2\pi\sqrt{LC}} \qquad (21.13)$$

REVIEW EXERCISES

Round all answers to two significant digits.

In exercises 1 through 10, using the given values for each series ac circuit, find the indicated quantity.

1. *RL* circuit: $L = 250$ mH, $f = 6.0$ kHz; X_L
2. *RL* circuit: $L = 750$ μH, $f = 55$ kHz; X_L
3. *RC* circuit: $C = 420$ nF, $f = 500$ Hz; X_C
4. *RC* circuit: $C = 700$ pF, $f = 100$ kHz; X_C
5. *RL* circuit: $X_L = 740$ Ω, $L = 60$ mH; f
6. *RL* circuit: $X_L = 1.1$ kΩ, $f = 900$ Hz; L
7. *RC* circuit: $X_C = 4.0$ kΩ, $f = 10$ kHz; C
8. *RC* circuit: $X_C = 850$ Ω, $C = 200$ nF; f
9. *RLC* circuit: $L = 200$ mH, $C = 500$ pF; f_r
10. *RLC* circuit: $L = 600$ mH, $C = 40$ nF; f_r

In exercises 11 through 22, using the given values for each series ac circuit, find the impedance Z in rectangular and polar form.

11. *RL* circuit: $R = 820$ Ω, $X_L = 1.0$ kΩ
12. *RL* circuit: $R = 3.3$ kΩ, $X_L = 1.8$ kΩ
13. *RL* circuit: $R = 150$ Ω, $X_L = 430$ Ω
14. *RL* circuit: $R = 1.0$ kΩ, $X_L = 640$ Ω
15. *RC* circuit: $R = 6.2$ kΩ, $X_C = 8.5$ kΩ
16. *RC* circuit: $R = 7.5$ kΩ, $X_C = 5.0$ kΩ
17. *RC* circuit: $R = 1.8$ kΩ, $X_C = 780$ Ω
18. *RC* circuit: $R = 330$ Ω, $X_C = 440$ Ω
19. *RLC* circuit: $R = 470$ Ω, $X_L = 910$ Ω, $X_C = 360$ Ω
20. *RLC* circuit: $R = 1.1$ kΩ, $X_L = 700$ Ω, $X_C = 1.3$ kΩ

21. *RLC* circuit: $R = 6.8\ \text{k}\Omega$, $X_L = 1.6\ \text{k}\Omega$, $X_C = 4.7\ \text{k}\Omega$

22. *RLC* circuit: $R = 560\ \Omega$, $X_L = 2.0\ \text{k}\Omega$, $X_C = 860\ \Omega$

In exercises 23 through 28, using the given values for a series ac circuit:

(a) Find Z in polar form.
(b) Find I and find V_T in rectangular and polar form.

23. *RL* circuit: $R = 390\ \Omega$, $X_L = 620\ \Omega$, $V_T = 15$ V

24. *RL* circuit: $R = 6.8\ \text{k}\Omega$, $X_L = 750\ \Omega$, $V_T = 14$ V

25. *RC* Circuit: $R = 6.2\ \text{k}\Omega$, $X_C = 5.6\ \text{k}\Omega$, $V_T = 24$ V

26. *RC* Circuit: $R = 4.3\ \text{k}\Omega$, $X_C = 13\ \text{k}\Omega$, $V_T = 8.5$ V

27. *RLC* Circuit: $R = 3.3\ \text{k}\Omega$, $X_L = 4.7\ \text{k}\Omega$, $X_C = 1.2\ \text{k}\Omega$, $V_T = 12$ V

28. *RLC* Circuit: $R = 820\ \Omega$, $X_L = 500\ \Omega$, $X_C = 1.1\ \text{k}\Omega$, $V_T = 36$ V

29. In an *RL* series circuit, $R = 1.8\ \text{k}\Omega$, $f = 900$ Hz, $L = 200$ mH, and $I = 10$ mA.
(a) Find X_L, V_R, and V_L.
(b) Find V_T and Z in polar and rectangular form.

30. In an *RL* series circuit, $R = 20\ \text{k}\Omega$, $f = 2.0$ kHz, $L = 1.6$ H, and $V_T = 10$ V.
(a) Find X_L, V_R, V_L, and I.
(b) Find V_T and Z in polar and rectangular form.

31. In an *RC* series circuit, $R = 3.3\ \text{k}\Omega$, $f = 2.3$ kHz, $C = 15$ nF, and $I = 800\ \mu$A.
(a) Find X_C, V_R, and V_C.
(b) Find V_T and Z in polar and rectangular form.

32. In an *RC* series circuit, $R = 510\ \Omega$, $f = 250$ Hz, $C = 1.0\ \mu$F, and $V_T = 10$ V.
(a) Find X_C, V_R, V_C, and I.
(b) Find V_T and Z in polar and rectangular form.

33. In an *RLC* series circuit, $R = 3.0\ \text{k}\Omega$, $L = 150$ mH, and $C = 10$ nF. The applied voltage $V_T = 32$ V with a frequency $f = 6.0$ kHz.
(a) Find X_L, X_C, V_R, V_L, V_C, and I.
(b) Find V_T and Z in rectangular and polar form.

34. In an *RLC* series circuit, $R = 6.2\ \text{k}\Omega$, $L = 50$ mH, and $C = 750$ pF. The current $I = 5.5$ mA with a frequency $f = 20$ kHz.
(a) Find X_L, X_C, V_R, V_L, and V_C.
(b) Find V_T and Z in rectangular and polar form.

35. In an *RLC* series circuit, $R = 750\ \Omega$, $L = 120$ mH, and $C = 7.5$ nF. The applied voltage $V_T = 30$ V with a frequency $f = 5.5$ kHz.
(a) Find X_L, X_C, V_R, V_L, V_C, and I.
(b) Find V_T and Z in rectangular and polar form.

36. In an *RLC* series circuit, $R = 3.9\ \text{k}\Omega$, $L = 10$ mH, and $C = 10$ nF. The current $I = 10$ mA with a frequency $f = 10$ kHz.
(a) Find X_L, X_C, V_R, V_L, and V_C.
(b) Find V_T and Z in rectangular and polar form.

37. Figure 21-17 shows an *RLC* series circuit containing two capacitors, an inductor, and a resistor connected to a voltage source. The applied voltage $V_T = 9.0$ V, $R = 680\ \Omega$, $X_{C_1} = 500\ \Omega$, $X_{C_2} = 1.6\ \text{k}\Omega$, and $X_L = 1.2\ \text{k}\Omega$.
(a) Find Z in rectangular and polar form and find the current I.

Note: $X_C = X_{C_1} + X_{C_2}$.

(b) Find V_T in rectangular and polar form and find the true power P.

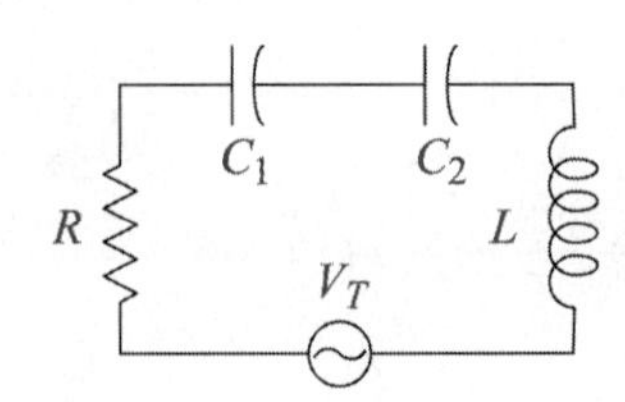

FIGURE 21-17 *RLC* series circuit for problems 37 and 38.

38. In the circuit in Figure 21-17, the current $I = 5.0$ mA, $R = 5.6\ \text{k}\Omega$, $X_{C_1} = 2.7\ \text{k}\Omega$, $X_{C_2} = 2.2\ \text{k}\Omega$, and $X_L = 8.2\ \text{k}\Omega$.
(a) Find V_T and Z in polar and rectangular form.
(b) Find the true power P.

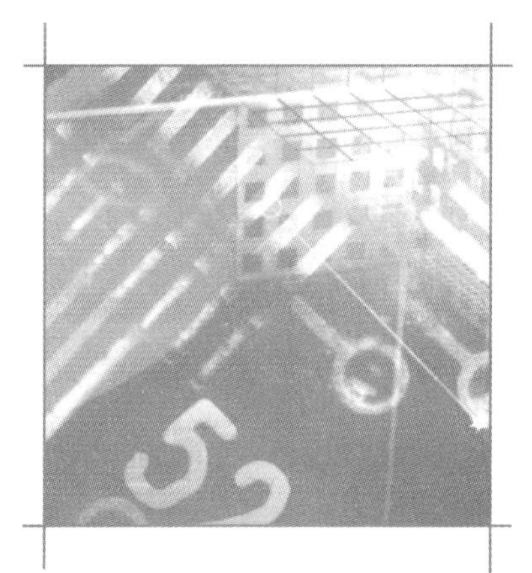

CHAPTER 22

Parallel AC Circuits

CHAPTER OBJECTIVES

In this chapter, you will learn:

- How to find the total current and impedance of an *RL* and an *RC* parallel circuit.
- How to change an *RL* and an *RC* parallel circuit to an equivalent series circuit.
- How to find the conductance, susceptance, and admittance of a parallel ac circuit.
- How to find the total current and impedance of an *RLC* parallel circuit.
- How to find the equivalent impedance of two impedances in parallel.
- How to find the total impedance of an ac network and reduce it to an equivalent series circuit.
- High, low and bandpass filters and how to analyze filter circuits.

Chapter 21 covers series ac circuits containing a resistor, an inductor, and a capacitor. This chapter studies parallel ac circuits containing the same components. The difference between series and parallel ac circuits is the same as the difference between series and parallel dc circuits. In a series ac circuit, the current is the *same* through each component but the voltage *divides* across each component. In a parallel ac circuit, the voltage is the *same* across each component but the current *divides* through each component. The analysis of parallel ac circuits focuses on the currents in the same way as the analysis of series ac circuits focuses on the voltages. When you combine both series and parallel components, you have an ac network. An ac network is analyzed in a similar way as a dc network and can be reduced to an equivalent series circuit.

• • •

22.1 *RL* Circuits

Figure 22-1(a) on page 436 shows an *RL* parallel circuit containing a resistor and an inductor in parallel. The voltage V across the resistor is the same as the voltage V across the inductor. Study the phasor diagram in Figure 22-1(b), which assigns a phase angle of 0° to the voltage V as a reference. The current through the resistor I_R is in phase with V and also has a phase angle of 0°. The voltage across the inductor V leads I_L by 90° ("ELI") which means that I_L lags V and I_R by 90°. I_L therefore has a phase angle of –90°. The total current I_T is the phasor sum of I_R and I_L . I_R is the real component and $-jI_L$ is the negative imaginary component:

Formula

RL Parallel Current

$$I_T = I_R - jI_L = \sqrt{I_R^2 + I_L^2}\ \angle \tan^{-1}\left(\frac{-I_L}{I_R}\right) \tag{22.1}$$

The total impedance of an ac parallel circuit is given by Ohm's law for both scalar and phasor quantities.

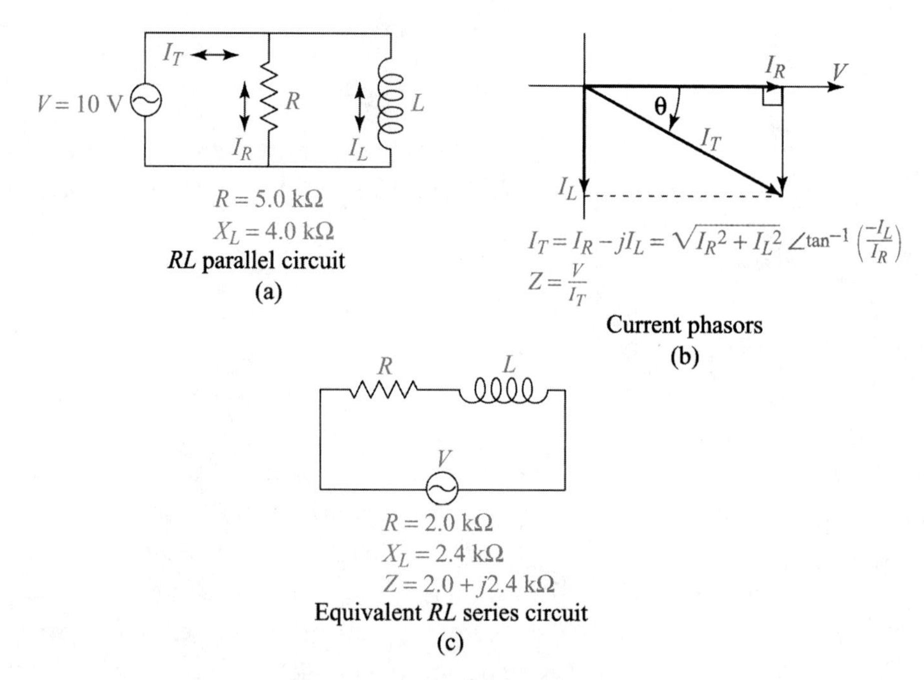

FIGURE 22-1 *RL* parallel circuit and equivalent *RL* series circuit for Example 22.1.

Law

Ohm's Law with I_T (Scalar and Phasor)

$$Z = \frac{V}{I_T} \tag{22.2}$$

EXAMPLE 22.1

In the *RL* parallel circuit in Figure 22-1(a), $V = 10$ V, $R = 5.0$ kΩ, and $X_L = 4.0$ kΩ.

(a) Find I_R and I_L. Find the total current I_T in polar and rectangular form.

(b) Find the impedance Z in polar and rectangular form.

Solution

(a) Find the magnitudes of I_R and I_L using Ohm's law:

$$I_R = \frac{V}{R} = \frac{10\text{ V}}{5.0\text{ k}\Omega} = 2.0\text{ mA}$$

$$I_L = \frac{V}{X_L} = \frac{10\text{ V}}{4.0\text{ k}\Omega} = 2.5\text{ mA}$$

Then apply formula (22.1) to find I_T, which is the phasor sum of I_R and $-I_L$:

$$I_T = 2.0 - j2.5\text{ mA} = \sqrt{2.0^2 + 2.5^2}\ \angle\tan^{-1}\left(\frac{-2.5}{2.0}\right)\text{ mA} \approx 3.2\ \angle{-51^\circ}\text{ mA}$$

Note that the total current I_T has a *negative* phase angle in an *RL* parallel circuit.

(b) Apply Ohm's law (22.2) and the division rule for complex phasors (20.6) to find the total impedance in polar form:

$$Z = \frac{V}{I_T} = \frac{10\ \angle 0^\circ\text{ V}}{3.2\ \angle{-51^\circ}\text{ mA}}$$

$$= \frac{10}{0.0032}\angle[0^\circ - (-51^\circ)]\ \Omega \approx 3.1\ \angle 51^\circ\text{ k}\Omega$$

Note that the phase angle of V is 0°, and the phase angle of the impedance (51°) is the negative of the phase angle of I_T (−51°). The impedance Z therefore has a *positive* phase angle in an *RL* parallel circuit. There is no impedance triangle in a parallel circuit, only in a series circuit. Impedances are linked to voltages, not currents.

To convert Z to rectangular form, use formula (20.4) for phasor components from Chapter 20:

$$Z = 3.1 \angle 51° \text{ k}\Omega = 3.1 \cos 51° + j3.1 \sin 51° \text{ k}\Omega \approx 2.0 + j2.4 \text{ k}\Omega$$

Observe that the imaginary component of Z is positive. This represents an inductive reactance and means that the *RL* parallel circuit in Figure 22-1(a) is equivalent to an *RL* series circuit where $R = 2.0$ kΩ and $X_L = 2.4$ kΩ. See Figure 22-1(c). Any *RL* parallel circuit can be changed to an equivalent *RL* series circuit by this procedure.

EXAMPLE 22.2 Given an *RL* parallel circuit with $R = 1.3$ kΩ, $L = 120$ mH, and $V = 22$ V with $f = 15$ kHz.

(a) Find I_T in rectangular and polar form.

(b) Find Z in polar and rectangular form and the equivalent *RL* series circuit.

Solution

(a) First find the reactance X_L by applying formula (20.1):

$$X_L = 2\pi(15 \text{ kHz})(120 \text{ mH}) \approx 11.3 \text{ k}\Omega$$

Find the currents I_R and I_L:

$$I_R = \frac{V}{R} = \frac{22 \text{ V}}{1.3 \text{ k}\Omega} \approx 16.9 \text{ mA}$$

$$I_L = \frac{V}{X_L} = \frac{22 \text{ V}}{11.3 \text{ k}\Omega} \approx 1.95 \text{ mA}$$

Then find I_T using formula (22.1):

$$I_T = 16.9 - j1.95 \text{ mA} = \sqrt{16.9^2 + 1.95^2} \angle \tan^{-1}\left(\frac{-1.95}{16.9}\right) \text{ mA}$$

$$\approx 17 \angle -6.6° \text{ mA}$$

(b) Find Z in polar form using formula (22.2):

$$Z = \frac{22 \text{ V} \angle 0°}{17 \angle -6.6° \text{ mA}} = \frac{22 \text{ V}}{0.017 \text{ mA}} \angle 0° - (-6.6°) = \approx 1.3 \angle 6.6° \text{ k}\Omega$$

Find Z in rectangular form using formula (20.4):

$$1.3 \angle 6.6° \text{ k}\Omega = 1.3 (\cos 6.6°) + j(1.3 \sin 6.6°) \text{ k}\Omega$$
$$\approx 1.3 + j0.15 \text{ k}\Omega = 1300 + j150 \ \Omega$$

The equivalent series circuit then has $R = 1.3$ kΩ and $X_L = 150$ Ω. Find the inductance of the series circuit using formula (21.1) solved for L:

$$L = \frac{X_L}{2\pi f} = \frac{150 \ \Omega}{2\pi(15 \text{ kHz})} \approx 1.6 \text{ mH}$$

EXERCISE 22.1

Round all answers to two significant digits.

In exercises 1 through 10, using the given values for the RL parallel circuit, find I_T and Z in rectangular and polar form.

1. $V = 12$ V, $R = 7.5$ kΩ, $X_L = 6.5$ kΩ
2. $V = 16$ V, $R = 1.5$ kΩ, $X_L = 3.9$ kΩ
3. $V = 5.0$ V, $R = 1.2$ kΩ, $X_L = 820$ Ω
4. $V = 9.0$ V, $R = 910$ Ω, $X_L = 1.3$ kΩ
5. $V = 30$ V, $R = 3.3$ kΩ, $X_L = 9.5$ kΩ
6. $V = 3.5$ V, $R = 1.2$ kΩ, $X_L = 1.5$ kΩ
7. $V = 20$ V, $R = 10$ kΩ, $X_L = 10$ kΩ
8. $V = 24$ V, $R = 450$ Ω, $X_L = 450$ Ω
9. $V = 13$ V, $R = 510$ Ω, $X_L = 730$ Ω
10. $V = 4.6$ V, $R = 820$ Ω, $X_L = 620$ Ω
11. Given an *RL* parallel circuit with $V = 6.0$ V, $R = 5.6$ kΩ, and $X_L = 4.3$ kΩ.
    (a) Find I_R, I_L, and I_T in polar and rectangular form.
    (b) Find Z in polar and rectangular form.
12. Given an *RL* parallel circuit with $R = 6.8$ kΩ, $X_L = 7.5$ kΩ, and $I_R = 800$ μA.
    (a) Find V, I_L, and I_T in polar and rectangular form.
    (b) Find Z in polar and rectangular form.
13. Given an *RL* parallel circuit with $R = 24$ kΩ, $L = 300$ mH, and $V = 12$ V with a frequency $f = 10$ kHz.
    (a) Find I_T in rectangular and polar form. See Example 22.2.
    (b) Find Z in polar and rectangular form and the equivalent *RL* series circuit.
14. Given an *RL* parallel circuit with $R = 1.8$ kΩ, $L = 100$ mH, and $V = 40$ V with a frequency $f = 2.5$ kHz.
    (a) Find I_T in rectangular and polar form. See Example 22.2.
    (b) Find Z in polar and rectangular form and the equivalent *RL* series circuit.
15. Given an *RL* parallel circuit with $R = 6.8$ kΩ, $L = 200$ mH, $f = 6.0$ kHz, and $I_L = 5.5$ mA.
    (a) Find I_T in rectangular and polar form.
    (b) Find Z in polar and rectangular form and the equivalent *RL* series circuit.
16. Given an *RL* parallel circuit with $R = 470$ Ω, $L = 10$ mH, $f = 12$ kHz, and $I_R = 6.2$ mA.
    (a) Find I_T in rectangular and polar form.
    (b) Find Z in polar and rectangular form and the equivalent *RL* series circuit.

22.2 *RC* Circuits

Figure 22-2(a) on page 439 shows an *RC* parallel circuit containing a resistor and a capacitor in parallel. The voltage V is the *same* across the resistor and the capacitor. The phase angle of V is assigned a reference angle of 0° for the phasor diagram in Figure 22-2(b). The current through the resistor I_R is in phase with V and also has a phase angle of 0°. The current through the capacitor I_C leads V by 90° ("ICE") which means that I_C also leads I_R by 90°. The total current I_T is the phasor sum of I_R and I_C where I_R is the real component and jI_C is the positive imaginary component:

Formula ***RC* Parallel Current**

$$I_T = I_R + jI_C = \sqrt{I_R^2 + I_C^2}\ \angle \tan^{-1}\left(\frac{I_C}{I_R}\right) \tag{22.3}$$

As in the parallel *RL* circuit, the phase angle of Z is the negative of the phase angle of I_T. Since I_T has a positive phase angle, Z_T *has a negative phase angle,* as in a series *RC* circuit.

EXAMPLE 22.3 In the *RC* parallel circuit in Figure 22-2(a), $V = 12$ V, $R = 750$ Ω, and $X_C = 800$ Ω.

(a) Find I_R and I_C. Find I_T in polar and rectangular form.

(b) Find Z in polar and rectangular form.

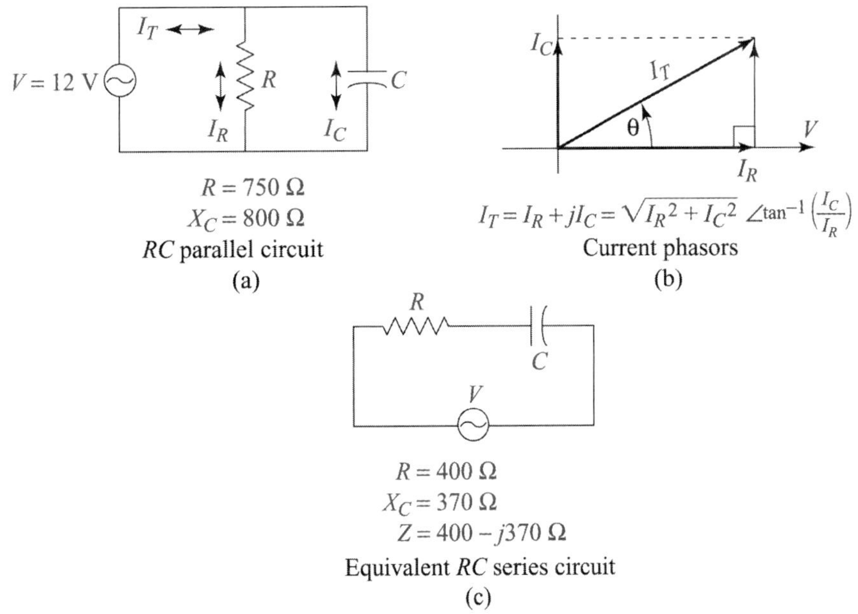

$R = 750\ \Omega$
$X_C = 800\ \Omega$
RC parallel circuit
(a)

$I_T = I_R + jI_C = \sqrt{I_R^2 + I_C^2}\ \angle \tan^{-1}\left(\frac{I_C}{I_R}\right)$
Current phasors
(b)

$R = 400\ \Omega$
$X_C = 370\ \Omega$
$Z = 400 - j370\ \Omega$
Equivalent *RC* series circuit
(c)

FIGURE 22-2 *RC* parallel circuit and equivalent *RC* series circuit for Example 22.3.

Solution

(a) To find the currents I_R and I_C, apply Ohm's law:

$$I_R = \frac{V}{R} = \frac{12\ \text{V}}{750\ \Omega} = 16\ \text{mA}$$

$$I_C = \frac{V}{X_C} = \frac{12\ \text{V}}{800\ \Omega} = 15\ \text{mA}$$

Apply formula (22.3) to find the total current:

$$I_T = 16 + j15\ \text{mA} = \sqrt{16^2 + 15^2}\ \angle \tan^{-1}\left(\frac{15}{16}\right)\ \text{mA} \approx 22\ \angle 43°\ \text{mA}$$

Note that the phase angle of I_T is a *positive* value for an *RC* parallel circuit.

(b) Use Ohm's law to find Z in polar form:

$$Z = \frac{V}{I_T} = \frac{12\ \angle 0°\ \text{V}}{22\ \angle 43°\ \text{mA}} = \frac{12}{0.022}\angle(0°–43°)\ \Omega \approx 545\ \angle{-43°}\ \Omega$$

Note that the phase angle of Z is a *negative* value for an *RC* parallel circuit. In rectangular form, Z is then:

$$Z = 545\cos(-43°) + j545\sin(-43°)\ \Omega \approx 400 - j370\ \Omega$$

Since the imaginary component of Z is negative, it represents a capacitive reactance. The *RC* parallel circuit in Figure 22-2(a) is equivalent to a series *RC* circuit with $R = 400\ \Omega$ and $X_C = 370\ \Omega$ as shown in Figure 22-2(c). Any *RC* parallel circuit can be changed to an equivalent *RC* series circuit.

EXAMPLE 22.4

Given an *RC* parallel circuit with $R = 1.3\ \text{k}\Omega$, $C = 10\ \text{nF}$, $I_C = 6.3\ \text{mA}$, and $f = 10\ \text{kHz}$.

(a) Find V and I_R. Find I_T in polar and rectangular form.

(b) Find Z and the equivalent *RC* series circuit.

Solution

(a) Using formula (21.6) find X_C:

$$X_C = \frac{1}{2\pi fC} = \frac{1}{2\pi(10 \text{ kHz})(10 \text{ nF})} \approx 1.6 \text{ k}\Omega$$

To find V and I_R, apply Ohm's law:

$$V = I_C X_C = (6.3 \text{ mA})(1.6 \text{ k}\Omega) \approx 10 \text{ V}$$

$$I_R = \frac{V}{R} = \frac{10 \text{ V}}{1.3 \text{ k}\Omega} \approx 7.7 \text{ mA}$$

The total current in rectangular and polar form is then:

$$I_T = 7.7 + j6.3 \text{ mA} = \sqrt{7.7^2 + 6.3^2} \angle \tan^{-1}\left(\frac{6.3}{7.7}\right) \text{ mA} \approx 9.9 \angle 39° \text{ mA}$$

(b) Find Z in polar form using Ohm's law:

$$Z = \frac{V}{I_T} = \frac{10 \angle 0° \text{ V}}{9.9 \angle 39° \text{ mA}} \approx 1.0 \angle -39° \text{ k}\Omega$$

The impedance in rectangular form is:

$$Z = 1.0 \cos(-39°) + j1.0 \sin(-39°) \text{ k}\Omega \approx 0.78 - j0.63 \text{ k}\Omega$$
$$= 780 - j630 \text{ }\Omega$$

The equivalent RC series circuit then has $R = 780$ Ω and $X_C = 630$ Ω. To find the equivalent value of the capacitance, use formula (22.6) solved for C:

$$C = \frac{1}{2\pi f X_C} = \frac{1}{2\pi(10 \text{ kHz})(630 \text{ }\Omega)} \approx 25 \text{ nF}$$

ERROR BOX

A common error when working with parallel circuits is confusing the phase angles for impedance and current. An inductive reactance X_L has a *positive* phase angle, but an inductive current has a *negative* phase angle. A capacitive reactance X_C has a *negative* phase angle, but a capacitive current has a *positive* phase angle. See if you can find the correct phase angles in the practice problems.

Practice Problems: An ac parallel circuit with an applied voltage of 12 V contains a resistance $R = 7.5$ kΩ in parallel with a reactance X. For each given value of X, find I_T and Z in polar form.

1. $X_L = 10$ kΩ **2.** $X_C = 10$ kΩ **3.** $X_L = 3.0$ kΩ **4.** $X_C = 3.0$ kΩ
5. $X_L = 960$ Ω **6.** $X_C = 960$ Ω **7.** $X_C = 22$ kΩ **8.** $X_L = 22$ kΩ

EXERCISE 22.2

Round all answers to two significant digits.

In exercises 1 through 10, using the given values for an RC *parallel circuit, find the total current* I_T *and the impedance* Z *in rectangular and polar form.*

1. $V = 25$ V, $R = 3.0$ kΩ, $X_C = 2.0$ kΩ
2. $V = 20$ V, $R = 5.6$ kΩ, $X_C = 8.2$ kΩ
3. $V = 30$ V, $R = 620$ Ω, $X_C = 470$ Ω
4. $V = 10$ V, $R = 560$ Ω, $X_C = 770$ Ω
5. $V = 6.0$ V, $R = 1.1$ kΩ, $X_C = 1.2$ kΩ
6. $V = 12$ V, $R = 11$ kΩ, $X_C = 10$ kΩ

7. $V = 24$ V, $R = 30$ kΩ, $X_C = 47$ kΩ
8. $V = 9.0$ V, $R = 13$ kΩ, $X_C = 11$ kΩ
9. $V = 15$ V, $R = 750$ Ω, $X_C = 1.0$ kΩ
10. $V = 16$ V, $R = 1.1$ kΩ, $X_C = 630$ Ω
11. Given an *RC* parallel circuit with $V = 15$ V, $R = 2.1$ kΩ, and $X_C = 1.2$ kΩ.
    (a) Find I_R and I_C. Find I_T in polar and rectangular form.
    (b) Find Z in polar and rectangular form.
12. Given an *RC* parallel circuit with $I_C = 2.2$ mA, $R = 8.2$ kΩ, and $X_C = 5.6$ kΩ.
    (a) Find V and I_R. Find I_T in polar and rectangular form.
    (b) Find Z in polar and rectangular form.
13. Given an *RC* parallel circuit with $R = 680$ Ω, C = 25 nF, $I_C = 9.5$ mA, and $f = 15$ kHz.
    (a) Find V and I_R. Find I_T in polar and rectangular form.
    (b) Find Z in polar and rectangular form and the equivalent *RC* series circuit.
14. Given an *RC* parallel circuit with $R = 6.8$ kΩ, C = 800 pF, $I_R = 900$ μA, and $f = 30$ kHz.
    (a) Find V and I_C. Find I_T in polar and rectangular form.
    (b) Find Z in polar and rectangular form and the equivalent *RC* series circuit.
15. Given an *RC* parallel circuit with $V = 18$ V, $R = 620$ Ω, C = 40 nF, and $f = 5.0$ kHz.
    (a) Find I_R and I_C. Find I_T in polar and rectangular form. See Example 22.4.
    (b) Find Z in polar and rectangular form and the equivalent *RC* series circuit.
16. Given an *RC* parallel circuit with $V = 6.0$ V, $R = 2.4$ kΩ, C = 500 pF, and $f = 20$ kHz.
    (a) Find I_R and I_C. Find I_T in polar and rectangular form. See Example 22.4.
    (b) Find Z in polar and rectangular form and the equivalent *RC* series circuit.

22.3 Susceptance and Admittance

In parallel dc circuits, conductance G is defined as the reciprocal of resistance and is measured in siemens (S):

$$G = \frac{1}{R}$$

Similarly, in parallel ac circuits, the phasors *susceptance B* and *admittance Y* are defined as reciprocals of reactance and impedance respectively, and are measured in siemens:

Formula **Susceptance (S)**

$$B = \frac{1}{X} \tag{22.4}$$

• • •

Formula **Admittance (S)**

$$Y = \frac{1}{Z} = \frac{1}{R} \pm j\frac{1}{X} = G \pm jB \tag{22.5}$$

Inductive $(-jB)$ Capacitive $(+jB)$

Because of the reciprocal relation for B in formula (22.4), the imaginary component jB in formula (22.5) is negative for an inductive reactance and positive for a capacitive reactance. Note that Y has the same phase angle as the current in a parallel circuit. Study the following examples, which consider the *RL* and *RC* circuits from Sections 22.1 and 22.2.

EXAMPLE 22.5 Given the *RL* parallel circuit of Example 22.1 where $R = 5.0$ kΩ and $X_L = 4.0$ kΩ.

(a) Find the conductance G and the magnitude of the susceptance B.

(b) Find the admittance Y in polar and rectangular form.

Solution

(a) To find G and B, find the reciprocals of R and X_L:

$$G = \frac{1}{R} = \frac{1}{5.0\ \text{k}\Omega} = 200\ \mu\text{S}$$

$$B = \frac{1}{X_L} = \frac{1}{4.0\ \text{k}\Omega} = 250\ \mu\text{S}$$

(b) To find the admittance Y, apply formula (22.5) making jB negative for the inductive reactance:

$$Y = G \pm jB = 200 - j250\ \mu\text{S}$$
$$= \left(\sqrt{250^2 + 200^2}\right) \angle \tan^{-1}\left(\frac{-250}{200}\right)\ \mu\text{S} \approx 320\ \angle{-51^\circ}\ \mu\text{S}$$

Note that the phase angle of Y is the same phase angle as I_T in Example 22.1. Also note that it is not necessary to know the voltage or currents in the ac circuit to find the admittance.

EXAMPLE 22.6 Given the RC parallel circuit of Example 22.3 where $R = 750\ \Omega$ and $X_C = 800\ \Omega$.

(a) Find the conductance G and the magnitude of the susceptance B.

(b) Find the admittance Y in polar and rectangular form.

Solution

(a) To find G and B, calculate the reciprocals of R and X_C:

$$G = \frac{1}{R} = \frac{1}{750\ \Omega} \approx 1.33\ \text{mS}$$

$$B = \frac{1}{X_C} = \frac{1}{800\ \Omega} = 1.25\ \text{mS}$$

(b) To find Y, apply formula (22.5) making jB positive for the capacitive reactance:

$$Y = G + jB = 1.33 + j1.25\ \text{mS}$$
$$= \left(\sqrt{1.33^2 + 1.25^2}\right) \angle \tan^{-1}\left(\frac{1.25}{1.33}\right)\ \text{mS} \approx 1.8\ \angle 43^\circ\ \text{mS}$$

Note that the phase angle of Y is the same phase angle as I_T in Example 22.3.

EXAMPLE 22.7 Given the RC parallel circuit of Example 22.4 where $R = 1.3\ \text{k}\Omega$, $C = 10$ nF, and $f =$ 10 kHz.

(a) Find Y and Z in polar and rectangular form.

(b) Find the equivalent RC series circuit.

Answers to Error Box Problems, page 440:

1. $I_T = 2.0\ \angle{-37^\circ}$ mA, $Z = 6.0\ \angle 37^\circ$ kΩ
2. $I_T = 2.0\ \angle 37^\circ$ mA, $Z = 6.0\ \angle{-37^\circ}$ kΩ
3. $I_T = 4.3\ \angle{-68^\circ}$ mA, $Z = 2.8\ \angle 68^\circ$ kΩ
4. $I_T = 4.3\ \angle 68^\circ$ mA, $Z = 2.8\ \angle{-68^\circ}$ kΩ
5. $I_T = 13\ \angle{-83^\circ}$ mA, $Z = 950\ \angle 83^\circ$ Ω
6. $I_T = 13\ \angle 83^\circ$ mA, $Z = 950\ \angle{-83^\circ}$ Ω
7. $I_T = 1.7\ \angle 19^\circ$ mA, $Z = 7.1\ \angle{-19^\circ}$ kΩ
8. $I_T = 1.7\ \angle{-19^\circ}$ mA, $Z = 7.1\ \angle 19^\circ$ kΩ

Solution In Example 22.4, Z is found by first calculating the currents. The solution using admittance is a more direct way to find the impedance, and it is not necessary to know the currents or voltage. To find Y, apply formula (22.5) where jB is positive, using $R = 1.3\ \text{k}\Omega$ and $X_C = 1.6\ \text{k}\Omega$ from Example 22.4:

$$Y = G + jB = \frac{1}{R} + j\frac{1}{X_C} = \frac{1}{1.3\ \text{k}\Omega} + j\frac{1}{1.6\ \text{k}\Omega} \approx 770 + j630\ \mu\text{S}$$

$$= \sqrt{770^2 + 630^2}\ \angle\tan^{-1}\left(\frac{630}{770}\right)\ \mu\text{S} \approx 995\ \angle 39°\ \mu\text{S}$$

(a) To find Z, note that since $Y = \frac{1}{Z}$, it follows that $Z = \frac{1}{Y}$. To perform the division in polar form, *the real number 1 is treated as a phasor with a phase angle of* 0°:

$$Z = \frac{1\ \angle 0°}{995\ \angle 39°\ \mu\text{S}} = \frac{1}{0.000995}\angle(0°–39°)\ \Omega \approx 1.0\ \angle{-39°}\ \text{k}\Omega$$

(b) Change Z to rectangular form to find the equivalent RC series circuit:

$$Z = 1.0\ \angle{-39°}\ \text{k}\Omega = 1000 \cos(-39°) + j1000 \sin(-39°)\ \Omega \approx 780 - j630\ \Omega$$

The equivalent RC series circuit has $R = 780\ \Omega$ and $X_C = 630\ \Omega$. This is the same result as Example 22.4.

EXERCISE 22.3

Round all answers to two significant digits.

In exercises 1 through 8, using the given values for an RL *parallel circuit, find* G *and the magnitude of* B. *Find the admittance* Y *in rectangular and polar form.*

1. $R = 7.5\ \text{k}\Omega, X_L = 6.5\ \text{k}\Omega$

2. $R = 1.5\ \text{k}\Omega, X_L = 3.9\ \text{k}\Omega$

3. $R = 1.2\ \text{k}\Omega, X_L = 820\ \Omega$

4. $R = 910\ \Omega, X_L = 1.3\ \text{k}\Omega$

5. $R = 10\ \text{k}\Omega, X_L = 10\ \text{k}\Omega$

6. $R = 450\ \Omega, X_L = 450\ \Omega$

7. $R = 510\ \Omega, X_L = 730\ \Omega$

8. $R = 820\ \Omega, X_L = 620\ \Omega$

For exercises 9 through 16, using the given values for an RC *parallel circuit, find* G *and the magnitude of* B. *Find the admittance* Y *in rectangular and polar form.*

9. $R = 3.0\ \text{k}\Omega, X_C = 2.0\ \text{k}\Omega$

10. $R = 5.6\ \text{k}\Omega, X_C = 8.2\ \text{k}\Omega$

11. $R = 620\ \Omega, X_C = 470\ \Omega$

12. $R = 560\ \Omega, X_C = 770\ \Omega$

13. $R = 30\ \text{k}\Omega, X_C = 47\ \text{k}\Omega$

14. $R = 13\ \text{k}\Omega, X_C = 11\ \text{k}\Omega$

15. $R = 750\ \Omega, X_C = 1.0\ \text{k}\Omega$

16. $R = 1.1\ \text{k}\Omega, X_C = 630\ \Omega$

17. Given an RL parallel circuit with $R = 24\ \text{k}\Omega$, $L = 300$ mH, and $f = 10$ kHz.
(a) Find Y and Z in polar and rectangular form.
(b) Find the equivalent RL series circuit.

18. Given an RL parallel circuit with $R = 6.8\ \text{k}\Omega$, $L = 200$ mH, and $f = 6.0$ kHz.
(a) Find Y and Z in polar and rectangular form.
(b) Find the equivalent RL series circuit.

19. Given an RC parallel circuit with $R = 680\ \Omega$, $C = 25$ nF, and $f = 15$ kHz.
(a) Find Y and Z in polar and rectangular form.
(b) Find the equivalent RC series circuit.

20. Given an RC parallel circuit with $R = 620\ \Omega$, $C = 40$ nF, and $f = 5.0$ kHz.
(a) Find Y and Z in polar and rectangular form.
(b) Find the equivalent RC series circuit.

22.4 *RLC* Circuits

When an inductor and a capacitor are connected in parallel, the current through the inductor tends to oppose the current through the capacitor. The currents oppose each other because the phase angle of I_L is –90° and the phase angle of I_C is 90°. The phasors therefore have opposite direction and subtract from each other. Consider the *RLC* parallel circuit in Figure 22-3.

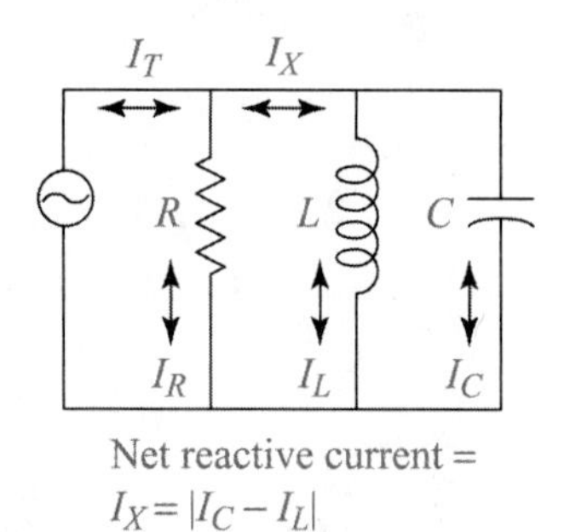

FIGURE 22-3 *RLC* parallel circuit.

The total current of the *RLC* circuit in rectangular form is:

Formula

RLC Parallel Current

$$I_T = I_R + j(I_C - I_L) = I_R \pm jI_X \quad I_X = |I_C - I_L| \tag{22.6}$$

Inductive ($-jI_X$) Capacitive ($+jI_X$)

I_X is the magnitude of the net reactive current. When $I_L > I_C$, I_X is inductive, and the imaginary term jI_X is negative. See Figure 22-4(a). When $I_C > I_L$, I_X is capacitive, and jI_X is positive. See Figure 22-4(b). The current I_T can be changed to polar form by using formula (22.1) when I_X is inductive and by using formula (22.3) when I_X is capacitive.

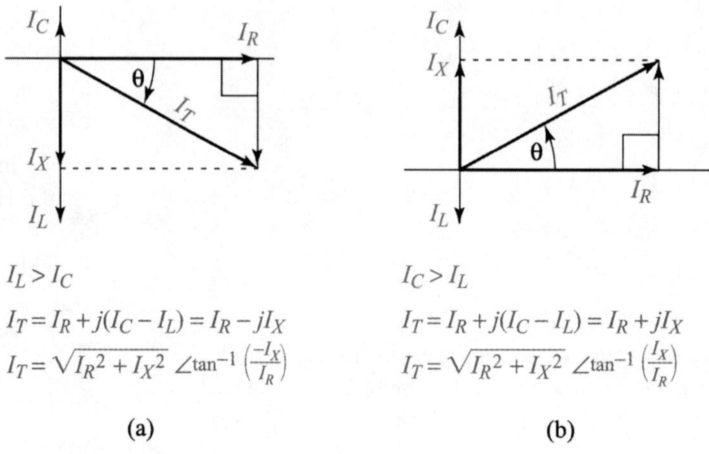

FIGURE 22-4 Current phasors in an *RLC* parallel circuit.

EXAMPLE 22.8 In the *RLC* circuit in Figure 22-3, I_R = 10 mA, I_L = 5.0 mA, and I_C = 15 mA. Find I_T in polar and rectangular form.

Solution Apply (22.6) to find I_T in rectangular form:

$$I_T = 10 + j(15 - 5.0) \text{ mA} = 10 + j10 \text{ mA}$$

The capacitive current I_C is greater than I_L so I_X is capacitive, and the imaginary term jI_X is positive. The *RLC* circuit is equivalent to an *RC* parallel circuit where $I_R = 10$ mA and $I_C = I_X = 10$ mA. Apply (22.3) to find the total current in polar form:

$$I_T = \sqrt{I_R^2 + I_C^2}\ \angle\tan^{-1}\left(\frac{I_C}{I_R}\right) = \sqrt{10^2 + 10^2}\ \angle\tan^{-1}\left(\frac{10}{10}\right)\text{ mA} \approx 14\ \angle 45°\text{ mA}$$

EXAMPLE 22.9 In the *RLC* parallel circuit in Figure 22-5(a), $R = 1.3$ kΩ, $L = 150$ mH, $C = 300$ nF, and the voltage $V = 30$ V with a frequency $f = 600$ Hz.

(a) Find I_T and Z in polar and rectangular form.

(b) Find the equivalent *RL* or *RC* series circuit.

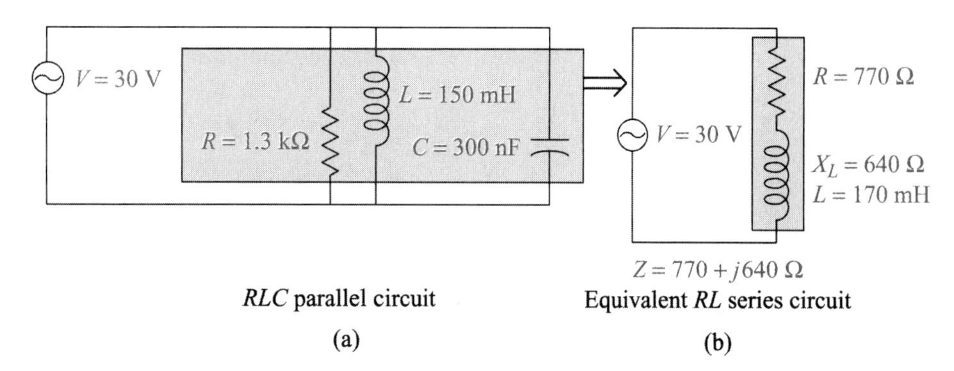

FIGURE 22-5 *RLC* parallel circuit and equivalent *RL* series circuit for Example 22.9.

Solution

(a) Find X_L and X_C using formulas (21.1) and (21.6):

$$X_L = 2\pi fL = 2\pi(600\text{ Hz})(150\text{ mH}) \approx 565\ \Omega$$

$$X_C = \frac{1}{2\pi fC} = \frac{1}{2\pi(600\text{ Hz})(300\text{ nF})} \approx 884\ \Omega$$

Apply Ohm's law to find the currents through the resistor, inductor, and capacitor:

$$I_R = \frac{V}{R} = \frac{30\text{ V}}{1.3\text{ k}\Omega} \approx 23\text{ mA}$$

$$I_L = \frac{V}{X_L} = \frac{30\text{ V}}{565\ \Omega} \approx 53\text{ mA}$$

$$I_C = \frac{V}{X_C} = \frac{30\text{ V}}{884\ \Omega} \approx 34\text{ mA}$$

Apply (22.6) to find the total current in rectangular form:

$$I_T = I_R + j(I_C - I_L) = 23 + j(34 - 53)\text{ mA} = 23 - j19\text{ mA}$$

Since $I_L > I_C$, I_X is inductive and jI_X is negative. The *RLC* parallel circuit is equivalent to an *RL* parallel circuit where $I_R = 23$ mA and $I_L = I_X = 19$ mA. Apply (22.1) to change I_T to polar form using $I_L = 19$ mA:

$$I_T = \sqrt{I_R^2 + I_L^2}\ \angle\tan^{-1}\left(\frac{-I_L}{I_R}\right)$$

$$= \sqrt{23^2 + 19^2}\ \angle\tan^{-1}\left(\frac{-19}{23}\right)\text{ mA} \approx 30\ \angle -40°\text{ mA}$$

To find Z use Ohm's law with phasors:

$$Z = \frac{V}{I_T} = \frac{30 \angle 0° \text{ V}}{30 \angle -40° \text{ mA}} = 1.0 \angle 40° \text{ k}\Omega$$
$$= 1000 \cos 40° + j1000 \sin 40° \ \Omega \approx 770 + j640 \ \Omega$$

(b) The equivalent series circuit is an RL series circuit where $R = 770\ \Omega$ and $X_L = 640\ \Omega$. See Figure 22-5(b). The value of the inductance L in the equivalent series circuit is found using formula (21.1) solved for L:

$$L = \frac{X_L}{2\pi f} = \frac{640\ \Omega}{2\pi(600 \text{ Hz})} \approx 170 \text{ mH}$$

In an RLC parallel circuit the admittance is given by:

Formula ***RLC* Admittance (S)**

$$Y = \frac{1}{R} + j\left(\frac{1}{X_C} - \frac{1}{X_L}\right) = G + j(B_C - B_L) = G \pm jB \qquad B = |B_C - B_L| \qquad (22.7)$$

Inductive ($-jB$) Capacitive ($+jB$)

B is the magnitude of the net susceptance. When $B_C > B_L$, the imaginary term jB is positive, and the net reactive current is capacitive. When $B_L > B_C$, the term jB is negative, and the net reactive current is inductive.

EXAMPLE 22.10 In Example 22.9, find I_T and Z in polar and rectangular form using the concept of admittance.

Solution Using X_L and X_C calculated in Example 22.9, apply (22.7) to find the admittance of the RLC circuit in Figure 22-5(a):

$$Y = \frac{1}{R} + j\left(\frac{1}{X_C} - \frac{1}{X_L}\right) = \frac{1}{1300\ \Omega} + j\left(\frac{1}{884\ \Omega} - \frac{1}{565\ \Omega}\right) \approx 770 - j640\ \mu\text{S}$$
$$= \sqrt{770^2 + 640^2}\ \angle \tan^{-1}\left(\frac{-640}{770}\right)\ \mu\text{S} \approx 1.0 \angle 40° \text{ mS}$$

The imaginary term and the phase angle are negative, which means the net reactive current is inductive. Find Z by taking the reciprocal of Y:

$$Z = \frac{1}{Y} = \frac{1 \angle 0°}{1.0 \angle -40° \text{ mS}} = 1.0 \angle 40° \text{ k}\Omega \approx 770 + j640\ \Omega$$

This is the same value of Z found by the method of branch currents in Example 22.9. To find I_T, note by Ohm's law that:

$$I = \frac{V}{Z} = V\left(\frac{1}{Z}\right) = VY$$

Therefore:

$$I_T = VY = (30 \text{ V})(770 - j640\ \mu\text{S}) \approx 23 - j19 \text{ mA}$$

Applying the procedures shown in the above examples, any *RLC* parallel circuit can be reduced to an *RL* or *RC* parallel circuit, and an equivalent *RL* or *RC* series circuit can be found for the parallel circuit.

EXERCISE 22.4

Round all answers to two significant digits.

In exercises 1 through 10, using the given values for an RLC *parallel circuit, find* I_T *in rectangular and polar form.*

1. $I_R = 30$ mA, $I_L = 80$ mA, $I_C = 40$ mA
2. $I_R = 50$ mA, $I_L = 100$ mA, $I_C = 60$ mA
3. $I_R = 55$ mA, $I_L = 30$ mA, $I_C = 90$ mA
4. $I_R = 65$ mA, $I_L = 25$ mA, $I_C = 55$ mA
5. $I_R = 700$ μA, $I_L = 400$ μA, $I_C = 800$ μA
6. $I_R = 500$ μA, $I_L = 600$ μA, $I_C = 300$ μA
7. $I_R = 1.1$ mA, $I_L = 7.5$ mA, $I_C = 6.6$ mA
8. $I_R = 8.5$ mA, $I_L = 5.0$ mA, $I_C = 10$ mA
9. $I_R = 5.0$ mA, $I_L = 45$ mA, $I_C = 45$ mA
10. $I_R = 12$ mA, $I_L = 32$ mA, $I_C = 32$ mA

In exercises 11 through 20, using the given values for an RLC *parallel circuit, find* I_T *and Z in rectangular and polar form.*

11. $V = 12$ V, $R = 390\ \Omega$, $X_L = 430\ \Omega$, $X_C = 750\ \Omega$
12. $V = 32$ V, $R = 1.5$ kΩ, $X_L = 3.6$ kΩ, $X_C = 6.5$ kΩ
13. $V = 36$ V, $R = 1.6$ kΩ, $X_L = 2.2$ kΩ, $X_C = 1.2$ kΩ
14. $V = 6.3$ V, $R = 11$ kΩ, $X_L = 24$ kΩ, $X_C = 13$ kΩ
15. $V = 3.5$ V, $R = 910\ \Omega$, $X_L = 1.5$ kΩ, $X_C = 1.5$ kΩ
16. $V = 40$ V, $R = 20\ \Omega$, $X_L = 15$ kΩ, $X_C = 15$ kΩ
17. $V = 24$ V, $R = 1.4$ kΩ, $X_L = 510\ \Omega$, $X_C = 1.5$ kΩ
18. $V = 5.5$ V, $R = 820\ \Omega$, $X_L = 660\ \Omega$, $X_C = 1.0$ kΩ
19. $V = 6.0$ V, $R = 10$ kΩ, $X_L = 20$ kΩ, $X_C = 5.6$ kΩ
20. $V = 14$ V, $R = 2.2$ kΩ, $X_L = 2.7$ kΩ, $X_C = 2.7$ kΩ

For exercises 21 through 30, find the admittance Y *in rectangular and polar form for each of the parallel circuits in exercises 11 through 20, respectively.*

31. Given an *RLC* parallel circuit where $V = 10$ V, $f = 20$ kHz, $R = 4.3$ kΩ, $L = 10$ mH, and $C = 5.0$ nF.
    (a) Find I_T and Z in polar and rectangular form. See Example 22.9.
    (b) Find the equivalent *RL* or *RC* series circuit.
32. In problem 31:
    (a) Find Y in polar and rectangular form.
    (b) Using Y, find the equivalent *RL* or *RC* series circuit. See Example 22.10.
33. Given an *RLC* parallel circuit where $V = 15$ V, $f = 50$ kHz, $R = 6.8$ kΩ, $L = 30$ mH, and $C = 500$ pF.
    (a) Find I_T and Z in polar and rectangular form. See Example 22.9.
    (b) Find the equivalent *RL* or *RC* series circuit.
34. In problem 33:
    (a) Find Y in polar and rectangular form.
    (b) Using Y, find the equivalent *RL* or *RC* series circuit. See Example 22.10.
35. Given the *RLC* parallel circuit in Figure 22-6(a) where $R_1 = 8.2$ kΩ, $R_2 = 15$ kΩ, $X_L = 5.6$ kΩ, $X_C = 10$ kΩ, and $f = 12$ kHz.
    (a) Find Y and Z in rectangular form. First find R_{eq} for the parallel resistors in Figure 22-6(b).
    (b) Find the equivalent *RL* or *RC* series circuit.

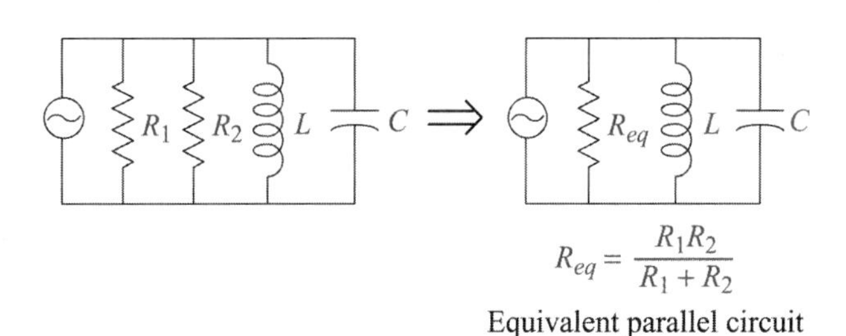

FIGURE 22-6 *RLC* parallel circuit for problems 35 and 36.

36. Given the *RLC* parallel circuit in Figure 22-6(a) where $R_1 = 1.2$ kΩ, $R_2 = 1.5$ kΩ, $L = 100$ mH, $C = 500$ nF, and $f = 1.0$ kHz.
    (a) Find Y and Z in rectangular form. First find R_{eq} for the parallel resistors in Figure 22-6(b).
    (b) Find the equivalent *RL* or *RC* series circuit.
37. In an *RLC* parallel circuit, $R = 1.5$ kΩ, $L = 50$ mH, $C = 10$ nF, $f = 5.0$ kHz, and the total current $I_T = 12$ mA.
    (a) Find Y and Z in polar form.
    (b) Find the applied voltage V.
38. In an *RLC* parallel circuit, $R = 3.3$ kΩ, $L = 60$ mH, $C = 100$ nF, $f = 2.5$ kHz, and the applied voltage $V = 25$ V.
    (a) Find Y and Z in polar form.
    (b) Find the total current I_T in polar form.

22.5 AC Networks

The formulas that apply to series-parallel ac circuits or ac networks are similar to those for a dc circuit. The total impedance of two or more impedances in series is equal to their sum:

Formula **Series Impedances**

$$Z_T = Z_1 + Z_2 + Z_3 + \cdots \tag{22.8}$$

The equivalent impedance of two impedances in parallel is equal to their product divided by their sum:

Formula **Parallel Impedances**

$$Z_{eq} = \frac{Z_1 Z_2}{Z_1 + Z_2} \tag{22.9}$$

EXAMPLE 22.11 Figure 22-7(a) shows an ac network consisting of a resistance in series with an inductance and a capacitance in parallel. If $R = 1.5$ kΩ, $X_L = 470$ Ω, and $X_C = 620$ Ω, find the total impedance of the circuit in rectangular and polar form.

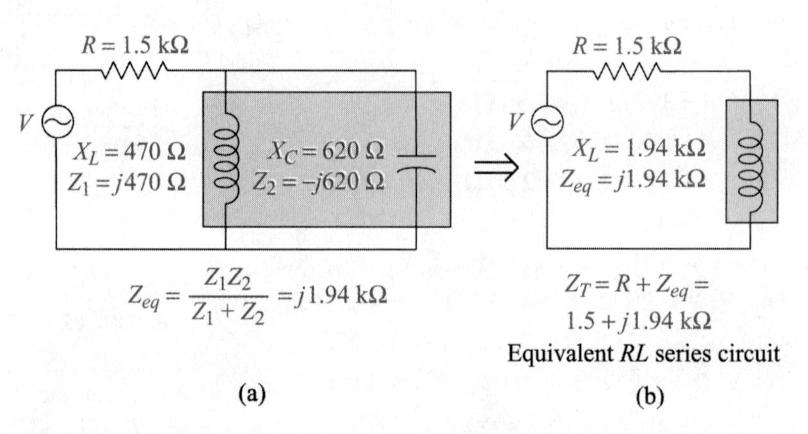

FIGURE 22-7 AC network for Example 22.11.

Solution First express each reactance as a complex phasor in rectangular form:

$$Z_1 = 0 + jX_L = 0 + j470\ \Omega = j470\ \Omega$$
$$Z_2 = 0 - jX_C = 0 - j620\ \Omega = -j620\ \Omega$$

Then apply (22.9) to Z_1 and Z_2 in parallel and find the equivalent impedance Z_{eq}:

$$Z_{eq} = \frac{(j470)(-j620)}{(j470) + (-j620)} = \frac{-j^2 291{,}400}{-j150}$$

You can simplify the fraction by dividing the 150 into 291,400, and the $-j$ into $-j^2$ as follows:

$$Z_{eq} = \frac{-j^2 291{,}400}{-j150} = \frac{(j)(\cancel{-j})291{,}400}{(\cancel{-j})150} \approx j1.94\ \text{k}\Omega$$

The equivalent impedance has a positive imaginary component and is inductive. Find the total impedance Z_T in rectangular form by adding R to Z_{eq}, and then change to polar form:

$$Z_T = R + Z_{eq} = 1.5 + j1.94 \text{ k}\Omega \approx 2.5 \angle 52° \text{ k}\Omega$$

The ac network is therefore equivalent to an RL series circuit where $R = 1.5$ kΩ and $X_L = 1.94$ kΩ. See Figure 22-7(b).

Observe that two phasors can be multiplied and divided in polar form but must be added in rectangular form. Therefore, when using formula (22.9), it may be necessary to work in both polar and rectangular forms as the next example shows.

EXAMPLE 22.12 Figure 22-8(a) shows an important ac network containing a capacitor in parallel with a resistor and an inductor in series. If $f = 10$ kHz, $R = 150$ Ω, $L = 10$ mH, and $C = 20$ nF, find the total impedance of the circuit in polar and rectangular form.

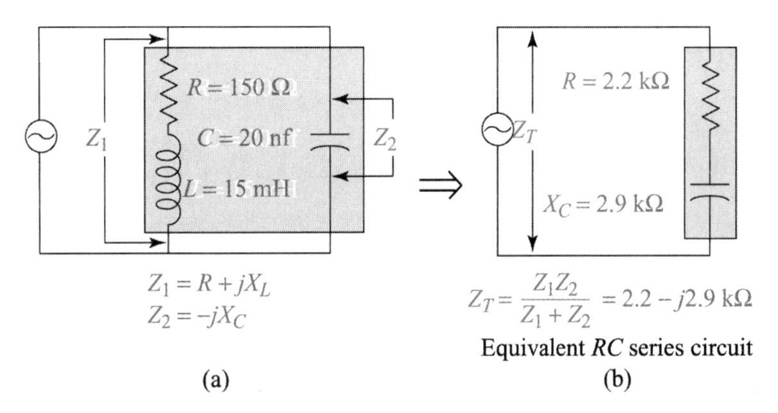

FIGURE 22-8 AC network for Example 22.12.

Solution Compute the reactances:

$$X_L = 2\pi fL = 2\pi(10 \text{ kHz})(15 \text{ mH}) \approx 942 \ \Omega$$

$$X_C = \frac{1}{2\pi fC} = \frac{1}{2\pi(10 \text{ kHz})(20 \text{ nF})} \approx 796 \ \Omega$$

Then find the impedance of the series string Z_1 by adding R and X_L in phasor form:

$$Z_1 = R + jX_L = 150 + j942 \ \Omega$$

To find the total impedance, apply formula (22.9) with $Z_1 = 150 + j942$ Ω and $Z_2 = -jX_C = -j796$ Ω:

$$Z_T = \frac{(150 + j942)(-j796)}{(150 + j942) + (-j796)} = \frac{(150 + j942)(-j796)}{150 - j146}$$

You can complete the calculation by changing each phasor to polar form and applying the rules for multiplying and dividing phasors:

$$Z_T = \frac{(954 \angle 81°)(796 \angle -90°)}{209 \angle 44°} = \frac{(954)(796)}{209} \angle(81° - 90° - 44°) \ \Omega$$

Z_T in polar and rectangular form is:

$$Z_T \approx 3.6 \angle{-53°}\ \text{k}\Omega \approx 2.2 - j2.9\ \text{k}\Omega$$

The ac network is equivalent to an *RC* series circuit where $R = 2.2\ \text{k}\Omega$ and $X_C = 2.9\ \text{k}\Omega$. See Figure 22-8(b).

EXERCISE 22.5

Round all answers to two significant digits.

In exercises 1 through 12, find Z_T *in polar and rectangular form.*

1. In Example 22.11, find Z_T when $R = 1.3\ \text{k}\Omega$, $X_L = 510\ \Omega$, and $X_C = 680\ \Omega$.

2. In Example 22.11, find Z_T when $R = 2.0\ \text{k}\Omega$, $X_L = 2.4\ \kappa\Omega$, and $X_C = 1.8\ \text{k}\Omega$.

3. In the ac network in Figure 22-9, find Z_T when $R = 750\ \Omega$, $X_L = 900\ \Omega$, and $X_C = 820\ \Omega$.

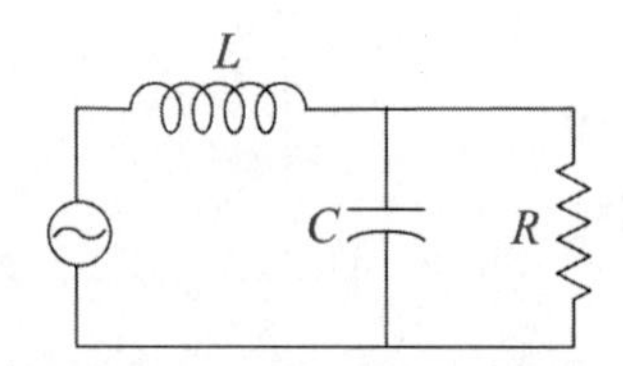

FIGURE 22-9 AC network for problems 3 and 4.

4. In the ac network in Figure 22-9, find Z_T when $R = 5.6\ \text{k}\Omega$, $X_L = 1.2\ \text{k}\Omega$, and $X_C = 3.3\ \text{k}\Omega$.

5. In Figure 22-8, find Z_T when $f = 2.5$ kHz, $R = 330\ \Omega$, $L = 30$ mH, and $C = 100$ nF.

6. In Figure 22-8, find Z_T when $f = 15$ kHz, $R = 470\ \Omega$, $L = 5.0$ mH, and $C = 30$ nF.

7. In the ac network in Figure 22-10, find Z_T when $X_C = 5.1\ \text{k}\Omega$, $R = 5.6\ \text{k}\Omega$, and $X_L = 4.7\ \text{k}\Omega$.

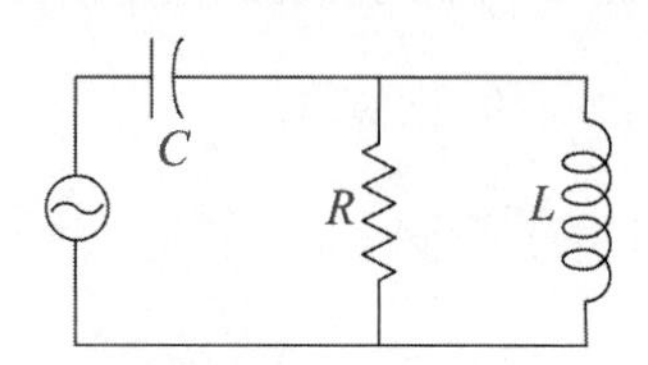

FIGURE 22-10 AC network for problems 7 and 8.

8. In the ac network in Figure 22-10, find Z_T when $X_C = 560\ \Omega$, $R = 430\ \Omega$, and $X_L = 750\ \Omega$.

9. In the ac network in Figure 22-11 find Z_T when $R = 10\ \text{k}\Omega$, $X_C = 20\ \text{k}\Omega$, and $X_L = 10\ \text{k}\Omega$.

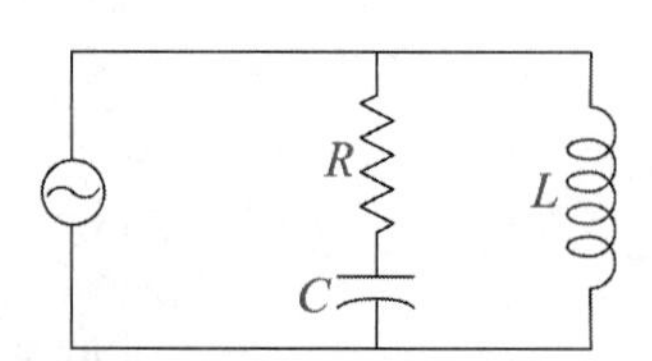

FIGURE 22-11 AC network for problems 9 and 10.

10. In the ac network of Figure 22-11, find Z_T when $R = 5.0\ \text{k}\Omega$, $X_C = 10\ \text{k}\Omega$, and $X_L = 10\ \text{k}\Omega$.

11. In the ac network of Figure 22-12, find Z_T when $R_S = 5.1\ \text{k}\Omega$, $R_1 = 100\ \Omega$, $R_2 = 200\ \Omega$, $X_L = 5.0\ \text{k}\Omega$, and $X_C = 6.0\ \text{k}\Omega$.

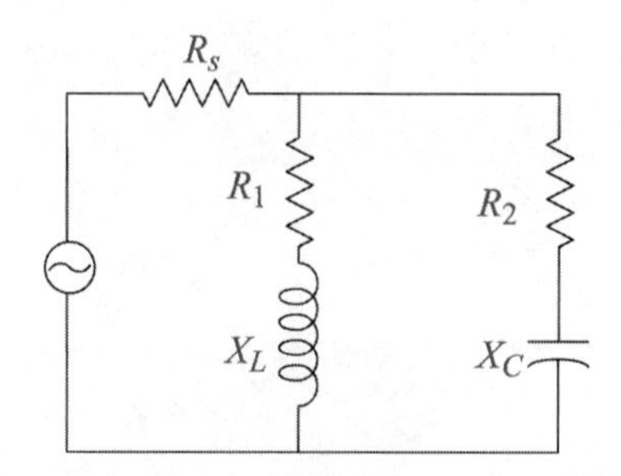

FIGURE 22-12 AC network for problems 11 and 12.

12. In the ac network of Figure 22-12, find Z_T when $R_S = 18\ \text{k}\Omega$, $R_1 = 500\ \Omega$, $R_2 = 200\ \Omega$, $X_L = 15\ \text{k}\Omega$, and $X_C = 12\ \text{k}\Omega$.

13. In Figure 22-7, given that $V_T = 15 \angle 0°$ V, find I_T, V_R, and V_L in rectangular and polar form.

Note: Kirchoff's voltage law applies to a closed loop in an ac circuit: $V_T = V_R + V_L$.

14. In Figure 22-8, given that $V_T = 10 \angle 0°$ V, find I_R, V_R, and V_L in rectangular and polar form.

Note: Kirchoff's voltage law applies to a closed loop in an ac circuit: $V_T = V_R + V_L$.

22.6 Filters

A mechanical filter separates fine particles and course particles. An electronic filter separates high frequencies and low frequencies. Capacitors and inductors are used for filtering because of their opposite response to frequencies. As the frequency increases, inductive reactance X_L increases but capacitive reactance X_C decreases.

Low-Pass Filters

A low-pass filter circuit allows low frequencies, such as audio frequencies, in the input voltage to pass to the load output voltage while higher frequencies are attenuated, or reduced, in the load. Two basic ways this can be done are:

1. Using a capacitor in parallel with the load to provide a bypass for the higher frequencies, allowing low frequencies to pass to the load. See Figure 22-13(a).
2. Using an inductor in series with the load to choke the higher frequencies, allowing low frequencies to pass to the load. See Figure 22-13(b).

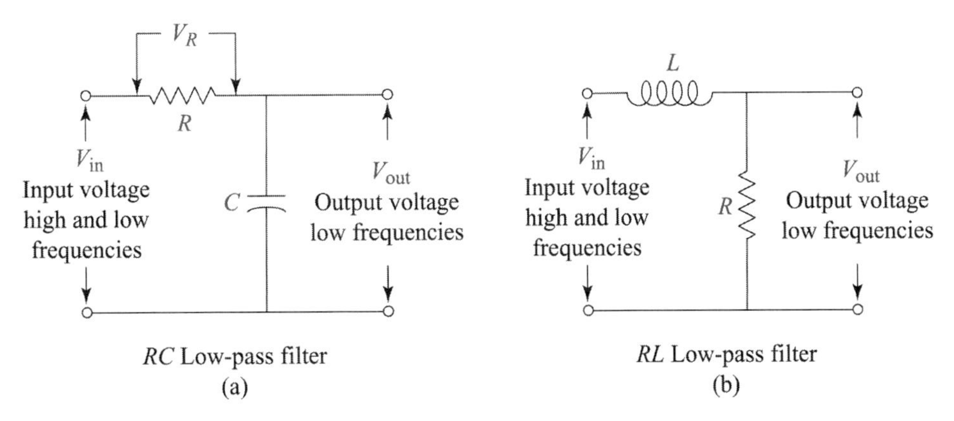

FIGURE 22-13 Low-pass filters.

EXAMPLE 22.13 In Figure 22-13(a) if $V_{in} = 20$ V, $R = 10$ kΩ, $C = 12$ nF, and $X_C = 10$ kΩ, find:

(a) The current I, V_R, and V_C.

(b) The frequency f.

Solution

(a) First find the impedance Z of the RC series branch, then find I and V_R:

$$Z = \sqrt{R^2 + X_C^2} = \sqrt{10^2 + 10^2} \approx 14.14 \text{ k}\Omega$$

$$I = \frac{V_{in}}{Z} = \frac{20 \text{ V}}{14.14 \text{ k}\Omega} = 1.41 \text{ mA}$$

$$V_R = IR = (1.41 \text{ mA})(10 \text{ k}\Omega) = 14.1 \text{ V}$$

The voltage across the capacitor V_C equals the output voltage V_{out}:

$$V_C = IX_C = (1.41 \text{ mA})(10 \text{ k}\Omega) = 14.1 \text{ V}$$

(b) Note in (a) that since $R = X_C$, $V_R = V_C$. When this condition exists in a low-pass filter, the output voltage equals 70.7% of the input voltage and the frequency is called the *cutoff frequency* f_c:

$$f = f_c = \frac{1}{2\pi RC} = \frac{1}{2\pi(10 \text{ k}\Omega)(12 \text{ nF})} \approx 1.33 \text{ kHz}$$

A low-pass filter attenuates frequencies above the cutoff frequency f_c and allows frequencies below f_c to produce output in the load.

High-Pass Filters

A high-pass filter circuit allows high frequencies, such as radio frequencies, in the input voltage to pass to the load output voltage while lower frequencies are attenuated, or reduced, in the load. Two basic ways this can be done are:

1. Using a coupling capacitor in series with the load to provide high reactance to the low frequencies, allowing high frequencies to pass to the load. See Figure 22-14(a).
2. Using an inductor in parallel with the load to provide a shunt for the low frequencies, allowing high frequencies to pass to the load. See Figure 22-14(b).

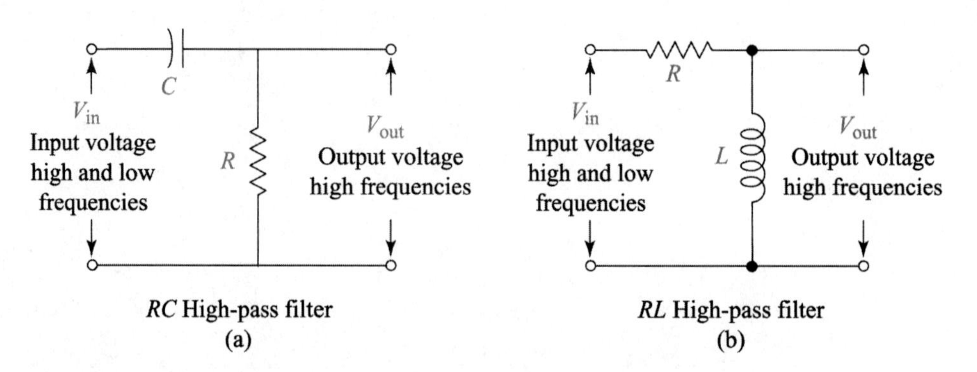

FIGURE 22-14 High-pass filters.

A high-pass filter allows frequencies above the cutoff frequency f_c to produce output in the load while frequencies lower than f_c cannot pass to the load. The formula for the cutoff frequency for the *RC* high-pass filter in Figure 22-14(a) is the same formula as that for the *RC* low-pass filter in Figure 22-13(a):

Formula **RC Cutoff Frequency**

$$f_c = \frac{1}{2\pi RC} \tag{22.10}$$

The formula for the cutoff frequency for the *RL* high-pass filter in Figure 22-14(b) is the same formula as that for the *RL* low-pass filter in Figure 22-13(b):

Formula **RL Cutoff Frequency**

$$f_c = \frac{R}{2\pi L} \tag{22.11}$$

EXAMPLE 22.14 In Figure 22-14(b), if $V_{in} = 10$ V, $R = 1.5$ kΩ, $L = 100$ mH, and $X_L = 2.0$ kΩ, find

(a) The current I, V_R, and V_{out}.

(b) The frequency of the output voltage and the cutoff frequency f_c.

Solution

(a) First find the impedance Z of the RL series branch, then find I and V_R:

$$Z = \sqrt{R^2 + X_L^2} = \sqrt{1.5^2 + 2.0^2} = 2.5 \text{ k}\Omega$$

$$I = \frac{V_{\text{in}}}{Z} = \frac{10 \text{ V}}{2.5 \text{ k}\Omega} = 4.0 \text{ mA}$$

$$V_R = IR = (4.0 \text{ mA})(1.5 \text{ k}\Omega) = 6.0 \text{ V}$$

The output voltage V_{out} is the same as the voltage across the inductor V_L:

$$V_{\text{out}} = V_L = IX_L = (4.0 \text{ mA})(2.0 \text{ k}\Omega) = 8.0 \text{ V}$$

(b) The frequency of the output voltage is the same as the frequency of the inductor voltage:

$$f = \frac{X_L}{2\pi L} = \frac{2.0 \text{ k}\Omega}{2\pi(100 \text{ mH})} \approx 3.18 \text{ kHz}$$

Use formula (22.11) to find the cutoff frequency:

$$f_c = \frac{R}{2\pi L} = \frac{1.5 \text{ k}\Omega}{2\pi(100 \text{ mH})} \approx 2.39 \text{ kHz}$$

Notice that the frequency of the output voltage 3.18 kHz is above the cutoff frequency 2.39 kHz.

Bandpass Filters

A bandpass filter combines a high-pass filter with a low-pass filter to allow only certain frequencies, between two cutoff frequencies, to pass to the output of the filter. All frequencies that do not fall within the range of the cutoff frequencies are attenuated and therefore produce little power at the output. Figure 22.15 shows an RC bandpass filter. The upper and lower cutoff frequencies are each calculated by applying formula (22.10) as shown in the next example.

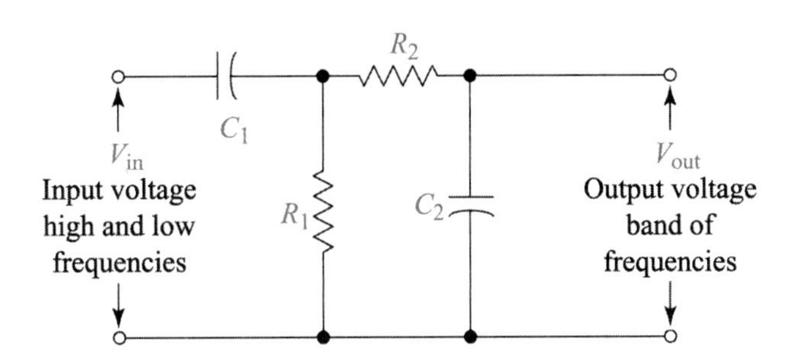

FIGURE 22-15 *RC* bandpass filter.

EXAMPLE 22.15 In Figure 22-15, if $R_1 = 3.0 \text{ k}\Omega$, $C_1 = 1.5\ \mu\text{F}$, $R_2 = 1.0 \text{ k}\Omega$, and C_2 =, 100 pF, find the upper and lower cutoff frequencies f_2 and f_1.

Solution To find the upper cutoff frequency f_2, apply formula (22.10) to the low-pass filter consisting of R_2 and C_2:

$$f_2 = \frac{1}{2\pi R_2 C_2} = \frac{1}{2\pi(1.0 \text{ k}\Omega)(100 \text{ pF})} \approx 1.59 \text{ kHz}$$

To find the lower cutoff frequency f_1, apply formula (22.10) to the high-pass filter consisting of R_1 and C_1:

$$f_1 = \frac{1}{2\pi R_1 C_1} = \frac{1}{2\pi(3.0\ \text{k}\Omega)(1.5\ \mu\text{F})} \approx 35.4\ \text{Hz}$$

The bandpass filter therefore passes frequencies above 35.4 Hz and below 1.59 kHz to the load.

When the cutoff frequencies differ by more than ten times, called a *decade*, the impedances of the high- and low-pass stages of the filter are far enough apart so that they can operate without interference. Assuming the frequencies are separated by at least a decade then the *bandwidth* can be defined as the difference between the two cutoff frequencies:

Formula

Bandwidth

$$\text{BW} = f_2 - f_1 \tag{22.12}$$

A bandpass filter can also be constructed using a series LC circuit and a resistor as shown in Figure 22-16. This combines the RL low-pass filter with the RC high-pass filter. The resonant frequency, f_r, of the LC circuit is when $X_L = X_C$ and the impedance is at a minimum. The resonant frequency therefore lies in the middle of the bandwidth:

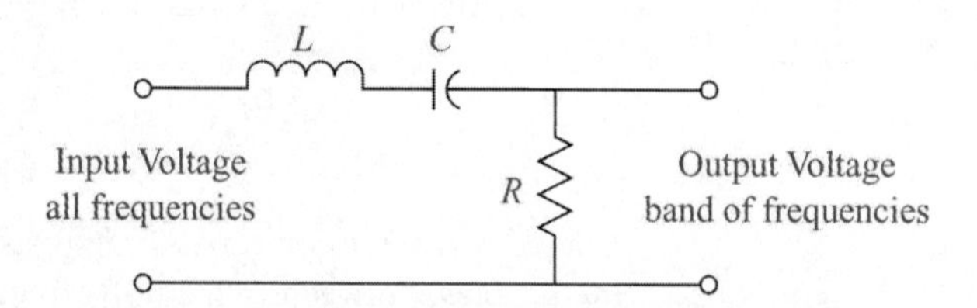

FIGURE 22-16 Series LC bandpass filter.

Formula

Resonant Frequency

$$f_r = f_1 + \frac{\text{BW}}{2} \tag{22.13}$$

The *quality*, Q, of a filter is the ratio of its resonant frequency to its bandwidth:

Formula

Quality

$$Q = \frac{f_r}{\text{BW}} \tag{22.14}$$

Therefore, the higher the quality, the narrower the bandwidth. Study the next example, which illustrates these ideas.

EXAMPLE 22.16

Given the bandpass filter in Figure 22-16 with $f_r = 2.0$ kHz and BW = 100 Hz. Find the cutoff frequencies f_1 and f_2, and the quality Q.

Solution Apply formula (22.13) solved for f_1 to find the lower cutoff frequency:

$$f_1 = f_r - \frac{\text{BW}}{2} = 2.0\ \text{kHz} - \frac{100\ \text{Hz}}{2} = 1950\ \text{Hz}$$

Apply (22.12) to find f_2:

$$f_2 = \text{BW} + f_1 = 100\ \text{Hz} + 1950\ \text{Hz} = 2050\ \text{Hz}$$

From (22.14) the quality is:

$$Q = \frac{fr}{\text{BW}} = \frac{2.0\ \text{kHz}}{100\ \text{Hz}} = 20$$

Band-Reject Filter

The band-reject filter functions opposite to that of a bandpass filter. It attenuates only a narrow band of frequencies and passes all others. Figure 22-17 shows a band-reject filter containing a series *LC* circuit. The output voltage is across the *LC* circuit, which produces a minimum impedance at the resonant frequency, f_r. The band-reject filter therefore prevents any signal equal to f_r, or close to f_r, from reaching the output, and is sometimes referred to as a *notch filter*.

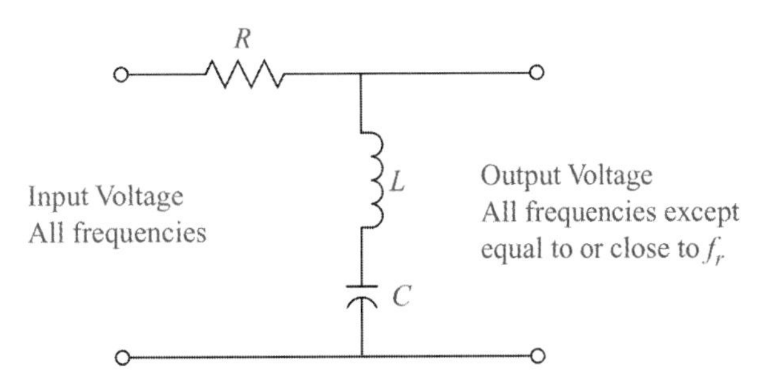

FIGURE 22-17 Band-reject filter.

EXERCISE 22.6

In exercises 1 through 4, refer to the RC *low-pass filter in Figure 22-13(a).*

1. Given $V_{in} = 20$ V, $R = 12$ kΩ, $C = 10$, and $X_C = 12$ kΩ. Find I, V_R, V_C, and f_c.
2. Given $V_{in} = 20$ V, $R = 5.0$ kΩ, $C = 10$, and $X_C = 5.0$ kΩ. Find I, V_R, V_C, and f_c.
3. Given $V_{in} = 10$ V, $R = 1.5$ kΩ, $C = 100$ nF.
   (a) Find f_c.
   (b) Find X_C, I, and V_0 at f_c.
4. Given $V_{in} = 10$ V, $R = 2.0$ kΩ, $C = 500$ nF.
   (a) Find f_c.
   (b) Find X_C, I, and V_0 at f_c.
5. For the *RL* low-pass filter in Figure 22-13(b), given $V_{in} = 10$ V, $R = 1.0$ kΩ, and $L = 50$ mH.
   (a) Find f_c.
   (b) Find X_L, I, and V_{out} at f_c.
6. For the *RL* low-pass filter in Figure 22-13(b), given $V_{in} =$ 10 V, $R = 1.0$ kΩ, and $L = 100$ mH.
   (a) Find f_c.
   (b) Find X_L, I, and V_{out} at f_c.
7. For the *RC* high-pass filter in Figure 22-14(a), given V_{in} $= 12$ V, $R = 1.5$ kΩ, and $C = 10$ nF.
   (a) Find f_c.
   (b) Find X_C, I, and V_{out} at f_c.
8. For the *RC* high-pass filter in Figure 22-14(a), given $V_{in} = 20$ V, $R = 10$ kΩ, and $C = 20$ nF.
   (a) Find f_c.
   (b) Find X_C, I, and V_{out} at f_c.

In exercises 9 through 12, refer to the RL *high-pass filter in Figure 22-14(b).*

9. Given $V_{in} = 10$ V, $R = 1.0$ kΩ, $L = 100$ mH, and $X_L =$ 2.0 kΩ.
   (a) Find I, V_R, and V_{out}.
   (b) Find the frequency of the output voltage and f_c.
10. Given $V_{in} = 10$ V, $R = 1.0$ kΩ, $L = 100$ mH, and $X_L =$ 1.5 kΩ.
    (a) Find I, V_R, and V_{out}.
    (b) Find the frequency of the output voltage and f_c.
11. Given $V_{in} = 20$ V, $R = 1.5$ kΩ, $L = 50$ mH.
    (a) Find f_c.
    (b) Find X_L, I, and V_{out} at f_c.
12. Given $V_{in} = 10$ V, $R = 1.5$ kΩ, $L = 20$ mH.
    (a) Find f_c.
    (b) Find X_L, I, and V_{out} at f_c.

13. In Figure 22-15, if $C_1 = 1.0\ \mu F$, $R_1 = 500\ \Omega$, $R_2 = 100\ k\Omega$, and $C_2 = 1.5$ nF, find the upper and lower cutoff frequencies f_2 and f_1.
14. In Figure 22-15, if $C_1 = 1.5\ \mu F$, $R_1 = 1.0\ k\Omega$, $R_2 = 50\ k\Omega$, and $C_2 = 2.0$ nF, find the upper and lower cutoff frequencies f_2 and f_1.
15. Given the bandpass filter in Figure 22-16 with f_r = 1.0 kHz and BW = 200 Hz. Find the cutoff frequencies f_1 and f_2, and the quality Q.
16. Given the bandpass filter in Figure 22-16 with f_r = 3.0 kHz and BW = 100 Hz. Find the cutoff frequencies f_1 and f_2, and the quality Q.
17. Given the band-reject filter in Figure 22-17 with f_r = 3.0 kHz and BW = 200 Hz. Find the cutoff frequencies f_1 and f_2, and the quality Q.
18. Given the band-reject filter in Figure 22-17 with f_r = 1.0 kHz and BW = 100 Hz. Find the cutoff frequencies f_1 and f_2, and the quality Q.

CHAPTER HIGHLIGHTS

22.1 *RL* CIRCUITS

In an *RL* parallel circuit, I_L lags V and I_R by 90° ("ELI") and has a negative phase angle of –90°.

RL Parallel Current

$$I_T = I_R - jI_L = \sqrt{I_R^2 + I_L^2}\ \angle\tan^{-1}\left(\frac{-I_L}{I_R}\right) \qquad (22.1)$$

• • •

Ohm's Law with I_T (Scalar and Phasor)

$$Z = \frac{V}{I_T} \qquad (22.2)$$

Key Example: For an *RL* parallel circuit with $V = 10$ V, $R = 5.0\ k\Omega$, and $X_L = 4.0\ k\Omega$.

$$I_R = \frac{10\ V}{5.0\ k\Omega} = 2.0\ mA$$

$$I_L = \frac{10\ V}{4.0\ k\Omega} = 2.5\ mA$$

$$I_T = I_R - jI_L = 2.0 - j2.5\ mA \approx 3.2\ \angle{-51^\circ}\ mA$$

$$Z = \frac{10\ \angle 0^\circ\ V}{3.2\ \angle{-51^\circ}\ mA} \approx 3.1\ \angle 51^\circ\ k\Omega \approx 2.0 + j2.4\ k\Omega$$

The equivalent *RL* series circuit is $R = 2.0\ k\Omega$ and $X_L = 2.4\ k\Omega$.

22.2 *RC* CIRCUITS

In an *RC* parallel circuit, I_C leads V and I_R by 90° ("ICE") and has a positive phase angle of 90°.

RC Parallel Current

$$I_T = I_R + jI_C = \sqrt{I_R^2 + I_C^2}\ \angle\tan^{-1}\left(\frac{I_C}{I_R}\right) \qquad (22.3)$$

Key Example: For an *RC* parallel circuit with $R = 1.3\ k\Omega$, $C = 10$ nF, $I_C = 6.3$ mA, and $f = 10$ kHz:

$$X_C = \frac{1}{2\pi(10\ kHz)(10\ nF)} \approx 1.6\ k\Omega$$

$$V = (6.3\ mA)(1.6\ k\Omega) \approx 10\ V$$

$$I_R = \frac{10\ V}{1.3\ k\Omega} \approx 7.7\ mA$$

$$I_T = I_R + jI_C = 7.7 + j6.3\ mA \approx 9.9\ \angle 39^\circ\ mA$$

$$Z = \frac{10\ \angle 0^\circ\ V}{9.9\ \angle 39^\circ\ mA} = 1.0\ \angle{-39^\circ}\ k\Omega \approx 780 - j630\ \Omega$$

$$C = \frac{1}{2\pi(10\ kHz)(630\ \Omega)} \approx 25\ nF$$

The equivalent *RC* series circuit is $R = 780\ \Omega$ and $C = 25$ nF.

22.3 SUSCEPTANCE AND ADMITTANCE

Susceptance (S)

$$B = \frac{1}{X} \qquad (22.4)$$

• • •

Admittance (S)

$$Y = \frac{1}{Z} = \frac{1}{R} \pm j\frac{1}{X} = G \pm jB \qquad (22.5)$$

Inductive ($-jB$) Capacitive ($+jB$)

Key Example: For the *RL* parallel circuit with $V = 10$ V, $R = 5.0\ k\Omega$, and $X_L = 4.0\ k\Omega$:

$$G = \frac{1}{R} = \frac{1}{5.0\ k\Omega} = 200\ \mu S$$

$$B = \frac{1}{X} = \frac{1}{4.0\ k\Omega} = 250\ \mu S$$

$$Y = G + jB = 200 - j250\ \mu S \approx 320\ \angle{-51^\circ}\ \mu S$$

22.4 *RLC* CIRCUITS

RLC Parallel Current

$$I_T = I_R + j(I_C - I_L) = I_R \pm jI_X \qquad (22.6)$$
$$I_X = |I_C - I_L|$$

Inductive ($-jI_X$) Capacitive ($+jI_X$)

Use formula (22.1) when I_X is inductive and formula (22.3) when I_X is capacitive.

RLC Admittance (S)

$$Y = \frac{1}{R} + j\left(\frac{1}{X_C} - \frac{1}{X_L}\right) = G + j(B_C - B_L)$$
$$= G \pm jB \qquad (22.7)$$
$$B = |B_C - B_L|$$

Inductive ($-jB$) Capacitive ($+jB$)

Key Example: For the *RLC* parallel circuit with $R = 1.3$ kΩ, $L = 150$ mH, $C = 300$ nF, $V = 30$ V, and $f = 600$ Hz:

$$X_L = 2\pi(600 \text{ Hz})(150 \text{ mH}) \approx 565\ \Omega$$
$$X_C = \frac{1}{2\pi(600 \text{ Hz})(300 \text{ nF})} \approx 884\ \Omega$$
$$I_R = \frac{30 \text{ V}}{1.3 \text{ k}\Omega} \approx 23 \text{ mA}$$
$$I_L = \frac{30 \text{ V}}{565\ \Omega} \approx 53 \text{ mA}$$
$$I_C = \frac{30 \text{ V}}{884\ \Omega} \approx 34 \text{ mA}$$
$$I_T = I_R + j(I_C - I_L) = 23 + j(34 - 53) \text{ mA}$$
$$= 23 - j19 \text{ mA} \approx 30\ \angle{-40°} \text{ mA}$$
$$Z = \frac{30\ \angle 0° \text{ V}}{30\ \angle{-40°} \text{ mA}} = 1.0\ \angle 40° \text{ k}\Omega \approx 770 + j640\ \Omega$$

The equivalent series circuit is an *RL* series circuit where $R = 770\ \Omega$ and $X_L = 640\ \Omega$.

22.5 AC NETWORKS

Series Impedances

$$Z_T = Z_1 + Z_2 + Z_3 + \cdots \qquad (22.8)$$

• • •

Parallel Impedances

$$Z_{eq} = \frac{Z_1 Z_2}{Z_1 + Z_2} \qquad (22.9)$$

Key Example: For the ac network in Figure 22-8(a), where $f = 10$ kHz, $R = 150\ \Omega$, $L = 15$ mH, and $C = 20$ nF:

$$X_L = 2\pi(10 \text{ kHz})(15 \text{ mH}) \approx 942\ \Omega$$
$$X_C = \frac{1}{2\pi(10 \text{ kHz})(20 \text{ nF})} \approx 796\ \Omega$$
$$Z_1 = 150 + j942\ \Omega$$
$$Z_2 = -jX_C = -j796\ \Omega$$
$$Z_T = \frac{(150 + j942)(-j796)}{(150 + j942) + (-j796)} = \frac{(150 + j942)(-j796)}{150 - j146}$$

Change to polar form to divide:

$$Z_T = \frac{(954\ \angle 81°)(796\ \angle{-90°})}{209\ \angle 44°} \approx 3.6\ \angle{-53°} \text{ k}\Omega$$
$$\approx 2.2 - j2.9 \text{ k}\Omega$$

The ac network is equivalent to an *RC* series circuit with $R = 2.2$ kΩ and $X_C = 2.9$ kΩ.

22.6 FILTERS

Figures 22-13 and 22-14 show basic *RC* and *RL* low-pass and high-pass filter circuits.

RC Cutoff Frequency

$$f_c = \frac{1}{2\pi RC} \qquad (22.10)$$

$$f_c = \frac{R}{2\pi L} \qquad (22.11)$$

Key Example: In Figure 22-14(b), if $V_{in} = 10$ V, $R = 1.5$ kΩ, $L = 100$ mH, and $X_L = 2.0$ kΩ:

$$Z = \sqrt{R^2 + X_L^2} = \sqrt{1.5^2 + 2.0^2} = 2.5 \text{ k}\Omega$$
$$I = \frac{V_{in}}{Z} = \frac{10 \text{ V}}{2.5 \text{ k}\Omega} = 4.0 \text{ mA}$$
$$V_R = IR = (4.0 \text{ mA})(1.5 \text{ k}\Omega) = 6.0 \text{ V}$$
$$V_{out} = V_L = IX_L = (4.0 \text{ mA})(2.0 \text{ k}\Omega) = 8.0 \text{ V}$$
$$f = \frac{X_L}{2\pi L} = \frac{2.0 \text{ k}\Omega}{2\pi(100 \text{ mH})} \approx 3.18 \text{ kHz}$$
$$f_c = \frac{R}{2\pi L} = \frac{1.5 \text{ k}\Omega}{2\pi(100 \text{ mH})} \approx 2.39 \text{ kHz}$$

Figures 22-16 and 22-17 show series *LC* bandpass and band-reject filters.

Bandwidth

$$\text{BW} = f_2 - f_1 \qquad (22.12)$$

Resonant Frequency

$$f_r = f_1 + \frac{BW}{2} \quad (22.13)$$

• • •

Quality

$$Q = \frac{f_r}{BW} \quad (22.14)$$

Key Example: In Figure 22-16 if $R = 300\ \Omega$, $f_r = 2.0$ kHz, and BW = 200 Hz.

$$f_1 = f_r - \frac{BW}{2} = 2.0\text{ kHz} - \frac{200\text{ Hz}}{2} = 1.9\text{ kHz}$$

$$f_2 = BW + f_1 = 200\text{ Hz} + 1.9\text{ kHz} = 2.1\text{ kHz}$$

$$Q = \frac{f_r}{BW} = \frac{2.0\text{ kHz}}{200\text{ Hz}} = 10$$

REVIEW EXERCISES

Round all answers to two significant digits.

In exercises 1 through 10, using the given values for the parallel ac circuit, find I_T *and* Z *in rectangular and polar form.*

1. RL circuit: $V = 9.0$ V, $R = 4.3$ kΩ, $X_L = 6.2$ kΩ
2. RL circuit: $V = 15$ V, $R = 5.6$ kΩ, $X_L = 1.8$ kΩ
3. RC circuit: $V = 12$ V, $R = 820\ \Omega$, $X_C = 1.8$ kΩ
4. RC circuit: $V = 6.4$ V, $R = 1.6$ kΩ, $X_C = 680\ \Omega$
5. RC circuit: $V = 40$ V, $R = 11$ kΩ, $X_C = 15$ kΩ
6. RLC circuit: $V = 30$ V, $R = 20$ kΩ, $X_L = 20$ kΩ, $X_C = 15$ kΩ
7. RLC circuit: $V = 8.0$ V, $R = 560\ \Omega$, $X_L = 350\ \Omega$, $X_C = 600\ \Omega$
8. RLC circuit: $V = 24$ V, $R = 1.8$ kΩ, $X_L = 750\ \Omega$, $X_C = 1.0$ kΩ
9. RLC circuit: $V = 20$ V, $R = 2.7$ kΩ, $X_L = 12$ kΩ, $X_C = 10$ kΩ
10. RLC circuit: $V = 5.5$ V, $R = 390\ \Omega$, $X_L = 470\ \Omega$, $X_C = 470\ \Omega$

In exercises 11 through 20, find the magnitude of B *and* Y *in rectangular and polar form for each of the parallel circuits in exercises 1 through 10.*

21. In an RL parallel circuit, $R = 560\ \Omega$, $L = 20$ mH, and the applied voltage $V = 6.0$ V with a frequency $f = 6.0$ kHz.
    (a) Find I_T in rectangular and polar form.
    (b) Find Z in polar and rectangular form and the equivalent RL series circuit.
22. In an RC parallel circuit, $R = 2.0$ kΩ, $C = 25$ nF, and the applied voltage $V = 20$ V with a frequency $f = 2.0$ kHz.
    (a) Find I_T in rectangular and polar form.
    (b) Find Z in polar and rectangular form and the equivalent RC series circuit.
23. In an RLC parallel circuit, $R = 1.5$ kΩ, $L = 50$ mH, $C = 10$ nF, and the applied voltage $V = 24$ V with a frequency $f = 10$ kHz.
    (a) Find I_T in rectangular and polar form.
    (b) Find Z in polar and rectangular form and the equivalent RL or RC series circuit.
24. In an RLC parallel circuit, $R = 750\ \Omega$, $L = 10$ mH, $C = 800$ pF, and the total current $I_T = 4.6$ mA with a frequency $f = 6.8$ kHz.
    (a) Find Y and Z in polar and rectangular form.
    (b) Find the applied voltage V and the equivalent RL or RC series circuit.
25. In an RLC parallel circuit, $R = 3.3$ kΩ, $L = 100$ mH, $C = 50$ nF, and $f = 1.5$ kHz.
    (a) Find Y and Z in polar and rectangular form.
    (b) Find the equivalent RL or RC series circuit.
26. In an RLC parallel circuit, $R = 15$ kΩ, $L = 430$ mH, $C = 12$ nF, and $f = 15$ kHz.
    (a) Find Y and Z in polar and rectangular form.
    (b) Find the equivalent RL or RC series circuit.
27. In the ac network in Figure 22-18, $R_1 = 150\ \Omega$, $X_L = 620\ \Omega$, $R_2 = 100\ \Omega$, and $X_C = 510\ \Omega$. Find the total impedance of the network in polar and rectangular form.

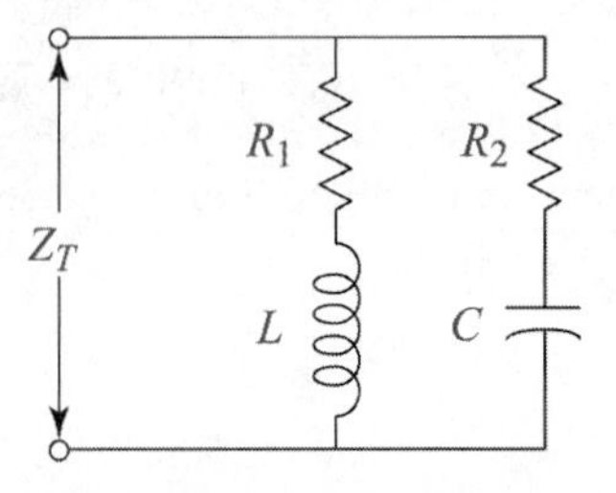

FIGURE 22-18 AC network for problems 27 and 28.

28. In the ac network in Figure 22-18, $R_1 = 1.0$ kΩ, $X_L = 3.0$ kΩ, $R_2 = 820\ \Omega$, and $X_C = 5.0$ kΩ. Find the total impedance of the network in polar and rectangular form.
29. In the ac network in Figure 22-19 on page 459, $R_1 = 1.0$ kΩ, $R_2 = 200\ \Omega$, $X_L = 2.0$ kΩ, and $X_C = 1.5$ kΩ. Find the total impedance of the network in polar and rectangular form.

30. In the ac network in Figure 22-19, $R_1 = 620\ \Omega$, $R_2 = 100\ \Omega$, $X_L = 900\ \Omega$, and $X_C = 1.2\ \text{k}\Omega$. Find the total impedance of the network in polar and rectangular form.

31. For the *RC* low-pass filter in Figure 22-13(a), given $V_{in} = 20$ V, $R = 1.0\ \text{k}\Omega$, and $C = 10$ nF.
(a) Find f_c.
(b) Find X_C, I, and V_{out} at f_c.

32. For the *RL* high-pass filter in Figure 22-14(b), given $V_{in} = 10$ V, $R = 1.0\ \text{k}\Omega$, $L = 50$ mH, and $X_L = 2.0\ \text{k}\Omega$.
(a) Find I, V_R, and V_{out}.
(b) Find the frequency of the output voltage and f_c.

33. Given the bandpass filter in Figure 22-16 with $f_r = 3.0$ kHz and BW = 200 Hz. Find the cutoff frequencies f_1 and f_2, and the quality Q.

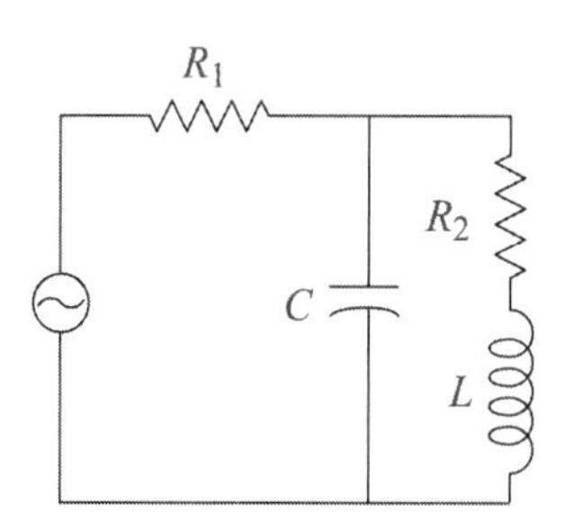

FIGURE 22-19 AC network for problems 29 and 30.

34. Given the band-reject filter in Figure 22-17 with $f_r = 3.6$ kHz and BW = 400 Hz. Find the cutoff frequencies f_1 and f_2, and the quality Q.

CHAPTER 23

Exponents and Logarithms

CHAPTER OBJECTIVES

In this chapter, you will learn:

- The definition of a fractional exponent and how it is applied.
- The definition of an exponential function.
- The meaning of the exponential number *e*.
- How to graph exponential growth and decay curves.
- The definition of a logarithm.
- How to find common and natural logarithms.
- How to solve exponential equations.

You have worked with exponents that are integers, that is, positive and negative whole numbers and zero. Exponents can also be defined as fractions or decimals. Fractional and decimal exponents allow you to express any number in terms of a number base, such as 10, and an exponent. This concept provides the basis for logarithms, which are the inverse of exponents. Logarithmic notation emphasizes the exponent in an expression, formula, or equation, and helps you to simplify calculations and solve equations with exponents. There are many applications of exponents and logarithms in electronics, particularly to current and voltage in *RC* and *RL* circuits and to power gain in ac networks.

• • •

23.1 Fractional Exponents

In Section 5.3, the five basic rules for exponents are presented where m and n are integers (positive or negative whole numbers or zero):

Rule **Product Rule for Exponents**

$$(x^m)(x^n) = x^{m+n} \tag{5.1}$$

Add exponents when multiplying like bases.

• • •

Rule **Quotient Rule for Exponents**

$$\frac{x^n}{x^m} = x^{n-m} \tag{5.2}$$

Subtract exponents when dividing like bases.

• • •

Rule **Power Rule for Exponents**

$$(x^m)^n = x^{mn} \tag{5.3}$$

Multiply exponents when raising to a power.

• • •

Rule **Factor Rule for Exponents**

$$(xy)^n = x^n y^n \text{ and } \left(\frac{x}{y}\right)^n = \frac{x^n}{y^n} \quad (5.4)$$

Raise each factor separately to the given power.

• • •

Rule **Radical Rule for Exponents**

$$\sqrt[n]{x^m} = x^{m/n} \quad (5.5)$$

Divide the index of the radical into the exponent of the radicand (number in the radical).

In rule (5.5) above, when n divides into m evenly, the result is an exponent that is an integer:

$$\sqrt[3]{x^6} = x^{6/3} = x^2$$

However, when n does not divide into m evenly, the result is a fractional exponent:

$$\sqrt[3]{x^2} = x^{2/3}$$

Rule (5.5) is therefore used to define a fractional exponent:

Formula **Fractional Exponent**

$$x^{m/n} = \sqrt[n]{x^m} \text{ or } \left(\sqrt[n]{x}\right)^m \qquad (x > 0) \quad (23.1)$$

The *denominator* of the fractional exponent is the *index of the root,* and the *numerator* is the *power.* For example, a denominator of 2 means square root, 3 means cube root, 4 means fourth root, and so on. When applying formula (23.1), you can take the root first and then raise to the power, or vice versa, but it is easier to take the root first.

EXAMPLE 23.1 Evaluate:

(a) $8^{2/3}$ **(b)** $25^{3/2}$ **(c)** $9^{1/2}$

Solution

(a) Apply formula (23.1). Take the cube root first and then raise to the second power:

$$8^{2/3} = \left(\sqrt[3]{8}\right)^2 = (2)^2 = 4$$

You can do fractional exponents on the calculator using parentheses:

8 [x^y] [(] 2 [÷] 3 [)] [=] → 4

(b) Take the square root first and then raise to the third power:

$$25^{3/2} = \left(\sqrt{25}\right)^3 = (5)^3 = 125$$

(c) The 1/2 power is the same as the square root:

$$9^{1/2} = \left(\sqrt{9}\right)^1 = 3$$

A special case of formula (23.1) is:

Formula **Fractional Exponent 1/*n***

$$x^{1/n} = \sqrt[n]{x} \qquad (x > 0) \qquad (23.2)$$

As a result of (23.2), any radical can be expressed as a fractional exponent, and vice versa. For example:

$$\sqrt[3]{10} = 10^{1/3}$$
$$\sqrt[4]{V} = V^{1/4}$$
$$\sqrt{PR} = (PR)^{1/2}$$

You can find the $\sqrt[3]{10}$ on the calculator by raising 10 to the 1/3 power using the power key and the reciprocal key [1/x] or [x⁻¹]

10 [xʸ] 3 [1/x] [=] → 2.15

Similarly, you can find $\sqrt[4]{2}$ by raising 2 to the 1/4 or 0.25 power:

2 [xʸ] 0.25 [=] → 1.19

EXAMPLE 23.2 Evaluate:

(a) $16^{-1/4}$ **(b)** $(-27)^{2/3}$ **(c)** $(0.04)^{-3/2}$ **(d)** $\left(\frac{1}{100}\right)^{-\frac{1}{2}}$

Solution

(a) A negative fractional exponent means the same as a negative integral exponent: invert and change to a positive exponent:

$$16^{-1/4} = \frac{1}{16^{1/4}} = \frac{1}{\sqrt[4]{16}} = \frac{1}{2}$$

(b) The cube root of a negative number is a real negative number:

$$(-27)^{2/3} = (\sqrt[3]{-27})^2 = (-3)^2 = 9$$

(c) The square root of a decimal has one-half the number of decimal places:

$$(0.04)^{-3/2} = \frac{1}{0.04^{3/2}} = \frac{1}{(\sqrt{0.04})^3} = \frac{1}{0.2^3} = \frac{1}{0.008} = 125$$

The calculator solution is, using 1.5 for 3/2:

DAL: 0.04 [∧] [(-)] 1.5 [=] → 125

Not DAL: 0.04 [xʸ] 1.5 [+/-] [=] → 125

(d) Invert the fraction and change to a positive exponent:

$$\left(\frac{1}{100}\right)^{-1/2} = (100)^{1/2} = \sqrt{100} = 10$$

Rules (5.1) to (5.5) for integral exponents also apply to fractional or decimal exponents as shown in the following examples.

EXAMPLE 23.3 Simplify:

(a) $(V_T^{0.5})(V_T^{2.5})$ **(b)** $\dfrac{X^{1.5}}{X^{2.5}}$

Solution

(a) Apply the product rule (5.1) and add the exponents:

$$(V_T^{0.5})(V_T^{2.5}) = V_T^{(0.5+2.5)} = V_T^3$$

(b) Apply the quotient rule (5.2) and subtract the exponents:

$$\frac{X^{1.5}}{X^{2.5}} = X^{(1.5-2.5)} = X^{-1} \text{ or } \frac{1}{X}$$

EXAMPLE 23.4 Evaluate without the calculator:

$$(4 \times 10^{-6})^{1.5}$$

Solution Use the factor rule (5.4) and apply the exponent to the number 4 and the power of 10 separately:

$$(4 \times 10^{-6})^{1.5} = (4)^{1.5} \times (10^{-6})^{1.5}$$

Apply the power rule (5.3) to the base 10 and use $\frac{3}{2} = 1.5$ for the base 4:

$$(4)^{1.5} \times (10^{-6})^{1.5} = 4^{3/2} \times 10^{(-6)(1.5)}$$
$$= \left(\sqrt{4}\right)^3 \times 10^{-9} = 8 \times 10^{-9}$$

This example can be checked on a calculator:

DAL: 4 [EE] [(-)] 6 [∧] 1.5 [=] → 8 E–9

Not DAL: 4 [EXP] 6 [+/-] [x^y] 1.5 [=] → 8 E–9

EXAMPLE 23.5 Simplify:

$$(0.25I^2R)^{1/2}$$

Solution Use the factor rule and apply the exponent 1/2 to each factor in the parentheses:

$$(0.25I^2R)^{1/2} = (0.25)^{1/2}\, I^{(2)(1/2)}\, R^{(1/2)}$$
$$= \sqrt{0.25}\, I^{(1)}\sqrt{R} = 0.5I\sqrt{R}$$

EXERCISE 23.1

In exercises 1 through 44, evaluate or simplify each expression. Try to do the exercises without the calculator to reinforce the ideas. Round decimal answers to three significant digits.

1. $25^{1/2}$

2. $16^{1/4}$

3. $8^{4/3}$

4. $9^{3/2}$

5. $(-27)^{2/3}$

6. $-81^{3/2}$

7. $4^{-3/2}$

8. $8^{-2/3}$

9. $100^{-1/2}$

10. $(1000)^{-2/3}$

11. $64^{-0.5}$

12. $9^{1.5}$

13. $0.027^{1/3}$

14. $0.16^{1/2}$

15. $\left(\dfrac{1}{8}\right)^{4/3}$

16. $\left(\dfrac{1}{16}\right)^{1/4}$

17. $3^{1/3}$

18. $5^{1/2}$

19. $10^{0.5}$

20. $10^{-0.5}$

21. $0.001^{-2/3}$

22. $0.0016^{3/4}$

23. $(25 \times 10^{-2})^{1.5}$

24. $(1.44 \times 10^{6})^{0.5}$

25. $(3.375 \times 10^{9})^{1/3}$

26. $(125 \times 10^{-12})^{2/3}$

27. $\left(\frac{25}{16}\right)^{3/2}$

28. $\left(\frac{81}{100}\right)^{1/2}$

29. $\left(\frac{8}{1000}\right)^{-2/3}$

30. $\left(\frac{-27}{8}\right)^{-1/3}$

31. $(100^{1/3})(100^{1/6})$

32. $(8^{1/2})(8^{-1/6})$

33. $(I_T^{-0.3})(I_T^{0.7})$

34. $(P_1^{0.3})(P_1^{0.2})$

35. $(-27e^{6})^{1/3}$

36. $(144X_C^4)^{0.5}$

37. $(0.04L^4X^2)^{-1/2}$

38. $0.001I^3R^3)^{2/3}$

39. $\left(\frac{V_x^2}{10^{-6}}\right)^{1/2}$

40. $\left(\frac{e^4}{10^{12}}\right)^{-1/4}$

41. $(3s^{1/2})^2(2s^{2/3})^3$

42. $(-5k^{1/3})^{-3}(10k^{3/2})^2$

43. $\left(\frac{9a^{-2}}{4b^4}\right)^{-1/2}$

44. $\left(\frac{64t^3}{t^6}\right)^{1/3}$

Applied Problems

In problems 45 through 50, solve each problem to two significant digits.

45. Find the maximum deflection δ (delta) of a beam in meters given by:

$$\delta = (3.1 \times 10^{-6})(20/25)(25^2 - 20^2)^{3/2}$$

46. Find the time t in minutes that it takes for the runoff of rainfall at a certain reservoir to reach a maximum given by:

$$t = 2.53\left(\frac{1000}{(0.20)(3.2)^2}\right)^{1/3}$$

Applications to Electronics

47. The bandwidth of a three-stage dc amplifier is given by:

$$B = B_1(2^{1/3} - 1)^{1/2}$$

where B_1 = bandwidth of one stage. Find B in kHz when $B_1 = 6.0$ kHz.

48. The current in amperes required to produce a magnetic flux density B in a circular coil is given by:

$$i = \frac{2rB}{(4\pi \times 10^{-7})(n)}$$

Find i when $B = 1.0 \times 10^{-3.5}$ T (tesla), $n = 2.0 \times 10^2$ turns and the radius $r = 0.10$ m.

49. The behavior of iron in a magnetic field gives rise to the equation:

$$W = nB^{1.6}$$

Calculate W when $n = 1.3$ and $B = 0.072$.

50. In calculating the heat dissipation from a circuit component, the following expression arises:

$$(QR)^{0.6}\left(\frac{Q}{R}\right)^{0.4}$$

Simplify this expression.

23.2 Exponential Functions

An exponential growth function is defined as:

Definition **Exponential Growth Function**

$$y = b^x \qquad b = \text{base} > 1 \tag{23.3}$$

EXAMPLE 23.6 Graph the exponential growth function $y = 2^x$ from $x = -4$ to $x = 4$.

Solution Substitute integral values of x from −4 to 4 and compute the powers of 2. For example, to calculate y when $x = -2$, a negative exponent means invert and change to a positive exponent:

$$y = 2^{-2} = \frac{1}{2^2} = \frac{1}{4} = 0.25$$

The table of values for x and y is as follows:

x	−4	−3	−2	−1	0	1	2	3	4
$y = 2^x$	0.0625	0.125	0.25	0.50	1	2	4	8	16

Plot the points on graph paper to produce the graph shown in Figure 23-1. Study the graph. Observe that as x increases, y increases at a faster and faster rate and becomes infinitely large. The graph gets steeper and steeper and is said to increase, or grow, exponentially. As x decreases, y decreases at a slower and slower rate and becomes infinitely small. The graph approaches the x axis as y gets closer and closer to zero but never touches it. The value of y never reaches zero and never becomes negative. The x axis is called an *asymptote* of the graph. By using the graph in Figure 23-1, a value of y corresponding to any value of the exponent x can be found. For example, $2^{1.7} \approx 3.2$ as shown on the graph by the arrows.

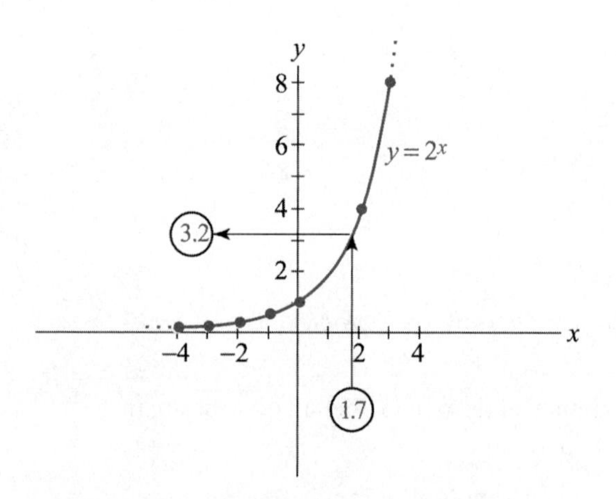

FIGURE 23-1 Exponential growth function for Example 23.6.

The calculator provides a more precise answer, which is an infinite decimal that is an irrational number:

2 [y^x] 1.7 [=] → 3.2490 …

The function in Figure 23-1 is called an exponential growth function because many natural patterns of growth behave in this way. Plants and animals in their early stages grow exponentially. Animal and human populations tend to increase exponentially when uncontrolled. The number of components on a computer chip tends to increase exponentially, and money in a compound interest bank account grows exponentially.

The opposite of exponential growth is called exponential decay:

Formula **Exponential Decay Function**

$$y = b^{-x} \quad b = \text{base} > 1 \tag{23.4}$$

EXAMPLE 23.7 Graph the exponential decay function $y = 2^{-x}$ from $x = -4$ to $x = 4$.

Solution Substitute integral values of x and compute the powers of 2:

x	−4	−3	−2	−1	0	1	2	3	4
$y = 2^{-x}$	16	8	4	2	1	0.50	0.25	0.125	0.0625

This table of values is the reverse of the table in Example 23.6 because when x becomes $-x$ it reverses the values of x, and therefore the values of y. The graph is shown in Figure 23-2 and is the mirror image of the graph in Figure 23-1 where the y axis is the mirror. Observe that as x increases, the value of y decreases at a slower and slower rate approaching zero but never reaching it. This process is called exponential decay because natural decay patterns, such as radioactive decay and biological decay, behave in this way.

The Exponential Function e^x

There are important applications of exponential decay in electronics that behave similar to the curve in Figure 23-2. When an inductive circuit is shorted, the current decays exponentially. When a capacitor is discharged, the voltage decays exponentially and, when a capacitor is charged *or* discharged, the current decays exponentially. In these examples, *the rate of decrease is proportional to the amount of current or voltage*. That is, as the current or voltage drops, the rate of decrease drops by the same ratio. For example, suppose the initial current is 10 mA, and it is decreasing at a rate of 2 mA per millisecond. When the current becomes half its value, 5 mA, the rate of decrease in the current will have slowed down to 1 mA per millisecond. The exponential change in these applications

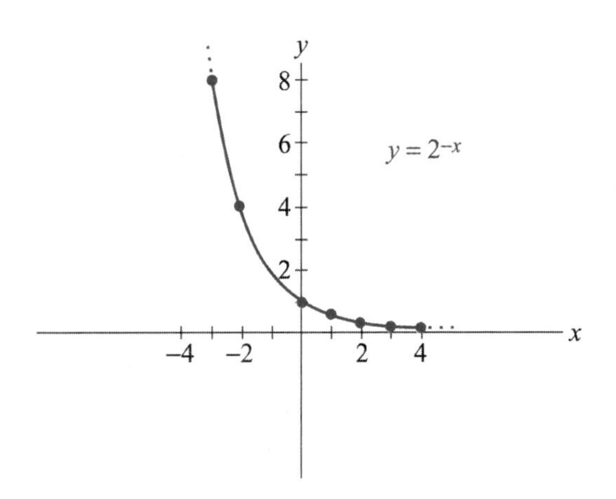

FIGURE 23-2 Exponential decay function for Example 23.7.

is best described mathematically by the exponential function e^x. The number e is an irrational number like π and is equal to the infinite decimal:

$$e = 2.718281828459\ \ldots$$

The value of e is found using the methods of calculus. The following expression approaches e as n increases toward infinity (∞):

$$\left(1 + \frac{1}{n}\right)^n \rightarrow e \text{ as } n \rightarrow \infty$$

You can approximate e on the calculator by letting n equal a large number such as 1,000,000:

$$\left(1 + \frac{1}{10^6}\right)^{1{,}000{,}000} = (1.000001)^{1{,}000{,}000} \approx 2.71828$$

Powers of e are done on the calculator with the key [e^x]. On most calculators this is done by pressing [2nd] or [SHIFT], and then [LN]. For example, e^3 is:

DAL: [2nd] [LN] 3 [=] → 20.086 ([2nd] [LN] = [e^x])

Not DAL: 3 [2nd] [LN] → 20.086

and $e^{-1.5}$ is:

DAL: [2nd] [LN] [(-)] 1.5 [=] → 0.2231

Not DAL: 1.5 [+/-] [2nd] [LN] → 0.2231

The letter e was chosen in honor of the Swiss mathematician Leonhard Euler (1707–1783), who was a great contributor to mathematics. Study the next example, which closes the circuit and shows an application of e to an RC circuit.

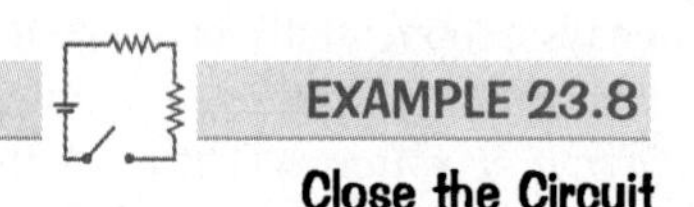

EXAMPLE 23.8

Close the Circuit

Figure 23-3 shows a battery connected to an RC series circuit with a switch S. The voltage across the resistance t seconds after the switch is closed is given by the exponential function:

$$v = Ve^{-t/RC}$$

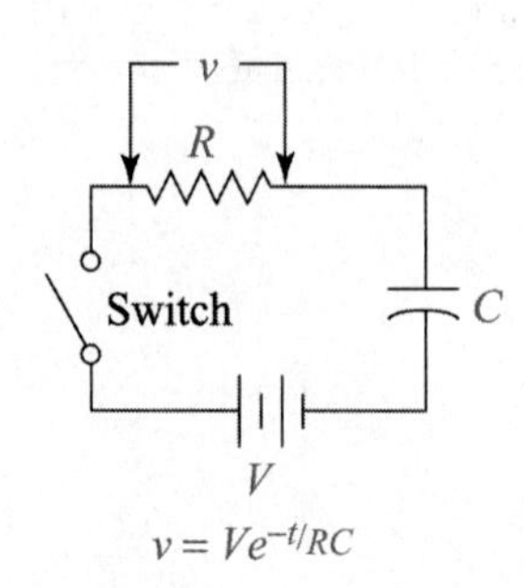

FIGURE 23-3 *RC* circuit for Example 23.8.

where V = battery voltage. If $V = 25$ V, $R = 200\ \Omega$, and $C = 100\ \mu$F:

(a) Calculate the value of RC in milliseconds.

(b) Find v when $t = 10$ ms.

(c) Find v when $t = 15$ ms.

Solution

(a) $RC = (200\ \Omega)(100 \times 10^{-6}\text{ F}) = 0.02\text{ s} = 20\text{ ms}$

(b) To find v when $t = 10$ ms, first compute the exponent of e by substituting the values of RC and t:

$$\frac{-t}{RC} = \frac{-0.010\text{ s}}{0.02\text{ s}} = -0.50$$

Then substitute the exponent and the value of V into the given formula and calculate $v = 25e^{-0.50}$:

DAL: 25 [×] [2nd] [LN] [(-)] 0.5 [=] → 15.2 V

Not DAL: 25 [×] 0.5 [+/-] [2nd] [LN] [=] → 15.2 V

(c) To find v when $t = 15$ ms, proceed the same as for $t = 10$ ms:

$$\frac{-t}{RC} = \frac{-0.015 \text{ s}}{0.02 \text{ s}} = -0.75$$

$$v = 25e^{-0.75} \approx 11.8 \text{ V}$$

EXERCISE 23.2

In exercises 1 through 20, graph each exponential function for the given range of values. Use an appropriate scale for the y axis.

1. $y = 3^x$; $x = -2$ to $x = 3$
2. $y = 1.5^x$; $x = -3$ to $x = 5$
3. $y = 2^{x-1}$; $x = -1$ to $x = 5$
4. $y = 2^{x+1}$; $x = -3$ to $x = 3$
5. $y = 10^{0.2x}$; $x = -2$ to $x = 5$
6. $y = 10^{0.3x}$; $x = -2$ to $x = 4$
7. $y = 1.5^{-x}$; $x = -5$ to $x = 3$
8. $y = 3^{-x}$; $x = -3$ to $x = 2$
9. $y = 2^{-x+1}$; $x = -3$ to $x = 3$
10. $y = 2^{-x-1}$; $x = -5$ to $x = 1$
11. $y = 1 - 2^x$; $x = -2$ to $x = 4$
12. $y = 1 - 2^{-x}$; $x = -4$ to $x = 2$
13. $y = e^t$; $t = -2$ to $t = 3$
14. $y = 2e^t$; $t = -3$ to $t = 2$
15. $y = 2e^{-t}$; $t = -2$ to $t = 3$
16. $y = e^{-t}$; $t = -3$ to $t = 2$
17. $y = 1 - e^{-t}$; $t = -3$ to $t = 2$
18. $y = 1 - e^t$; $t = -2$ to $t = 3$
19. $y = 1 - e^{0.5t}$; $t = -3$ to $t = 5$
20. $y = 1 - e^{-0.5t}$; $t = -5$ to $t = 3$

Applied Problems

21. The world population was growing in the year 2000 at the rate of approximately 1.5% per year. This is expressed by the exponential function:

$$A = A_0(1.015)^n$$

where A = population n years from 2000 and $A_0 = 6.0 \times 10^9$ (2000 estimate).
(a) Graph the population growth curve from 2000 to 2050 ($n = 0$ to $n = 50$). Use appropriate scales.
(b) In what year will the population reach 10 billion?

22. Inflation causes prices to increase according to the exponential function:

$$P = P_0(1 + r)^n$$

where P = price n years from now, P_0 = present price and r = rate of inflation per year.
(a) If $P_0 = \$10.00$ is the present price of a movie and $r = 6\%$, graph P versus n for $n = 0$ to $n = 16$.
(b) After how many years will the price double?

23. The radioactive decay of one gram of Strontium 90, found in nuclear power plants, is given by the exponential function:

$$m = e^{-0.03t}$$

where m = mass after t years. Graph mass vs. time for $t = 0$ to $t = 80$ yrs and determine the half life of Sr 90 (when $m = 0.50$ g) from the graph.

24. Depreciation of an item can be expressed by the exponential function:

$$P = P_0(1 - r)^n$$

where P = value n years from now, P_0 = present value, and r = rate of depreciation.
(a) If a \$20,000 car depreciates at a rate of 20% per year, graph P vs. n for $n = 0$ to $n = 10$ yr.
(b) After how many years will the value decrease by 50%?

Applications to Electronics

25. In the RC circuit in Figure 23-3, given the exponential function:

$$v = V_e^{-t/RC}$$

with $V = 25$ V, $R = 200\ \Omega$, and $C = 100\ \mu$f, find the voltage v when t equals:
(a) 5 ms.
(b) 20 ms.
(c) 0 ms.

26. In the RC circuit in Figure 23.3, given the exponential function:

$$v = V_e^{-t/RC}$$

with $V = 12$ V, $R = 100\ \Omega$, and $C = 200\ \mu$f.
(a) Calculate RC in milliseconds. Find v when t is:
(b) 20 ms.
(c) 30 ms.

27. In the RC circuit in Figure 23.3, the current t seconds after the switch is closed is given by:

$$i = \frac{V}{R}e^{-t/RC}$$

Given $V = 100$ V, $R = 2.0$ MΩ, and $C = 200$ nF.
(a) Calculate RC in milliseconds. Find i when t is:
(b) 200 ms.
(c) 400 ms.

28. In the RC circuit in Figure 23.3, the voltage across the capacitor t seconds after the switch is closed is given by:

$$v = V(1 - e^{-t/RC})$$

Given $V = 10$ V, $R = 1.0$ kΩ, and $C = 50$ μF.
(a) Calculate RC in millisconds. Find v when t is:
(b) 50 ms.
(c) 100 ms.

23.3 Common and Natural Logarithms

A logarithm is another name for an exponent. Logarithms were discovered in 1614 by John Napier of Scotland to simplify calculations by using exponents. A logarithmic function is the inverse of an exponential function in the same way that $\sqrt{x}$ is the inverse of x^2 and $\tan^{-1}$ is the inverse of tan:

Formula **Logarithmic Function**

$$y = \log_b x \text{ means } b^y = x \quad (b > 1) \tag{23.5}$$

Definition (23.5) says: "The logarithm of x to the base b is the exponent y," or simply, "log x to the base $b = y$". A logarithm is another way of looking at an exponent. For example:

$$\log_2 16 = 4 \text{ means } 2^4 = 16$$

$$\log_5 25 = 2 \text{ means } 5^2 = 25$$

These statements are examples of logarithmic form and exponential form. Study the following example, which provides more examples.

EXAMPLE 23.9 Give some examples of logarithmic and exponential form.

Solution

	Exponential Form	Logarithmic Form
(a)	$5^3 = 125$	$\log_5 125 = 3$
(b)	$4^{1.5} = 8$	$\log_4 8 = 1.5$
(c)	$10^3 = 1000$	log 1000 = 3 (base 10 is understood)
(d)	$10^{-6} = 0.000001$	log 0.000001 = −6 (base 10 is understood)
(e)	$e^2 = 7.39$	ln 7.39 = 2 (ln means $\log_e$)
(f)	$e^{-1} = 0.368$	ln 0.368 = −1 (ln means $\log_e$)

Note in examples (c) and (d) in the table that when the base is not written it is understood to be 10. Note also in examples (e) and (f) that ln means $\log_e$.

EXAMPLE 23.10

Find x or y without the calculator:

(a) $\log_2 x = 3$
(b) $\log x = -1$
(c) $y = \log_8 64$
(d) $y = \log 1000$

Solution Write each in exponential form:

(a) $\log_2 x = 3 \rightarrow 2^3 = x \rightarrow x = 8$
(b) $\log x = -1 \rightarrow 10^{-1} = x \rightarrow x = 0.1$
(c) $y = \log_8 64 \rightarrow 8^y = 64 \rightarrow y = 2$
(d) $y = \log 1000 \rightarrow 10^y = 1000 \rightarrow y = 3$

Common Logarithms

The number 10 is used as a base for logarithms because our number system is based on 10, making it easy to work with. Base 10 logarithms are called *common logarithms* and the base is not written.

EXAMPLE 23.11

Construct a table of common logs for the following powers of 10: –6, –3, –2, –1, 0, 1, 2, 3, 6.

TABLE 23-1 COMMON LOGARITHMS

$\log 10^{-6}$	$= \log 0.000001$	$= -6$
$\log 10^{-3}$	$= \log 0.001$	$= -3$
$\log 10^{-2}$	$= \log 0.01$	$= -2$
$\log 10^{-1}$	$= \log 0.1$	$= -1$
$\log 10^{0}$	$= \log 1.0$	$= 0$
$\log 10^{1}$	$= \log 10$	$= 1$
$\log 10^{2}$	$= \log 100$	$= 2$
$\log 10^{3}$	$= \log 1000$	$= 3$
$\log 10^{6}$	$= \log 1{,}000{,}000$	$= 6$

Solution Observe in Table 23-1 that the power of 10 *is* the logarithm. That is:

$$\log (10^x) = x$$

The logarithm of a number between two numbers in the table lies between the two powers. For example, $\log 50 = 1.70$, which is between 1 and 2, and $\log 0.5 = -0.301$, which is between –1 and 0. You should memorize this table. It will give you a better understanding of logarithms and provide a quick check on results.

EXAMPLE 23.12

Find x:

(a) $10^x = 30.5$ **(b)** $10^x = 0.411$ **(c)** $10^x = 23 \times 10^6$

Solution

(a) Write $10^x = 30.5$ in logarithmic form: $\log 30.5 = x$. Then find log 30.5 on the calculator:

DAL: [log] 30.5 [=] → 1.48

Not DAL: 30.5 [log] → 1.48

Similarly, for (b) and (c):

(b) *DAL:* [log] 0.411 [=] → −0.386

Not DAL: 0.411 [log] → −0.386

Observe in (b) that for a number between 0 and 1 the logarithm is negative.

(c) *DAL:* [log] 23 [EE] 6 [=] → 7.36

Not DAL: 23 [EE] 6 [log] → 7.36

Note that $\log 1 = 0$ because $10^0 = 1$. Also note that logarithms are not defined for zero or negative numbers.

The exponential key [10^x] is the inverse of [log]. For example, if $\log x = 0.502$, then $10^{0.502} = x$, and you use the key [10^x] to find x:

DAL: [10^x] 0.502 [=] → 3.18

Not DAL: 0.502 [10^x] → 3.18

Natural Logarithms

The exponential number e is also used as a base for logarithms. The reason for this is because many exponential functions are more simply expressed in terms of e, such as the exponential function in Example 23.8 for an *RC* circuit. Base e logarithms are called *natural logs*. When e is the base, we use "ln" instead of "log" and do not write the base e as shown in Example 23.9. In particular, note that $\ln e = 1$ because $e^1 = e$, and $\ln 1 = 0$ because $e^0 = 1$. Use [LN] on the calculator to find the natural log of a number. For example ln 0.587 is:

DAL: [LN] 0.587 [=] → −0.533

Not DAL: 0.587 [LN] → −0.533

The key [e^x] is the inverse of [LN]. If $\ln x = 0.868$, then $e^{0.868} = x$. Use the [e^x] key to find x:

DAL: [e^x] 0.868 [=] → 2.38

Not DAL: 0.868 [e^x] → −2.38

EXAMPLE 23.13 Find x or y to three significant digits:

(a) $y = \log 0.521$ **(b)** $\ln x = 0.664$

Solution

(a) To find y use the [log] key:

DAL: [log] 0.521 [=] → −0.283

Not DAL: 0.521 [log] → −0.283

(b) To find x use the [e^x] key

DAL: [e^x] 0.664 [=] → 1.94

Not DAL: 0.664 [e^x] → 1.94

ERROR BOX

A common error when working with logarithms is not seeing a simple relationship between a logarithm and an exponent. As shown with the powers of 10 in Example 23.11, $\log(10^x) = x$. Similarly $\ln(e^x) = x$ because in exponential form this says that $e^{(x)} = (e^x)$. To reinforce the concepts, see if you can apply these ideas and do the practice problems without the calculator.

Practice Problems: Find the value of x or y *without the calculator*. Give the answers to 9 and 10 in terms of e.

1. $\log 10^3 = y$ **2.** $\ln e^2 = y$ **3.** $\log(1/10) = y$ **4.** $\ln(1/e^2) = y$
5. $\log 10^{1.5} = y$ **6.** $\ln e^{0.5} = y$ **7.** $\log x = 2$ **8.** $\log x = -1$
9. $\ln x = 1$ **10.** $\ln x = 0.5$

It is possible to find the log of a number to any base b with the calculator using the formula:

Formula

Log of a Number *a* to any Base *b*:

$$\log_b a = \frac{\log a}{\log b} \tag{23.6}$$

For example, the $\log_2 23$ on the calculator is:

DAL: [log] 23 [÷] [log] 2 [=] → 4.52
Not DAL: 23 [log] [÷] 2 [log] [=] → 4.52

Rules for Logarithms

The rules for logarithms for any base b follow from the rules for exponents:

Rule

Product Rule for Logs

$$\log_b(xy) = \log_b x + \log_b y \tag{23.7}$$

• • •

Quotient Rule for Logs

$$\log_b\left(\frac{x}{y}\right) = \log_b x - \log_b y \tag{23.8}$$

• • •

Power Rule for Logs

$$\log_b(y^x) = x(\log_b y) \tag{23.9}$$

The rules for logarithms help you to simplify expressions with logarithms and solve equations with exponents.

EXAMPLE 23.14 Simplify each expression by applying the rules for logarithms.

(a) $\log (0.1x)$ **(b)** $\log 500R - \log 5R$ **(c)** $\ln e^{2t}$ **(d)** $\ln e^{0.5t} + \ln e^{-0.1t}$

Solution

(a) Apply the product rule (23.6) and note that $\log 0.1 = -1$:

$$\log(0.1x) = \log 0.1 + \log x = -1 + \log x = \log x - 1$$

(b) Apply the quotient rule (23.7):

$$\log 500R - \log 5R = \log\left(\frac{500R}{5R}\right) = \log 100 = 2$$

(c) Apply the power rule (23.8) and note that $\ln e = 1$:

$$\ln e^{2t} = (2t) \ln e = 2t(1) = 2t$$

(d) Apply the power rule:

$$\ln e^{0.5t} + \ln e^{-0.1t} = (0.5t) \ln e + (-0.1t) \ln e = 0.5t\,(1) - 0.1t\,(1) = 0.4t$$

EXERCISE 23.3

In exercises 1 through 14, change each from exponential form to logarithmic form.

1. $2^3 = 8$
2. $5^4 = 625$
3. $5^{-1} = 0.2$
4. $2^{-5} = 0.03125$
5. $10^4 = 10{,}000$
6. $10^{1.6} = 39.8$
7. $10^{-0.2} = 0.631$
8. $10^{-2} = 0.01$
9. $e^2 = 7.39$
10. $e^{0.5} = 1.649$
11. $e^{-1.5} = 0.223$
12. $e^{-1} = 0.368$
13. $e^{-2.3} = 0.100$
14. $e^{-0.625} = 0.535$

In exercises 15 through 28, change each from logarithmic form to exponential form.

15. $\log_5 25 = 2$
16. $\log_2 16 = 4$
17. $\log_2 0.125 = -3$
18. $\log_5 0.04 = -2$
19. $\log 1000 = 3$
20. $\log 200 = 2.3$
21. $\log 0.316 = -0.5$
22. $\log 0.1 = -1$
23. $\ln 3.0 = 1.1$
24. $\ln 54.6 = 4.0$
25. $\ln 0.145 = -1.93$
26. $\ln 0.60 = -0.511$
27. $\ln 0.0523 = -2.95$
28. $\ln 0.0288 = -3.55$

In exercises 29 through 44, find x *or* y *without the calculator. Check with the calculator.*

29. $\log_2 x = 16$
30. $\log_3 x = 4$
31. $\log x = 2$
32. $\log x = 3$
33. $\log_5 x = -1$
34. $\log_2 x = -2$
35. $\log x = -3$
36. $\log x = -6$
37. $y = \log_3 27$
38. $y = \log_5 125$
39. $y = \log_5 0.04$
40. $y = \log_2 0.25$
41. $y = \log 10{,}000$
42. $y = \log 1$
43. $y = \log 0.001$
44. $y = \log 0.1$

In exercises 45 through 58, find each logarithm to three significant digits.

45. $\log 86.1$
46. $\log 5.33$
47. $\log 0.105$
48. $\log 0.0484$
49. $\log (5.23 \times 10^3)$
50. $\log (1.23 \times 10^6)$
51. $\log (2.55 \times 10^{-6})$
52. $\log (34.1 \times 10^{-9})$
53. $\ln 25$
54. $\ln 1.43$
55. $\ln 0.124$
56. $\ln 0.559$
57. $\ln (6.52 \times 10^{-3})$
58. $\ln (45 \times 10^3)$

In exercises 59 through 74, find the value of x to three significant digits.

59. $\log x = 2.32$
60. $\log x = 5.89$
61. $\log x = 0.708$
62. $\log x = 0.445$
63. $\ln x = 6.06$
64. $\ln x = 8.56$
65. $\ln x = 1.35$
66. $\ln x = 1.98$
67. $\ln x = 0.383$
68. $\ln x = 0.132$
69. $\log x = -3.12$
70. $\log x = -0.831$
71. $\ln x = -1.35$
72. $\ln x = -2.12$
73. $\ln x = -0.845$
74. $\ln x = -0.603$

Answers to Error Box Problems, page 473:

1. 3 **2.** 2 **3.** −1 **4.** −2 **5.** 1.5 **6.** 0.5 **7.** 100 **8.** 1/10 **9.** e **10.** $\sqrt{e}$

In exercises 75 through 86, simplify each expression by applying the rules for logarithms.

75. $\log 100x$

76. $\log 0.01y$

77. $\log 0.5y + \log 2y$

78. $\log 10a + \log 10x$

79. $\ln e^{2t} - \ln e^{t}$

80. $\ln e^{-t} - \ln e^{-2t}$

81. $\ln\left(\frac{e^2}{t}\right)$

82. $\ln\left(\frac{5}{e^t}\right)$

83. $2\log 10^x - \log 10^2 x$

84. $\log 15x^2 - \log 3x$

85. $\ln 20e^{-2t} + \ln 5e^{t}$

86. $\ln e^{-0.2t} + \ln e^{0.5t}$

Applied Problems

Solve each applied problem to two significant digits.

87. At an inflation rate of r percent a year, prices will double in approximately n years where:

$$n = \frac{\ln 2}{r}$$

Find n when r is:

(a) 3% **(b)** 5% **(c)** 10%

88. When interest is compounded daily in a bank account at a rate of r percent a year, the principal (initial amount) will double in approximately n years given by:

$$n = \frac{\log 2}{\log(1 + r)}$$

Find n when r is:

(a) 3% **(b)** 5% **(c)** 6%

89. The German lightning calculator Zacharias Dase (1823–1861) was able to mentally calculate the product of two n-digit numbers in an incredibly small amount of time t given approximately by the formula:

$$\log t = 2.72 \log n + \log 0.102$$

where t is in seconds. For example, he could multiply two 5-digit numbers in his head in less than nine seconds. Find how long it took him to mentally multiply two:

(a) 7-digit numbers.
(b) 10-digit numbers.

90. Because of pollution the fish in a certain river are decreasing according to the formula:

$$\ln(P/P_0) = -0.075t$$

where P = population t years from now and P_0 = original population.

(a) After how many years will there be only 50% left? (Let $(P/P_0) = 0.5$)
(b) What percentage will die in the first year of pollution?

Applications to Electronics

91. The current in an RC series circuit is given by the exponential function:

$$i = I_0 e^{-t/RC}$$

where t = time and I_0 = current when $t = 0$. Express this function in logarithmic form by first solving for $\frac{i}{I_0}$.

92. The voltage in an RL series circuit is given by the exponential function:

$$v = V_0 e^{-Rt/L}$$

where t = time and V_0 = voltage when $t = 0$. Express this function in logarithmic form by first solving for $\frac{v}{V_0}$.

93. Find the decibel gain of an amplifier is given by:

$$\text{dB} = 10 \log\left(\frac{15}{0.40}\right)$$

94. The capacitance of a cylindrical capacitor is given by:

$$C = \frac{cl}{(18 \times 10^9)\ln(R_2/R_1)}$$

where c = dielectric constant, l = length, R_1 = inside radius, and R_2 = outside radius. Find C when $c = 5.8$, $l = 0.15$ m, $R_2 = 1.1$ cm, and $R_1 = 1.0$ cm.

CHAPTER HIGHLIGHTS

23.1 FRACTIONAL EXPONENTS

Fractional Exponent

$$x^{m/n} = \sqrt[n]{x^m} \text{ or } \left(\sqrt[n]{x}\right)^m \quad (x > 0) \qquad (23.1)$$

• • •

Fractional Exponent 1/n

$$x^{1/n} = \sqrt[n]{x} \quad (x > 0) \qquad (23.2)$$

Key Example:

(a) $100^{3/2} = \left(\sqrt{100}\right)^3 = 1000$

(b) $(-8)^{2/3} = (\sqrt[3]{-8})^2 = 4$

(c) $(4 \times 10^{-6})^{1/2} = 4^{1/2} \times 10^{(-6)(1/2)} = 2 \times 10^{-3}$

23.2 EXPONENTIAL FUNCTIONS

Exponential Growth Function

$$y = b^x \quad b = \text{base} > 1 \tag{23.3}$$

Key Example: See Figure 23-4.

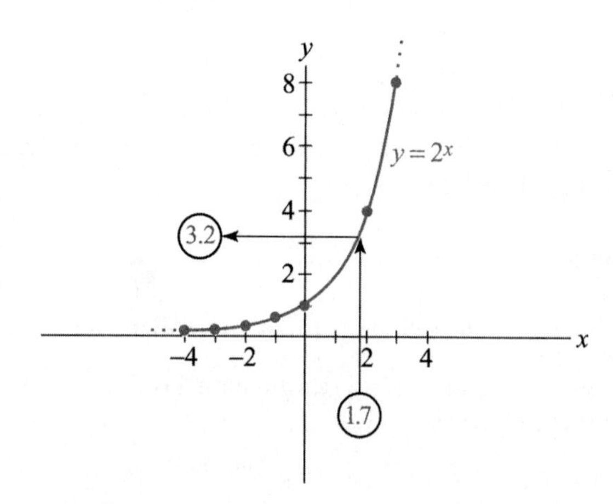

FIGURE 23-4 Key Example: Exponential growth function.

Exponential Decay Function

$$y = b^{-x} \quad b = \text{base} > 1 \tag{23.4}$$

Key Example: See Figure 23-5.

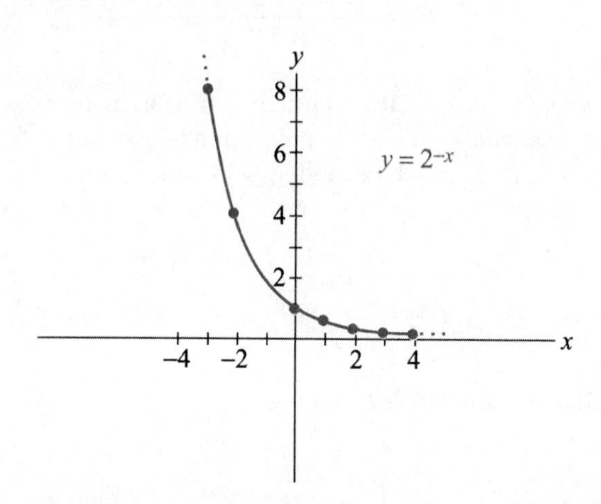

FIGURE 23-5 Key Example: Exponential decay function.

23.3 COMMON AND NATURAL LOGARITHMS

Logarithmic Function

$$y = \log_b x \text{ means } b^y = x \quad (b > 1) \tag{23.5}$$

Key Example:

	Exponential Form	Logarithmic Form
(a)	$5^3 = 125$	$\log_5 125 = 3$
(b)	$4^{1.5} = 8$	$\log_4 8 = 1.5$
(c)	$10^3 = 1000$	$\log 1000 = 3$ (base 10 is understood)
(d)	$10^{-6} = 0.000001$	$\log 0.000001 = -6$ (base 10 is understood)
(e)	$e^2 = 7.39$	$\ln 7.39 = 2$ (ln means $\log_e$)
(f)	$e^{-1} = 0.368$	$\ln 0.368 = -1$ (ln means $\log_e$)

Key Example:

(a) $y = \log 0.521$; y: [log] 0.521 [=] → −0.283

(b) $y = \ln 15.2$; y: [LN] 15.2 [=] → 2.72

(c) $\log x = -1.31$; x: [10ˣ] [(−)] 1.31 [=] → 0.049

(d) $\ln x = 0.664$; x: [eˣ] 0.664 [=] → 1.94

Log of a Number *a* to any Base *b*:

$$\log_b a = \frac{\log a}{\log b} \tag{23.6}$$

• • •

Product Rule for Logs

$$\log_b (xy) = \log_b x + \log_b y \tag{23.7}$$

• • •

Quotient Rule for Logs

$$\log_b \left(\frac{x}{y}\right) = \log_b x - \log_b y \tag{23.8}$$

• • •

Power Rule for Logs

$$\log_b (y^x) = x(\log_b y) \tag{23.9}$$

Key Example:

$$\ln 5e^{2t} - \ln 5e^t = \ln\left(\frac{5e^{2t}}{5e^t}\right) = \ln(e^{2t-t}) = \ln e^t = t$$

REVIEW EXERCISES

In exercises 1 through 12, evaluate or simplify each expression.

1. $27^{2/3}$
2. $16^{3/4}$
3. $10{,}000^{-1/2}$
4. $0.001^{-1/3}$
5. $-9^{3/2}$
6. $(-8)^{2/3}$
7. $(64 \times 10^{-3})^{2/3}$
8. $(0.25 \times 10^{6})^{1.5}$
9. $(100^{-1/3})(100^{4/3})$
10. $(e^{0.5t})^{-2}$
11. $\left(\dfrac{10^4}{I^2}\right)^{1.5}$
12. $\left(\dfrac{v^3}{0.001r^3}\right)^{-2/3}$

In exercises 13 through 16, graph each exponential function for the given range of values. Use an appropriate scale for the y axis.

13. $y = 2^{1.2x}$; $x = -2$ to $x = 3$
14. $y = 5^{-x/2}$; $x = -3$ to $x = 2$
15. $y = e^t - 1$; $t = -2$ to $t = 3$
16. $y = 2 - e^{-t}$; $t = -3$ to $t = 2$

In exercises 17 through 24, change each to logarithmic form.

17. $10^3 = 1000$
18. $2^6 = 64$
19. $5^{-0.54} = 0.419$
20. $10^{-2.3} = 0.00501$
21. $e^{-0.75} = 0.472$
22. $e^{-3.5} = 0.0302$
23. $e^4 = 54.6$
24. $e^{0.28} = 1.32$

In exercises 25 through 32, change each to exponential form.

25. $\log 0.0776 = -1.11$
26. $\log 0.443 = -0.354$
27. $\log 5.33 = 0.727$
28. $\log 12.7 = 1.1$
29. $\ln 2.69 = 0.99$
30. $\ln 23 = 3.14$
31. $\ln (4.55 \times 10^{-3}) = -5.39$
32. $\ln 0.972 = -0.0284$

In exercises 33 through 40, find x or y without the calculator. Check with the calculator.

33. $\log_2 x = -1$
34. $\log_5 x = -2$
35. $\log x = 6$
36 $\log x = -1$
37. $y = \log_5 25$
38. $y = \log_2 0.5$
39. $y = \log 0.01$
40. $y = \log 100$

In exercises 41 through 48, find each logarithm to three significant digits.

41. $\log 25$
42. $\log 6.7$
43. $\log (5.23 \times 10^{-6})$
44. $\log 0.932$
45. $\ln 0.675$
46. $\ln (33 \times 10^{-3})$
47. $\ln 36.5$
48. $\ln 12.2$

In exercises 49 through 56, find x *to three significant digits.*

49. $\log x = 0.859$
50. $\log x = 0.0338$
51. $\log x = -0.336$
52. $\log x = -3.25$
53. $\ln x = 4.21$
54. $\ln x = 1.61$
55. $\ln x = -0.434$
56. $\ln x = -5.52$

In exercises 57 through 60, simplify each expression by applying the rules for logarithms.

57. $\log (0.001V)$
58. $\log 25q^2 - \log 2.5q$
59. $2\ln e^{10t} + \ln e^{-20t}$
60. $\ln\left(\dfrac{e^3}{t^2}\right)$

Applied Problems

Solve each applied problem to two significant digits.

61. A power cable supported only at the ends will lie in a curve called a *catenary,* which is given by the function:

$$y = \frac{e^x + e^{-x}}{2}$$

 (a) Find the value of y when $x = 0.5$.
 (b) Find the value of y when $x = -0.5$.

62. An epidemic spreads according to the exponential equation:

$$N = N_0\,(2)^{0.20t}$$

 where the initial number of infected people at $t = 0$ is $N_0 = 10$, and N = number infected after t days.
 (a) Graph N vs t for $t = 2$ to $t = 30$.
 (b) How many days does it take for the number of infected people to double?

63. The half-life H_L of radioactive carbon 14 (C14), which is present in all living things at a constant level, is given by:

$$H_L = (\ln 2)/k$$

 where $k = 12.4 \times 10^{-3}$. C14 decays slowly after death and age can therefore be determined by the amount of C14 still present. Find the half-life of C14.

64. When radiation penetrates a substance, it diminishes in intensity as the depth increases expressed by:

$$\ln(I/I_0) = -kx$$

 where I_0 = initial intensity, I = intensity at depth x, and k = coefficient of absorption. If $k = 2.6$ per mm for X-rays:
 (a) At what depth will I be reduced to 1% of I_0?
 (b) What percent will I be reduced to at $x = 1.0$ mm?

Applications to Electronics

65. A discharged battery is being charged by a solar cell. After t hours the voltage is given by:

$$V = 13.2(1 - e^{-0.1t})$$

Find the voltage when t is:
(a) 5 h **(b)** 10 h

66. The formula for the decibel (dB) gain of an amplifier in terms of power is:

$$A_p = 10 \log \frac{P_{out}}{P_{in}}$$

Where P_{out} = output power and P_{in} = input power. Find the dB gain A_p when P_{in} = 0.50 W and P_{out} is:
(a) 5.0 W **(b)** 50 W

67. In the RC circuit in Example 23.8, the current t seconds after the switch is closed is given by:

$$i = \frac{V}{R} e^{-t/RC}$$

If $V = 10$ V, $R = 1.2$ kΩ, and $C = 5.5$ μF, find i when t is:
(a) 10 ms. **(b)** 20 ms.

68. A circular coil of $n = 5.0 \times 10^2$ turns and radius $r = 0.30$ m has a magnetic flux density of $B = 1.0 \times 10^{-3.3}$ T (tesla). Find the current in amperes required to produce this flux density in a vacuum given by:

$$i = \frac{2rB}{(4\pi \times 10^{-7})(n)}$$

69. The input voltage in a network is given by:

$$\log 18 - \log V_1 = 1.2$$

(a) Solve for $\log V_1$ and change to exponential form.
(b) Find the value of V_1.

70. In an RL series circuit, the current t seconds after the power is turned on is given by:

$$i = I(1 - e^{-Rt/L})$$

Given $I = 100$ mA, $R = 200$ Ω, and $L = 400$ mH.
(a) Calculate L/R. Find i when t is:
(b) 1.5 ms.
(c) 3.5 ms.

CHAPTER 24

Applications of Logarithms

CHAPTER OBJECTIVES

In this chapter, you will learn:

- How to solve exponential equations.
- How to find time constants in *RC* and *RL* circuits.
- How to use universal time constant curves.
- How to find instantaneous values of current and voltage in *RC* and *RL* circuits.
- How to compute power, voltage, and current gain in electrical networks.
- How to use Bode plots on a semilogarithmic graph.

In electronics, exponential equations describe current and voltage changes in *RC* and *RL* series circuits. Logarithms enable you to understand and solve these equations in a direct way. Logarithms are also useful when working with power changes in amplifiers and antennas because human senses are not as precise as electronic measurements. When these changes are defined in terms of logarithms, they more accurately reflect human responses to the changes. Sound intensity levels and power gain are therefore set up on a logarithmic scale, and logarithms are used to find values of power, voltage, and current.

• • •

24.1 Exponential Equations

The power rule for logarithms (23.8) enables you to solve equations with exponents as shown in the following example.

EXAMPLE 24.1 Solve the exponential equation: $2^x = 50$

Solution Take the common logarithm (or natural logarithm) of both sides of the equation:

$$\log(2^x) = \log 50$$

Apply the multiplication rule (23.8) and make the exponent x a coefficient of the log function:

$$(x)(\log 2) = \log 50$$

Then divide both sides by (log 2) to solve for x:

$$\frac{(x)\cancel{(\log 2)}}{\cancel{(\log 2)}} = \frac{\log 50}{(\log 2)}$$

$$x = \frac{\log 50}{(\log 2)} = \frac{1.70}{0.301} \approx 5.64$$

This can be done in one series of steps on the calculator:

DAL: [log] 50 [÷] [log] 2 [=] → 5.64

Not DAL: 50 [log] [÷] 2 [log] [=] → 5.64

The result means that $2^{5.64} = 50$. You can quickly check whether the answer is approximately correct without the calculator. Observe that the value of x should be between 5 and 6 since $2^5 = 32$ and $2^6 = 64$.

EXAMPLE 24.2

Solve the exponential equation: $1 - e^{-10t} = 0.70$

Solution First solve for the exponential term by moving e^{-10t} and 0.70 to opposite sides of the equation:

$$e^{-10t} = 1 - 0.70 = 0.30$$

Express the equation in logarithmic form using natural logarithms:

$$\ln 0.30 = -10t$$

You can also take the natural logarithm of both sides, which leads to the same result. Divide both sides by −10 to solve for t:

$$t = \frac{\ln 0.30}{-10} \approx 0.12$$

ERROR BOX

When solving an exponential equation, it is easy to enter the wrong number or press the wrong key on the calculator. You should therefore check that the answer makes sense by estimating the result when possible. For example, for the exponential equation: $5^n = 92$, consider that $5^2 = 25$ and $5^3 = 125$. Therefore, $2 < n < 3$.

For the exponential equation: $e^t = 0.5$, you can estimate $e \approx 3$. Then $3^{-1} = 1/3 \approx 0.3$ and $3^0 = 1$. Therefore, since $0.3 < 0.5 < 1$ it follows that $-1 < t < 0$. See if you can apply these ideas to the practice problems.

Practice Problems: Estimate each answer between two integers, without the calculator, for each exponential equation:

1. $3^x = 25$ **2.** $4^n = 70$ **3.** $2^x = 0.6$ **4.** $10^n = 0.05$

5. $e^t = 10$ **6.** $e^t = 0.2$ **7.** $2^n = 100$ **8.** $10^k = 2500$

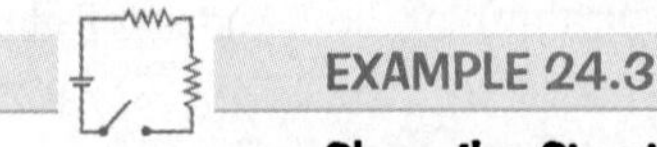

EXAMPLE 24.3

Close the Circuit

The charging current for the RC circuit in Figure 24-1 is given by the exponential equation:

$$i_C = \frac{V}{R} e^{-(t/RC)}$$

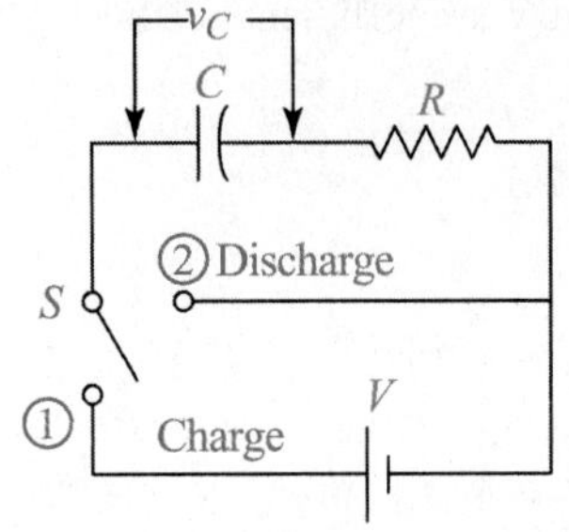

If $V = 12$ V, $R = 100\ \Omega$, and $C = 5.0\ \mu$F, find the time t when i_C is:

(a) Find the value of RC

Find the time t when i_C is:

(b) 5.0 mA

(c) 10 mA

FIGURE 24-1 *RC* circuit, charge and discharge current, for Example 24.3.

Solution

(a) $RC = (100\ \Omega)(5.0 \times 10^{-6}\text{F}) = 0.0005$

(b) To find t when $i_C = 5.0$ mA, substitute the given values into the formula, using $RC = 0.0005$:

$$5.0 \text{ mA} = \frac{12 \text{ V}}{100 \ \Omega} e^{-t/(0.0005)}$$

Simplify: $0.005 \text{ A} = 0.12e^{-t/0.0005} = 0.12e^{-2000t}$

or $0.12e^{-2000t} = 0.005 \text{ A}$

Divide both sides by 0.12: $e^{-2000t} = \frac{0.005}{0.12} \approx 0.0417$

Write the equation in logarithmic form and solve for t:

$$\ln 0.0417 = -2000t$$

$$t = \frac{\ln 0.0417}{-2000} \approx 1.59 \text{ ms}$$

(c) To find t when $i_C = 10$ mA, proceed the same as in (b). The equation and solution are:

$$0.010 = 0.12e^{-2000t}$$

$$e^{-2000t} = \frac{0.010}{0.12} \approx 0.0833$$

$$\ln 0.0833 = -2000t$$

$$t = \frac{\ln 0.0833}{-2000} \approx 1.24 \text{ ms}$$

EXERCISE 24.1

In exercises 1 through 22, solve each exponential equation to three significant digits.

1. $2^x = 20$

2. $2^x = 1.3$

3. $5^n = 0.22$

4. $4^n = 23$

5. $3^{a+1} = 50$

6. $5^{b-1} = 100$

7. $10^x = 20$

8. $10^{2x} = 0.15$

9. $10^{x-1} = 0.372$

10. $10^{x+1} = 0.977$

11. $10^{-3x} - 10 = 40$

12. $3 + 10^{2x} = 14$

13. $e^{5t} = 5.2$

14. $e^{1.7t} = 0.36$

15. $e^{-1.3t} = 0.135$

16. $e^{-0.356t} = 4.32$

17. $1 - e^{-300t} = 0.90$

18. $1 - e^{-0.75t} = 0.50$

19. $1 - e^{0.25t} = -1.62$

20. $1 - e^{5.34t} = 0.98$

21. $e^{-20t} - 1 = -0.55$

22. $e^{-10t} - 1 = -0.95$

Applied Problems

In problems 23 through 30, round each answer to two significant digits.

23. Interest on savings accounts is compounded daily in most banks. A formula for interest compounded daily is:

$$A = Pe^{(r)(n)}$$

where A = final amount, P = principal, n = number of years, and r = annual interest rate. For a principal of \$1000 at an interest rate $r = 6\%$, find how many years it will take for the amount A to grow to:
(a) \$1500. **(b)** \$2000. **(c)** \$3000.

24. The world population was growing in the year 2000 at the rate of approximately 1.5% per year. This is expressed by the exponential function:

$$A = A_0(1.015)^n$$

where A = population n years from 2000 and $A_0 = 6.0 \times 10^9$ (2000 estimate). Based on this exponential growth rate, what year will the population:
(a) reach 7 billion?
(b) double to 12 billion?

Applications to Electronics

25. In Example 24.3, find the time t when the current i is:
(a) 8.0 mA. **(b)** 20 mA.

26. In the RC circuit in Figure 24-1, after the capacitor is charged, the switch is moved to discharge. The capacitor discharges through the resistance and the voltage across the capacitor is given by:

$$v_C = Ve^{-t/RC}$$

If $V = 60$ V, $R = 1.0$ kΩ, and $C = 5.0$ μF, find the time t when the capacitor voltage is:
(a) $v_C = 15$ V. **(b)** $v_C = 5.0$ V.

27. A discharged battery is being charged by a solar cell. The voltage after t hours is given by:

$$v = 13.2(1 - e^{-0.10t})$$

Find t when:
(a) $v = 12.0$ V. **(b)** $v = 12.8$ V.

28. In problem 27, if 13.2 V is the maximum voltage of the fully charged battery, find how long it takes for the voltage to reach:
(a) 95% of the maximum.
(b) 99% of the maximum.

29. In the RL circuit in Figure 24-2, the current after the switch S is closed is given by:

$$i_L = \frac{V}{R}(1 - e^{-Rt/L})$$

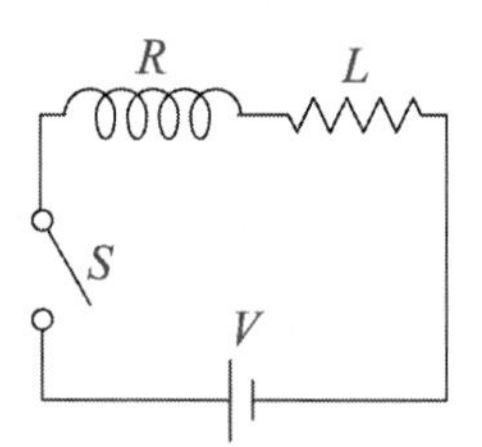

FIGURE 24-2 *RL* circuit for problems 29 and 30.

If $V = 50$ V, $R = 100\ \Omega$, and $L = 0.5$ H, find the time t when i_L is:
(a) 100 mA. **(b)** 400 mA.

30. In problem 29, if $V = 2.4$ V, $R = 75\ \Omega$, and $L = 800$ mH, find the time t when i_L is:
(a) 5.0 mA. **(b)** 15 mA.

24.2 *RC* and *RL* Circuits

The current and voltage changes in RC and RL circuits are exponential functions, one of which is seen in Example 24.3 on page 480. These functions can be described by universal curves, which can be applied to any RC or RL circuit as discussed below.

RC Circuit

Figure 24-3 shows an RC circuit, which contains a constant dc voltage source V and a single pole double throw (SPDT) switch S. Associated with this circuit is a time constant τ (tau), which provides a standard unit for comparing different circuits:

Formula ***RC* Time Constant**

$$\tau = RC \tag{24.1}$$

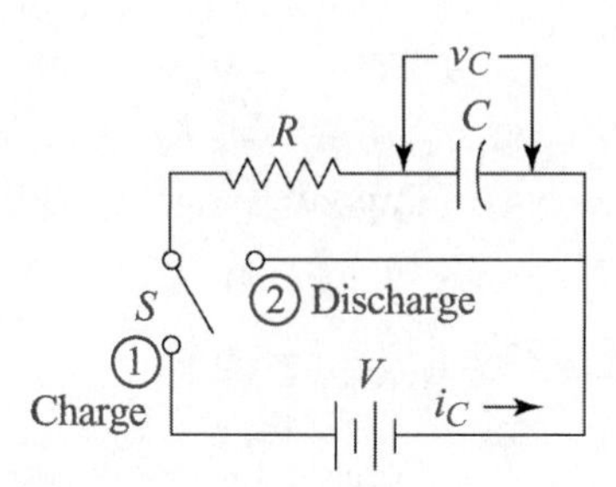

FIGURE 24-3 Exponential change in *RC* circuit for Example 24.4.

When S is switched to position 1, the capacitor charges and the current t seconds after the switch is closed is given by:

Answers to Error Box Problems, page 480:
1. $2 < x < 3$ **2.** $3 < n < 4$ **3.** $-1 < x < 0$ **4.** $-2 < n < -1$ **5.** $2 < t < 3$ **6.** $-2 < t < -1$
7. $6 < n < 7$ **8.** $3 < k < 4$

Formula

Capacitor Charge and Discharge Current

$$i_C = \frac{V}{R} e^{-t/\tau} \tag{24.2}$$

When the capacitor is fully charged, the capacitor voltage v_C equals the applied voltage V. If S is then switched to position 2, the capacitor will discharge through the resistance. The current t seconds after S is switched to position 2 is also given by formula (24.2). When the capacitor is charging or discharging, the initial current at $t = 0$ is $I = \frac{V}{R}$. The current then exponentially decays to practically zero in a time interval equal to five time constants (5τ). See Figure 24-4, which shows the universal curves for *RC* and *RL* circuits.

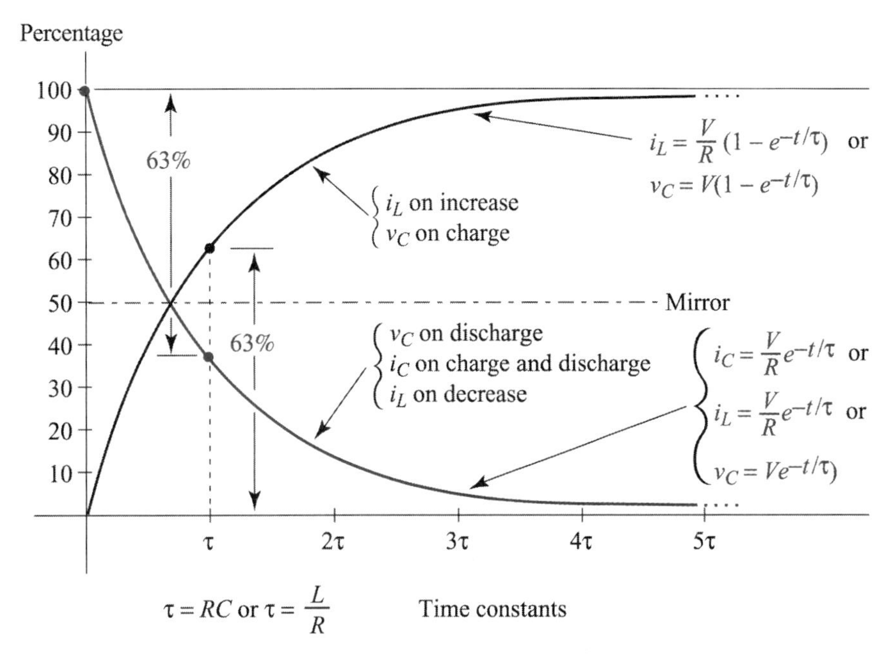

FIGURE 24-4 Universal time constant curves for *RC* and *RL* circuits.

When the capacitor is discharging, the voltage across the capacitor t seconds after the switch is closed is given by:

Formula

Capacitor Discharge Voltage

$$v_C = Ve^{-t/\tau} \tag{24.3}$$

Formula (24.3) is a curve of exponential decay like capacitor current. When $t = 0$, the initial voltage across the capacitor $= V$, and it exponentially decays to practically zero in a time interval equal to five time constants.

When the capacitor is charging, the voltage across the capacitor t seconds after the switch is closed is given by:

Formula

Capacitor Charge Voltage

$$v_C = V(1 - e^{-t/\tau}) \tag{24.4}$$

Formula (24.4) is a rising exponential curve, which starts at zero and increases at a slower and slower rate toward a maximum value $= V$. For all practical purposes it reaches the maximum or steady-state value in a time interval equal to five time constants. See Figure 24-4. This rising curve is the mirror image of the decay curve. The horizontal line passing through the point of intersection of the curves represents the mirror.

RL Circuit

Figure 24-5 shows an *RL* circuit, which contains a constant dc voltage source *V* and a single pole double throw (SPDT) switch *S*. The time constant for this circuit is:

Formula **_RL_ Time Constant**

$$\tau = \frac{L}{R} \tag{24.5}$$

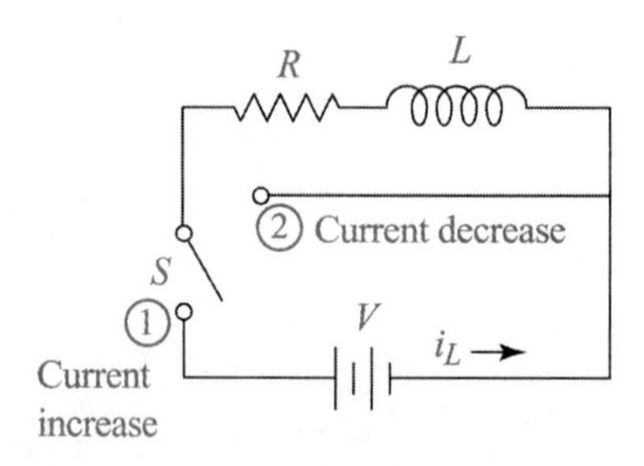

FIGURE 24-5 Exponential change in *RL* circuit for Example 24.5.

When *S* is switched to position 1, the inductor current i_L, *t* seconds after the switch is closed, is given by:

Formula **Inductor Current Increase**

$$i_L = \frac{V}{R}(1 - e^{-t/\tau}) \tag{24.6}$$

Formula (24.6) is the same rising exponential curve as formula (24.4), which starts at zero and increases at a slower and slower rate toward the steady state or maximum value $I = \frac{V}{R}$. For all practical purposes it reaches the maximum value in a time interval equal to five time constants. See Figure 24-4. After the current reaches its maximum value, if *S* is switched to position 2, the current decays exponentially and, *t* seconds after the switch is closed, is given by:

Formula **Inductor Current Decrease**

$$i_L = \frac{V}{R}e^{-t/\tau} \tag{24.7}$$

Formula (24.7) is the same curve as capacitor current (24.2) and capacitor discharge voltage (24.3). At $t = 0$, the inductor current $I = \frac{V}{R}$ and decays practically to zero in five time constants.

The two types of exponential curves in Figure 24-4 are based on multiples of the time constant and the percentage of current or voltage. Since the curves have the same shape, *the decrease, as a percent of the initial value for the decay curve, equals the increase, as a percent of the final value, for the rising curve.* This percent change is given by the formula:

Formula

Current and Voltage Change in *RC* and *RL* Circuits

$$\text{Percent Change} = (1 - e^{-t/\tau}) \times 100\% \qquad \textbf{(24.8)}$$

$$t / \tau = \text{number of time constants}$$

The values for t / τ and the percent change are shown in Table 24-1. After 0.70 time constants the percent change is 50%, and after 1.0 time constant the percent change is 63%. After 5.0 time constants, the percent change is 99%, and the circuit reaches steady state for all practical purposes.

TABLE 24-1 TIME CONSTANTS

Number of Time Constants t/τ	Percent Change
0.25	22%
0.50	39%
0.70	50%
1.0	63%
2.0	86%
3.0	95%
4.0	98%
5.0	99%

Study the following examples which show how to apply the above ideas.

EXAMPLE 24.4

For the *RC* circuit in Figure 24-3, $V = 20$ V, $R = 270\ \Omega$, and $C = 10\ \mu$F. If S is switched to position 1 to charge the capacitor:

(a) Find i_C and v_C after 1.5 ms.

(b) Find t when $i_C = 20$ mA.

(c) Find t when v_C has increased to 95% of its final value.

Solution

(a) To find values of i_C and v_C, first find the time constant using formula (24.1):

$$\tau = RC = (270\ \Omega)(10\ \mu\text{F}) = (270)(10 \times 10^{-6}) = 0.0027\ \text{s} = 2.7\ \text{ms}$$

Then to find i_C, substitute $t = 1.5$ ms, $V = 20$ V, $R = 270\ \Omega$, and $\tau = 2.7$ ms in formula (24.2):

$$i_C = \frac{V}{R} e^{-t/\tau} = \frac{20}{270} e^{-1.5/2.7} = 0.074(0.5738)\ \text{A} \approx 43\ \text{mA}$$

Note that all calculations are done using milliseconds for t and τ.

To find v_C after 1.5 ms, substitute $t = 1.5$ ms, $V = 20$ V, and $\tau = -2.7$ ms in formula (24.4):

$$v_C = V(1 - e^{-t/\tau}) = 20(1 - e^{-1.5/2.7}) = 20(0.4262)\text{ V} \approx 8.5\text{ V}$$

(b) To find t when $i_C = 20$ mA, substitute the given values in formula (24.2) and solve for the exponential term:

$$0.020 = \frac{20}{270} e^{-t/2.7}$$
$$0.020 = 0.074e^{-t/2.7}$$
$$e^{-t/2.7} = \frac{0.020}{0.074} \approx 0.27$$

Rewrite in logarithmic form and solve for t:

$$\ln 0.27 = \frac{-t}{2.7}$$
$$t = -2.7(\ln 0.27) \approx 3.5\text{ ms}$$

(c) To find how long it takes for v_C to increase to 95% of its final value, use Table 24-1 and note that this percent change corresponds to 3.0 time constants. Therefore the time is:

$$t = 3.0\tau = 3.0(2.7\text{ ms}) = 8.1\text{ ms}$$

EXAMPLE 24.5

In the RL circuit in Figure 24-5, $R = 330\ \Omega$, $L = 500$ mH, and $V = 30$ V. If S is switched to position 1 causing the current to increase:

(a) Find i_L after 3.2 ms.

(b) Find t when $i_L = 25$ mA.

(c) Find t when the current has increased to 30% of its final value.

Solution

(a) First find the time constant. Apply formula (24.5):

$$\tau = \frac{L}{R} = \frac{500\text{ mH}}{330\ \Omega} \approx 1.52\text{ ms}$$

Substitute $t = 3.2$ ms in formula (24.6) and calculate i_L:

$$i_L = \frac{V}{R}(1 - e^{-t/\tau}) = \frac{30\text{ V}}{330\ \Omega}(1 - e^{-3.2/1.52}) \approx 80\text{ mA}$$

(b) Substitute $i_L + 25$ mA in formula (24.6). Solve for the exponential term and use natural logarithms to find t:

$$0.025 = \frac{30\text{ V}}{330\ \Omega}(1 - e^{-t/1.52}) = 0.0909(1 - e^{-t/1.52})$$

Divide both sides by 0.0909: $$1 - e^{-t/1.52} = \frac{0.025}{0.0909} = 0.275$$

Solve for $e^{-t/1.52}$: $$e^{-t/1.52} = 1 - 0.275 = 0.725$$

Change to logarithmic form: $$\ln 0.725 = \frac{-t}{1.52}$$

Solve for t: $$t = (-1.52)(\ln 0.725) \approx 0.490\text{ ms}$$
$$= 490\ \mu\text{s}$$

(c) To find t when the current has increased to 30% of its value, you cannot use Table 24-1. Use formula (24.8) for percent change to find t:

$$\text{Percent Change} = (1 - e^{-t/\tau}) \times 100\%$$
$$30\% = (1 - e^{-t/1.52}) \times 100\%$$

Divide both sides by 100%: $1 - e^{-t/1.52} = \dfrac{30\%}{100\%} = 0.30$

Solve for $e^{-t/1.52}$: $e^{-t/1.52} = 1 - 0.30 = 0.70$

Change to logarithmic form: $\ln 0.70 = \dfrac{-t}{1.52}$

Solve for t: $t = (-1.52)(\ln 0.70) \approx 0.540$ ms $= 540\ \mu$s

EXERCISE 24.2

Round all answers to two significant digits.

In exercises 1 through 16, find the time constant of the given circuit.

1. *RC* Circuit: $R = 200\ \Omega$, $C = 5.0\ \mu$F
2. *RC* Circuit: $R = 560\ \Omega$, $C = 3.2\ \mu$F
3. *RC* Circuit: $R = 1.3$ kΩ, $C = 2.0\ \mu$F
4. *RC* Circuit: $R = 1.8$ kΩ, $C = 5.1\ \mu$F
5. *RC* Circuit: $R = 10$ kΩ, $C = 400$ nF
6. *RC* Circuit: $R = 22$ kΩ, $C = 150$ nF
7. *RC* Circuit: $R = 2.7$ MΩ, $C = 200$ pF
8. *RC* Circuit: $R = 1.6$ MΩ, $C = 400$ pF
9. *RL* Circuit: $R = 30\ \Omega$, $L = 60$ mH
10. *RL* Circuit: $R = 430\ \Omega$, $L = 680$ mH
11. *RL* Circuit: $R = 75\ \Omega$, $L = 500$ mH
12. *RL* Circuit: $R = 47\ \Omega$, $L = 300$ mH
13. *RL* Circuit: $R = 1.0$ kΩ, $L = 750$ mH
14. *RL* Circuit: $R = 1.3$ kΩ, $L = 1.0$ H
15. *RL* Circuit: $R = 4.3\ \Omega$, $L = 900\ \mu$H
16. *RL* Circuit: $R = 33\ \Omega$, $L = 10$ mH
17. In the *RC* circuit in Figure 24-3, $V = 50$ V, $R = 200\ \Omega$, and $C = 10\ \mu$F.
    (a) Find the time constant of the circuit.
    (b) Graph the exponential decay of current when S is switched to position 2 after the capacitor is charged. Show points for 1, 2, 3, and 4 time constants. See Figure 24-4.
18. Do the same as problem 17 for $V = 100$ V, $R = 2.0$ kΩ, and $C = 5.0\ \mu$F.
19. Given the *RL* circuit in Figure 24-5 with $V = 60$ V, $R = 75\ \Omega$, and $L = 300$ mH.
    (a) Find the time constant of the circuit.
    (b) Graph the exponential curve of increasing current when S is switched to position 1, showing points for 1, 2, 3, and 4 time constants. See Figure 24-4.
20. Do the same as problem 19 when $V = 20$ V, $R = 1.0$ kΩ, and $L = 800$ mH.
21. Given the *RC* circuit in Figure 24-3 with $V = 60$ V, $R = 750\ \Omega$, and $C = 20\ \mu$F. If S is switched to position 1 to charge the capacitor:
    (a) Find i_C and v_C after 20 ms.
    (b) Find t when $i_C = 50$ mA.
    (c) Find t when $v_C = 25$ V.
22. In problem 21, after the capacitor is fully charged, S is switched to position 2 discharging the capacitor.
    (a) Find i_C and v_C after 10 ms.
    (b) Find t when $i_C = 10$ mA.
    (c) Find t when $v_C = 10$ V.
23. In problem 21, find the time for:
    (a) i_C to decrease 50% of its initial value.
    (b) v_C to increase to 90% of its final value.
24. In problem 22, find the time for:
    (a) v_C to decrease 39% of its initial value.
    (b) i_C to decrease 80% of its initial value.
25. In the *RL* circuit in Figure 24-5, $V = 100$ V, $R = 100\ \Omega$, and $L = 500$ mH. If S is switched to position 1 causing the current to increase:
    (a) Find i_L after 12 ms.
    (b) Find t when $i_L = 500$ mA.
    (c) Find t when the current has increased to 22% of its final value.
26. In problem 25, after the circuit reaches steady state, S is switched to position 2 and the current decreases.
    (a) Find i_L after 4 ms.
    (b) Find t when $i_L = 250$ mA.
    (c) Find t when the current has decreased 95% of its initial value.

27. In Figure 24-3, when S is switched to position 1 the charge on the capacitor in coulombs (C) is given by the exponential equation:

$$q_C = CV(1 - e^{-t/RC})$$

This equation is the same as formula (24.4) where CV represents 100% charge. If $V = 40$ V, $C = 20$ μF, and $R = 1.0$ kΩ:

(a) Find q_C after 25 ms.
(b) Find t when $q_C = 300$ μC.
(c) Find t when the capacitor is 95% charged.

28. In Figure 24-3, when S is switched to position 1, the voltage across the resistance is given by the exponential equation:

$$v_R = Ve^{-t/RC}$$

This equation is the same as formula (24.3). If $V = 20$ V, $C = 500$ nF, and $R = 2.0$ kΩ:

(a) Find v_R after 500 μs.
(b) Find t when $v_R = 15$ V.
(c) Find t when v_R has decreased 63% of its initial value.

24.3 Power and Gain

Human senses detect changes in sound levels based on a *ratio* of the power increase and not based on the actual power increase. For example, suppose the power output of an amplifier increases from 5 W to 10 W. This is an *actual* increase of 5 W, but the *ratio* of the power increase from 5 W to 10 W is 10/5 = 2/1 or two times. If the power increases another 5 W to 15 W, the increase in sound that the human ear detects will appear less than the first increase of 5 W. But, if the power increases by the *same* ratio from 10 W to 20 W, the increase in sound *will* appear to be the same as the first increase. That is, the increase in sound when the power doubles from 5 W to 10 W will sound similar to when the power doubles from 10 W to 20 W, or from 20 W to 40 W, and so on.

As a result of the above phenomena, sound levels are defined in terms of logarithms of power ratios. The quotient rule for logarithms (23.7) shows that logarithms of the above power values: 5 W, 10 W, 20 W, and 40 W, are equally apart on a logarithmic scale:

$$\log 10 - \log 5 = \log (10/5) = \log 2$$
$$\log 20 - \log 10 = \log (20/10) = \log 2$$
$$\log 40 - \log 20 = \log (40/20) = \log 2$$

When the ratio of two values are the same, the difference in the logarithms of the two values is also the same; therefore, a logarithmic scale better represents the response of a human ear. The original logarithmic unit for a power increase of ten times is called the bel in honor of Alexander Graham Bell:

$$\text{bel} = \log\left(\frac{P_{out}}{P_{in}}\right)$$

where P_{out} = output power and P_{in} = input power. The bel turned out to be too large a unit for practical use. The smallest change in sound intensity that a human ear can detect is the decibel (dB), which is one-tenth of a bel. The dB is therefore the unit used to measure logarithmic change in power levels:

Formula **Power Gain (Decibels)**

$$A_p = 10 \log\left(\frac{P_{out}}{P_{in}}\right) \tag{24.9}$$

EXAMPLE 24.6

The power input to an amplifier $P_{in} = 0.25$ W, and the power output $P_{out} = 4.0$ W. Find the power gain in decibels.

Solution Apply formula (24.9):

$$A_p = 10 \log\left(\frac{4.0}{0.25}\right) = 10 \log 16$$

$$A_p = 10(1.204) \approx 12.0 \text{ dB gain}$$

EXAMPLE 24.7

The power gain of a circuit is $A_p = -20$ dB. If $P_{in} = 500$ mW, find P_{out}.

Solution Substitute the given values in formula (24.9):

$$-20 = 10 \log\left(\frac{P_{out}}{500}\right)$$

Divide both sides by 10. Then change to exponential form and solve for P_{out}:

$$\log\left(\frac{P_{out}}{500}\right) = -2$$

$$10^2 = \frac{P_{out}}{500}$$

$$P_{out} = 5.0 \text{ mW}$$

In an amplifier, when the input resistance R is the same as the output resistance R, you can apply the power formula $P = V^2/R$ to P_{out} and P_{in} and write formula (24.9) in terms of voltage:

$$A_p = 10 \log\left(\frac{V_{out}^2}{R} \div \frac{V_{in}^2}{R}\right)$$

where V_{in} = input voltage and V_{out} = output voltage.

When inverting and multiplying the fractions, the R's divide out, and the fraction becomes:

$$A_p = 10 \log\left(\frac{V_{out}}{V_{in}}\right)^2$$

Applying the power rule (23.8) makes the exponent 2 a coefficient, which multiplies by 10 to give the formula:

Formula

Voltage Gain (Decibels)

$$A_v = 20 \log\left(\frac{V_{out}}{V_{in}}\right) \qquad (24.10)$$

EXAMPLE 24.8

The input voltage of a circuit is $V_{in} = 40$ mV, and the output voltage is $V_{out} = 5.5$ V. If the input and output resistances are the same, find the power gain.

Solution Apply formula (24.10) using 0.040 V for 40 mV:

$$A_p = 20 \log\left(\frac{5.5}{0.040}\right) = 20\ (2.14) \approx 43 \text{ dB gain}$$

EXAMPLE 24.9

The voltage gain in a network is $A_p = -20\ dB$. If the input and output resistances are the same and $V_{out} = 1.6$ V, what is the input voltage?

Solution Substitute in formula (24.10) and solve for the logarithmic term:

$$-20 = 20 \log\left(\frac{1.6}{V_{in}}\right)$$

Divide by 20:

$$\log\left(\frac{1.6}{V_{in}}\right) = \frac{-20}{20} = -1.0$$

Change to exponential form:

$$10^{-1.0} = \frac{1.6}{V_{in}}$$

Solve for V_{in}:

$$V_{in} = \frac{1.6}{10^{-1.0}} = 16 \text{ V}$$

Notice in this example that when the voltage changes by ten times, the decibel change is 20 dB. This is referred to as a *decade* change and is discussed further in the next section.

Decibel Levels

Decibel levels must be in reference to some standard level that corresponds to 0 dB. One of the common references used in telecommunications is 1 mW. For this reference, the symbol dBm is used and it means $P_{in} = 1$ mW:

Formula

Standard Power Gain

$$A_{p(\text{dBm})} = 10 \log\left(\frac{P_{out}}{1 \text{ mW}}\right) \tag{24.11}$$

EXAMPLE 24.10

The power of a circuit is 2.0 mW. Find the power gain in dBm.

Solution Apply formula (24.11):

$$A_{p(\text{dBm})} = 10 \log\left(\frac{2.0}{1}\right) = 3.0 \text{ dBm}$$

Antenna Gain

The gain of an antenna compares the power supplied to the antenna with the power required by a standard reference antenna to produce the same signal strength. If P_{in} is the power supplied to the given antenna and P_{out} is the power supplied to the standard antenna, the power gain of the given antenna is given by formula (24.9).

EXAMPLE 24.11

A multielement transmitting antenna produces a 150-mV/m (millivolts per meter) signal at a receiving station when supplied with 500 W. A standard dipole antenna requires 2 kW to produce the same signal strength at the receiving station. What is the power gain of the multielement antenna?

Solution Substitute $P_{out} = 2$ kW $= 2000$ W and $P_{in} = 500$ W in (24.9) and compute the power gain:

$$A_p = 10 \log \frac{2000}{500} = 6.0 \text{ dB}$$

EXERCISE 24.3

Round all answers to two significant digits. In exercises 1 through 10, find the power gain A_p *in decibels.*

1. $P_{in} = 50$ mW, $P_{out} = 5.0$ W
2. $P_{in} = 1.5$ W, $P_{out} = 15$ mW
3. $P_{in} = 100$ mW, $P_{out} = 25$ mW
4. $P_{in} = 1.0$ W, $P_{out} = 4.0$ W
5. $P_{in} = 10$ W, $P_{out} = 20$ W
6. $P_{in} = 24$ W, $P_{out} = 12$ W
7. $P_{in} = 1.0$ mW, $P_{out} = 2.5$ W
8. $P_{in} = 1.0$ mW, $P_{out} = 1.6$ W
9. $P_{in} = 400$ μW, $P_{out} = 3.6$ W
10. $P_{in} = 500$ mW, $P_{out} = 500$ μW

In exercises 11 through 20, find the voltage gain A_V *in decibels.*

11. $V_{in} = 40$ mV, $V_{out} = 400$ mV
12. $V_{in} = 1.0$ V, $V_{out} = 100$ mV
13. $V_{in} = 90$ mV, $V_{out} = 45$ mV
14. $V_{in} = 750$ mV, $V_{out} = 1.5$ V
15. $V_{in} = 800$ μV, $V_{out} = 33$ mV
16. $V_{in} = 750$ μV, $V_{out} = 41$ mV
17. $V_{in} = 70$ μV, $V_{out} = 6.0$ mV
18. $V_{in} = 150$ μV, $V_{out} = 4.3$ mV
19. $V_{in} = 1.0$ mV, $V_{out} = 30$ mV
20. $V_{in} = 1.0$ mV, $V_{out} = 15$ mV

In exercises 21 through 26, find the standard power gain A_p *in dBm for the given power output.*

21. $P_{out} = 20$ mW
22. $P_{out} = 10$ mW
23. $P_{out} = 1.0$ W
24. $P_{out} = 2.0$ W
25. $P_{out} = 300$ μW
26. $P_{out} = 100$ μW

In exercises 27 through 58 given the power or voltage gain, find the indicated value.

27. $P_{in} = 1.0$ mW, $A_p = 20$ dB; P_{out}
28. $P_{in} = 3.0$ mW, $A_p = 6$ dB; P_{out}
29. $P_{in} = 10$ mW, $A_p = 3$ dB; P_{out}
30. $P_{in} = 20$ mW, $A_p = 20$ dB; P_{out}
31. $P_{in} = 8.0$ W, $A_p = -6$ dB; P_{out}
32. $P_{in} = 12$ mW, $A_p = -3$ dB; P_{out}
33. $P_{in} = 7.0$ mW, $A_p = 30$ dB; P_{out}
34. $P_{in} = 10$ mW, $A_p = 25$ dB; P_{out}
35. $P_{out} = 10$ mW, $A_p = -6$ dB; P_{in}
36. $P_{out} = 2.0$ W, $A_p = -30$ dB; P_{in}
37. $P_{out} = 4.0$ W, $A_p = 6$ dB; P_{in}
38. $P_{out} = 100$ μW, $A_p = -20$ dB; P_{in}
39. $P_{out} = 100$ μW, $A_p = -20$ dB; P_{in}
40. $P_{out} = 10.0$ mW, $A_p = -6$ dB; P_{in}
41. $P_{in} = 10$ mW, $A_p = 10$ dB; P_{out}
42. $P_{out} = 100$ μW, $A_p = -10$ dB; P_{in}
43. $V_{in} = 1.0$ V, $A_v = 6$ dB; V_{out}
44. $V_{in} = 1.0$ V, $A_v = 20$ dB; V_{out}
45. $V_{in} = 4.0$ V, $A_v = -20$ dB; V_{out}
46. $V_{in} = 71$ mV, $A_v = -3$ dB; V_{out}
47. $V_{in} = 2.5$ V, $A_v = 6$ dB; V_{out}
48. $V_{in} = 30$ mV, $A_v = 3$ dB; V_{out}
49. $V_{in} = 1.5$ V, $A_v = 10$ dB; V_{out}
50. $V_{in} = 2.5$ V, $A_v = -10$ dB; V_{out}
51. $V_{out} = 4.5$ mV, $A_v = -20$ dB; V_{in}
52. $V_{out} = 100$ mV, $A_v = -6$ dB; V_{in}
53. $V_{out} = 7.6$ V, $A_v = 6$ dB; V_{in}
54. $V_{out} = 650$ mV, $A_v = 20$ dB; V_{in}
55. $V_{out} = 2.8$ V, $A_v = -3$ dB; V_{in}
56. $V_{out} = 5.6$ V, $A_v = 3$ dB; V_{in}
57. $V_{out} = 12$ V, $A_v = 30$ dB; V_{in}
58. $V_{out} = 9.0$ mV, $A_v = 16$ dB; V_{in}
59. A medium frequency directional antenna produces a signal of 110 mV/m at a receiving station when supplied with 1.0 kW of power. A standard vertical antenna requires 2.0 kW of power to produce the same signal strength at the receiving station. What is the power gain of the directional antenna?
60. A medium frequency directional antenna produces a signal of 900 mV/m at a receiving station when supplied with 300 W of power. A standard vertical antenna requires 1.0 kW of power to produce the same signal strength at the receiving station. What is the power gain of the directional antenna?
61. A multielement antenna produces 25 μV/m at a receiving station when supplied with 75 W of power. A standard dipole half wave antenna requires 500 W of power to produce the same signal strength at the receiving station. What is the power gain of the multielement antenna?
62. A multielement antenna produces 50 μV/m at a receiving station when supplied with 100 W of power. A standard dipole half-wave antenna requires 600 W of power to produce the same signal strength at the receiving station. What is the power gain of the multielement antenna?

24.4 Bode Plots

A bode plot is a semilogarithmic graph showing the frequency response of a circuit and is very helpful in understanding the frequency characteristics of the circuit. A semilogarithmic graph uses one linear scale and one logarithmic scale. The logarithmic scale is drawn so that distances are proportional to the logarithms of the plotted values. For example, consider the values 10, 100, and 1000. Their logarithms are log 10 = 1, log 100 = 2, and log 1000 = 3. Since their logarithms all differ by the same amount, the values 10, 100, and 1000 are plotted equidistant on a logarithmic scale. The distance between 1 and 100 is twice the distance between 1 and 10. See Figures 24-6 and 24-7. A Bode plot shows the change in decibels compared with the change in frequency for a circuit. Study the following examples, which explain more thoroughly the Bode plot.

EXAMPLE 24.12 Given the *RC* low-pass filter in Figure 22-13(a) with $R = 2.0\ \text{k}\Omega$ and $C = 800$ nF. Construct a Bode plot of the frequency response on *three cycle* semilogarithmic paper.

Solution First calculate the cutoff frequency applying formula (22.10):

$$f_c = \frac{1}{2\pi(2.0\ \text{k}\Omega)(800\ \text{nF})} \approx 100\ \text{Hz}$$

Figure 24-6 shows the Bode plot of the frequency response. The frequencies are plotted on the top horizontal scale, which is logarithmic. The frequencies start at 10 Hz on the left and increase by three decades (cycles), which are powers of 10, to 100 Hz, 1.0 kHz, and 10 kHz on the right.

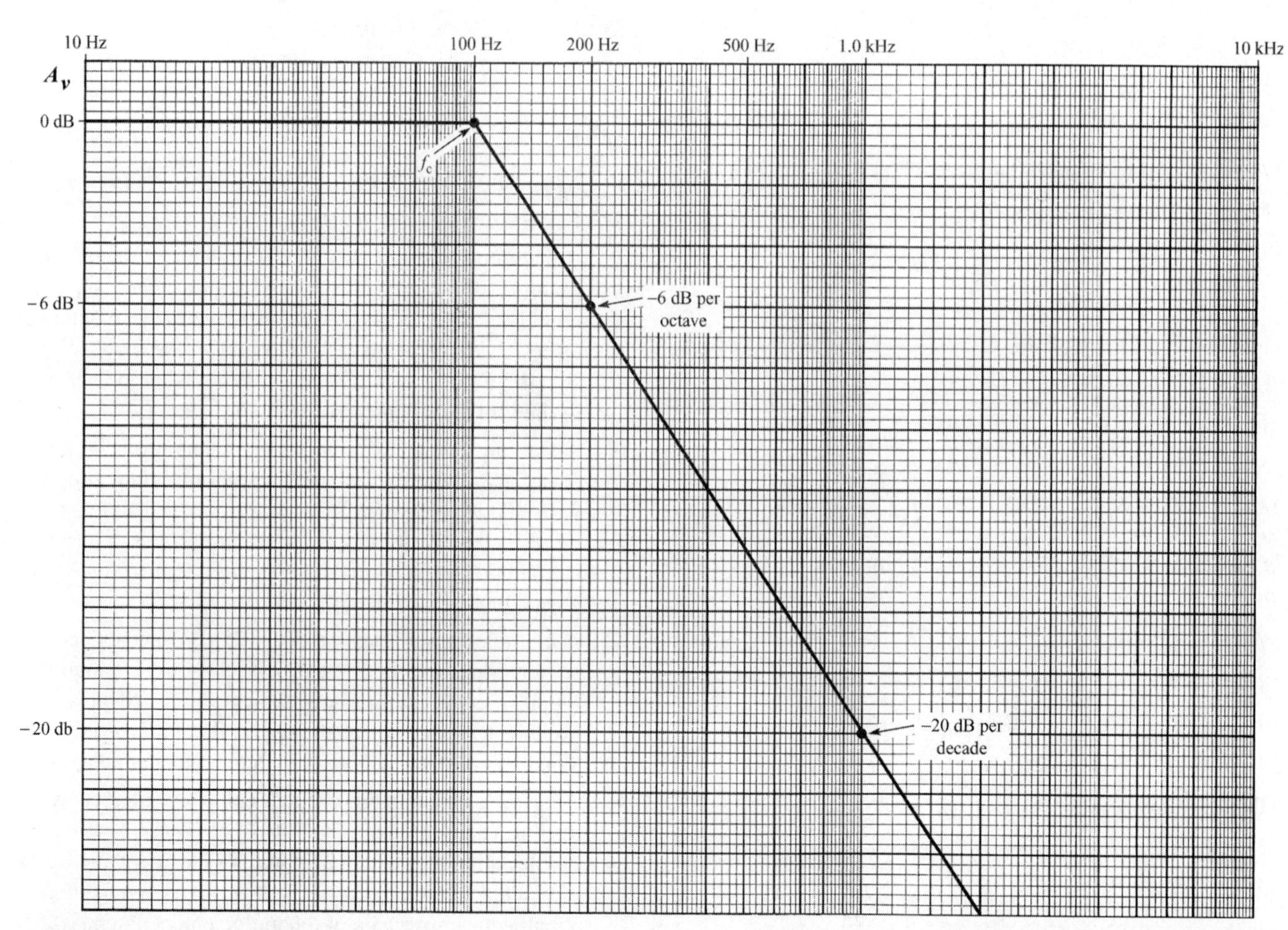

FIGURE 24-6 Bode plot of frequency response of *RC* low-pass filter for Example 24.12. Frequency decreases 20 dB per decade and 6 dB per octave.

Notice that because of the logarithmic scale these decades are equally spaced. The voltage gains in decibels are plotted on the left vertical scale, which is linear and proportional.

All frequencies below f_c are shown on the 0-dB line, which corresponds to no attenuation of these frequencies by the filter. Above f_c the frequency response decreases. In Example 24.9 it is shown that when the voltage changes by ten times, called a decade change, the decibel change is 20 dB. Similarly, in the Bode plot a decade change in the frequency corresponds to a 20 dB change in A_v. This is shown in the Bode plot as a straight line where A_v changes 20 dB to the right of f_c when the frequency increases a decade from 100 Hz to 1.0 kHz. Observe also on the Bode plot that when the frequency doubles (or halves) from 100 Hz to 200 Hz, known as an *octave* change, the frequency changes by 6 dB. These relationships of 20 dB per decade and 6 dB per octave are properties of all Bode plots.

EXAMPLE 24.13 Consider an amplifier whose frequency response is the same as that of a bandpass filter, which is discussed in Section 22.6 on Filters. The mid-frequency gain of the amplifier $A_v = 30$ dB. The lower cutoff frequency $f_1 = 100$ Hz and the upper cutoff frequency $f_2 =$ 1.2 kHz. Construct a Bode plot of the frequency response on *five-cycle* semilog paper.

Solution Figure 24-7 shows the Bode plot of the frequency response. The frequencies are plotted on the top horizontal scale, which is logarithmic; and the voltage gains in decibels are plotted on the left vertical scale, which is linear. The frequencies start at 1.0 Hz on the left and increase by five decades (cycles), which are powers of 10, to 100 kHz, on

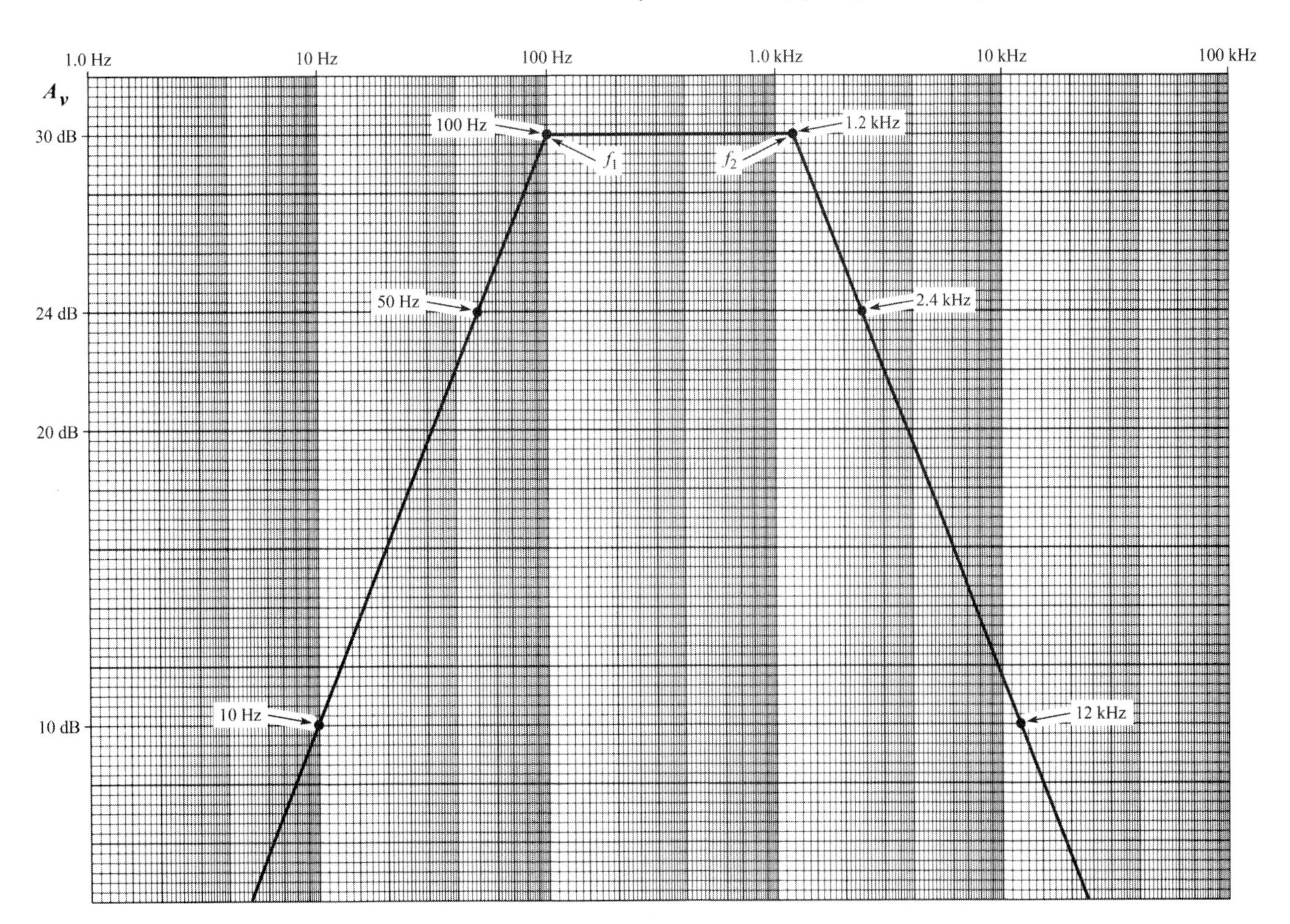

FIGURE 24-7 Bode Plot of voltage gain vs. frequency response of amplifier for Example 24.13.

the right. All frequencies between the cutoff frequencies are shown on the 30-dB line, which is the mid-frequency gain. The frequency response decreases to the left of f_1, and to the right of f_2. This is shown in the Bode plot as a straight line where A_v decreases 20 dB (from 30 dB to 10 dB) to the right and to the left of the cutoff frequencies, when the frequency changes by a decade from the cutoff frequency. On the left, 10 dB corresponds to a decade decrease from 100 Hz to 10 Hz. On the right, 10 dB corresponds to a decade increase from 1.2 kHz to 12 kHz. Observe also on the Bode plot that when the frequency decreases by half from 100 Hz to 50 Hz, which is an octave change, the frequency changes by 6 dB from 30 dB to 24 dB and similarity on the right.

The Bode plot is an approximation of the actual frequency response. The actual frequency response tends to curve away from the line near the cutoff frequency but approaches the line as as you move farther away from the cutoff frequency.

EXERCISE 24.4

Round all answers to two significant digits. In exercises 1 through 6, given R *and* C *for the low-pass filter circuit in Figure 22-13(a), find the cutoff frequency* f_c *and construct a Bode plot on three-cycle semilog graph paper using a voltage gain* $A_v = 0$ *dB. Show the gain changes of 6 dB per octave and 20 dB per decade. See Example 24.12.*

1. $R = 11\text{ k}\Omega$, $C = 1.2\text{ nF}$

2. $R = 2.0\text{ k}\Omega$, $C = 500\text{ nF}$

3. $R = 6.8\text{ k}\Omega$, $C = 260\text{ nF}$

4. $R = 16\text{ k}\Omega$, $C = 20\text{ nF}$

5. $R = 5.1\text{ k}\Omega$, $C = 2.1\text{ nF}$

6. $R = 750\ \Omega$, $C = 8.5\text{ nF}$

In exercises 7 through 12 given R *and* C *for the high-pass filter circuit in Figure 22-14(a). Find the cutoff frequency* f_c *and construct a Bode plot on three cycle semilog graph paper using a voltage gain* $A_v = 0$ *dB. Show the gain changes of 6 dB per octave and 20 dB per decade.*

7. $C = 260\text{ nF}$, $R = 6.8\text{ k}\Omega$

8. $C = 1.0\ \mu\text{F}$, $R = 820\ \Omega$

9. $C = 50\text{ nF}$, $R = 3.3\text{ k}\Omega$

10. $C = 800\text{ nF}$, $R = 2.0\text{ k}\Omega$

11. $C = 160\text{ nF}$, $R = 1.5\text{ k}\Omega$

12. $C = 300\text{ nF}$, $R = 330\ \Omega$

In exercises 13 through 18, given the mid-frequency gain of an amplifier A_v*, the lower cutoff frequency* f_1*, and the upper cutoff frequency* f_2*, construct a Bode plot of the frequency response on five-cycle semilog paper. See Example 24.13.*

13. $A_v = 30$ dB, $f_1 = 100$ Hz, $f_2 = 1.5$ kHz

14. $A_v = 10$ dB, $f_1 = 50$ Hz, $f_2 = 3.0$ kHz

15. $A_v = 20$ dB, $f_1 = 90$ Hz, $f_2 = 5.0$ kHz

16. $A_v = -10$ dB, $f_1 = 100$ Hz, $f_2 = 4.0$ kHz

17. $A_v = -25$ dB, $f_1 = 60$ Hz, $f_2 = 1.0$ kHz

18. $A_v = 35$ dB, $f_1 = 80$ Hz, $f_2 = 1.5$ kHz

CHAPTER HIGHLIGHTS

24.1 EXPONENTIAL EQUATIONS

To solve an exponential equation, isolate the exponential term and apply the power rule (23.8) or change to exponential form.

Key Example:

$$1 - e^{-10t} = 0.70$$

$$e^{-10t} = 1 - 0.70 = 0.30$$

$$\ln 0.30 = -10t$$

$$t = \frac{\ln 0.30}{-10} \approx 0.12$$

24.2 *RC* AND *RL* CIRCUITS

Study the universal time constant curves in Figure 24-4 and Table 24-1.

RC Time Constant

$$\tau = RC \qquad (24.1)$$

• • •

Capacitor Charge and Discharge Current

$$i_C = \frac{V}{R} e^{-t/\tau} \qquad (24.2)$$

• • •

Capacitor Discharge Voltage

$$v_C = Ve^{-t/\tau} \quad (24.3)$$

• • •

Capacitor Charge Voltage

$$v_C = V(1 - e^{-t/\tau}) \quad (24.4)$$

Key Example: For the *RC* circuit in Figure 24-3, $V = 20$ V, $R = 270\ \Omega$, and $C = 10\ \mu$F. When *S* is switched to position 1 to charge the capacitor:

$$\tau = RC = (270\ \Omega)(10\ \mu\text{F}) = 2.7\text{ ms}$$

After 1.5 ms:

$$i_C = \frac{V}{R}e^{-t/\tau} = \frac{20}{270}e^{-1.5/2.7} = 0.074(0.5738) \approx 43\text{ mA}$$

$$v_C = V(1 - e^{-t/\tau}) = 20(1 - e^{-1.5/2.7}) \approx 8.5\text{ V}$$

When $i_C = 20$ mA, the value of t is:

$$0.020 = \frac{20}{270}e^{-t/2.7}$$

$$e^{-t/2.7} = 0.27 \rightarrow \ln 0.27 = \frac{-t}{2.7}$$

$$t = -2.7(\ln 0.27) \approx 3.5\text{ ms}$$

From Table 24-1, v_C increases to 95% of its final value in 3.0 time constants:

$$t = 3.0\tau = 3.0(2.7\text{ ms}) = 8.1\text{ ms}$$

***RL* Time Constant**

$$\tau = \frac{L}{R} \quad (24.5)$$

• • •

Inductor Current Increase

$$i_L = \frac{V}{R}(1 - e^{-t/\tau}) \quad (24.6)$$

• • •

Inductor Current Decrease

$$i_L = \frac{V}{R}e^{-t/\tau} \quad (24.7)$$

• • •

Current and Voltage Change in *RC* and *RL* Circuits

$$\text{Percent Change} = (1 - e^{-t/\tau}) \times 100\% \quad (24.8)$$

t/τ = number of time constants

Key Example: For the *RL* circuit in Figure 24-5, $R = 330\ \Omega$, $L = 500$ mH, and $V = 30$ V. If *S* is switched to position 1 causing the current to increase:

$$\tau = \frac{L}{R} = \frac{500\text{ mH}}{330\ \Omega} \approx 1.52\text{ ms}$$

When $t = 3.2$ ms:

$$i_L = \frac{V}{R}(1 - e^{-t/\tau}) = \frac{30\text{ V}}{330\ \Omega}(1 - e^{-3.2/1.52}) \approx 80\text{ mA}$$

When $i_L = 25$ mA, the value of t is:

$$0.025 = \frac{30\text{ V}}{330\ \Omega}(1 - e^{-t/1.52})$$

$$e^{-t/1.52} = 0.725 \rightarrow \ln 0.725 = \frac{-t}{1.52}$$

$$t = (-1.52)(\ln 0.725) \approx 0.490\text{ ms} = 490\ \mu\text{s}$$

To find t when the current increases to 70% of its final value:

$$70\% = (1 - e^{-t/1.52}) \times 100\%$$

$$e^{-t/1.52} = 0.30 \rightarrow \ln 0.30 = \frac{-t}{1.52}$$

$$t = -1.52(\ln 0.30) \approx 1.8\text{ ms}$$

24.3 POWER AND GAIN

Power Gain (Decibels)

$$A_p = 10\log\left(\frac{P_{out}}{P_{in}}\right) \quad (24.9)$$

• • •

Voltage Gain (Decibels)

$$A_v = 20\log\left(\frac{V_{out}}{V_{in}}\right) \quad (24.10)$$

• • •

Standard Power Gain

$$A_{p(\text{dBm})} = 10\log\left(\frac{P_{out}}{1\text{ mW}}\right) \quad (24.11)$$

Key Example: If the input power $P_{in} = 6.0$ mW, for a power gain of 30 dB the output power P_{out} is:

$$30 = 10\log\left(\frac{P_{out}}{6.0}\right)$$

$$\log\left(\frac{P_{out}}{6.0}\right) = 3 \rightarrow 10^3 = \frac{P_{out}}{6.0}$$

$$P_{out} = 6000\text{ mW} = 6\text{ W}$$

24.4 BODE PLOTS

A Bode plot is a semilogarithmic graph of the frequency response of a circuit. The frequencies are plotted on the horizontal logarithmic scale and A_v is plotted on the vertical linear scale. A property of Bode plots is that A_v changes 20 dB per decade (ten times), and 6 dB per octave (two times) change in the frequency. See Figures 24-6 and 24-7.

REVIEW EXERCISES

In exercises 1 through 6, solve each exponential equation to three significant digits.

1. $2^x = 37$

2. $10^{3x} = 30$

3. $e^{2t} = 8.8$

4. $e^{-3.3t} = 5.6$

5. $1 - e^{-10t} = 0.50$

6. $e^{-5t} - 1 = -0.6$

In exercises 7 through 14, find the time constant of the given circuit to two significant digits.

7. RC Circuit: $R = 240\ \Omega$, $C = 6.0\ \mu F$

8. RC Circuit: $R = 1.2\ k\Omega$, $C = 5.6\ \mu F$

9. RC Circuit: $R = 11\ k\Omega$, $C = 360$ nF

10. RC Circuit: $R = 9.1\ M\Omega$, $C = 150$ pF

11. RL Circuit: $R = 13\ \Omega$, $L = 150$ mH

12. RL Circuit: $R = 51\ \Omega$, $L = 400$ mH

13. RL Circuit: $R = 4.7\ \Omega$, $L = 2.5$ mH

14. RL Circuit: $R = 1.6\ k\Omega$, $L = 1.0$ H

In exercises 15 through 22, find A_p *or* A_v *in decibels to two significant digits.*

15. $P_{in} = 40$ mW, $P_{out} = 10$ mW

16. $P_{in} = 6.0$ W, $P_{out} = 12$ W

17. $P_{in} = 1.0$ W, $P_{out} = 60$ W

18. $P_{in} = 14$, $P_{out} = 7.0$ mW

19. $V_{in} = 15$ V, $V_{out} = 1.5$ V

20. $V_{in} = 220$ mV, $V_{out} = 2.2$ V

21. $V_{in} = 1.0$ mV, $V_{out} = 25$ V

22. $V_{in} = 1.6$ mV, $V_{out} = 800\ \mu V$

In exercises 23 through 30, given A_p *or* A_v*, find the indicated value.*

23. $P_{in} = 8.0$ mW, $A_p = 25$ dB; P_{out}

24. $P_{in} = 8.0$ mW, $A_p = -20$ dB gain; P_{out}

25. $P_{out} = 50$ W, $A_p = -6$ dB; P_{in}

26. $P_{out} = 1.0$ W, $A_p = 20$ dB; P_{in}

27. $V_{in} = 10$ mV, $A_v = 40$ dB; V_{out}

28. $V_{in} = 1.0$ mV, $A_v = -12$ dB; V_{out}

29. $V_{out} = 9.0$ V, $A_v = -20$ dB; V_{in}

30. $V_{out} = 900$ mV, $A_v = 6$ dB; V_{in}

In problems 31 through 44, solve each applied problem to two significant digits.

31. In the RC circuit in Figure 24-3, $V = 40$ V, $R = 1.5\ k\Omega$, and $C = 10\ \mu F$. If S is switched to position 1 to charge the capacitor:

(a) Find the time constant of the circuit.

(b) Give the equation for the exponential decay of current i_C in milliamps.

(c) Give the equation for the increase in capacitor voltage v_C.

(d) Find i_C and v_C after 10 ms.

(e) Find t when $i_C = 5.0$ mA.

(f) Find t when $v_C = 20$ V.

32. In problem 31, after the capacitor is fully charged, S is switched to position 2 discharging the capacitor.

(a) Give the equation for the exponential decay of current i_C in milliamps.

(b) Give the equation for the decrease in capacitor voltage.

(c) Find i_C and v_C after 6.0 ms.

(d) Find t when $i_C = 12$ mA.

(e) Find t when $v_C = 30$ V.

33. In problem 31, find the time for:

(a) i_C to decrease 39% of its initial value.

(b) v_C to increase to 80% of its final value.

34. In problem 32, find the time for:

(a) v_C to decrease 63% of its initial value.

(b) i_C to decrease 70% of its initial value.

35. In the RL circuit in Figure 24-5, $V = 100$ V, $R = 100\ \Omega$, and $L = 500$ mH. If S is switched to position 1 causing the current to increase:

(a) Find the time constant of the circuit.

(b) Give the exponential equation for the increase of current i_L.

(c) Find i_L after 8 ms.

(d) Find t when $i_L = 650$ mA.

(e) Find t when the current has increased to 95% of its final value.

36. In problem 35, after the circuit reaches steady state, S is switched to position 2 and the current decreases.

(a) Give the exponential equation for the decrease of current i_L.

(b) Find i_L after 12 ms.
(c) Find t when $i_L = 300$ mA.
(d) Find t when the current has decreased 50% of its initial value.

37. In Figure 24-3, when S is switched to position 1, the voltage across the resistance v_R is given by the voltage decay curve in Figure 24-4. If $V = 9.0$ V, $C = 10$ μF, and $R = 1.2$ kΩ:
(a) Give the exponential equation for the decay of voltage.
(b) Find v_R after 20 ms.
(c) Find the time for v_R to decay 40% of its initial value.

38. In Figure 24-5, when S is switched to position 1, the voltage across the resistance v_R is given by the rising exponential curve in Figure 24-4. If $V = 60$ V, $R = 200$ Ω, and $L = 800$ mH:
(a) Give the exponential equation for the rise of voltage.
(b) Find v_R after 1.2 ms.
(c) Find the time for v_R to increase to 98% of its final value.

39. An audio frequency amplifier has an input power of 2 mW and an input voltage of 1.5 V. If the decibel gain is 30 dB and the input and output resistances are the same, what are the output power and output voltage?

40. A directional antenna produces a signal of 100 μV/m at a receiving station when supplied with 50 W of power. A standard vertical antenna requires 250 W of power to produce the same signal strength at the receiving station. What is the power gain of the directional antenna?

41. Given the RC low-pass filter in Figure 22-13(a) with $R = 1.5$ kΩ and $C = 750$ nF. Construct a Bode plot of the frequency response on three-cycle semilogarithmic paper.

42. Given the RC high-pass filter in Figure 22-14(a) with $R = 2.7$ kΩ and $C = 40$ nF. Construct a Bode plot of the frequency response on three-cycle semilogarithmic paper.

43. Given an amplifier whose mid-frequency gain $A_v = 6.0$ dB. The lower cutoff frequency $f_1 = 110$ Hz and the upper cutoff frequency $f_2 = 5.0$ kHz. Construct a Bode plot of the frequency response on five-cycle semilog paper.

44. Given an amplifier whose mid-frequency gain $A_v = -4.0$ dB. The lower cutoff frequency $f_1 = 40$ Hz and the upper cutoff frequency $f_2 = 7.5$ kHz. Construct a Bode plot of the frequency response on five-cycle semilog paper.

CHAPTER 25
Computer Number Systems

CHAPTER OBJECTIVES

In this chapter, you will learn:

- The binary, octal, and hexadecimal number systems.
- How to add and subtract binary numbers.
- How to subtract decimal numbers using complements.
- How to subtract binary numbers using complements.
- How to convert between decimal, binary, octal, and hexadecimal numbers.
- The binary and hexadecimal codes used in computers.

Our number system is called the decimal number system because it is based on the number 10. It evolved from the use of our fingers for counting. However, a number system can be based on any number (except 1 or 0) and used just as effectively as our system. In the electronics of a microprocessor or computer, switches are either open or closed, current is either flowing in one direction or the other, magnetic cores are either magnetized clockwise or counterclockwise. Because of this *bistable* nature of electronic states, the base 2 or binary system is the primary system used in computers. Data in a computer is represented by a series of binary digits or *bits,* and calculations are done by use of binary arithmetic. Other systems compatible with the binary system, the base 8 or octal system, and the base 16 or hexadecimal system, are also used in computers. To work with computers and their electronics it is therefore necessary to understand the binary, octal, and hexadecimal number systems. This chapter studies these three computer number systems and the computer codes that use these systems to represent data.

• • •

25.1 Binary Number System

The binary or base 2 system uses only two digits, 0 and 1, to express all numbers. To better understand the base 2 system, and the base 8 and base 16 systems, it is helpful to first take a closer look at our base 10 decimal system and how it works.

Decimal Numbers

The decimal number system contains 10 digits:

$$0, 1, 2, 3, 4, 5, 6, 7, 8, 9$$

which represent the first 10 positive integers in the system. The *base* of the decimal system is therefore 10. A number larger than nine is expressed using these digits, where each digit represents a multiple of a *power of ten.* The powers of ten increase from left to right. For example, the number 3789 in expanded form is equal to:

$$\begin{aligned} 3789 &= 3(10^3) + 7(10^2) + 8(10^1) + 9(10^0) \\ &= 3(1000) + 7(100) + 8(10) + 9(1) = 3789 \end{aligned}$$

The powers of ten are called the place values of the digits and are shown in Table 25-1.

TABLE 25-1 PLACE VALUES OF THE DECIMAL DIGITS

Power	10^3	10^2	10^1	10^0	.	10^{-1}	10^{-2}	10^{-3}
Value	1000	100	10	1	Decimal Point	0.1	0.01	0.001

Observe in Table 25-1 that digits to the right of the decimal point correspond to negative powers of ten.

EXAMPLE 25.1 Write the decimal number 52.34 in expanded form

Solution Using the place values of the digits from Table 25-1, the number in expanded form is:

$$\begin{aligned} 52.34 &= 5(10^1) + 2(10^0) + 3(10^{-1}) + 4(10^{-2}) \\ &= 5(10) + 2(1) + 3(0.1) + 4(0.01) \end{aligned}$$

Binary Numbers

The base of the binary system is 2. It contains only two digits, 0 and 1, which represent the first two positive integers. A number larger than one is expressed using the digits 0 and 1, as shown in Table 25-2. Each digit represents a power of two in the same way that each digit in the decimal system represents a power of ten. For example, the binary number 1101_2 in expanded form is:

$$\begin{aligned} 1101_2 &= 1(2^3) + 1(2^2) + 0(2^1) + 1(2^0) \\ &= 1(8) + 1(4) + 0(2) + 1(1) = 13 \end{aligned}$$

TABLE 25-2 BINARY AND DECIMAL NUMBERS

Binary	Decimal
0	0
1	1
10	2
11	3
100	4
101	5
110	6
111	7
1000	8
1001	9
1010	10
1011	11
1100	12
1101	13
1110	14
1111	15
10000	16

Therefore the binary number $1101_2 = 13$ in the decimal system. The subscript 2 is used to identify the base when working with numbers from different bases. If there is no subscript, the base is understood to be 10. The place values of the binary digits, including negative powers of two to the right of the *binary point,* are shown in Table 25-3.

TABLE 25-3 PLACE VALUES OF THE BINARY DIGITS

Power	2^6	2^5	2^4	2^3	2^2	2^1	2^0	.	2^{-1}	2^{-2}	2^{-3}
Value	64	32	16	8	4	2	1	Binary Point	$\frac{1}{2} = 0.5$	$\frac{1}{4} = 0.25$	$\frac{1}{8} = 0.125$

Notice in Table 25-3 that the negative powers of two are fractions whose numerators are 1 and whose denominators are 2, 4, 8, 16, and so on.

EXAMPLE 25.2

Convert to decimal numbers:

(a) 10010_2 **(b)** 110100_2 **(c)** 10.101_2

Solution

(a) To convert a binary number to a decimal number, multiply each digit by its place value in Table 25-3 and add:

$$\begin{aligned} 10010_2 &= 1(2^4) + 0(2^3) + 0(2^2) + 1(2^1) + 0(2^0) \\ &= 16 + 0 + 0 + 2 + 0 = 18 \end{aligned}$$

(b) Observe in (a) that you only need to add the place values that have digits of 1. You do not need to consider the zero digits:

$$110100_2 = 1(25) + 1(2^4) + 1(2^2) = 32 + 16 + 4 = 52$$

(c) The binary number 10.101 contains a binary point and is not an integer. It contains a binary fraction, and you need to add negative powers of 2:

$$10.101 = 1(2^1) + 1(2^{-1}) + 1(2^{-3}) = 2 + 0.5 + 0.125 = 2.625$$

To convert a decimal integer to a binary integer, apply the following algorithm called the *method of remainders:*

Algorithm

To Convert Decimal Integer to Binary Integer

1. Divide the decimal integer by 2.
2. Record the remainder (0 or 1).
3. Divide the quotient by 2.
4. Record the remainder again.
5. Repeat steps 3 and 4 until the quotient is zero.
6. The remainders in reverse order represent the binary integer.

The flowchart for the method of remainders algorithm is shown in Figure 25-1 on page 502.

EXAMPLE 25.3 Convert the decimal integers to binary integers:

(a) 46 **(b)** 108

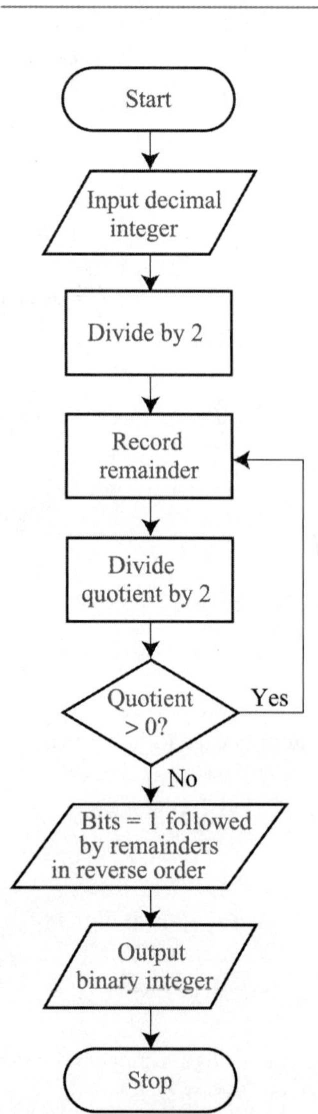

FIGURE 25-1 Flowchart for converting a decimal integer to a binary integer for Example 25.3.

Solution

(a) To convert 46 to a binary integer, apply the method of remainders. Divide 46 by 2 and record the remainder:

$$2\overline{)46}^{\;23} \quad \text{Remainder} = \mathbf{0}$$

Continue dividing the quotients and writing down the remainders until the quotient is zero. Work from the top down as follows:

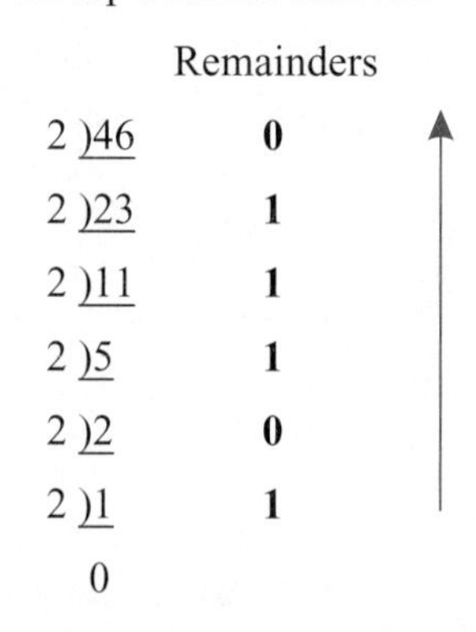

The binary digits or bits are the remainders, reading from the bottom up:

$$46 = 101110_2$$

(b) To convert 108 to a binary integer, apply the method of remainders:

	Remainders
2)108	**0**
2)54	**0**
2)27	**1**
2)13	**1**
2)6	**0**
2)3	**1**
2)1	**1**
0	

Therefore: $108 = 1101100_2$

Algorithm

To Convert Decimal Fraction to Binary Fraction

1. Multiply the fraction by 2.
2. Record the integral digit to the left of the decimal point (0 or 1).
3. Multiply the *fractional part of the result by 2.*
4. Record the integral digit again.
5. Continue steps 3 and 4 until the fractional part of the result is zero. The integral digits, in order, represent the binary fraction.

The flowchart for converting a decimal fraction to a binary fraction is shown in Figure 25-2.

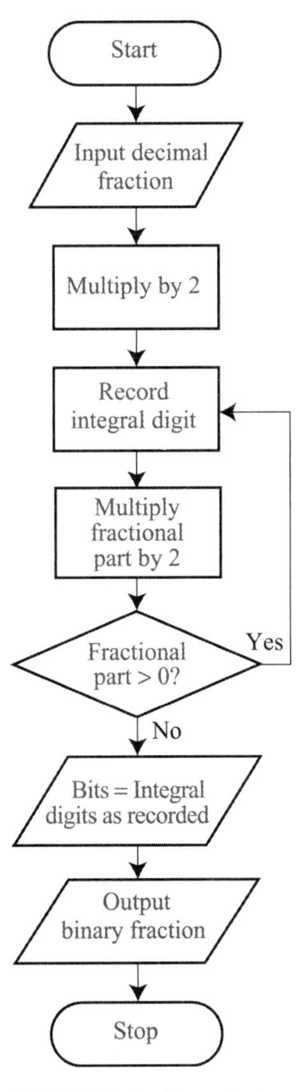

FIGURE 25-2 Flowchart for converting a decimal fraction to a binary fraction for Example 25.4.

EXAMPLE 25.4 Convert decimal fraction 0.625 to a binary fraction.

Solution Apply the above algorithm. Multiply 0.625 and subsequent fractional parts by 2. Write down the integral digits until the fractional part is zero:

Fractional Part ↓	*Integral Digits*
$2(0.625) = \mathbf{1}.250$	**1**
$2(0.250) = \mathbf{0}.50$	**0**
$2(0.50) = \mathbf{1}.00$	**1** ↓

The binary fraction is the integral digits reading from the top down:

$$0.625 = 0.101_2$$

The conversion of a decimal fraction can result in a repeating binary fraction. For example, the decimal fraction 0.8 = 0.1100 1100 1100 . . . . See exercise 46.

EXAMPLE 25.5 Convert 44.25 to a binary number.

Solution Convert the integral part, 44, and the fractional part, 0.25, separately, applying the preceding algorithms:

	Remainders ↑
2)44	**0**
2)22	**0**
2)11	**1**
2)5	**1**
2)2	**0**
2)1	**1**
0	

	Integral digits ↓
$2(0.25) = \mathbf{0}.50$	**0**
$2(0.50) = \mathbf{1}.00$	**1**

Reading the remainders backward and the integral digits forward, the conversion is:

$$44.25 = 101100.01_2$$

EXERCISE 25.1

In exercises 1 through 8, write each decimal number in expanded form.

1. 1328
2. 5586
3. 99,150
4. 45,217
5. 38.96
6. 21.152
7. 10.01
8. 101.11

In exercises 9 through 24, convert each binary number to a decimal number.

9. 11
10. 10
11. 100
12. 111
13. 1110
14. 1001
15. 10001
16. 10101

17. 110011

18. 100100

19. 10.11

20. 11.01

21. 1.101

22. 10.001

23. 11.1011

24. 1.1111

In exercises 25 through 42, convert each decimal number to a binary number.

25. 4

26. 6

27. 11

28. 27

29. 57

30. 63

31. 99

32. 111

33. 128

34. 200

35. 0.5

36. 0.75

37. 0.875

38. 0.375

39. 2.25

40. 7.125

41. 33.625

42. 5.5625

43. Complete Table 25-2 for the decimal numbers 17 through 32.

44. How many binary digits are needed to write the following decimal numbers?
(a) 100
(b) 1000
(c) 10,000
(d) 1,000,000

45. One way a computer stores data is by a binary-coded decimal (BCD) where each four bits represent a decimal digit. For example, 0010 1001 is a BCD for the decimal number 29. What decimal numbers do each of the following BCD's represent?
(a) 0100 1001
(b) 0100 0000
(c) 0000 0111
(d) 0001 0010

46. Show that the decimal fraction 0.8 is equal to the repeating binary number 0.1100 1100 1100 … . Apply the method shown in Example 25.4 and show that the integral digits repeat.

25.2 Binary Arithmetic

The laws of binary arithmetic are essentially the same as those for decimal numbers. However, since there are only two digits in the binary system, addition is much simpler. The subscript 2 for a binary number is omitted in this section, and all numbers are understood to be binary.

Binary Addition

Table 25-4 shows the four basic binary sums plus two other useful sums for adding binary numbers. The table tells the digit to record and the digit to carry to the next column. Study the following examples, which show how to add binary numbers.

TABLE 25-4 BINARY SUMS

Sum	
$0 + 0 = 0$	
$0 + 1 = 1$	
$1 + 1 = 10$	(0 and carry *1*)
$1 + 1 + 1 = 10 + 1 = 11$	(1 and carry *1*)
$1 + 1 + 1 + 1 = 11 + 1 = 100$	(0 and carry *10*)

EXAMPLE 25.6 Add the binary numbers:

(a) $110 + 111$ **(b)** $111.1 + 101.0$

Solution

(a) Add each column starting with the units column and working to the left, applying the sums in Table 25-4 and carrying digits:

$$\begin{array}{r} \\ 110 \\ +111 \\ \hline 1 \end{array} \Rightarrow \begin{array}{r} \mathit{1} \\ 110 \\ +111 \\ \hline 01 \end{array} \Rightarrow \begin{array}{r} \mathit{1} \\ 110 \\ +111 \\ \hline 1101 \end{array} \quad \text{(carried digit)}$$

Column two adds up to 10. Therefore 1 is carried to the third column as shown on the top. The decimal equivalent of the addition is:

$$\begin{array}{r} 6 \\ +7 \\ \hline 13 \end{array}$$

(b) To add numbers with binary points, align the binary points, as in decimal arithmetic, and add the same way as shown in (a):

$$\begin{array}{r} \\ 111.1 \\ +101.0 \\ \hline .1 \end{array} \Rightarrow \begin{array}{r} \mathit{1} \\ 111.1 \\ +101.0 \\ \hline 0.1 \end{array} \Rightarrow \begin{array}{r} \mathit{11} \\ 111.1 \\ +101.0 \\ \hline 00.1 \end{array} \Rightarrow \begin{array}{r} \mathit{11} \\ 111.1 \\ +101.0 \\ \hline 1100.1 \end{array} \quad \text{(carried digits)}$$

EXAMPLE 25.7

Add the binary numbers:

$$\begin{array}{r} 10 \\ 11 \\ 111 \\ 101 \\ \hline \end{array}$$

Solution The addition is as follows with the carried digits on the top:

$$\begin{array}{r} \mathit{1} \\ \mathit{101} \\ 10 \\ 11 \\ 111 \\ 101 \\ \hline 10001 \end{array} \quad \text{(carried digits)}$$

The first column adds up to 11 and carry 1. The second column adds up to 100 with a carry of 0 to the third column and 1 to the fourth column. The third column adds up to 10 with another carry of 1 to the fourth column. You can also do this addition by adding partial sums of two numbers at a time. The method of partial sums is used in computers.

It is common to add zeros in front of a binary number so that all columns being used are represented. For example, if four columns are used, 11 is written 0011 and 101 is written 0101. Most computers store numbers using a fixed number of digits and supply leading zeros if necessary.

Binary Subtraction Using Complements

Subtraction in a computer is usually done by adding complements since it simplifies the microprocessor circuitry. Binary complements are defined as follows:

Definitions

1's Complement

The number obtained by subtracting each digit from 1. This is done directly by changing each 0 to 1 and each 1 to 0.

2's Complement

The number obtained by adding 1 to its 1's complement.

EXAMPLE 25.8 Find the 1's complement and the 2's complement for the four binary numbers: 101, 110, 1011, and 111.

Solution Apply the above rules for the 1's complement and the 2's complement:

Binary Number	1's Complement	2's Complement
101	010	010 + 1 = 011
110	001	001 + 1 = 010
1011	0100	0100 + 1 = 0101
111→ 0111 *	1000	1000 + 1 = 1001

*One or more leading zeros do not affect the results when performing operations on binary numbers. Note in the table, for the binary number 111, a leading zero should be added to avoid having all zeros in the 1's complement. The next two examples show how binary subraction is done by adding complements.

EXAMPLE 25.9 Subtract the following binary numbers by:

(a) Adding the 1's complement.

(b) Adding the 2's complement.

$$\begin{array}{r} 1011 \\ -1001 \\ \hline \end{array}$$

Solution

(a) To subtract by adding the 1's complement, first obtain the 1's complement of the subtrahend by changing 0's to 1's and 1's to 0's: 1001 → 0110. Then use the following addition algorithm:

$$\begin{array}{rl} \mathit{11} & \text{(carried digits)} \\ 1011 & \\ +0110 & \\ \hline \textcircled{1}0001 & \\ +\mathit{1} & \text{(end-around carry)} \\ \hline 0010 & = 10 \end{array}$$

Take the leading one of the sum and add it to the remaining digits to obtain the answer to the subtraction. This is called *end-around carry* in computer terminology.

(b) To subtract by adding the 2's complement, first obtain the 2's complement of the subtrahend:

$$0110 + 1 = 0111$$

Then use the following addition algorithm:

$$\begin{array}{rl} \mathit{111} & \text{(carried digits)} \\ 1011 & \\ +0111 & \\ \hline 10010 & = 0010 = 10 \quad \text{(delete leading one)} \end{array}$$

Delete the leading one of the sum to obtain the result of the subtraction.

EXAMPLE 25.10 Subtract the binary numbers:

$$\begin{array}{r} 10110 \\ \underline{-1101} \end{array}$$

Solution To add by using either the 1's complement or the 2's complement *it is first necessary to add leading zeros to the subtrahend (number being subtracted) so both numbers have the same number of digits. Subtraction using the 1's complement is then as follows:*

$$\begin{array}{r} 10110 \\ \underline{-1101} \end{array} \Rightarrow \begin{array}{r} 10110 \\ \underline{-01101} \end{array} \Rightarrow \begin{array}{r} 10110 \\ \underline{+10010} \end{array} \Rightarrow \begin{array}{r} \mathit{1\,1} \\ 10110 \\ \underline{+10010} \\ (1)01000 \\ \underline{+\ \mathit{1}} \\ 1001 \end{array} \quad \begin{array}{l} \text{(Carried digits} \\ \\ \\ \\ \text{(end-around carry)} \\ \\ \end{array}$$

See exercise 31, which asks you to check this example using the 2's complement.

EXERCISE 25.2

In exercises 1 through 14, add the binary numbers.

1. $\begin{array}{r} 11 \\ \underline{10} \end{array}$

2. $\begin{array}{r} 11 \\ \underline{100} \end{array}$

3. $\begin{array}{r} 100 \\ \underline{111} \end{array}$

4. $\begin{array}{r} 111 \\ \underline{111} \end{array}$

5. $\begin{array}{r} 110 \\ 011 \\ \underline{110} \end{array}$

6. $\begin{array}{r} 011 \\ 011 \\ \underline{100} \end{array}$

7. $\begin{array}{r} 11.1 \\ 10.1 \\ \underline{10.1} \end{array}$

8. $\begin{array}{r} 1.10 \\ 1.10 \\ \underline{11.11} \end{array}$

9. $\begin{array}{r} 1 \\ 10 \\ 101 \\ \underline{111} \end{array}$

10. $\begin{array}{r} 1 \\ 111 \\ 101 \\ \underline{101} \end{array}$

11. $\begin{array}{r} 010 \\ 101 \\ 101 \\ \underline{111} \end{array}$

12. $\begin{array}{r} 001 \\ 100 \\ 011 \\ \underline{110} \end{array}$

13. $\begin{array}{r} 101.01 \\ 11.01 \\ \underline{111.11} \end{array}$

14. $\begin{array}{r} 111.10 \\ 100.11 \\ \underline{110.01} \end{array}$

In exercises 15 through 30, subtract the binary numbers by adding the 1's complement or the 2's complement.

15. $\begin{array}{r} 11 \\ \underline{-10} \end{array}$

16. $\begin{array}{r} 111 \\ \underline{-10} \end{array}$

17. $\begin{array}{r} 111 \\ \underline{-11} \end{array}$

18. $\begin{array}{r} 101 \\ \underline{-\ 10} \end{array}$

19. $\begin{array}{r} 111 \\ \underline{-101} \end{array}$

20. $\begin{array}{r} 1101 \\ \underline{-1010} \end{array}$

21. $\begin{array}{r} 1111 \\ \underline{-110} \end{array}$

22. $\begin{array}{r} 10110 \\ \underline{-101} \end{array}$

23. $\begin{array}{r} 110 \\ \underline{-101} \end{array}$

24. $\begin{array}{r} 111 \\ \underline{-110} \end{array}$

25. $\begin{array}{r} 1101 \\ \underline{-1010} \end{array}$

26. $\begin{array}{r} 1100 \\ \underline{-111} \end{array}$

27. $\begin{array}{r} 11000 \\ \underline{-110} \end{array}$

28. $\begin{array}{r} 10111 \\ \underline{-1111} \end{array}$

29. $\begin{array}{r} 100.01 \\ \underline{-1.11} \end{array}$

30. $\begin{array}{r} 1110.1 \\ \underline{-101.1} \end{array}$

31. Subtract the binary numbers in Example 25.10 by adding the 2's complement, and check that you get the same answer.

32. Many computers use storage registers that are set for a fixed number of digits. Consider a computer that stores the first 10 binary digits. A binary number such as 10011 will be stored as 0 000 010 011. Show that if this number is subtracted from 0 000 110 111 by using all 10 digits and adding the 1's complement, the computer will obtain the correct answer, which is 100 100.

25.3 Octal Number System

After studying binary numbers, you realize that they can be awkward to work with unless you are a computer. A binary number is generally more than three times as long as a decimal number. A long string of 0's and 1's can be difficult to recognize or remember. Number systems with a larger base that can easily be converted to the binary system are therefore used in computers also. Octal (base 8) and hexadecimal (base 16) numbers are used to identify storage locations in a computer memory and to code and represent data. The octal number system contains the eight digits:

$$0, 1, 2, 3, 4, 5, 6, 7$$

The place values of the octal digits are powers of eight and are shown in Table 25-5. Table 25-6 shows the first sixteen octal numbers with their decimal and binary equivalents. Note that after the number 7 in the octal system, the next place must be used to represent higher numbers so that the decimal number 8 equals 10_8 in the octal system.

TABLE 25-5 PLACE VALUES OF THE OCTAL DIGITS

Power	8^4	8^3	8^2	8^1	8^0	.	8^{-1}	8^{-2}
Value	4096	512	64	8	1	Octal Point	$\frac{1}{8} = 0.125$	$\frac{1}{64} = 0.015625$

TABLE 25-6 DECIMAL, OCTAL, AND BINARY NUMBERS

Decimal	Octal	Binary
0	0	0
1	1	1
2	2	10
3	3	11
4	4	100
5	5	101
6	6	110
7	7	111
8	10	1000
9	11	1001
10	12	1010
11	13	1011
12	14	1100
13	15	1101
14	16	1110
15	17	1111
16	20	10000

Conversions Among Octal, Decimal, and Binary Numbers

Conversions between octal and decimal numbers use procedures that are similar to those for binary and decimal numbers. Study the following examples.

EXAMPLE 25.11 Convert the octal number 372_8 to a decimal number.

Solution Multiply each digit by its place value in Table 25-5 and add:

$$372_8 = 3(64) + 7(8) + 2(1) = 192 + 56 + 2 = 250$$

To convert a decimal number to an octal number, you use the method of remainders similar to that for binary integers:

Algorithm

To Convert Decimal Number to Octal Number

1. Divide the decimal integer by 8.
2. Record the remainder.
3. Divide the quotient by 8.
4. Record the remainder.
5. Repeat steps 3 and 4 until the quotient is zero.
6. The remainders in reverse order represent the octal number.

EXAMPLE 25.12 Convert the decimal number 281 to an octal number.

Solution Apply the above method of remainders:

	Remainders
8)281	**1** ↑
8)35	**3**
8)4	**4**
0	

The digits of the octal number are the remainders reading from the bottom up: $281 = 431_8$.

Conversions between octal and binary numbers can be done very directly. Since $2^3 = 8$, every three binary bits is equivalent to one octal digit.

Algorithm

To Convert Octal Number to Binary Number

Change each digit into its 3-bit binary equivalent, adding leading zeros if necessary.

EXAMPLE 25.13 Convert the octal number 561_8 to a binary number.

Solution Write the three bits corresponding to each digit:

$$\underbrace{5}_{101} \quad \underbrace{6}_{110} \quad \underbrace{1}_{001}$$

Then: $$561_8 = 101\ 110\ 001_2$$

Algorithm **To Convert Binary Number to Octal Number**

Arrange the binary number in groups of three bits. Add leading zeros if necessary. Change each three bits into an octal digit.

EXAMPLE 25.14 Convert the binary number 1111010_2 to an octal number.

Solution Arrange the binary number in groups of three bits. Add two leading zeros and write the corresponding octal digits:

$$\underbrace{001}_{1} \quad \underbrace{111}_{7} \quad \underbrace{010}_{2}$$

Therefore: $$1\ 111\ 010_2 = 172_8$$

EXAMPLE 25.15 Express each number as a binary, octal and decimal number:

(a) 763_8 **(b)** $11\ 101\ 100_2$

Solution

(a) To express 763_8 as a binary number, change each digit into its 3-bit binary equivalent:

$$\overbrace{111}^{7} \quad \overbrace{110}^{6} \quad \overbrace{011}^{3}$$

Then: $$763_8 = 111\ 110\ 011_2$$

To express as a decimal number, multiply each digit by its place value:

$$763_8 = 7(8^2) + 6(8) + 3(1) = 448 + 48 + 3 = 499$$

(b) To express 11101100_2 as an octal number, change each three bits to an octal digit:

$$\underbrace{011}_{3} \quad \underbrace{101}_{5} \quad \underbrace{100}_{4}$$

Then: $$11101100_2 = 354_8$$

To change to a decimal number, add up the place values of the 1's:

$$11101100_2 = 2^7 + 2^6 + 2^5 + 2^3 + 2^2 = 128 + 64 + 32 + 8 + 4 = 236$$

If your calculator has a BASE key, you can do the above conversions on the calculator. One possible calculator solution for (b) is to first set the calculator for binary mode (BIN) by pressing the BASE or MODE key and then entering the following:

111011100 OCT → 354*o* DEC → 236

The octal number 354 appears with an '*o*' (octal) or other symbol to identify it.

EXERCISE 25.3

In exercises 1 through 8, convert each octal number to a decimal number.

1. 12
2. 17
3. 65
4. 70
5. 324
6. 635
7. 7777
8. 1001

In exercises 9 through 16, convert each decimal number to an octal number.

9. 9
10. 13
11. 38
12. 60
13. 212
14. 481
15. 1100
16. 3030

In exercises 17 through 24, convert each octal number to a binary number.

17. 10
18. 15
19. 23
20. 34
21. 761
22. 520
23. 1054
24. 4076

In exercises 25 through 32, convert each binary number to an octal number.

25. 1100
26. 1011
27. 11011
28. 10111
29. 101111
30. 101000
31. 11001010
32. 1110100

33. **(a)** Express 623_8 as a binary and as a decimal number.

(b) Express 10100101_2 as an octal and as a decimal number.

34. Express each decimal number as an octal and as a binary number:
(a) 400
(b) 100

35. Complete Table 25-6 for the decimal numbers 17 through 32.

36. In a computer, data is represented by use of a binary code. For example, one code, which uses six bits for each character, represents the letter A as the binary number 110001. Find the octal and decimal equivalents of this number.

25.4 Hexadecimal Number System

The hexadecimal number system, which has a base of 16, is compatible with the binary number system. It is very important in computer applications and has replaced the octal system in almost all such applications. The hexadecimal or hex number system contains the following 16 digits:

0, 1, 2, 3, 4, 5, 6, 7, 8, 9, A, B, C, D, E, F

The letters A through F are equal to the decimal numbers 10 through 15, respectively. Table 25-7 shows the first 21 hexadecimal numbers with their decimal and binary equivalents. Note that after the number F (15) in the hexadecimal system, the next place must be used to represent higher numbers. The place values of the hexadecimal digits are powers of 16 and are shown in Table 25-8 on page 512.

TABLE 25-7 DECIMAL, HEXADECIMAL, AND BINARY NUMBERS

Decimal	Hexadecimal	Binary
0	0	0
1	1	1
2	2	10
3	3	11
4	4	100
5	5	101
6	6	110

TABLE 25-7 *(continued)*

Decimal	Hexadecimal	Binary
7	7	111
8	8	1000
9	9	1001
10	A	1010
11	B	1011
12	C	1100
13	D	1101
14	E	1110
15	F	1111
16	10	10000
17	11	10001
18	12	10010
19	13	10011
20	14	10100

TABLE 25-8 PLACE VALUES OF THE HEXADECIMAL DIGITS

Powers	16^3	16^2	16^1	16^0	.	16^{-1}
Values	4096	256	16	1	Hex Point	$\frac{1}{16} = 0.0625$

Conversions Among Hexadecimal, Decimal, and Binary Numbers

Conversions between hexadecimal and decimal numbers use the same procedures as those shown for binary and octal numbers except that the hexadecimal numbers A through F must be substituted for the decimal numbers 10 through 15, respectively. Study the following examples that illustrate these conversions:

EXAMPLE 25.16 Convert each hexadecimal number to its decimal equivalent:

(a) E5 **(b)** 2B.4 **(c)** FAD

Solution

(a) To convert E5 to a decimal number, multiply each digit by its place value shown in Table 25-8 and add the products. Use 14 for the digit E:

$$E5_{16} = E(16) + 5(1) = 14(16) + 5(1) = 229$$

(b) To convert 2B.4 to a decimal number, do the same as in (a) using 11 for B:

$$2B.4_{16} = 2(16) + B(1) + 4(0.0625) = 2(16) + 11(1) + 4(0.0625) = 43.25$$

(c) To convert FAD to a decimal number, use 15 for F, 10 for A, and 13 for D:

$$FAD_{16} = F(256) + A(16) + D(1) = 15(256) + 10(16) + 13(1) = 4013$$

To convert a decimal number to a hexadecimal number use the method of remainders:

Algorithm

To Convert Decimal Number to Hexadecimal Number

1. Divide the decimal integer by 16.
2. Record the remainder.
3. Divide the quotient by 16.
4. Record the remainder.
5. Repeat steps 3 and 4 until the quotient is zero.
6. The remainders in reverse order represent the hexadecimal number.

EXAMPLE 25.17

Convert 380 to a hexadecimal number.

Solution Apply the method of remainders algorithm above:

	Remainders
16)380	12 = **C** ↑
16)23	**7**
16)1	**1**
0	

Note that a remainder of 12 is changed to a C. You must change remainders from 10 through 15 to A through F, respectively. The hexadecimal number is given by the remainders reading from the bottom up:

$$380 = 17C_{16}$$

Division by 16 can be done on a calculator and the remainder found by multiplying the decimal part of the quotient by 16. For example, the remainders when dividing 380 by 16 are found as follows:

380 [÷] 16 [=] → 23.75 → 0.75 [×] 16 [=] → **12**

23 [÷] 16 [=] → 1.4375 → 0.4375 [×] 16 [=] → **7**

1 [÷] 16 [=] → 0.0625 → 0.0625 [×] 16 [=] → **1**

Figure 25-3 shows a flowchart for converting a decimal number to a hexadecimal number as done in Example 25.17. Conversions between hexadecimal and binary numbers are done directly because each hexadecimal digit is equivalent to four binary bits.

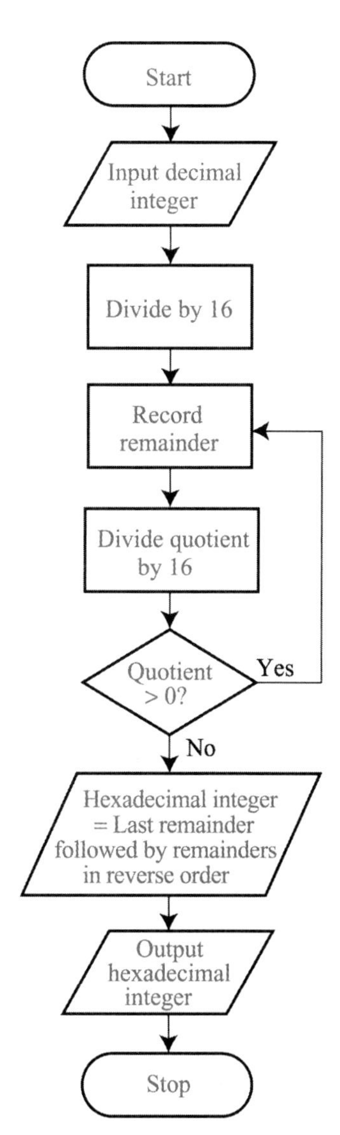

FIGURE 25-3 Flowchart for converting a decimal number to a hexadecimal number for Example 25.17.

Algorithm

To Convert Hexadecimal Number to Binary Number

Change each digit into its 4 bit binary equivalent, adding leading zeros if necessary.

EXAMPLE 25.18 Convert the hexadecimal number 7D to a binary number.

Solution Write the number as follows using Table 25-7:

$$\overbrace{0111}^{7} \quad \overbrace{1101}^{D}$$

Then: $7D_{16} = 1111101_2$

Algorithm

To Convert Binary Number to Hexadecimal Number

Arrange the binary number in groups of four bits. Add leading zeros if necessary. Change each four bits into a hexadecimal digit.

EXAMPLE 25.19 Convert the binary number 1011010 to a hexadecimal digit.

Solution Write the number as follows using Table 25-7:

$$\underbrace{0101}_{5} \quad \underbrace{1010}_{A}$$

Then: $1011010_2 = 5A_{16}$

Conversions Among Decimal, Binary, and Hexadecimal Numbers

Applying the ideas presented in this chapter, you should now be able to express a number in any of the three bases 2, 10, and 16 as the next example illustrates.

EXAMPLE 25.20 Express the decimal number 252 in bases 2 and 16.

Solution You can start with base 2 or 16. Starting with base 2, use the method of remainders to convert 252:

	Remainders
2)252	**0** ↑
2)126	**0**
2)63	**1**
2)31	**1**
2)15	**1**
2)7	**1**
2)3	**1**
2)1	**1**
0	

Then: $252 = 11111100_2$

Change the binary number to a hexadecimal number by using 4-bit equivalents:

$$\underbrace{1111}_{F} \quad \underbrace{1100}_{C}$$

Then: $252 = FC_{16} = 374_8 = 11111100_2$

One way to do this example on a calculator with a BASE mode is:

252 BIN → 1111 1100 *b* HEX → FC *h*

The *b* stands for binary and *h* for hexadecimal.

ERROR BOX

A common error when working with hexadecimal numbers is incorrectly using or identifying the digits A through F for 10 through 15. See if you can do the practice problems *without a table or calculator* to better learn the hexadecimal system.

Practice Problems: Change each hexadecimal number to a decimal number.

1. 1A **2.** B2 **3.** 3C **4.** D4

5. 5E **6.** F6 **7.** E0 **8.** 1F

Change each decimal number to a hexadecimal number.

9. 161 **10.** 43 **11.** 195 **12.** 77

13. 229 **14.** 111 **15.** 989 **16.** 666

EXERCISE 25.4

In exercises 1 through 14, convert each hexadecimal number to a decimal number.

1. 1B
2. 2E
3. A0
4. C3
5. CD
6. FF
7. 123
8. 285
9. A0B
10. BE7
11. 10.2
12. 2C.A
13. 1001
14. 10000

In exercises 15 through 26, convert each decimal number to a hexadecimal number.

15. 21
16. 24
17. 40
18. 35
19. 85
20. 72
21. 142
22. 116
23. 343
24. 421
25. 1000
26. 1010

In exercises 27 through 36, convert each hexadecimal number to a binary number.

27. 32
28. 89
29. AB
30. FE
31. 6C7
32. B81
33. 1E0B
34. 2A2A
35. C0D0
36. F111

In exercises 37 through 46, convert each binary number to a hexadecimal number.

37. 1110
38. 1011
39. 11011
40. 11101
41. 101111
42. 111111
43. 1011100
44. 1011010
45. 11101001
46. 10000000

47. Express the decimal number 160 in bases 2 and 16.

48. Express the decimal number 341 in bases 2 and 16.

49. In a computer, data is represented by use of a binary code. For example, the XS-3 code represents the decimal number 267 as the binary number 0101 1001 1010. The XS-3 is a 4-bit code since four bits represent each character. Find the hexadecimal and decimal equivalents of this binary number.

50. One standard 6-bit binary computer code encodes the word GO as the binary number 110111 100110. Find the hexadecimal and decimal equivalents of this number.

25.5 Hexadecimal Arithmetic

Addition of hexadecimal numbers can be done directly as with decimal numbers. Subtraction can be done by adding complements, which is the method used for binary numbers on a computer.

Hexadecimal Addition

When adding hexadecimal numbers you have to think in terms of 16's and remember that the letters A, B, C, D, E, and F represent the decimal numbers 10, 11, 12, 13, 14, and 15, respectively. Study the steps in the next example, which illustrates the procedure.

EXAMPLE 25. 21 Add the hexadecimal numbers:
(a) A3 + 87 **(b)** E6A + BD **(c)** C6C0 + F45

Solution

(a) When you add each column carry a 1 to the next column when the sum is 16 or greater. Then write down the difference from 16 at the bottom:

$$\begin{array}{r} \\ \mathrm{A}\ 3 \\ +\ \ 8\ 7 \\ \hline \mathrm{A} \\ (10) \end{array} \Rightarrow \begin{array}{r} \mathit{1} \\ \mathrm{A}\ 3 \\ +\ \ 8\ 7 \\ \hline 2\ \mathrm{A} \\ (18) \end{array} \Rightarrow \begin{array}{r} \mathit{1}\phantom{\ \mathrm{A}\ 3} \\ \mathrm{A}\ 3 \\ +\ \ 8\ 7 \\ \hline 1\ 2\ \mathrm{A} \\ \\ \end{array} \quad \text{(carried digits)}$$

Add the first column: 3 + 7 to get (10), which you write down as A. Add the second column: A + 8 to get (18). Since (18) is greater than 16, take the difference of (18) from 16, which is 2. Write down the 2 and carry 1 to the third column. Then bring down the 1.

(b)

$$\begin{array}{r} \mathit{1}\phantom{\ \mathrm{A}} \\ \mathrm{E}\ 6\ \mathrm{A} \\ +\ \ \mathrm{B}\ \mathrm{D} \\ \hline 7 \\ (23) \end{array} \Rightarrow \begin{array}{r} \mathit{1}\ \mathit{1}\phantom{\ \mathrm{A}} \\ \mathrm{E}\ 6\ \mathrm{A} \\ +\ \ \mathrm{B}\ \mathrm{D} \\ \hline 2\ 7 \\ (18) \end{array} \Rightarrow \begin{array}{r} \mathit{1}\ \mathit{1}\phantom{\ \mathrm{A}} \\ \mathrm{E}\ 6\ \mathrm{A} \\ +\ \ \mathrm{B}\ \mathrm{D} \\ \hline \mathrm{F}\ 2\ 7 \\ (15) \end{array} \quad \text{(carried digits)}$$

Add the first column: A(10) + D(13) to get (23). Write down the difference from 16, which is 7, and carry 1. Add the second column: 1 + 6 + B(11) to get (18). Write down the difference from 16, which is 2, and carry 1. Add the third column E(14) + 1 to get (15), which is F.

(c) See if you can follow the third example:

$$\begin{array}{r} \\ \mathrm{C}\ 6\ \mathrm{C}\ 0 \\ +\ \ \mathrm{F}\ 4\ 5 \\ \hline 5 \\ \\ \end{array} \Rightarrow \begin{array}{r} \mathit{1}\phantom{\ \mathrm{C}\ 0} \\ \mathrm{C}\ 6\ \mathrm{C}\ 0 \\ +\ \ \mathrm{F}\ 4\ 5 \\ \hline 0\ 5 \\ (16) \end{array} \Rightarrow \begin{array}{r} \mathit{1}\ \mathit{1}\phantom{\ \mathrm{C}\ 0} \\ \mathrm{C}\ 6\ \mathrm{C}\ 0 \\ +\ \ \mathrm{F}\ 4\ 5 \\ \hline 6\ 0\ 5 \\ (22) \end{array} \Rightarrow \begin{array}{r} \mathit{1}\ \mathit{1}\phantom{\ \mathrm{C}\ 0} \\ \mathrm{C}\ 6\ \mathrm{C}\ 0 \\ +\ \ \mathrm{F}\ 4\ 5 \\ \hline \mathrm{D}\ 6\ 0\ 5 \\ (13) \end{array} \quad \text{(carried digits)}$$

Hexadecimal Subtraction Using Complements

Hexadecimal complements are defined as follows:

Definitions

15's Complement:

The number obtained by subtracting each digit from 15.

16's Complement:

The number obtained by adding 1 to its 15's complement.

Answers to Error Box Problems, page 515:
1. 26 **2.** 178 **3.** 60 **4.** 212 **5.** 94 **6.** 246 **7.** 14 **8.** 32 **9.** A1 **10.** 2B **11.** C3 **12.** 4D **13.** E5 **14.** 6F **15.** 3DD **16.** 29A

Hexadecimal subtraction can be done by the 16's complement method by first changing the subtrahend (number to be subtracted) to its 16's complement, then adding the numbers and deleting the leading 1. This is the same procedure as subtracting binary numbers by adding the 2's complement. Study the next example.

EXAMPLE 25.22 Subtract the hexadecimal numbers:

(a) $\begin{array}{r} 4\ C \\ -1\ B \\ \hline \end{array}$ **(b)** $\begin{array}{r} 3\ F\ F \\ -2\ A\ 9 \\ \hline \end{array}$

Solution

(a) Change the subtrahend 1B to its 16's complement by subtracting each digit from 15 and adding 1:

$$15 - 1 = E(14) \quad \text{and} \quad 15 - B(11) = 4 \rightarrow E4 + 1 = E5$$

Add 4C to E5 and delete the leading one:

$$\begin{array}{rrrl} & \mathit{1} & & \text{(carried digits)} \\ & 4 & C & \\ + & E & 5 & \\ \hline \not{1} & 3 & 1 & \text{(delete leading one)} \\ & (19) & (17) & \end{array}$$

(b) Change 2A9 to its 16's complement: 2A9 → D56 + 1 → D57. Add 3FF to D56 and delete the leading one:

$$\begin{array}{rrrrl} & \mathit{1} & \mathit{1} & & \text{(carried digits)} \\ & 3 & F & F & \\ + & D & 5 & 7 & \\ \hline \not{1} & 1 & 5 & 6 & \text{(delete leading one)} \\ & & (21) & (22) & \end{array}$$

EXERCISE 25.5

In exercises 1 through 20, add the hexadecimal numbers.

1. $\begin{array}{r} 3B \\ A5 \\ \hline \end{array}$
2. $\begin{array}{r} 4C \\ 56 \\ \hline \end{array}$
3. $\begin{array}{r} FA \\ 4C \\ \hline \end{array}$
4. $\begin{array}{r} AD \\ F2 \\ \hline \end{array}$
5. $\begin{array}{r} 78 \\ 44 \\ \hline \end{array}$
6. $\begin{array}{r} 94 \\ 88 \\ \hline \end{array}$
7. $\begin{array}{r} 37C \\ 5E4 \\ \hline \end{array}$
8. $\begin{array}{r} 2DD \\ 11A \\ \hline \end{array}$
9. $\begin{array}{r} DFE \\ 708 \\ \hline \end{array}$
10. $\begin{array}{r} 984 \\ A0B \\ \hline \end{array}$
11. $\begin{array}{r} 793 \\ 934 \\ \hline \end{array}$
12. $\begin{array}{r} 826 \\ 885 \\ \hline \end{array}$
13. $\begin{array}{r} 648 \\ ADA \\ \hline \end{array}$
14. $\begin{array}{r} CAB \\ 90F \\ \hline \end{array}$
15. $\begin{array}{r} 12F8 \\ C4A \\ \hline \end{array}$
16. $\begin{array}{r} B7BB \\ F49 \\ \hline \end{array}$
17. $\begin{array}{r} 1A2B \\ 5CC7 \\ \hline \end{array}$
18. $\begin{array}{r} D2DE \\ AF54 \\ \hline \end{array}$
19. $\begin{array}{r} 8361 \\ 9504 \\ \hline \end{array}$
20. $\begin{array}{r} A0E5 \\ ABAB \\ \hline \end{array}$

In the exercises 21 through 40, subtract the hexadecimal numbers.

21. $\begin{array}{r} C3 \\ 20 \\ \hline \end{array}$
22. $\begin{array}{r} 9E \\ 4A \\ \hline \end{array}$
23. $\begin{array}{r} FE \\ D1 \\ \hline \end{array}$
24. $\begin{array}{r} A2 \\ 67 \\ \hline \end{array}$
25. $\begin{array}{r} 98 \\ 49 \\ \hline \end{array}$
26. $\begin{array}{r} 94 \\ 88 \\ \hline \end{array}$
27. $\begin{array}{r} 8CA \\ 690 \\ \hline \end{array}$
28. $\begin{array}{r} FAF \\ 1B1 \\ \hline \end{array}$
29. $\begin{array}{r} DEF \\ CBA \\ \hline \end{array}$
30. $\begin{array}{r} 90E \\ 50F \\ \hline \end{array}$
31. $\begin{array}{r} 793 \\ 734 \\ \hline \end{array}$
32. $\begin{array}{r} 672 \\ 583 \\ \hline \end{array}$

33. $\begin{array}{r} \text{DAD} \\ \text{ACA} \\ \hline \end{array}$ **35.** $\begin{array}{r} \text{F24E} \\ \text{B4A} \\ \hline \end{array}$ **37.** $\begin{array}{r} \text{8A2B} \\ \text{5C7C} \\ \hline \end{array}$ **39.** $\begin{array}{r} 5371 \\ 3594 \\ \hline \end{array}$

34. $\begin{array}{r} \text{CAB} \\ \text{90F} \\ \hline \end{array}$ **36.** $\begin{array}{r} \text{9AAA} \\ \text{94E} \\ \hline \end{array}$ **38.** $\begin{array}{r} \text{D2DE} \\ \text{8F54} \\ \hline \end{array}$ **40.** $\begin{array}{r} \text{F0E0} \\ \text{AABB} \\ \hline \end{array}$

25.6 Computer Codes

Alphanumeric data consists of letters or numbers or both, such as words, addresses, equations, and tables. Such data is represented in a computer using a binary code. Several binary codes have been used throughout the development of computers, but today the one that is primarily used is ASCII (pronounced "AS-KEY"), which stands for American Standard Code for Information Interchange. Another code, less frequently used, is EBCDIC (pronounced "EBB-SE-DIC"), which stands for Extended Binary-Coded Decimal Interchange Code. Both these codes are extensions of a 4-bit binary-coded decimal (BCD) where four numeric bits are used to represent the digits 0 to 9. For example, the BCD representation of the decimal number 278 uses 12 bits:

$$\overbrace{0010}^{2}\quad\overbrace{0111}^{7}\quad\overbrace{1000}^{8}$$

Each group of four bits represents one numeric *character* and is the binary equivalent of the digit. This is different from the straight binary representation of 278 which uses 9 bits:

$$278 = 1\ 0001\ 0110_2$$

Straight binary representation of numbers is used in computers when performing arithmetic operations. Four-bit binary-coded decimal representation is used for information processing when only numeric characters are involved. When alphanumeric characters are used, more than four bits are necessary, and codes have been developed that use six, seven, and eight bits to represent each character. The computer codes ASCII and EBCDIC use eight bits to code a character: four *numeric bits* (least significant bits) and four leading *zone bits* (more significant bits). For example, in ASCII, A is coded as:

Zone Bits	*Numeric Bits*
0100	0001

In EBCDIC, A is coded as:

Zone Bits	*Numeric Bits*
1100	0001

Observe that the numeric bits are the same for A in ASCII and EBCDIC.

A code of n bits contains 2^n combinations of bits. An 8-bit code such as ASCII can therefore accommodate $2^8 = 256$ characters. This includes the basic 36-character set consisting of the 10 digits and 26 letters and as many as 220 other characters such as %, =, ?, and so on.

ASCII

The ASCII code was developed by the American National Standards Institute (ANSI) and is used in microcomputers, and in non-IBM mainframe systems. Table 25-9 shows the ASCII code for the 36 basic characters and 12 other characters. The hexadecimal or

TABLE 25-9 AMERICAN STANDARD CODE FOR INFORMATION INTERCHANGE (ASCII)

			Zone Bits		
	Hexadecimal		3	4	5
		Binary	0011	0100	0101
Numeric Bits	0	0000	0	@	P
	1	0001	1	A	Q
	2	0010	2	B	R
	3	0011	3	C	S
	4	0100	4	D	T
	5	0101	5	E	U
	6	0110	6	F	V
	7	0111	7	G	W
	8	1000	8	H	X
	9	1001	9	I	Y
	A	1010	:	J	Z
	B	1011	;	K	[
	C	1100	<	L	\
	D	1101	=	M	]
	E	1110	>	N	^
	F	1111	?	O	_

hex equivalents of each binary representation are also given and are used more often than the binary representation. They are a convenient way of referring to a character since eight bits can readily be converted to two hex digits and vice versa. For example, the character K is coded as 0100 1011, which in hex is 4B. Observe in Table 25-9 that the numeric bits for the integers 0 to 9 are the binary equivalents of the integers.

A computer usually stores an extra bit for each character called a *check bit* or *parity bit* that precedes the zone bits. The check bit is chosen so that the sum of the 1-bits is either even or odd depending on whether the computer operates on even parity or odd parity. For example, if the computer operates on even parity, the check bit for K would be 0 since the sum of the 1-bits is an even number 4. If the computer operates on odd parity, the check bit for K would be 1:

Check bit
↓

K	even parity	0	0100	1011
K	odd parity	1	0100	1011

The check bit may be used to detect an error when a character is transmitted. If a bit has been erased or changed, the parity changes, and the computer, detecting the wrong parity, transmits the data again. An odd-parity bit has the advantage that all zeros (0 0000 0000) will never be transmitted and thus will not be confused with “no information.”

EXAMPLE 25.23 Given the name JEAN:

(a) What is the ASCII code for this name in binary, hexadecimal, and decimal code?

(b) If the computer uses an odd-parity check. What would the binary code for JEAN be?

Solution

(a) Using Table 25-9, the binary and hexadecimal ASCII codes are:

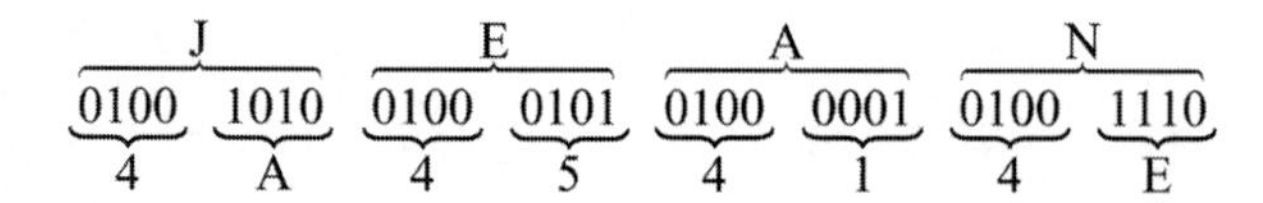

You can find the decimal code by converting the hexadecimal (or binary) numbers:

$$\begin{aligned}
\text{J: } 4A_{16} &= 4(16) + 10(1) = 74\\
\text{E: } 45_{16} &= 4(16) + 5(1) = 69\\
\text{A: } 41_{16} &= 4(16) + 1(1) = 65\\
\text{N: } 4E_{16} &= 4(16) + 14(1) = 78
\end{aligned}$$

(b) An odd-parity check means the bits in each byte must add up to an odd number. The 1-bits for J and E add up to odd numbers and the 1-bits for A and N add up to even numbers. Therefore, a 0-bit is added to J and E, and a 1-bit is added to A and N:

J	E	A	N
0 0100 1010	0 0100 0101	1 0100 0001	1 0100 1110

The number of bits used to represent a character is called a *byte*. A byte in ASCII contains eight bits, or nine bits including the check bit. A computer word is a set of bytes that are processed as a group. Words vary in length. For example, in certain computers, a word consists of four 8-bit bytes or 32 bits.

EXAMPLE 25.24 What is the following computer word in ASCII using four 8-bit bytes?

00110001 00111001 00111001 00110101 ?

Solution Partition the bits into 8-bit bytes and use Table 25-9 to obtain:

0011 0010	0011 0000	0011 0000	0011 0001
2	0	0	1

EXERCISE 25.6

In exercises 1 through 6, express each number in:

(a) 4-bit BCD code
(b) straight binary representation

1. 18
2. 37
3. 122
4. 253
5. 349
6. 356

In exercises 7 ythrough 14, express each computer word using ASCII code in:

(a) binary code
(b) hexadecimal code
(c) decimal code

7. X7
8. 3M
9. CIA
10. FBI
11. SAM
12. PAT
13. ROSE
14. JOSE

15. **(a)** What is a byte?
(b) How long is a byte in ASCII including a parity bit?

16. **(a)** What is a computer word?
(b) A word containing four characters with 16-bit bytes contains how many bits?

17. What would be the binary ASCII code for the word NYC if the computer uses an even-parity check?

18. What would be the binary ASCII code for the word LAX if the computer uses an odd-parity check?

19. What is the following computer word in ASCII?

0011 0111 0011 0000 0011 0111

20. What is the following computer word in ASCII?

0100 0010 0100 1001 0101 0100

21. The ASCII code for the character "$" in octal code is 44. Express this in:

(a) 8-bit binary code

(b) hexadecimal code

22. Complete the following word in ASCII, where each byte has a check bit added for odd parity, by supplying the last byte:

1 0100 0001
1 0101 0011
0 0100 0011
0 0100 1001

23. A computer using an even-parity check receives the following binary ASCII code for the mathematical equation: $1 - 1 = 0$:

1 0011 0001 1 0110 1111 1 0011 0001
1 0011 1101 0 0011 0000

Using Table 25-9, troubleshoot and correct the error in this transmission.

24. A computer using an odd-parity check receives the following binary ASCII code for the mathematical relationship: $5 > 3 > 2$:

1 0011 0101 0 0011 1110 1 0011 0010
0 0011 1110 0 0011 0010

Using Table 25-9, troubleshoot and correct the error in this transmission.

CHAPTER HIGHLIGHTS

25.1 BINARY NUMBER SYSTEM

TABLE 25-3 PLACE VALUES OF THE BINARY DIGITS

Power	2^6	2^5	2^4	2^3	2^2	2^1	2^0	.	2^{-1}	2^{-2}	2^{-3}
Value	64	32	16	8	4	2	1	Binary Point	$\frac{1}{2} = 0.5$	$\frac{1}{4} = 0.25$	$\frac{1}{8} = 0.125$

Key Example:

$$110100_2 = (1 \times 2^5) + (1 \times 2^4) + (1 \times 2^2) = 32 + 16 + 4 = 52$$

To Convert Decimal Integer to Binary Integer

1. Divide the decimal integer by 2.
2. Record the remainder (0 or 1).
3. Divide the quotient by 2.
4. Record the remainder again.
5. Repeat steps 3 and 4 until the quotient is zero.
6. The remainders in reverse order represent the binary integer.

Key Example: The decimal number $46 = 101110_2$ as follows:

	Remainders
2)46	**0**
2)23	**1**
2)11	**1**
2)5	**1**
2)2	**0**
2)1	**1**
0	

25.2 BINARY ARITHMETIC

TABLE 25-4 BINARY SUMS

0 + 0 = 0	
0 + 1 = 1	
1 + 1 = 10	(0 and carry *1*)
1 + 1 + 1 = 10 + 1 = 11	(1 and carry *1*)
1 + 1 + 1 + 1 = 11 + 1 = 100	(0 and carry *10*)

Key Example:

$$\begin{array}{r} 110 \\ +111 \\ \hline 1 \end{array} \Rightarrow \begin{array}{r} \mathit{1} \\ 110 \\ +111 \\ \hline 01 \end{array} \Rightarrow \begin{array}{r} \mathit{1} \\ 110 \\ +111 \\ \hline 1101 \end{array} \quad \text{(carried digit)}$$

1's Complement:

The number obtained by subtracting each digit from 1. This is done directly by changing each 0 to 1 and each 1 to 0.

2's Complement:

The number obtained by adding 1 to its 1's complement.

Key Example: Subtraction adding the 1's complement:

$$\begin{array}{r}1011\\-1001\\\hline\end{array} \Rightarrow \begin{array}{r}\mathit{11}\\1011\\+0110\\\hline 10001\end{array} \text{(carried digits)} \Rightarrow \begin{array}{r}\textcircled{1}0001\\ \rightarrow +\mathit{1}\\\hline 0010\end{array} = 10 \text{ (end-around carry)}$$

Subtraction adding the 2's complement:

$$\begin{array}{r}1011\\-1001\\\hline\end{array} \Rightarrow \begin{array}{r}\mathit{111}\\1011\\+0111\\\hline \not{1}0010\end{array} \text{(carried digits)} = 0010 = 10 \text{ (delete leading one)}$$

25.3 OCTAL NUMBER SYSTEM

See Table 25-5 below.

Key Example:

$$372_8 = 3(64) + 7(8) + 2(1) = 192 + 56 + 2 = 250$$

To Convert Decimal Number to Octal Number

1. Divide the decimal integer by 8.
2. Record the remainder.
3. Divide the quotient by 8.
4. Record the remainder.
5. Repeat steps 3 and 4 until the quotient is zero.
6. The remainders in reverse order represent the octal number.

Key Example: The decimal number 281 = 431_8 as follows:

	Remainders
8)281	**1** ↑
8)35	**3**
8)4	**4**
0	

To Convert Octal Number to Binary Number

Change each digit into its 3-bit binary equivalent, adding leading zeros if necessary.

Key Example: 561_8 = 101 110 001_2 as follows:

$$\underbrace{101}_{5}\quad\underbrace{101}_{6}\quad\underbrace{001}_{1}$$

To Convert Binary Number to Octal Number

Arrange the binary numbers in groups of three bits. Add leading zeros if necessary. Change each three bits into an octal digit.

Key Example: The binary number 1 111 010_2 = 172_8 as follows:

$$\underbrace{001}_{1}\quad\underbrace{111}_{7}\quad\underbrace{010}_{2}$$

25.4 HEXADECIMAL NUMBER SYSTEM

See Table 25-8 below.

The hexadecimal number system contains the following sixteen digits:

0, 1, 2, 3, 4, 5, 6, 7, 8, 9, A, B, C, D, E, F

The letters A to F are equal to the decimal numbers 10 to 15, respectively.

Key Example: E5 = $(14 \times 16) + (5 \times 1) = 224 + 5 = 229$

To Convert Decimal Number to Hexadecimal Number

1. Divide the decimal integer by 16.
2. Record the remainder.
3. Divide the quotient by 16.

TABLE 25-5 PLACE VALUES OF THE OCTAL DIGITS

Power	8^4	8^3	8^2	8^1	8^0	.	8^{-1}	8^{-2}
Value	4096	512	64	8	1	Octal Point	$\frac{1}{8} = 0.125$	$\frac{1}{64} = 0.015625$

TABLE 25-8 PLACE VALUES OF THE HEXADECIMAL DIGITS

Powers	16^3	16^2	16^1	16^0	.	16^{-1}
Values	4096	256	16	1	Hex Point	$\frac{1}{16} = 0.0625$

4. Record the remainder.
5. Repeat steps 3 and 4 until the quotient is zero.
6. The remainders in reverse order represent the hexadecimal number.

Key Example: The decimal number $380 = 17C_{16}$ as follows:

	Remainders
16)380	12 = **C** ↑
16)23	**7**
16)1	**1**
0	

To Convert Hexadecimal Number to Binary Number

Change each digit into its 4-bit binary equivalent, adding leading zeros if necessary.

Key Example: $7D_{16} = 1111101_2$ as follows:

7	D
0111	1101

To Convert Binary Number to Hexadecimal Number

Arrange the binary number in groups of four bits. Add leading zeros if necessary. Change each four bits into a hexidecimal digit.

Key Example: $1011010_2 = 5A_{16}$ as follows:

0101	1010
5	A

25.5 HEXADECIMAL ARITHMETIC

Key Example:

```
                  1 1              1 1   (carried digits)
  E 6 A           E 6 A            E 6 A
+   B D   ⇒ +       B D   ⇒ +        B D
-------         -------          -------
      7             2 7            F 2 7
   (23)          (18)           (15)
```

15's Complement:

The number obtained by subtracting each digit from 15.

16's Complement:

The number obtained by adding 1 to its 15's complement.

Key Example: Subtraction adding the 16's complement:

```
                 1   1         (carried digits)
  4C                 4   C
- 1B    ⇒    +       B   5     (delete leading 1)
-----        -----------
             1̸   3   1
               (19) (17)
```

25.6 COMPUTER CODES

ASCII is the binary code used in microcomputers and non-IBM mainframe systems. ASCII uses eight bits for alphanumeric characters and may or may not use an extra parity bit. A byte is the number of bits used to represent a character. See Table 25-9.

Key Example: The binary and hexadecimal code for JEAN is:

J		E		A		N	
0100	1010	0100	0101	0100	0001	0100	1110
4	A	4	5	4	1	4	E

REVIEW EXERCISES

In exercises 1 through 6, convert each binary number to a decimal number.

1. 1111 **4.** 111110

2. 1101 **5.** 100.11

3. 11011 **6.** 101.101

In exercises 7 through 12, convert each decimal number to a binary number.

7. 19 **10.** 125

8. 34 **11.** 0.75

9. 88 **12.** 5.5

In exercises 13 through 16, calculate each binary sum.

```
13.  101        15.   11
     111             111
     ---             100
                     101
                     ---

14.   11        16.  1.01
     110             0.11
     101            11.01
     ---            -----
```

In exercises 17 through 20 perform each binary subtraction.

```
17.  1010       18.   1100
     -101            -1001
     ----            -----

19.  10000      20.  11001
     -1111           -1110
     -----           -----
```

In exercises 21 through 26 convert each number to a decimal number.

21. 75_8
22. 167_8
23. $3A_{16}$
24. EB_{16}
25. $17A_{16}$
26. DAD_{16}

In exercises 27 through 32, convert each number to a binary number.

27. 46_8
28. 103_8
29. $2A_{16}$
30. $8B_{16}$
31. $9CD_{16}$
32. $FE5_{16}$

In exercises 33 through 38, convert each decimal number to:

(a) an octal number
(b) a hexadecimal number

33. 27
34. 63
35. 111
36. 347
37. 1056
38. 2112

In exercises 39 through 44, convert each binary number to:

(a) an octal number
(b) a hexadecimal number

39. 111
40. 1101
41. 10011
42. 11110
43. 1010101
44. 11001100

In exercises 45 through 48, add the hexadecimal numbers.

45. $\begin{array}{r} B4 \\ +7E \\ \hline \end{array}$

46. $\begin{array}{r} A83 \\ +\ \ 56 \\ \hline \end{array}$

47. $\begin{array}{r} DFE \\ +70A \\ \hline \end{array}$

48. $\begin{array}{r} C984 \\ +\ \ A0B \\ \hline \end{array}$

In exercises 49 through 52, subtract the hexadecimal numbers.

49. $\begin{array}{r} C5 \\ -62 \\ \hline \end{array}$

50. $\begin{array}{r} 4E7 \\ -\ \ B5 \\ \hline \end{array}$

51. $\begin{array}{r} CAB \\ -2EE \\ \hline \end{array}$

52. $\begin{array}{r} 406B \\ -\ \ 91F \\ \hline \end{array}$

In exercises 53 through 56, express each computer word in ASCII code.

53. GO
54. 101
55. XS3
56. SEA

57. Complete the computer message in hexadecimal ASCII code by supplying the last word:

4841535445

4D414B4553

58. Complete the following message in binary ASCIII code:

0100 1100
0100 1111
0101 0110

59. A computer using an even-parity check receives the following binary ASCII code for the mathematical equation: $1 + 1 = 2$:

1 0011 0001 0 0010 1011 1 0011 0001
0 0011 1101 1 0011 0010

If the ASCII code for the character "+" in hex is 2B, using Table 25-9 troubleshoot and correct the error in this transmission.

60. Given a binary number, the 1's complement of the 1's complement is the original number, since each digit is reversed twice back to the original. Show that the 2's complement of the 2's complement is also the original number for the binary numbers:
(a) 1100
(b) 1011000

CHAPTER 26

Boolean Algebra

CHAPTER OBJECTIVES

In this chapter, you will learn:

- The basic operations, laws, and theorems of Boolean algebra.
- How to use truth tables to prove Boolean theorems.
- The three basic logic gates AND, OR, and NOT.
- How to construct truth tables for logic circuits.
- How to express the output of a logic circuit using Boolean algebra.
- How to simplify a logic circuit using Karnaugh maps.

In the mid-nineteenth century the English mathematician George Boole developed a mathematical system called Boolean algebra or Boolean logic. A century later, in 1938, Claude Shannon applied these ideas to switching circuit problems. His mathematical methods of working with switches and relays proved very useful in the simplification of circuits and are used today in the design of logic networks and computers. The laws of Boolean algebra are similar in some ways to those of arithmetic and algebra, but in other ways they are quite different. These laws provide a mathematical model for the basic logic gates of digital circuits. Useful visual tools for simplifying Boolean expressions and logic circuits are diagrams called Karnaugh maps, which are introduced in this chapter.

• • •

26.1 Boolean Algebra

Boolean algebra consists of a set of algebraic elements, or variables, that are combined with the following basic Boolean operations to form statements:

Operations

Boolean Operations

1. The OR operation "+"
2. The AND operation "·"
3. The *complement* of an element A, called NOT A, and written $\overline{A}$. (26.1)

The operations are defined in terms of *truth tables*, where each element is assigned a truth value of 1 ("true") or 0 ("false"). These are shown in Table 26-1 for any two elements A and B. Observe in the table that $A \cdot B$ is written in shortcut form as AB, similar to multiplication in algebra. Also notice that there are four possible combinations of the truth values 0 and 1 for two elements.

Table 26-1 shows that:

TABLE 26-1 TRUTH TABLES FOR AND, OR, AND NOT

		AND	OR	NOT
A	B	AB	$A+B$	$\overline{A}$
0	0	0	0	1
0	1	0	1	1
1	0	0	1	0
1	1	1	1	0

1. A AND B is true *only if both* A and B are true.
2. A OR B is true *if either* A or B is true.
3. NOT A is true when A is false and vice versa.

The truth values for AND, OR, and NOT come from their application in verbal language. For example, suppose A and B represent the following statements:

A: Computers use binary numbers.

B: $1 + 1 = 3$

Here, A is a true statement, and B is not true.

The statement AB is: Computers use binary numbers and $1 + 1 = 3$. This statement is *not* true because B is not true, which agrees with the truth table.

The statement $A + B$ is: Computers use binary numbers or $1 + 1 = 3$. This statement *is* true because A is true, which also agrees with the truth table.

The statement $\overline{A}$ is: Computers do not use binary numbers. This statement is *not* true because A is true. The statement $\overline{B}$ is: $1 + 1 \neq 3$. This statement *is* true because B is not true.

These last two statements also agree with the truth table.

If two Boolean expressions have the same truth values in a truth table, they are equivalent and are a Boolean theorem. The truth values are based on the defined truth tables for the operations of AND, OR, and NOT shown in Table 26-1. Study the following example of a proof using a truth table.

EXAMPLE 26.1 Prove the Boolean theorem:

$$A + AB = A$$

Solution To prove a Boolean theorem, construct a truth table containing each of the elements and statements and show that the truth values for each side of the equality are the same. That is, show that the statement A, and the statement $A + AB$ have the same truth values. The table is as follows:

A	B	AB	$A + AB$
0	0	0	0
0	1	0	0
1	0	0	1
1	1	1	1

↑ SAME ↑

The first three columns in the truth table are the same as in the truth table for AB in Table 26-1. The last column compares the column for A and the column for AB, and applies the truth values for the operation of OR in Table 26-1. Here A is treated as "A", and AB is treated as "B" in Table 26-1. Since the column for A and the column for A + AB have the same truth values, the two statements are equivalent and the theorem is true.

The set of elements with the operations of AND, OR, and NOT satisfy laws like those of arithmetic:

Laws

Commutative Laws

$$A + B = B + A \qquad (26.2)$$
$$AB = BA$$

• • •

Laws **Associative Laws**

$$(A + B) + C = A + (B + C)$$
$$(AB)C = A(BC) \qquad (26.3)$$

• • •

Laws **Distributive Laws**

$$A(B + C) = AB + AC$$
$$A + BC = (A + B)(A + C) \qquad (26.4)$$

Note in the distributive laws (26.4) that the AND operation is done before OR. The OR operation needs to be enclosed in parentheses when it is to be done before AND. The first distributive law is similar to the one in arithmetic, where + is addition and · is multiplication, while the second distributive law does not apply in arithmetic.

The 0 and 1 shown in the truth tables are also basic elements in Boolean algebra and satisfy the following laws:

Laws **Identity Laws**

$$A + 0 = A$$
$$A \cdot 1 = A \qquad (26.5)$$

• • •

Laws **Complement Laws**

$$A + \overline{A} = 1$$
$$A \cdot \overline{A} = 0 \qquad (26.6)$$
$$\overline{\overline{A}} = A$$

The *dual* of a Boolean statement is the statement obtained by interchanging the operations of + and ·, and by interchanging 0 and 1. Note that the distributive laws (26.4) and the laws (26.5) and (26.6) above contain dual statements. Dual statements have the following property:

Rule **Dual Property**

If a statement is true, the dual statement is also true.

Several basic theorems follow from the above laws and the dual property. They are shown in Table 26-2, where each statement is written next to its dual.

TABLE 26-2 BASIC THEOREMS IN BOOLEAN ALGEBRA

Statement	Dual
$\overline{0} = 1$	$\overline{1} = 0$
$A + 1 = 1$	$A \cdot 0 = 0$
$A + A = A$	$A \cdot A = A$
$A + \overline{A} = 1$	$A \cdot \overline{A} = 0$

Observe in Table 26-2 that:

1. The complement of 1 is 0 and vice versa.
2. An element operated on itself results in the same element. This is very different from arithmetic.
3. An element operated on by its complement results in 1 for '+' and 0 for '·'.

It is also possible to prove a theorem by applying the laws directly without a truth table, as the next example shows.

EXAMPLE 26.2

Prove the Boolean theorem of Example 26.1: $A + AB = A$, by applying the above laws.

Solution In Example 26.1, this theorem is proved by a truth table. Start with the left side and reduce it to the right side as follows.

Apply the distributive law: $A + AB = A(1 + B)$

Apply the commutative law: $A(1 + B) = A(B + 1)$

In Table 26-2, the second theorem says that any element $A + 1 = 1$. Therefore, $B + 1 = 1$ and:

$$A(B + 1) = A(1) = A \cdot 1$$

By the identity law: $A \cdot 1 = A$

The left side now equals the right side and the theorem is proved.

EXAMPLE 26.3

Prove the Boolean theorem:

$$\overline{AB} = \overline{A} + \overline{B}$$

Solution This statement says: NOT (AB) is equivalent to (NOT A) OR (NOT B). To show this equivalence, construct a truth table containing columns for A and B, columns for each new element, $\overline{A}$ and $\overline{B}$, and each statement, AB, $\overline{AB}$, and $\overline{A} + \overline{B}$:

A	B	AB	$\overline{A}$	$\overline{B}$	$\overline{AB}$	$\overline{A}+\overline{B}$
0	0	0	1	1	1	1
0	1	0	1	0	1	1
1	0	0	0	1	1	1
1	1	1	0	0	0	0

SAME (last two columns)

To obtain the truth values for each column, apply the defined truth values for AND, OR, and NOT in Table 26-1. Since the last two columns have the same truth values, they are equivalent, and the theorem is true.

EXAMPLE 26.4

Prove the Associative law:

$$(AB)C = A(BC)$$

Solution There are *eight* possible combinations of truth values for three elements. The truth table, shown on page 529, therefore has eight rows.

The last two columns are the same, which proves the theorem. Note that the 0's and 1's in the columns for A, B, and C represent the first eight binary numbers. There is a direct relationship between binary numbers that have two digits and Boolean algebra that has two elements, 0 and 1.

A	B	C	AB	BC	$(AB)C$	$A(BC)$
0	0	0	0	0	0	0
0	0	1	0	0	0	0
0	1	0	0	0	0	0
0	1	1	0	1	0	0
1	0	0	0	0	0	0
1	0	1	0	0	0	0
1	1	0	1	0	0	0
1	1	1	1	1	1	1

SAME

The theorem in Example 26.3 is one of two important laws in Boolean algebra known as De Morgan's laws, which are duals of each other:

Laws

De Morgan's Laws

$$\overline{AB} = \overline{A} + \overline{B}$$
$$\overline{A + B} = \overline{A} \cdot \overline{B} \qquad (26.7)$$

EXAMPLE 26.5

Prove De Morgan's law for three variables: $\overline{A + B + C} = \overline{A}\,\overline{B}\,\overline{C}$ using a truth table.

Solution

Note in the table that the AND and OR operations for three variables are extensions of the definitions for two variables. The columns that represent each side of DeMorgan's law are the same, proving the theorem.

A	B	C	$A+B+C$	$\overline{A+B+C}$	$\overline{A}$	$\overline{B}$	$\overline{C}$	$\overline{A}\,\overline{B}\,\overline{C}$
0	0	0	0	1	1	1	1	1
0	0	1	1	0	1	1	0	0
0	1	0	1	0	1	0	1	0
0	1	1	1	0	1	0	0	0
1	0	0	1	0	0	1	1	0
1	0	1	1	0	0	1	0	0
1	1	0	1	0	0	0	1	0
1	1	1	1	0	0	0	0	0

SAME

EXERCISE 26.1

1. Given the following statements:

   A: Copper is a poor conductor.
   B: $(0.1)^2 = 0.01$.

   Write the statements *A* AND *B*, *A* OR *B*, NOT *A*, and NOT *B*, and tell whether each is true, or not true, based on the truth of ***A*** and ***B***.

2. Do the same as exercise 1 for the following statements:

   A: $V = IR$

   B: New York is in California.

3. Which of the laws for Boolean algebra also apply in arithmetic and algebra?

4. Explain why the complement laws for Boolean algebra cannot apply for addition or multiplication in algebra.

In exercises 5 through 16, give the dual of each Boolean statement.

5. $A + B = B + A$
6. $(A + B) + C = A + (B + C)$
7. $A \cdot 1 = A$
8. $A \cdot \overline{A} = 0$
9. $A + 1 = 1$
10. $A + \overline{A} = 1$
11. $A + 1 = A + \overline{A}$
12. $A + 0 = A \cdot A$
13. $A(B + 1) = A + 0$
14. $AB + BC = (A + C)B$
15. $A + \overline{A}B = A + B$
16. $\overline{A + B} = \overline{A} \cdot \overline{B}$

In exercises 17 through 28, prove each Boolean theorem by the use of truth tables.

17. $A + A = A$
18. $\overline{AA} = \overline{A}$
19. $A(A + B) = A$
20. $(A + B)(A + B) = A + B$
21. $\overline{\overline{AB}} = AB$
22. $A + \overline{A}B = A + B$
23. $\overline{A} + \overline{A} \cdot \overline{B} = \overline{A}$
24. $\overline{A + B} = \overline{A} \cdot \overline{B}$
25. $(A + B) + C = A + (B + C)$
26. $(A + B)(A + C) = A + BC$
27. $\overline{A}(\overline{B}\,\overline{C}) = \overline{A}(\overline{B + C})$
28. $\overline{A} + \overline{B} + \overline{C} = \overline{ABC}$
29. Prove $A(A + B) = A$ by applying the laws of Boolean algebra.
30. Prove $B(A + AB) = AB$ by applying the laws of Boolean algebra.

26.2 Logic Circuits

Series and Parallel Switching Circuits

Figure 26-1 shows the two basic kinds of switching circuits. A series circuit contains two or more switches connected end to end in the same path. In a series circuit, current will flow only if all switches are closed. A parallel circuit contains two or more switches connected in parallel paths. In a parallel circuit, current will flow if either switch is closed. The properties of each circuit are shown in a truth table where 0 = open and 1 = closed.

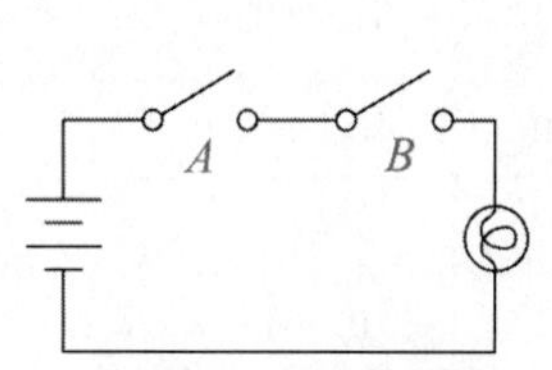

Series switching circuit

A	B	Output A AND B
0	0	0
0	1	0
1	0	0
1	1	1

(a)

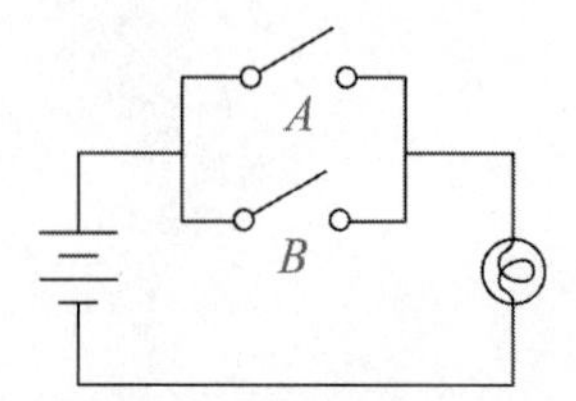

Parallel switching circuit

A	B	Output A OR B
0	0	0
0	1	1
1	0	1
1	1	1

(b)

FIGURE 26-1 Basic switching circuits.

Note the following about the truth tables:

1. The truth table for the series circuit is the same as the truth table for the Boolean operation AND.
2. The truth table for the parallel circuit is the same as the truth table for the Boolean operation OR.

Because of this, the concepts of Boolean algebra can be applied to the analysis of switching circuits, or logic circuits, as follows.

Logic Gates

Logic gates control the transmission of input pulses and are the building blocks of electronic circuits. There are three basic logic gates corresponding to the Boolean operations of AND, OR, and NOT.

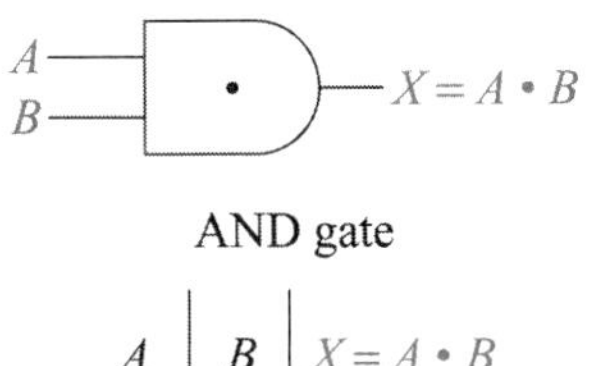

A	B	$X = A \cdot B$
0	0	0
0	1	0
1	0	0
1	1	1

FIGURE 26-2 AND gate.

The AND Gate An AND gate corresponds to the series switching circuit shown in Figure 26-1. The output X is expressed in terms of the inputs A and B using Boolean symbols: $X = A \cdot B$. The AND gate and its truth table are shown symbolically in Figure 26-2. It will allow a pulse or signal to be transmitted, denoted by $X = 1$, *only* when $A = 1$ and $B = 1$, that is, only when there is an input to both A and B. When there is an output from a gate, it is said to be *enabled.*

The OR Gate An OR gate corresponds to the *parallel* switching circuit shown in Figure 26-1. The output X is expressed using Boolean symbols: $X = A + B$. The OR gate and its truth table are shown symbolically in Figure 26-3. It allows a pulse or signal to be transmitted, $X = 1$, when either $A = 1$ or $B = 1$; that is, when there is an input at either A or B.

The AND gate and the OR gate can have more than two inputs. An AND gate with three inputs has output $X = A \cdot B \cdot C$, and an OR gate with three inputs has output $X = A + B + C$.

The NOT Gate A NOT gate corresponds to the complement operation and is called an inverter. The NOT gate has only one input A. Its output is expressed in Boolean symbols: $X = \overline{A}$. The NOT gate and its truth table are shown symbolically in Figure 26-4(a).

A NOT gate can be thought of as an electrical switch that allows current to flow through a circuit when it is open and does not allow current to flow when it is closed, as shown in Figure 26-4(b). The current flows through the lamp when the switch is open.

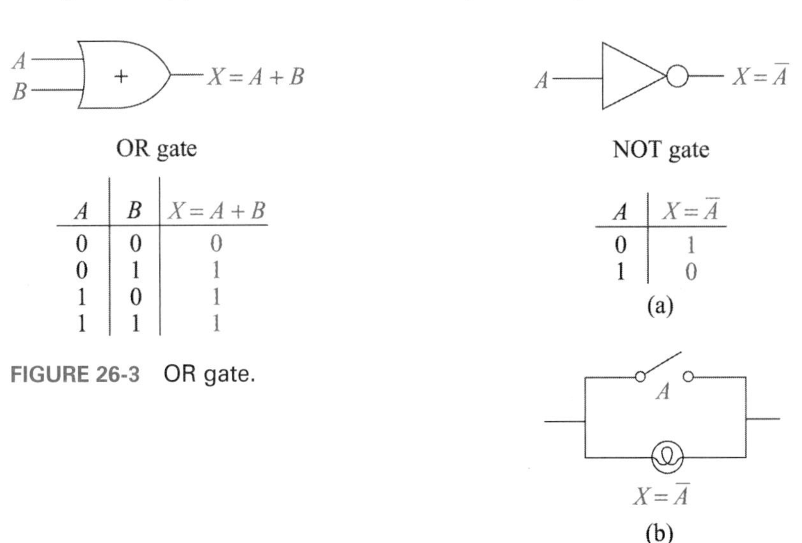

A	B	$X = A + B$
0	0	0
0	1	1
1	0	1
1	1	1

FIGURE 26-3 OR gate.

A	$X = \overline{A}$
0	1
1	0

FIGURE 26-4 NOT gate.

When the switch is closed, the lamp is short circuited and no current flows through it. A NOT gate can also be thought of as a voltage inverter that changes high voltage to low voltage and vice versa.

A *logic circuit* is a circuit containing one or more of the three basic logic gates. Since the three basic logic gates have the same truth tables as the corresponding Boolean operations, *logic circuits have the properties of Boolean algebra.* The following example shows a common AND-OR logic circuit.

EXAMPLE 26.6 Figure 26-5 shows a logic circuit consisting of two AND gates feeding into an OR gate.

(a) Find the Boolean expression for the output X in terms of the inputs A and B.

(b) Construct the truth table for the circuit and show that the output X is equivalent to $A \cdot B$.

(c) Show that the output X is equivalent to $A \cdot B$ by applying the laws of Boolean algebra.

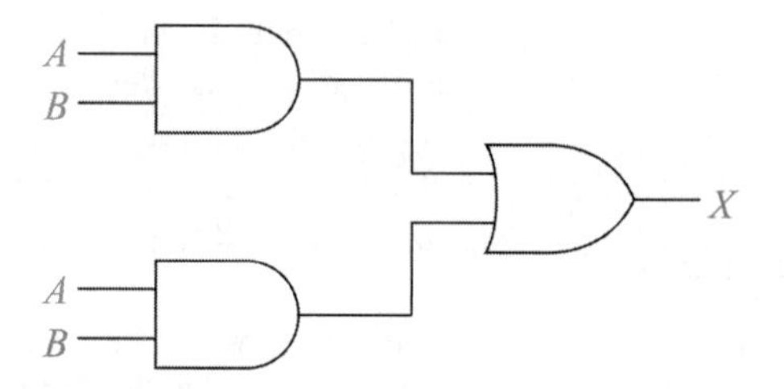

FIGURE 26-5 AND-OR logic circuit for Example 26.6.

Solution

(a) To find the output X, find the output of each gate separately, moving in the direction of the signal from left to right. The output of each AND gate is $A \cdot B$. See Figure 26-6. These become the inputs of the OR gate, and the output of the OR gate is then:

$$X = A \cdot B + A \cdot B$$

This is known as a *Boolean sum of products.*

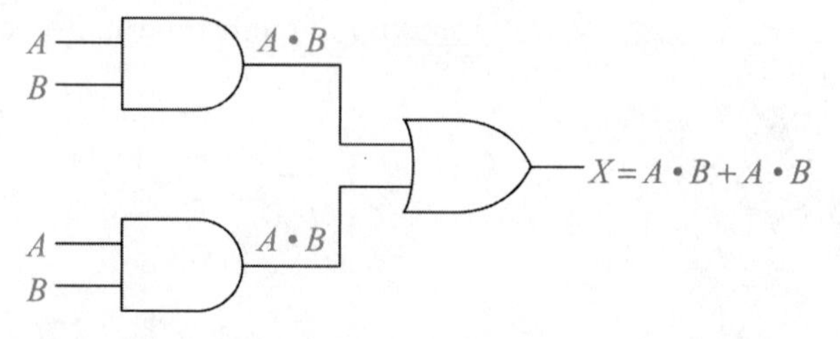

FIGURE 26-6 Output of AND-OR logic circuit for Example 26.6.

(b) To construct the truth table shown on page 533, work with the Boolean expression for the output X and proceed as in Section 26.1 with the four combinations of the inputs A and B:

In the table, the output X is equivalent to the product $A \cdot B$ since the last two columns are the same.

(c) To show this equivalence using the laws of Boolean algebra, apply the theorem $A + A = A$ from Table 26-2 where A represents the product $A \cdot B$:

$$X = A \cdot B + A \cdot B = A \cdot B$$

A	B	$A \cdot B$	$X = A \cdot B + A \cdot B$
0	0	0	0
0	1	0	0
1	0	0	0
1	1	1	1

↑——— **SAME** ———↑

This means that the logic circuit in Figure 26-5 can be simplified to a basic AND gate. Simplifying switching circuits is one of the important applications of Boolean algebra.

ERROR BOX

A common error when working with logic circuits is confusing the AND and OR gates. An AND gate has an output *only when all* inputs are 1. The OR gate has an output *when any* input is 1. See if you can do the practice problems without looking at the truth tables.

Practice Problems: Given the following inputs for an AND gate and an OR gate. Tell the output for each gate.

1. $A = 0, B = 0$ **2.** $A = 1, B = 0$ **3.** $A = 0, B = 1$

4. $A = 1, B = 1$ **5.** $A = 1, B = 1, C = 0$ **6.** $A = 1, B = 1, C = 1$

7. $A = 1, B = 0, C = 0$ **8.** $A = 1, B = 0, C = 1$ **9.** $A = 0, B = 0, C = 0$

10. $A = 0, B = 1, C = 0$

EXAMPLE 26.7

Given the logic circuit in Figure 26-7 with inputs A, B, and C:

(a) Find the output in terms of A, B, and C.

(b) Construct the truth table for the circuit and show that the output X is equivalent to $B \cdot C$.

(c) Show that the output X is equivalent to $B \cdot C$ by applying the laws of Boolean algebra.

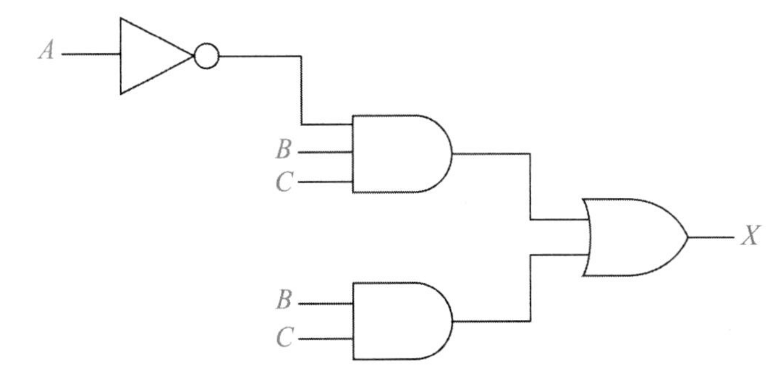

FIGURE 26-7 Logic circuit for Example 26.7.

Solution

(a) Moving from left to right, the output of each gate is shown in Figure 26-8 on page 534. The output X is made up of the two AND gates feeding into the OR gate:

$$X = \bar{A}BC + BC$$

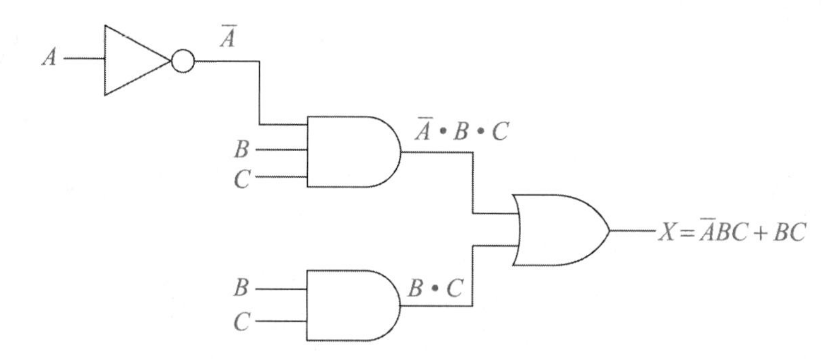

FIGURE 26-8 Output of the logic circuit for Example 26.7.

(b) The truth table for the circuit contains eight possible combinations because of the three inputs:

A	B	C	$\overline{A}$	$\overline{A}B$	$\overline{A}BC$	BC	$X = \overline{A}BC + BC$
0	0	0	1	0	0	0	0
0	0	1	1	0	0	0	0
0	1	0	1	1	0	0	0
0	1	1	1	1	1	1	1
1	0	0	0	0	0	0	0
1	0	1	0	0	0	0	0
1	1	0	0	0	0	0	0
1	1	1	0	0	0	1	1

SAME

The last two columns in the truth table are identical, which shows that:

$$X = \overline{A}BC + BC = BC$$

(c) To show this equivalence using the laws of Boolean algebra, first apply the distributive law to the term BC in X:

$$X = \overline{A}BC + BC = BC(\overline{A} + 1)$$

Then from Table 26-2, $A + 1 = 1$ for any element A. Applying this to $\overline{A}$ you have:

$$X = BC(\overline{A} + 1) = BC(1) = BC$$

NAND and NOR Gates

Two useful logic circuits are the NAND and NOR gates. A NAND gate is equivalent to an AND gate followed by a NOT gate. A NOR gate is equivalent to an OR gate followed by a NOT gate. Figure 26-9 on page 535 shows the two gates and their truth tables. Note that the bubble in the NAND and NOR gates represents the NOT operation. The NAND gate has a high output when any input is low, whereas the NOR gate has a low output when any input is high.

Answers (AND,OR) *to Error Box Problems, page 533:*
1. 0,0 **2.** 0,1 **3.** 0,1 **4.** 1,1 **5.** 0,1 **6.** 1,1 **7.** 0,1 **8.** 0,1 **9.** 0,0 **10.** 0,1

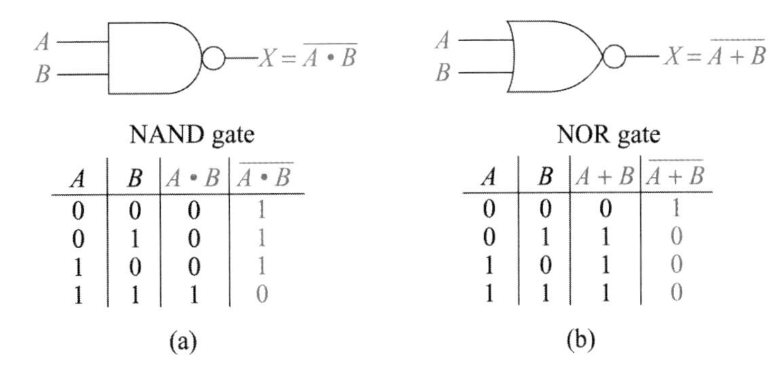

A	B	$A \bullet B$	$\overline{A \bullet B}$
0	0	0	1
0	1	0	1
1	0	0	1
1	1	1	0

A	B	$A + B$	$\overline{A + B}$
0	0	0	1
0	1	1	0
1	0	1	0
1	1	1	0

FIGURE 26-9 NAND and NOR gates.

EXAMPLE 26.8

For the logic circuit in Figure 26-10 showing two NAND gates with inputs A and B feeding into a NOR gate:

(a) Find the output in terms of A and B.

(b) Construct the truth table for the circuit and show that the output X is equivalent to AB.

(c) Show that the output X is equivalent to AB by applying the laws of Boolean algebra.

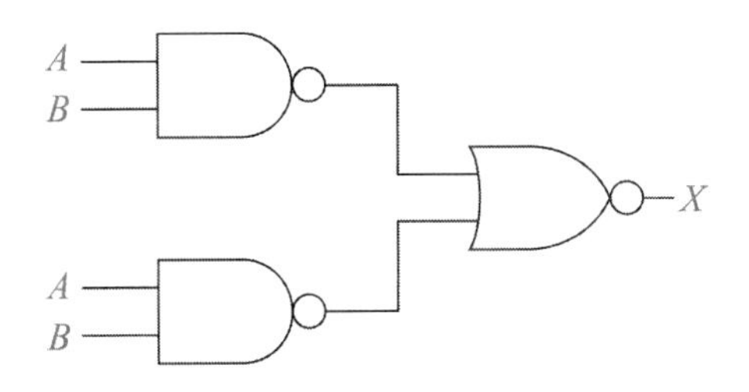

FIGURE 26-10 Logic circuit for Example 26.8.

Solution

(a) The output of each NAND gate is $\overline{AB}$, which inputs into the NOR gate producing the output:

$$X = \overline{\overline{AB} + \overline{AB}}$$

(b) The truth table for the output X and AB is:

A	B	$\overline{AB}$	$\overline{AB} + \overline{AB}$	$X = \overline{\overline{AB} + \overline{AB}}$	AB
0	0	1	1	0	0
0	1	1	1	0	0
1	0	1	1	0	0
1	1	0	0	1	1

SAME (last two columns)

The last two columns of the truth table are the same, which shows that the output X is equivalent to AB or the output of an AND gate.

(c) To show this equivalence using the laws of Boolean algebra, apply the second of De Morgan's laws (26.7) to change the output as follows:

$$\overline{\overline{AB} + \overline{AB}} = \overline{\overline{AB}} \cdot \overline{\overline{AB}}$$

Then by the complement law (26.6), $\overline{\overline{A}} = A$, and the basic theorem in Table 26-2, $A \cdot A = A$:

$$\overline{\overline{AB}} \cdot \overline{\overline{AB}} = AB \cdot AB = AB$$

EXERCISE 26.2

1. Figure 26-11 shows a combination of a series and parallel switching circuit with three switches.
   (a) What combination of closed switches will produce an output at the lamp?
   (b) What combination of open switches will not produce an output?

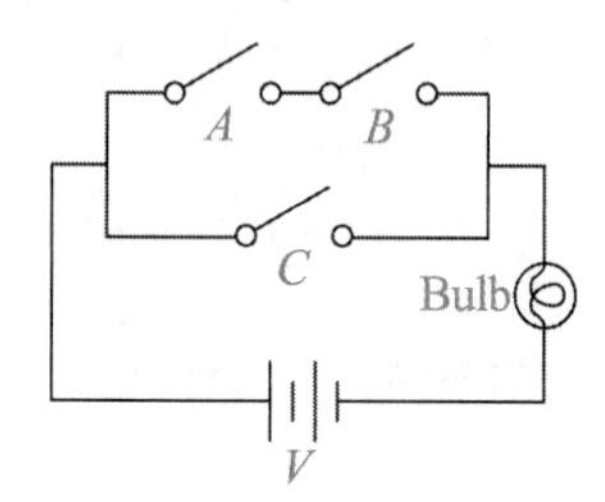

FIGURE 26-11 Switching circuit for problems 1 and 2.

2. Express the output of the circuit in Figure 26-11 in Boolean algebra when all three switches are closed.

In exercises 3 through 24, give the output of each logic gate for the given inputs.

3. AND gate: $A = 1, B = 0$
4. AND gate: $A = 1, B = 1$
5. AND gate: $A = 0, B = 1$
6. AND gate: $A = 0, B = 0$
7. OR gate: $A = 0, B = 0$
8. OR gate: $A = 0, B = 1$
9. OR gate: $A = 1, B = 0$
10. OR gate: $A = 1, B = 1$
11. NOT gate: $A = 1$
12. NOT gate: $A = 0$
13. NAND gate: $A = 0, B = 1$
14. NAND gate: $A = 1, B = 1$
15. NAND gate: $A = 0, B = 0$
16. NAND gate: $A = 1, B = 0$
17. NOR gate: $A = 1, B = 1$
18. NOR gate: $A = 0, B = 0$
19. NOR gate: $A = 0, B = 1$
20. NOR gate: $A = 1, B = 0$
21. AND gate: $A = 1, B = 0, C = 1$
22. AND gate: $A = 1, B = 1, C = 1$
23. OR gate: $A = 1, B = 1, C = 0$
24. OR gate: $A = 0, B = 0, C = 0$

For each logic circuit in exercises 25 through 28:

(a) Determine the output X.
(b) Construct the truth table showing the output X.

25.

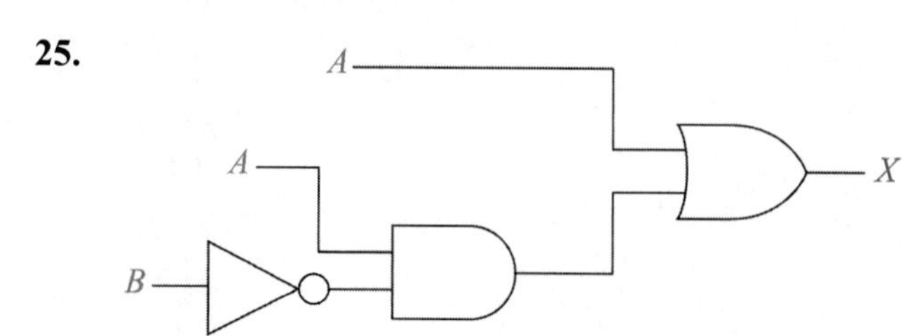

FIGURE 26-12 Logic circuit for problems 25 and 29.

26.

A
A
B
X

FIGURE 26-13 Logic circuit for problems 26 and 30.

27.

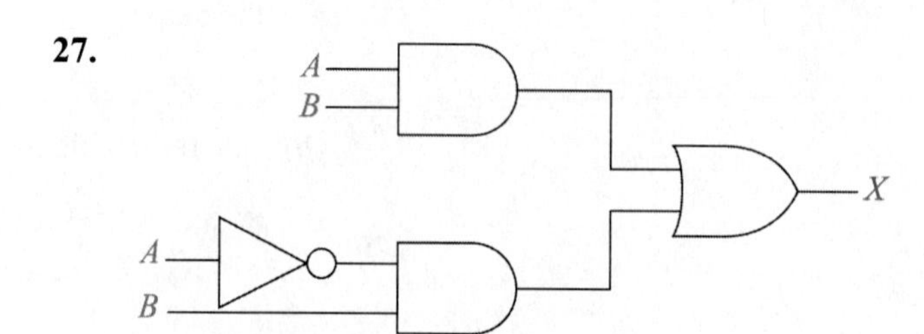

FIGURE 26-14 Logic circuit for problems 27 and 31.

28.

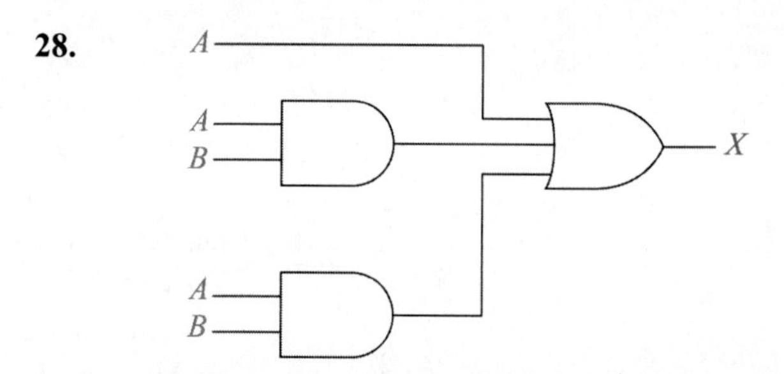

FIGURE 26-15 Logic circuit for problems 28 and 32.

In exercises 29 through 32, refer to the figure and show that the output is equivalent to the given output for X by applying the laws of Boolean algebra. See Examples 26.6 and 26.7.

29. Figure 26-12; $X = A$

30. Figure 26-13; $X = \overline{A}$

31. Figure 26-14; $X = B$

32. Figure 26-15; $X = A$

For each logic circuit in exercises 33 through 36:

(a) Determine the output X.
(b) Construct the truth table and show that the output is equivalent to the given output for X.
(c) Show that the output is equivalent to the given output for X by applying the laws of Boolean algebra.

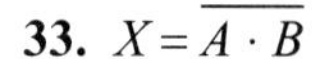

33. $X = \overline{A \cdot B}$

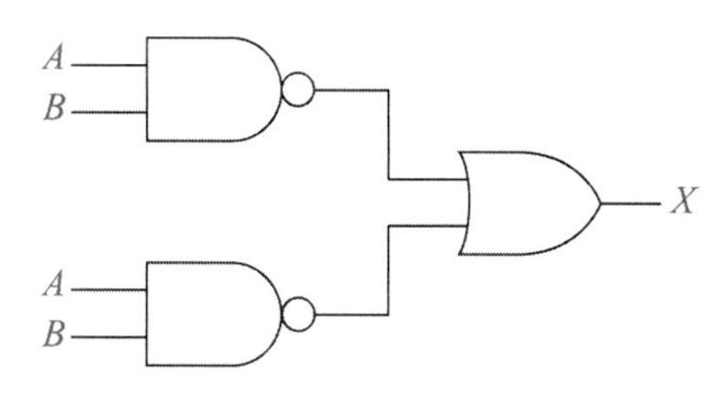

FIGURE 26-16 Logic circuit for problem 33.

34. $X = A + B$

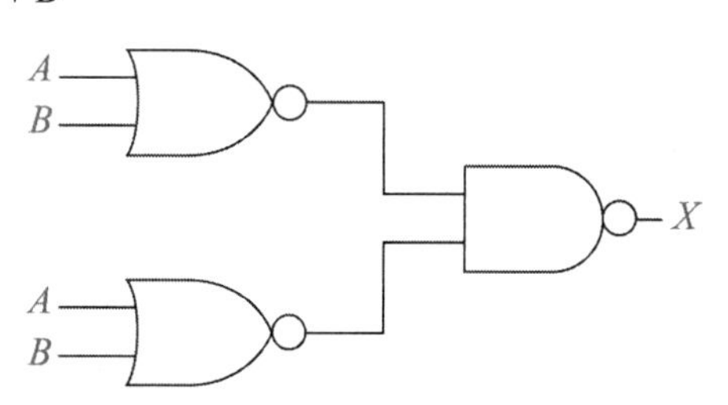

FIGURE 26-17 Logic circuit for problem 34.

35. $X = BC$

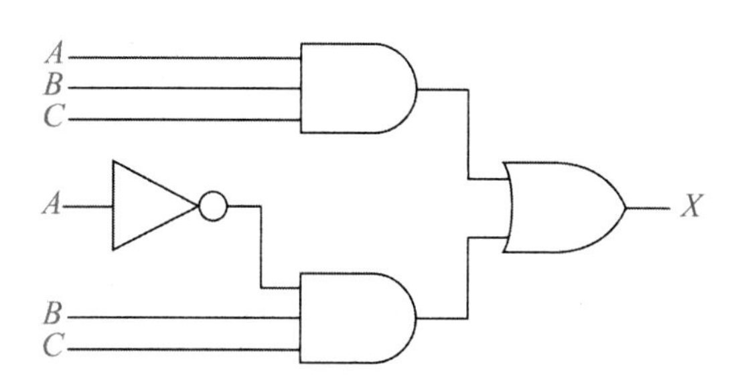

FIGURE 26-18 Logic circuit for problem 35.

36. $X = A$

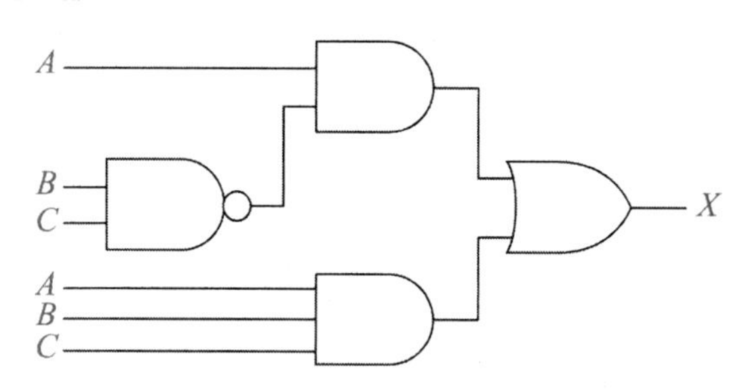

FIGURE 26-19 Logic circuit for problem 36.

26.3 Karnaugh Maps

In the Sections 26.1 and 26.2, it is shown that certain Boolean sums of products are equivalent to simpler expressions. Example 26.6 shows that: $AB + AB = AB$, and Example 26.7 shows that: $\overline{A}BC + BC = BC$. A useful way to simplify Boolean sums of products is by using a Karnaugh map. These are checkerboard-like diagrams in which squares represent products of Boolean elements. Figure 26-20 shows a Karnaugh map for two variables *A and B*.

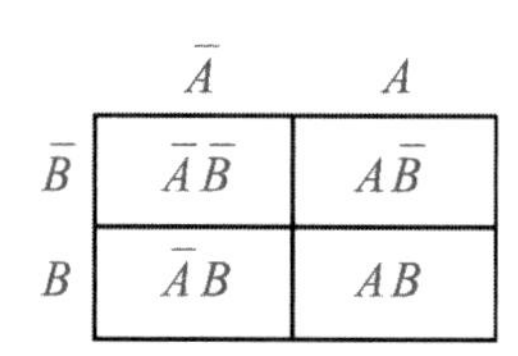

	$\overline{A}$	A
$\overline{B}$	$\overline{A}\,\overline{B}$	$A\overline{B}$
B	$\overline{A}B$	AB

FIGURE 26-20 Karnaugh map for two variables.

In a truth table, each variable has two states: 0 or 1. In a Karnaugh map, each variable appears as itself and its complement. The complement is like the other state of the variable. If A is 0, then $\overline{A}$ is 1, and vice versa. Each row and each column in the two-variable map represents one variable. The first column represents $\overline{A}$, the first row $\overline{B}$; the second column represents A and the second row B. Each small square in the map represents the product of the variables in that column and that row. For example, the upper left square is $\overline{A}\,\overline{B}$, the upper right $A\overline{B}$, the lower left $\overline{A}B$, and the lower right AB. These four products are the four fundamental Boolean products for two variables.

EXAMPLE 26.9 Simplify the Boolean sum of products by using a Karnaugh map:

$$AB + A\overline{B}$$

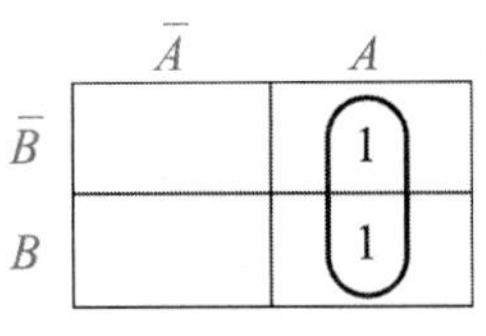

FIGURE 26-21 Adjacent squares in a two-variable Karnaugh map for Example 26.9.

Solution Place a 1 in each square of a two-variable Karnaugh map corresponding to one of these products as shown in Figure 26-21.

When two adjacent squares have 1's, circle them together as shown. The sum of all the products in a row, or a column, is equivalent to the variable that represents that row or column. Since the circled squares constitute the column A, it follows that:

$$AB + A\overline{B} = A$$

This result can also be shown using truth tables or the theorems of Boolean algebra.

EXAMPLE 26.10 Simplify the Boolean sum of products by using a Karnaugh map:

$$AB + \overline{A}B + \overline{A}\,\overline{B}$$

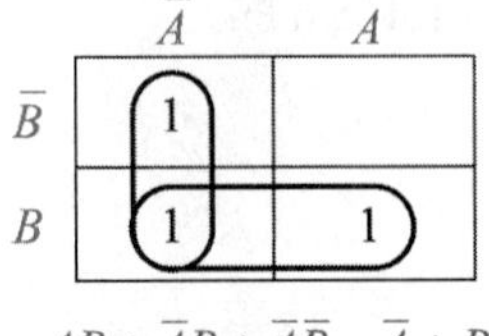

FIGURE 26-22 Adjacent squares in a Karnaugh map for Example 26.10.

Solution Place a 1 in each square corresponding to one of the products as shown in Figure 26-22.

Circle adjacent squares containing 1's. One of the circles represents the variable $\overline{A}$ and the other circle the variable B. The sum of these two variables then represents the equivalent expression:

$$AB + \overline{A}B + \overline{A}\,\overline{B} = \overline{A} + B$$

Three Variables

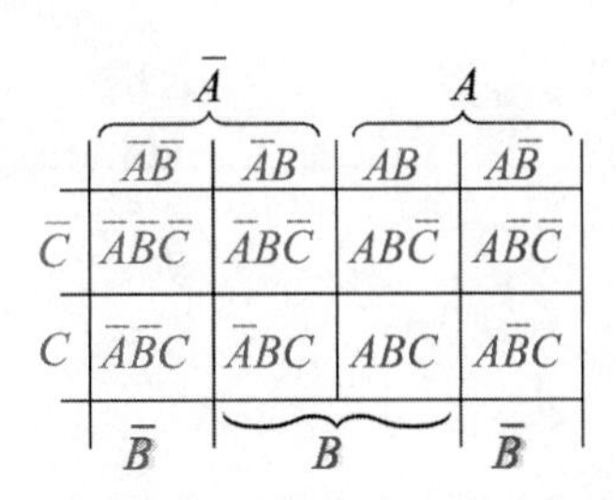

FIGURE 26-23 Karnaugh map for three variables.

When there are two variables, there are four possible products in the Karnaugh map corresponding to the combinations of 0's and 1's in the truth table. Similarly, when there are three variables, there are eight possible products to consider in a Karnaugh map. Figure 26-23 shows the Karnaugh map for three variables. The columns represent the four products from the Karnaugh map for two variables, A and B. The rows represent the variables $\overline{C}$ and C. Each square represents the product of three variables, and there are eight possible combinations of products. Note also that the first two columns constitute $\overline{A}$, the last two columns A, the middle two columns B, and the first and last columns $\overline{B}$.

EXAMPLE 26.11 Simplify the Boolean sum of products using a Karnaugh map:

$$ABC + A\overline{B}C + AB\overline{C}$$

Solution Place a 1 in each square corresponding to a product in a three-variable Karnaugh map. Circle together any 1's that are adjacent to each other as shown in Figure 26-24 on page 539.

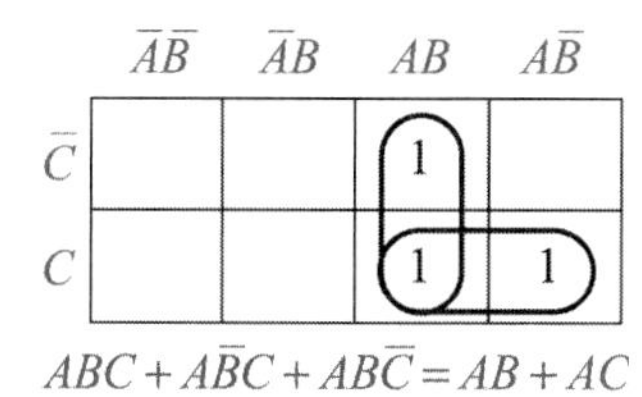

FIGURE 26-24 Adjacent squares in a three-variable Karnaugh map for Example 26.11.

Each circled group of products can now be replaced by the product common to the group. AB is common to ABC and $AB\overline{C}$, and AC is common to ABC and $A\overline{B}C$. Therefore, the original expression is equivalent to the sum of two products:

$$ABC + A\overline{B}C + AB\overline{C} = AB + AC$$

The first column and the last column are also considered adjacent columns in a Karnaugh map, as they contain common products. You can think of the map as being cut out and wrapped onto a cylinder so that the first and last columns are touching. Study the next example, which illustrates this situation.

EXAMPLE 26.12

Simplify the following Boolean sum of products:

$$\overline{A}BC + A\overline{B}C + \overline{A}\,\overline{B}C$$

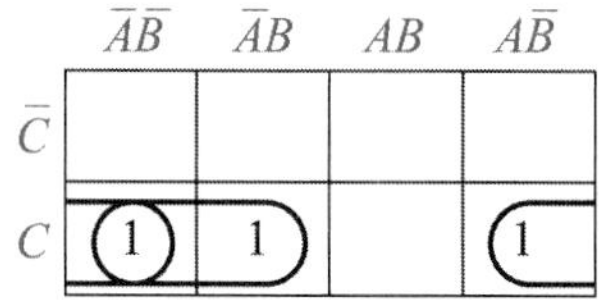

	$\overline{A}\overline{B}$	$\overline{A}B$	AB	$A\overline{B}$
$\overline{C}$				
C	1	1		1

$\overline{A}BC + A\overline{B}C + \overline{A}\overline{B}C = \overline{A}C + \overline{B}C$

FIGURE 26-25 Adjacent squares on the edges of a Karnaugh map for Example 26.12.

Solution Place a 1 in each square corresponding to a product. One of the products is in the first column and one is in the last column, but they are adjacent as both are in the bottom row. See Figure 26-25.

Therefore, there are two groups of 1's that should be circled. $\overline{A}C$ is common to $\overline{A}BC$ and $\overline{A}\,\overline{B}C$. Also $\overline{B}C$ is common to $A\overline{B}C$ and $\overline{A}\,\overline{B}C$. The Boolean expression is therefore equivalent to:

$$\overline{A}BC + A\overline{B}C + \overline{A}\,\overline{B}C = \overline{A}C + \overline{B}C$$

EXAMPLE 26.13

Simplify the Boolean expression:

$$\overline{A}\,\overline{B}\,\overline{C} + \overline{A}\,\overline{B}C + A\overline{B}\,\overline{C} + A\overline{B}C$$

Solution As shown in Figure 26-23, four adjacent squares constitute each of the single variables A, B, $\overline{A}$, and $\overline{B}$. Also each row of four squares constitutes C or $\overline{C}$. This means that if any four adjacent squares or any row are circled, they correspond to one of these single variables. The given four products represent four adjacent squares, two on the left edge and two on the right edge, and can therefore be replaced by the single variable $\overline{B}$:

$$\overline{A}\,\overline{B}\,\overline{C} + \overline{A}\,\overline{B}\,C + A\,\overline{B}\,\overline{C} + A\,\overline{B}\,C = \overline{B}$$

Four Variables

A Karnaugh map of four variables contains 16 products, as shown in Figure 26-26. Each column, each row, or any four adjacent squares correspond to a product of two variables. Two adjacent squares correspond to a product of three variables.

	$\overline{A}\overline{B}$	$\overline{A}B$	AB	$A\overline{B}$
$\overline{C}\overline{D}$	$\overline{A}\overline{B}\overline{C}\overline{D}$	$\overline{A}B\overline{C}\overline{D}$	$AB\overline{C}\overline{D}$	$A\overline{B}\overline{C}\overline{D}$
$\overline{C}D$	$\overline{A}\overline{B}\overline{C}D$	$\overline{A}B\overline{C}D$	$AB\overline{C}D$	$A\overline{B}\overline{C}D$
CD	$\overline{A}\overline{B}CD$	$\overline{A}BCD$	$ABCD$	$A\overline{B}CD$
$C\overline{D}$	$\overline{A}\overline{B}C\overline{D}$	$\overline{A}BC\overline{D}$	$ABC\overline{D}$	$A\overline{B}C\overline{D}$

FIGURE 26-26 Karnaugh map for four variables.

EXAMPLE 26.14 Simplify the Boolean expression:

$$ABCD + \overline{A}BCD + AB\overline{C}D + \overline{A}B\overline{C}D$$

Solution To simplify this expression using the Karnaugh map, place a 1 in each square corresponding to a product, as shown in Figure 26-27.

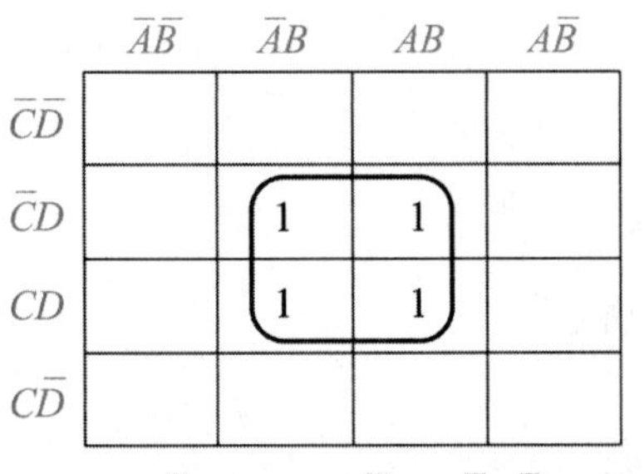

$$ABCD + \overline{A}BCD + AB\overline{C}D + \overline{A}B\overline{C}D = BD$$

FIGURE 26-27 Four adjacent squares in a four-variable Karnaugh map for Example 26.14.

Circle adjacent squares that form a group of four adjacent squares. The expression is then equivalent to the product of the two variables common to each square, which is BD:

$$ABCD + \overline{A}BCD + AB\overline{C}D + \overline{A}B\overline{C}D = BD$$

In a four-variable map, not only are the first and last columns considered adjacent but also the first and last rows are considered adjacent. Study the next example, which illustrates this situation.

EXAMPLE 26.15 Simplify the Boolean expression:

$$ABC\overline{D} + AB\overline{C}\,\overline{D} + A\overline{B}C\overline{D} + \overline{A}\,\overline{B}C\overline{D}$$

Solution Place a 1 in each square corresponding to a product as shown in Figure 26-28. There are then *two* pairs of adjacent squares in the last row and one pair of adjacent squares in the third column.

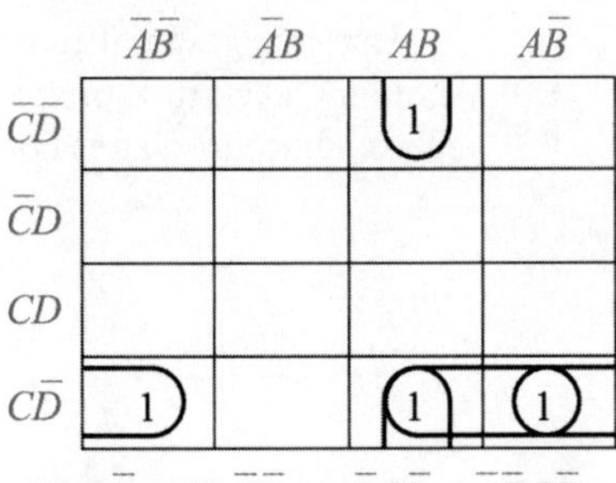

$$ABC\overline{D} + AB\overline{C}\overline{D} + A\overline{B}C\overline{D} + \overline{A}\overline{B}C\overline{D} =$$

$$AB\overline{D} + \overline{B}C\overline{D}$$

FIGURE 26-28 Adjacent squares on the edges of a Karnaugh map for Example 26.15.

Each pair of adjacent squares corresponds to the product of three variables that is common to the two squares. However, the pair of adjacent squares in the third column ($AB\overline{C}\,\overline{D}$ and $ABC\overline{D}$) and one pair of adjacent squares in the last row ($A\overline{B}C\overline{D}$ and $\overline{A}\,\overline{B}C\overline{D}$) include the other two adjacent squares in the last row ($ABC\overline{D}$ and $A\overline{B}C\overline{D}$). Therefore, the three pairs of adjacent squares can be replaced by only two products of three variables resulting in the Boolean expression:

$$ABC\overline{D} + AB\overline{C}\,\overline{D} + A\overline{B}C\overline{D} + \overline{A}\,\overline{B}C\overline{D} = AB\overline{D} + \overline{B}C\overline{D}$$

Karnaugh maps can also be drawn for five or more variables and are very useful for simplifying Boolean expressions that represent complex logic circuits.

EXERCISE 26.3

In exercises 1 through 14, simplify each Boolean sum of products by using a Karnaugh map.

1. $AB + \overline{A}B$
2. $A\overline{B} + \overline{A}\,\overline{B}$
3. $\overline{A}\,\overline{B} + \overline{A}B$
4. $AB + A\overline{B}$
5. $\overline{A}B + A\overline{B} + \overline{A}\,\overline{B}$
6. $AB + A\overline{B} + \overline{A}B$
7. $\overline{A}B + AB + \overline{A}\,\overline{B}$
8. $AB + A\overline{B} + \overline{A}\,\overline{B}$
9. $ABC + \overline{A}BC + AB\overline{C}$
10. $\overline{A}BC + \overline{A}B\overline{C} + \overline{A}\,\overline{B}\,\overline{C}$
11. $\overline{A}\,\overline{B}C + A\overline{B}C + \overline{A}\,\overline{B}\,\overline{C}$
12. $\overline{A}\,\overline{B}\,\overline{C} + AB\overline{C} + A\overline{B}\,\overline{C}$
13. $A\overline{B}\,\overline{C} + \overline{A}\,\overline{B}\,\overline{C} + \overline{A}BC$
14. $ABC + \overline{A}\,\overline{B}C + \overline{A}\,\overline{B}\,\overline{C}$
15. Show that the Boolean expression: $ABC + A\overline{B}C + AB\overline{C} + A\overline{B}\,\overline{C}$ can be simplified to a single variable.
16. Show that the Boolean expression: $\overline{A}B\overline{C} + \overline{A}BC + \overline{A}\,\overline{B}C + \overline{A}\,\overline{B}\,\overline{C}$ can be simplified to a single variable.
17. Show that the Boolean expression: $A\overline{B} + \overline{A}B$ cannot be simplified.
18. Show that the Boolean expression: $ABC + \overline{A}B\overline{C} + A\overline{B}\,\overline{C}$ cannot be simplified.

In exercises 19 through 22, simplify each Boolean expression containing four variables:

19. $\overline{A}\,\overline{B}\,\overline{C}\,\overline{D} + \overline{A}B\overline{C}\,\overline{D} + AB\overline{C}\,\overline{D} + A\overline{B}\,\overline{C}\,\overline{D}$
20. $ABCD + A\overline{B}CD + AB\overline{C}D + A\overline{B}\,\overline{C}D$
21. $\overline{A}\,\overline{B}CD + \overline{A}\,\overline{B}C\overline{D} + A\overline{B}CD$
22. $AB\overline{C}\,\overline{D} + A\overline{B}C\overline{D} + ABC\overline{D}$
23. Show that the Boolean expression: $AB + A\overline{B}C + A\overline{B}\,\overline{C}$ can be simplified to a single variable.
24. Show that the Boolean expression: $\overline{A}BC + ABCD + ABC\overline{D}$ can be simplified to an expression containing two variables.

CHAPTER HIGHLIGHTS

26.1 BOOLEAN ALGEBRA

Boolean Operations

1. The OR operation "+"
2. The AND operation "·"
3. The *complement* of an element A, called NOT A, and written $\overline{A}$. (26.1)

TABLE 26-1 TRUTH TABLES FOR AND, OR, AND NOT

		AND	OR	NOT
A	B	AB	$A+B$	$\overline{A}$
0	0	0	0	1
0	1	0	1	1
1	0	0	1	0
1	1	1	1	0

Commutative Laws

$$A + B = B + A \qquad AB = BA \tag{26.2}$$

• • •

Associative Laws

$$(A + B) + C = A + (B + C) \qquad (AB)C = A(BC) \tag{26.3}$$

• • •

Distributive Laws

$$A + BC = (A + B)(A + C) \qquad A(B + C) = AB + AC \tag{26.4}$$

• • •

Identity Laws

$$A + 0 = A \qquad A \cdot 1 = A \tag{26.5}$$

• • •

Complement Laws

$$A + \overline{A} = 1$$
$$A \cdot \overline{A} = 0 \qquad (26.6)$$
$$\overline{\overline{A}} = A$$

• • •

Dual Property

If a statement is true, the dual statement is also true.

TABLE 26-2 BASIC THEOREMS IN BOOLEAN ALGEBRA

Statement	Dual
$\overline{0} = 1$	$\overline{1} = 0$
$A + 1 = 1$	$A \cdot 0 = 0$
$A + A = A$	$A \cdot A = A$
$A + \overline{A} = 1$	$A \cdot \overline{A} = 0$

De Morgan's Laws

$$\overline{AB} = \overline{A} + \overline{B} \qquad (26.7)$$
$$\overline{A + B} = \overline{A} \cdot \overline{B}$$

Key Example: To prove the Boolean theorem: $\overline{AB} = \overline{A} + \overline{B}$, construct a truth table:

A	B	AB	$\overline{A}$	$\overline{B}$	$\overline{AB}$	$\overline{A} + \overline{B}$
0	0	0	1	1	1	1
0	1	0	1	0	1	1
1	0	0	0	1	1	1
1	1	1	0	0	0	0

26.2 LOGIC CIRCUITS

The truth tables for the three basic logic gates AND, OR, and NOT are the same as the corresponding operations in Boolean algebra shown in Table 26-1.

Key Example: For the AND-OR logic circuit in Figure 26-29, the truth table is:

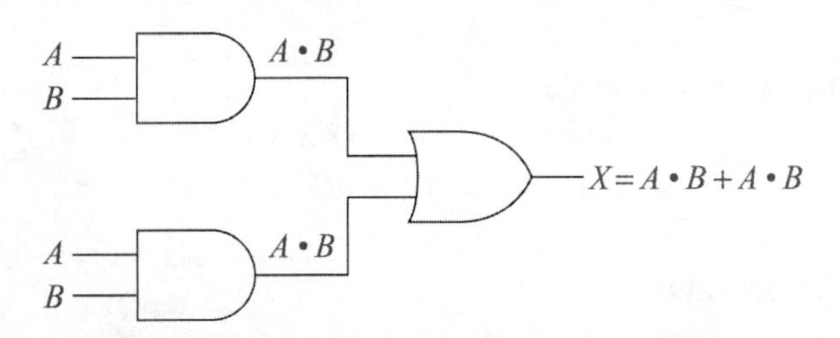

FIGURE 26-29 Key Example: Output of AND-OR circuit.

A	B	$A \cdot B$	$X = A \cdot B + A \cdot B$
0	0	0	0
0	1	0	0
1	0	0	0
1	1	1	1

The output $X = AB$ since the last two columns are the same. Apply the theorem $A + A = A$ from Table 26-2 to the term $A \cdot B$ to show the equivalence using Boolean algebra:

$$X = A \cdot B + A \cdot B = A \cdot B$$

26.3 KARNAUGH MAPS

Two adjacent squares can be replaced by a term containing one less variable. Four adjacent squares can be replaced by a term containing two less variables. Refer to Figures 26-30 and 26-31.

	$\overline{A}$	A
$\overline{B}$	$\overline{A}\,\overline{B}$	$A\overline{B}$
B	$\overline{A}B$	AB

FIGURE 26-30 Karnaugh map for two variables.

	$\overline{A}\,\overline{B}$	$\overline{A}B$	AB	$A\overline{B}$
$\overline{C}$	$\overline{A}\,\overline{B}\,\overline{C}$	$\overline{A}B\overline{C}$	$AB\overline{C}$	$A\overline{B}\,\overline{C}$
C	$\overline{A}\,\overline{B}C$	$\overline{A}BC$	ABC	$A\overline{B}C$
	$\overline{B}$	B		$\overline{B}$

FIGURE 26-31 Karnaugh map for three variables.

Key Example: See Figure 26-32.

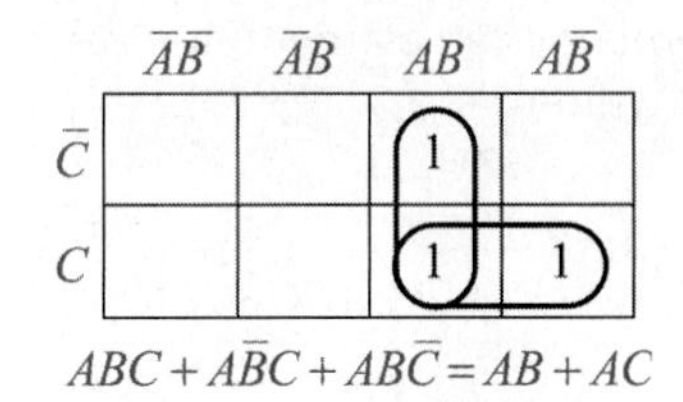

FIGURE 26-32 Adjacent squares in a three-variable Karnaugh map.

REVIEW EXERCISES

In exercises 1 through 6:

(a) Prove each Boolean statement by the use of truth tables.
(b) Prove each Boolean statement by applying Boolean algebra.
(c) Give the dual of each Boolean statement.

1. $A + AB = A$
2. $A(A + BC) = A$
3. $\overline{A}B + \overline{A}\,\overline{B} = \overline{A}$
4. $A(\overline{A} + B) = AB$
5. $(A + B)(B + C) = AC + B$
6. $\overline{A}\,\overline{(B + C)} = \overline{A}\,\overline{B}\,\overline{C}$

In exercises 7 through 12, give the output of each logic gate for the given inputs:

7. AND gate; $A = 0, B = 1$
8. OR gate; $A = 1, B = 0$
9. NOR gate; $A = 1, B = 0$
10. NAND gate; $A = 0, B = 0$
11. AND gate; $A = 0, B = 1, C = 1$
12. OR gate; $A = 0, B = 0, C = 1$

For each logic circuit in exercises 13 through 16:

(a) Determine the output X.
(b) Construct the truth table and show that the output is equivalent to the given output for X.
(c) Show that the output is equivalent to the given output for X by applying the laws of Boolean algebra.

13. $X = A + B$

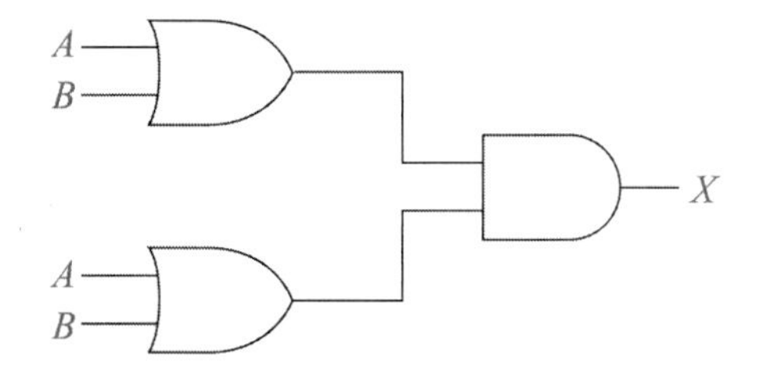

FIGURE 26-33 Logic circuit for problem 13.

14. $X = A+B$

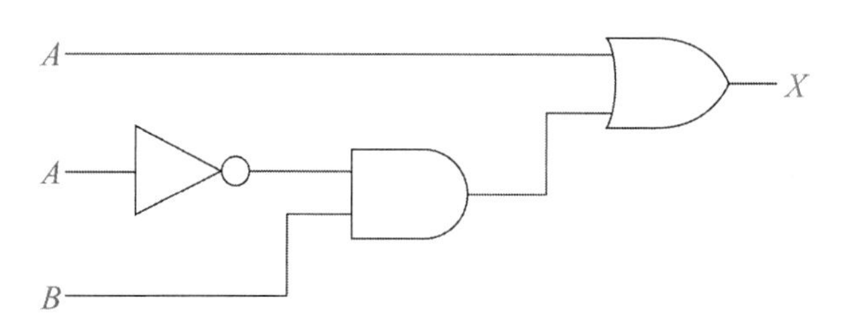

FIGURE 26-34 Logic circuit for problem 14.

15. $X = AB$

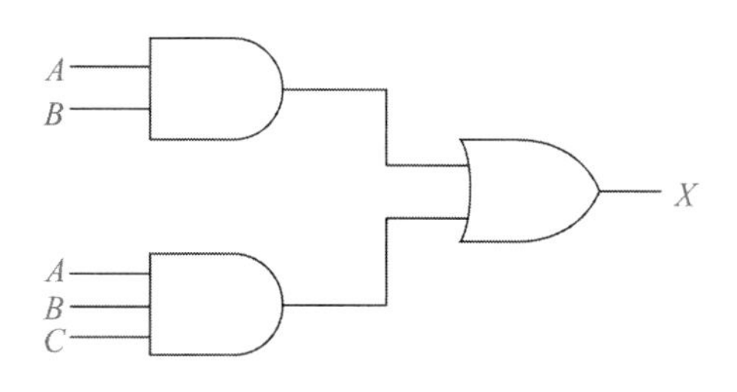

FIGURE 26-35 Logic circuit for problem 15.

16. $X = \overline{A} + \overline{B}$

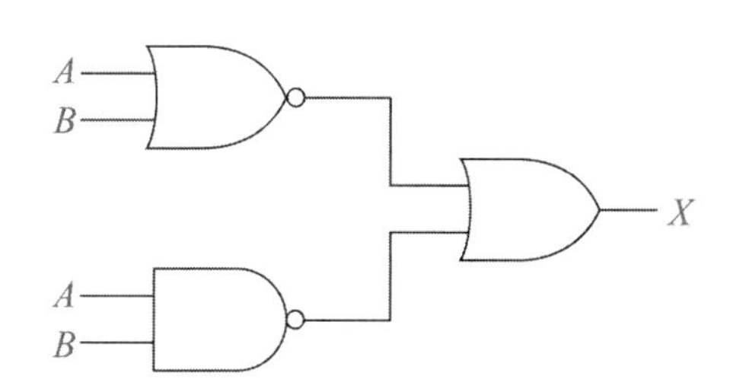

FIGURE 26-36 Logic circuit for problem 16.

In exercises 17 through 22, simplify each Boolean expression using a Karnaugh map.

17. $A + \overline{A}B + \overline{A}\,\overline{B}$
18. $ABC + AB\overline{C} + AB$
19. $\overline{A}\,\overline{B}\,\overline{C} + \overline{A}\,\overline{B}C + \overline{A}B\overline{C}$
20. $ABC + A\overline{B}C + \overline{A}\,\overline{B}C$
21. $\overline{A}B\overline{C} + AB\overline{C} + \overline{A}BC + ABC$
22. $ABCD + ABC\overline{D} + AB\overline{C}\,\overline{D}$

CHAPTER 27
Statistics

CHAPTER OBJECTIVES

In this chapter, you will learn:

- How to calculate the mean, median, and mode of a distribution.
- How to set up a frequency distribution.
- How to calculate the standard deviation of a distribution.
- How to graph a histogram of a frequency distribution.
- The normal curve and how to interpret it.

Many problems in electronics cannot be solved by theoretical methods alone. Empirical or experimental methods are often needed to find the best solution to a problem. Empirical methods involve gathering data, constructing tables and graphs to represent the data, and applying statistical formulas to study and analyze the results. For example, the design of a new electronic component, circuit, or computer system may require constant testing and modification, based on experimental data and results, before the system is put into final production. The technician must have the knowledge of how data is collected and analyzed to be able to improve the quality of a system and judge whether components, such as capacitors, resistors, or inductors, are functioning properly. This chapter introduces basic statistical measures that apply to experimental data including the mean, median, mode and standard deviation, and studies their application to small data distributions and larger frequency distributions.

• • •

27.1 Mean, Median, and Mode

Given a set of data or a distribution, there are three measures that are used to represent the "center" or "middle" of the distribution. These are called *measures of central tendency.* The one used most often is the mean $\bar{x}$ (x bar), which is commonly referred to as the "average."

Definition

Mean of a Distribution

The arithmetic sum of the terms x divided by the number of terms:

$$\bar{x} = \frac{\Sigma x}{n} \text{ where } \Sigma = \text{sum and } n = \text{number of terms} \tag{27.1}$$

The Greek capital sigma, Σ, represents sum, and Σx means that the terms, x, are to be summed up.

EXAMPLE 27.1

Given the distribution: X: 3, 5, 8, 8, 11, 12, 13, 20, 22; calculate the mean.

Solution Count the number of terms $n = 9$. Apply formula (27.1):

$$\bar{x} = \frac{\Sigma x}{n} = \frac{3 + 5 + 8 + 8 + 11 + 12 + 13 + 20 + 22}{9} = \frac{102}{9} \approx 11.3$$

Example 27.1 can be done on a calculator that has statistics keys [$\bar{x}$] and [Σ+] or [M+]. Set the calculator for statistics by pressing the [MODE] or an equivalent key, if necessary, and then enter each term followed by [Σ+] or [M+]:

3 [Σ+] 5 [Σ+] 8 [Σ+] 8 [Σ+] 11 [Σ+] 12 [Σ+] 13 [Σ+] 20 [Σ+] 22 [Σ+]

The display will show the number of terms entered, which should be 9. Then press [$\bar{x}$] and [=], if necessary, to display the mean:

[$\bar{x}$] [=] → 11.3

The second measure of central tendency is the median:

Definition

Median of a Distribution

The middle term when the terms are *arranged in size order.* For an even number of terms, it is the mean of the two middle terms. (27.2)

The median divides the distribution in half such that 50% of the terms are equal to or greater than the median, and 50% are equal to or less than the median.

EXAMPLE 27.2 Given the distribution:

X: 53, 26, 44, 35, 40, 50, 33, 53

Find the median.

Solution Arrange the terms in size order:

26, 33, 35, 40, 44, 50, 53, 53

Find the mean of the two middle terms, since the number of terms $n = 8$ is an even number:

$$\text{Median} = \frac{40 + 44}{2} = 42$$

Note that one-half of the terms are greater than 42 and one-half are less than 42.

The third measure of central tendency is the mode:

Definition

Mode of a Distribution

The term or terms which occur most often with the highest frequency. (27.3)

A distribution can have one mode, more than one mode, or no mode. See the next example.

EXAMPLE 27.3 Given the distribution:

X: 6.3, 6.4, 6.5, 6.6, 6.7, 6.8, 6.8, 6.8, 7.0, 7.0

Find the mode.

Solution The mode is 6.8, since it occurs three times or has the highest frequency = 3. The term with the next highest frequency = 2, is 7.0. If two or more terms have the same highest frequency, there is more than one mode. If all the terms have the same frequency, there is no mode.

EXAMPLE 27.4

Given the following distribution of grades on an examination:

95, 83, 73, 73, 75, 66, 60, 77, 89, 90, 60, 72, 55

Find the mean, median, and mode of the distribution.

Solution The number of terms $n = 13$. Apply formula (27.1) to find the mean:

$$\bar{x} = \frac{\Sigma x}{n} = \frac{95 + 83 + 73 + 73 + 75 + 66 + 60 + 77 + 89 + 90 + 60 + 72 + 55}{13}$$
$$= \frac{968}{13} \approx 74.5$$

Arrange the terms in size order to find the median:

55, 60, 60, 66, 72, 73, 73, 75, 77, 83, 89, 90, 95

Since there is an odd number of terms, $n = 13$, the median is the middle or seventh term = 73.

Two terms have the highest frequency of 2, therefore there are two modes: 60 and 73.

The next example closes the circuit and introduces the idea of a frequency distribution.

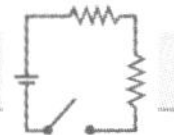

EXAMPLE 27.5

Close the Circuit

A number of students measure the current in amps for a circuit with the following results:

1.2 A, 1.2 A, 1.1 A, 1.3 A, 1.1 A, 1.4 A, 1.0 A,
1.3 A, 1.1 A, 1.2 A, 1.1 A, 1.4 A, 1.0 A, 1.3 A

(a) Find the mean, median, and mode for this distribution.

(b) Which of the three measures best represents the "average" current?

Solution

(a) When a distribution contains terms with various frequencies, do a tally by counting the number of times each term appears and make a frequency distribution table:

x (amps)	Tally	Frequency (*f*)
1.0	||	2
1.1	||||	4
1.2	|||	3
1.3	|||	3
1.4	||	2

The numbers in the frequency column tell how many times each term appears in the distribution. The mean is calculated by multiplying each term x by its frequency f and summing up the f and fx columns:

x (amps)	f	fx
1.0	2	2.0
1.1	4	4.4
1.2	3	3.6
1.3	3	3.9
1.4	2	2.8
	$\Sigma f = 14$	$\Sigma fx = 16.7$

The sum of the frequencies, $\Sigma f = 14$, represents the total number of terms n, and the sum, $\Sigma fx = 16.7$, represents the sum of all the terms Σx. The mean is therefore given by:

$$\bar{x} = \frac{\Sigma x}{n} = \frac{\Sigma fx}{\Sigma f} = \frac{16.7}{14} \approx 1.19 \text{ A}$$

The mean can be done on a calculator that has statistics keys [$\bar{x}$], [Σ+] or [M+], and [FRQ]. Set the calculator for statistics mode and enter each term and its frequency followed by the key [Σ+] or [M+] :

1.0 [FRQ] 2 [Σ+]
1.1 [FRQ] 4 [Σ+]
1.2 [FRQ] 3 [Σ+]
1.3 [FRQ] 3 [Σ+]
1.4 [FRQ] 2 [Σ+]

Then press [$\bar{x}$] and [=] , if necessary, to display the mean:

[$\bar{x}$] [=] → 1.19 A

To find the median arrange the terms in size order:

1.0, 1.0, 1.1, 1.1, 1.1, 1.1, 1.2, 1.2, 1.2, 1.3, 1.3, 1.3, 1.4, 1.4

The two middle terms are both 1.2, therefore the median is 1.2 A. The mode is 1.1 A since it has the highest frequency = 4.

(b) The mean is the best measure of the average current because the terms of the distribution do not vary much, and it takes into account the size of every term in the calculation.

The mean is not always the best measure of central tendency. For example, for an ac voltage or current wave, the "mean" ac voltage or current is actually zero, which is one of the reasons why rms values are used. The median is a better measure of central tendency when the terms of a distribution vary greatly and the mean is too high or too low a value to be representative of many of the terms in the distribution. The mode is used when matters of style or popularity are important.

EXERCISE 27.1

In exercises 1 through 8, find the mean, median and mode for each distribution.

1. 1, 5, 8, 4, 7, 6, 11, 7, 9, 12, 15
2. 82, 68, 76, 74, 57, 63, 75, 82, 83
3. 5.3, 7.2, 9.1, 5.5, 6.2, 8.8, 5.4
4. 3.2, 3.3, 4.1, 2.2, 4.2, 1.0, 3.3, 4.1, 1.5
5. 100, 105, 108, 110, 101, 105, 102, 104
6. 135, 136, 132, 132, 133, 134, 133, 134
7. 5.0, 1.0, 0, −0.8, −1.0, 0.4, −0.2, 0
8. 5, 7, −4, 3, −2, −3, −1, 1, 5, −5
9. In exercise 1, add 10 to each of the terms in the distribution to form a new distribution.
   (a) Find the mean, median, and mode of the new distribution.
   (b) What appears to be true about the mean, median, and mode of a distribution when all of the terms are increased by the same amount?
10. In exercise 1, multiply each of the terms in the distribution by 10 to form a new distribution.
    (a) Find the mean, median, and mode of the new distribution.
    (b) What appears to be true about the mean, median, and mode of a distribution when all of the terms are multiplied by the same amount?
11. Consider the two distributions:

    X: 3, 3, 3, 3, 3
    Y: 1, 2, 3, 4, 5

    (a) Show that the mean and median are the same for both distributions.
    (b) Are both distributions the same? If not, explain how they differ.
12. Consider the two distributions:

    X: 21, 22, 23, 24, 25, 26
    Y: 1, 7, 20, 27, 32, 54

    (a) Show that the mean and median are the same for both distributions.
    (b) Are both distributions the same? If not, explain how they differ.

Applied Problems

13. Your college assigns the following point values to letter grades: A = 4, B = 3, C = 2, and D = 1.
    (a) Using these point values, calculate the average or mean of the grades: A, B, A, C, C, B, D, A.
    (b) What letter grade is the mean closest to?
14. Do the same as in exercise 11 for the grades: A, B, B, A, C, C, B, D, C
15. The air temperature in degrees Fahrenheit is measured every hour during 12 hours of a summer day in Chicago:

    75°, 77°, 79°, 80°, 80°, 82°,
    83°, 85°, 82°, 80°, 77°, 76°

    Find the mean, median, and modal temperatures for the 12 hours.
16. During a 13-day period, the number of fire alarms in a city for each day were:

    14, 20, 19, 20, 17, 18, 10, 12, 12, 20, 19, 14, 15

    Find the mean, median, and modal number of fire alarms.
17. The annual percentage rates for a 30-year fixed-rate home mortgage offered by a number of banks are:

    7.85%, 7.80%, 7.75%, 7.80%, 7.90%,
    8.30%, 7.85%, 7.85%, 8.00%, 7.75%,
    7.80%, 8.50%, 7.90%, 7.85%, 7.85%

    (a) Find the mean, median, and mode of this data.
    (b) Which of the three measures best represents the "average" rate?
18. The annual salaries of the employees of a small company are:

    $25,000, $28,000, $27,000, $18,000,
    $13,000, $35,000, $34,000, $32,000,
    $33,000, $44,000, $100,000, $150,000

    (a) Find the mean, median, and mode of the salaries.
    (b) Which of the three measures best represents the "average" salary?

Applications to Electronics

19. A number of students measure the voltage of a battery with the following results:

    9.1 V, 9.1 V, 9.2 V, 9.0 V, 9.3 V,
    9.0 V, 9.1 V, 9.0 V, 9.1 V, 8.9 V, 9.2 V,
    9.1 V, 9.3 V, 8.9 V, 9.4 V, 9.1 V

    (a) Make a frequency distribution table for this data and calculate the mean, median, and mode. See Example 27.5.
    (b) Which of the three measures best represents the voltage of the battery?
20. An electrical technician measures the power output every hour of a generating station in kilowatts over a 16-hour period:

    1200 kW, 1300 kW, 1200 kW, 1100 kW,
    1000 kW, 1400 kW, 1300 kW, 1500 kW,
    1400 kW, 1400 kW, 1200 kW, 1200 kW,
    1300 kW, 1100 kW, 1100 kW, 1000 kW

(a) Make a frequency distribution table for this data and calculate the mean, median, and mode. See Example 27.5.

(b) Which of the three measures best represents the "average" power output of the generating station?

21. A company that produces inductors tests a sample of one type of inductor and finds the following values for the inductances in the sample:

27.3 mH, 27.4 mH, 27.4 mH, 27.5 mH, 27.3 mH,
27.2 mH, 27.3 mH, 27.4 mH, 27.5 mH, 27.5 mH,
27.4 mH, 27.3 mH, 27.2 mH, 27.3 mH, 27.4 mH,
27.4 mH, 27.5 mH, 27.6 mH

(a) Make a frequency distribution table for this data and calculate the mean, median, and mode.

(b) Which of the three measures best represents the value of the inductance?

22. A company which manufactures computer circuits finds that the impedance of the circuit varies. The impedances in ohms of a sample of the circuits are measured to be:

1510 Ω, 1500 Ω, 1520 Ω, 1500 Ω, 1490 Ω, 1490 Ω,
1510 Ω, 1500 Ω, 1520 Ω, 1510 Ω 1490 Ω, 1500 Ω,
1500 Ω, 1510 Ω, 1510 Ω, 1500 Ω, 1490 Ω, 1520 Ω,
1500 Ω, 1510 Ω, 1510 Ω

(a) Make a frequency distribution table for this data and calculate the mean, median, and mode.

(b) Which of the three measures best represents the impedance of the circuit?

27.2 Standard Deviation

The mean, median, and mode provide only a measure of the central tendency of a distribution. Distributions may have the same mean, median, and mode but be quite different. Consider the three distributions:

51, 52, 53, 53, 54, 55
0, 45, 53, 53, 83, 84
53, 53, 53, 53, 53, 53

Each of these distributions has the same value for the mean, median, and mode = 53. However, what is different about them is how the terms vary. The terms in the first distribution vary uniformly while the terms in the second vary a lot, and the terms in the third do not vary at all. The statistical measure of variation is called the *standard deviation.* It is denoted by the Greek letter σ (sigma) and is defined:

Formula **Sample Standard Deviation**

$$\sigma_{n-1} = \sqrt{\frac{\Sigma(x - \bar{x})^2}{n - 1}} \qquad (27.4)$$

The formula for the standard deviation is arrived at by theoretical considerations and is an effective measure of the dispersion of the terms about the mean for a sample of data. In general, the majority of the terms in any distribution lie within one standard deviation from the mean. Study the next example, which shows how to apply formula (27.4)

EXAMPLE 27.6 Find the mean and standard deviation of the distribution:

51, 52, 53, 53, 54, 55

Solution This is the first of the three distributions shown above. Formula (27.4) requires the following steps:

1. Calculate the mean $\bar{x}$.
2. Calculate the difference of each term from the mean: $(x - \bar{x})$.
3. Square each difference from the mean: $(x - \bar{x})^2$.

4. Sum the squares of the differences from the mean: $\Sigma(x - \bar{x})^2$.

5. Divide this sum by $n - 1$: $\dfrac{\Sigma(x - \bar{x})^2}{n - 1}$

6. Take the square root of the result: $\sqrt{\dfrac{\Sigma(x - \bar{x})^2}{n - 1}}$

The calculation for the standard deviation is best done in a table as follows:

x	$\bar{x}$	$x - \bar{x}$	$(x - \bar{x})^2$
51	53	−2	4
52	53	−1	1
53	53	0	0
53	53	0	0
54	53	1	1
55	53	2	4
$\Sigma x = 318$			$\Sigma(x - \bar{x})^2 = 10$

Enter the terms in the first column and sum this column. Use this sum to calculate the mean:

$$\bar{x} = \frac{\Sigma x}{n} = \frac{318}{6} = 53$$

Enter the mean in the second column and calculate the algebraic differences from the mean in the third column. Square these differences, which are all positive, in the fourth column. Sum the fourth column and use this sum to calculate the standard deviation:

$$\sigma_{n-1} = \sqrt{\frac{\Sigma(x - \bar{x})^2}{n - 1}} = \sqrt{\frac{10}{5}} = \sqrt{2} \approx 1.41$$

The value for $\sigma_{n-1} = 1.41$ tells us that the majority (about two-thirds) of the terms are within 1.41 units from the mean. The standard deviation can be done on a statistical calculator by entering the data the same way as for the mean and then pressing [σxn−1] and [=], if necessary, for the standard deviation:

51 [Σ+] 52 [Σ+] 53 [Σ+] 53 [Σ+] 54 [Σ+] 55 [Σ+] [x̄] [=] → 53 [σxn−1] [=] → 1.41

The standard deviation reveals how the terms spread out around the mean. Compare the three distributions given at the beginning of the section. Even though the mean, median, and mode are the same for each distribution, the standard deviation for the second distribution:

0, 45, 53, 53, 83, 84

is 30.8, which is much greater than 1.41 for the first distribution. This tells you that the terms are more spread out from the mean. The standard deviation for the third distribution above:

53, 53, 53, 53, 53, 53

is 0, which tells you that the terms do not vary from the mean at all.

The next two examples close the circuit and show applications of statistics to electronics.

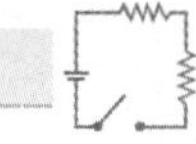

EXAMPLE 27.7

Close the Circuit

A number of students measure the resistance of a coil with the following results:

100 Ω, 102 Ω, 104 Ω, 105 Ω, 100 Ω, 102 Ω, 105 Ω, 106 Ω, 103 Ω

Find the mean and standard deviation of the distribution.

Solution Arrange the terms in order and set up columns for x, $\bar{x}$, $x - \bar{x}$, and $(x - \bar{x})^2$.

x (Ω)	$\bar{x}$	$x - \bar{x}$	$(x - \bar{x})^2$
100	103	−3	9
100	103	−3	9
102	103	−1	1
102	103	−1	1
103	103	0	0
104	103	1	1
105	103	2	4
105	103	2	4
106	103	3	9

$\Sigma x = 927$ $\Sigma(x - \bar{x})^2 = 38$

Sum the first column and calculate the mean:

$$\bar{x} = \frac{\Sigma x}{n} = \frac{927}{9} = 103\ \Omega$$

Sum the last column and calculate the standard deviation:

$$\sigma_{n-1} = \sqrt{\frac{\Sigma(x - \bar{x})^2}{n - 1}} = \sqrt{\frac{38}{8}} = \sqrt{4.75} \approx 2.18\ \Omega$$

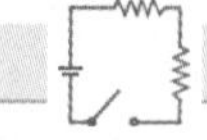

EXAMPLE 27.8

Close the Circuit

An electronics technician accurately measures the capacitance of twenty samples of a capacitor whose nominal value is 100 nF with 1% tolerance, with the following results:

97 nF, 98 nF, 98 nF, 99 nF, 99 nF, 99 nF, 100 nF, 100 nF, 100 nF, 100 nF, 100 nF, 100 nF, 101 nF, 101 nF, 101 nF, 101 nF, 102 nF, 102 nF, 103 nF, 104 nF

(a) Construct a frequency distribution table for this data.
(b) Find the mean and standard deviation of this set of data.

Solution

(a) Counting the number of times each term appears produces the frequency table:

x (nF)	f
97	1
98	2
99	3
100	6
101	4
102	2
103	1
104	1

(b) The mean and standard deviation can be calculated by listing all twenty values in a table like the one in Example 27.7. They can also be found with a statistical calculator by using the [FRQ] key. Enter each term and its frequency, and then press [$\bar{x}$] and [σ_{xn-1}] as follows:

97 [FRQ] 1 [Σ+]
98 [FRQ] 2 [Σ+]
99 [FRQ] 3 [Σ+]
100 [FRQ] 6 [Σ+]
101 [FRQ] 4 [Σ+]
102 [FRQ] 2 [Σ+]
103 [FRQ] 1 [Σ+]
104 [FRQ] 1 [Σ+]

[$\bar{x}$] [=] → 100.25 nF [σ_{xn-1}] [=] → 1.71 nF

EXERCISE 27.2

In exercises 1 through 10, find the mean and standard deviation for each distribution. Round answers to three significant digits.

1. 1, 3, 4, 5, 8, 10

2. 2, 4, 5, 7, 8, 11

3. 28, 29, 30, 35, 39, 42, 50

4. 14, 16, 18, 23, 29, 30, 32

5. 10, 20, −10, 0, −20, −10, 20, 10

6. 1.2, −2.4, −1.6, 0.8, −2.2, 1.0, 2.0, −0.4

7. 5.5, 5.3, 5.1, 5.6, 5.9, 5.5, 5.8, 5.0, 5.6

8. 1.3, 1.6, 1.7, 1.7, 1.8, 1.9, 1.9, 2.0, 2.0

9. 103, 110, 104, 111, 101, 100, 105, 105, 106, 105

10. 150, 155, 155, 158, 160, 162, 162, 168, 170, 170

11. In problem 1, add 10 to each term in the distribution to form a new distribution.
(a) Find the mean and standard deviation of the new distribution.
(b) What can you say about the mean and standard deviation of a distribution when all the terms are increased by the same amount?

12. In problem 2, multiply each term by 10 to form a new distribution.
(a) Find the mean and standard deviation of the new distribution.
(b) What can you say about the mean and standard deviation of a distribution when all the terms are multiplied by the same amount?

In exercises 13 through 16, set up a frequency table and find the mean and standard deviation for each distribution. See Example 27.7.

13. 8, 9, 6, 10, 5, 5, 6, 8, 8, 9, 7, 10, 7, 7, 8

14. 3, 3, 7, 1, 5, 6, 10, 3, 5, 6, 7, 5, 3, 4, 6

15. 1.5, 1.5, 1.4, 1.1, 1.2, 1.3, 1.2, 1.3, 1.2, 1.4, 1.4, 1.2, 1.5, 1.1, 1.3, 1.1

16. 2.2, 2.6, 2.1, 2.4, 2.4, 2.5, 2.6, 2.3, 2.1, 2.2, 2.6, 2.8, 2.2, 2.4, 2.3, 2.5

Applied Problems

17. During a fifteen-day period, the number of computers sold in a department store each day were:

10, 22, 18, 20, 15, 16, 11, 12, 12, 22, 15, 16, 16

Find the mean and standard deviation for this data.

18. The air temperature in degrees Celsius measured every hour during 12 hours of a summer day in Toronto is found to be:

25°, 25°, 26°, 27°, 27°, 28°,
28°, 29°, 29°, 28°, 27°, 26°

Find the mean and standard deviation of the temperatures for this day.

19. The amounts of money earned by a number of students during the summer are:

$3,000, $3,400, $2,500, $2,800, $3,300,
$2,600, $3,200, $3,800, $3,300, $4,100,
$2,800, $3,400, $3,100

Find the mean and standard deviation of this distribution.

20. The interest rates offered by a number of savings banks are:

3.80%, 3.80%, 3.75%, 3.80%, 3.90%,
4.00%, 3.80%, 3.85%, 3.90%, 3.75%,
3.80%, 3.90%, 4.00%, 3.85%, 3.85%

Find the mean and standard deviation for this data.

Applications to Electronics

21. A number of students measure the current in a circuit with the following results:

42 mA, 42 mA, 40 mA, 41 mA,
40 mA, 39 mA, 40 mA, 41 mA

Calculate the mean and standard deviation of this distribution.

22. A electrical technician measures the resistance of ten 100-Ω marked resistors produced by a company to be:

101 Ω, 98 Ω, 101 Ω, 100 Ω, 99 Ω,
100 Ω, 99 Ω, 102 Ω, 101 Ω, 103 Ω

Calculate the mean and standard deviation of this distribution.

23. A company that manufactures batteries finds that the voltage of the batteries vary. The voltages of a sample of batteries are found to be:

1.5 V, 1.5 V, 1.4 V, 1.5 V, 1.6 V, 1.6 V, 1.7 V,
1.4 V, 1.4 V, 1.5 V, 1.5 V, 1.8 V, 1.7 V, 1.6 V,
1.6 V, 1.5 V, 1.5 V, 1.6 V, 1.7 V, 1.7 V, 1.5 V

(a) Make a frequency distribution table for this data.
(b) Calculate the mean and standard deviation. See Example 27.7.

24. A electronics technician measures the impedance of a sample of computer circuits produced in an assembly line to be:

1.5 kΩ, 1.5 kΩ, 1.3 kΩ, 1.2 kΩ, 1.4 kΩ, 1.4 kΩ,
1.3 kΩ, 1.5 kΩ, 1.4 kΩ, 1.4 kΩ, 1.4 kΩ, 1.2 kΩ,
1.3 kΩ, 1.3 kΩ, 1.2 kΩ, 1.4 kΩ

(a) Make a frequency distribution table for this data.
(b) Calculate the mean and standard deviation. See Example 27.7.

27.3 Histograms and the Normal Curve

When a distribution contains a large number of terms, a *histogram* is generally used to illustrate the data. A histogram is a graph of a frequency distribution, similar to a bar graph, that helps you to visualize the distribution at a glance. See Figure 27-1.

EXAMPLE 27.9

A sample of 50 resistors, whose nominal value is 100 Ω, are measured and the results tabulated in the following frequency distribution table. Draw a histogram for this frequency table.

x (Ω)	*f*
97	2
98	4
99	6
100	16
101	15
102	3
103	3
104	1

Solution Figure 27-1 on page 537 shows the histogram of the fifty samples. The values of the resistances are plotted on the horizontal axis and the frequencies on the vertical axis. Unlike a bar graph, there are no gaps between the bars. The bars are drawn touching each other to give a feeling of continuity for the data. The height of a bar shows the frequency of the term, and the shape of the graph shows you how the terms are distributed. The terms are concentrated in the center at 100 Ω and 101 Ω and decrease to the right and left of the center.

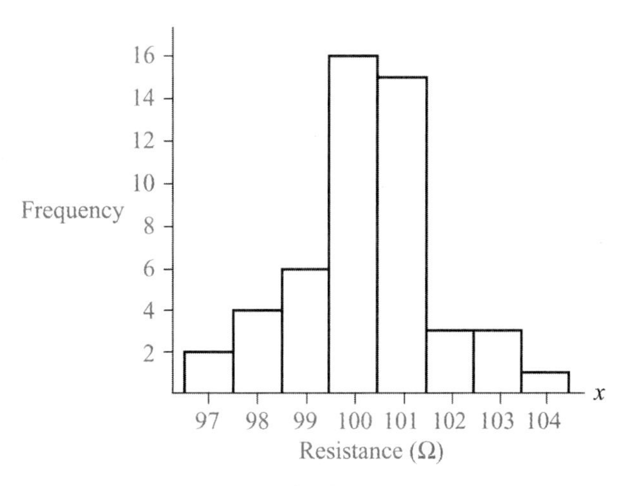

FIGURE 27-1 Histogram for Example 27.9.

As the number of terms in a distribution increases, it becomes necessary to group them into intervals in a frequency table to limit the size of the table and the histogram.

EXAMPLE 27.10 A manufacturer of computer DVD disks finds that the total usable memory in megabytes varies because of defects in the recording track. A random sample of 100 disks is tested, and the results are tabulated using intervals in the following frequency distribution table. Draw a histogram of the data.

Interval (MB)	Frequency
505–509	4
510–514	10
515–519	12
520–524	23
525–529	35
530–534	11
535–539	5

Solution Each interval must have the same width and the number of intervals cannot be too large or too small to show the data effectively. This table contains seven intervals and each interval contains five terms. The histogram is shown in Figure 27-2. Each bar represents one interval, and each interval is associated with its midpoint. For example, the midpoint of the first interval 505–509 is:

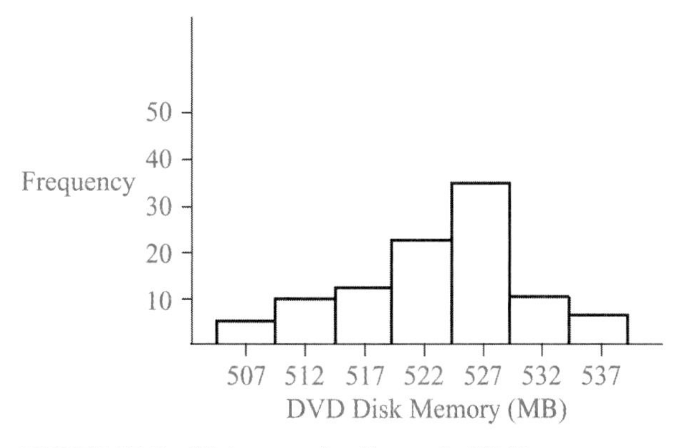

FIGURE 27-2 Histogram for Example 27.10.

$$\text{Midpoint} = \frac{505 + 509}{2} = 507 \text{ MB}$$

The *range* of the data is the difference between the largest and smallest term:

$$\text{Range} = 539 - 505 = 34 \text{ MB}$$

Normal Curve

When the data in a distribution is very large (several hundred or more), a curve rather than a histogram is used to portray the data. A very useful and important frequency distribution for a large set of random data is the *normal distribution* or *normal curve*. The normal curve is sometimes referred to as the bell-shaped curve because of its shape. See Figure 27-3.

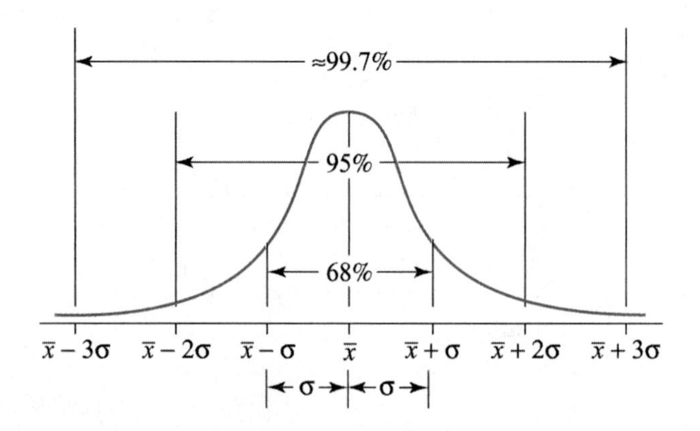

FIGURE 27-3 Normal curve.

Properties of the Normal Curve

1. The curve is symmetrical about the center, which is the highest point.
2. The center is equal to the mean, the median, and the mode.
3. Fifty percent of the terms lie above the mean and 50% lie below the mean.
4. Sixty-eight percent of the terms lie within one standard deviation of the mean.
5. Ninety-five percent of the terms lie within two standard deviations of the mean.
6. Approximately 100% (99.7%) of the terms lie within three standard deviations of the mean.

When the mean = 0 and the standard deviation = 1, the equation of the normal curve is:

$$y = \frac{1}{\sqrt{2\pi}} e^{-x^2/2}$$

where e is the base of natural logarithms introduced in Chapter 24.

EXAMPLE 27.11 List some examples of large random distributions that tend toward normal distributions.

Solution

1. Human and animal traits such as height, weight, intelligence, shoe size, and life span.
2. Specifications of electrical and electronic parts produced in large quantities such as resistor, capacitor, and inductor values, power gains of amplifiers, voltages of batteries, and power outputs of transformers.
3. Lives of electrical and electronic equipment produced in large quantities such as batteries, resistors, motors, computer monitors, and disk drives.

4. Sizes of mechanical parts produced in large quantity such as bolts, nuts, rods, gears, and beams.
5. Scores on aptitude tests administered in large numbers.

EXAMPLE 27.12

The power gain of 1000 amplifiers is found to be normally distributed with a mean of $\bar{x} = 30$ dB and a standard deviation of $\sigma = 2$ dB. What percentage of the amplifiers have a power gain:

(a) Greater than 30 dB?
(b) Between 28 dB and 32 dB?
(c) Between 26 dB and 34 dB?
(d) Less than 24 dB or more than 36 dB?

Solution Compute the range of values for one, two, and three standard deviations from the mean and tabulate them as shown below. Since $\bar{x} = 30$ dB and $\sigma = 2$ dB, the ranges are:

Standard Deviations	Range of Values	Percent of Terms
$\bar{x} \pm \sigma = 30$ dB $\pm$ 2 dB	28 dB–32 dB	68%
$\bar{x} \pm 2\sigma = 30$ dB $\pm$ 4 dB	26 dB–34 dB	95%
$\bar{x} \pm 3\sigma = 30$ dB $\pm$ 6 dB	24 dB–36 dB	99.7%

(a) The mean $\bar{x} = 30$ dB divides the curve in half. Therefore, 50% of the amplifiers have a power gain greater than 30 dB.

(b) From the properties of the normal curve, 68% of the terms lie within one standard deviation of the mean. Therefore, from the first range of values shown above, 68% of the amplifiers have a power gain between 28 dB and 32 dB.

(c) From the properties of the normal curve, 95% of the terms lie within two standard deviations of the mean. Therefore, from the second range of values shown above, 95% of the amplifiers have a power gain between 26 dB and 34 dB.

(d) The third range of values above shows that 99.7% of the amplifiers have a power gain between 24 dB and 36 dB. The amplifiers that have a power gain less than 24 dB or greater than 36 dB represent the remaining amplifiers that are outside this interval. Since the entire curve represents 100% of the amplifiers, it follows that 100% − 99.7% = 0.3% of the amplifiers have a power gain less than 24 dB or greater than 36 dB.

EXAMPLE 27.13

A manufacturer of automobile batteries finds that the life of the batteries is normally distributed with a mean of $\bar{x} = 60$ months and a standard deviation of $\sigma = 5$ months. Out of 1000 batteries, how many will last:

(a) Between 60 and 65 months?
(b) More than 65 months?
(c) Less than 50 months?

Solution

(a) Since $\sigma = 5$ months, 65 months is one standard deviation above the mean. From Figure 27-3, half of 68% = 34% of the distribution lies between $\bar{x}$ and $\bar{x} + \sigma$. Since 1000 batteries represents 100% of the distribution:

$$(1000)(34\%) = 340 \text{ Batteries will last more than 65 months}$$

(b) Since 50 % of the normal distribution is above the mean and 34% lies between $\bar{x}$ and $\bar{x} + \sigma$, the percentage of the distribution above one standard distribution (65 months) is: 50% − 34% = 16%. Therefore:

$$(1000)(16\%) = 160 \text{ Batteries will last more than 65 months}$$

(c) 50 months is two standard deviations below the mean. Similar to the reasoning in (a), Figure 27-3 shows that 50% − 1/2(95%) = 2.5% of the distribution is below two standard deviations from the mean. Therefore:

$$(1000)(2.5\%) = 25 \text{ Batteries will last less than 50 months}$$

EXERCISE 27.3

In exercises 1 through 6, draw a histogram for each frequency distribution.

1.

x	f
5	2
6	3
7	5
8	8
9	7
10	2

2.

x	f
3	3
4	6
5	7
6	12
7	10
8	5

3.

x	f
1–5	5
6–10	12
11–15	15
16–20	20
21–25	10
26–30	6

4.

x	f
21–25	4
26–30	8
31–35	16
36–40	13
41–45	9
46–50	4

5.

x	f
9–19	12
20–30	20
31–41	26
42–52	32
53–63	28
64–74	14
75–85	8

6.

x	f
50–54	5
55–59	10
60–64	22
65–69	28
70–74	20
75–79	11
80–84	4

Applied Problems

7. Assuming that the scores on the college scholastic aptitude test are normally distributed with a mean $\bar{x} = 400$ and a standard deviation $\sigma = 100$, what percentage of scores are:

(a) Less than 400?
(b) Between 300 and 500?
(c) Between 200 and 600?
(d) Less than 200 or greater than 600?

8. The weights of 1000 people are found to be normally distributed with a mean of 120 lb and a standard deviation of 20lb. What percentage of people weigh:

(a) More than 120 lb?
(b) Between 100 lb and 140 lb?
(c) Between 60 lb and 180 lb?
(d) Less than 60 lb or more than 180 lb?

9. In Example 27-13:

(a) How many batteries will last between 55 and 70 months?
(b) How many batteries will last more than 75 months?

10. Out of 10,000 adults, it is found that their weights are normally distributed with $\bar{x} = 160$ lbs. and $\sigma = 10$ lbs. What percent of these adults:

(a) Weigh more than 180 lbs?
(b) Weigh less than 130 lbs?
(c) How many weigh between 150 and 170 lbs?

Applications to Electronics

11. Draw a histogram for the data in problem 21, Exercise 27.2.

12. Draw a histogram for the data in problem 22, Exercise 27.2.

13. A manufacturer of computer hard drives finds that the total usable memory in gigabytes varies because of defects in the drive. A random sample of 100 hard drives is tested and the results tabulated in the following frequency distribution table. Draw a histogram of the data.

Interval (GB)	Frequency
7.75–7.79	5
7.80–7.84	24
7.85–7.89	32
7.90–7.94	18
7.95–7.99	14
8.00–8.04	7

14. A manufacturer of induction coils tests a random sample of 200 coils and tabulates the results in the following frequency distribution table. Draw a histogram of the data.

Interval (mH)	Frequency
16–18	10
19–21	20
22–24	55
25–27	65
28–30	22
31–33	15
34–36	13

15. The life of a computer monitor cathode ray tube is normally distributed with a mean $\bar{x}$ = 10,000 hours and a standard deviation (a)
Less than 10,000 h?
(b) Between 9800 h and 10,200 h?
(c) Between 9400 h and 10,600 h?
(d) Less than 9800 h or more than 10,200 h?

16. The voltage gain of a large number of amplifiers is found to be normally distributed with a mean of $\bar{x}$ = 25 dB and a standard deviation of σ = 1 dB. What percentage of the amplifiers have a voltage gain:
(a) Greater than 25 dB?
(b) Between 24 dB and 26 dB?
(c) Between 23 dB and 27 dB?
(d) Less than 23 dB or more than 27 dB?

17. A manufacturer of resistors finds that the range of actual values of its 1-kΩ marked resistor is normally distributed with $\bar{x}$ = 1.0 kΩ and σ = 50Ω. Out of 5000 resistors how many have actual values:
(a) Between 900 Ω and 1.1 kΩ?
(b) Between 950 Ω and 1.0 kΩ?
(c) Greater than 1.15 kΩ?
(d) Less than 1.10 kΩ?

18. A manufacturer of capacitors finds that the range of actual values of its 1-μF marked capacitor is normally distributed with $\bar{x}$ = 1.0 μF and σ = 100 nF. Out of 1000 capacitors how many have actual values:
(a) Between 900 nF and 1.1 μF?
(b) Between 800 nF and 1.0 μF?
(c) Greater than 1.30 μF?
(d) Less than 1.20 μF?

19. In exercise 15, what percentage of tubes will last:
(a) Between 9800 h and 10,000 h?
(b) Less than 9800 h?
(c) Less than 10,200 h?

20. In exercise 16, what percentage of the amplifiers have a voltage gain:
(a) Between 23 dB and 25 dB?
(b) More than 23 dB?
(c) More than 27 dB?

CHAPTER HIGHLIGHTS

27.1 MEAN, MEDIAN, AND MODE

Mean of a Distribution

The arithmetic sum of the terms x divided by the number of terms:

$$\bar{x} = \frac{\Sigma x}{n} \text{ where } \Sigma = \text{sum and} \quad (27.1)$$

n = number of terms

• • •

Median of a Distribution

The middle term when the terms are *arranged in size order.* For an even number of terms, it is the mean of the two middle terms. (27.2)

• • •

Mode of a Distribution

The term or terms which occur most often with the highest frequency. (27.3)

Key Example: Given the distribution:

55, 60, 60, 66, 72, 73, 73, 75, 77, 83, 89, 90, 95

$$\bar{x} = \frac{\Sigma x}{n} = \frac{968}{13} \approx 74.5, \text{ Median} = 73, \text{ Modes} = 60, 73$$

27.2 STANDARD DEVIATION

Sample Standard Deviation

$$\sigma_{n-1} = \sqrt{\frac{\Sigma(x - \bar{x})^2}{n - 1}} \quad (27.4)$$

Key Example: The standard deviation of the distribution: 51, 52, 53, 54, 55 is calculated:

x	$\bar{x}$	$x - \bar{x}$	$(x - \bar{x})^2$
51	53	−2	4
52	53	−1	1
53	53	0	0
53	53	0	0
54	53	1	1
55	53	2	4

$\Sigma x = 318$ $\Sigma(x - \bar{x})^2 = 10$

$$\bar{x} = \frac{\Sigma x}{n} = \frac{318}{6} = 53$$

$$\sigma_{n-1} = \sqrt{\frac{\Sigma(x - \bar{x})^2}{n - 1}} = \sqrt{\frac{10}{5}} = \sqrt{2} \approx 1.41$$

27.3 HISTOGRAMS AND THE NORMAL CURVE

Key Example: For the frequency distribution below, its accompanying histogram is shown in Figure 27-4.

Interval (MB)	Frequency
505–509	4
510–514	10
515–519	12
520–524	23
525–529	35
530–534	11
535–539	5

The normal curve for large random distributions is shown in Figure 27-5.

Key Example: The power gain of 1000 amplifiers is normally distributed with a mean of $\bar{x}$ = 30 dB and a standard deviation of σ = 2 dB:

Standard Deviations	Range of Values	Percent of Terms
$\bar{x} \pm \sigma$ = 30 dB ± 2 dB	28 dB–32 dB	68%
$\bar{x} \pm 2\sigma$ = 30 dB ± 4 dB	26 dB–34 dB	95%
$\bar{x} \pm 3\sigma$ = 30 dB ± 6 dB	24 dB–36 dB	99.7%

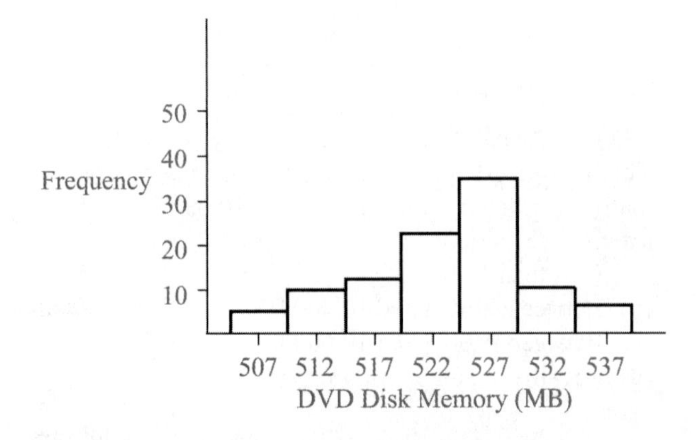

FIGURE 27-4 Key Example: Histogram.

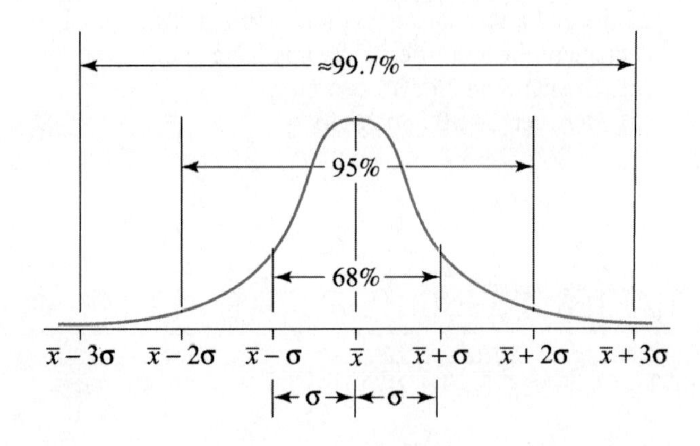

FIGURE 27-5 Key Example: Normal curve.

REVIEW EXERCISES

In exercises 1 through 4, find the mean, median, mode, and standard deviation for each distribution.

1. 1, 5, 8, 13, 15, 20 , 22
2. 3, 25, 55, 15, 17, 49, 8, 45, 17, 10
3. 1.5, 1.6, 1.4, 1.0, 1.9, 1.0, 1.7, 1.5, 1.3
4. 208, 202, 200, 210, 205, 200, 204, 203

In exercises 5 and 6, draw a histogram for each frequency distribution.

5.

x	f
5	3
7	4
9	6
11	8
13	8
15	2

6.

x	f
1–5	6
6–10	10
11–15	15
16–20	18
21–25	9
26–30	5

Applied Problems

7. The stopping distances of 10 different cars traveling at 55 mi/h are found to be:

50 ft, 52 ft, 49 ft, 48 ft, 49 ft,
50 ft, 51 ft, 50 ft, 51 ft, 52 ft

(a) Find the mean, median, and mode of the data.
(b) Find the standard deviation of the data.

8. The grades in a course are distributed as follows:

B, D, B, D, C, C, A, C, C, C, B, C, D, A, C, F

Letting A = 4.0, B = 3.0, C = 2.0, D = 1.0 and F = 0.0:

(a) Construct a frequency distribution table for this data.
(b) Find the mean and standard deviation of the data.
(c) Draw the histogram for the data.

Applications to Electronics

9. Eight of the same type induction coils manufactured by a company are found to have the following inductances:

1.12 mH, 1.11 mH, 1.10 mH, 1.12 mH,
1.11 mH, 1.09 mH, 1.12 mH, 1.10 mH

(a) Find the mean, median, and mode of the data.
(b) Find the standard deviation of the data.

10. The actual power consumption of 18 "100-watt bulbs" is measured to be:

98 W, 101 W, 103 W, 100 W, 99 W, 98 W, 100 W,
101 W, 97 W, 103 W, 102 W, 100 W, 99 W, 99 W,
99 W, 101 W, 103 W, 97 W

(a) Construct a frequency distribution table for this data.
(b) Find the mean and standard deviation of the data.
(c) Draw the histogram for the data.

11. The life of a computer disk drive is normally distributed with a mean $\bar{x}$ = 5,000 hours and a standard deviation σ = 100 hours. Out 10,000 manufactured disk drives, what percentage will last:
(a) Less than 5,000 hours?
(b) Between 4900 h and 5100 h?
(c) Between 4800 h and 5200 h?
(d) Less than 4700 h or more than 5300 h?
How many will last:
(e) More than 4900 h?
(f) Less than 5200 h?

12. The power gain of an amplifier is found to be normally distributed with a mean of $\bar{x}$ = 30 dB and a standard deviation of σ = 2 dB. Out of 1000 manufactured amplifiers what percentage will have a power gain:
(a) Greater than 30 dB?
(b) Between 28 dB and 32 dB?
(c) Between 24 dB and 36 dB?

How many will have a power gain:
(d) Less than 26 dB or more than 34 dB?
(e) More than 26 dB?

APPENDIX A

Calculator Functions

Basic scientific calculator operations are shown in Sections 1.4, 2.4, 3.4, and 4.4. Other functions and operations are shown in examples throughout the text. Below is a listing of basic key operations found on scientific calculators with references in brackets to examples in the text that illustrate the operation.

Scientific Calculator Functions

Key	Function	Examples
( )	Sets parentheses	[1.2, 1.3, 2.7]
STO or M_{in}	Stores a number in memory	[1.23]
RCL, MR, or RM	Recalls a number from memory	[1.23]
$a^{b/c}$ or >Frac	Enters a fraction	[1.8, 1.11]
FIX	Fixes the number of significant digits	[2.21, 2.23]
$1/x$ or x^{-1}	Calculates the reciprocal of a number	[1.20, 3.31, 9.19]
+/-	Changes the sign of a number	[3.19, 4.5, 4.6]
(–)	Minus key for a number	[3.19, 4.5, 4.6]
x^2	Squares a number	[3.2, 3.3, 3.4, 3.32]
$\sqrt{\ }$	Calculates the square root of a number	[3.22, 3.32, 3.33]
y^x, x^y, or ∧	Raises to a power	[3.1, 3.2, 3.3]
$\sqrt[3]{\ }$	Finds the cube root of a number	[3.34]
$\sqrt[x]{y}$ or $x^{\frac{1}{y}}$	Finds any root of a number	[3.34]
MODE	Sets the operational mode	[4.5, 4.20, 25.17]
SCI or ENG	Sets for scientific or engineering notation	[4.5, 4.17, 4.21]
EXP or EE	Sets the display to enter a power of ten for scientific notation	[3.17, 3.18, 3.19]
m, μ, k, M	Electrical units on engineering calculator	[4.13, 4.21]

Key	Function	Examples
[π]	Enters the number π	[5.35, 16.1]
[DRG]	Sets the angle measurement for degrees, radians, or gradient	[16.2, 16.11, 17.4]
[sin], [cos], [tan]	Finds the sine, cosine, or tangent of an angle	[16.11, 17.4]
[$\sin^{-1}$], [$\cos^{-1}$], [$\tan^{-1}$]	Finds the inverse sine, inverse cosine, or inverse tangent of a number	[16.13, 16.14, 17.6]
[CPLX]	Sets the calculator to complex mode	[20.10]
[i]	The imaginary unit of a complex number	[20.10]
[R→P]	Changes from rectangular to polar form	[20.17, 21.5]
[P→R]	Changes from polar to rectangular form	[20.17, 21.5]
[e^x]	Raises e to the x power	[23.8]
[log]	Finds the logarithm of a number to the base 10	[23.12, 23.13, 24.1]
[LN] or [lnx]	Finds the logarithm of a number to the base e	[23.13]
[BASE]	Sets the calculator to a number base mode	[25.17, 25.22]
[BIN], [OCT], [HEX]	Used to change a number to binary, octal, or hexadecimal form	[25.17, 25.22]
[Σ+] or [M+]	Sums the numbers entered	[27.1, 27.5]
[$\bar{x}$]	Calculates the mean	[27.1, 27.5]
[FRQ]	Enters the frequency of a term	[27.5, 27.8]
[σ_{xn-1}]	Calculates the standard deviation	[27.6, 27.8]

Graphing Calculator Functions

Graphing calculators use direct algebraic logic where the function key is pressed before a value is entered. For example, the calculations below are done as follows:

Calculation	Keystrokes
5^2	5 [x^2] [ENTER]
$\sqrt[3]{64}$	64 [∧] 3 [x^{-1}] [ENTER]
tan 40°	[TAN] 40 [ENTER]
ln 10	[LN] 10 [ENTER]

Graphing calculators have many computer like capabilities. This section introduces some of the basic graphing functions of the popular TI-83.

Before drawing a graph press the MODE key to set the conditions for the graph. There are eight choices as follows:

1. Notation: normal, scientific or engineering
2. Accuracy: floating decimal or number of decimal places
3. Angle measure: radians or degrees
4. Type of graph: function, parametric, polar, or sequence
5. Type of display: connected line or dot
6. Type of plot: sequential or simultaneous plotting
7. Type of number real or complex numbers
8. Type of screen: full, horizontal split with functions, or graph with table

To choose the mode, move the cursor using the arrow keys, <, >, ∧, ∨ for left, right, up, and down and press ENTER to select. The following examples use normal notation, floating decimal, radian measure, function mode, connected line, sequential plotting, real numbers, and full or split screen.

To set the range of values on the x and y axes you can use the ZOOM key. Some of the choices are as follows:

ZOOM 5: Sets a square window without distortion; x ranges from −4.7 to 4.7 and y from −3.1 to 3.1.

ZOOM 6: Sets a standard window; x and y range from −10 to 10.

ZOOM 7: Sets a window for trigonometric functions; x ranges from -2π to 2π and y from −4 to 4.

ZOOM 0: Sets a window to best fit the curve.

You can set any values you want for the x and y axes by using the WINDOW key.

To enter the equation of a graph, press Y= and enter the expression using the X, T, Θ key for the variable. The variable is X when in function mode, T in parametric mode and θ in polar mode. Use the CLEAR key to clear a line or the entire display. For example, to enter $y = \sin x$, key in the following:

Y= CLEAR SIN X, T, Θ ENTER

The display shows the first function and the cursor moves to the second function:

$$Y_1 = \sin x$$
$$Y_2 =$$

Press ZOOM 7 to set the window and display the graph. The calculator graphs $y = \sin x$ from -2π to 2π. If you set your own window, press GRAPH to display the graph.

You can enter a second function by pressing Y= again. It is possible to enter and graph up to ten functions. For example, to graph the voltage wave $y = 3 \cos x$ and the current wave $y = 0.5 \sin x$ on the same set of axes, key in:

Y= CLEAR 3 × COS X, T, Θ ENTER CLEAR 0.5 × SIN X, T, Θ ZOOM 7

You can select which functions you want the calculator to graph by moving the cursor to the equals sign and pressing ENTER to highlight the sign. If you do not want the function to be graphed, press ENTER again to remove the highlight.

You can draw a horizontal or vertical line on a graph by pressing DRAW 3 or 4, then positioning the cursor where you want the line.

Another useful feature is to see the values of x and y for any point on the graph. Press TRACE to locate the cursor on a graph. The coordinates of the point are shown at the bottom of the display. Use the left and right arrow keys to move the cursor along the graph and to see the coordinates of the points. If there are two graphs, use the up and down arrow keys to move the point between the graphs. You can also see a table of values of x and y by pressing TABLE, and you can set the conditions for the table by using TBL SET.

APPENDIX B

Notation and Formulas

Electrical Notation

Quantity	Letter(s) Used	Unit	Symbol
Admittance	Y	Siemen	S
Angular velocity	ω (omega)	Radians/Second	rad/s
Capacitance	C	Farad	F
Charge	Q	Coulomb	C
Conductance	G	Siemen	S
Current	I, i	Ampere	A
Energy	E	Kilowatt-hour	kWh
Frequency	f	Hertz	Hz
Impedance	Z	Ohm	Ω (omega)
Inductance	L	Henry	H
Period	T	Second	s
Power	P	Watt	W
Power gain	dB	Decibel	dB
Reactance	X	Ohm	Ω
Resistance	R	Ohm	Ω
Resonant frequency	f_r	Hertz	Hz
Susceptance	B	Siemen	S
Time Constant	τ (tau)	Second	s
Voltage	V, v	Volt	V

Metric or SI Prefixes

Power of 10	Prefix	Symbol
10^{-15}	femto	f
10^{-12}	pico	p
10^{-9}	nano	n
10^{-6}	micro	μ (mu)
10^{-3}	milli	m
10^{-2}	centi	c
10^{3}	kilo	k
10^{6}	mega	M
10	giga	G
10^{12}	tera	T

Mathematical Formulas

Exponent Rules

$$(x^m)(x^n) = x^{m+n} \tag{5.1}$$

$$\frac{x^n}{x^m} x^{n-m} \tag{5.2}$$

$$(x^m)^n = x^{mn} \tag{5.3}$$

$$(xy)^n = x^n y^n \text{ and } \left(\frac{x}{y}\right)^n = \frac{x^n}{y^n} \tag{5.4}$$

$$\sqrt[n]{x^m} = x^{m/n} \tag{5.5}$$

$$x^0 = 1 \quad (x \neq 0) \tag{5.6}$$

$$x^{-n} = \left(\frac{1}{x}\right)^n = \frac{1}{x^n} \text{ and } \left(\frac{x}{y}\right)^{-n} = \left(\frac{y}{x}\right)^n \quad (x, y \neq 0) \tag{5.7}$$

$$x^{-1} = \frac{1}{x} \text{ and } \left(\frac{x}{y}\right)^{-1} = \frac{y}{x} \quad (x, y \neq 0) \tag{5.8}$$

$$x^{m/n} = \sqrt[n]{x^m} \text{ or } \left(\sqrt[n]{x}\right)^m \quad (x > 0) \tag{23.1}$$

$$x^{1/n} = \sqrt[n]{x} \quad (x > 0) \tag{23.2}$$

Quadratic Formula

$$x = \frac{-b \pm \sqrt{b^2 - 4ac}}{2a} \tag{8.4}$$

Angles

$$\pi \text{ radians} = 180^0$$

$$1 \text{ rad} = \frac{180°}{\pi} \approx \frac{180°}{3.142} \approx 57.3°$$

For a point (x,y) on a circle of radius r:

$$\begin{aligned} x^2 + y^2 &= r^2 \\ \sin\theta &= \frac{y}{r} \\ \cos\theta &= \frac{x}{r} \\ \tan\theta &= \frac{y}{x} \end{aligned} \tag{17.1}$$

$$-90° \le \sin^{-1} x \le 90°$$
$$0° \le \cos^{-1} x \le 180° \qquad (17.2)$$
$$-90° < \tan^{-1} x < 90°$$

Adding Vector Components F_x and F_y

$$\text{Magnitude of the Resultant } F = \sqrt{F_x^2 + F_y^2}$$
$$\text{Reference Angle} = \tan^{-1}\left|\frac{F_y}{F_x}\right| \qquad 0^0 \le \theta < 360° \qquad (19.1)$$

Adding Phasor Components V_x and V_y

$$\text{Magnitude of the Resultant } V = \sqrt{V_x^2 + V_y^2}$$
$$\text{Angle } \theta = \tan^{-1}\left|\frac{V_y}{V_x}\right| \qquad -90^0 \le \theta < 90° \qquad (19.2)$$

Vector or Phasor Components

$$\cos\theta = \frac{V_x}{V} \rightarrow V_x = V\cos\theta \ (x \text{ component})$$
$$\sin\theta = \frac{V_y}{V} \rightarrow V_y = V\sin\theta \ (y \text{ component}) \qquad (19.3)$$

Law of Sines

$$\frac{a}{\sin A} = \frac{b}{\sin B} = \frac{c}{\sin C} \qquad (19.4)$$

Law of Cosines

$$c^2 = a^2 + b^2 - 2ab(\cos C) \qquad (19.5)$$

Complex Phasors

$$j = \sqrt{-1} \text{ and } j^2 = \left(\sqrt{-1}\right)^2 = -1 \qquad (20.1)$$

$$\sqrt{-x} = j\sqrt{x} \quad (x = \text{positive number}) \qquad (20.2)$$

$$x + jy = \sqrt{x^2 + y^2}\ \angle\tan^{-1}\left(\frac{y}{x}\right) = Z\ \angle\theta \quad (x \ne 0) \qquad (20.3)$$

$$Z \angle\theta = Z \cos\theta + j(Z \sin\theta) = x + jy \quad (20.4)$$

$$(Z_1 \angle\theta_1)(Z_2 \angle\theta_2) = Z_1 Z_2 \angle(\theta_1 + \theta_2) \quad (20.5)$$

$$\frac{Z_2 \angle\theta_2}{Z_1 \angle\theta_1} = \frac{Z_2}{Z_1} \angle(\theta_2 - \theta_1) \quad (20.6)$$

Logarithms

$y = \log_b x$ means $b^y = x \quad (b > 1)$ (23.5)

$y = \log x$ means $10^y = x$

$y = \ln x$ means $e^y = x$ where $e \approx 2.718$

Log of a Number a to any Base b:

$$\log_b a = \frac{\log a}{\log b} \quad (23.6)$$

$$\log_b (xy) = \log_b x + \log_b y \quad (23.7)$$

$$\log_b \left(\frac{x}{y}\right) = \log_b x - \log_b y \quad (23.8)$$

$$\log_b (y^x) = x (\log_b y) \quad (23.9)$$

Electrical Formulas

DC Circuit

Ohm's law:

$I = \frac{V}{R}$ (7.1) $\quad$ $V = IR$ (7.1a) $\quad$ $R = \frac{V}{I}$ (7.1b)

Power formulas:

$P = VI$ (7.2) $\quad$ $I = \frac{P}{V}$ (7.2a) $\quad$ $V = \frac{P}{I}$ (7.2b)

$P = I^2R$ (7.3a) $\quad$ $P = \frac{V^2}{R}$ (7.3b) $\quad$ $R = \frac{P}{I^2}$ (7.4a)

$R = \frac{V^2}{P}$ (7.4b) $\quad$ $I = \sqrt{\frac{P}{R}}$ (7.5a) $\quad$ $V = \sqrt{PR}$ (7.5b)

Series DC Circuit

$$R_T = R_1 + R_2 + R_3 + \ldots + R_n \quad (11.1)$$

$$V_T = V_1 + V_2 + V_3 + \ldots + V_n \quad (11.2)$$

$$P_T = P_1 + P_2 + P_3 + \ldots + P_n \quad (11.5)$$

Voltage division:

$$V_x = \frac{R_x}{R_T} V_T \tag{14.1}$$

Parallel DC Circuit

$$I_T = I_1 + I_2 + I_3 + \ldots + I_n \tag{11.6}$$

$$\frac{1}{R_T} = \frac{1}{R_1} + \frac{1}{R_2} + \frac{1}{R_3} + \ldots + \frac{1}{R_n} \tag{11.7}$$

Two parallel resistances:

$$R_T = \frac{R_1 R_2}{R_1 + R_2} \tag{11.9}$$

$$R_1 = \frac{R_2 R_T}{R_2 - R_T} \tag{11.10}$$

Conductance:

$$G = \frac{1}{R} \tag{11.11}$$

$$G_T = G_1 + G_2 + G_3 + \ldots + G_n \tag{11.12}$$

Kirchhoff's Laws

The algebraic sum of all currents entering and leaving any branch point in a circuit equals zero. (14.2)

The algebraic sum of the voltages around any closed path equals zero. (14.3)

AC Circuit

$$V_{\text{rms}} = \left(\frac{1}{\sqrt{2}}\right) V_P \approx (0.707) V_P$$

$$I_{\text{rms}} = \left(\frac{1}{\sqrt{2}}\right) I_P \approx (0.707) I_P \tag{18.3}$$

$$V_P = \left(\sqrt{2}\right) V_{\text{rms}} \approx (1.41) V_{\text{rms}}$$

$$I_P = \left(\sqrt{2}\right) I_{\text{rms}} \approx (1.41) I_{\text{rms}} \tag{18.4}$$

Resistive circuit only:

$$I_{\text{rms}} = \frac{V_{\text{rms}}}{R} \tag{18.5}$$

$$P = (V_{\text{rms}})(I_{\text{rms}})$$

Angular velocity (rad/s):

$$\omega = \frac{\theta}{t} \text{ or } \theta = \omega t \tag{18.6}$$

Frequency:

$$f = \frac{\omega}{2\pi} \text{ or } \omega = 2\pi f \quad (18.7)$$

Angle (θ):

$$\theta = \omega t = 2\pi ft \quad (18.8)$$

AC voltage and current waves:

$$v = V_P \sin \omega t \text{ or } v = V_P \sin 2\pi ft$$

$$i = I_P \sin \omega t \text{ or } i = I_P \sin 2\pi ft \quad (18.9)$$

Period (sec):

$$T = \frac{1}{f} = \frac{2\pi}{\omega} \quad (18.10)$$

AC Series Circuit

***RL* circuit:**

$$X_L = 2\pi fL \quad (21.1)$$

$$V_T = V_R + jV_L = \left(\sqrt{V_R^2 + V_L^2}\right) \angle \tan^{-1}\left(\frac{V_L}{V_R}\right) \quad (21.3)$$

$$Z = R + jX_L = \left(\sqrt{R^2 + X_L^2}\right) \angle \tan^{-1}\left(\frac{X_L}{R}\right) \quad (21.4)$$

$$I = \frac{V}{Z} \quad (21.5)$$

***RC* circuit:**

$$X_C = \frac{1}{2\pi fC} \quad (21.6)$$

$$V_T = V_R - jV_C = \left(\sqrt{V_R^2 + V_C^2}\right) \angle \tan^{-1}\left(\frac{-V_C}{V_R}\right) \quad (21.8)$$

$$Z = R - jX_C = \left(\sqrt{R^2 + X_C^2}\right) \angle \tan^{-1}\left(\frac{-X_C}{R}\right) \quad (21.9)$$

***RLC* circuit:**

$$Z = R + j(X_L - X_C) = R \pm jX \quad (21.10)$$

Use (21.4) when $X_L > X_C$. Use (21.9) when $X_L < X_C$.

$$V_T = V_R + j(V_L - V_C) = V_R \pm jV_X \quad (21.11)$$

(Use 21.3) when $X_L > X_C$. Use (21.8) when $X_L < X_C$.

True power (W):

$$P = I^2R = VI \cos \theta \quad (21.12a)$$

Reactive power (VAR):

$$Q = VI \sin \theta \quad (21.12b)$$

Apparent power (VA):

$$S = VI \tag{21.12c}$$

Resonant frequency:

$$f_r = \frac{1}{2\pi\sqrt{LC}} \tag{21.13}$$

AC Parallel Circuit

***RL* circuit:**

$$I_T = I_R - jI_L = \sqrt{I_R^2 + I_L^2}\ \angle\tan^{-1}\left(\frac{-I_L}{I_R}\right) \tag{22.1}$$

***RC* circuit:**

$$I_T = I_R - jI_C = \sqrt{I_R^2 + I_C^2}\ \angle\tan^{-1}\left(\frac{I_C}{I_R}\right) \tag{22.3}$$

Susceptance:

$$B = \frac{1}{X} \tag{22.4}$$

Admittance:

$$Y = \frac{1}{Z} = \frac{1}{R} \pm j\frac{1}{X} = G \pm jB \tag{22.5}$$

Inductive $(-jB)$ Capacitive $(+jB)$

***RLC* circuit:**

$$I_T = I_R + j(I_C - I_L) = I_R \pm jI_X \tag{22.6}$$

Inductive $(-jI_X)$ Capacitive $(+jI_X)$

$$Y = \frac{1}{R} + j\left(\frac{1}{X_C} - \frac{1}{X_L}\right) = G + j(B_C - B_L) = G \pm jB \tag{22.7}$$

Inductive $(-jB)$ Capacitive $(+jB)$

AC Network

Series impedances:

$$Z_T = Z_1 + Z_2 + Z_3 + \ldots \tag{22.8}$$

Parallel impedances:

$$Z_{eq} = \frac{Z_1 Z_2}{Z_1 + Z_2} \tag{22.9}$$

Filters

***RC* cutoff frequency:**

$$f_c = \frac{1}{2\pi RC} \tag{22.10}$$

***RL* cutoff frequency:**

$$f_c = \frac{R}{2\pi L} \tag{22.11}$$

Bandwidth:

$$\text{BW} = f_2 - f_1 \tag{22.12}$$

Resonant frequency:

$$f_r = f_1 + \frac{\text{BW}}{2} \tag{22.13}$$

Quality:

$$Q = \frac{f_r}{\text{BW}} \tag{22.14}$$

RC and *RL* DC Circuits

***RC* circuit:**

$$\tau = RC \tag{24.1}$$

Capacitor charge and discharge:

$$i_C = \frac{V}{R} e^{-t/\tau} \tag{24.2}$$

Capacitor discharge:

$$v_C = Ve^{-t/\tau} \tag{24.3}$$

Capacitor charge:

$$v_C = V(1 - e^{-t/\tau}) \tag{24.4}$$

***RL* circuit:**

$$\tau = \frac{L}{R} \tag{24.5}$$

Inductor current increase:

$$i_L = \frac{V}{R}(1 - e^{-t/\tau}) \tag{24.6}$$

Inductor current decrease:

$$i_L = \frac{V}{R} e^{-t/\tau} \tag{24.7}$$

Current and voltage change in *RC* and *RL* circuits:

$$\text{Percent change} = (1 - e^{-t/\tau}) \times 100\% \tag{24.8}$$

Power Gain

$$A_p = 10 \log\left(\frac{P_{\text{out}}}{P_{\text{in}}}\right) \tag{24.9}$$

$$A_v = 20 \log\left(\frac{V_{\text{out}}}{V_{\text{in}}}\right) \tag{24.10}$$

Standard power gain:

$$A_{p(\text{dBm})} = 10 \log\left(\frac{P_{\text{out}}}{1 \text{ mW}}\right) \tag{24.11}$$

Logic Gates

		AND	OR	NOT
A	*B*	*AB*	*A* + *B*	$\overline{A}$
0	0	0	0	1
0	1	0	1	1
1	0	0	1	0
1	1	1	1	0

APPENDIX C

Answers to Odd-Numbered Problems

CHAPTER 1

Exercise 1.1

1. 16
3. 120
5. 4
7. 36
9. 40
11. 50
13. 26
15. 77
17. 45
19. 10
21. 2
23. 120
25. 5
27. 25
29. 1
31. 10
33. 55
35. 1500
37. 1
39. 3
41. 20 mi/gal; 21 mi/gal
43. 54,000,000 mi
45. 1 A
47. 16 V
49. 7

Exercise 1.2

1. $\frac{3}{5}$
3. $\frac{4}{5}$
5. $\frac{3}{4}$
7. $\frac{2}{9}$
9. $\frac{2}{11}$
11. $\frac{9}{35}$
13. $\frac{2}{15}$
15. $\frac{4}{15}$
17. $\frac{4}{3}$
19. 2
21. $\frac{2}{5}$
23. $\frac{28}{45}$
25. 2
27. $\frac{5}{16}$
29. $\frac{5}{6}$
31. $\frac{2}{21}$
33. 6
35. $\frac{1}{3}$
37. $\frac{1}{2}$
39. 8
41. 6
43. 1
45. $\frac{3}{10}$ ft
47. $300
49. 27 V
51. $\frac{1}{16}$ A
53. $\frac{9}{25}$ W

Exercise 1.3

1. $\frac{7}{7} = 1$
3. $\frac{5}{8}$
5. $\frac{13}{20}$
7. $\frac{19}{24}$
9. $\frac{14}{15}$
11. $\frac{31}{60}$
13. $\frac{57}{200}$
15. $\frac{1}{4}$
17. $\frac{1}{6}$
19. $\frac{11}{20}$
21. $\frac{85}{24}$
23. 5
25. $\frac{33}{80}$
27. $\frac{127}{200}$
29. $\frac{7}{20}$
31. $\frac{63}{100}$
33. $\frac{31}{60}$
35. $\frac{2}{3}$
37. $\frac{7}{12}$
39. $\frac{1}{80}$
41. $\frac{13}{12}$
43. $\frac{14}{15}$
45. $\frac{41}{100}$
47. $\frac{11}{20}$
49. $\frac{3}{8}$
51. $\frac{33}{40}$
53. 1 lb
55. $\frac{13}{4}$ in
57. 10 Ω
59. 20 Ω
61. $\frac{20}{3}$ Ω
63. 4

Exercise 1.4

1. 4
3. 47
5. 130
7. 412
9. 2
11. 147
13. 60
15. 40
17. 1
19. 1
21. 4
23. 12
25. 710 (OIL)
27. 666.7 Ω

Review Exercises

1. 22
3. 12
5. 27
7. $\frac{15}{4}$
9. $\frac{3}{2}$
11. $\frac{15}{2}$
13. 4
15. $\frac{5}{2}$
17. $\frac{11}{16}$
19. $\frac{1}{5}$
21. 2
23. $\frac{7}{24}$
25. $\frac{5}{36}$
27. $\frac{27}{100}$
29. $\frac{8}{7}$
31. 200
33. 147
35. 2
37. 25 mi/h

39. 250 ft
41. \$1250
43. $\frac{1}{2}$ A
45. 10 V
47. $\frac{18}{5}$ W
49. 27 V
51. 20 Ω
53. 20 Ω
55. $\frac{25}{9}$

CHAPTER 2

Exercise 2.1

1. 10.09
3. 0.91
5. 22.71
7. 4.05
9. 0.0062
11. 12
13. 6.25
15. 0.00001
17. 30
19. 0.12
21. 10
23. 0.08
25. 0.8
27. 12.3
29. 0.02
31. 4
33. 10
35. 40
37. 7
39. 0.08
41. 0.2083 mm
43. 0.5555 cm^2
45. 8.4 ft
47. 40.1 V
49. 17.0 cm

Exercise 2.2

1. 0.75, 75%
3. 0.4, 40%
5. 0.375, 37.5%
7. 0.04, 4%
9. 15%, $\frac{3}{20}$
11. 10%, $\frac{1}{10}$
13. 12%, $\frac{3}{25}$
15. 28%, $\frac{7}{25}$
17. $\frac{1}{5}$, 0.2
19. $\frac{3}{20}$, 0.15
21. $\frac{43}{50}$, 0.86
23. $\frac{99}{100}$, 0.99
25. 0.05, 5%
27. 1.25, 125%
29. $\frac{1}{20}$, 5%
31. $\frac{3}{2}$, 150%
33. $\frac{6}{5}$, 1.2
35. $\frac{7}{125}$, 0.056
37. 33%, 0.33
39. $\frac{3}{16}$, 18.75%
41. $\frac{17}{200}$, 0.085
43. $\frac{1}{1000}$, 0.1%
45. 3
47. 15
49. 1.35
51. 75
53. 1.65
55. 2.5
57. \$13.20
59. 0.95 ft^3
61. \$38.00, \$42.75
63. \$64.00, \$68.80
65. \$42.00, \$44.10
67. 12%
69. 110%, $\frac{1120}{1000}$, $\frac{9}{8}$, $1\frac{1}{6}$, 1.19
71. 165 Ω
73. 0.11 A, 1.21 A
75. 40% decrease
77. 20% increase

Exercise 2.3

1. 3
3. 2
5. 3
7. 3
9. 4
11. 4
13. 4
15. 4
17. 4
19. 4
21. 5, 5.2, 5.15
23. 300, 320, 318
25. 0.4, 0.44, 0.445
27. 80,000, 82,000, 81,900
29. 100, $1\bar{0}0$, 101
31. 20, 19, 18.9
33. 20, 20, 20.0
35. 300, 340, 336
37. 300, 260, $26\bar{0}$
39. 1000, $1\bar{0}00$, 999
41. 461 m^2
43. 152 lb
45. 18 V
47. 203 in^2
49. 120 V
51. 1480 W
53. (a) \$43
    (b) \$46

Exercise 2.4

1. 2
3. 21.1
5. 2.0
7. 0.5
9. 147
11. 2.4
13. 0.68
15. 0.143
17. 0.860
19. 35.5
21. 1.90
23. 110
25. 8.27
27. 0.00482
29. 0.00266
31. 197,000,000 mi^2
33. (a) 4.67%
    (b) 19.1%
35. 33.0 V
37. 0.415 A

Review Exercises

1. 31.26
3. 4.04
5. 220.1
7. 0.25
9. 0.006
11. 20
13. 44
15. 0.8, 80%
17. 0.15, 15%
19. $\frac{4}{25}$, 16%
21. $\frac{7}{20}$, 0.35
23. $\frac{11}{200}$, 0.055
25. 0.005, 0.5%
27. 8.4
29. 39
31. 82.5
33. 0.5
35. 33
37. 0.95
39. 31.0
41. 66.9
43. 25.3 in, 26.7 in^2
45. \$12,000
47. \$59.84, \$64.33
49. 76
51. (a) 0.41 A
    (b) 2.5 A
53. 10%
55. 1200 Ω
57. 21 mA

CHAPTER 3

Exercise 3.1

1. 81
3. 128
5. 100,000
7. 121
9. $\frac{1}{16}$, 0.0625
11. $\frac{9}{25}$, 0.36
13. $\frac{1}{10,000}$, 0.0001
15. $\frac{25}{16}$, 1.5625
17. 0.49, $\frac{49}{100}$
19. $\frac{8}{125}$, 0.064
21. 0.0001, $\frac{1}{10,000}$
23. 0.0000001
25. $\frac{1}{9}$, 0.111 …
27. 1
29. $\frac{1}{40}$, 0.025

31. $\frac{6}{125}$, 0.048
33. $\frac{1}{8}$, 0.125
35. $\frac{1}{20}$, 0.05
37. $\frac{25}{8}$, 3.125
39. $\frac{9}{50}$, 0.18
41. $\frac{1}{500}$, 0.002
43. $\frac{5}{64}$, 0.078125
45. $\frac{17}{250}$, 0.068
47. $\frac{75}{4}$, 18.75
49. $\frac{1}{9}$, 0.111 ...
51. 0.512 cm^3
53. \$26,620
55. 2.25 mm^2
57. 20 Ω
59. 17 W
61. 0.06 W

Exercise 3.2

1. 10^3
3. 10^0
5. 10^{-6}
7. 10^{-4}
9. 10^{-2}
11. 10^{-1}
13. 100,000
15. 1,000,000
17. 0.01, $\frac{1}{100}$
19. 0.1, $\frac{1}{10}$
21. 0.000001, $\frac{1}{1,000,000}$
23. 10
25. 10^6
27. 10^{-5}
29. 10^{-6}
31. 10^9
33. 10^{-9}
35. 10^3
37. 10^{-2}
39. $10^1 = 10$
41. 10^{-2}
43. 10^2
45. 10^{-1}
47. 10^4
49. 10^2
51. 10^{-3}
53. 10^9
55. 10^{-1}
57. 12×10^2
59. $24 \times 10^1 = 24 \times 10$
61. 6×10^{-12}
63. 12×10^{-1}
65. 3×10^3
67. 2.5×10^7
69. 2.5×10^{-2}
71. 0.75×10^6
73. 16×10^3, 16,000
75. 7×10^{-9}, 0.000000007
77. 6.5×10^4, 65×10^3, 65,000
79. 29×10^{-4}, 2.9×10^{-3}, 0.0029
81. $4.5 \times 10 = 45$
83. 480 W
85. 3.0 V
87. 0.05 A
89. 2000 Ω

Exercise 3.3

1. 4
3. $\frac{5}{2}$
5. $\frac{2}{3}$
7. 0.8
9. 0.1
11. 200
13. 0.09
15. 3
17. $\frac{1}{2}$
19. 0.1
21. 10
23. 14
25. 28
27. 1.4
29. 2.1
31. 12
33. $0.6 = \frac{3}{5}$
35. 2
37. $0.1 = \frac{1}{10}$
39. 64×10^9
41. 16×10^8
43. 0.64×10^{10}
45. 0.001×10^9
47. 9×10^{11}
49. 150×10^6
51. 4×10^3
53. 12×10
55. 0.5×10^2
57. 2×10^3
59. 6×10^3
61. 0.2×10^1
63. 1.5×10^{-5}
65. 0.75×10^5
67. 70 ft
69. 2.5
71. 20 Ω
73. 60 V
75. 0.2 A
77. 6.4 W
79. 0.05 A
81. 2

Exercise 3.4

1. 58.1
3. 0.0101
5. 0.655
7. 79.7
9. 0.242
11. 57.0
13. 15.0
15. 10.1
17. 57.4
19. 97.9
21. 148 cm^3
23. 0.794 W
25. 4830 Ω

Review Exercises

1. 343
3. $\frac{1}{16}$, 0.0625
5. 0.216, $\frac{27}{125}$
7. 1.21, $\frac{121}{100}$
9. 0.01, $\frac{1}{100}$
11. $\frac{2}{5}$, 0.4
13. 1
15. 0.29, $\frac{29}{100}$
17. 7
19. $\frac{2}{5} = 0.4$
21. $0.3 = \frac{3}{10}$
23. 4
25. 1
27. $0.06 = \frac{3}{50}$
29. 7
31. 100,000,000
33. $\frac{1}{10,000}$, 0.0001
35. $10^2 = 100$
37. $10^4 = 10,000$
39. $10^{-5} = 0.00001$, $\frac{1}{100,000}$
41. $10^9 = 1,000,000,000$
43. $10^6 = 1,000,000$
45. 14×10^2
47. 2×10^{-3}
49. $0.10 \times 10^{-9} = 1.0 \times 10^{-10}$
51. 1×10^8
53. 16×10^5
55. $6.5 \times 10^3 = 65 \times 10^2$
57. 81×10^8
59. 9×10^3
61. 4×10^3
63. 7×10^3
65. 0.1×10^7
67. 0.0140
69. 277
71. 51.7
73. 9.48
75. 56 ft
77. 50 mm = 5 cm
79. 50 Ω
81. 0.04 A
83. 6.5 Ω
85. $0.075 \times 10^3 = 75$ W
87. 0.0031 Ω

CHAPTER 4

Exercise 4.1

1. 4.26×10^{10}, 42.6×10^9
3. 9.30×10^{-4}, 930×10^{-6}
5. 2.35×10^0
7. 1.17×10^{-3}
9. 3.34×10^2, 334×10^0
11. 1.12×10^{-1}, 112×10^{-3}
13. 1×10^3
15. 1.62×10^5, 162×10^3
17. 56,400,000, 5.64×10^7, 56.4×10^6

19. 1,200,000, 1.2×10^{6}
21. 0.000000114, 1.14×10^{-7}, 114×10^{-9}
23. 52.8, 5.28×10^{1}, 52.8×10^{0}
25. 0.00569, 5.69×10^{-3}
27. 5,660,000, 5.66×10^{6}
29. 0.000552, 5.52×10^{-4}, 552×10^{-6}
31. 26,400,000, 2.64×10^{7}, 26.4×10^{6}
33. 0.00000390, 3.90×10^{-6}
35. 1.45, 1.45×10^{0}
37. 0.0101, 1.01×10^{-2}, 10.1×10^{-3}
39. 112,000, 1.12×10^{5}, 112×10^{3}
41. 0.0000105, 1.05×10^{-5}, 10.5×10^{-6}
43. 871,000,000, 8.71×10^{8}, 871×10^{6}
45. 1.23, 1.23×10^{0}
47. 125,000, 1.25×10^{5}, 125×10^{3}
49. 0.503, 5.03×10^{-1}, 503×10^{-3}
51. 1.29×10^{7}, 12.9×10^{6}, 12,900,000
53. 1.35×10^{7}, 13.5×10^{6}, 13,500,000
55. 3.41×10^{6}, 3,410,000
57. 7.18×10^{-8}, 71.8×10^{-9}, 0.0000000718
59. 1.49×10^{5}, 149×10^{3}, 149,000
61. 9.42×10^{-3}
63. 7.88×10^{12}
65. 53.3×10^{-3}
67. 52.9×10^{3}
69. 7.28×10^{-9}
71. 57.0×10^{0}
73. 899×10^{3}
75. 5.00×10^{-6}
77. 40.0×10^{6}
79. 250×10^{9}
81. $3490.57
83. 2.40×10^{-6} m
85. 439×10^{9} kWh
87. 183×10^{-3} V
89. 92.3×10^{-3} A
91. 5.41×10^{3} Ω

Exercise 4.2

1. 2300 V
3. 0.031 A
5. 1500 Ω
7. 1500 kV
9. 78 mC
11. 6800 kW
13. 7.4 MV
15. 1300 pF
17. 3.9 MHz
19. 0.045 mH
21. 0.40 kg
23. 0.23 mm
25. 5.5 mC
27. 0.040 MV
29. 0.0034 A
31. 7500 mΩ
33. 880 pF
35. 0.001 kV
37. 4.1 m
39. 600 cm^2
41. 0.35 kHz
43. 5500 μS
45. 0.056 MΩ
47. 0.860 m^2
49. 0.000505 mF
51. 3.5 km
53. 0.56 ms, 560 μS
55. 300 kV, 0.30 MV
57. 1.3 V
59. 0.16 A, 160 mA
61. 0.48 MΩ, 480 kΩ
63. 0.77 W, 770 mW
65. 300 m, 30,000 cm

Exercise 4.3

1. 180 ft
3. 54 kg
5. 2.6 m
7. 77 lb
9. 1.3 hp
11. 190 W
13. 1100 W
15. 2600 J
17. 250 ft^2
19. 1.9 m^2
21. 0.78 in^2
23. 110 ft/s
25. 40 cm
27. 3.1 mi
29. 14 cm
31. 35°C
33. 4900 ft
35. 160 hp
37. 9500 Btu
39. 250 g
41. 130 kg
43. 0.055 in
45. 1.85 km
47. 54 J, 40 ft · lb
49. 0.025 mm
51. 13,000 hp
53. 0.55 cm^2, 0.085 in^2

Exercise 4.4

1. 539
3. 43.3×10^{12}
5. 25×10^{6}
7. 215×10^{-3}
9. 430×10^{6}
11. 2.08×10^{6}
13. 1.24×10^{-3}
15. 78.9 J
17. 4.20 V
19. 747 Ω
21. 30.3 kΩ

Review Exercises

1. 8.76×10^{7}, 87.6×10^{6}
3. 1.12×10^{-1}, 112×10^{-3}
5. 3.47×10^{5}, 347×10^{3}
7. 330,000,000
9. 124
11. 0.000339
13. 1,260,000, 1.26×10^{6}
15. 5.55, 5.55×10^{0}
17. 733,000, 7.33×10^{5}, 733×10^{3}
19. 3.66, 3.66×10^{0}
21. 0.594 W
23. 45.3 μA
25. 0.670 nF
27. 56.0 cm
29. 32.0 kV
31. 10.4 cm^2
33. 26.5 lb
35. 0.762 m
37. 7.11 hp
39. 4.47 MW
41. 6.21 mi
43. 2.72×10^{7}, 27.2×10^{6}, 27,200,000
45. 4.13×10^{-6}, 0.00000413
47. 5.03×10^{-6}, 0.00000503
49. 33.0×10^{3}
51. 91.3×10^{12}
53. 200×10^{-6}
55. 2.05×10^{-3}
57. 3.18

59. 500×10^9

61. 6,370 km = 6.37×10^3 km = 3960 mi = 3.960×10^3 mi

63. 30 caliber

65. 150,000 kW = 201×10^3 hp

67. 1.47 kΩ = 1470 Ω

69. 56.4 V

71. 0.011 A = 11.0 mA

73. 2220 Ω = 2.22 kΩ

75. 8.33×10^{-3} H

77. 50.4×10^{-6} H = 50.4 μH

CHAPTER 5

Exercise 5.1

1. −3

3. −10

5. 4

7. 5

9. 5.3

11. 5

13. −0.10

15. $\frac{1}{6}$ (0.17)

17. $\frac{5}{8}$ (0.625)

19. −56

21. 3

23. −7

25. −2

27. $-\frac{1}{10}$

29. −75

31. 0.06

33. $\frac{3}{2}$ (1.5)

35. −2

37. −0.11

39. 0

41. $-\frac{5}{2}$ (−2.5)

43. −27

45. −0.6

47. 48

49. $-\frac{1}{5}$ (−0.2)

51. $-\frac{1}{2}$ (−0.5)

53. $\frac{23}{6}$ (3.83)

55. $\frac{3}{2}$ (1.5)

57. −22

59. 8848 m − (−11,034 m) = 19,882 m; 29,028 ft − (−36,201 ft) = 65,229 ft

61. 0° − 45° + 90° = + 45°

63. positive

65. −158°C

67. 1.3 + 0.8 − 1.7 − 0.4 = 0

Exercise 5.2

1. 451

3. 390

5. 1.33

7. 140

9. 677

11. $-2x$

13. $11q - 4p$

15. $2IR + 16V$

17. $2de - d$

19. $x^2 - 3x$

21. $-1.5V_1 - 2.3V_2$

23. $12Q_a - 2Q_b$

25. $-\left(\frac{1}{2}\right)P - \frac{1}{20}$

27. $-by - 1$

29. $7V - I$

31. $5x^2 - x + 1$

33. $4h^2 + 5h + 5$

35. $3rs^3 - 3rs^2$

37. $\frac{R_x}{3}$

39. $-4.5P_2 + 8.9P_1$

41. $5\pi + 1$

43. $5V_0 - 2V_1$

45. $n^2 - 3n$

47. $8H + 1$

49. $1.3v - 1.1r$

51. $3\pi rh - 2\pi r^2$

53. $7d^2 + d - 4$

55. $3.8IR + 0.4$

57. $\left(\frac{1}{12}\right)W + 2$

Exercise 5.3

1. x^7

3. −1

5. y^9

7. R^2

9. −6

11. v^6

13. $(-2)^6 = 64$

15. $27p^3v^6$

17. 10^9

19. $\frac{x^2}{9b^4}$

21. I^2

23. a^6

25. $\frac{1}{8^2} = \frac{1}{64}$

27. 1

29. $\frac{1}{(-3^3)} = -\frac{1}{27}$

31. 1

33. 8

35. $\frac{9}{2} = 4.5$

37. $(-2)^3 = -8$

39. $9^2 = 81$

41. I^2

43. $\frac{9}{T^2}$

45. $-\frac{v}{4}$

47. $\frac{3y}{x}$

49. $10^3 = 1000$

51. $2x^2$

53. $0.006a^2b^2$

55. $5m^3 - 5m^2 + 15m$

57. $30y^3$

59. $2a - b$

61. $3f - 2$

63. $x^3y + xy^3$

65. $2I^2RV$

67. $\left(\frac{1}{25}\right) \times 10^6 = 40 \times 10^3$

69. 4×10^3

71. 720×10^{-9}

73. 2.03×10^{-3}

75. 2.00×10^{-3}

77. 12.0 V

79. 3.63 mA

81. 43.7 mW

83. 3.63 mA

85. 11 kΩ

87. 120 Ω

89. 1.74×10^{-3} Ω = 1.74 mΩ

91. 982 kHz

93. 2 * X^3 − 6 * X^2 + 10 * X

Review Exercises

1. 2
3. −8.1
5. $-\frac{1}{8}$
7. $-\frac{1}{20}$
9. 7
11. −0.6
13. $\frac{3}{4}$
15. −10
17. $-\frac{4}{5}$
19. −3.8
21. $3x - 2x^2$
23. $0.4V_1 - 1.1V_2$
25. $\left(\frac{1}{10}\right)F - \left(\frac{2}{5}\right)G$
27. $-4 + 5CV$
29. $3X_C^2 - X_L^2$
31. $-\dfrac{R_x}{3}$
33. $5.5I_1 - 2.7$
35. m^9
37. $(-1)^9 = -1$
39. q^{12}
41. $64x^6y^2$
43. $-\dfrac{1}{3d^2e^2}$
45. $\dfrac{1}{f^3}$
47. 1
49. $\frac{3}{4}$
51. $-21X_C^4$
53. $180I^5R^5$
55. $\dfrac{20x}{w}$
57. $42a^3b^3 - 42a^2b^4$
59. 2.94×10^9
61. $\left(\dfrac{4}{5}\right) \times 10^{-2} = 8 \times 10^{-3}$
63. 207 mV
65. $V = 19.2$ V
67. 13.9 kΩ
69. 13.3 mA
71. 320.5°C
73. $-2.6e^2 + 12.6e + 0.2$
75. −3.0 V
77. **(a)** 48.5 mA
    (b) 776 mW
79. 12.0 Ω
81. 6.45 kΩ
83. 4.77 A

CHAPTER 6

Exercise 6.1

1. 4
3. 25
5. 4
7. −8.6
9. 52
11. 55
13. 1
15. −1
17. 16
19. 0.5
21. −11
23. 3
25. 1.5
27. −6
29. 8.0
31. 0
33. 1.6
35. 5.5
37. −2
39. $\frac{2}{3} = 0.67$
41. 5
43. 68
45. 115
47. 12
49. 5
51. −25°C
53. 3.5 sec
55. 100 V
57. 24 W

Exercise 6.2

1. −4
3. −6
5. 6
7. 2.3
9. 6
11. $\frac{1}{3} = 0.33$
13. −2
15. −2
17. 1.5
19. −1
21. −2.3
23. −8
25. 3
27. 12
29. 8
31. 24
33. 4.8
35. 0.8
37. −16
39. 17
41. 2
43. $24.80
45. 500 V
47. 15 V
49. 33 Ω

Review Exercises

1. 6
3. 3
5. 1.5
7. −5
9. −13
11. $-\frac{3}{5} = -0.60$
13. 2
15. 0.4
17. 2
19. 50
21. $\frac{1}{2} = 0.5$
23. 100
25. −12
27. −1
29. $-\frac{3}{4} = -0.75$
31. 62.5 mi/h
33. $1000
35. 21 V
37. 7.5 Ω

CHAPTER 7

Exercise 7.1

1. 3.3 A
3. 120 V
5. 7.5 Ω

7. 1.5 V
9. 3.3 mA
11. 17 V
13. 3.0 kΩ
15. 2.2 mA
17. 24 kΩ
19. 22 kΩ
21. (a) 3.6 A
(b) 1.8 A
(c) 3.2 A
23. (a) 20 V
(b) 30 V
(c) $\frac{10(100\%)}{20} = 50\%$
25. (a) 160 mA
(b) 120 mA
(c) 80 mA
27. (a) 5.6 A
(b) 3.5 A
29. (a) 150 mA
(b) 100 mA
31. (a) 60 mΩ
(b) 50 mΩ
33. (a) V increases 50%
(b) V decreases 10%
35. (a) R increases $\frac{2}{1}$
(b) R decreases $\frac{2}{3}$

Exercise 7.2

1. 5.5 W, 22 Ω
3. 1.5 A, 5.3 Ω
5. 13 V, 8.9 Ω
7. 14 W, 1.2 A
9. 6.4 W, 32 V
11. 10 Ω, 3.1 A
13. 22 Ω, 18 V
15. 2.0 A, 600 V
17. 37 V, 670 mA
19. 1.7 mW, 120 kΩ
21. 3.0 W, 68 V
23. 160 Ω, 240 V
25. 19 mA, 2.9 V
27. 3.3 kW, 600 mA
29. 12 W, 4.7 V
31. (a) 1.7 A, 5.0 A
(b) 140 Ω, 48 Ω
33. (a) 880 mA
(b) 330 mA
(c) 220 mA
35. (a) 15 kW
(b) 810 Ω
37. (a) 170 mW
(b) 1.1 V
39. 3.1 W, 2.4 W, 5.5 W
41. (a) 81 V
(b) 1.2 kWh = 4.3 MJ
43. 360 W, 160 W, 520 W
45. $P = VI = V\left(\frac{V}{R}\right) = \frac{V^2}{R}$

Review Exercises

1. 300 mA, 11 W
3. 90 Ω, 900 mW
5. 180 V, 400 W
7. 80 Ω, 4.5 mW
9. 1.3 A, 64 Ω
11. 33 V, 74 kΩ
13. 2.2 kV, 3.0 A
15. 360 mV, 140 Ω
17. (a) 80 mA
(b) 40 mA
(c) Half, I is directly proportional to V_s
19. (a) 150 mA
(b) 300 mA
(c) 100%, I is inversely proportional to R
21. (a) 9.0 V
(b) 4.5 V
23. 18 Ω
25. (a) 200 mA
(b) 3.6 kWh=13 MJ
27. (a) 670 mΩ
(b) 60 A
29. $\frac{3}{1}$

CHAPTER 8

Exercise 8.1

1. $x^2 + 6x + 8$
3. $C^2 + 3C - 4$
5. $I^2 - 100$
7. $2V^2 - V - 10$
9. $3x^2 + 7x + 4$
11. $t^2 - 8t + 16$
13. $4P^2 - 2P - 20$
15. $V_1^2 - 9$
17. $6X^2 - 7XR + 2R^2$
19. $9x^2 + 18x + 9$
21. $25L^2 - 4C^2$
23. $16y^2 - 9$
25. $3a^2 - 3b^2$
27. $0.8p^2 + pq + 0.3q^2$
29. $2C^2 + 4CL + 2L^2$
31. $E^3 - E^2 - E + 1$
33. $L^3 - 3L^2 + 5L - 6$
35. $T^3 - 1$
37. $(n + 3)(n - 2) = n^2 + n - 6$
39. $(l - 5)^2 = l^2 - 10l + 25$
41. $(t + 4)(2t + 2) = 2t^2 + 10t + 8$
43. $0.03T^2 + 0.28T - 24$

Exercise 8.2

1. (2)(2)(3)(3)
3. $(2)(2)(7)(x)(y)(y)$
5. $(-2)(2)(17)(a)(a)(b)(b)$
7. $(-3)(13)(R_1)(R_2)$
9. $(2)(5)(5)(I_1)(I_1)(R_1)$
11. $(2)(3)\left(\frac{1}{x}\right)$
13. $5(R + 3)$
15. $x(3 - a)$
17. $a(a - 2)$
19. $p^2(3p + 1)$
21. $0.5V(I_1 - 0.5I_2)$
23. $0.2r(s^2 + 2t^2)$
25. $3.3XY(2Y + 1)$
27. $2I^2(2R_1 + R_2)$
29. $\left(\frac{4}{b}\right)(3a + 4)$
31. $\left(\frac{4}{I}\right)(P_1 + 2P_2)$
33. $5(2s^2 + 4s - 3)$
35. $4v(6v^2 - 4v + 3)$
37. $6ab(4a^2 - 5ab + 7b^2)$
39. $xyz(x + y + z)$
41. $4w(w^2 + 7w + 10)$
43. $14R(4I_1^2 + 3I_2^2)$
45. $\left(\frac{e^{-1}}{R}\right)(V - V_0)$

Exercise 8.3

1. $(a + 2b)(a - 2b)$
3. $(X + 0.4)(X - 0.4)$
5. $(0.5X + Y)(0.5X - Y)$
7. $(Z + 1)(Z + 5)$
9. $(v - 1)(v - 3)$
11. $(C_1 + 3)^2$
13. prime
15. $(A - 4)^2$

17. $(n+1)(n-7)$
19. $50(m+2)(m-2)$
21. $(3G+1)(G+1)$
23. $(4e-1)(e-1)$
25. $(2x+1)(x-3)$
27. $(5x+6y)(x-y)$
29. prime
31. $4(2R-1)(R+4)$
33. $2(x+2y)(x-2y)$
35. $5P(2V+5)(V-1)$
37. $(S-0.3)^2$
39. $(L^2+1)^2$
41. $\left(\frac{1}{2}\right)(T+5)(T-5)$
43. $4(t+3)(t+4)$
45. $(0.5t-1.0)^2$
47. $R(I_1+I_2)(I_1-I_2)$
49. $X^2Y - X_0^2Y = Y(X+X_0)(X-X_0)$

Exercise 8.4

1. ± 2
3. $\pm\frac{3}{7}$
5. $5, -2$
7. $1, -\frac{1}{3} = -0.333$
9. $\frac{1}{2}, \frac{1}{2} = 0.50$
11. $2, \frac{2}{3} = 0.667$
13. $0, \frac{1}{7} = 0.143$
15. $\pm\frac{8}{5} = \pm 1.60$
17. ± 0.740
19. $-1, \frac{1}{2} = 0.50$
21. $2, -\frac{5}{3} = -1.67$
23. $0.193, -5.19$
25. $1.62, -0.618$
27. $2.41, -0.414$
29. $1.28, -0.781$
31. $3.65, -1.65$
33. $1.31, 0.191$
35. $1.58, 0.423$
37. 3.6 ft, 7.2 ft
39. 6 ft, 16 ft
41. $3500\ \Omega = 3.5\ \text{k}\Omega$
43. 5.0 A, 8.0 A
45. 0.89 A, 0.11 A

Review Exercises

1. $W^2 - 2W - 8$
3. $0.02k^2 + 0.03k - 0.02$
5. $4m^2 - 1$
7. $3X^2 - 3Y^2$
9. $I^3 - 2I^2 - 4I + 8$
11. $2x(x-2)$
13. $12C(A^2+B^2)$
15. $3z(7z^2+5z-9)$
17. $\left(\frac{3}{R}\right)(V_1+3V_2)$
19. $rst(r+s+t)$
21. prime
23. $(4x+5y)(4x-5y)$
25. $\left(10+\frac{A}{2}\right)\left(10-\frac{A}{2}\right)$
27. $(P+7)(P-3)$
29. $(X+0.1)(X+0.2)$
31. $(2q-3)(q-1)$
33. $(2u-3v)(3u-2v)$
35. $5(5X-2)(X+1)$
37. $\left(\frac{1}{10}\right)(s+3)(s-3)$
39. ± 4
41. $5, -1$
43. $0, \frac{7}{8} = 0.875$
45. ± 2.10
47. $5, -10$
49. $2, 2$
51. $3.73, 0.268$
53. $2.62, 0.382$
55. $0.809, -0.309$
57. $1.70, -2.68$
59. $(3l-2)(2l+3) = 6l^2 + 5l - 6$
61. $6(t+3)(t+4)$
63. 22 cm, 27 cm
65. $(2t-3)(t+5) = 2t^2 + 7t - 15$
67. $25R(3I_1 + 5I_2)$
69. 5.0 A
71. 3.1 kΩ
73. 2 A, 4 A

CHAPTER 9

Exercise 9.1

1. $\frac{5}{2x}$
3. $\frac{7a^2}{9}$
5. $\frac{a+b}{a}$
7. $\frac{L}{M-N}$
9. $\frac{b+d}{c}$
11. n
13. $\frac{R_1-R_2}{5}$
15. 3
17. not reducible
19. $\frac{2V+1}{2}$
21. $\frac{t+3}{3t-1}$
23. not reducible
25. $\frac{2a+x}{2a-x}$
27. $\frac{7X_L-1}{9X_L-4}$
29. $\frac{2(2f-3)}{f-3}$
31. $\frac{3F_x^2-3F_y^2}{1.5F_x-1.5F_y} = 2(F_x+F_y)$
33. $2t+1$

Exercise 9.2

1. $\frac{6}{x}$
3. $\frac{1}{y}$
5. $\frac{1}{2}$
7. $\frac{5b}{a}$
9. $\frac{5g(g-h)}{14h}$
11. $\frac{s-t}{s}$
13. 4
15. $\frac{x^2+y^2}{2}$
17. $\frac{I_2}{5(I_1-I_2)}$
19. $2(a+1)$
21. $5(2D-1)$
23. $\frac{R_2}{R_1^2}$

25. $\frac{2x(y+2)}{3}$

27. $\frac{10^3}{K}$

29. 1

31. $\frac{4r}{3\pi}$

33. $P = VI = V\left(\frac{V}{R}\right) = \frac{V^2}{R}$

35. $\frac{3R_1}{4}$

Exercise 9.3

1. $\frac{4x}{5}$
3. $\frac{1}{3a}$
5. $\frac{5R+9}{60}$
7. $\frac{35-12I}{40}$
9. $\frac{17}{63z}$
11. $\frac{37p}{45}$
13. $\frac{5t}{2s}$
15. $\frac{6-8w}{rw^2}$
17. $\frac{4V_1^2+3V_2^2}{V_1V_2}$
19. $\frac{2X-3XY}{12Y^2}$
21. $\frac{27+25d}{30cd^2}$
23. $\frac{25}{4\times10^4} = \frac{1}{16\times10^2}$
25. $\frac{x-a}{a(a+x)}$
27. $\frac{5x+2}{x(x+1)}$
29. $\frac{4+3h-h^2}{h(h+2)}$
31. $\frac{2d^2+6cd-5c^2}{2d^2}$
33. $\frac{5xy-2}{10y^2}$
35. $\frac{8.1+2.3V}{I(1+V)}$
37. $\frac{4V+1}{2(V+1)}$
39. $\frac{-2}{(e+1)(e-1)}$
41. $\frac{p^2-5}{2(p-1)}$
43. $\frac{4-t}{(t+1)(t-1)}$
45. $\frac{2}{3}$
47. $\frac{-1}{r-1} = \frac{1}{1-r}$
49. $\frac{7I_1^2+9I_1I_2-I_2^2}{I_1I_2(I_1+I_2)}$
51. $\frac{a^2+b^2}{b^2} = \frac{c^2}{b^2}$
53. $\frac{R_1R_2}{R_1+R_2}$
55. $\frac{6R_2+3R_1}{R_1R_2}$
57. $\frac{R_1R_2R_3}{R_1R_2+R_1R_3+R_2R_3}$

Exercise 9.4

1. 12
3. 20
5. 2
7. 1
9. 5
11. 0.5
13. 100
15. 2.5
17. $\frac{1}{5} = 0.2$
19. −4
21. 6
23. 12
25. 0.4
27. $\frac{1}{4} = 0.25$
29. 30
31. 4.5
33. 4
35. 3.2
37. −4
39. 3
41. $\frac{1}{2} = 0.5$
43. 7
45. $100
47. 260 Ω
49. 1.3 kΩ = 1300 Ω
51. 25

Review Exercises

1. $5x^2$
3. $\frac{5}{a+4b}$
5. not reducible
7. $\frac{2f-1}{2f+1}$
9. $\frac{16x}{45}$
11. $\frac{9V}{10}$
13. 1
15. 3
17. $\frac{R_1+2R_2}{2R_1+R_2}$
19. 5
21. $\frac{21}{5x}$
23. $\frac{23}{15r}$
25. $\frac{15b-6a}{14a^2b^2}$
27. $\frac{9x-1}{2x(x-1)}$
29. $\frac{15V-25}{(V+5)(V-5)}$
31. $\frac{0.66-0.44r}{t(1-r)}$
33. $\frac{-11m-15}{10m}$
35. 8
37. 150
39. 7
41. 6.7
43. −2
45. 2
47. 4.5h
49. t^2
51. $\frac{R_2(2R_2+3R_3)}{R_2+R_3}$

53. 60 μF

55. 2.0 mH

57. 100 Ω

CHAPTER 10

Exercise 10.1

1. $I = \frac{P}{V}$

3. $r = \frac{C}{2\pi}$

5. $P = I^2R$

7. $m = \frac{F}{a}$

9. $f = \frac{1}{2\pi X_C C}$

11. $V = \sqrt{PR}$

13. $r = \sqrt{\frac{A}{\pi}}$

15. $r_g = \frac{V - IR}{I}$

17. $w = \frac{P - 2l}{2}$

19. $I = \frac{V - V_0}{R}$

21. $V = \frac{P}{I_1 + I_2}$

23. $w = \frac{A}{l_1 + l_2}$

25. $I = \frac{V_T}{R_1 + R_2 + R_3}$

27. $\beta = \frac{I_C}{I_E - I_C}$

29. $R_2 = \frac{R_{eq}R_1}{R_1 - R_{eq}}$

31. $L_2 = \frac{L_T L_1}{L_1 - L_T}$

33. $L = \frac{Z}{\omega(1 + \omega CZ)}$

35. (a) $h = 116{,}600 – 550T$
(b) 1100 ft

37. (a) $C = \left(\frac{5}{9}\right)(F - 32)$

(b) 25°C

39. $P = (IR)I = I^2R$

41. $P = V\left(\frac{V}{R}\right) = \frac{V^2}{R} \rightarrow \left(\frac{R}{P}\right)P$
$= \left(\frac{R}{P}\right)\left(\frac{V^2}{R}\right) \rightarrow R = \frac{V^2}{P}$

43. (a) $R_1 = \frac{V - V_0}{I}$

(b) 6.0 Ω

45. $R_x = \frac{R_1 R_{eq}}{R_1 - R_{eq}}$

47. (a) $\alpha = \frac{R_t - R_0}{R_0 t}$

(b) $\frac{0.0031}{°C}$

49. (a) $\rho = \frac{RA}{l}$

(b) 2.1×10^{-8} Ω–m

51. $f = \frac{1}{2\pi\sqrt{LC}}$

53. (a) $-16V$
(b) $\Delta V = R(\Delta I)$
(c) $\Delta V = 80(-0.2) = -16$ V
(d) $\Delta V = 80(0.4) = 32$ V

55. (a) $\Delta V = IR$
(b) $\Delta V = (-0.1)IR$

Exercise 10.2

1. $42,500, $53,500

3. $53,182

5. $42

7. 89

9. (a) $2608.70
(b) $260.87

11. 1925 lines/min

13. 8 mi/h

15. 1.5 h

17. 2.5 h

19. 5.2 mH, 5.8 mH

21. 50 mA, 75 mA, 150 mA

23. 6.0 V, 5.7 V

25. 200 cm, 150 cm

27. 21 V, 42 V, 7.0 V

29. 4 A, 2 A, 1 A

31. $V_R = 4.8$ V, $V_B = 7.2$ V

33. 36

Review Exercises

1. $Q = \frac{Er^2}{k}$

3. $r = \sqrt{\frac{Q_1Q_2}{F}}$

5. $a = \frac{2(d - v_0 t)}{t^2}$

7. $R_1 = \frac{V - IR_2}{I}$

9. $L_1 = \frac{L_T L_2}{L_2 - L_T}$

11. (a) $r = \sqrt{\frac{V}{\pi h}}$

(b) 2

13. $500

15. 2.0 h

17. 2.6×10^8 m/s

19. (a) $I = \frac{V_2 - V_1}{R_1 + R_2}$

(b) 0.80 A = 800 mA

21. (a) $R_2 = 2(R_T - R_1)$
(b) 60 Ω

23. 42 V

25. 4.8 ft, 2.9 ft, 2.4 ft

27. (a) $\Delta I_T = \frac{\Delta V}{R_0}$

(b) 30 mA

CHAPTER 11

Exercise 11.1

1. 300 Ω, 80 mA, 8 V, 16 V

3. 4.5 kΩ, 33 mA, 50 V, 100 V

5. 60 Ω, 40 Ω, 20 Ω, 80 V, 40 V

7. 960 V, 480 V, 480 V, 1.5 kΩ, 1.5 kΩ

9. 100 mA, 120 Ω, 180 Ω, 12 V, 18 V

11. 70 Ω, 140 mA, 2.1 V, 3.1 V, 4.7 V

13. 100 mA, 200 mW, 240 mW, 360 mW, 800 mW

15. 7.2 V, 1.6 mW, 2.4 mW, 2.4 mW, 6.5 mW

17. 4.0 V, 8.0 V, 16 V, 430 Ω

19. 20 W, 20 W, 20 W, 1.4 A

21. 240 Ω

23. 160 Ω

25. 15 V, 30 V, 27 W

27. 45 V or 15 V

29. 2.7 kΩ

Exercise 11.2

1. 7.5 Ω

3. 230 Ω

5. 28 kΩ

7. 12 Ω

9. 460 Ω

11. 150 Ω

13. 30 Ω

15. 130 Ω

17. 1.9 kΩ

19. 1.2 kΩ

21. 130 mS, 7.5 Ω

23. 9.1 mS, 110 Ω

25. 170 mS, 6.0 Ω

27. 11 mS, 94 Ω

29. 2.5 mS, 390 Ω

31. 300 Ω

33. 1.0 kΩ

35. 11 Ω

37. 20 Ω

39. 150 Ω

41. $I_1 = 500$ mA, $I_2 = 400$ mA, $I_T = 900$ mA, $R_T = 13\ \Omega$

43. $V = 10$ V, $R_1 = 5\ \Omega$, $R_2 = 10\ \Omega$, $R_T = 3.3\ \Omega$, $I_1 = 2.0$ A, $I_T = 3.0$ A

45. $I_1 = 330$ mA, $I_2 = 250$ mA, $I_3 = 280$ mA, $I_T = 860$ mA, $G_T = 17$ mS, $R_T = 58\ \Omega$

47. $R_1R_2 = R_TR_2 + R_TR_1 \rightarrow R_1R_2 = R_T(R_1 + R_2) \rightarrow$

$$R_T = \frac{R_1R_2}{R_1 + R_2}$$

49. $R_{eq} = \frac{(100)(150)}{250} = 60\ \Omega;$

$$\frac{60}{100} = 60\%, \quad \frac{60}{150} = 40\%$$

Exercise 11.3

1. 35 Ω

3. 5.2 kΩ

5. 370 Ω

7. 30 Ω

9. 13 Ω

11. $R_T = \frac{R_1R_2}{R_1 + R_2} + R_3;\ R_T = 180\ \Omega$

13. $I_T = 26$ mA, $V_2 = 6.2$ V, $V_4 = 2.6$ V, $I_4 = 8.7$ mA

15. $R_T = 240\ \Omega$

17. $R_1 = 20\ \Omega$

19. $R_T = 50\ \Omega$

21. 46 mA, $V_1 = 4.6$ V, $V_2 = 1.4$ V

23. 2.9 W

Review Exercises

1. 51 Ω, 470 mA, 16 V, 8.5 V

3. 150 Ω, 100 Ω, 50 Ω, 4 V, 2 V, 160 mW, 80 mW, 240 mW

5. 1.1 kV, 690 W, 1.0 kW, 1.1 kW, 2.8 kW

7. 13 Ω, 26 Ω, 52 Ω

9. **(a)** $R_D = 3.4\ \Omega$ **(b)** 15 Ω

11. 69 mS, 15 Ω

13. 16 mS, 62 Ω

15. 290 Ω

17. 110 mS, 9.0 Ω

19. 18 mS, 55 Ω

21. 450 Ω

23. 110 Ω

25. 1.2 A, 590 mA, 400 mA, 2.2 A, 18 Ω, 55 mS

27. **(a)** $R_T = \frac{(R_1 + R_2)R_3}{(R_1 + R_2) + R_3}$ **(b)** 94 Ω

29. $R_T = 50\ \Omega$, $V_1 = 25$ V, $V_3 = 14$ V, $P_T = 13$ W

31. **(a)** $R_T = \frac{(R_2 + R_3)R_4}{R_2 + R_3 + R_4}$ **(b)** $R_T = 1.1$ kΩ, $V_T = 39$ V, $V_1 = 8.4$ V

33. $\frac{1}{R_T} = \frac{2}{R_1} + \frac{1}{2R_1} = \frac{5}{2R_1} \rightarrow R_T = \frac{2R_1}{5}$

CHAPTER 12

Exercise 12.1

1. through 16.

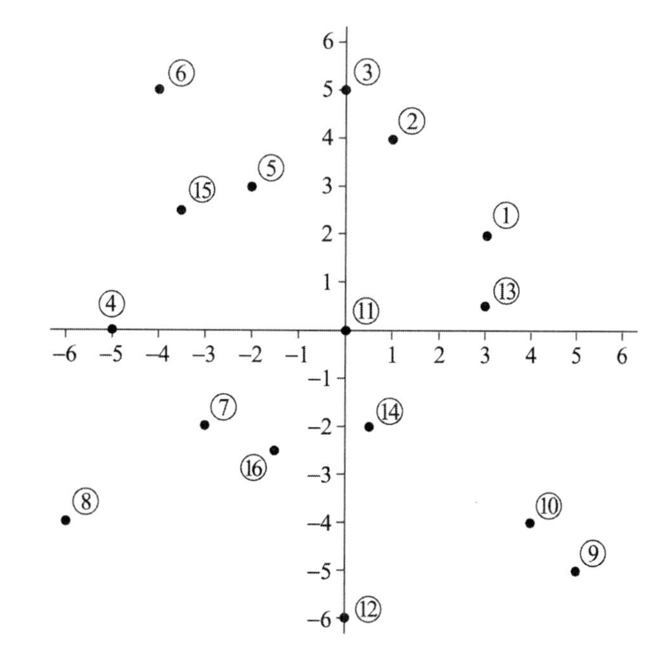

17. $x - y = 3$

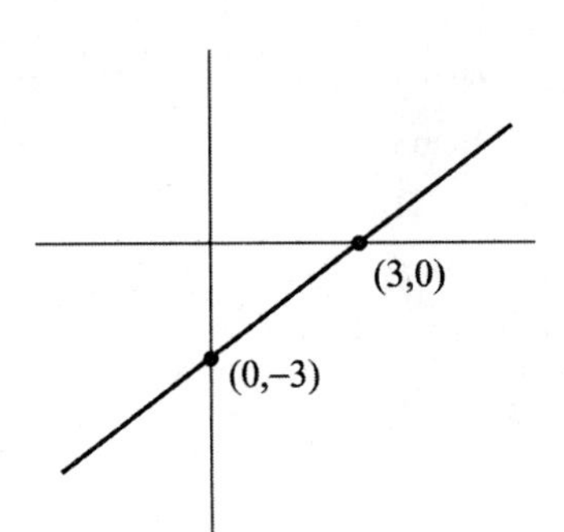

19. $x - 2y = 4$

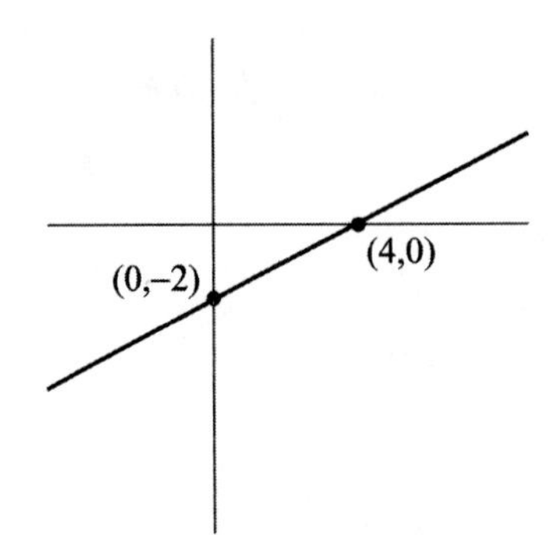

21. $x + y = 0$

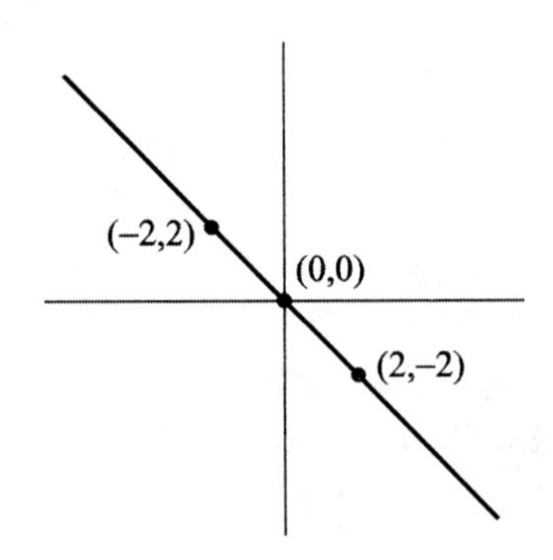

23. $2x - 3y = 6$

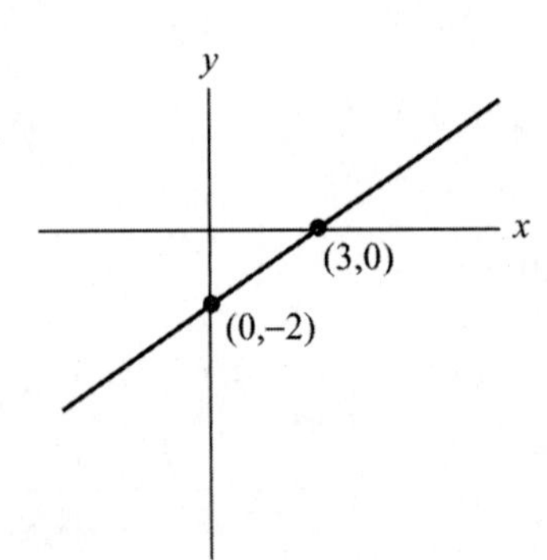

25. $3x + y = 6$

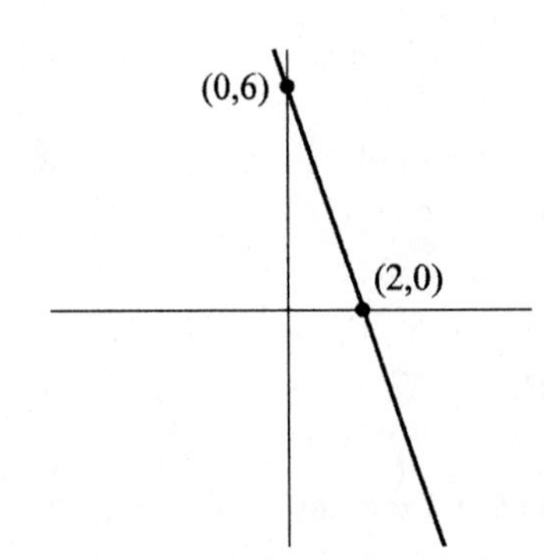

27. $x + 2y = 3$

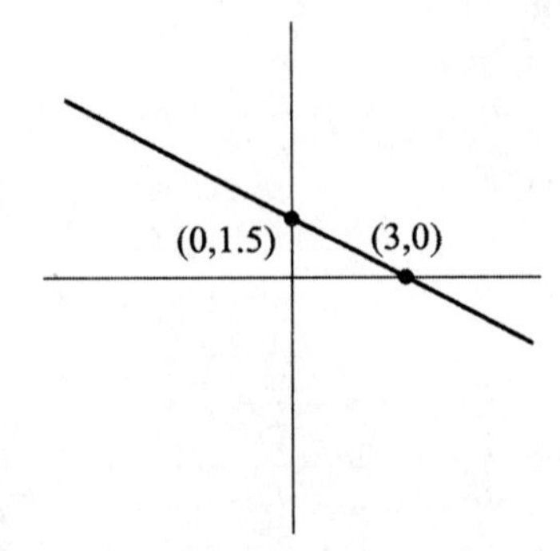

29. $y = 2x - 2$

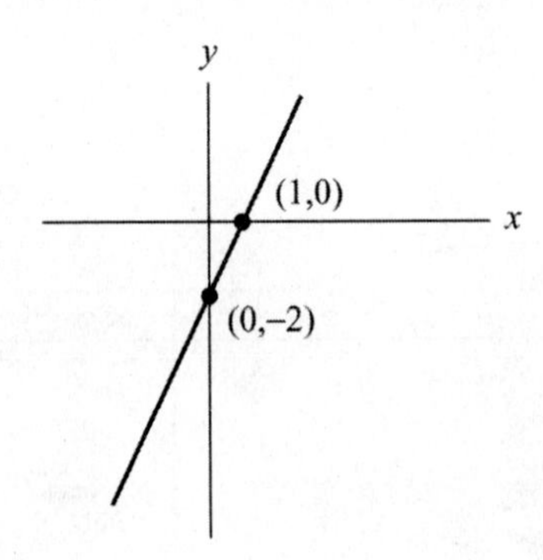

31. $2a + b = 4$

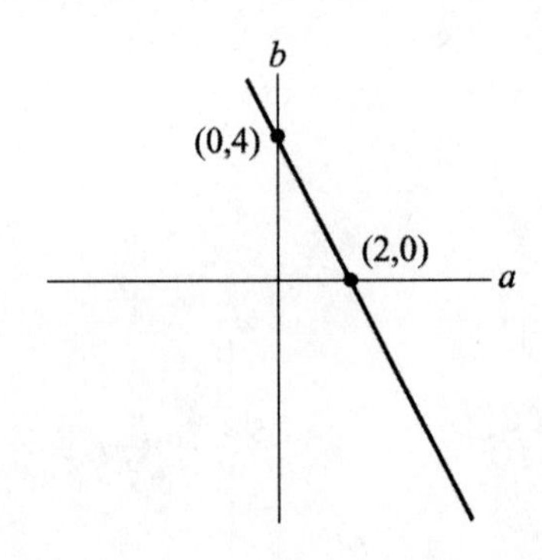

33. $5r - 2t = 10$

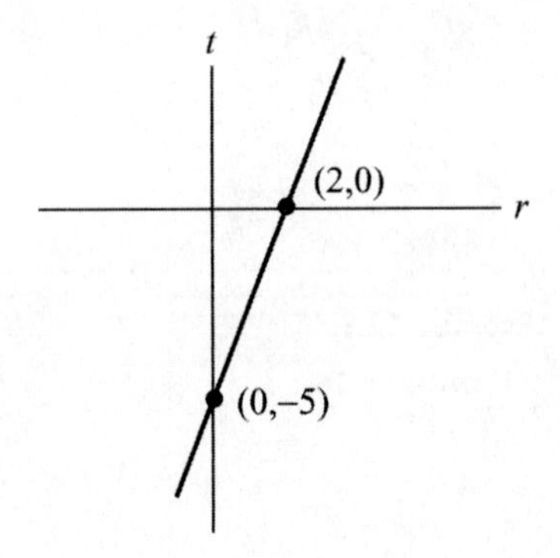

35. $I = \frac{V}{10}$

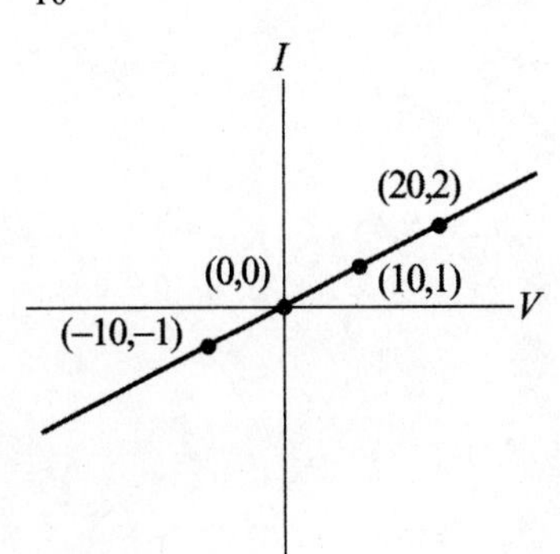

37. $P - 20I = 0$

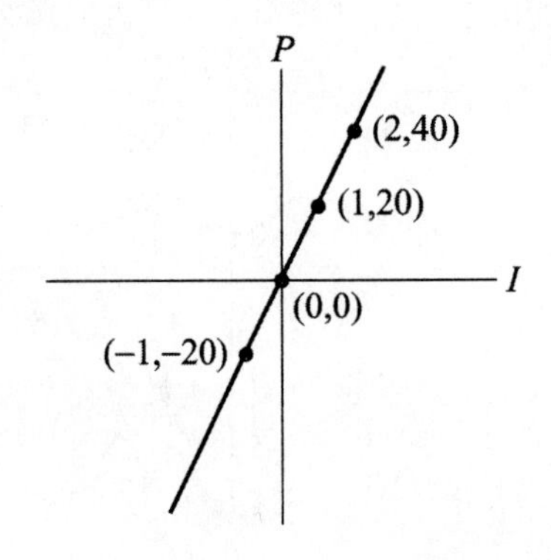

39. $2V_1 + V_2 = 5$

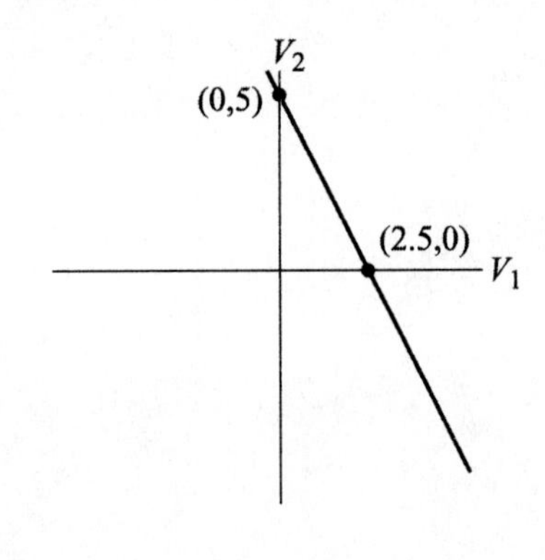

41. $I = \dfrac{V}{12\ \Omega}$

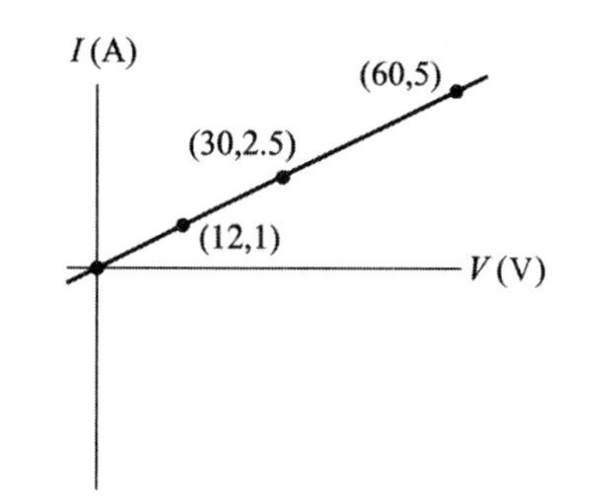

43. $P = 120I$

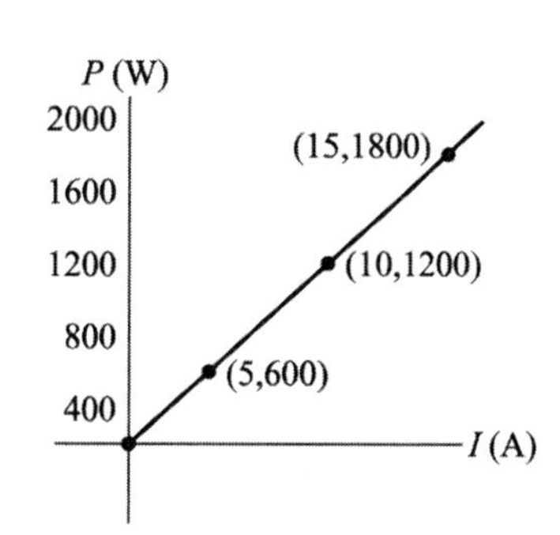

45. $P = 200$ V

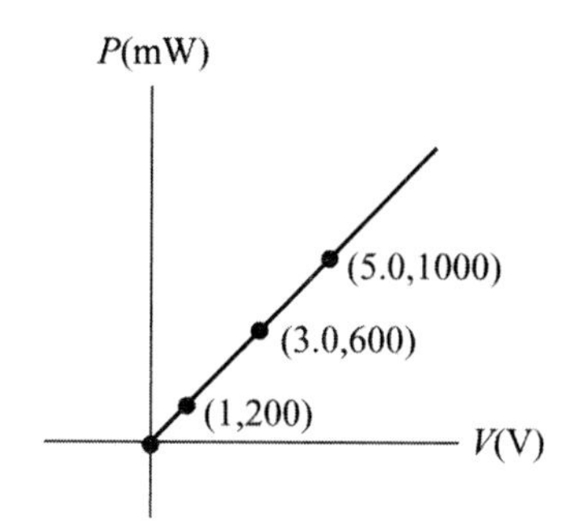

Exercise 12.2

1. $y = x$

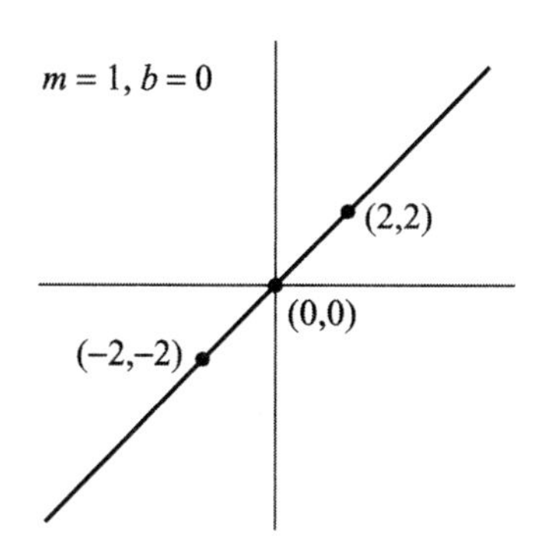

3. $y = -3x$

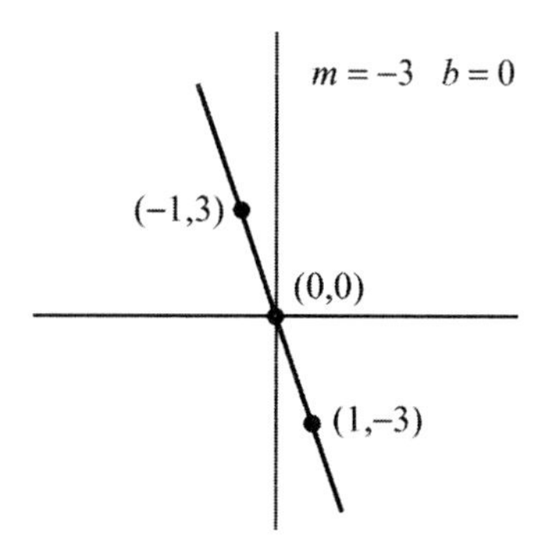

5. $y = 2x - 2$

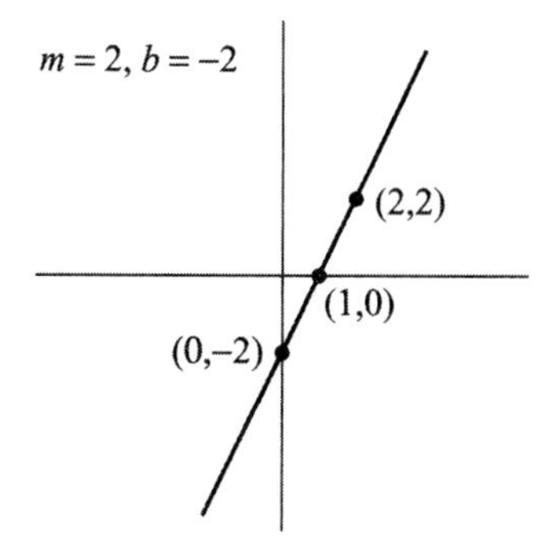

7. $y = -2x - 2$

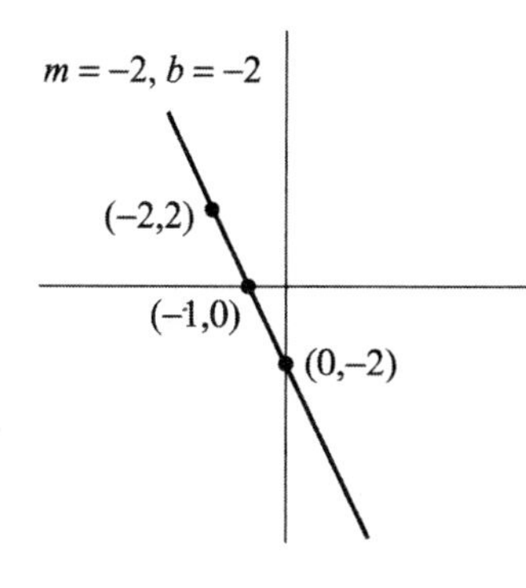

9. $y = x - 3$

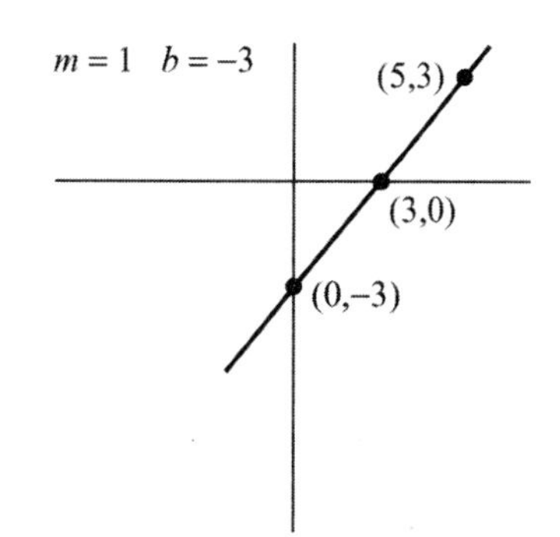

11. $y = -x - 4$

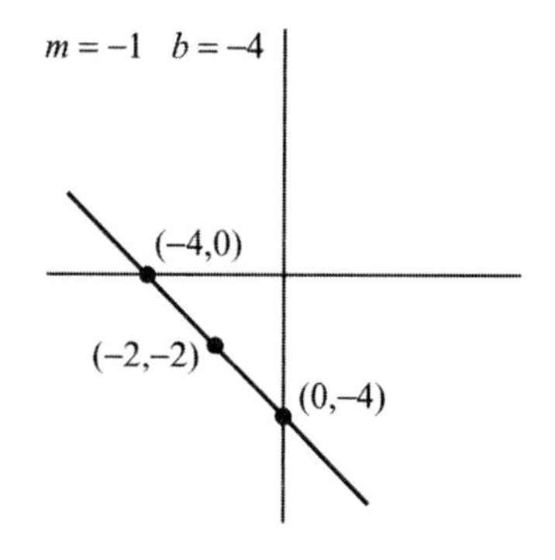

13. $y = -4x - 4$

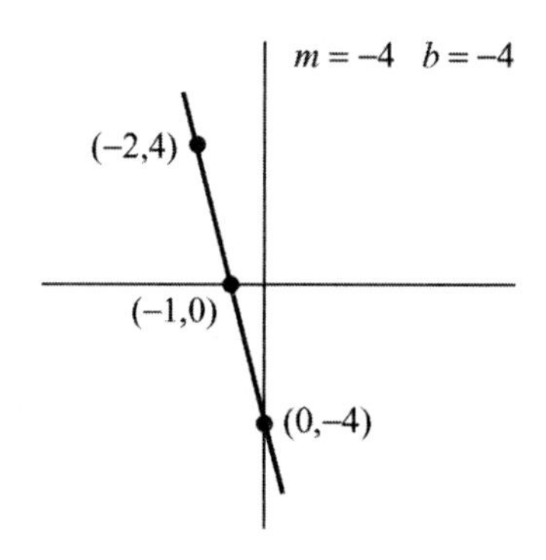

15. $y = 3x - 4$

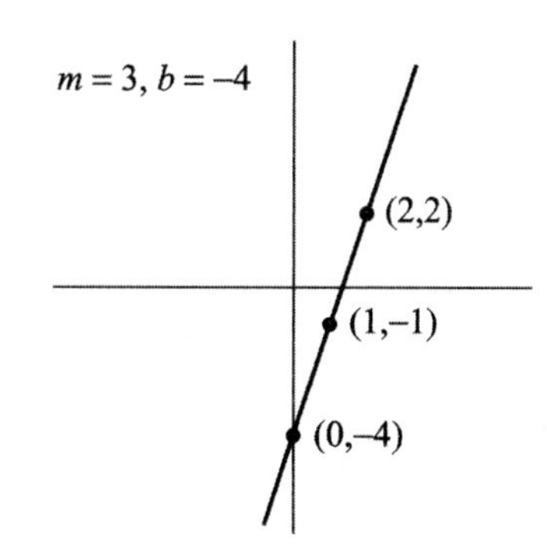

17. $y = \frac{3}{2}x$

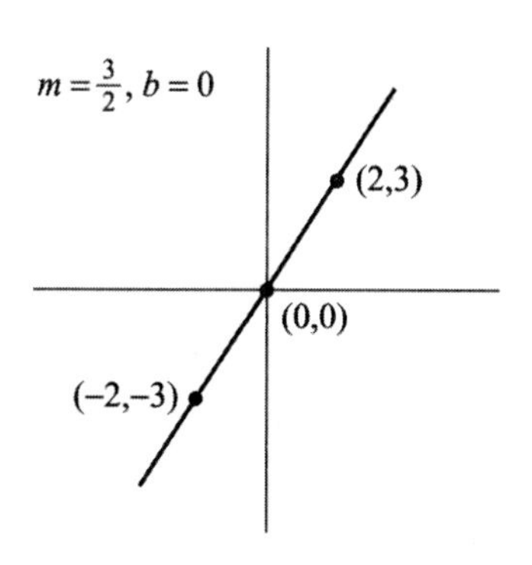

19. $y = -\frac{1}{2}x$

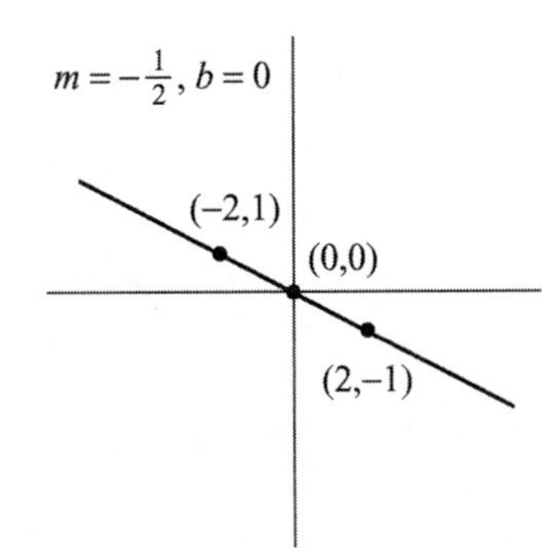

21. $y = x + \frac{3}{2}$

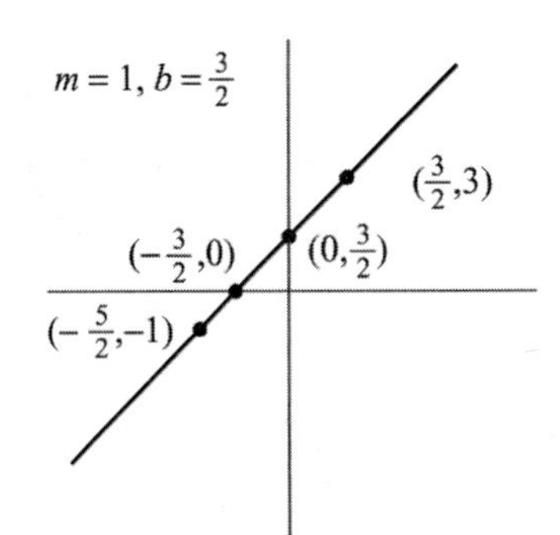

23. $y = \frac{3}{2}x - 3$

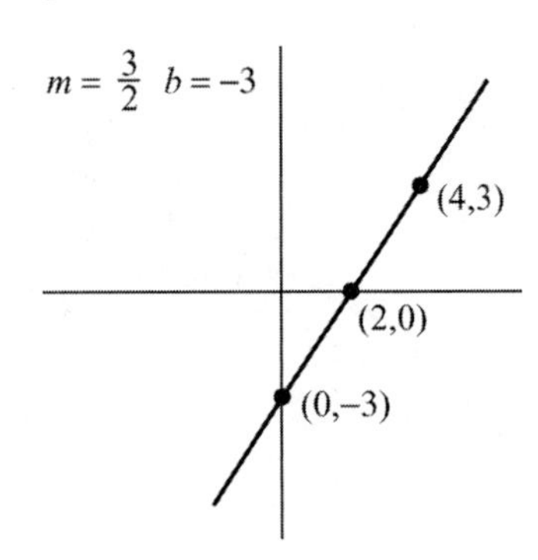

25. $P = 3.5I$

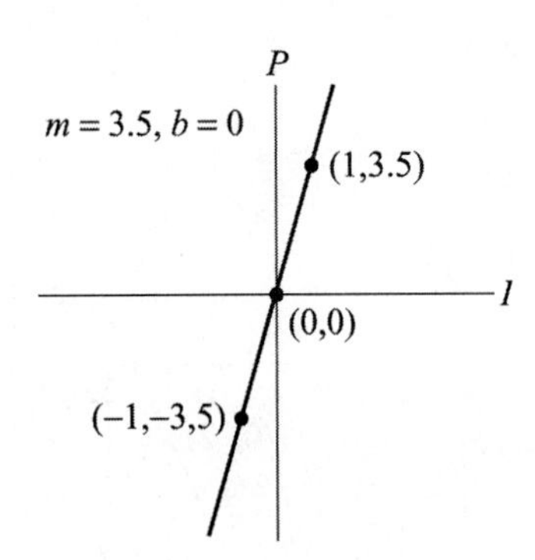

27. $I = 0.9V$

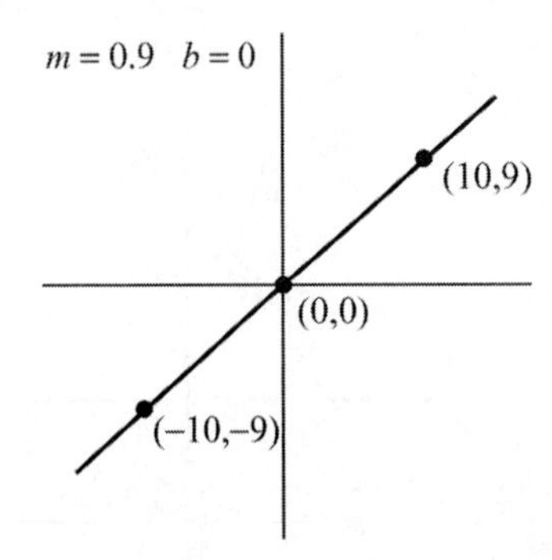

29. $V = 3.2R + 2.3$

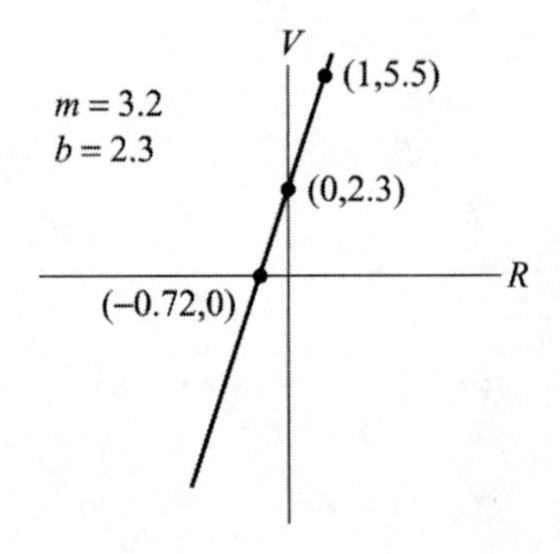

31. $v = 4.0z - 3.0$

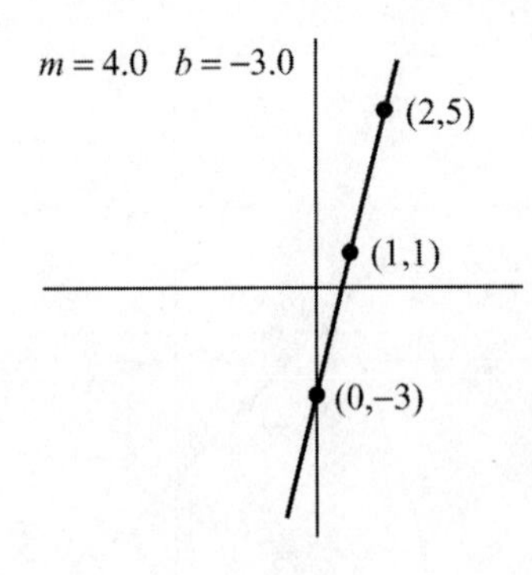

33. $R = -1.6t + 5.0$

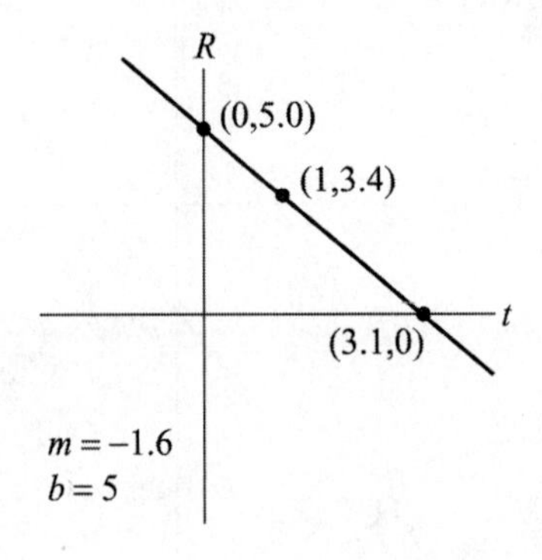

35. $I_2 = 0.5I_1 + 1$

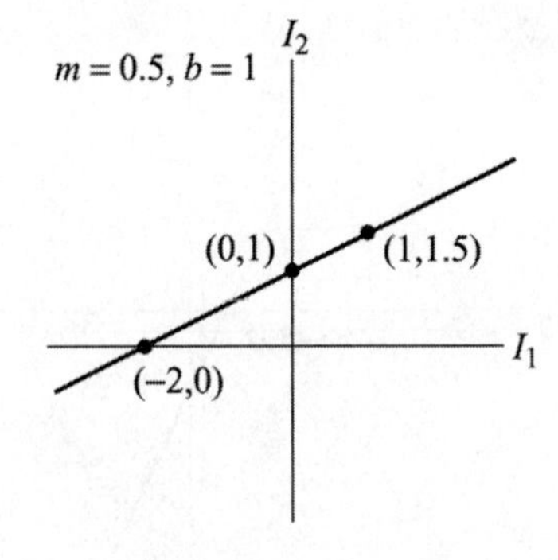

37. $v = 100 + 50t$

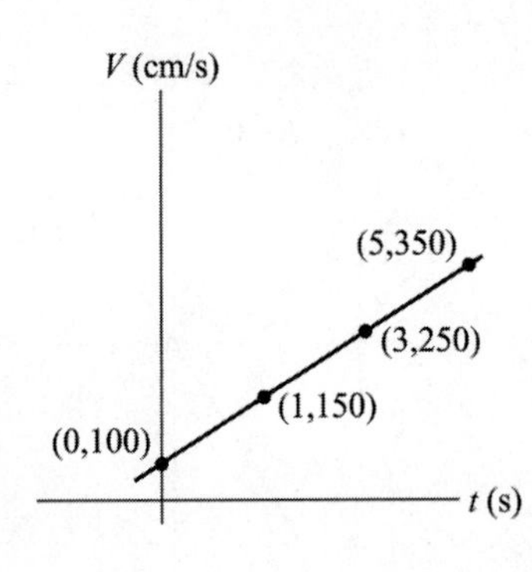

39. **(a)**

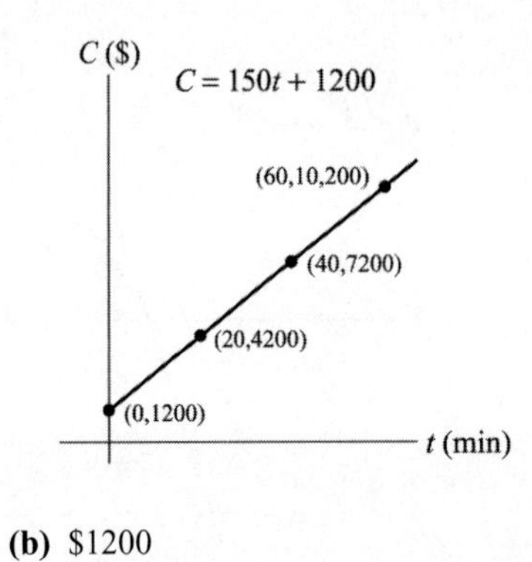

(b) $1200

41. Slope $= \dfrac{I}{R} = 0.0067$

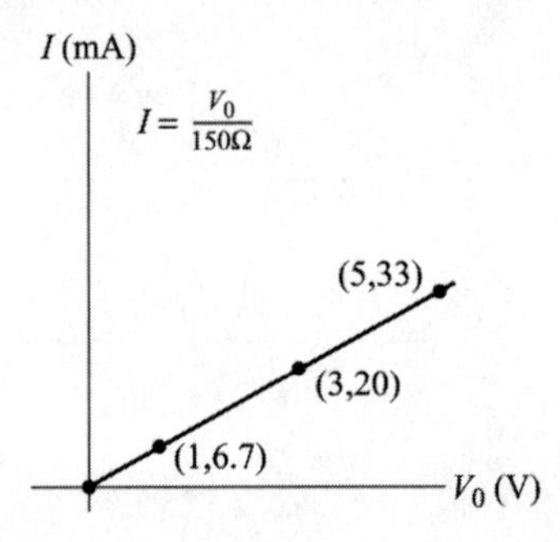

43. $V_L = 9.0 - V_S$

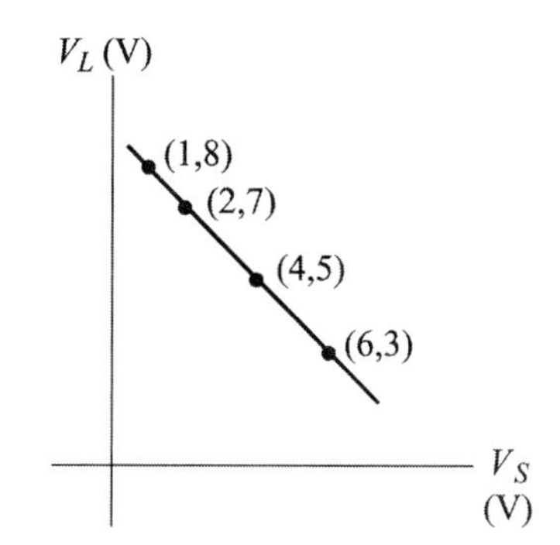

45. $R_t = 0.03t + 15$

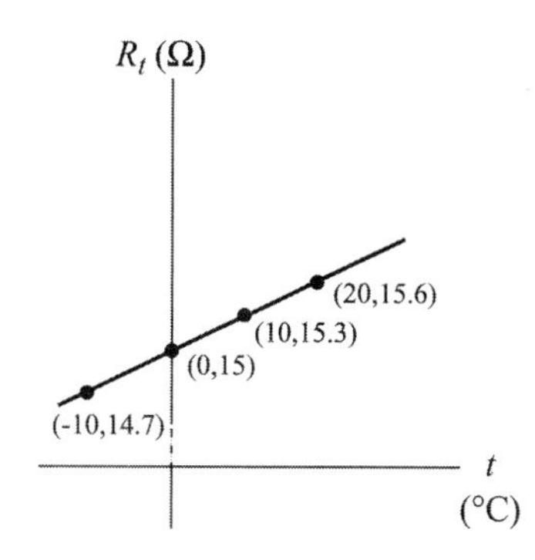

Exercise 12.3

1. $y = x^2$

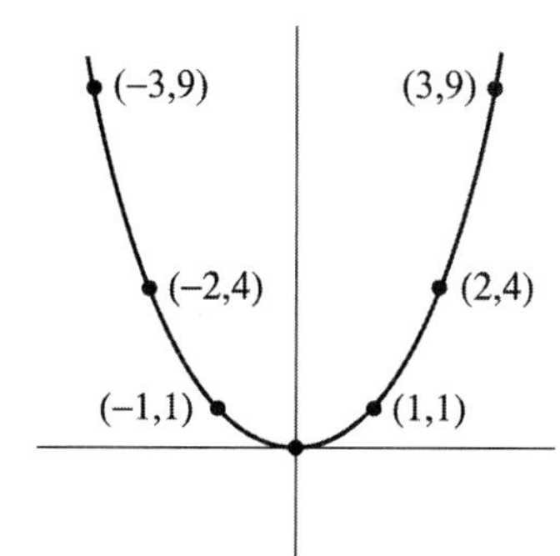

3. $y = 0.5x^2$

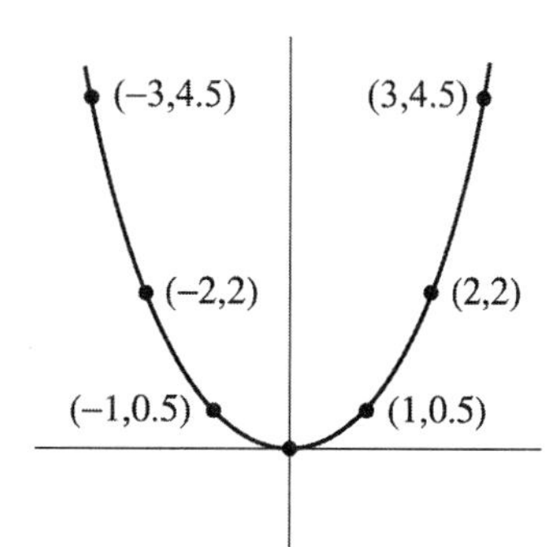

5. $y = 1.2x^2$

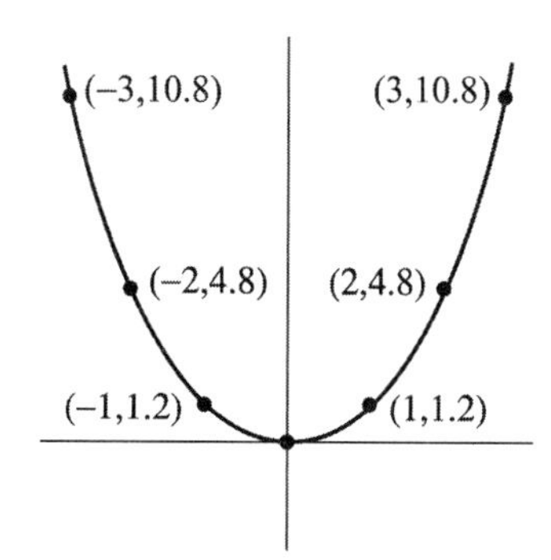

7. $y = 1.5x^2$

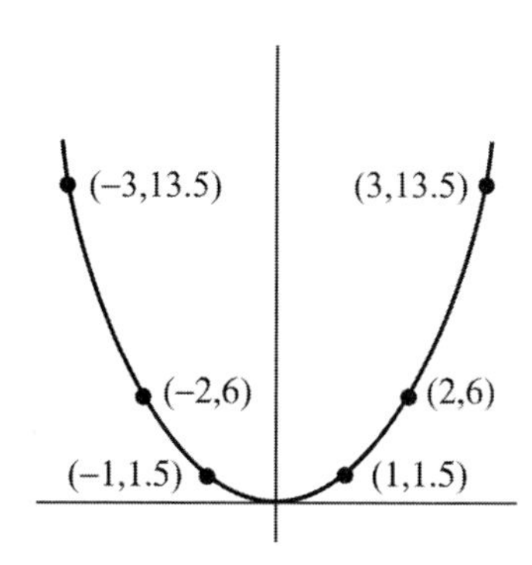

9. $y = -2x^2$

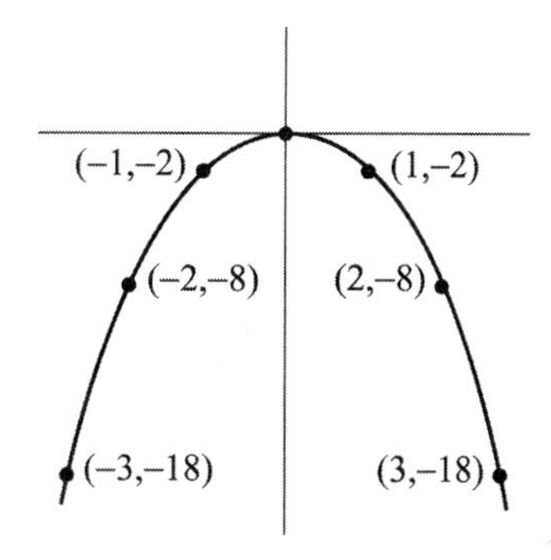

11. $y = -0.6x^2$

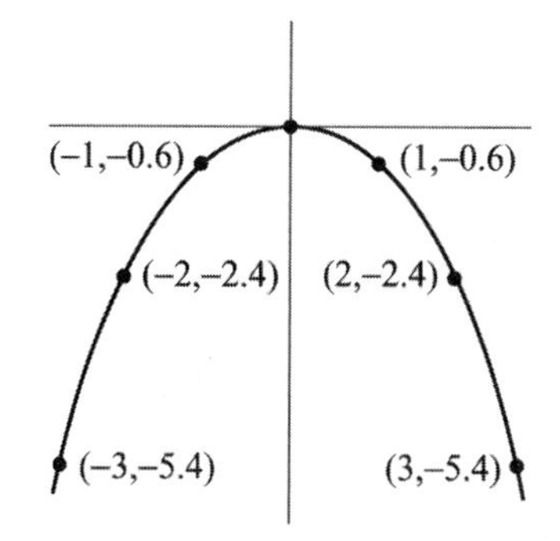

13. $P = 10I^2$

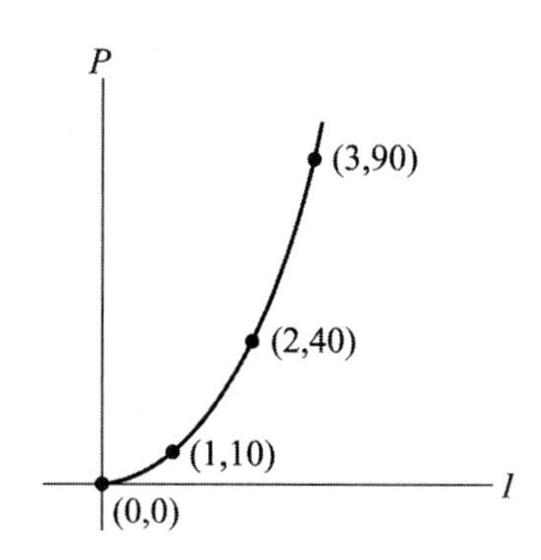

15. $P = 75I^2$

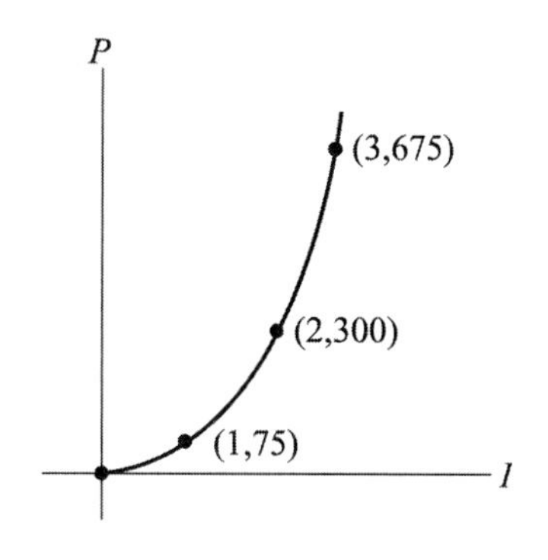

17. $P = 0.1V^2$

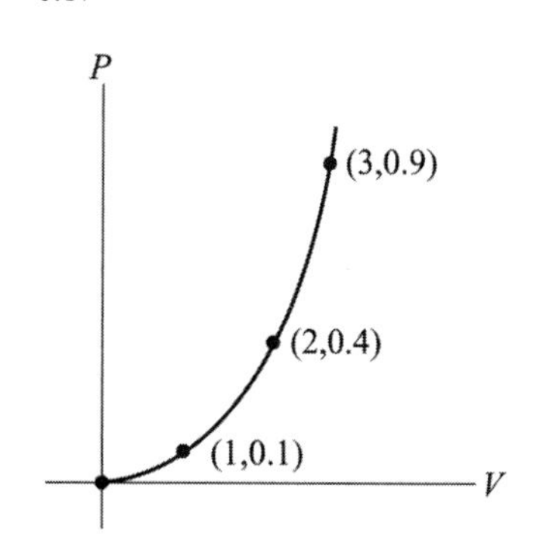

19. $P = 0.05V^2$

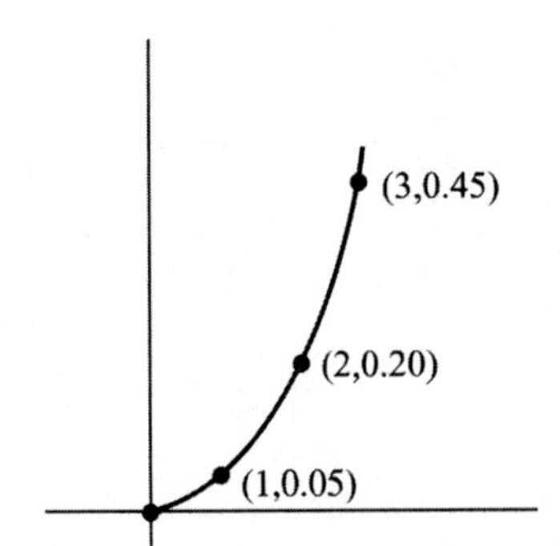

21. $I = \frac{12}{R}$

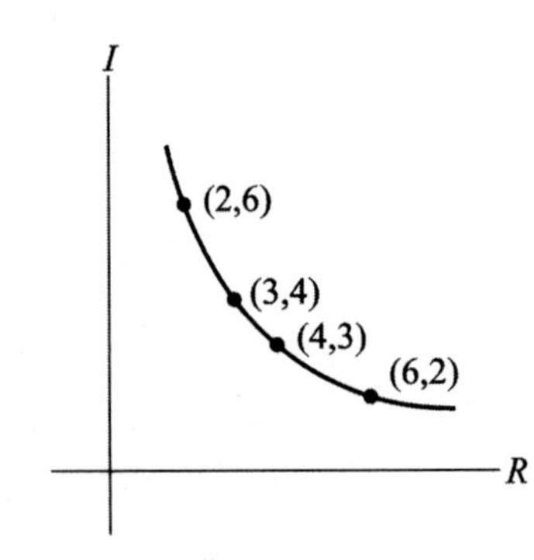

23. $I = \frac{220}{R}$

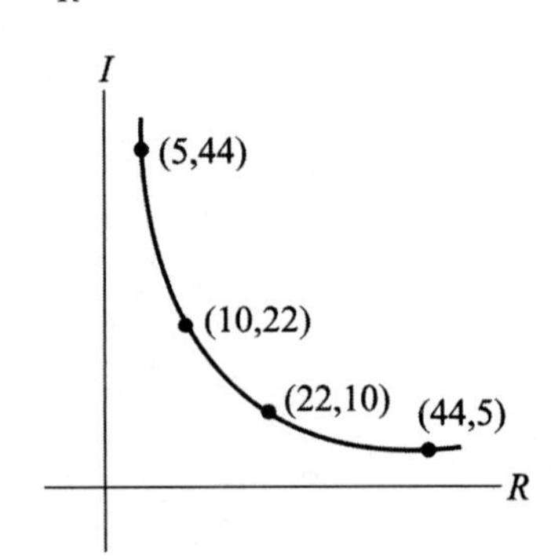

25. $I = \frac{9.2}{R}$

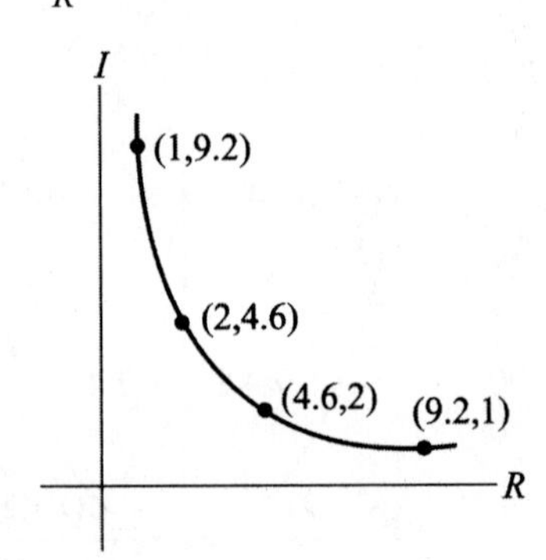

27. $I = \frac{5.5}{V}$

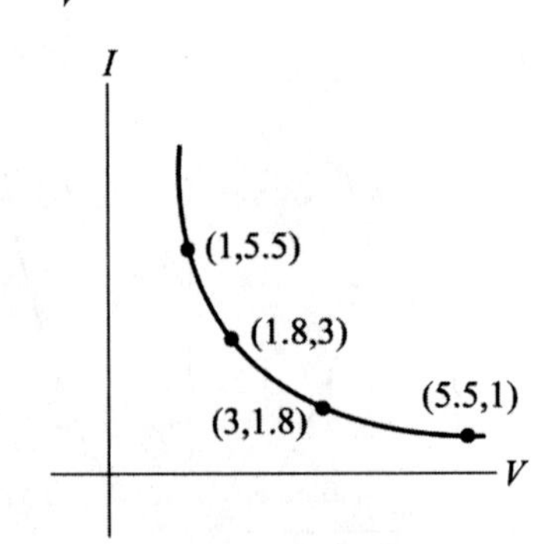

29. $d = 16t^2$

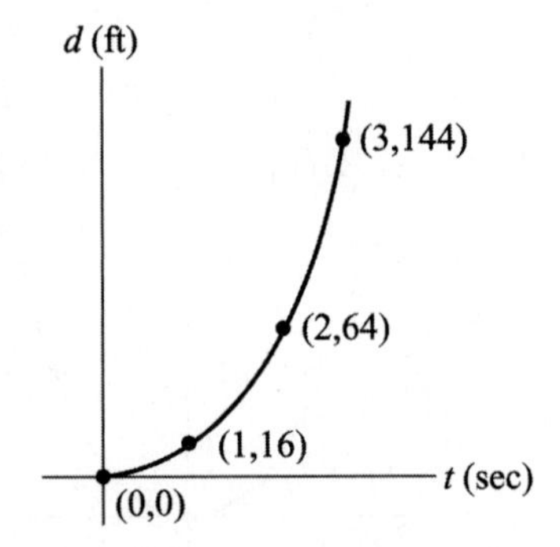

31. $P = 22I^2$

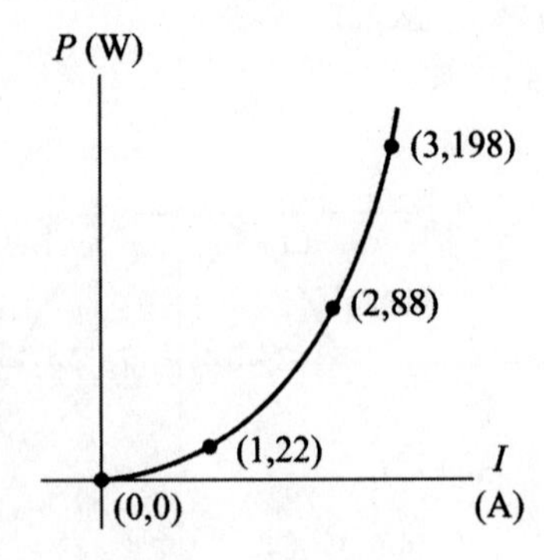

33. $P = \frac{V^2}{200}$

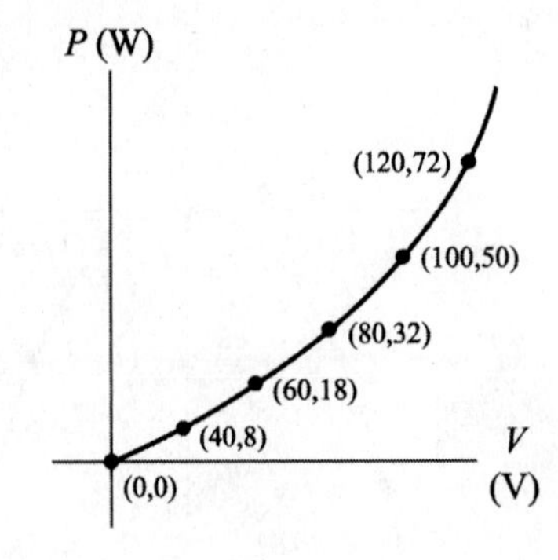

35. $IR = 9.0$

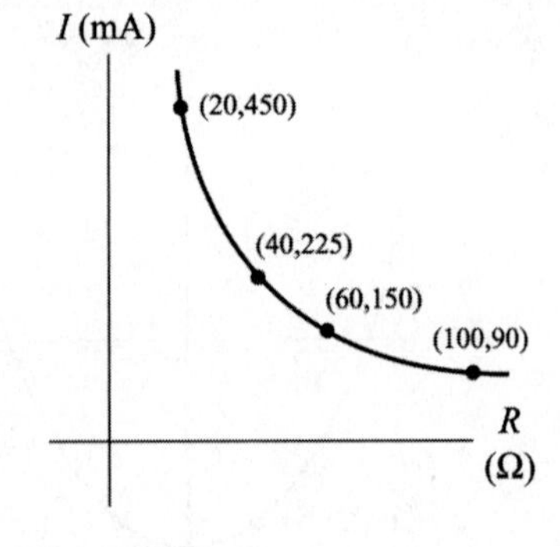

37. $PR = 6400$

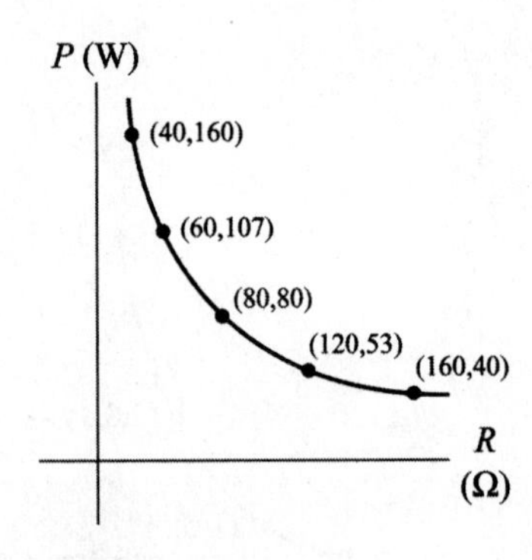

39. (a)

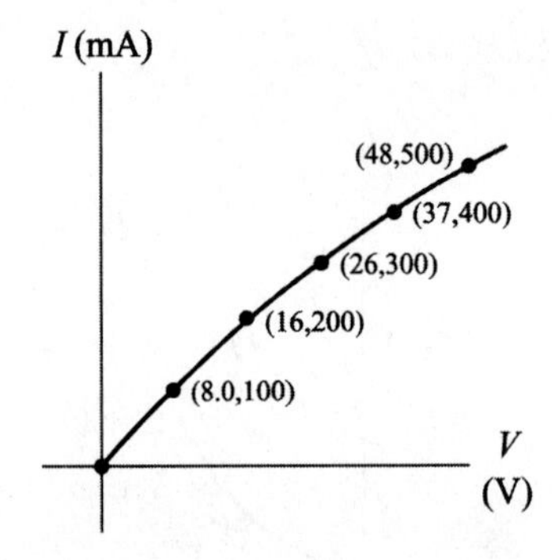

(b) 80Ω, 80 Ω, 87 Ω, 93 Ω, 96 Ω

41. (a)

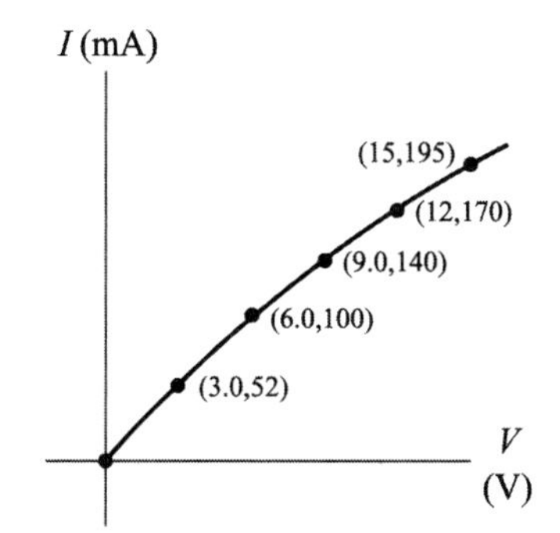

(b) 58 Ω, 60 Ω, 64 Ω, 71 Ω, 77 Ω

43.

R_t (Ω)
(20,15.6)
(10,13.5)
(15,14.5)
(5,12.7)
(0,12.0)
(−5,11.4)
t
(°C)

Review Exercises

1. $2x + y = 2$

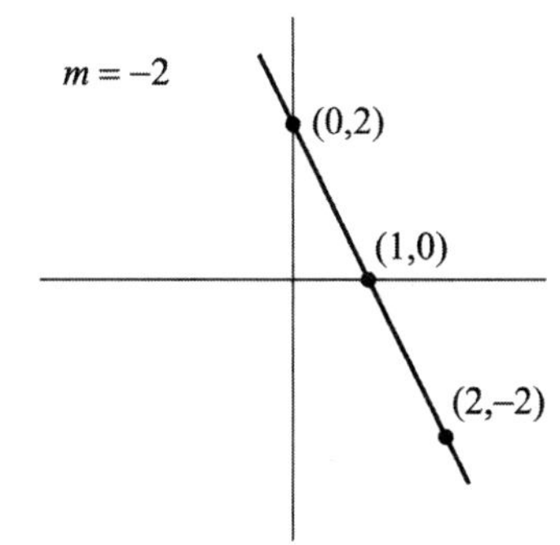

3. $2x - 5y = 5$

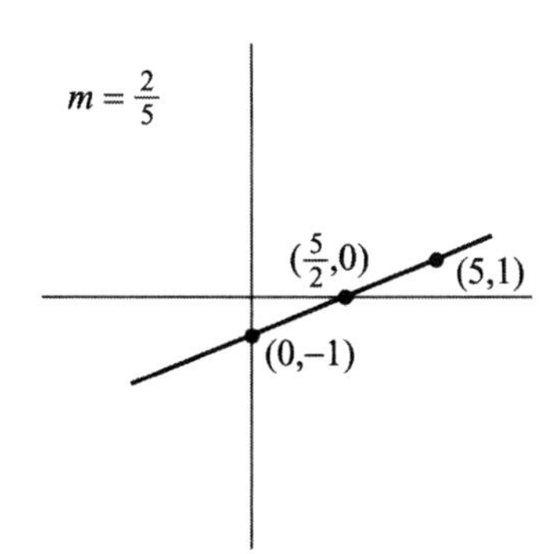

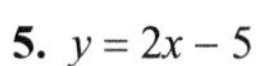

5. $y = 2x - 5$

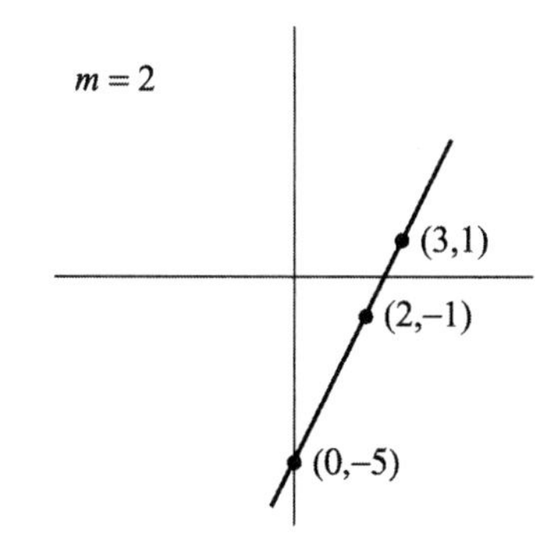

7. $y = -x + 2$
$m = -1$

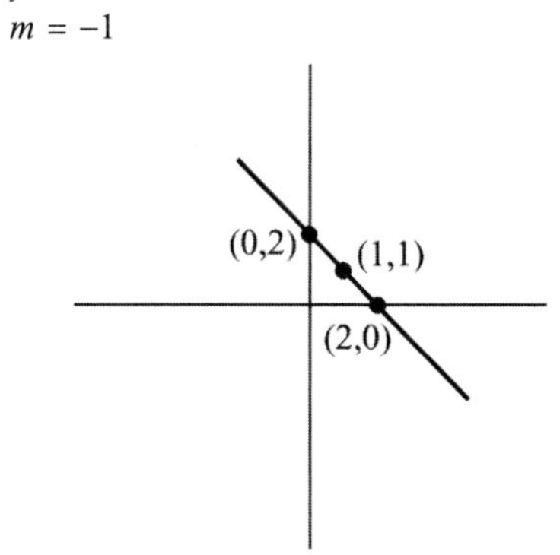

9. $5x + 2y = 0$

$m = -\frac{5}{2}$
(−2,5)
(0,0)
(2,−5)

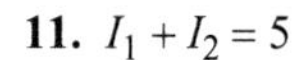

11. $I_1 + I_2 = 5$

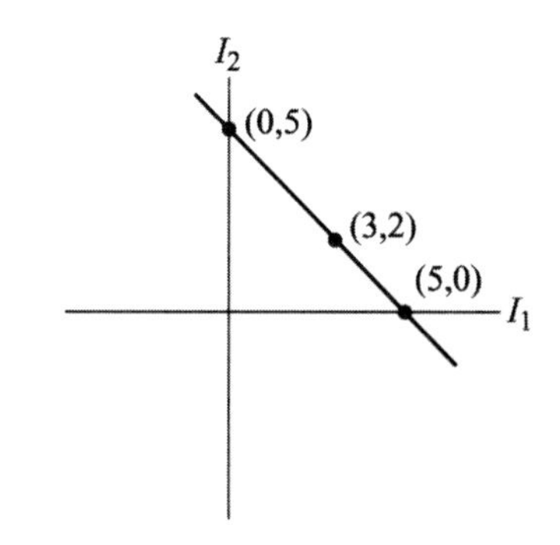

13. $P - 10V = 0$

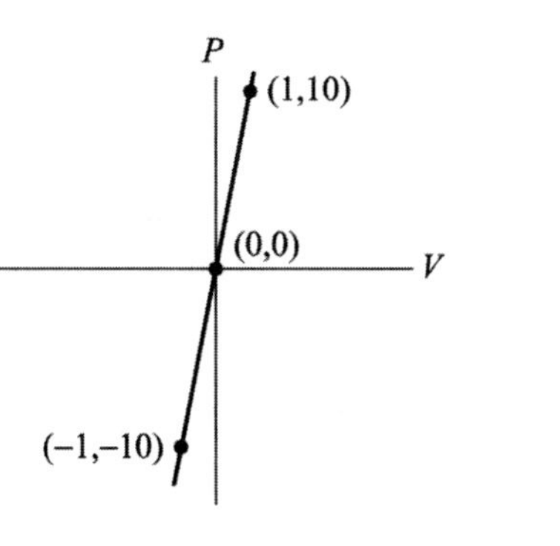

15. $I = 0.04\ V$

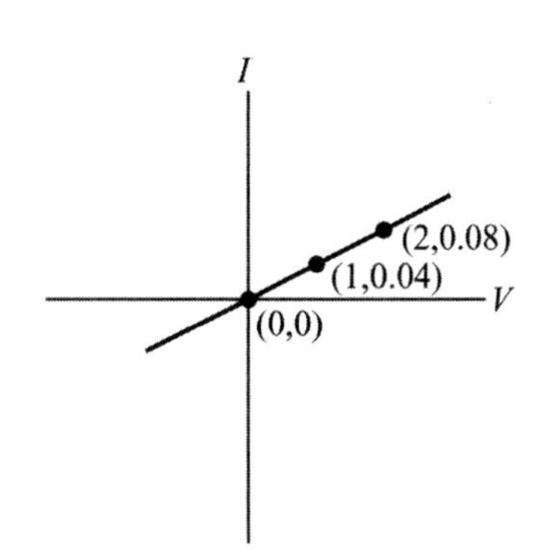

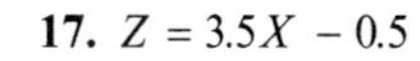

17. $Z = 3.5X - 0.5$

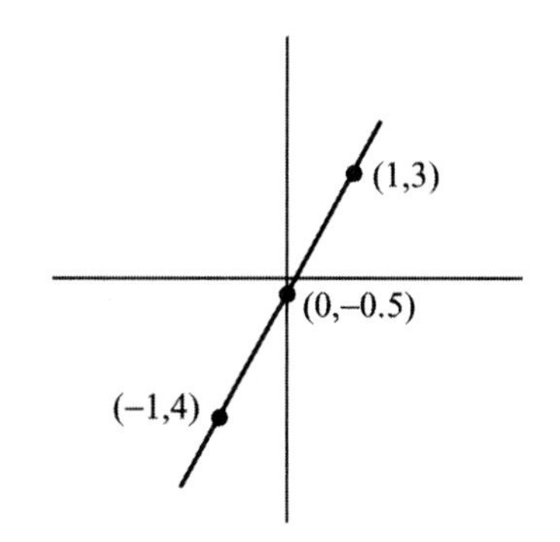

19. $V = 1.6R + 3.2$

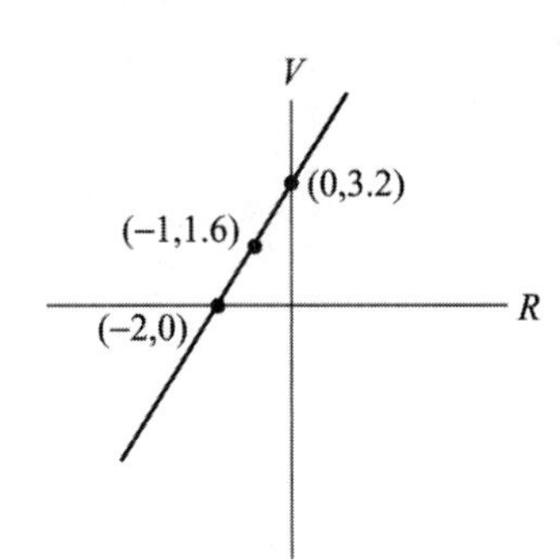

21. $y = 1.5x^2$

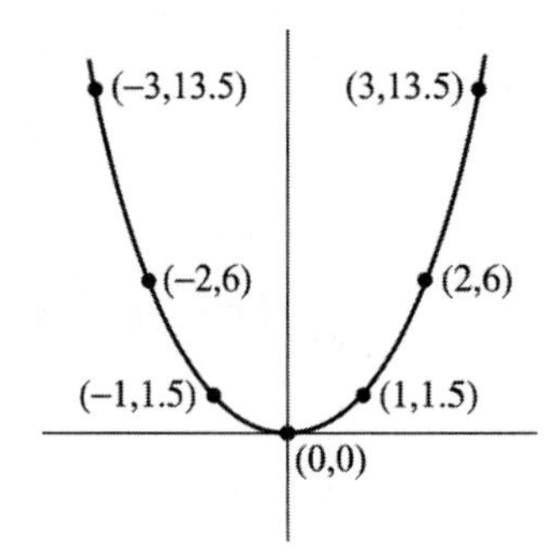

23. $y = 2.2x^2$

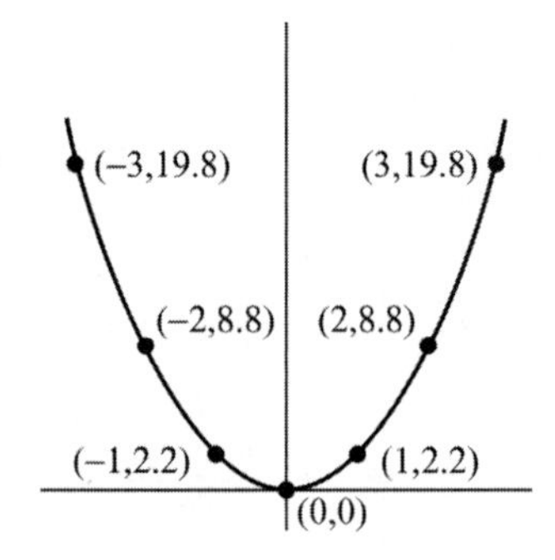

25. $y = -3x^2$

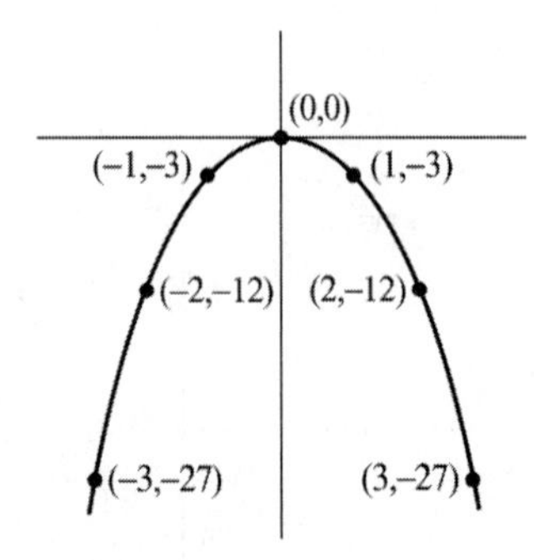

27. $P = 12I^2$

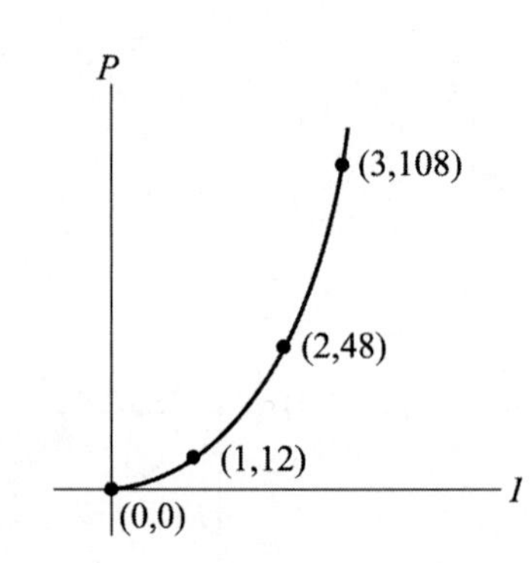

29. $P = 1.4V^2$

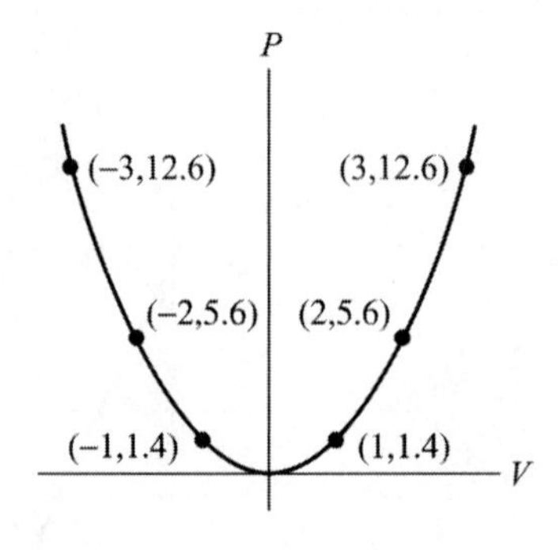

31. $I = \frac{28}{R}$

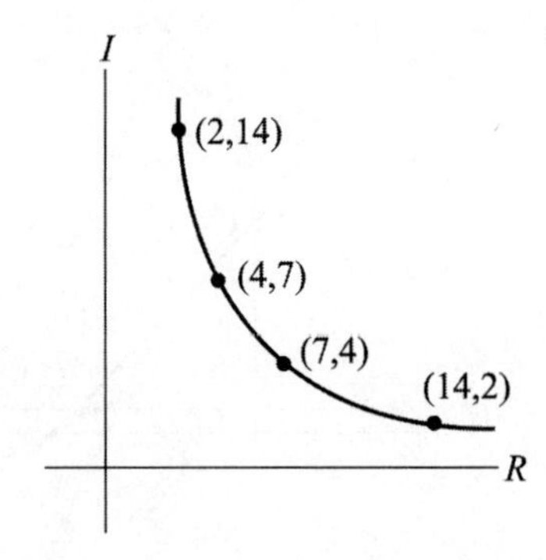

33. $I = \frac{60}{V}$

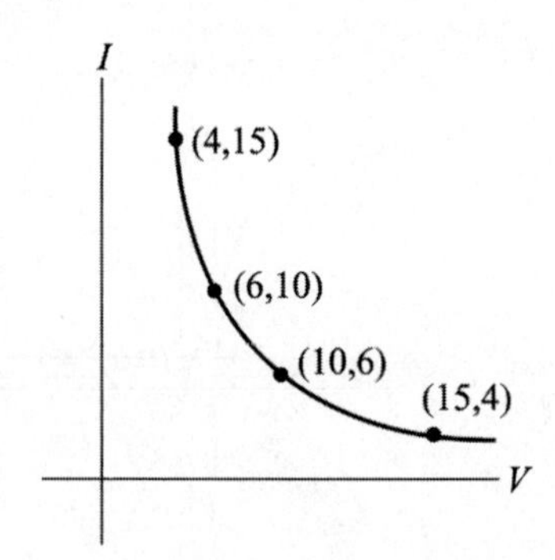

35. $F = 1.5x$

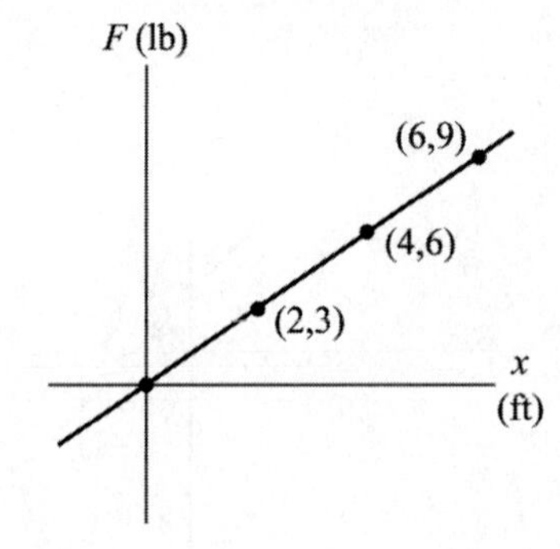

37. $C = 2\pi r$

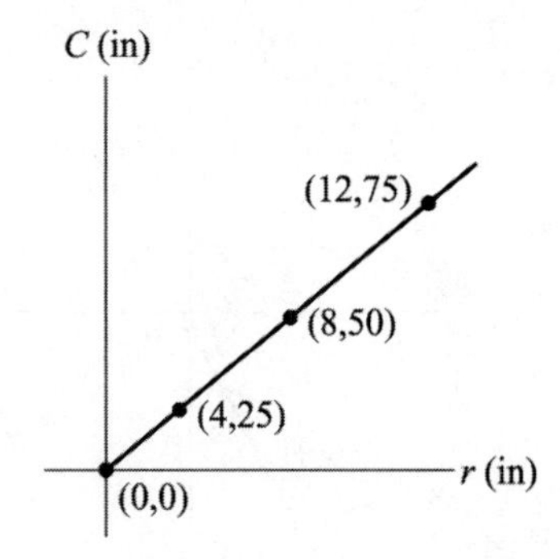

39. $I = \frac{V}{3.6\ \text{k}\Omega}$

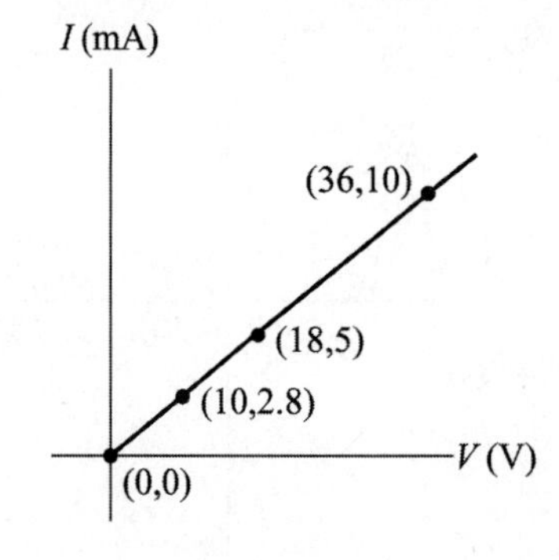

41. $Q = 2.76 + 1.22t$

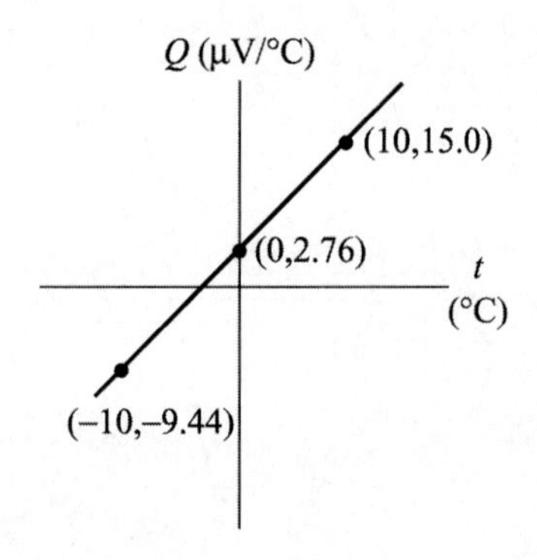

43. $P = 120I^2$

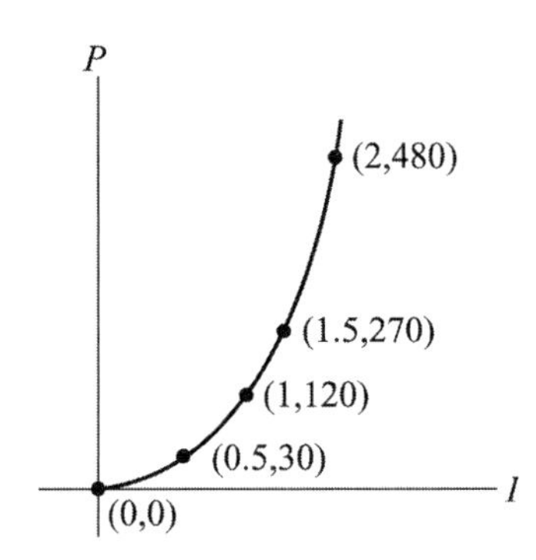

45. **(a)**

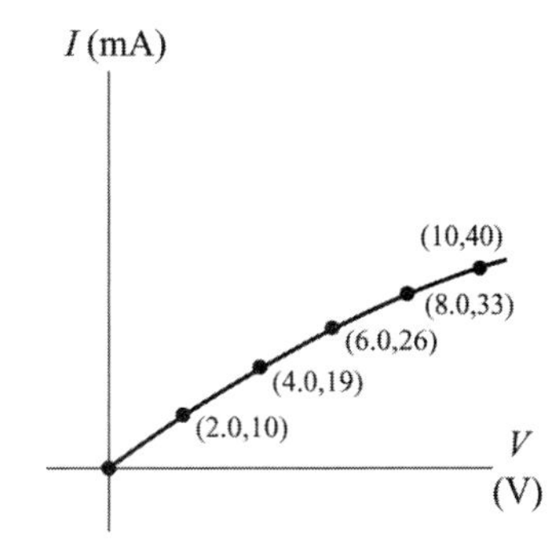

(b) 200 Ω, 211 Ω, 231 Ω, 242 Ω, 250 Ω

CHAPTER 13

Exercise 13.1

1.

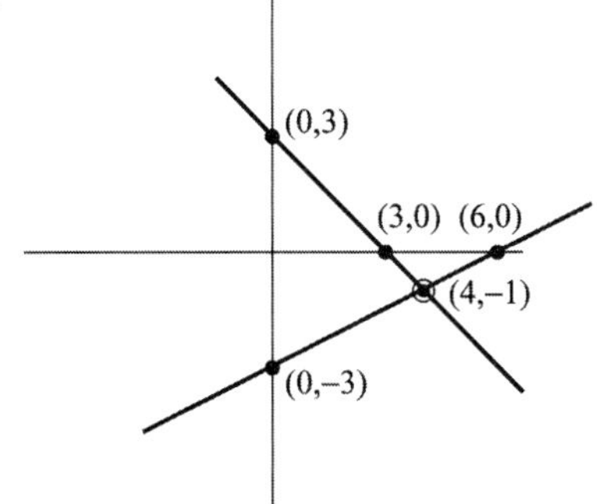

3.

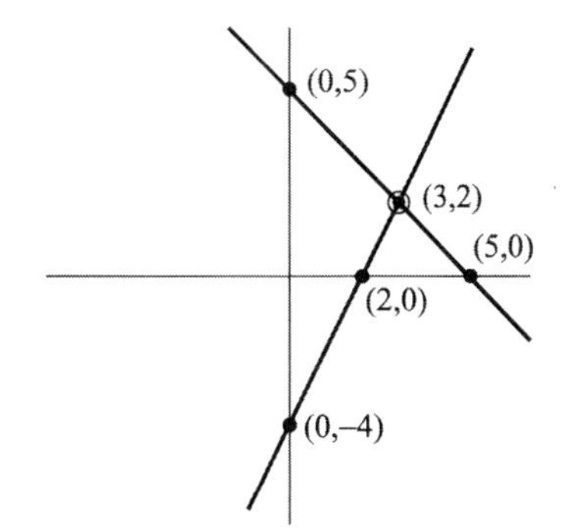

5.

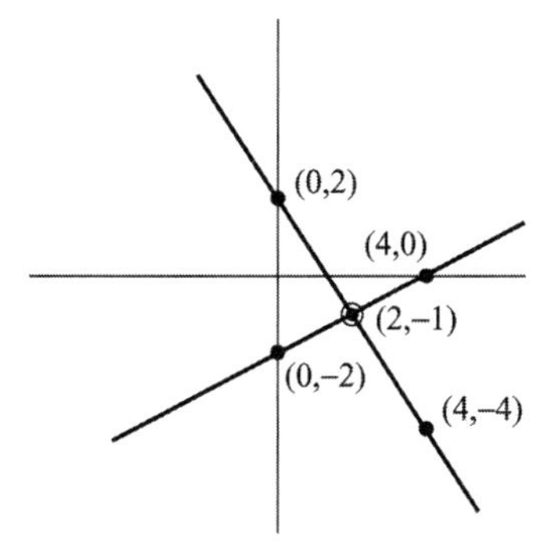

7.

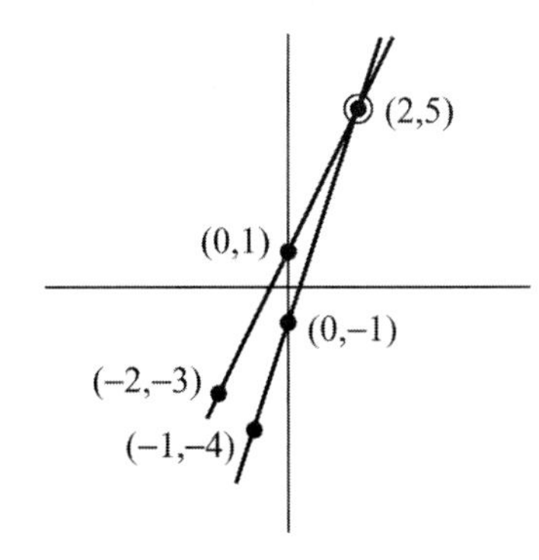

9.

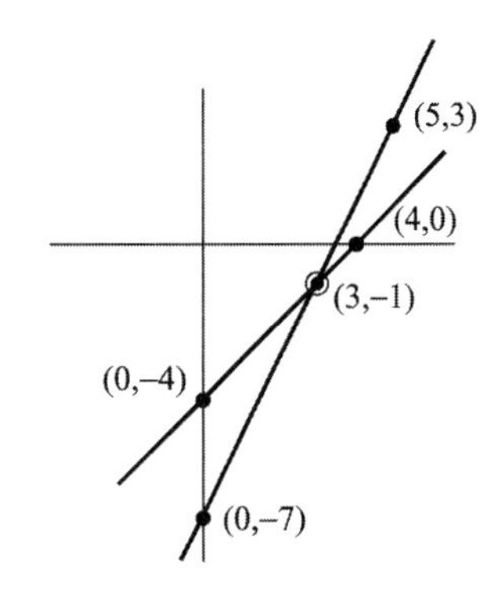

11.

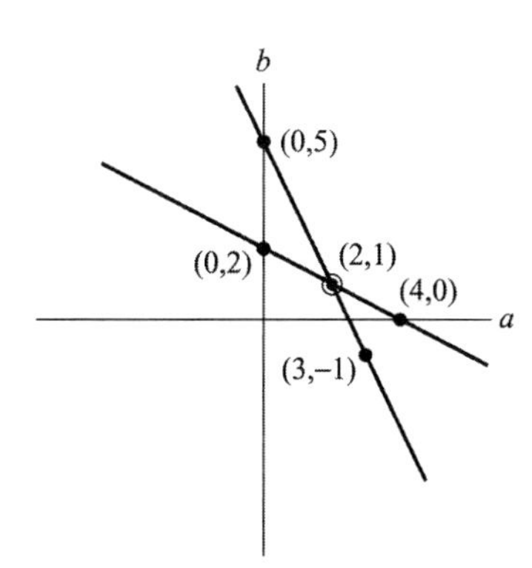

13.

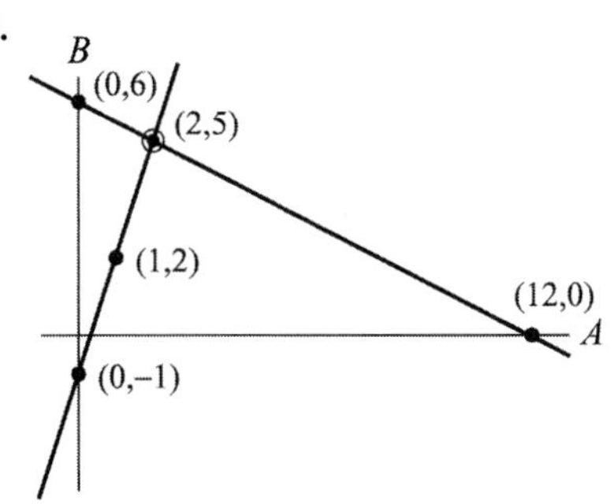

15.

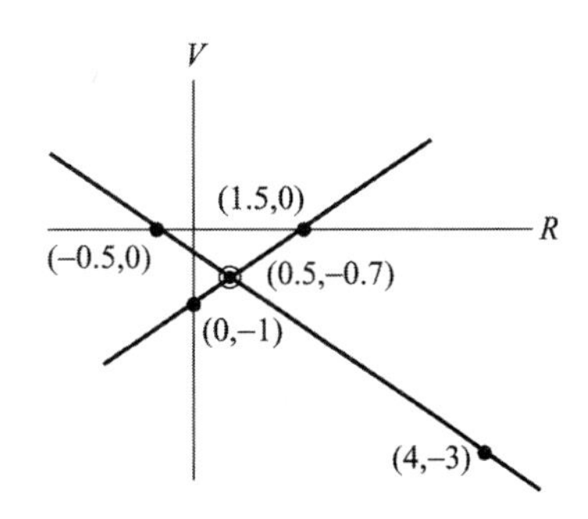

17.

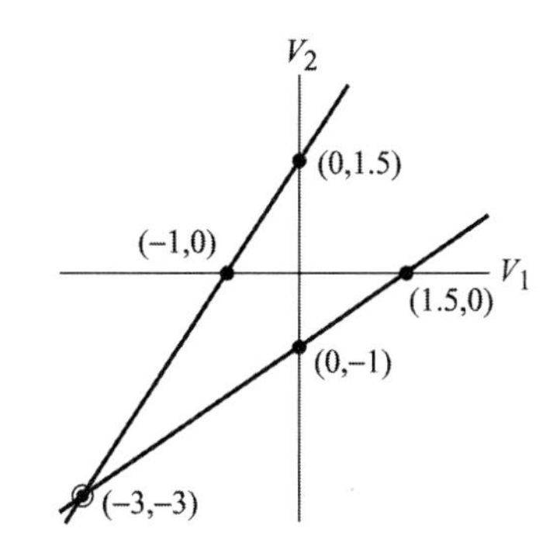

19.

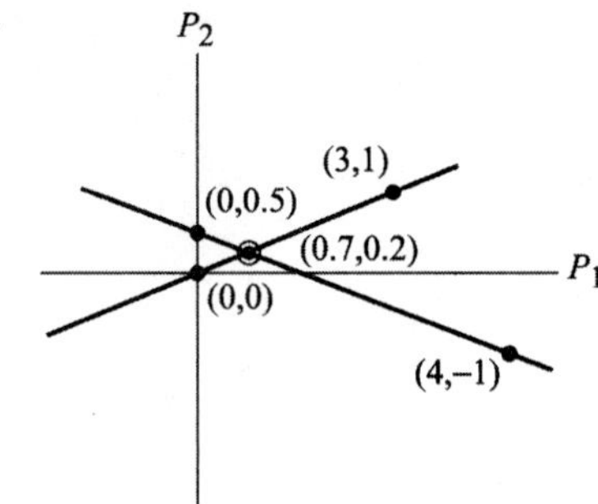

25.

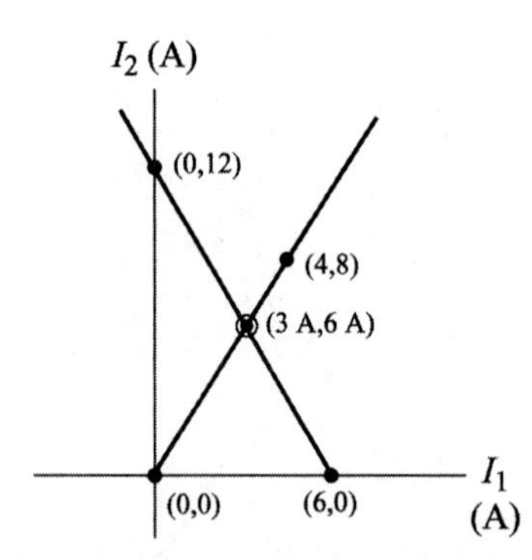

27. $V_1 + V_2 = 20$
$V_1 - V_2 = 10$

Exercise 13.2

1. (4,−1)
3. (3,2)
5. (2,−1)
7. (2,5)
9. (3,−1)
11. (2,1)
13. (2,5)
15. (2,0)
17. (−0.5,0.25)
19. (−3,−3)
21. (12,4)
23. (1.9,−0.2)
25. (6,1)
27. (−2.8,−3.2)

21.

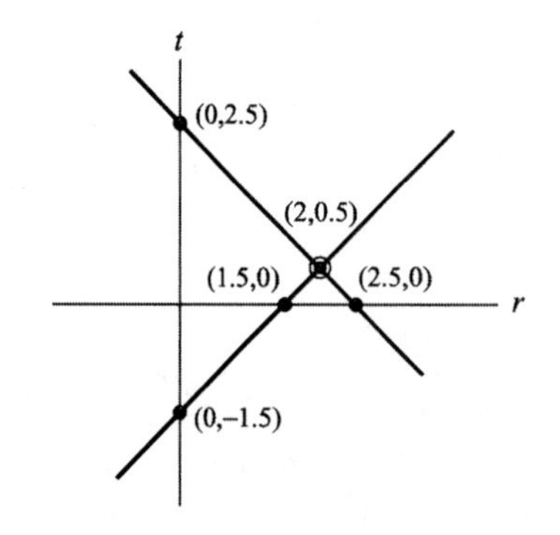

29. (−2.4,5.2)
31. 18 m = 59 ft, 36 m = 120 ft
33. Mr \$300, Ms \$200
35. 3.3 h, 1.7 h
37. 100 Ω, 200 Ω
39. 22 V, 12 V
41. 110 mA, 74 mA
43. (a) 40 mS, 50 mS
(b) 25 Ω, 20 Ω
45. 3 ns, 5 ns
47. $R_x = 80\ \Omega$, $R_y = 40\ \Omega$

Exercise 13.3

1. 5
3. 21
5. −7.8
7. (2,5)
9. (−0.4,0.5)
11. (3,1)
13. (−0.5,0.33)
15. (−2.8,−3.2)
17. (−3,−3)
19. (0.71,0.24)
21. (−2,−4)
23. (0.6,0.4)
25. (1.3,1.1)
27. (3.4,1.6)
29. (−2.2,−4.4)
31. (−2.4,5.2)
33. (1.9,−0.23)
35. 1000, 1500
37. 1.3 h, 2.7 h
39. 190 mA, 100 mA
41. 32 V, 24 V
43. $R_0 = 20\ \Omega$, $\alpha = 0.0039/°C$

Exercise 13.4

1. 2

23.

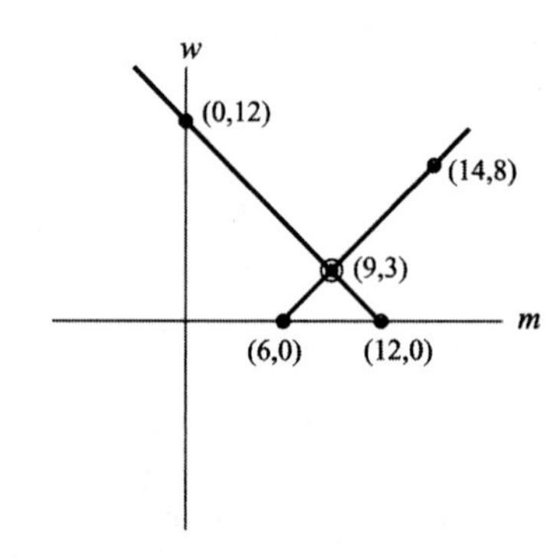

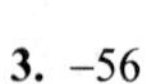

3. −56
5. 13
7. (1, 2, 3)
9. (4, −1, 3)
11. (2, 1, 3)
13. (0.1, 0.5, −0.3)
15. (3, −5, −1)
17. (0.5, 0.67, 0.6)
19. (0.5, 1.0, −0.5)
21. 0.25 gal, 0.75 gal, 0.50 gal
23. 50 V, 200 V, 150 V
25. 720 mA, 880 mA, 160 mA
27. 1.8 A, 1.1 A, 1.6 A

Review Exercises

1.

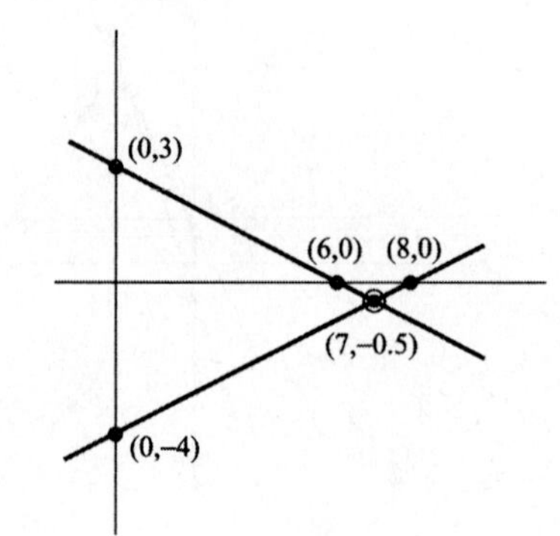

3.

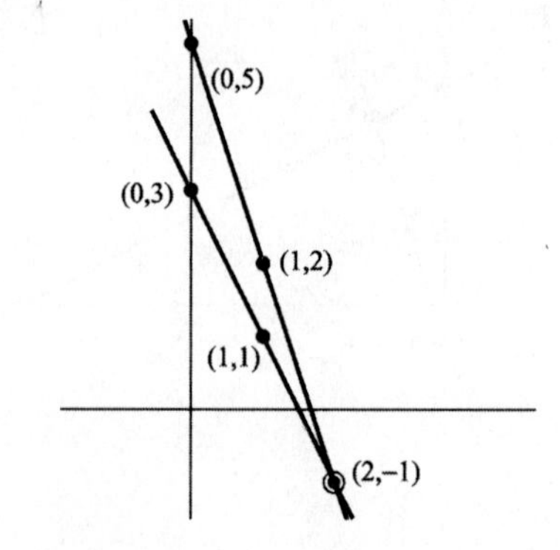

5.

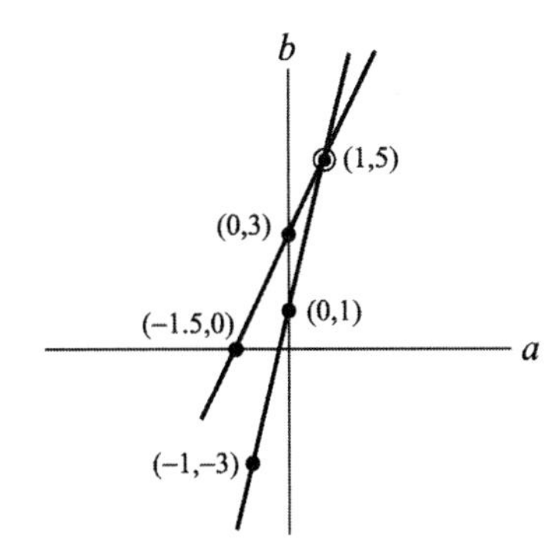

7.

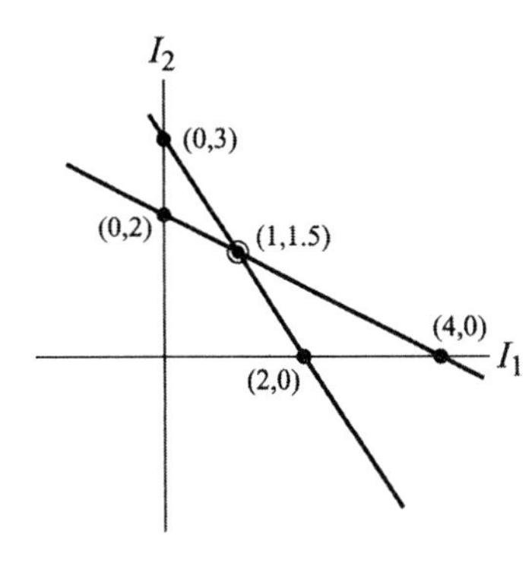

9.

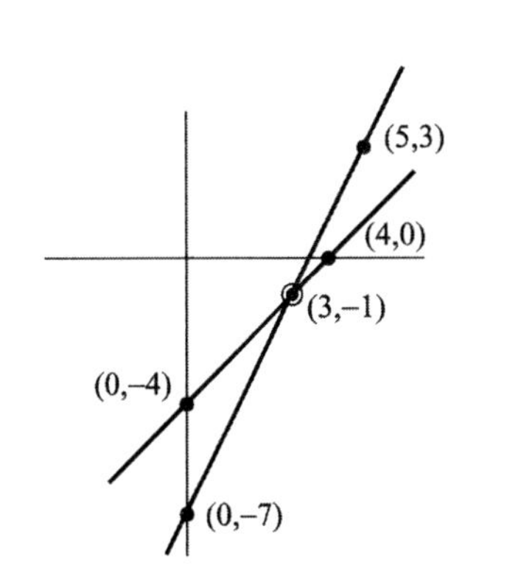

11. (4,1)

13. (2,–1)

15. (7,2)

17. (–0.75,1.8)

19. (2,0)

21. (5,–3)

23. (0.4,0.9)

25. –28

27. –0.44

29. 0

31. 250

33. (3,0.33)

35. (2,0.5)

37. (0.4,0.9)

39. (2.5,3.5)

41. (10,–8)

43. (3,–2)

45. (2, –1, 3)

47. (1.5, 1, 0.5)

49. (3, –1, 1)

51. (1.5, 2, 3.5)

53. 31 m, 46.5 m

55. $11,500

57. 20 V, 5 V

59. 10 mS, 18 mS; 100 Ω, 56 Ω

61. 0.60, 0.40, 0.20

63. 200 mA, 100 mA

CHAPTER 14

Exercise 14.1

1. 3.2 V, 1.9 V, 3.9 V

3. 23 Ω, 6.4 V, 4.3 V, 8.6 V, 4.3 V, 640 mA, 640 mA, 210 mA, 430 mA, 210 mA

5. 12 V, 8.0 V, 4.0 V

7. 16 Ω, 1.5 A, 900 mA, 600 mA, 9.0 V, 4.5 V, 4.5 V, 15 V

9. 2.9 kΩ, 70 mA, 20 mA, 10 mA, 40 mA, 200 V, 200 V, 80 V, 120 V

11. 25 Ω

13. 24 V, 9 V, 27 V

Exercise 14.2

1. See statement in text.

3. $I_1 - I_2 - I_3 = 0$

5. $V_0 - I_1R_1 - I_2R_2 = 0$

7. $I_2R_2 - I_3R_3 = 0$

9. $I_1 + I_2 - I_3 = 0$

11. $V_1 - I_3R_3 - I_1R_1 = 0$

13. $I_1 - I_2 - I_3 + I_4 = 0$

15. $V_1 - I_1R_1 - V_2 + I_4R_4 = 0$

17. $V_2 - I_3R_3 - I_4R_4 = 0$

19. $V_1 - I_1R_1 - I_3R_3 = 0$

Exercise 14.3

1. 1.0 A, 500 mA, 500 mA

3. 280 mA, 190 mA, 90 mA

5. 100 mA, 700 mA, 800 mA

7. 100 mA, 300 mA, 400 mA

9. 1.7 A, 1.9 A, 230 mA

11. 4.0 V, 10 Ω

13. 200 mA, 500 mA, 500 mA, 800 mA

15. 14 mA, 8.0 mA, 6.0 mA

Review Exercises

1. 26 V, 6.0 V, 8.0 V, 12 V

3. 57 Ω, 230 mA, 13 V, 3.0 V, 4.0 V, 6.0 V, I = 130 mA, $I_2 = I_3 = I_4$ = 100 mA

5. **X**: $I_1 + I_3 - I_2 = 0 \leftrightarrow$ **Y**: $I_2 - I_1 - I_3 = 0$

7. 600 mA, 1.2 A, 600 mA

9. 280 mA, 240 mA, 520 mA, 14 V, 24 V, 36 V

11. 1.4 A, 400 mA, 900 mA, 900 mA

13. 100 mA, 75 mA, 25 mA

CHAPTER 15

Exercise 15.1

1. 190 mA, 4.6 V, 3.5 V, 6.9 V

3. 100 mA, 13 mA, 20 V, 4.0 V

5. 600 mA, 600 mA, 1.2 A, 6.0 V, 12 V, 6.0 V

7. 500 mA, 100 mA, 400 mA, 5 V, 1 V, 4 V

9. 4.6 mA, 4.3 mA, 380 μA, 9.3 V, 13 V, 750 mV

Exercise 15.2

1. V_{TH} = 4.9V, R_{TH} = 5.7 Ω, I_L = 330 mA, V_L = 3.0 V

3. V_{TH} = 80 V, R_{TH} = 670 Ω, I_L = 22 mA, V_L = 65 V

5. V_{TH} = 13 V, R_{TH} = 2.2 Ω, I_L = 1.6 A, V_L = 9.7 V

7. V_{TH} = 60 V, R_{TH} = 50 Ω, I_L = 240 mA, V_L = 48 V

9. V_{TH} = 33 V, R_{TH} = 1.2 kΩ, I_L = 12 mA, V_L = 18 V

11. V_{TH} = 6.8 V, R_{TH} = 12 Ω, I_L = 180 mA, V_L = 4.6 V

Exercise 15.3

1. I_N = 2.3 A, R_N = 3.4 Ω, I_L = 590 mA, V_L = 5.9 V

3. I_N = 100 mA, R_N = 170 Ω, I_L = 36 mA, V_L = 11 V

5. **(a)** 600 mA, 12 V
(b) 600 mA, 12 V

7. I_L = 400 mA, V_L = 2.0 V

Review Exercises

1. 830 mA, 8.3 V, 4.2 V, 2.5 V

3. 280 mA, 240 mA, 520 mA, 14 V, 24 V, 36 V

5. $V_{TH} = 10$ V, $R_{TH} = 2.0$ kΩ, $I_3 = 2.5$ mA, $V_3 = 5$ V

7. $V_{TH} = 30$ V, $R_{TH} = 0$ Ω $I_2 = 30$ mA, $V_2 = 30$ V

9. (a) $V_{TH} = 56$ V, $R_{TH} = 100$ Ω, $I_L = 340$ mA, $V_L = 20$ V
   (b) $R_T = 130$ Ω, $I_T = 740$ mA, $V_L = 20$ V, $I_L = 340$ mA

11. $I_N = 83$ mA, $R_N = 880$ Ω, $I_3 = 31$ mA, $V_3 = 46$ V

13. $I_N = 5.0$ mA, $R_N = 2.0$ kΩ, $I_3 = 2.5$ mA, $V_3 = 5$ V

CHAPTER 16

Exercise 16.1

1. 30°
3. 270°
5. 135°
7. 18°
9. 63°
11. 29°
13. 92°
15. 5.2°
17. $\frac{\pi}{3}$
19. 2π
21. $\frac{2\pi}{3}$
23. $\frac{\pi}{18}$
25. 0.96 rad
27. 1.7 rad
29. 0.14 rad
31. 3.0 rad
33. 0.0175 rad
35. $A = B = 70°$, $C = 110°$
37. $A = 60°$, $B = 30°$
39. $A = 50°$, $B = 100°$, $C = 30°$
41. $A = 35°$, $B = 55°$
43. $A = 0.97$ rad, $B = 2.2$ rad
45. $30° = \frac{\pi}{6}$
47. (a) 36°
    (b) 1.2 rad

Exercise 16.2

1. 13
3. 22
5. 3.2
7. 5.2
9. 0.85
11. 600
13. $x = 3.6$, $y = 4.8$
15. $x = 3.5$, $y = 3$
17. $x = 1.4$, $y = 0.69$
19. $x = 4.5$, $y = 7.5$
21. 6.4 ft
23. 15 mi
25. 90 ft
27. 6.4 kΩ
29. 29 kΩ
31. 16 V
33. 170 mm

Exercise 16.3

1. $\sin A = \cos B = 0.471$
   $\cos A = \sin B = 0.882$
   $\tan A = 0.533$, $\tan B = 1.88$
3. $\sin A = \cos B = 0.800$
   $\cos A = \sin B = 0.600$
   $\tan A = 1.33$, $\tan B = 0.750$
5. $\sin A = \cos B = 0.581$
   $\cos A = \sin B = 0.814$
   $\tan A = 0.714$, $\tan B = 1.4$
7. $\sin A = \cos B = 0.640$
   $\cos A = \sin B = 0.768$
   $\tan A = 0.833$, $\tan B = 1.2$
9. $\sin A = \cos B = 0.836$
   $\cos A = \sin B = 0.549$
   $\tan A = 1.52$, $\tan B = 0.656$
11. $\sin A = \cos B = 0.333$
    $\cos A = \sin B = 0.943$
    $\tan A = 0.354$, $\tan B = 2.83$
13. $\sin A = \cos B = 0.707$
    $\cos A = \sin B = 0.707$
    $\tan A = 1.00$, $\tan B = 1.00$
15. $\sin A = \cos B = 0.816$
    $\cos A = \sin B = 0.577$
    $\tan A = 1.41$, $\tan B = 0.707$
17. $\sin A = \cos B = 0.750$
    $\cos A = \sin B = 0.265$
    $\tan A = 2.80$, $\tan B = 0.354$
19. $\sin A = \cos B = 0.500$
    $\cos A = \sin B = 0.866$
    $\tan A = 0.577$, $\tan B = 1.73$
21. 0.500
23. 0.176
25. 0.163
27. 0.0612
29. 1.96
31. 0.853
33. 0.866
35. 1.00
37. 60.0°
39. 19.3°
41. 45.0°
43. 65.0°
45. 13.1°
47. 22.6°
49. 60.0°
51. $B = 40.0°$, $a = 2.30$, $b = 1.93$
53. $A = 64.5°$, $b = 4.91$, $c = 11.4$
55. $A = 38.7°$, $B = 51.3°$, $b = 6.24$
57. $A = 22.0°$, $B = 68.0°$, $c = 6.69$
59. $A = 0.841$ rad, $b = 4.03$, $c = 6.04$
61. $B = 45.0°$, $b = 2.00$, $c = 2.83$
63. 89 m
65. 770 ft
67. $Z = 4.5$ kΩ, $\theta = 35°$
69. $v = 212$ V
71. $56° = 0.98$ rad

Review Exercises

1. 60.0°
3. 12.0°
5. 86.0°
7. 0.802°
9. $\frac{\pi}{6} = 0.524$ rad
11. $\frac{\pi}{9} = 0.349$ rad
13. 5.24 rad
15. $\frac{7\pi}{9} = 2.44$ rad
17. $A = C = 70.0°$, $B = 110°$
19. 2.50
21. 62.4
23. 200
25. 18.4
27. $x = 4.00$, $y = 7.21$
29. $\sin A = \cos B = 0.800$
    $\cos A = \sin B = 0.600$
    $\tan A = 1.33$, $\tan B = 0.750$

31. $\sin A = \cos B = 0.385$
$\cos A = \sin B = 0.923$
$\tan A = 0.417$, $\tan B = 2.40$

33. $\sin A = \cos B = 0.680$
$\sin B = \cos A = 0.733$
$\tan A = 0.927$, $\tan B = 1.08$

35. $\sin A = \cos B = 0.600$
$\sin B = \cos A = 0.800$
$\tan A = 0.750$, $\tan B = 1.33$

37. $\sin A = \cos B = 0.555$
$\cos A = \sin B = 0.832$
$\tan A = 0.333$, $\tan B = 1.50$

39. 0.866

41. 1.52

43. 0.0175

45. 0.718

47. 0.839

49. 11.5°

51. 30.0°

53. 17.3°

55. 58.0°

57. 60°, 2.89, 5.77

59. 34°, 18.5, 27.4

61. 48.6°, 41.4°, 1.32

63. 33.7°, 56.3°, 18.0

65. 36 mi

67. 75 m

69. **(a)** 8.5 kΩ
(b) 16°

71. 3.6 mm × 7.2 mm

73. **(a)** 0.029°
(b) 0.14°

CHAPTER 17

Exercise 17.1

[Sin, Cos, Tan]

1. 0.800, 0.600, 1.33

3. 0.882, −0.471, −1.88

5. −0.385, −0.923, 0.417

7. −0.600, 0.800, −0.75

9. −0.707, 0.707, −1.00

11. 0.793, 0.609, 1.30

13. 0.809, −0.588, −1.38

15. −0.625, −0.781, 0.800

17. −0.903, 0.430, −2.10

19. 0.800, 0.600, 1.33

21. 0.800, 0.750

23. −0.923, 0.385

25. −0.943, 2.83

27. 0.832, 0.555

29. −0.714, −1.02

31. 0.707, −0.707

33. −0.800, 0.750

35. 0.422, −0.906, −0.467

37. **(a)** $V_T = 18.4$ V
(b) −0.652, 0.759, −0.857

39. 1.42, −1.42

Exercise 17.2

1. sin 50° = 0.766

3. −tan 30° = −0.577

5. cos 15° = 0.966

7. −tan 80° = −5.67

9. −sin 30° = −0.500

11. −cos 45° = −0.707

13. −tan 50.6° = −1.22

15. tan 63.0° = 1.97

17. cos 90° = 0

19. tan 180° = 0

21. sin 270° = −1

23. −tan 53°= −1.33

25. −sin 80° = −0.985

27. cos 0° = 1

29. −tan 45° = −1.00

31. −tan 22.5° = −0.415

33. 9.00 knots

35. **(a)** −0.424
(b) $X_C = 1.95$ kΩ, $Z = 5.00$ kΩ

37. **(a)** 850 mA
(b) 736 mA
(c) 0 mA
(d) −601 mA

39. **(a)** 114 V
(b) −70.5 V
(c) 0 V

Exercise 17.3

1. 30.0°

3. 50.0°

5. 72.7°

7. 142.0°

9. −41.9°

11. −60.0°

13. 25.0°, 335.0°

15. 40.2°, 220.2°

17. 213.8°, 326.2°

19. 153.3°, 333.3°

21. 99.0°, 279.0°

23. 57.3°

25. −33.8°

27. −77.4°

29. 0.981 rad, 5.30 rad

31. 2.36 rad, 5.50 rad

33. 3.67 rad, 5.76 rad

35. 1.19 rad, 4.33 rad

37. 0.785 rad, 2.36 rad, 3.93 rad, 5.50 rad

39. 1.06 rad

41. −0.703 rad

43. −1.40, 125°

45. 8.24°

47. −1.50, −56.3°

49. 333°, −27.0°

51. **(a)** 0.629 rad, 2.51 rad
(b) 1.67 ms, 6.67 ms

Review Exercises

[Sin, Cos, Tan]

1. 0.923, 0.385, 2.40

3. 0.882, −0.471, −1.88

5. −0.300, 0.954, −0.314

7. 0.741, 0.671, 1.11

9. −sin 50° = −0.766

11. tan 65° = 2.14

13. cos 40° = 0.766

15. −tan 88.3° = −34.1

17. 77.5° = 1.35 rad

19. 65.5° = 1.14 rad

21. 60° = 1.05 rad, 300° = 5.24 rad

23. 136° = 2.37 rad, 316° = 5.51 rad

25. 41.5° = 0.725 rad, 138° = 2.42 rad

27. 36.9° = 0.644 rad, 143° = 2.50 rad

29. 108° = 1.89 rad, 288° = 5.03 rad

31. 64.7° = 1.13 rad

33. 36.9° = 0.644 rad

35. −30.5° = −0.532 rad

37. 18.8 cm^2

39. **(a)** −4.00 A
(b) −1.70 A
(c) −0.351 A = −351 mA

41. 40.5 mA, −32.9°

CHAPTER 18

Exercise 18.1

1.

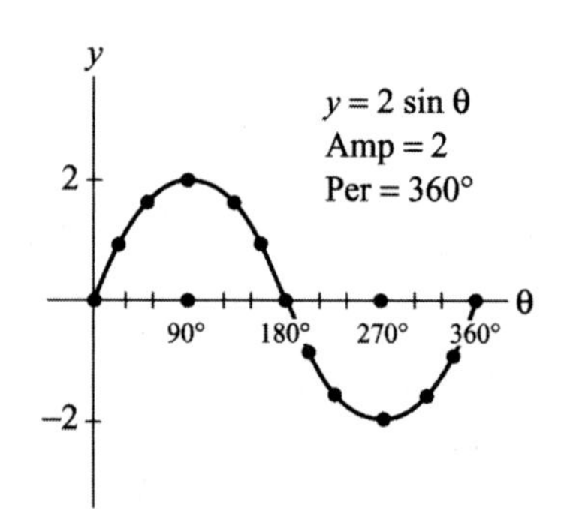

3.

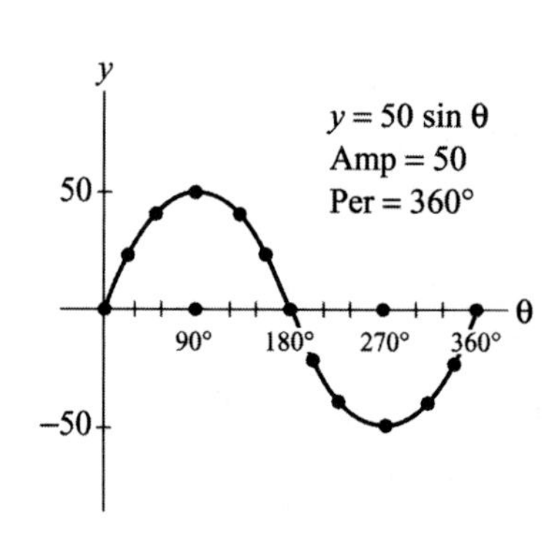

5.

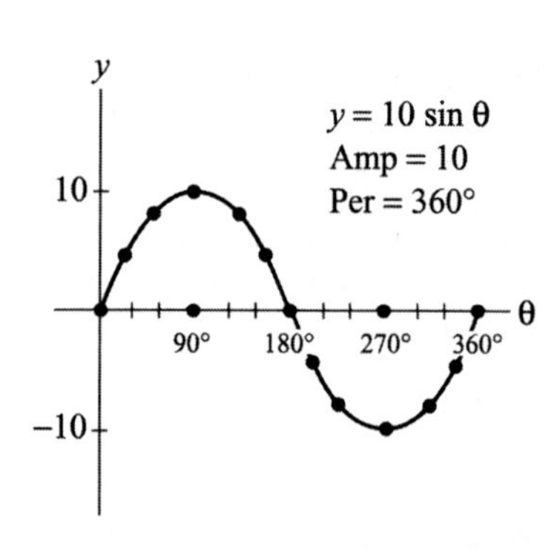

7.

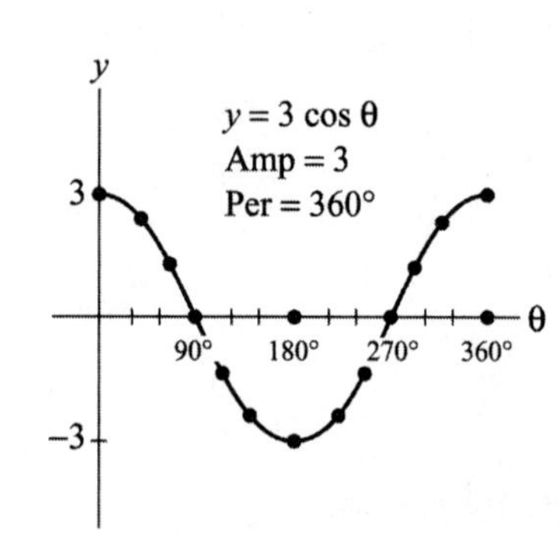

9.

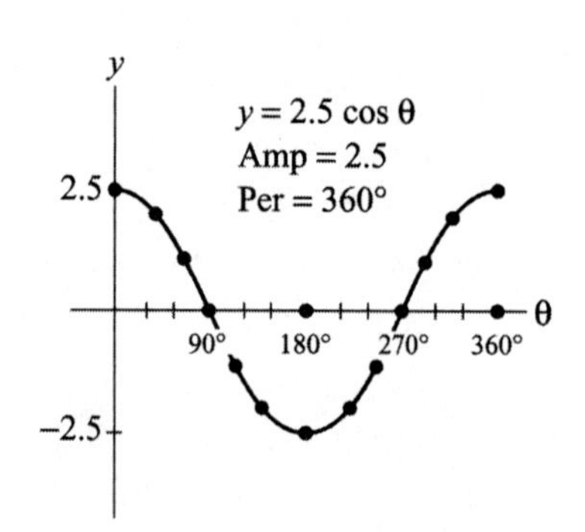

11.

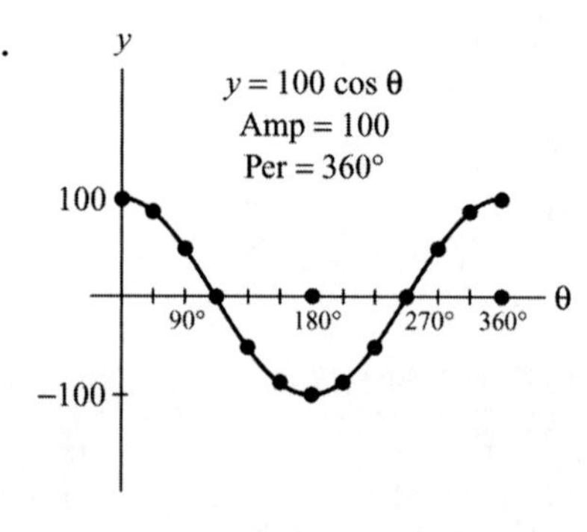

13.

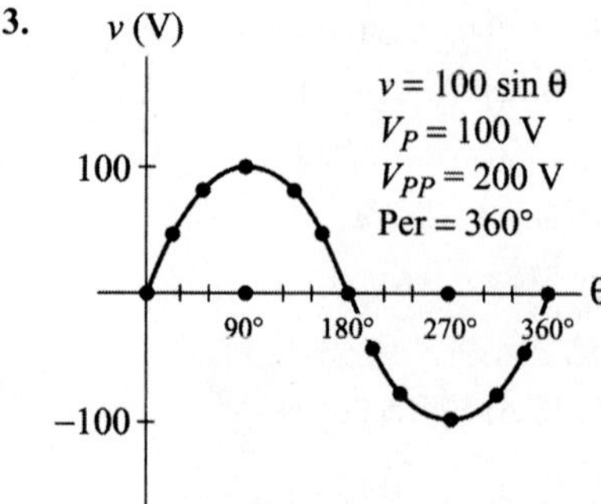

15.

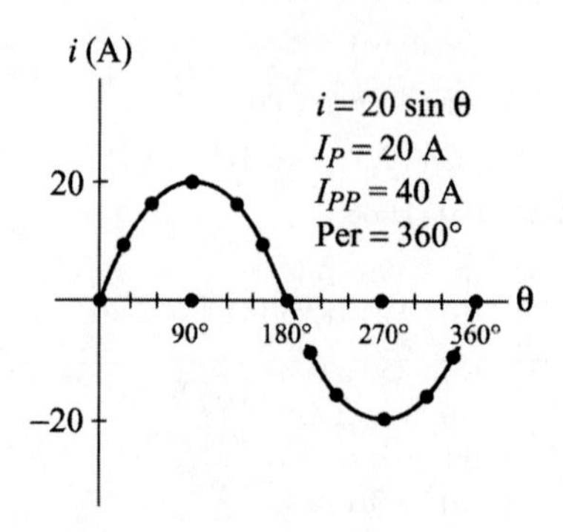

17.

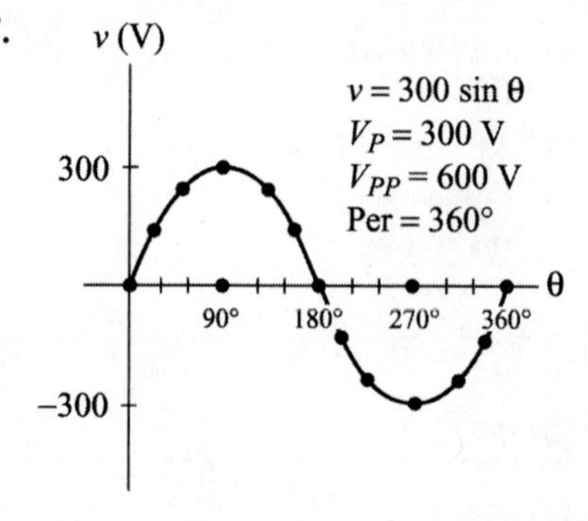

19.

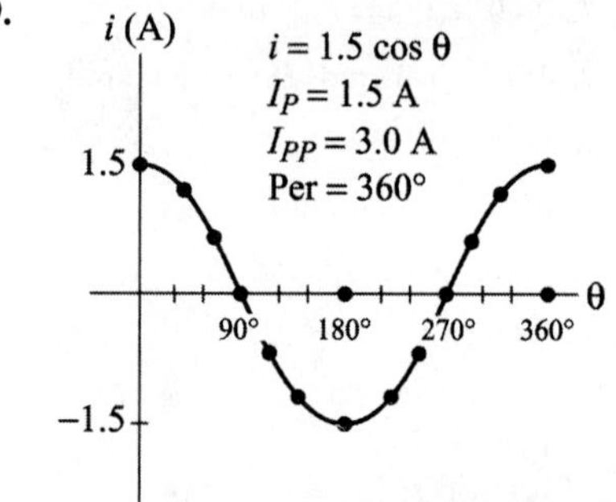

21.

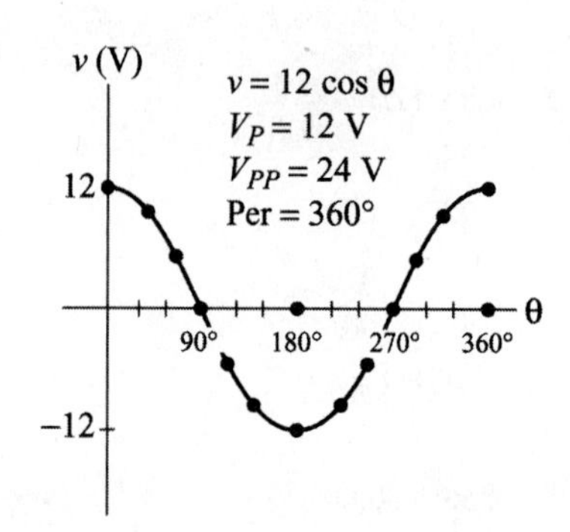

23.

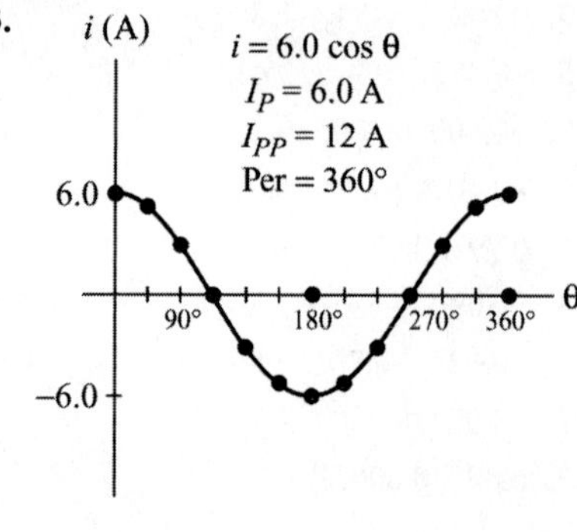

25.

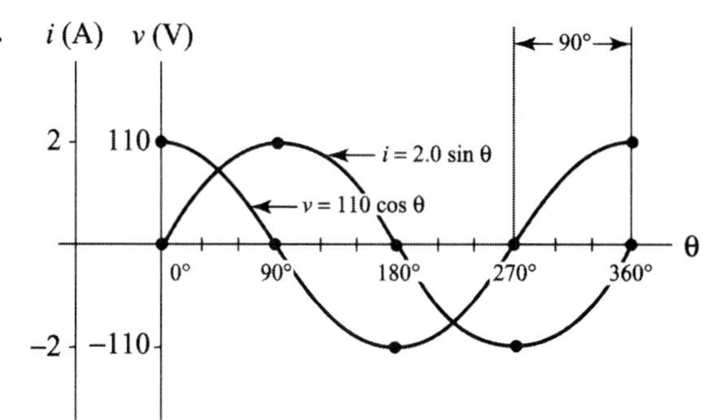

27. **(a)** 39 V
(b) 0.52 rad, 2.6 rad

Exercise 18.2

1. 300 V, 210 V
3. 4.4 A, 1.6 A
5. 280 V, 99 V
7. 300 mA, 110 mA
9. 11 A, 4.0 A
11. 400 V, 140 V
13. 160 V, 320 V
15. 1.7 A, 3.4 A
17. 850 mA, 1.7 A
19. 110 V, 220 V
21. 110 mA, 220 mA
23. 17 V, 34 V
25. 110 V, 160 V, 2.2 A, 3.1 A, 240 W
27. **(a)** 33 V, 11 V, 17 V, 5.5 V
(b) 610 mW, 910 mW, 300 mW, 1.8 W
29. **(a)** 67 mA, 53 mA, 120 mA
(b) 5.3 W, 4.3 W, 9.6 W

Exercise 18.3

1. 55 Hz, 18 ms
3. 70 Hz, 14 ms
5. 60 Hz, 17 ms
7. 48 Hz, 21 ms
9. 25 Hz, 40 ms
11. 840 kHz, 1.2 μs
13. 12 kHz, 84 μs
15. 17 ms, 120π rad/s
17. 22 ms, 90π rad/s
19. 2.0 ms, 1000π rad/s
21. 5.0 ms, 400π rad/s
23. 1.1 μs, 5.7×10^6 rad/s
25. 14 ns, 440×10^6 rad/s
27. 4.0 A, 120π rad/s, 60 Hz, 17 ms
29. 110 V, 140π rad/s, 70 Hz, 14 ms
31. 310 V, 300 rad/s, 48 Hz, 21 ms
33. 450 mA, 500 rad/s, 80 Hz, 13 ms
35. 1.2 A, 200π rad/s, 100 Hz, 10 ms
37. 72 V, 3.0×10^6 rad/s, 480 kHz, 2.1 μs
39. 390 rad/s, 63 Hz, 16 ms
41. $v = 180 \sin 120\pi t$
43. $i = 2.5 \sin 100\pi t$
45. $i = 0.45 \sin 420t$
47. $v = 160 \sin 290t$
49.

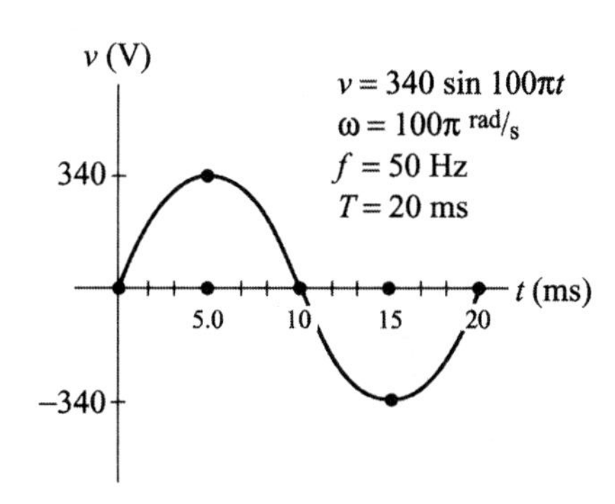

51.

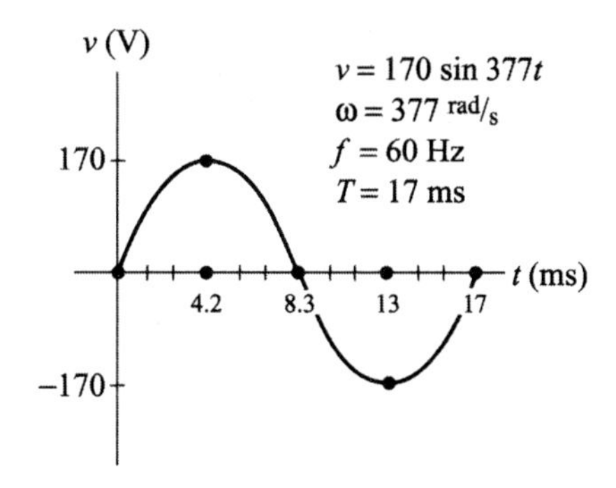

53.

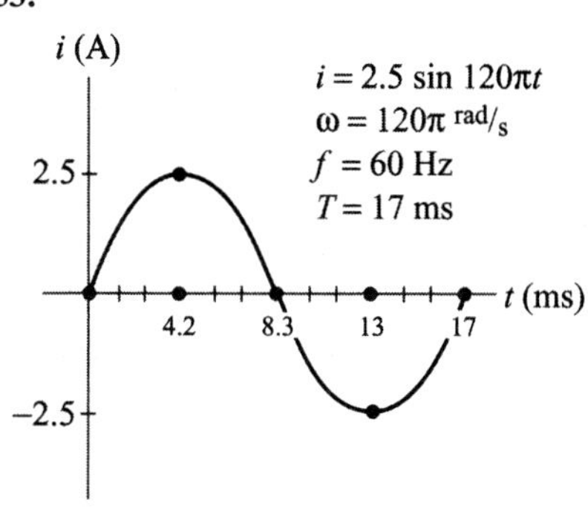

55.

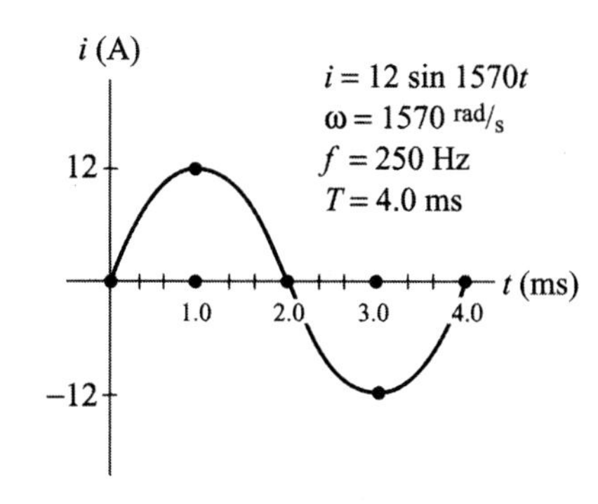

57. **(a)** 77 V, 86 V
(b) 1.8 ms, 6.5 ms
59. **(a)** 1.2 A, 910 mA
(b) 12 ms, 18 ms

Exercise 18.4

1. 1.5 A, 50 Hz, 20 ms, $\frac{\pi}{2}$
3. 300 V, 100 Hz, 10 ms, $\frac{\pi}{2}$
5. 60 V, 60 Hz, 17 ms, $-\frac{\pi}{2}$
7. 18 A, 50 Hz, 20 ms, $\frac{\pi}{2}$
9. 5.5 A, 100 Hz, 10 ms, $-\frac{\pi}{2}$
11. $i = 3.3 \sin\left(120\pi t + \frac{\pi}{2}\right)$
13. $v = 400 \sin 100\pi t$
15. $v = 350 \sin\left(100\pi t + \frac{\pi}{2}\right)$
17. $i = 12 \sin 120\pi t$
19. $v = 170 \sin\left(350t + \frac{\pi}{2}\right)$
21. $i = 0.50 \sin\left(100\pi t - \frac{\pi}{2}\right)$

23. (a)

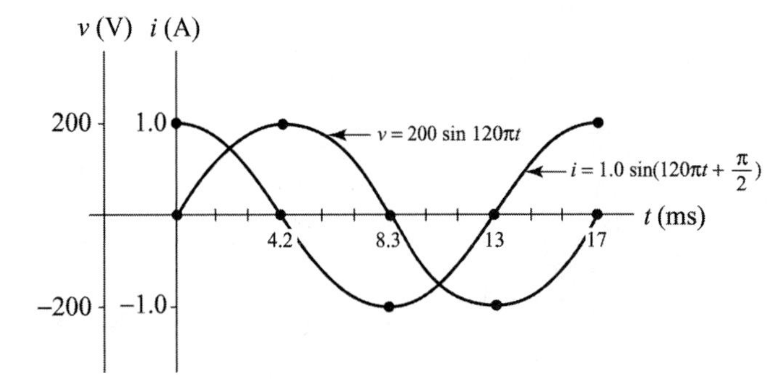

(b) t difference = 4.2 ms

25. (a) Both: 50 Hz, 20 ms; v: $\frac{\pi}{2}$ rad, i: 0 rad

(b) 5 ms, v leads i by $\frac{\pi}{2}$

27. (a) − 2.0 A, 1.7 A

(b) 4.2 ms, 13 ms

Review Exercises

1.

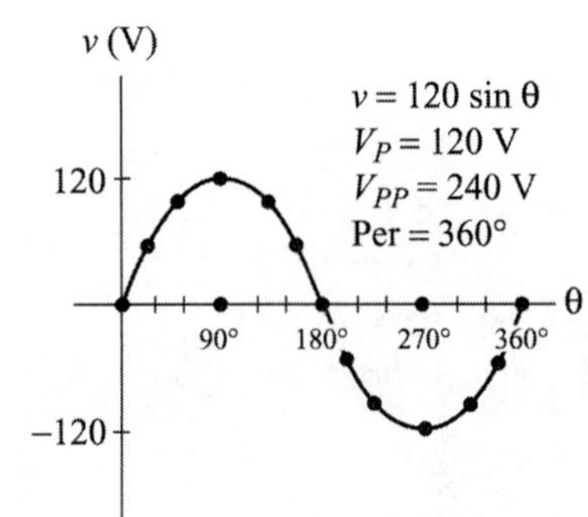

3.

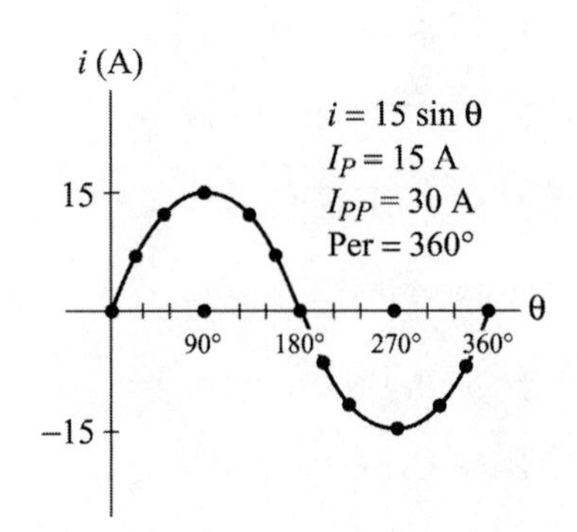

5. 160 V, 113 V, 120π rad/s, 60 Hz, 17 ms, 0 rad

7. 3.0 A, 2.1 A, 200π rad/s, 100 Hz, 10 ms, 0 rad

9. 170 V, 120 V, 350 rad/s, 56 Hz, 18 ms, 0 rad

11. 1.6 A, 1.1 A, 400 rad/s, 64 Hz, 16 ms, 0 rad

13. 1.2 A, 850 mA, 140π rad/s, 70 Hz, 14 ms, $\frac{\pi}{2}$

15. 320 V, 230 V, 377 rad/s, 60 Hz, 17 ms, $\frac{\pi}{2}$

17. $i = 2.4 \sin 120\pi t$

19. $v = 110 \sin 100\pi t$

21. $v = 120 \sin 370t$

23. $i = 13 \sin 200\pi t$

25. $i = 14 \sin\left(100\pi t + \frac{\pi}{2}\right)$

27. $v = 220 \sin\left(120\pi t + \frac{\pi}{2}\right)$

29. 50 mA, 71 mA, 75 V, 110 V, 3.8 W

31. (a) 83 Hz, 12 ms, 520 rad/s

(b) $v = 120 \sin 520t$

33.

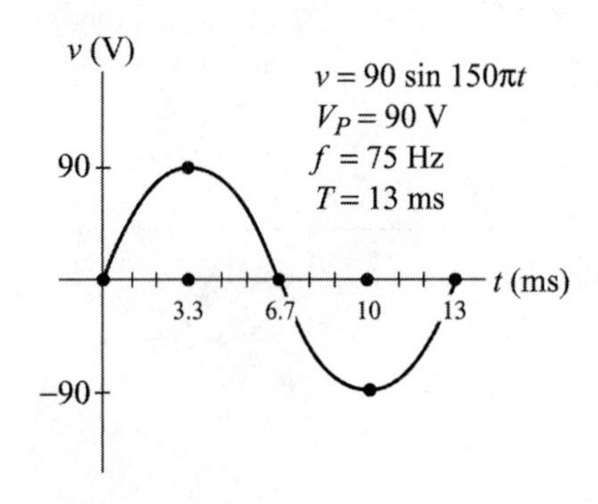

35. (a) 400 mA, −500 mA

(b) 1.1 ms, 5.6 ms

(c) 0 ms, 6.7 ms

CHAPTER 19

Exercise 19.1

1. 26 N, 67°

3. 59 ft/s, 154°

5. 6.4×10^{-6} N, 231°

7. 49 m/s/s, 59°

9. 40 ∠34° kΩ

11. 5.3 ∠−63° kΩ

13. 12 ∠90° V

15. 1.3 ∠−39° kΩ

17. 300 ∠57° μA

19. 7.3 ∠9.7° kΩ

21. 20 ∠−21° kΩ

23. 46 N, 300°

25. 7.1 mi, 10°

27. 71 ∠−62° V

29. 86 ∠54° mA

31. 18 ∠34° kΩ

33. 30 ∠−29° kΩ

35. 3.4×10^3 ∠169° N/C

Exercise 19.2

1. $C = 40°$, $a = 7.5$, $c = 5.3$

3. $C = 83°$, $b = 29$, $c = 31$

5. $B = 45°$, $b = 6.5$, $c = 5.2$

7. $B = 75°$, $a = 34$, $b = 39$

9. $A = 42°$, $b = 21$, $c = 13$

11. $C = 75°$, $a = 2.2$, $b = 1.9$

13. $B = 63°$, $a = 4.2$, $c = 7.2$

15. $A = 20°$, $b = 68$, $c = 51$

17. $B = 47°$, $C = 66°$, $c = 5.0$

19. $A = 47°$, $C = 46°$, $a = 630$

21. $A = 51°$, $B = 24°$, $b = 0.24$

23. $A = 19°$, $C = 10°$, $c = 5.7$

25. $AB = 1500$ ft, $AC = 1200$ ft

27. 15×10^{-6} N

Exercise 19.3

1. $c = 7.1$, $A = 42°$, $B = 68°$

3. $c = 81$, $A = 19°$, $B = 31°$

5. $c = 2.6$, $A = 25°$, $B = 137°$

7. $b = 0.42$, $A = 10°$, $C = 25°$

9. $a = 0.44$, $B = 55°$, $C = 109°$

11. $A = 29°$, $B = 47°$, $C = 104°$

13. $A = 62°, B = 73°, C = 45°$

15. $A = 130°, B = 26°, C = 24°$

17. $A = 67°, B = 67°, C = 46°$

19. $A = 39°, B = 58°, C = 83°$

21. 450 m

23. 113°, 41°, 26°

25. **(a)** 43 A
(b) 23 A
(c) $I_T = I_P\sqrt{3} = 1.73I_P$

27. 17 kΩ, 30°

Review Exercises

1. 20 N, 37°

3. 3.4 m/s, 199°

5. 11 ft/s/s, 309°

7. $13 \angle -37°$ kΩ

9. $150 \angle -90°$ mA

11. $13 \angle -21°$ kΩ

13. $C = 20°, a = 7.2, c = 2.5$

15. $C = 66°, a = 19, b = 24$

17. $B = 49°, C = 91°, c = 17$

19. $A = 29°, C = 54°, a = 4.1$

21. $A = 43°, B = 27°, c = 25$

23. $B = 42°, C = 78°, a = 13$

25. $A = 39°, B = 57°, C = 84°$

27. $A = 122°, B = 18°, C = 40°$

29. 9.3 lb, 213°

31. No: 86°, 50°, 44°

33. **(a)** $26 \angle 51°$ kΩ
(b) $7.1 \angle -45°$ V
(c) $29 \angle 27°$ mA

35. **(a)** 9.5 A
(b) 9.2 A

37. 1.2 kΩ, 134°

CHAPTER 20

Exercise 20.1

1. $j4$

3. $j6$

5. $j0.2$

7. $j1.2$

9. $j0.07$

11. $j2$

13. $-j0.3$

15. $j2$

17. $j\sqrt{3}$

19. $j2\sqrt{2}$

21. $j(2 \times 10^3)$

23. −12

25. j

27. 10

29. −2

31. $-j0.6$

33. $-j800$

35. $j(4 \times 10^3)$

37. $-j48$

39. −1

41. $-j5$

43. $-j3$

45. $-j0.5$

47. $j0.5$

49. $-j3$

51. 0.5

53. $-j$

55. $j150$

57. 35 V

59. $-j220$ Ω

Exercise 20.2

1. $4 + j2$

3. $0 + j5$

5. $8.4 + j2.7$

7. $3 - j4$

9. $17 - j7$

11. $3 + j4$

13. $0 + j8$

15. $8.6 - j4.6$

17. $61 + j0$

19. $39 + j54$

21. $6 + j18$

23. $-4 + j0$

25. $0.1 - j0.2$

27. $-1 + j3$

29. $1 - j$

31. $1 - j2$

33. $1.4 - j1.8$

35. $4.5 + j0.5$

37. $1 - j0.5$

39. $0 + j2.5$

41. $x^2 - j2x + j2x - j^2 4$
$= x^2 - (-1)4 = x^2 + 4$

43. $13.7 + j3.2$ kΩ

45. $1.4 - j0.20$ kΩ

47. $100 - j50$ V

Exercise 20.3

1.

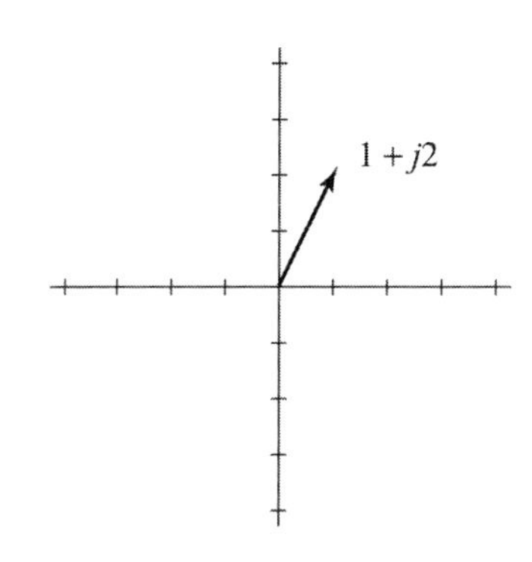

3.

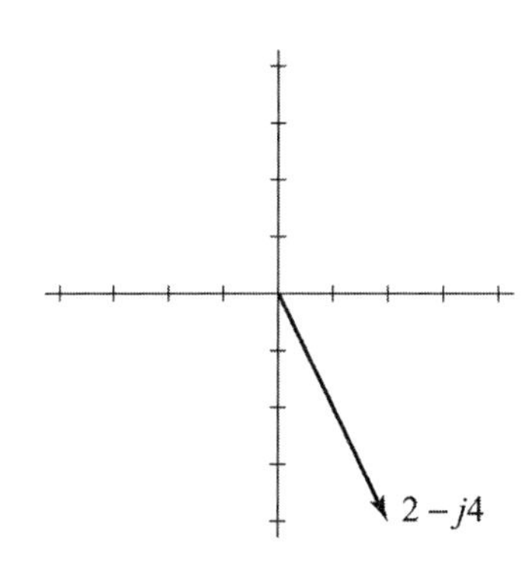

5.

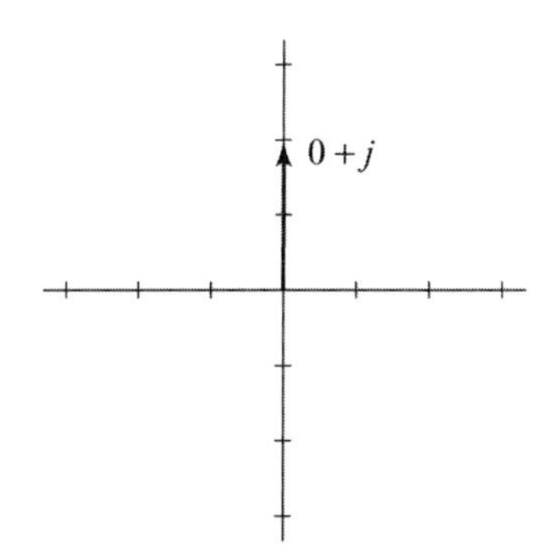

7.

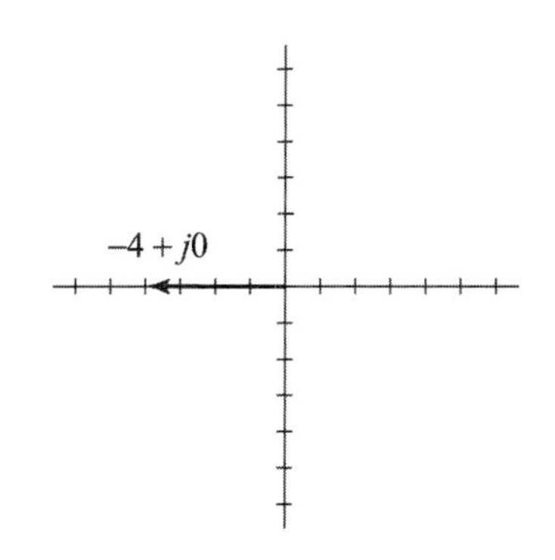

9.

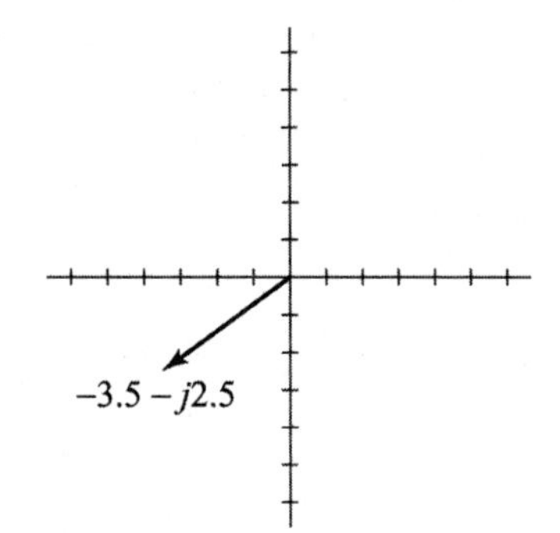

11.

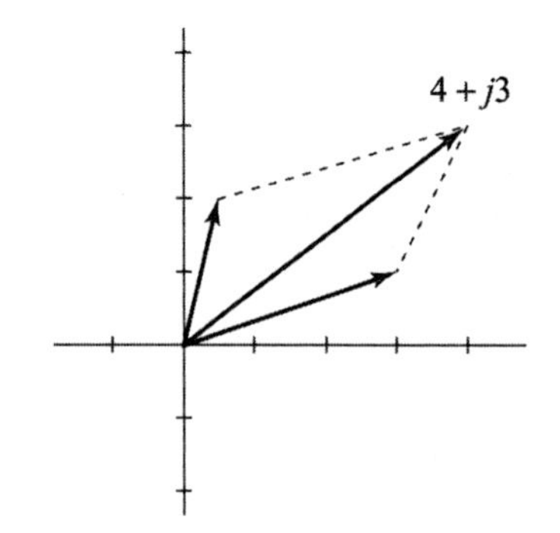

13.

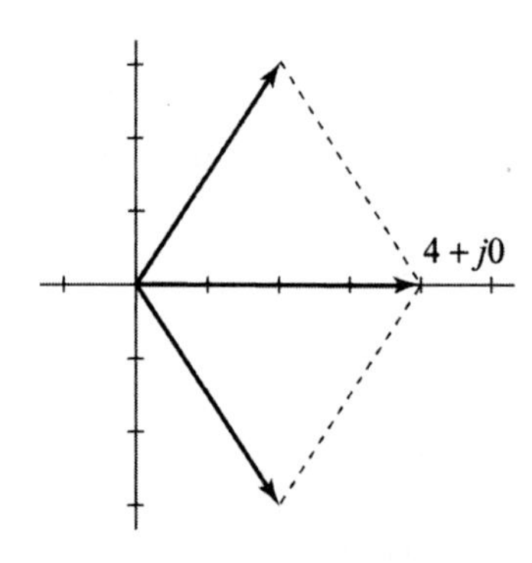

15.

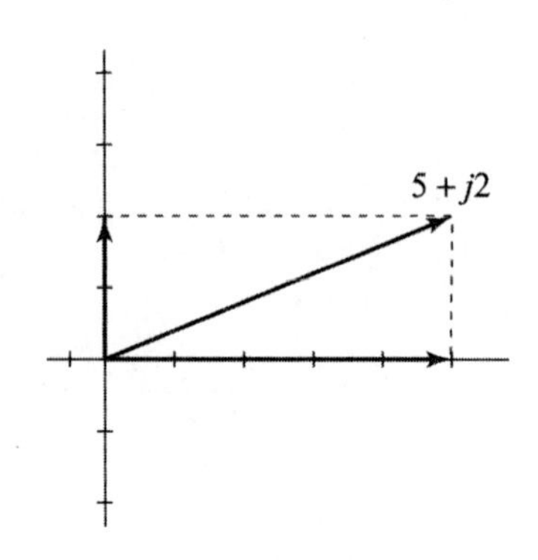

17.

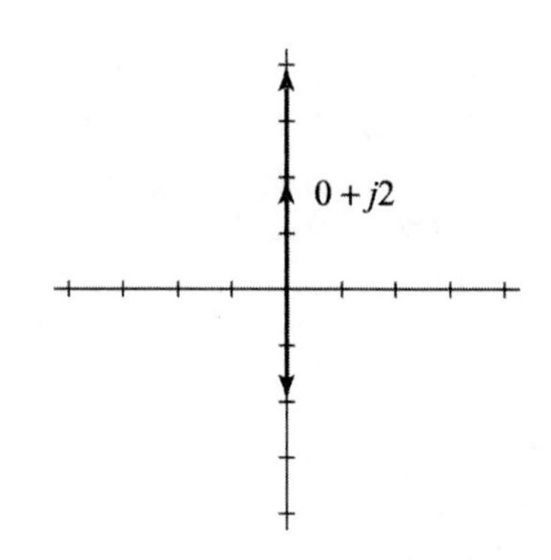

19.

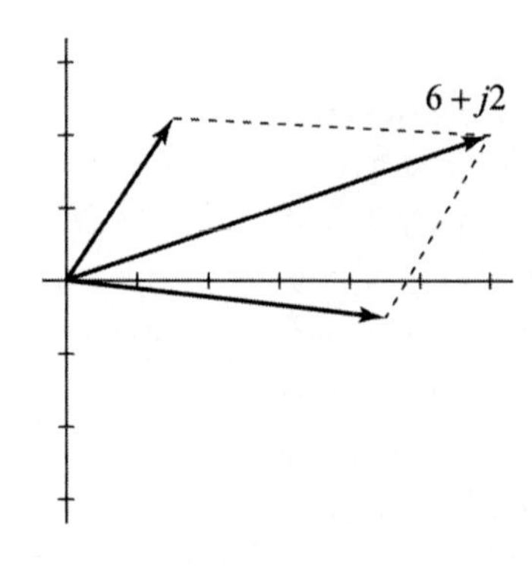

21. $2.8 \angle 45^\circ$
23. $4.5 \angle -63^\circ$
25. $1.4 \angle -45^\circ$
27. $4.5 \angle 27^\circ$
29. $55 \angle -90^\circ$
31. $75 \angle 0^\circ$
33. $25 \angle 53^\circ$
35. $1.3 \angle -67^\circ$
37. $4.3 \angle 65^\circ$
39. $2.1 + j2.1$
41. $43 - j25$
43. $100 + j0$
45. $0 - j10$
47. $2.9 - j1.4$
49. $0.60 + j4.3$
51. $0 + j15$
53. $7.1 - j7.1$
55. **(a)** $5.5 + j2.5\ \text{k}\Omega = 6.0 \angle 24^\circ\ \text{k}\Omega$
(b) $3.3 - j4.7\ \text{k}\Omega = 5.7 \angle -55^\circ\ \text{k}\Omega$

Exercise 20.4

1. $89 \angle 40^\circ$
3. $32 \angle 64^\circ$
5. $97 \angle -8.1^\circ$
7. $110 \angle 90^\circ$
9. $10 \angle -90^\circ$
11. $6.4 \angle -30^\circ$
13. $5.7 \angle -76^\circ$
15. $8 \angle 60^\circ$
17. $660 \angle -45^\circ$
19. $200 \angle -50^\circ$
21. $180 \angle 90^\circ$
23. $18 \angle 0^\circ$
25. $3.0 \angle 50^\circ$
27. $0.5 \angle -68^\circ$
29. $4.0 \angle 90^\circ$
31. $1.7 \angle 40^\circ$
33. $0.55 \angle 90^\circ$
35. $4 \angle 0^\circ$
37. $26 \angle -90^\circ$
39. $1.5 \angle 90^\circ$
41. $2.2 \angle -27^\circ$
43. $0.88 \angle 38^\circ$
45. $4.8 \angle 48^\circ$ mA
47. $3.6 \angle 65^\circ$ mA
49. $31 \angle 63^\circ\ \text{k}\Omega$

Review Exercises

1. $j9$
3. $-j$
5. $-j1.6$
7. -32
9. 96
11. $-j72$
13. $-j2.5$
15. 5
17. $6 + j2$
19. $22 - j7$
21. $-2.1 - j4.1$
23. $0.2 + j0.4$
25. $2 - j4$
27. $1 + j0$
29. $3.2 \angle 72^\circ$
31. $45 \angle -90^\circ$
33. $5 \angle 0^\circ$
35. $87 + j50$
37. $0 + j5.5$
39. $8.9 \angle 39^\circ$
41. $40 \angle -31^\circ$
43. $3.0 \angle 90^\circ$
45. $14 \angle 40^\circ$
47. $60 \angle -70^\circ$
49. $60 \angle 30^\circ$
51. $3 \angle 20^\circ$
53. $3.0 \angle -8.0^\circ$
55. $1 + j0$
57. **(a)** $22 \angle 56^\circ$ V
(b) $6.4 \angle -34^\circ$ V
59. **(a)** $75 \angle 60^\circ$ V
(b) $1.1 \angle -28^\circ$ V
61. $8.3 \angle 46^\circ\ \text{k}\Omega = 5.8 + j6.0\ \text{k}\Omega$

CHAPTER 21

Exercise 21.1

1. 3.8 kΩ

3. 630 Ω

5. 10 kΩ

7. 110 mH

9. 40 mH

11. 640 Hz

13. 7.4 kHz

15. 800 Ω

17. 1.4 kΩ

19. 2.5 V

21. 2.5 mA

23. $0.75 + j1.5$ kΩ = 1.7 ∠63° kΩ

25. $3.3 + j4.5$ kΩ = 5.6 ∠54° Ω

27. $2 + j1.2$ kΩ = 2.3 ∠31° kΩ

29. $910 + j750$ Ω = 1.2 ∠39° kΩ

31. **(a)** $V_R = 12$ V, $V_L = 16$ V
(b) $V_T = 20$ ∠52° V, $Z = 1.0$ ∠52° kΩ

33. **(a)** $V_R = 5$ V, $V_L = 7.5$ V
(b) $V_T = 9.0$ ∠56° V, $Z = 1.8$ ∠56° kΩ

35. **(a)** $V_R = 17$ V, $V_L = 14$ V
(b) $V_T = 21$ ∠39° V, $Z = 28$ ∠39° kΩ

37. **(a)** $V_R = 41$ V, $V_L = 33$ V
(b) $V_T = 52$ ∠39° V, $Z = 3.5$ ∠39° kΩ

39. **(a)** 700 ∠28° Ω
(b) $I = 21$ mA, $V_R = 13$ V, $V_L = 7.1$ V

41. **(a)** 3.2 ∠18° kΩ
(b) $I = 3.8$ mA, $V_R = 11$ V, $V_L = 3.8$ V

43. **(a)** 4.9 ∠29° kΩ
(b) $I = 12$ mA, $V_R = 52$ V, $V_L = 29$ V

45. **(a)** 16 ∠45° kΩ
(b) $I = 3.2$ mA, $V_R = 35$ V, $V_L = 35$ V

47. **(a)** $X_L = 1.4$ kΩ, $I = 5.3$ mA, $R = 1.8$ kΩ, $V_R = 9.6$ V
(b) $V_T = 12$ ∠38° V

49. **(a)** $X_L = 280$ Ω, $I = 400$ mA, $V_R = 40$ V, $V_L = 110$ V
(b) $Z = 300$ ∠71° Ω

Exercise 21.2

1. 1.3 kΩ

3. 21 kΩ

5. 530 Ω

7. 1.6 nF

9. 99 nF

11. 1.3 kHz

13. 13 kHz

15. 1.7 kΩ

17. 8.0 kΩ

19. 4.8 V

21. 400 μA

23. $1.6 - j1.2$ kΩ = 2.0 ∠−37° kΩ

25. $680 - j910$ Ω = 1.1 ∠−53° kΩ

27. $550 - j550$ Ω = 780 ∠−45° Ω

29. $0.82 - j1.0$ kΩ = 1.3 ∠−51° kΩ

31. **(a)** $V_R = 24$ V, $V_C = 13$ V
(b) $V_T = 27$ ∠−28° V, $Z = 8.5$ ∠−28° kΩ

33. **(a)** $V_R = 5.0$ V, $V_C = 13$ V
(b) $V_T = 13$ ∠−68° V, $Z = 27$ ∠−68° kΩ

35. **(a)** $V_R = 2.0$ V, $V_C = 3.2$ V
(b) $V_T = 3.8$ ∠−58° V, $Z = 940$ ∠−58° Ω

37. **(a)** $V_R = 28$ V, $V_C = 18$ V
(b) $V_T = 33$ ∠−32° V, $Z = 5.1$ ∠−32° kΩ

39. **(a)** $Z = 920$ ∠−42° Ω
(b) $I = 22$ mA, $V_R = 15$ V, $V_C = 13$ V

41. **(a)** $Z = 1.5$ ∠−35° Ω
(b) $I = 16$ mA, $V_R = 20$ V, $V_C = 14$ V

43. **(a)** $Z = 15$ ∠−58° kΩ
(b) $I = 780$ μA, $V_R = 6.4$ V, $V_C = 10$ V

45. **(a)** $Z = 1.1$ ∠−45° Ω
(b) $I = 9.1$ mA, $V_R = 6.9$ V, $V_C = 6.9$ V

47. **(a)** $X_C = 1.6$ kΩ, $I = 4.1$ mA, $V_C = 6.5$ V, $V_R = 14$ V
(b) $Z = 3.7$ ∠−26° kΩ

49. **(a)** $X_C = 18$ kΩ, $V_R = 94$ V, $V_C = 76$ V, $I = 4.3$ mA
(b) $Z = 28$ ∠−39° kΩ

Exercise 21.3

1. $X = 300$ Ω, $Z = 750 + j300$ Ω = 810 ∠22° Ω

3. $X = 2.2$ kΩ, $Z = 1.8 - j2.2$ kΩ = 2.8 ∠−51° kΩ

5. $X = 4.8$ kΩ, $Z = 3.9 + j4.8$ kΩ = 6.2 ∠51° kΩ

7. $X = 450$ Ω, $Z = 1000 - j450$ Ω = 1100 ∠−24° Ω

9. $X = 350$ Ω, $Z = 200 + j350$ Ω = 400 ∠60° Ω

11. $I = 2.5$ mA, $V_R = 8.3$ V, $V_L = 3.0$ V, $V_C = 12$ V, $V_T = 12$ ∠−47° V

13. $I = 72$ mA, $V_R = 20$ V, $V_L = 37$ V, $V_C = 33$ V, $V_T = 20$ ∠13° V

15. $I = 4.8$ mA, $V_R = 4.3$ V, $V_L = 20$ V, $V_C = 7.1$ V, $V_T = 14$ ∠72° V

17. $I = 43$ mA, $V_R = 26$ V, $V_L = 51$ V, $V_C = 36$ V, $V_T = 30$ ∠31° V

19. 3.6 kHz

21. 28 kHz

23. 1.5 μF

25. 70 mH

27. **(a)** $X_L = 820$ Ω, $X_C = 610$ Ω, $Z = 330 + j210$ Ω = 390 ∠32° Ω
(b) $I = 31$ mA, $V_R = 10$ V, $V_L = 25$ V, $V_C = 19$ V, $V_T = 10 + j6$ V = 12 ∠32° V

29. **(a)** $X_L = 3.8$ kΩ, $X_C = 6.6$ kΩ
$Z = 2.2 - j2.9$ kΩ
= 3.6 ∠−52° kΩ
(b) $I = 5.5$ mA,
$V_C = 21$ V,
$V_R = 12$ V, $V_L = 21$ V
$V_T = 12 - j15.8$ V
= 20 ∠−52° V

31. **(a)** $Z = 600 + j450$ Ω = 750 ∠37° Ω
(b) $I = 27$ mA, $P = 430$ mW, $Q = 320$ mVAR, $S = 540$ mVI, p.f. = 0.80

33. **(a)** $V_R = 18$ V, $V_L = 36$ V, $V_C = 16$ V, $V_T = 18 + j20$ V = 27 ∠48° V
(b) $Z = 1.8 + j2.0$ kΩ = 2.7 ∠48° kΩ

35. **(a)** $X_L = 1.9$ kΩ, $X_C = 3.2$ kΩ, $Z = 3.9 - j1.3$ kΩ = 4.1 ∠−18° kΩ
(b) $I = 2.4$ mA, $V_{R_1} = 6.6$ V,
$V_{R_2} = 2.9$ V,
$V_L = 4.6$ V,
$V_C = 7.7$ V,
$V_T = 9.5 - j3.2$ V = 10 ∠−18° V

Review Exercises

1. 9.4 kΩ

3. 760 Ω

5. 2.0 kHz

7. 4.0 nF

9. 16 kHz

11. $820 + j1000$ Ω = 1.3 ∠51° kΩ

13. $150 + j430$ Ω = 460 ∠71° Ω

15. $6.2 - j8.5$ kΩ = 11 ∠−54° kΩ

17. $1800 - j780\ \Omega = 2.0\ \angle{-23^\circ}\ \text{k}\Omega$

19. $470 + j550\ \Omega = 720\ \angle 49^\circ\ \Omega$

21. $6.8 - j3.1\ \text{k}\Omega = 7.5\ \angle{-25^\circ}$

23. **(a)** $Z = 730\ \angle 58^\circ\ \Omega$
(b) $I = 20$ mA,
$V_T = 8.0 + j13\ \text{V} = 15\ \angle 58^\circ\ \text{V}$

25. **(a)** $Z = 8.4\ \angle{-42^\circ}\ \Omega$
(b) $I = 2.9$ mA,
$V_T = 18 - j16\ \text{V} = 24\ \angle{-42^\circ}\ \text{V}$

27. **(a)** $Z = 4.8\ \angle 47^\circ\ \text{k}\Omega$
(b) $I = 2.5$ mA,
$V_T = 8.2 + j8.7\ \text{V} = 12\ \angle 47^\circ\ \text{V}$

29. **(a)** $X_L = 1.1\ \text{k}\Omega$, $V_R = 18$ V,
$V_L = 11$ V
(b) $V_T = 18 + j11\ \text{V} = 21\ \angle 31^\circ$ V,
$Z = 1.8 + j1.1\ \text{k}\Omega = 2.1\ \angle 31^\circ\ \text{k}\Omega$

31. **(a)** $X_C = 4.6\ \text{k}\Omega$, $V_R = 2.6$ V,
$V_C = 3.7$ V
(b) $V_T = 2.6 - j3.7\ \text{V} = 4.5\ \angle{-54^\circ}$ V,
$Z = 3.3 - j4.6\ \text{k}\Omega = 5.7\ \angle{-54^\circ}\ \text{k}\Omega$

33. **(a)** $X_L = 5.7\ \text{k}\Omega$, $X_C = 2.7\ \text{k}\Omega$,
$V_R = 23$ V, $V_L = 43$ V,
$V_C = 20$ V, $I = 7.5$ mA
(b) $V_T = 23 + j23\ \text{V} = 32\ \angle 45^\circ$ V,
$Z = 3.0 + j3.0\ \text{k}\Omega = 4.2\ \angle 45^\circ\ \text{k}\Omega$

35. **(a)** $X_L = 4.1\ \text{k}\Omega$, $X_C = 3.9\ \text{k}\Omega$,
$V_R = 28$ V, $V_L = 155$ V,
$V_C = 145$ V, $I = 38$ mA
(b) $V_T = 28 + j11\ \text{V} = 30\ \angle 21^\circ$ V,
$Z = 750 + j280\ \Omega = 800\ \angle 21^\circ\ \Omega$

37. **(a)** $Z = 680 - j900\ \Omega = 1.1\ \angle{-53^\circ}\ \Omega$,
$I = 8.0$ mA
(b) $V_T = 5.4 - j7.2\ \text{V} = 9.0\ \angle{-53^\circ}$ V,
$P = 43$ mW

CHAPTER 22

Exercise 22.1

1. $1.6 - j1.8\ \text{mA} = 2.4\ \angle{-49^\circ}$ mA,
$3.2 + j3.7\ \text{k}\Omega = 4.9\ \angle 49^\circ\ \text{k}\Omega$

3. $4.2 - j6.1\ \text{mA} = 7.4\ \angle{-56^\circ}$ mA,
$380 + j560\ \Omega = 680\ \angle 56^\circ\ \Omega$

5. $I_T = 9.1 - j3.2\ \text{mA} = 9.6\ \angle{-19^\circ}$ mA,
$Z = 2.9 + j1.0\ \text{k}\Omega = 3.1\ \angle{-19^\circ}\ \text{k}\Omega$

7. $2.0 - j2.0\ \text{mA} = 2.8\ \angle{-45^\circ}$ mA,
$5.0 + j5.0\ \text{k}\Omega = 7.1\ \angle 45^\circ\ \text{k}\Omega$

9. $25 - j18\ \text{mA} = 31\ \angle{-35^\circ}$ mA,
$340 + j240\ \Omega = 420\ \angle 35^\circ\ \Omega$

11. **(a)** $I_R = 1.1$ mA, $I_L = 1.4$ mA
(b) $I_T = 1.1 - j1.4\ \text{mA} = 1.8\ \angle{-52^\circ}$ mA,
$Z = 2.1 + j2.7\ \text{k}\Omega = 3.4\ \angle 52^\circ\text{k}\Omega$

13. **(a)** $I_T = 500 - j640\ \mu\text{A} = 810\ \angle{-52^\circ}\ \mu\text{A}$,
$Z = 9.1 + j12\ \text{k}\Omega = 15\ \angle 52^\circ\ \text{k}\Omega$
(b) $R = 9.1\ \text{k}\Omega$, $L = 190$ mH

15. **(a)** $I_T = 6.1 - j5.5\ \text{mA} = 8.2\ \angle{-42^\circ}$ mA
(b) $Z = 3.8 + j3.4\ \text{k}\Omega = 5.0\ \angle 42^\circ\ \text{k}\Omega$,
$R = 3.8\ \text{k}\Omega$, $L = 90$ mH

Exercise 22.2

1. $8.3 + j13\ \text{mA} = 15\ \angle 56^\circ$ mA,
$920 - j1400\ \Omega = 1.7\ \angle{-56^\circ}\ \text{k}\Omega$

3. $48 + j64\ \text{mA} = 80\ \angle 53^\circ$ mA,
$230 - j300\ \Omega = 370\ \angle{-53^\circ}\ \Omega$

5. $I_T = 5.4 + j5.0\ \text{mA} = 7.4\ \angle 43^\circ$ mA
$Z = 600 - j550\ \Omega = 810\ \angle{-43^\circ}\ \Omega$

7. $800 + j510\ \mu\text{A} = 950\ \angle 33^\circ\ \mu\text{A}$,
$21 - j14\ \text{k}\Omega = 25\ \angle{-33^\circ}\ \text{k}\Omega$

9. $20 + j15\ \text{mA} = 25\ \angle 37^\circ$ mA,
$480 - j360\ \Omega = 600\ \angle{-37^\circ}\ \Omega$

11. **(a)** $I_R = 7.1$ mA, $I_C = 13$ mA,
$I_T = 7.1 + j13\ \text{mA} = 14\ \angle 60^\circ$ mA
(b) $Z = 520 - j900\ \Omega = 1.0\ \angle{-60^\circ}\ \text{k}\Omega$

13. **(a)** $V = 4.0$ V, $I_R = 5.9$ mA,
$I_T = 5.9 + j9.5\ \text{mA} = 11\ \angle 58^\circ$ mA
(b) $Z = 190 - j310\ \Omega = 360\ \angle{-58^\circ}\ \Omega$,
$R = 190\ \Omega$, $C = 35$ nF

15. **(a)** $I_R = 29$ mA, $I_C = 23$ mA,
$I_T = 29 + j23\ \text{mA} = 37\ \angle 38^\circ$ mA
(b) $Z = 390 - j300\ \Omega = 490\ \angle{-38^\circ}\ \Omega$,
$R = 390\ \Omega$, $C = 100$ nF

Exercise 22.3

1. $G = 130\ \mu\text{S}$, $B = 150\ \mu\text{S}$, $Y = 130 - j150\ \mu\text{S} = 200\ \angle{-49^\circ}\ \mu\text{S}$

3. $G = 830\ \mu\text{S}$, $B = 1.2$ mS, $Y = 0.83 - j1.2\ \text{mS} = 1.5\ \angle{-56^\circ}$ mS

5. $G = 100\ \mu\text{S}$, $B = 100\ \mu\text{S}$, $Y = 100 - j100\ \mu\text{S} = 140\ \angle{-45^\circ}\ \mu\text{S}$

7. $G = 2.0$ mS, $B = 1.4$ mS, $Y = 2.0 - j1.4\ \text{mS} = 2.4\ \angle{-35^\circ}$ mS

9. $G = 330\ \mu\text{S}$, $B = 500\ \mu\text{S}$, $Y = 330 + j500\ \mu\text{S} = 600\ \angle 56^\circ\ \mu\text{S}$

11. $G = 1.6$ mS, $B = 2.1$ mS, $Y = 1.6 + j2.1\ \text{mS} = 2.7\ \angle 53^\circ$ mS

13. $G = 33\ \mu\text{S}$, $B = 21\ \mu\text{S}$, $Y = 33 + j21\ \mu\text{S} = 40\ \angle 33^\circ\ \mu\text{S}$

15. $G = 1.3$ mS, $B = 1.0$ mS, $Y = 1.3 + j1.0\ \text{mS} = 1.7\ \angle 37^\circ\ \mu\text{S}$

17. **(a)** $Y = 42 - j53\ \mu\text{S} = 67\ \angle{-52^\circ}\ \mu\text{S}$,
$Z = 9.2 + j12\ \text{k}\Omega = 15\ \angle 52^\circ\ \text{k}\Omega$
(b) $R = 9.2\ \text{k}\Omega$, $L = 180$ mH

19. **(a)** $Y = 1.5 + j2.4\ \text{mS} = 2.8\ \angle 58^\circ$ mS,
$Z = 190 - j310\ \Omega = 360\ \angle{-58^\circ}\ \Omega$
(b) $R = 190\ \Omega$, $C = 35$ pF

Exercise 22.4

1. $30 - j40\ \text{mA} = 50\ \angle{-53^\circ}$ mA

3. $55 + j60\ \text{mA} = 81\ \angle 47^\circ$ mA

5. $I_T = 700 + j400\ \mu\text{A} = 810\ \angle 30^\circ\ \mu\text{A}$

7. $1.1 - j0.90\ \text{mA} = 1.4\ \angle{-39^\circ}$ mA

9. $5.0 + j0\ \text{mA} = 5.0\ \angle 0^\circ$ mA

11. $I_T = 31 - j12\ \text{mA} = 33\ \angle{-21^\circ}$ mA,
$Z_T = 340 + j130\ \Omega = 360\ \angle 21^\circ\ \Omega$

13. $I_T = 23 + j14$ mA $= 26 \angle 31°$ mA,
$Z_T = 1.2 - j0.71$ kΩ $= 1.4 \angle -31°$ kΩ

15. $I_T = 3.8 + j0$ mA $= 3.8 \angle 0°$ mA
$Z = 910 + j0$ Ω $= 910 \angle 0°$ Ω

17. $I_T = 17 - j31$ mA $= 35 \angle -61°$ mA,
$Z_T = 330 + j590$ Ω $= 680 \angle 61°$ Ω

19. $I_T = 600 + j770$ μA $= 1.0 \angle 52°$ mA,
$Z_T = 3.8 - j4.8$ kΩ $= 6.1 \angle -52°$ kΩ

21. $2.6 - j0.99$ mS $= 2.7 \angle -21°$ mS

23. $630 + j380$ μS $= 730 \angle 31°$ μS

25. $1.1 + j0$ mS $= 1.1 \angle 0°$ mS

27. $0.71 - j1.3$ mS $= 1.5 \angle -61°$ mS

29. $100 + j130$ μS $= 160 \angle 52°$ μS

31. **(a)** $I_T = 2.3 - j1.7$ mA $= 2.9 \angle -36°$ mA,
$Z = 2.8 + j2.0$ kΩ $= 3.5 \angle 36°$ kΩ
(b) $R = 2.8$ kΩ, $L = 16$ mH

33. **(a)** $I_T = 2.2 + 0.77$ mA $= 2.3 \angle 19°$ mA,
$Z = 6.1 - j2.1$ kΩ $= 6.4 \angle -19°$ kΩ
(b) $R = 6.1$ kΩ, $C = 1.5$ nF

35. **(a)** $Y = 190 - j79$ μS, $Z = 4.5 + j1.9$ kΩ
(b) $R = 4.5$ kΩ, $L = 25$ mH

37. **(a)** $Y = 740 \angle -26°$ μS, $Z = 1.3 \angle 26°$ kΩ
(b) $V = 16$ V

Exercise 22.5

1. $1.3 + j2.0$ kΩ $= 2.4 \angle 57°$ kΩ

3. $410 + j530$ Ω $= 670 \angle 52°$ Ω

5. $990 \angle -8.4°$ Ω $= 980 - j140$ Ω

7. $2.3 - j2.3$ kΩ $= 3.3 \angle -45°$ kΩ

9. $5.0 + j15$ kΩ $= 16 \angle 72°$ kΩ

11. $13 + j28$ kΩ $= 31 \angle 65°$ kΩ

13. $I_T = 6.1 \angle -52°$ mA $= 3.7 - j4.8$ mA,
$V_R = 9.1 \angle -52°$ V $= 5.6 - j7.2$ V,
$V_L = 12 \angle 37°$ V $= 9.4 + j7.2$ V

Exercise 22.6

1. $I = 1.18$ mA,
$V_R = 14.1$ V,
$V_C = 14.1$ V,
$f_c = 1.33$ kHz

3. **(a)** $f_c = 1.06$ kHz
(b) $X_C = 1.5$ kΩ,
$I = 4.71$ mA,
$V_{OUT} = 7.07$ V

5. **(a)** $f_c = 3.18$ kHz
(b) $X_L = 1.0$ kΩ, $I = 7.07$ mA,
$V_{OUT} = 7.07$ V

7. **(a)** $f_c = 10.6$ kHz
(b) $X_C = 1.5$ kΩ,
$I = 5.66$ mA,
$V_{OUT} = 8.49$ V

9. **(a)** $I = 4.47$ mA,
$V_R = 4.47$ V,
$V_{OUT} = 8.94$ V
(b) $f = 3.18$ kHz,
$f_c = 1.59$ kHz

11. **(a)** $f_c = 4.77$ kHz
(b) $X_L = 1.5$ kΩ,
$I = 9.42$ mA,
$V_{OUT} = 14.1$ V

13. $f_{c_2} = 1.06$ kHz,
$f_{c_1} = 318$ Hz

15. 900 Hz–1100 Hz, 5

17. 2.9 kHz–3.1 kHz, 15

Review Exercises

1. $I_T = 2.1 - j1.5$ mA $= 2.5 \angle -35°$ mA,
$Z = 2.9 + j2.0$ kΩ $= 3.5 \angle 35°$ kΩ

3. $I_T = 15 + j6.7$ mA $= 16 \angle 24°$ mA,
$Z = 680 - j310$ Ω $= 750 \angle -24°$ Ω

5. $I_T = 3.6 + j2.7$ mA $= 4.5 \angle 36°$ mA,
$Z = 7.2 - j5.2$ kΩ $= 8.9 \angle -36°$ kΩ

7. $I_T = 14 - j9.5$ mA $= 17 \angle -34°$ mA,
$Z = 390 + j260$ Ω $= 470 \angle 34°$ Ω

9. $I_T = 7.4 + j0.33$ mA $= 7.4 \angle 2.6°$ mA,
$Z = 2.7 - j0.12$ kΩ $= 2.7 \angle -2.6°$ kΩ

11. $B = 160$ μS, $Y = 230 - j160$ μS $= 280 \angle -35°$ μS

13. $B = 560$ μS, $Y = 1.2 + j0.56$ mS $= 1.3 \angle 24°$ mS

15. $B = 67$ μS, $Y = 91 + j67$ μS $= 110 \angle 36°$ μS

17. $B = 1.2$ mS, $Y = 1.8 - j1.2$ mS $= 2.1 \angle -34°$ mS

19. $B = 17$ μS, $Y = 370 + j17$ μS $= 370 \angle 2.6°$ μS

21. **(a)** $I_T = 11 - j8.0$ mA $= 13 \angle -37°$ mA
(b) $Z = 360 + j270$ Ω $= 450 \angle 37°$ Ω;
$R = 360$ Ω,
$L = 7.1$ mH

23. **(a)** $I_T = 16 + j7.5$ mA $= 18 \angle 25°$ mA
(b) $Z = 1.2 - j0.57$ kΩ $= 1.4 \angle -25°$ kΩ;
$R = 1.2$ kΩ,
$C = 28$ nF

25. **(a)** $Y = 300 - j590$ μS $= 660 \angle -63°$ μS,
$Z = 0.69 + j1.3$ kΩ $= 1.5 \angle 63°$ kΩ
(b) $R = 690$ Ω, $L = 140$ mH

27. $Z = 1.2 \angle -26°$ kΩ $= 1.1 - j0.54$ kΩ

29. $Z = 6.0 \angle -65°$ kΩ $= 2.6 - j5.4$ kΩ

31. **(a)** $f_c = 15.9$ kHz
(b) $X_C = 1.0$ kΩ,
$I = 14.1$ mA,
$V_{OUT} = 14.1$ V

33. 2.9 kHz–3.1 kHz, 15

CHAPTER 23

Exercise 23.1

1. 5
3. 16
5. 9
7. $\frac{1}{8} = 0.125$
9. $\frac{1}{10} = 0.1$
11. $\frac{1}{8} = 0.125$
13. 0.3
15. $\frac{1}{16} = 0.0625$
17. 1.44
19. 3.16
21. 100
23. 125×10^{-3}
25. 1.5×10^{3}
27. $\frac{125}{64} = 1.95$
29. 10
31. 10
33. $I_T^{0.4}$
35. $-3e^2$
37. $\dfrac{5}{L^2 X}$
39. $\dfrac{V_x}{10^{-3}}$
41. $72s^3$
43. $\dfrac{2ab^2}{3}$
45. 8.4 mm
47. 3.1 kHz
49. 0.019

Exercise 23.2

1. $y = 3^x$

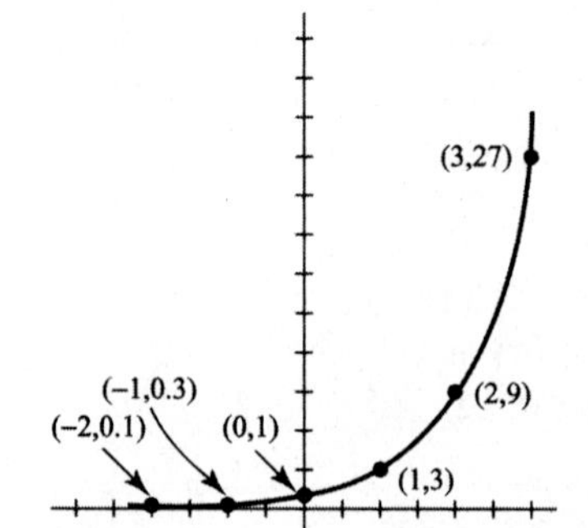

3. $y = 2^{x-1}$

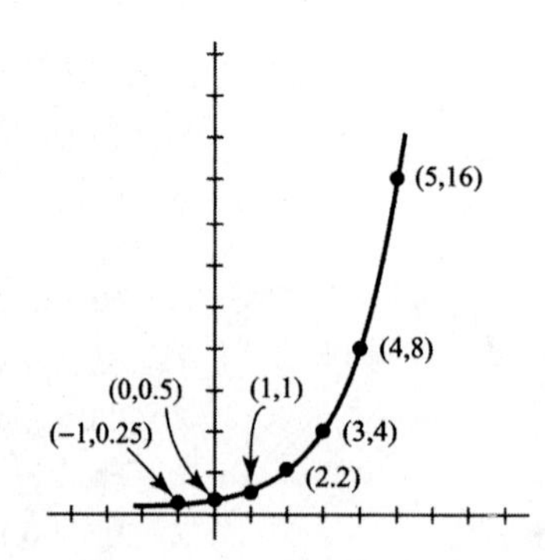

5. $y = 10^{0.2x}$

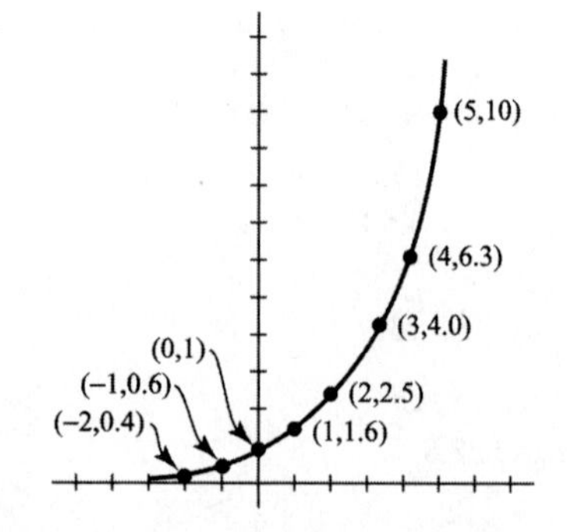

7. $y = 1.5^{-x}$

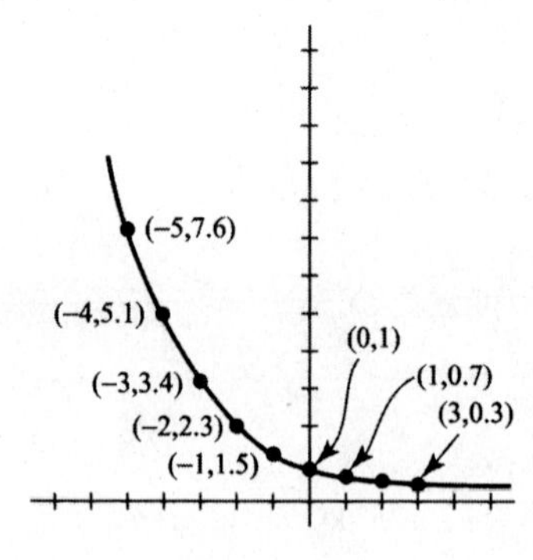

9. $y = 2^{-x+1}$

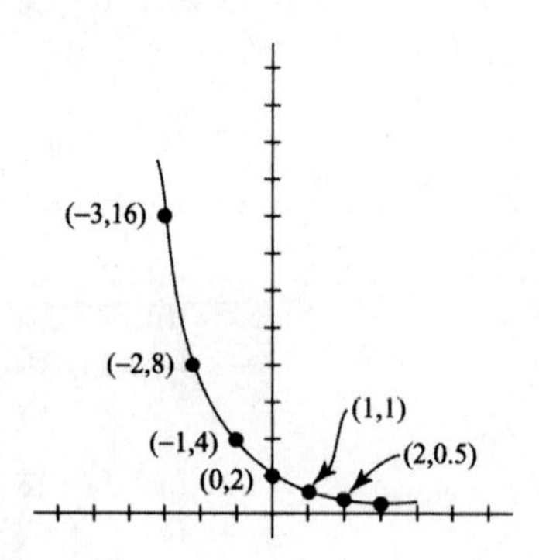

11. $y = 1 - 2^x$

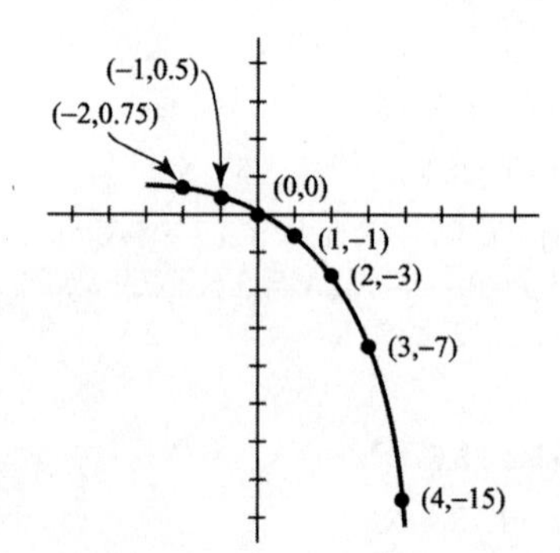

13. $y = e$

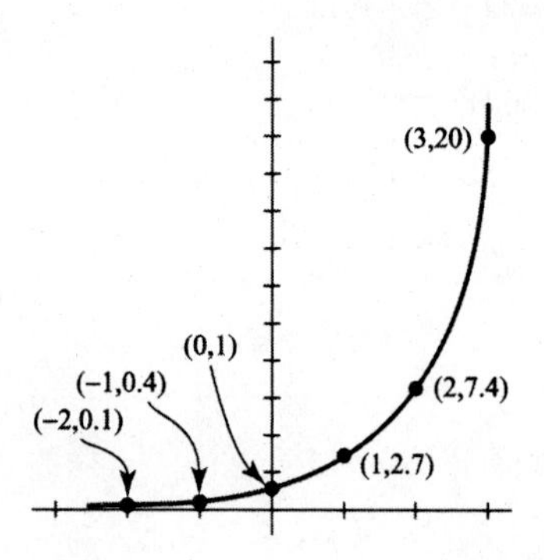

15. $y = 2e^{-t}$

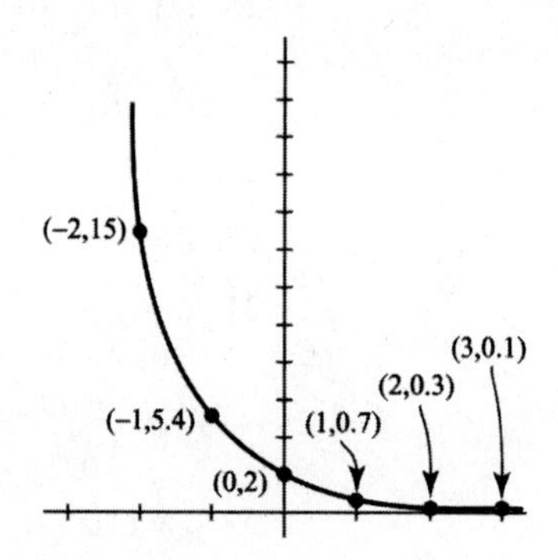

17. $y = 1 - e^{-t}$

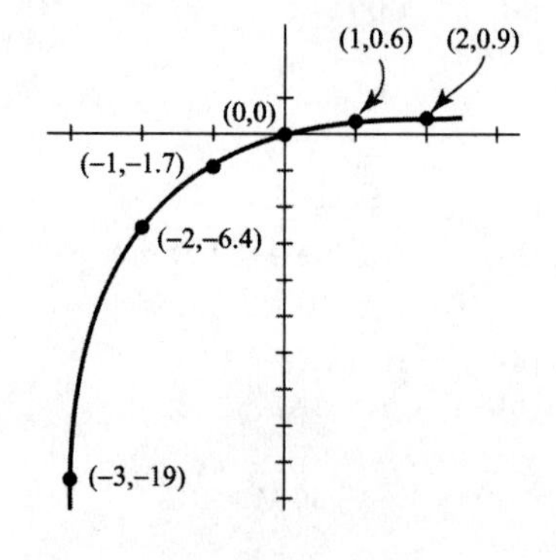

19. $y = 1 - e^{0.5t}$

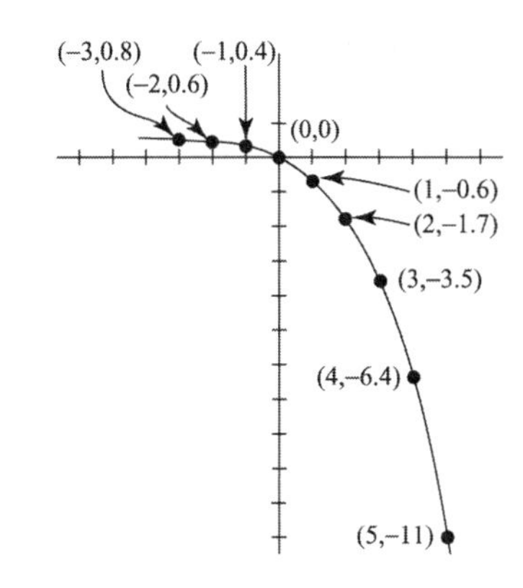

21. (a) $A = 6.0 \times 10^9(1.015)^n$

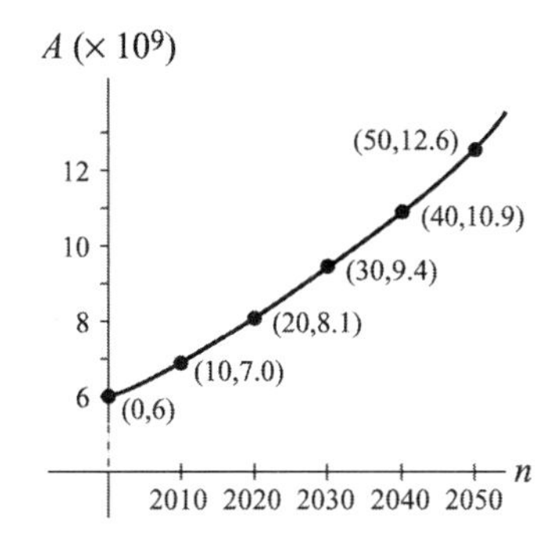

(b) ≈ 2035

23.

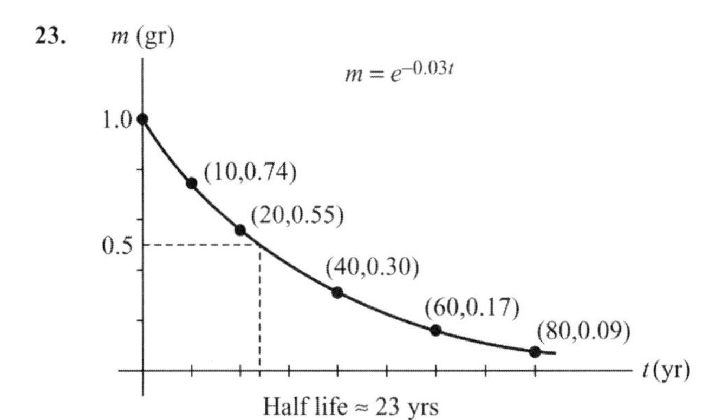

25. **(a)** 19.5 V
(b) 9.2 V
(c) 25 V

27. **(a)** 400 ms
(b) 30 μA
(c) 18 μA

Exercise 23.3

1. $\log_2 8 = 3$

3. $\log_5 0.2 = -1$

5. $\log 10{,}000 = 4$

7. $\log 0.631 = -0.2$

9. $\ln 7.39 = 2$

11. $\ln 0.223 = -1.5$

13. $\ln 0.100 = -2.3$

15. $5^2 = 25$

17. $2^{-3} = 0.125$

19. $10^3 = 1000$

21. $10^{-0.5} = 0.316$

23. $e^{1.1} = 3.0$

25. $e^{-1.93} = 0.145$

27. $e^{-2.95} = 0.0523$

29. 4

31. 100

33. $1/5 = 0.2$

35. 0.001

37. 3

39. −2

41. 4

43. −3

45. 1.94

47. −0.979

49. 3.72

51. −5.59

53. 3.22

55. −2.09

57. −5.03

59. 209

61. 5.11

63. 428

65. 3.86

67. 1.47

69. 759×10^{-6}

71. 0.259

73. 0.430

75. $2 + \log x$

77. $2\log y$

79. t

81. $2 - \ln t$

83. 0

85. $\ln 100 - t$

87. **(a)** 23 yrs
(b) 14 yrs
(c) 12 yrs

89. **(a)** 20 s
(b) 54 s

91. $\ln\left(\frac{i}{I_0}\right) = -\frac{t}{RC}$

93. 16 dB

Review Exercises

1. 9

3. 0.01

5. −27

7. 16×10^{-2}

9. 100

11. $\frac{10^6}{I^3}$

13. $y = 2^{1.2x}$

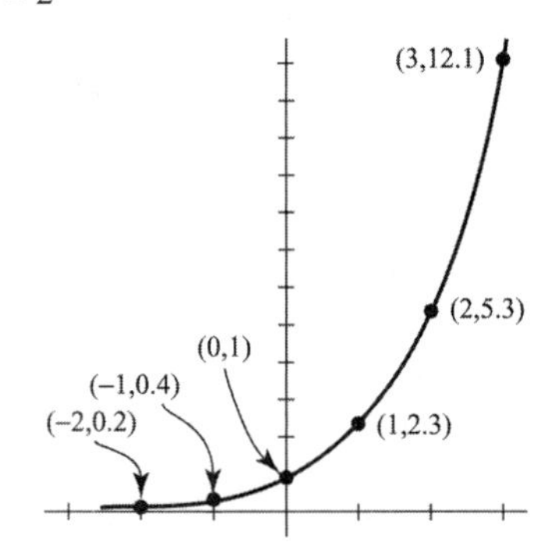

15. $y = e^t - 1$

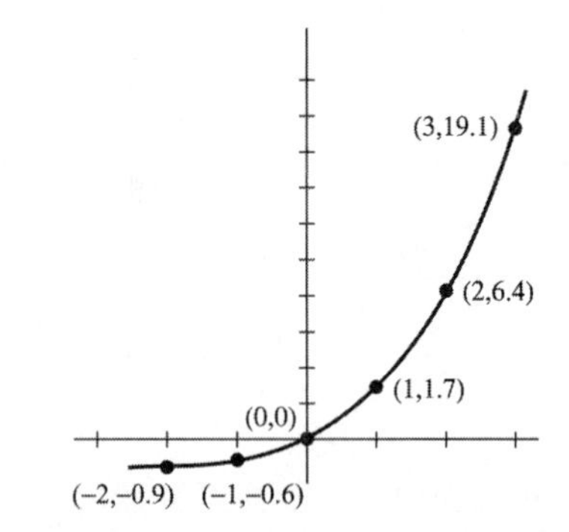

17. $\log 1000 = 3$

19. $\log_5 0.419 = -0.54$

21. $\ln 0.472 = -0.75$

23. $\ln 54.6 = 4$

25. $10^{-1.11} = 0.0776$

27. $10^{0.727} - 5.33$

29. $e^{0.99} = 2.69$

31. $e^{-5.39} = 4.55 \times 10^{-3}$

33. $\frac{1}{2} = 0.5$

35. 1,000,000

37. 2

39. –2

41. 1.40

43. –5.28

45. –0.393

47. 3.60

49. 7.23

51. 0.461

53. 67.4

55. 0.648

57. $\log V - 3$

59. 0

61. **(a)** 1.1
(b) 1.1

63. 56 yrs

65. **(a)** 5.2 h
(b) 8.3 h

67. **(a)** 1.8 mA
(b) 400 μA

69. **(a)** $V_1 = 10^{0.0553}$
(b) 1.1 V

CHAPTER 24

Exercise 24.1

1. 4.32

3. –0.941

5. 2.56

7. 1.30

9. 0.571

11. –0.566

13. 0.330

15. 1.54

17. 0.00768

19. 3.85

21. 0.0399

23. **(a)** 6.8 yr
(b) 12 yr
(c) 18 yr

25. **(a)** 1.4 ms
(b) 900 μS

27. **(a)** 24 h
(b) 35 h

29. **(a)** 1.1 ms
(b) 8.0 ms

Exercise 24.2

1. 1.0 ms

3. 2.6 ms

5. 4.0 ms

7. 540 μs

9, 2.0 ms

11. 6.7 ms

13. 750 μs

15. 210 μs

17. **(a)** 2 ms

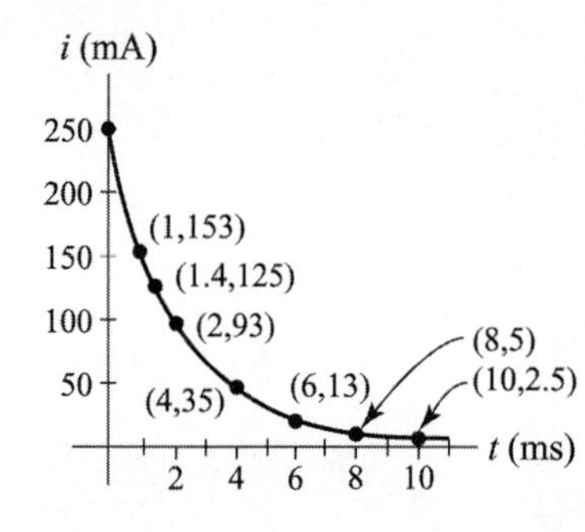

(b) $i = 250e^{-t/2}$

19. **(a)** 4 ms

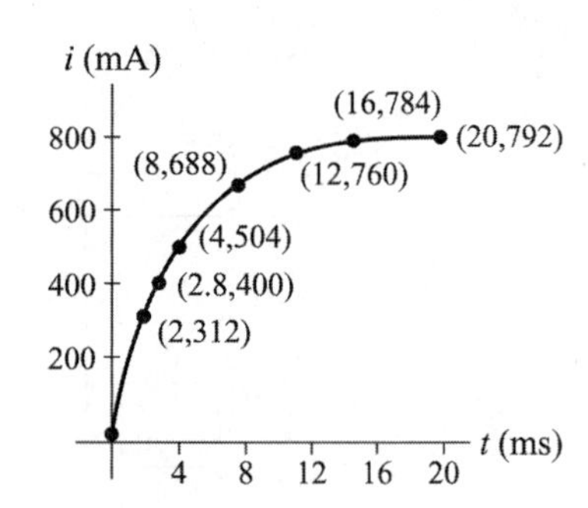

(b) $i = 800(1 - e^{-t/4})$

21. **(a)** 21 mA, 44 V
(b) 7.1 ms
(c) 8.1 ms

23. **(a)** 10 ms
(b) 35 ms

25. **(a)** 910 mA
(b) 3.5 ms
(c) 1.2 ms

27. **(a)** 570 μC
(b) 9.4 ms
(c) 60 ms

Exercise 24.3

1. 20 dB

3. –6.0 dB

5. 3.0 dB

7. 34 dB

9. 40 dB

11. 20 dB

13. –6.0 dB

15. 32 dB

17. 39 dB

19. 30 dB

21. 13 dBm

23. 30 dBm

25. –5.2 dBm

27. 100 mW

29. 20 mW

31. 2.0 W

33. 7.0 W

35. 40 mW

37. 1.0 W

39. 1.0 W

41. 1 mW

43. 2.0 V

45. 400 mV

47. 5.0 V

49. 4.7 V

51. 450 mV

53. 15 V

55. 4.0 V

57. 380 mV

59. 3.0 dB

61. 8.2 dB

Exercise 24.4

1.

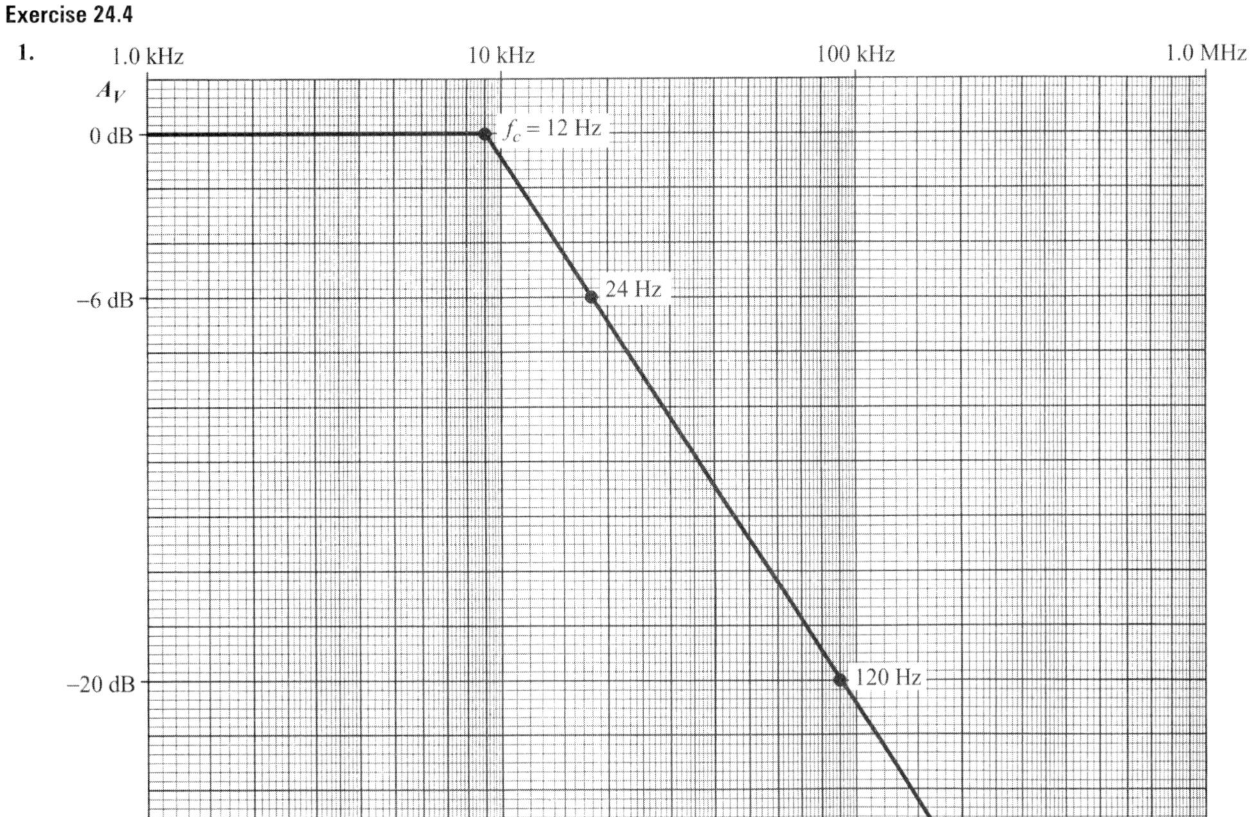

3.

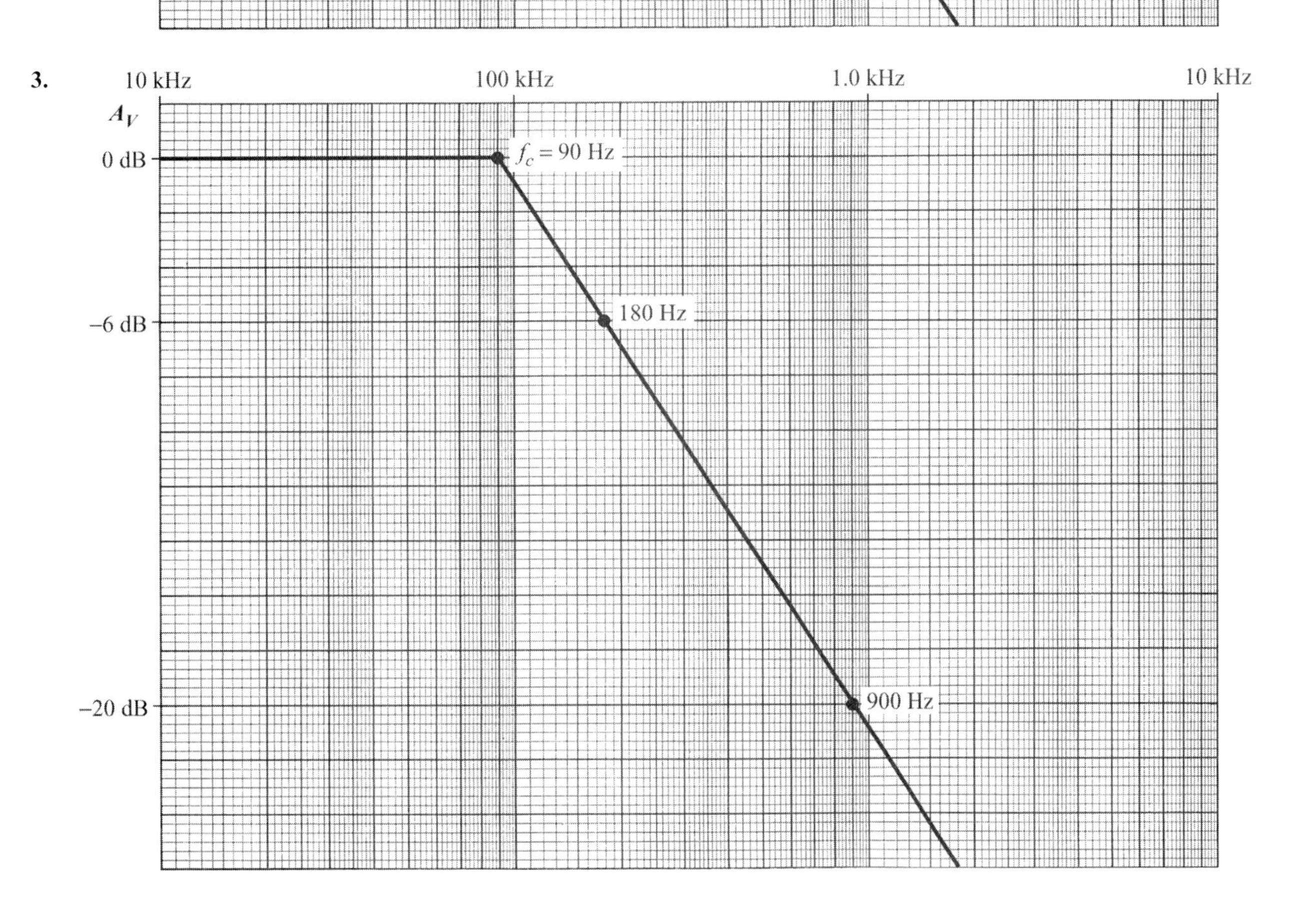

5.

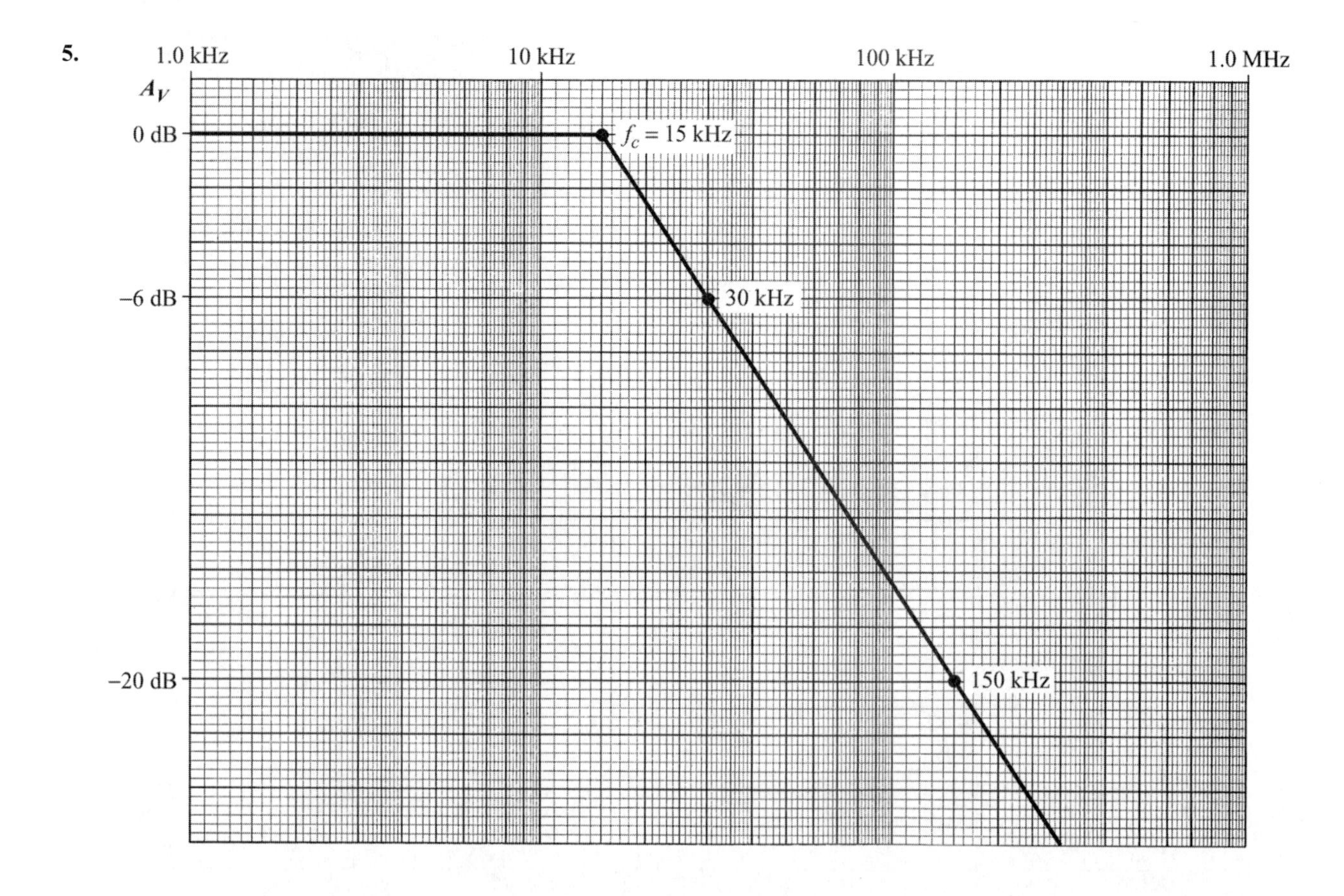
1.0 kHz
10 kHz
100 kHz
1.0 MHz
A_V
0 dB
−6 dB
−20 dB
f_c = 15 kHz
30 kHz
150 kHz

7.

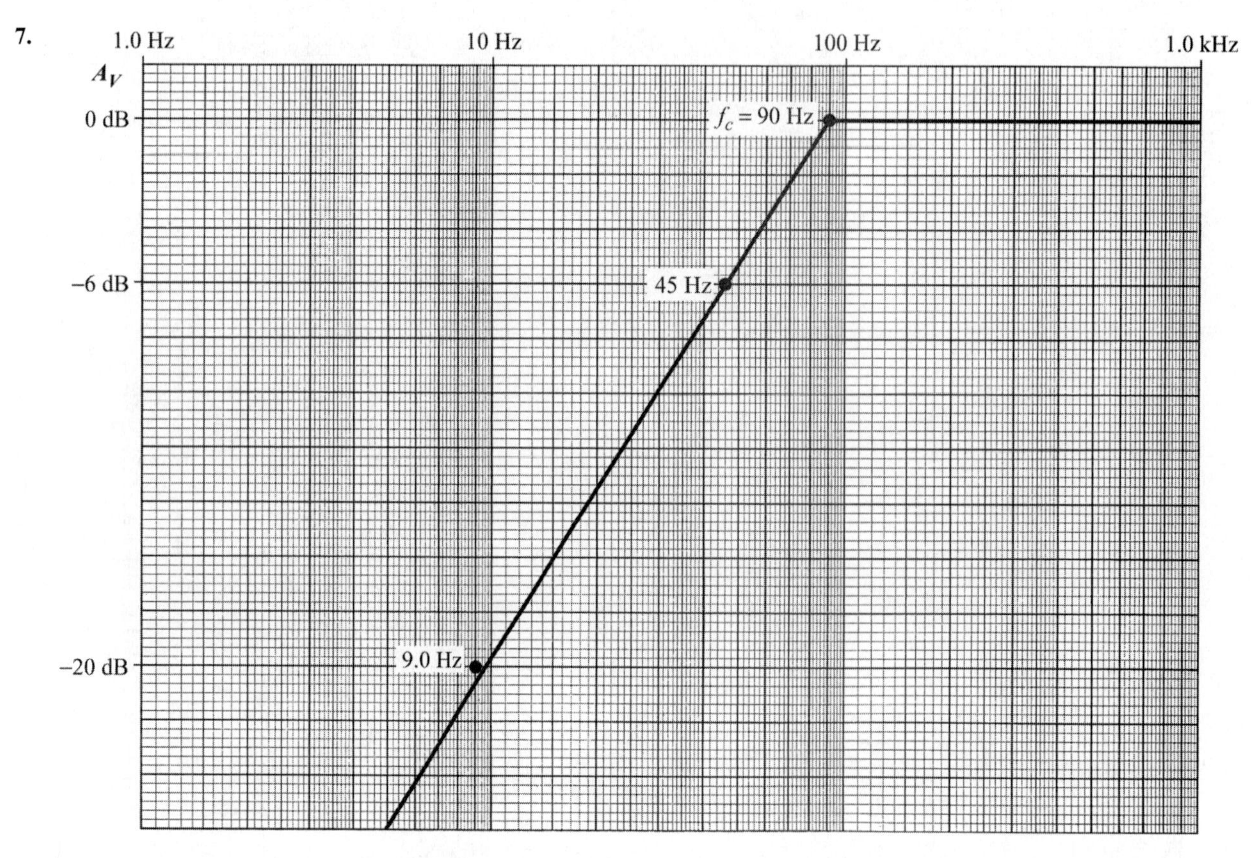
1.0 Hz
10 Hz
100 Hz
1.0 kHz
A_V
0 dB
−6 dB
−20 dB
f_c = 90 Hz
45 Hz
9.0 Hz

9.

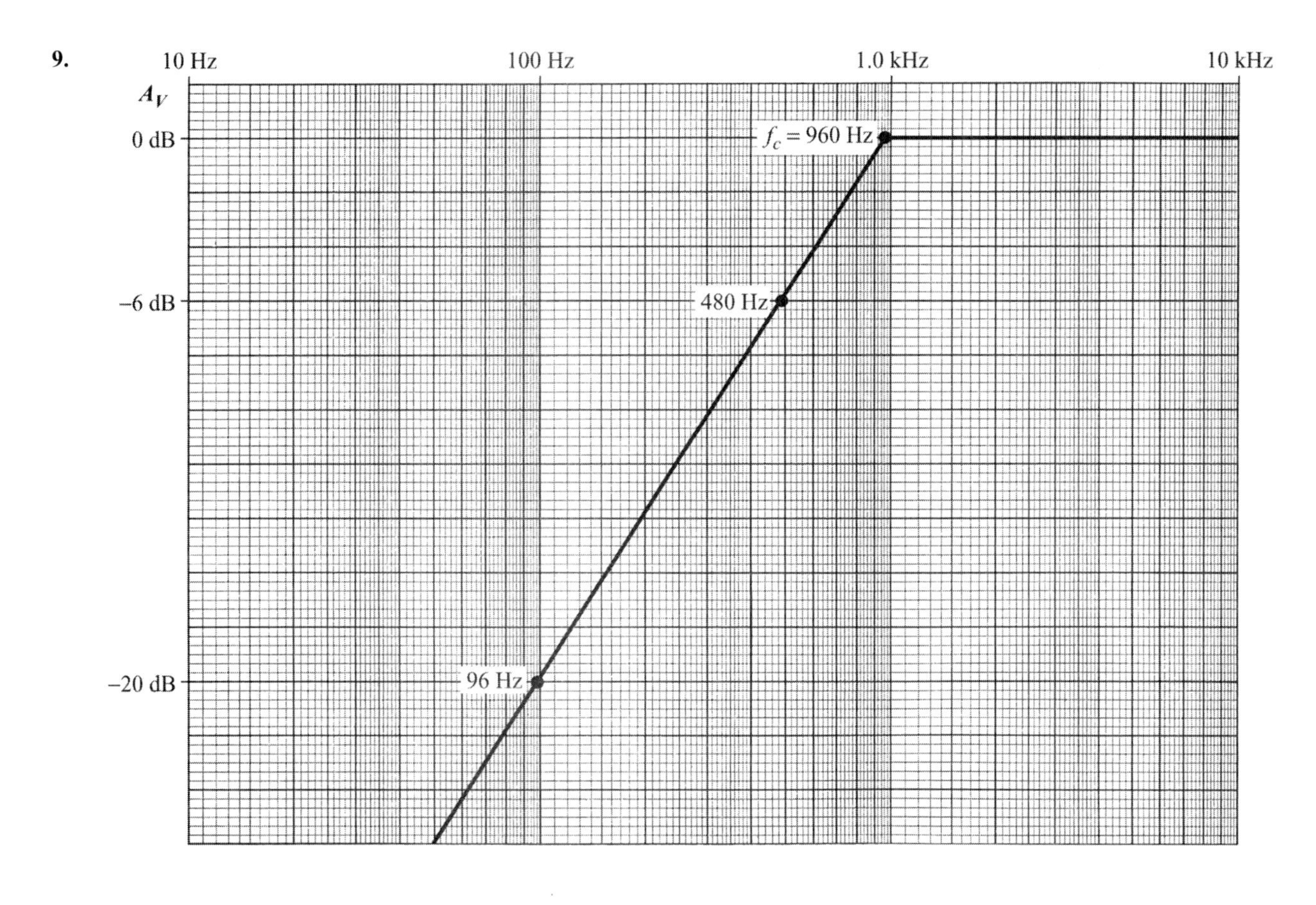
10 Hz
100 Hz
1.0 kHz
10 kHz
A_V
0 dB
−6 dB
−20 dB
f_c = 960 Hz
480 Hz
96 Hz

11.

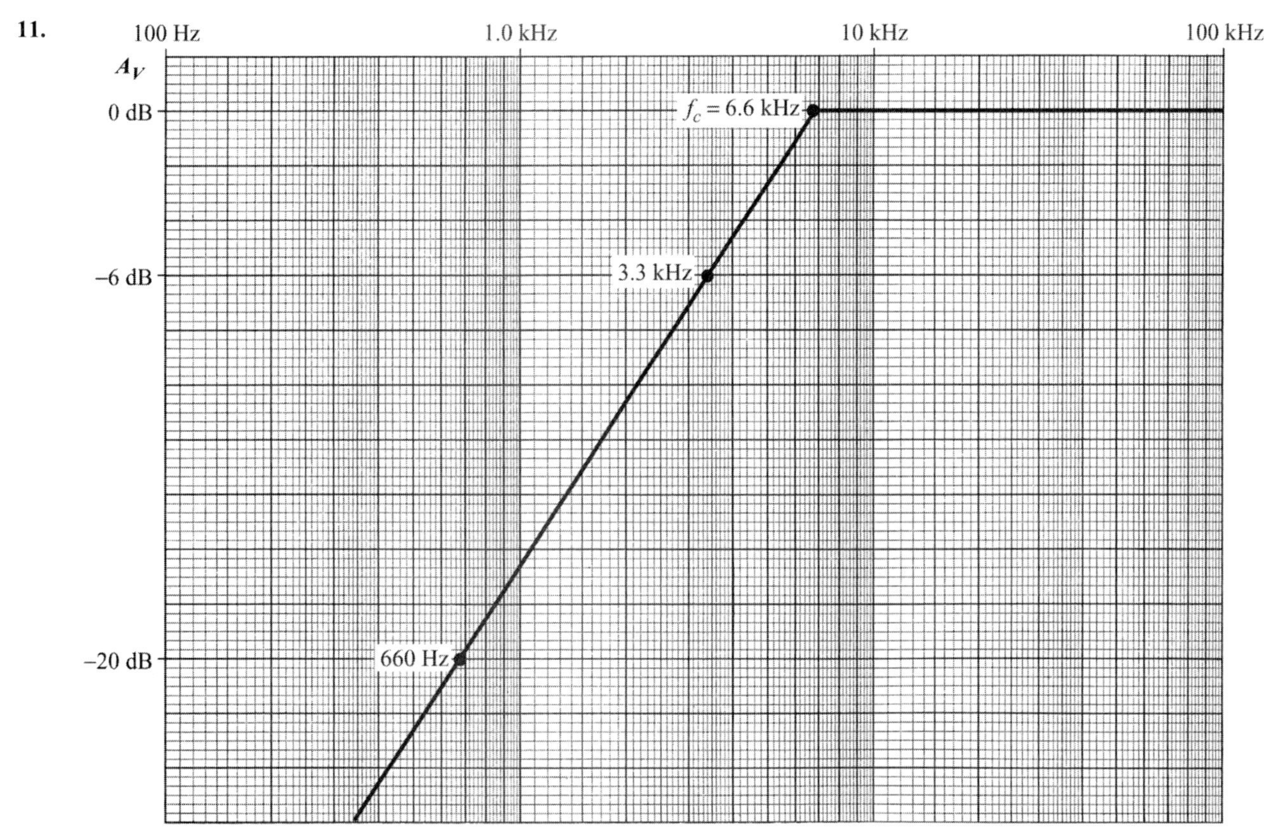
100 Hz
1.0 kHz
10 kHz
100 kHz
A_V
0 dB
−6 dB
−20 dB
f_c = 6.6 kHz
3.3 kHz
660 Hz

13.

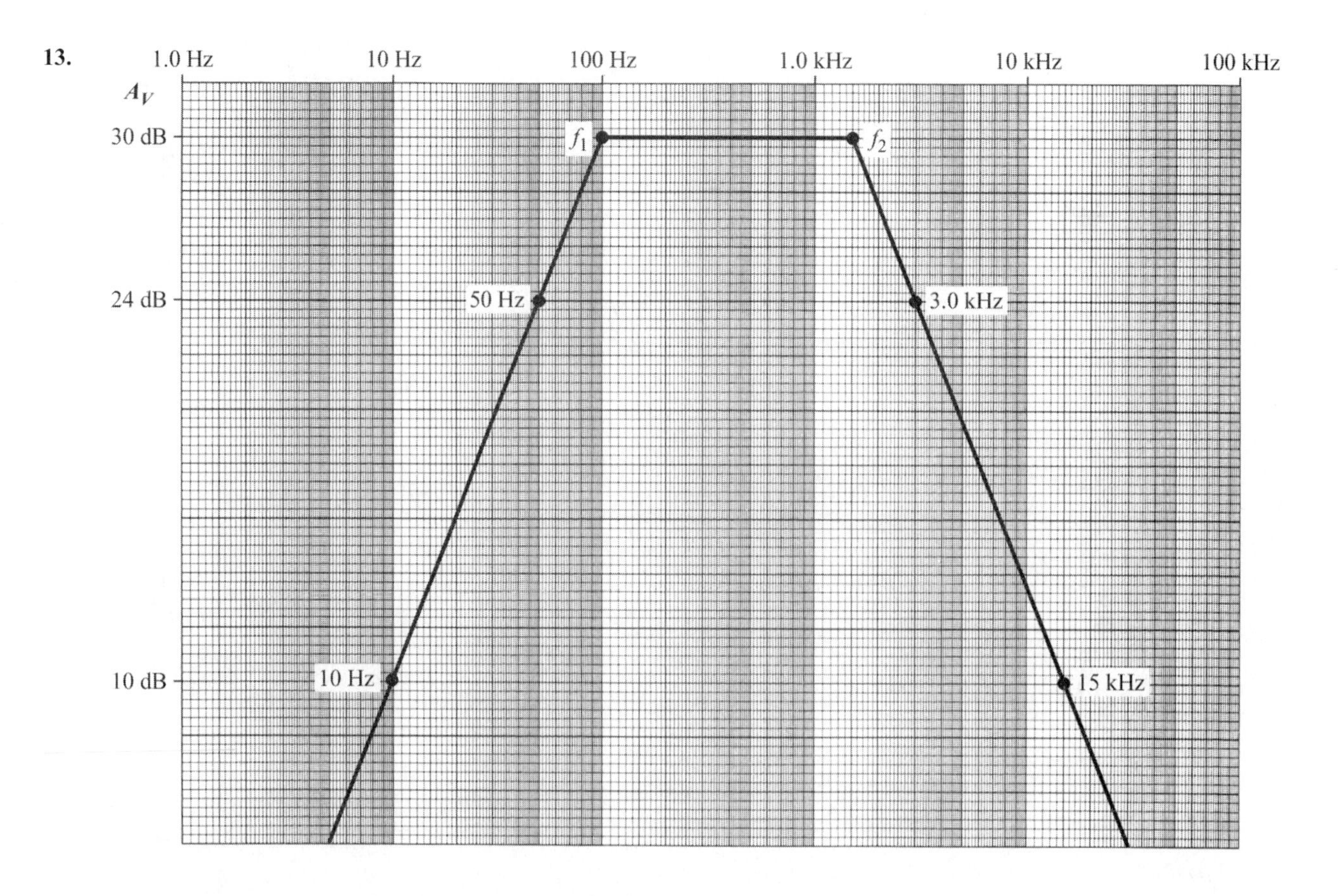
1.0 Hz
10 Hz
100 Hz
1.0 kHz
10 kHz
100 kHz
A_V
30 dB
24 dB
10 dB
f_1
f_2
50 Hz
3.0 kHz
10 Hz
15 kHz

15.

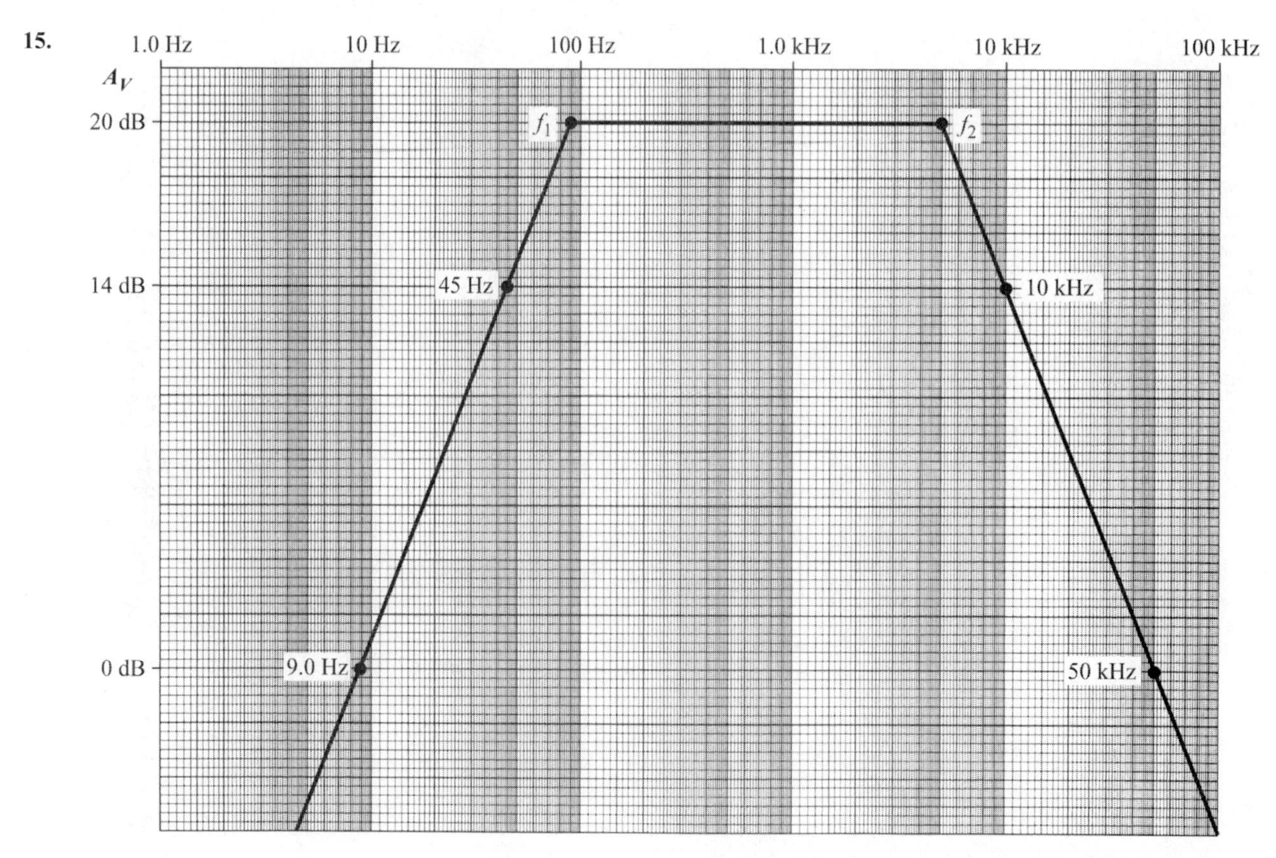
1.0 Hz
10 Hz
100 Hz
1.0 kHz
10 kHz
100 kHz
A_V
20 dB
14 dB
0 dB
f_1
f_2
45 Hz
10 kHz
9.0 Hz
50 kHz

17.

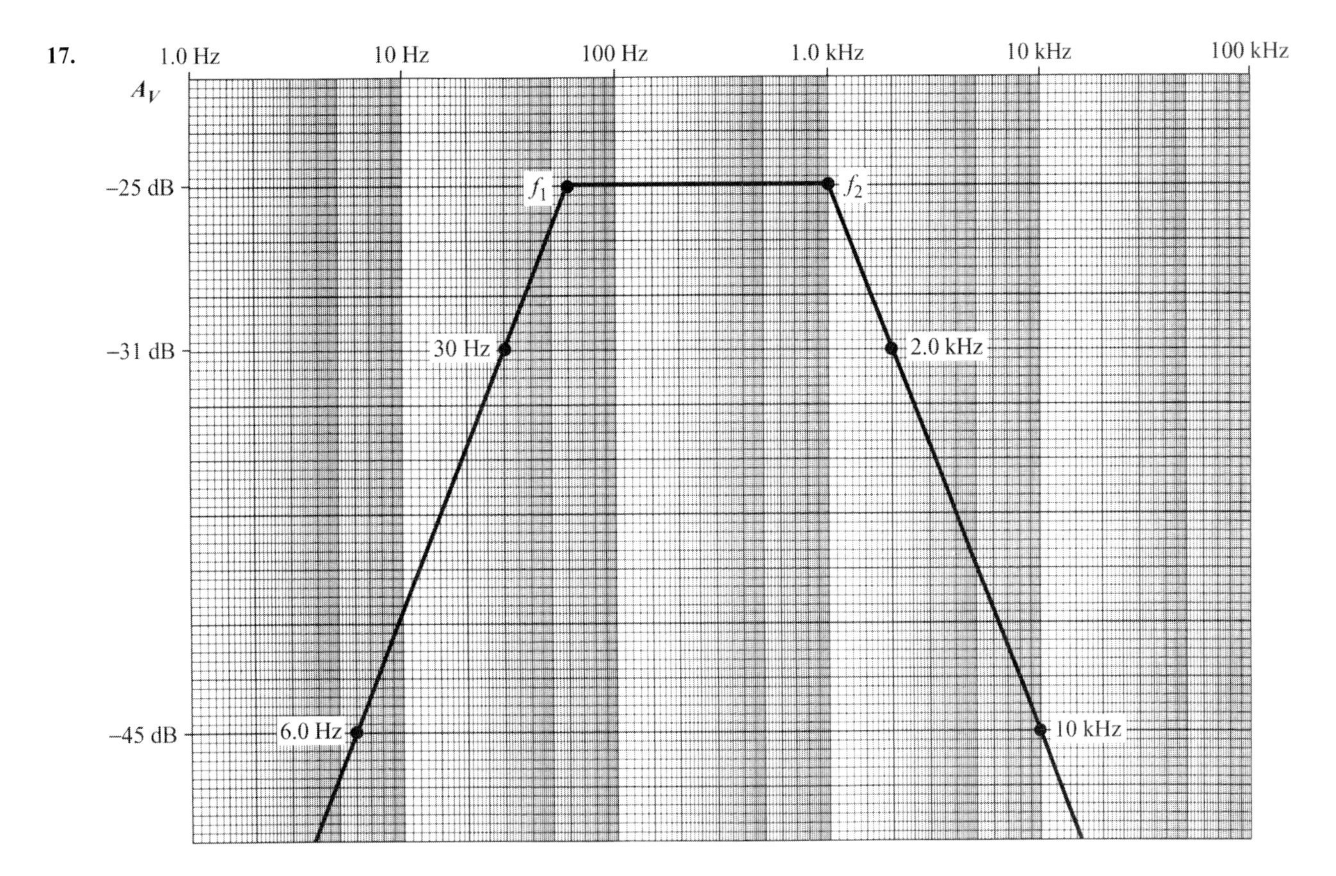

Review Exercises

1. 5.21

3. 1.09

5. 0.0693

7. 1.4 ms

9. 4.0 ms

11. 12 ms

13. 530 μs

15. −6.0 dB

17. 18 dB

19. −20 dB

21. 88 dB

23. 2.5 W

25. 200 W

27. 1.0 V

29. 90 V

31. **(a)** 15 ms
(b) $i_C = 27e^{-67t}$
(c) $v_C = 40(1 - e^{-67t})$
(d) 14 mA, 20 V
(e) 25 ms
(f) 10 ms

33. **(a)** 7.5 ms
(b) 24 ms

35. **(a)** 5 ms
(b) $i_L = 1.0(1 - e^{-200t})$
(c) 800 mA
(d) 5.2 ms
(e) 15 ms

37. **(a)** $v_R = 9.0e^{-83t}$
(b) 1.7 V
(c) 6.1 ms

39. 2.0 W, 47 V

41. *See graph on page 616.*

43. *See graph on page 616.*

CHAPTER 25

Exercise 25.1

1. 1(1000) + 3(100) + 2(10) + 8(1)

3. 9(10,000) + 9(1000) + 1(100) + 5(10) + 0(1)

5. 3(10) + 8(1) + 9(0.1) + 6(0.01)

7. 1(10) + 0(1) + 0(0.1) + 1(0.01)

9. 3

11. 4

13. 14

15. 17

17. 51

19. 2.75

21. 1.625

23. 3.6875

25. 100

27. 1011

29. 111 001

31. 1 100 011

33. 10 000 000

35. 0.1

37. 0.111

39. 10.01

41. 100 001.101

43.

17	10 001	25	11 001
18	19 919	26	11 010
19	19 011	27	11 011
20	10 100	28	11 100
21	10 101	29	11 101
22	10 110	30	11 110
23	10 111	31	11 111
24	11 000	32	100 000

45. **(a)** 49
(b) 40
(c) 7
(d) 12

Exercise 25.2

1. 101

3. 1011

5. 1111

7. 1000.1

9. 1111

Graph for Chapter 24, Review Exercise number 41.

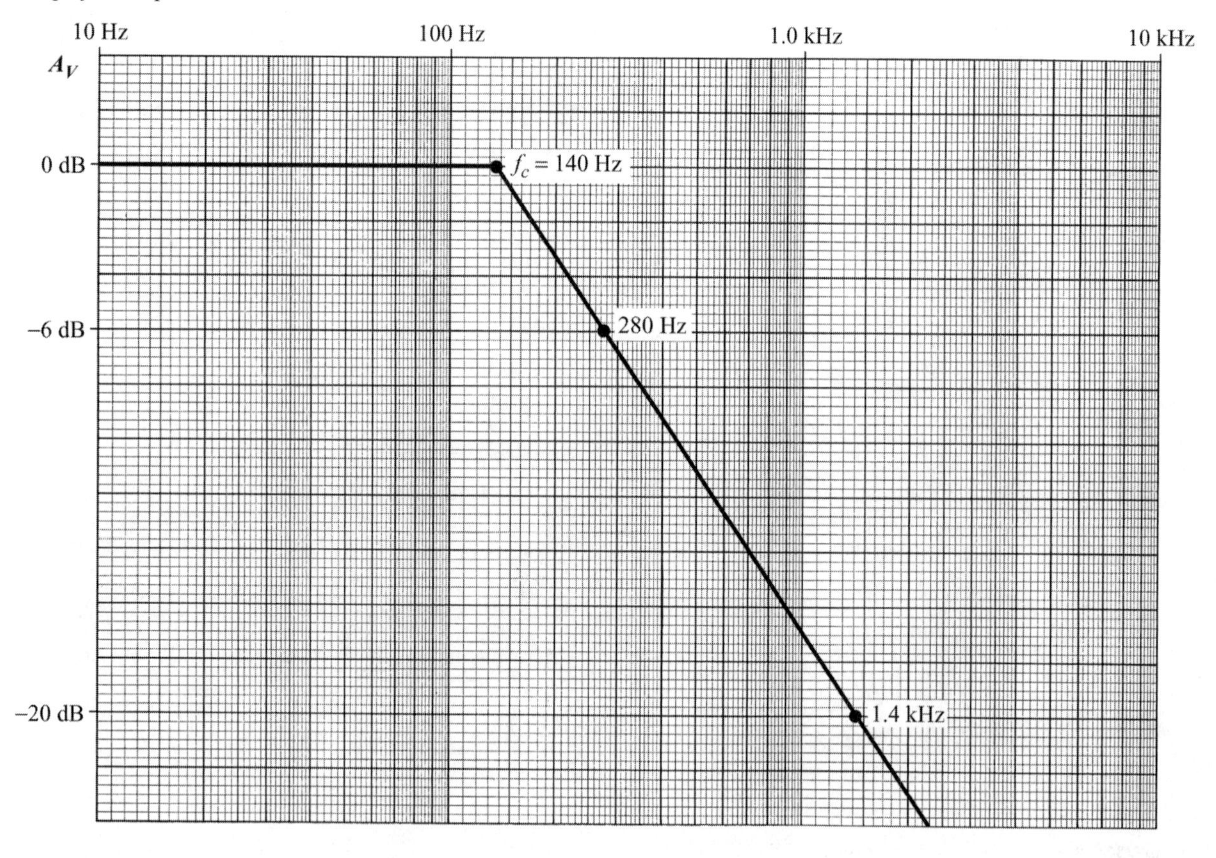

Graph for Chapter 24, Review Exercise number 43.

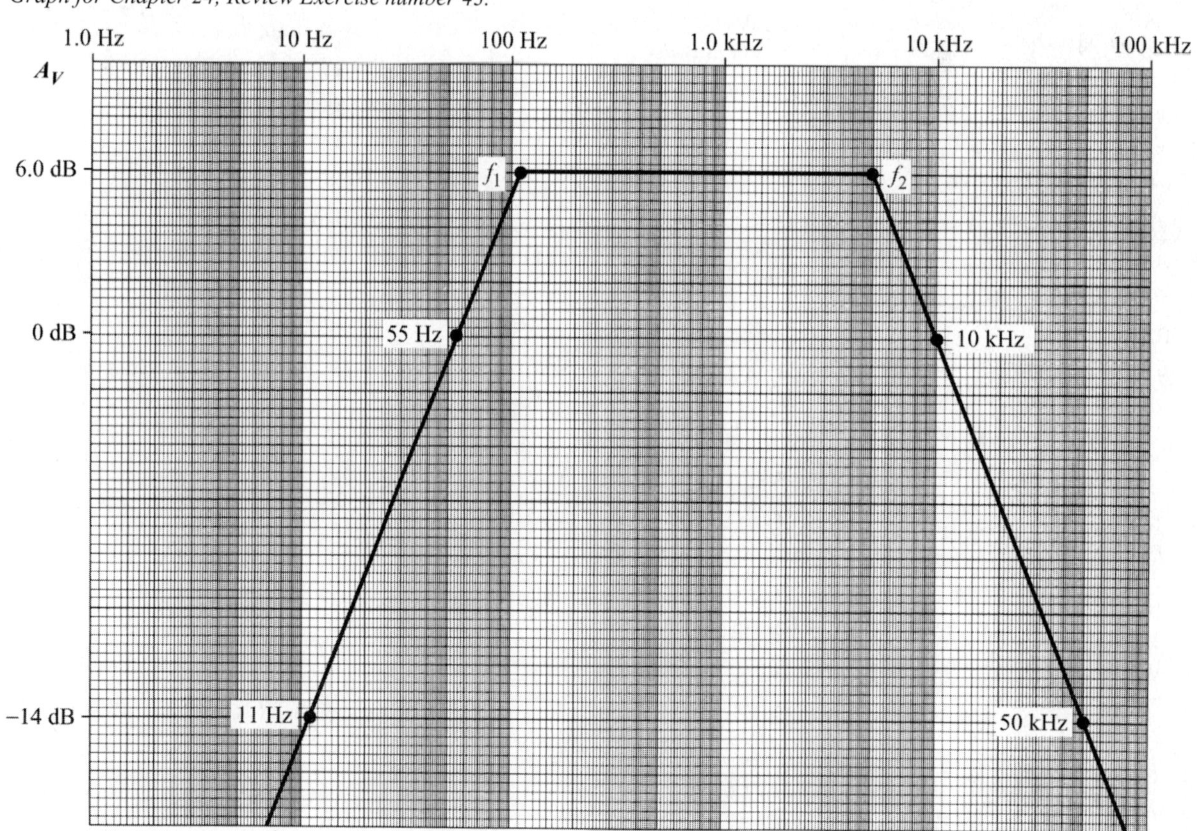

11. 10011

13. 10 000.01

15. 1

17. 100

19. 10
10110

31. 10010
$\overline{1\,01001} \rightarrow 1001$

21. 1001

23. 1

25. 11

27. 100 10

29. 10.10

Exercise 25.3

1. 10

3. 53

5. 212

7. 4095

9. 11

11. 46

13. 324

15. 2114

17. 1000

19. 10022

21. 111 110 001

23. 1 000 101 100

25. 14

27. 33

29. 57

31. 312

33. **(a)** $110\ 010\ 011_2 = 403$
(b) $245_8 = 165$

35.

17	21	1001	25	31	11001
18	22	10010	26	32	11010
19	23	10011	27	33	11011
20	24	10100	28	34	11100
21	25	10101	29	35	11101
22	26	10110	30	36	11110
23	27	10111	31	37	11111
24	30	11000	32	40	100000

Exercise 25.4

1. 27

3. 160

5. 205

7. 291

9. 2571

11. 16.125

13. 4097

15. 15

17. 28

19. 55

21. 8E

23. 157

25. 3E8

27. 11 0010

29. 1010 1011

31. 110 1100 0111

33. 1 1110 0000 1011

35. 1100 0000 1101 0000

37. E

39. 1B

41. 2F

43. 5C

45. E9

47. $1010\ 0000_2 = A0_{16}$

49. $59A_{16} = 1434$

Exercise 25.5

1. E0

3. 146

5. BC

7. 960

9. 1506

11. 10C7

13. 1122

15. 1F42

17. 76F2

19. 11865

21. A3

23. 2D

25. 4F

27. 23A

29. 135

31. 5F

33. 2E3

35. E7O4

37. 2DAF

37. 1DDD

Exercise 25.6

1. **(a)** 0001 1000
(b) 10010

3. **(a)** 0001 0010 0010
(b) 111 1010

5. **(a)** 0011 0100 1001
(b) 1 0101 1101

7. **(a)** 0101 1000 0011 0111
(b) 58 37
(c) 88 55

9. **(a)** 0100 0011 0100 1001 0100 0001
(b) 43 49 41
(c) 67 73 65

11. **(a)** 0101 0011 0100 0001 0100 1101
(b) 53 41 4D
(c) 83 65 77

13. **(a)** 0101 0010 0100 1111 0101 0011 0100 0101
(b) 52 4F 53 45
(c) 82 79 83 69

15. **(a)** Number of bits used to denote a character
(b) 9 bits

17. 0 0100 1110 0 0101 1001 1 0100 0011

19. 707

21. **(a)** 0010 0100
(b) 24

23. "–" should be 0 0110 1111

Review Exercises

1. 15

3. 27

5. 4.75

7. 10011

9. 101 1000

11. 0.11

13. 1100

15. 10011

17. 101

19. 1

21. 61

23. 58

25. 378

27. 100 110

29. 10 1010

31. 1001 1100 1101

33. **(a)** 33
(b) 1B

35. **(a)** 157
(b) 6F

37. **(a)** 2040
(b) 420

39. **(a)** 7
(b) 7

41. **(a)** 23
(b) 13

43. **(a)** 125
(b) 55

45. 132

47. 1508

49. 63

51. 9BD

53. 0100 0111 0100 1111

55. 0101 1000 0101 0011 0011 0011

57. 57 41 53 54 45 [WASTE]

59. "=" should be: 1 0011 1101

CHAPTER 26

Exercise 26.1

1. Copper is a poor conductor and $(0.1)^2 = 0.01$. Not True.
Copper is a poor conductor or $(0.1)^2 = 0.01$. True.
Copper is not a poor conductor. True
$(0.1)^2 \neq 0.01$. Not True

3. Commutative laws, Identity laws and the Distributive law: $A(B + C) = AB + AC$

5. $AB = BA$

7. $A + 0 = A$

9. $A \cdot 0 = 0$

11. $A \cdot 0 = A \cdot \overline{A}$

13. $A + (B \cdot 0) = A \cdot 1$

15. $A(\overline{A} + B) = AB$

17.

A	$A + A$
1	1
0	0

19.

A	B	$A + B$	$A(A + B)$
0	0	0	0
0	1	1	0
1	0	1	1
1	1	1	1

21.

A	B	AB	$\overline{\overline{AB}}$
0	0	0	0
0	1	0	0
1	0	0	0
1	1	1	1

23.

A	B	$\overline{A}$	$\overline{B}$	$\overline{A}\,\overline{B}$	$\overline{A} + \overline{A}\,\overline{B}$
0	0	1	1	1	1
0	1	1	0	0	1
1	0	0	1	0	0
1	1	0	0	0	0

25.

A	B	C	$A + B$	$B + C$	$(A + B) + C$	$A + (B + C)$
0	0	0	0	0	0	0
0	0	1	0	1	1	1
0	1	0	1	1	1	1
0	1	1	1	1	1	1
1	0	0	1	0	1	1
1	0	1	1	1	1	1
1	1	0	1	1	1	1
1	1	1	1	1	1	1

27.

A	B	C	$\overline{A}$	$\overline{B}$	$\overline{C}$	$\overline{B}\,\overline{C}$	$\overline{B + C}$	$\overline{A}(\overline{B}\,\overline{C})$	$\overline{A}(\overline{B + C})$
0	0	0	1	1	1	1	1	1	1
0	0	1	1	1	0	0	0	0	0
0	1	0	1	0	1	0	0	0	0
0	1	1	1	0	0	0	0	0	0
1	0	0	0	1	1	1	1	0	0
1	0	1	0	1	0	0	0	0	0
1	1	0	0	0	1	0	0	0	0
1	1	1	0	0	0	0	0	0	0

29. $A(A + B) = AA + AB = A + AB = A(1 + B) = A(1) = A$

Exercise 26.2

1. **(a)** (A and B) or C
(b) C and (A or B)

3. 0

5. 0

7. 0

9. 1

11. 0

13. 1

15. 1

17. 0

19. 0

21. 0

23. 1

25. **(a)** $A + A\overline{B}$

(b)

A	B	$\overline{B}$	$A\overline{B}$	$A + A\overline{B}$
0	0	1	0	0
0	1	0	0	0
1	0	1	1	1
1	1	0	0	1

27. **(a)** $AB + \overline{A}B$

(b)

A	B	$\overline{A}$	AB	$\overline{A}B$	$AB + \overline{A}B$
0	0	1	0	0	0
0	1	1	0	1	1
1	0	0	0	0	0
1	1	0	1	0	1

29. $A + A\overline{B} = A(1 + \overline{B}) = A(1) = A$

31. $AB + \overline{A}B = B(A + \overline{A}) = B(1) = B$

33. **(a)** $\overline{AB} + \overline{AB}$

(b)

A	B	$\overline{AB}$	$\overline{AB} + \overline{AB}$
0	0	1	1
0	1	1	1
1	0	1	1
1	1	0	0

(c) $\overline{AB} + \overline{AB} = \overline{AB}$

35. **(a)** $ABC + \overline{A}BC$

(b)

A	B	C	$\overline{A}$	BC	ABC	$\overline{A}BC$	$ABC + \overline{A}BC$
0	0	0	1	0	0	0	0
0	0	1	1	0	0	0	0
0	1	0	1	0	0	0	0
0	1	1	1	1	0	1	1
1	0	0	0	0	0	0	0
1	0	1	0	0	0	0	0
1	1	0	0	0	0	0	0
1	1	1	0	1	1	0	1

(c) $ABC + \overline{A}BC = BC(A + \overline{A}) = BC(1) = BC$

Exercise 26.3

1. B

3. $\overline{A}$

5. $\overline{A} + \overline{B}$

7. $\overline{A} + B$

9. $AB + BC$

11. $\overline{A}\,\overline{B} + \overline{B}C$

13. $\overline{B}\,\overline{C} + \overline{A}BC$

15. A

17. no adjacent squares

19. $\overline{C}\,\overline{D}$

21. $\overline{A}\,\overline{B}C + \overline{B}CD$

23. A

Review Exercises

1. **(a)**

A	B	AB	$A + AB$
0	0	0	0
0	1	0	0
1	0	0	1
1	1	1	1

(b) $A + AB = A(1 + B) = A(1) = A$

(c) $A(A + B) = A$

3. **(a)**

A	B	$\overline{A}$	$\overline{B}$	$\overline{A}B$	$\overline{A}\,\overline{B}$	$\overline{A}B + \overline{A}\,\overline{B}$
0	0	1	1	0	1	1
0	1	1	0	1	0	1
1	0	0	1	0	0	0
1	1	0	0	0	0	0

(b) $\overline{A}B + \overline{A}\,\overline{B} = \overline{A}(B + \overline{B}) = \overline{A}(1) = \overline{A}$

(c) $(\overline{A} + B)(\overline{A} + \overline{B}) = \overline{A}$

5. **(a)**

A	B	C	$A + B$	$B + C$	AC	$AC + B$	$(A + B)(B + C)$
0	0	0	0	0	0	0	0
0	0	1	0	1	0	0	0
0	1	0	1	1	0	1	1
0	1	1	1	1	0	1	1
1	0	0	1	0	0	0	0
1	0	1	1	1	1	1	1
1	1	0	1	1	0	1	1
1	1	1	1	1	1	1	1

(b)
$$(A + B)(B + C) = AB + AC + BB + BC$$
$$= AB + AC + B + BC = AB + AC + B(1 + C)$$
$$= AB + AC + B = AC + B(1 + A) = AC + B$$

(c) $AB + BC = (A + C)B$

7. 0

9. 0

11. 0

13. **(a)** $(A + B)(A + B) = A + B$

(b)

A	B	$A + B$	$(A + B)(A + B)$
0	0	0	0
0	1	1	1
1	0	1	1
1	1	1	1

(c) $(A + B)(A + B) = A + B$

15. **(a)** $AB + ABC$

(b)

A	B	C	AB	ABC	$AB + ABC$
0	0	0	0	0	0
0	0	1	0	0	0
0	1	0	0	0	0
0	1	1	0	0	0
1	0	0	0	0	0
1	0	1	0	0	0
1	1	0	1	0	1
1	1	1	1	1	1

(c) $AB + ABC = AB(1 + C) = AB(1) = AB$

17. 1

19. $\overline{A}\,\overline{B} + \overline{A}\,\overline{C}$

21. B

CHAPTER 27

Exercise 27.1

1. $\bar{x} = 7.73$, Median = 7, Mode = 7
3. $\bar{x} = 6.79$, Median = 6.2, No mode
5. $\bar{x} = 104$, Median = 104.5, Mode = 105
7. $\bar{x} = -0.0125$, Median = 0, Mode = 0
9. (a) $\bar{x} = 17.73$, Median = 17, Mode = 17
   (b) Mean, median and mode increase by same amount
11. (a) Mean = median = 3
    (b) **X:** Terms do not vary; **Y:** Terms increase uniformly
13. (a) 2.875
    (b) *B*
15. $\bar{x} = 79.7$, Median = 80, Mode = 80
17. (a) $\bar{x} = 7.92\%$, Median = 7.85%, Mode = 7.85%
    (b) Mean best represents the average rate
19. (a) $\bar{x} = 9.11$ V, Median = 9.1 V, Mode = 9.1 V

x (V)	*f*
8.9	2
9.0	3
9.1	6
9.2	2
9.3	2
9.4	1

   (b) Mean best represents actual voltage

21. (a) $\bar{x} = 27.4$ mH, Median = 27.4 mH, Mode = 27.4 mH

x (mH)3	*f*
27.2	2
27.3	5
27.4	6
27.5	4
27.6	1

   (b) All equally represent the inductance

Exercise 27.2

1. $\bar{x} = 5.17$, $\sigma_{n-1} = 3.31$
3. $\bar{x} = 36.1$, $\sigma_{n-1} = 8.07$
5. $\bar{x} = 2.5$, $\sigma_{n-1} = 14.9$
7. $\bar{x} = 5.48$, $\sigma_{n-1} = 0.299$
9. $\bar{x} = 105$, $\sigma_{n-1} = 3.46$
11. (a) $\bar{x} = 15.17$, $\sigma_{n-1} = 3.31$
    (b) Mean is increased by the same amount; standard deviation does not change.
13. $\bar{x} = 7.53$, $\sigma_{n-1} = 1.60$

x	*f*
5	2
6	2
7	3
8	4
9	2
10	2

15. $\bar{x} = 1.29$, $\sigma_{n-1} = 0.144$

x	*f*
1.1	3
1.2	4
1.3	3
1.4	3
1.5	3

17. $\bar{x} = 15.8$, $\sigma_{n-1} = 3.94$
19. $\bar{x} = \$3180$, $\sigma_{n-1} = \$455$
21. $\bar{x} = 40.6$ mA, $\sigma_{n-1} = 1.06$ mA
23. (a)

x(V)	*f*
1.4	3
1.5	8
1.6	5
1.7	4
1.8	1

   (b) $\bar{x} = 1.56$ V, $\sigma_{n-1} = 0.112$ V

Exercise 27.3

1.

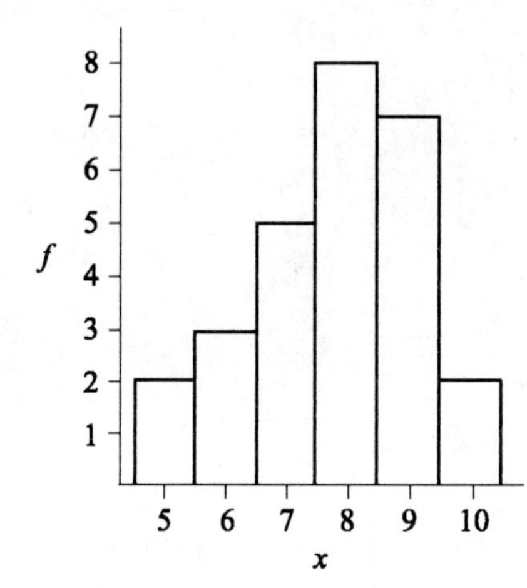

3.

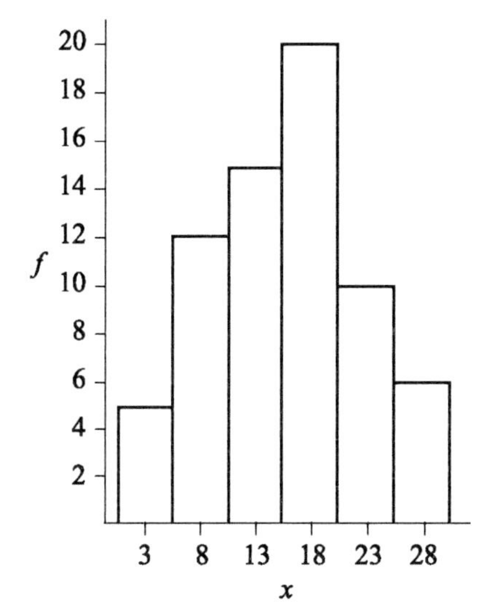

5.

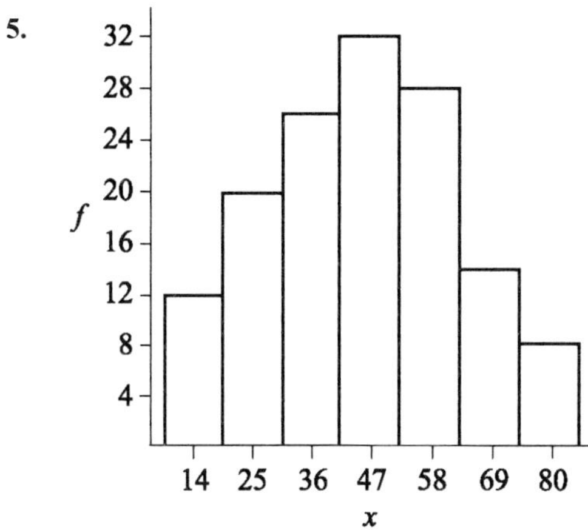

7. **(a)** 50%
(b) 68%
(c) 95%
(d) 5%

9. **(a)** 815
(b) 2

11.

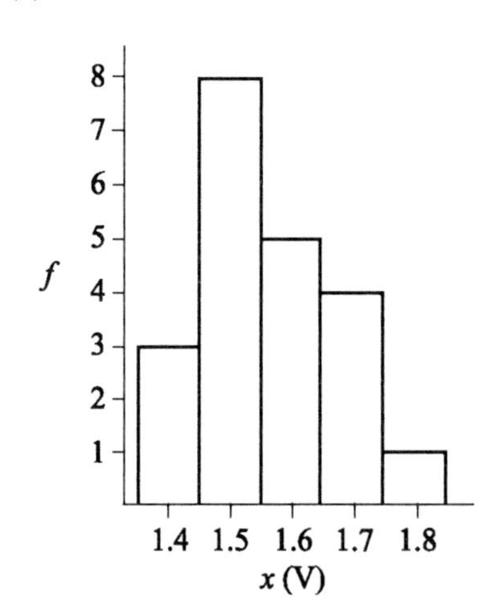

13.

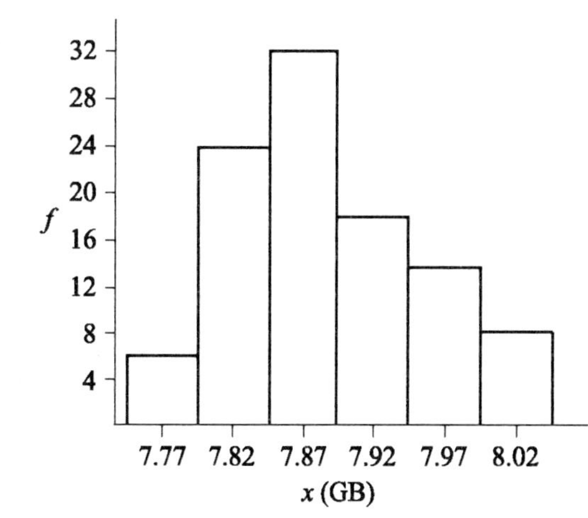

15. **(a)** 50%
(b) 68%
(c) 99.7%
(d) 32%

17. **(a)** 4750
(b) 1700
(c) 8
(d) 125

19. **(a)** 34%
(b) 16%
(c) 84%

Review Exercises

1. $\bar{x} = 12$, Median = 13, No Mode, $\sigma_{n-1} = 7.75$

3. $\bar{x} = 1.43$, Median = 1.5, Mode = 1.5, $\sigma_{n-1} = 0.30$

5.

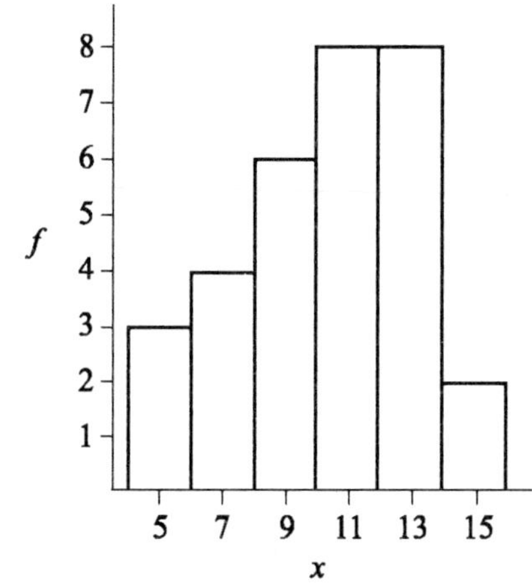

7. **(a)** $\bar{x} = 50.2$ ft, Median = 50 ft, Mode = 50 ft
(b) $\sigma_{n-1} = 1.32$ ft

9. **(a)** $\bar{x} = 1.11$ mH, Median = 1.11 mH, Mode = 1.12 mH
(b) $\sigma_{n-1} = 0.0113$ mH

11. **(a)** 50%
(b) 68%
(c) 95%
(d) 0.3%
(e) 8400
(f) 9750

INDEX

NOTE: Page numbers followed by f refer to figures, page numbers followed by t refer to tables.

O

P

S

T

U

V

W

X

Y

Z